Lecture Notes in Computer Science 1491
Edited by G. Goos, J. Hartmanis and J. van Leeuwen

Springer

Berlin
Heidelberg
New York
Barcelona
Hong Kong
London
Milan
Paris
Singapore
Tokyo

Wolfgang Reisig Grzegorz Rozenberg (Eds.)

Lectures on Petri Nets I: Basic Models

Advances in Petri Nets

Springer

Volume Editors

Wolfgang Reisig
Humboldt Universität zu Berlin, Institut für Informatik
Unter den Linden 6, D-10099 Berlin, Germany
E-mail: reisig@informatik.hu-berlin.de

Grzegorz Rozenberg
Leiden University, Department of Computer Science
P.O. Box 9512, 2300 RA Leiden, The Netherlands
E-mail: rozenber@wi.leidenuniv.nl

Cataloging-in-Publication data applied for

Die Deutsche Bibliothek - CIP-Einheitsaufnahme

Lectures on Petri nets : advances in Petri nets / Wolfgang Reisig ; Grzegorz Rozenberg (ed.). - Berlin ; Heidelberg ; New York ; Barcelona ; Hong Kong ; London ; Milan ; Paris ; Singapore ; Tokyo : Springer
1. Basic models. - 1998
(Lecture notes in computer science ; 1491)
ISBN 3-540-65306-6

CR Subject Classification (1998): F.1-3, D.2, C.1-2, G.2, D.4, J.4

ISSN 0302-9743
ISBN 3-540-65306-6 Springer-Verlag Berlin Heidelberg New York

Typesetting: Camera-ready by author
SPIN 10638863 06/3142 – 5 4 3 2 1 0 Printed on acid-free paper

Preface Volume One

Since Petri nets were introduced by C. A. Petri in his seminal PhD thesis in 1964, both the theory and the applications of this model have been flourishing. Although many other models of concurrent and distributed systems have been developed since then, Petri nets are still a central model for concurrent systems with respect to both the theory and the applications. They are often used as a yardstick for other models of concurrency.

The main attraction of Petri nets is the way in which the basic aspects of concurrent systems are captured both conceptually and mathematically. The intuitively appealing graphical notation makes Petri nets the model of choice in many applications. The natural way in which Petri nets allow one to formally capture many of the basic notions and issues of concurrent systems has contributed greatly to the development of a rich theory of concurrent systems based on Petri nets.

"Petri nets" is actually a generic name for a whole class of models which can be divided into three main layers. The first layer is the layer of fundamental research — the basic model here is that of *elementary net systems.* For modeling real-life systems of nontrivial size, elementary net systems may explode in size and become much too large to be managed effectively. The second layer allows one to collapse the repetitive features of elementary net systems in order to get more compact representations. The basic model here is *place/transition systems.* Finally, the third layer is that of high level nets, where one uses essentially algebra and logic to yield compact nets suitable for real-life applications. *Coloured Petri nets* and *predicate/transition nets* are the best known high level models.

For many practical applications, one needs to consider the execution time and/or stochastic processes. This leads to *timed and stochastic Petri nets.*

The two volumes are based on the Advanced Course on Petri Nets, held in Dagstuhl (Germany) in September 1996. Some material not presented in Dagstuhl is also included, in order to give a more balanced presentation of the area of Petri nets. The two volumes address those who are

- interested in systems design and would like to learn to use Petri nets,
- familiar with subareas of the theory or the applications of nets and wish to become acquainted with the whole area,

- interested in learning about recent results presented within a unified framework,
- going to learn how to apply Petri nets in various practical situations,
- interested in the relationship of Petri nets to other models of concurrent systems.

Previous Advanced Courses on Petri Nets were held in Hamburg (1978) and in Bad Honnef (1986).

The bulk of current research is presented annually at the International Conference on Application and Theory of Petri nets (ICATPN). This conference is oriented towards a broad audience. It is preceded by two days of tutorials and workshops, covering a wide variety of topics on both elementary and advanced level.

As a further service to the community, the news group on Petri nets provides a forum for news and views about all aspects of Petri nets (PetriNets–request@daimi.aau.dk). The *Petri Net Newsletter*, a communication forum for researchers in Petri nets, can be ordered through Gesellschaft für Informatik, Wissenschaftszentrum, Ahrstr. 45, 53175 Bonn, Germany.

This volume presents basic classes of Petri nets, some central research issues concerning these classes, and the relationship to other models of concurrency. Together with the companion volume on applications, it provides a broad and balanced picture of current trends in the theory and the applications of Petri nets.

Acknowledgements

The editors are very much indebted to all the authors, and also to the participants at the Advanced Course in Dagstuhl for their constructive criticism. The editors are grateful to the board of directors of the Dagstuhl conference center for allowing two weeks for the course, and to the Deutsche Forschungsgemeinschaft for its financial support. Many thanks to Springer–Verlag, and in particular to Mr. A. Hoffmann for his efficient handling of this book. Finally, the editors once again express their admiration of C. A. Petri, without whom the current understanding of concurrent systems would be much poorer.

Berlin — Wolfgang Reisig
Leiden — Grzegorz Rozenberg

June 1998

Table of Contents Volume One

Table of Contents Volume Two

Informal Introduction to Petri Nets

W. Reisig[1] and G. Rozenberg[2]

[1] Humboldt-Universität zu Berlin, Germany
[2] Leiden University, Netherlands

Introduction

"Petri Nets" is a collective term that, in the course of time, has come to designate a large number of system models, analysis techniques, graphical representations and notational conventions that are based on specific assumptions about the world of information processing.

The most basic are the notions of *local states* and *local actions* and their mutual relationships. In general, a local action is influenced by a *subset* of local states only. Moreover, in Petri nets local states and local actions are considered to be on an equal footing. The fundamental role of local states and local actions is also expressed in the standardized graphical convention, where local states are represented by circles and local actions are represented by boxes.

In this paper we give two examples which illustrate the process of constructing a formal model from an informal description using Petri nets. In the process of doing this we obtain examples of basic classes of Petri nets that are adressed in this volume.

1 Example 1: A Library

The following example has been taken from the initial phases of a software development process, where the initial system requirements are formulated with the impreciseness and incompleteness which is typical for the everyday use of natural languages. Then we gradually formalize, clarify and complete these requirements. The goal is to obtain a formal description of actions of a public library. We begin with the very basic description of the structure of a library: there is a *stock of books* that can be accessed by *users*. Thus we are dealing here with *two* components. It seems very natural to describe the stock of books as being *passive* and the users as being *active*. If we represent the passive component by a circle and the active component by a box, then we get the configuration shown in Fig. 1.1. The arrows between the components indicate the flow of objects and information.

Figure 1.1. A simple meaningful description of a library

In bigger libraries a user may not have free access to the books. Instead, there may be one (or several) counters where the library staff is available to the users. Hence the organizational structure is as given in Fig. 1.2.

Figure 1.2. Organisation of a large library

A large volume of book borrowing requires a more "sophisticated" organization. First of all, we need to distinguish the lending (outflow) and return (inflow) of books. In Fig. 1.3 we assume that a book has to be ordered by the user before the library staff provides it. In other words, there has to be a flow of information

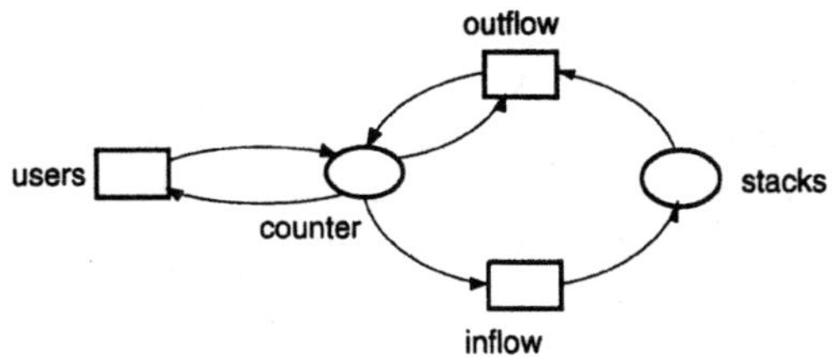

Figure 1.3. Organisational distinction between outflow and inflow of books

from the user to the provider before the required book is given to the user.

We refine the picture even further by identifying different interactions of the user with the library as illustrated in Fig. 1.4. In order to be able to deal with

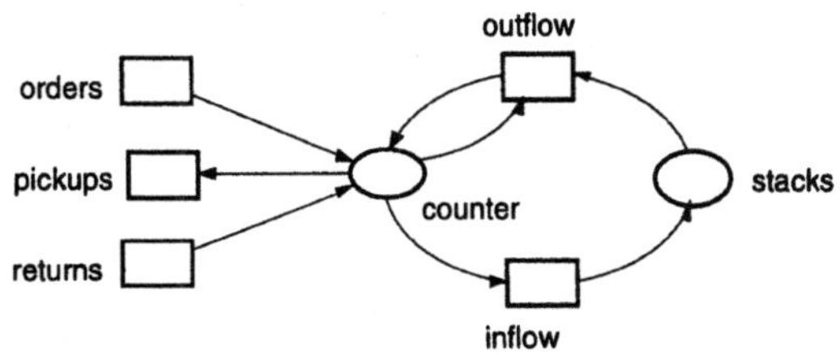

Figure 1.4. Decompostion of user interactions with the library into the acts of ordering, picking up and returning books

large numbers of people at the same time, it seems reasonable to establish separate counters for ordering, picking up and returning books (Fig. 1.5). The library

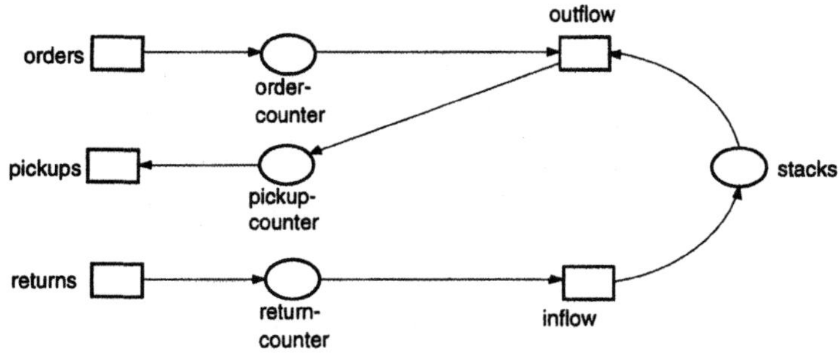

Figure 1.5. Introduction of separate counters

keeps a record of the borrowed books, in the simplest case just by using index cards that are stored with the books in the stacks. When a book is borrowed, its card is placed in a borrowed-book file, and when the book is returned, its card is taken out from this file and returned with the book to the stacks. Figure 1.6 shows the addition of the borrowed-book file to the system.

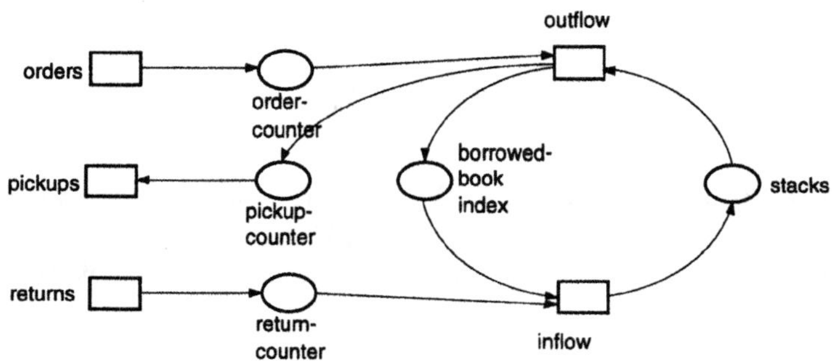

Figure 1.6. Introduction of a borrowed-book index

1.1 Passive and Active Components

The library example has been developed far enough to demonstrate a number of some central principles of system design using Petri nets. First of all, a system is partitioned into various components, and every component is designated as either passive or active. In our example this was quite obvious. The passive components (i.e., various counters, the stacks, and the borrowed-books index cards) are able to store things or to make them "visible"(traceable). The active components (in our example the users in their three roles of ordering, picking up or returning a book as well as the actions of retrieving a book from the stacks and returning a book back to the stacks) are able to produce, transport or change things.

It goes without saying that the arrows are important. An arrow does not represent a system component, but rather an abstract relationship between components, e.g., logical connections, access rights, physical proximity, or direct links.

As a matter of fact, there are no arrows connecting two passive or two active components. Instead, each arrow connects two components of different sorts. This represents the separation of components that we have mentioned already.

Whenever this principle is violated in a modeled system, it can be assumed that the criteria that led to a separation of individual components from their environment were wrongly assessed. Let us illustrate this with an example. Figure 1.7 shows a wrong and a right way (in terms of Petri nets) of model-

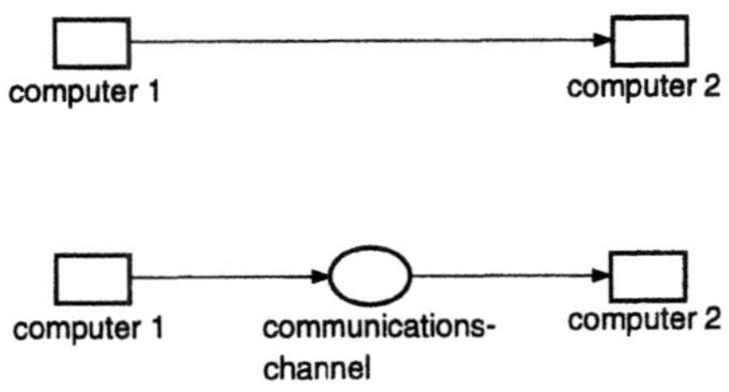

Figure 1.7. Representations of a simple computer link

ing a communication channel in a computer network. In the upper drawing the communication channel is either not modeled as a separate component, or it is (wrongly) modeled as an arrow. This would apply that a communication channel *always* links exactly two components. This may prove to be not realistic, since a communication channel is in fact quite complex: data may get corrupted or lost in the channel. In modeling this situation in a right way, a communication channel becomes a passive component. In this way we may model corrupting or losing data - see Fig. 1.8. The advantage of "right modeling" is very clearly seen, when a communication channel links several computers, as illustrated in Fig. 1.9.

1.2 Dynamic Behavior

What we have obtained until now, is a static representation of passive and active components, and the relationship between them. We move now to model the dynamic behavior of the represented system. To this aim, let us return to our library example, as represented in Fig. 1.6. We will inscribe now passive

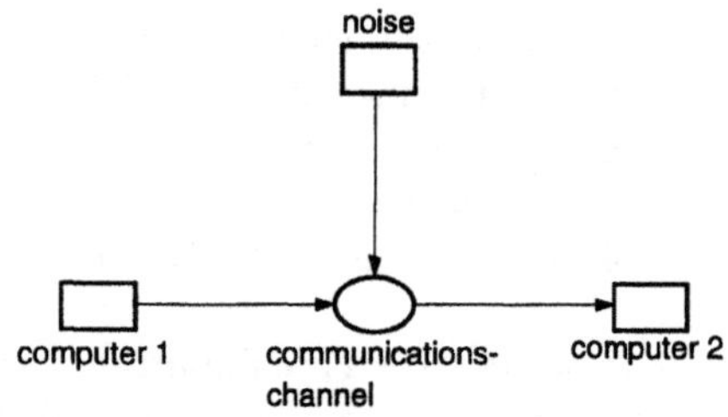

Figure 1.8. Extension of Fig. 1.7

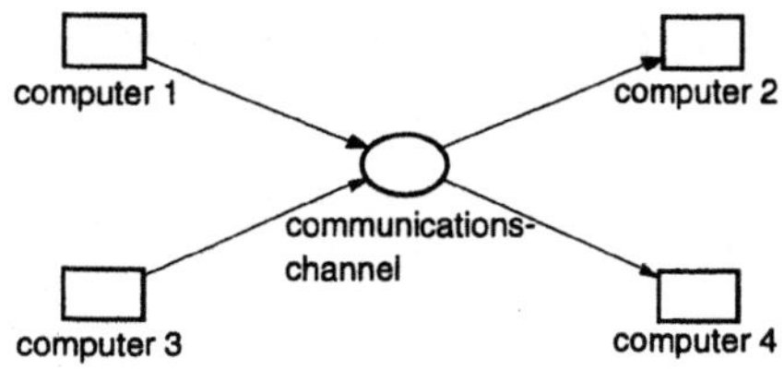

Figure 1.9. Several computers linked via one channel

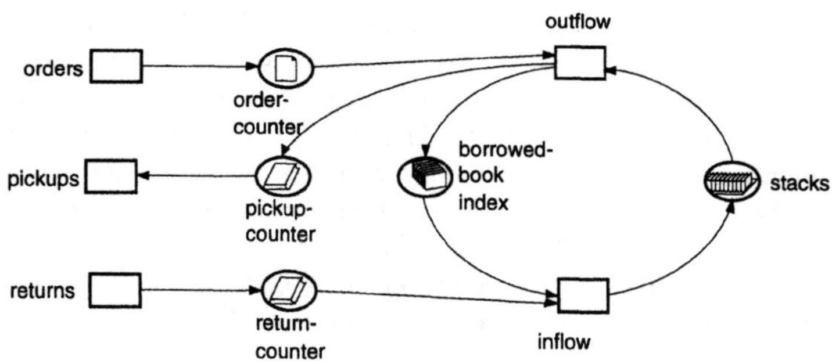

Figure 1.10. Order forms, books, index cards

components by concrete objects. Thus a (filled out) order form is at the order counter, a book is at the pickup counter waiting for the person who ordered it, and at the return counter there is a book that has been returned but not yet put back to the stacks. The books in the stacks and the cards in the borrowed book file are also shown.

The active components can now redistribute these objects.

Typically, one would expect the following rules. The *outflow* component accepts an order form, retrieves the corresponding book together with its index card from the stacks, brings the book to the pickup counter, and the index card to the borrowed-book index. The *inflow* component accepts a book from the return counter, and returns it to the stacks, together with the corresponding index card. The *orders* and *returns* components deliver orders and books to the respective counters, and the *pickups* component removes books from the pickup counter.

The Petri Net approach provides formal, yet intuitive concepts to describe such a behavior. To this end, the arcs of Fig. 1.10 are inscribed by *terms*, as shown in Fig. 1.11. The variable x stands for books; *order*(x) and *card*(x) stand for an

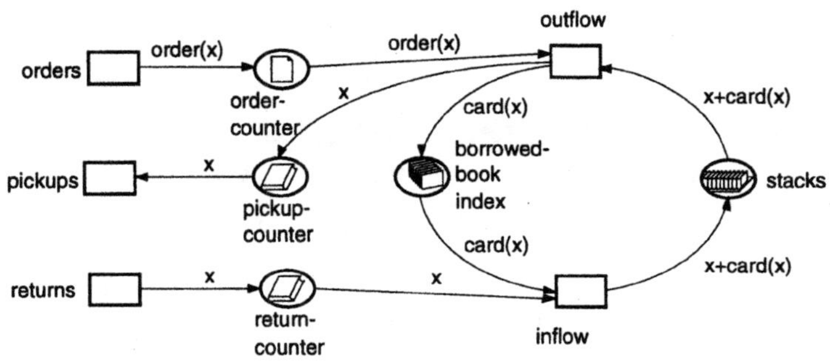

Figure 1.11. The library as a predicate/transition net

order of the book x and the index card of the book x, respectively. There is a universal *occurrence rule* for active components, which yields the above described behavior for the active components of Fig. 1.11. Formulated in a general setting, the occurrence rule assumes an interpretation of the variable x, by a concrete book; for example, $x =$ "...". Occurrence of an active component moves then the corresponding items as indicated by the inscriptions and directions of arcs.

Thus we have arrived, through a sequence of modifications, at a formal description of our library system by a Petri net. The change from Fig. 1.2 to Fig. 1.3 yields a more detailed description of the functions exercised by library staff. A similar relationship exists in the change from Fig. 1.3 to Fig. 1.4, and in the change from Fig. 1.4 to Fig. 1.5, where more detailed descriptions were given of the roles played by library users and the counters of the library staff. On the other hand, the change from Fig. 1.5 to Fig. 1.6 is different. In this case, a new component is added to the model, viz., the *borrowed-book file*. Thus, Petri net models of real world systems are systematically developed through two kinds of modifications: replacing a component with a more detailed subnet (a *refinement step*), or by adding components to the system, i.e., specifying an *environment* for a given net.

2 Example 2: A Container Interchange Station

Heavy freight is frequently transported in *containers* on top of train cars. Such train cars form *container trains*. They can be driven to *container interchange stations*, where the containers are *rearranged*, thus forming new container trains. A container interchange station has *container cranes* which can pick up containers from train cars and drop them at other train cars.

Figure 2.1 shows a typical container interchange station.

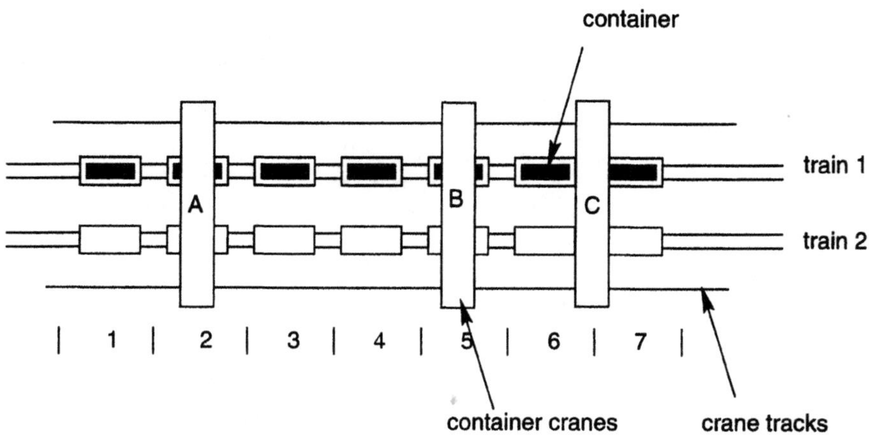

Figure 2.1. A container interchange station

Two trains may be positioned side by side, bridged by three container cranes. For the sake of simplicity, the range of the cranes is divided into seven sections, with each section covering one car of each train.

The situation outlined in Fig. 2.1 shows container cranes A and B over the second and fifth section, respectively. Crane C is moving in between sections 6 and 7.

Figure 2.2 represents the behavior of a single container crane.

In the depicted state, the crane is idle, loaded by a container, and its left neighboring section is free, whereas its right neighboring section is occupied by its right neighboring crane.

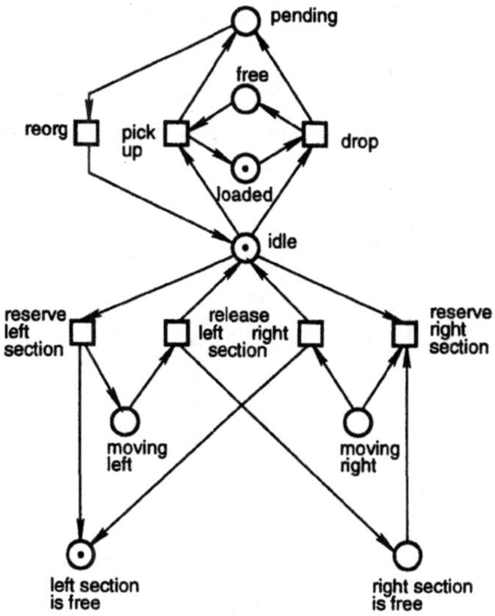

Figure 2.2. Behavior of a single container crane

Two actions are *enabled* in this state, *drop* and *reserve left section.* Both actions have two *preconditions*, (*idle* and *loaded*, and *idle* and *left section is free*, respectively), which are *fulfilled* in Fig. 2.2. Occurrence of *drop* leads to *free* and *pending*, and *reorg* is the only enabled action then. Alternatively to *drop*, the action *reserve left section* may occur in the initial state of Fig. 2.2. This leads to *moving left*, which in turn enables *release right.* To sum up, an action may occur if all of its preconditions and none of its postconditions are fulfilled. Its occurrence leads to a state where none of its preconditions and all of its postconditions are fulfilled. Graphically, this is depicted by Fig. 2.3.

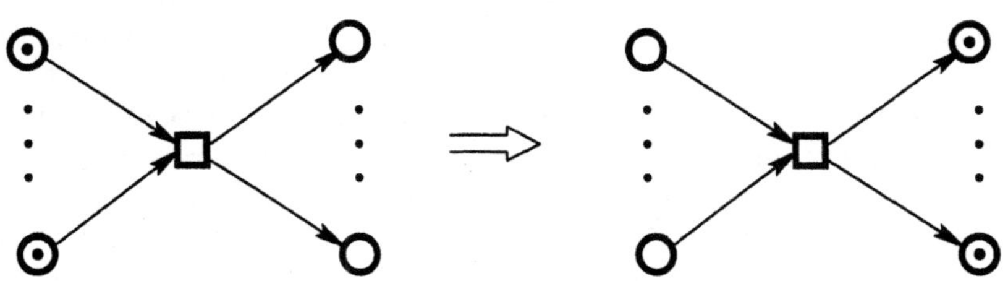

Figure 2.3. Occurrence of an event

Here, actions and conditions are represented as squares and circles, respectively. Arrows ending and starting at the square indicate preconditions and post-conditions of the action. A dot ("token") in a circle indicates that the corresponding condition is fulfilled.

Petri Nets of this kind are called *Elementary Net Systems.*

Figure 2.4 extends the environment of the above described container crane.

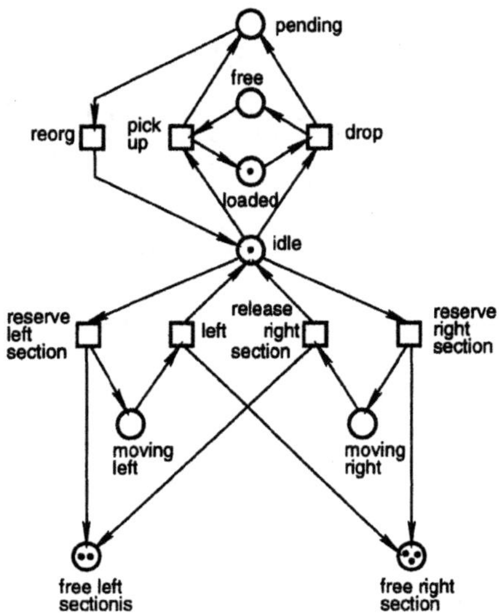

Figure 2.4. Counting the free sections in a container crane's environment

Not only the most adjacent left and right sections are represented, but the number of *all* adjacent free sections is given: each dot in *free left sections* and in *free right sections* represents a free section. Boxes and circles in Fig. 2.4 are called *places* and *transitions*, respectively. Each circle may carry *any* number of tokens (not at most one, as is the case with Elementary Net Systems). *Preplaces* and *postplaces* of transitions are defined analogously to pre- and postconditions, as above. A transition is *enabled* whenever each preplace carries at least one token. Occurrence of a transition can graphically be depicted as e.g. in Fig. 2.5.

The initial state of Fig. 2.4 enables three transitions: *drop*, *reserve left section*, and *reserve right section*, respectively. Occurrence of *reserve left section* would decrease the token load of *free left sections* to one token, and brings one token to *moving left.* This enables *release left section*, which increases the token load of *free right sections* to four. Petri Nets of this kind are denoted as *place/transition nets.*

Figure 2.6 shows three container cranes, with neighboring cranes sharing their right or left free sections.

The depicted situation describes the state of the container interchange station, as in Fig. 2.1. The occurrence rule for transitions described above reflex,

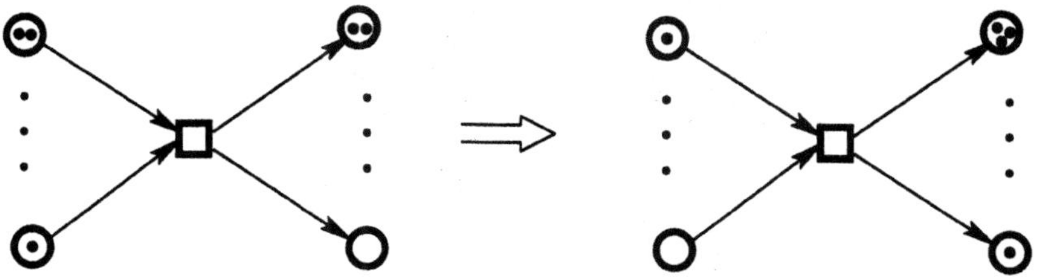

Figure 2.5. Occurrence of a transition

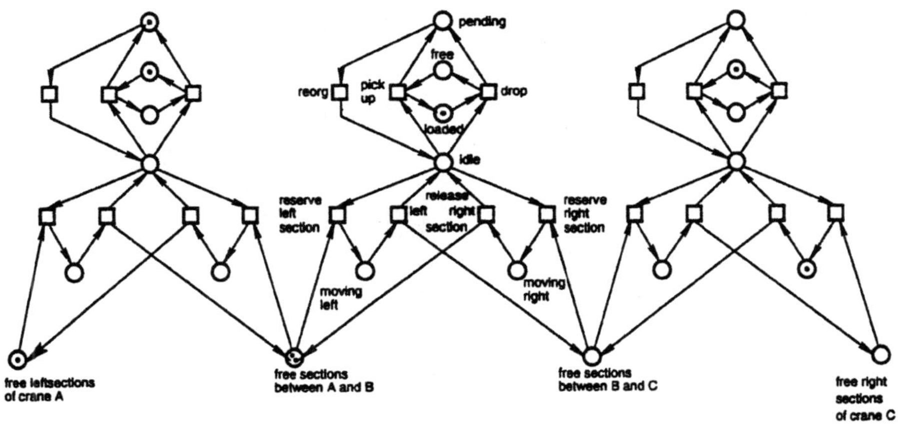

Figure 2.6. Model of the container exchange station. Fig. 2.1

in precise way, the behavior of the station. Now, Fig. 2.6 is an example of a place/transition net.

The regular structure of Fig. 2.6 suggests that one can represent the generic structure of container cranes only once, while all the details of Fig. 2.6 are given by proper inscriptions. This, in fact, is done in Fig. 2.7.

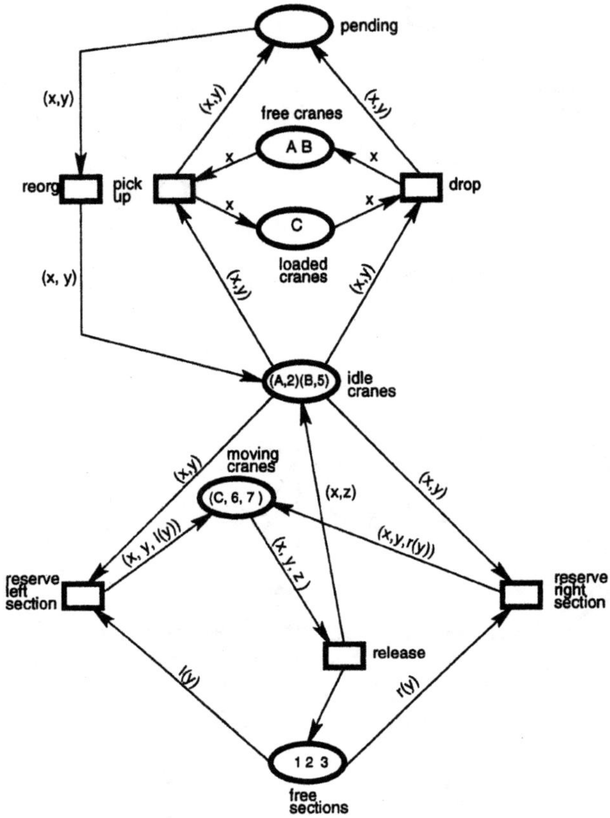

Figure 2.7. A high level representation, corresponding to Fig. 2.4

This figure is an example of a *high level net* which is a kind of nets that has been considered in the first example already.

References

[Reisig 92] W. Reisig: *A Primer in Petri Net Design.* Springer-Verlag (1992)

Elementary Net Systems

Grzegorz Rozenberg and Joost Engelfriet

Department of Computer Science, Leiden University
P.O.Box 9512, 2300 RA Leiden, The Netherlands
e-mail: {rozenber | engelfri}@wi.leidenuniv.nl

1 Introduction

The area of Petri Nets was initiated by C.A.Petri in the early sixties ([Pet62]). Since then this area has been developed tremendously in both the theory and the applications. Although many other models of concurrent and distributed systems have been developed in the meantime, the Petri net model is still a central model for such systems. It is also often used as a yardstick for other models.

One of the main attractions of Petri nets is the way in which the basic aspects of concurrent systems are identified both conceptually and mathematically. The ease of conceptual modelling (based also on a natural graphical notation) makes Petri nets the model of choice in many applications. The natural way in which Petri nets allow to formally capture many of the basic notions and issues of concurrent systems contributed greatly to the development of a rich theory of concurrent systems based on Petri nets.

Petri nets is actually a generic name for a whole class of net-based models which can be divided into three main layers. The first layer is the most fundamental and is especially well suited for a thorough investigation of foundational issues of concurrent systems. The basic model here is that of Elementary Net Systems, or EN systems (introduced in [RozThi86, Thi87, Roz87]). This model is not suitable for practical applications because the size of the model explodes even for simple but nontrivial applications. The second layer is an "intermediate" model where one folds some repetitive features of EN systems in order to get more compact representations. The basic model here is Place/Transition Systems, or P/T systems (see for example [Pet81, Rei82]). Finally, the third layer is that of high-level nets, where one uses essentially algebraic and logical tools to yield "compact nets" that are suited to real-life applications. Predicate/Transition Nets (see, e.g., [Gen87]) and Coloured Petri Nets (see, e.g., [Jen92]) are the best known high-level models.

In the framework of EN systems a concurrent system is seen as consisting of local states, local transitions (between local states), and the neighbourhood relationship between the local transitions and the local states. The global state of a system (its configuration) is simply the collection of all local states that (con)currently hold. The extent of change caused by a (local) transition is fixed and is restricted to the neighbourhood of the transition; it does not depend on the part of the global state that is outside that neighbourhood. This simple and elegant setup lends itself to a nice graphical representation of both the static structure of the system and its dynamic behaviour.

The EN system model has resulted from a number of modifications of the basic system model called Condition/Event Systems, or C/E systems, introduced by Petri (see, e.g., [Rei82, Thi87]). Perhaps the most significant difference is that in C/E systems transitions can also be reversed, recovering in this way the history of the system. An EN system can also be viewed as a special case of a P/T system.

In this chapter we present a comprehensive introduction to the theory of EN systems, covering both their structure and behaviour. The chapter is organized as follows. It consists of eight sections, of which this is the first. The second section recalls some standard mathematical notions needed in this chapter. Section 3 is a basic introduction to EN systems, both informally and formally. It begins with an informal introduction of EN systems. This is followed by the formalization of the notion of a net which represents the structure of an EN system. It consists of places (i.e., local states) and transitions (i.e., local transitions), connected by the neighbourhood relation. Then the dynamic execution of a net in both a sequential and a concurrent setting is discussed. An EN system is a net with an initial state, where all its executions begin. The sequential executions of the system that start from the initial state are called its firing sequences. Considering all states that are reachable by such a firing sequence leads to the state space of an EN system, formalized by the notion of a sequential configuration graph. The sequential configuration graph is extended to the (full) configuration graph of the system, in which also concurrent execution steps are represented. It is shown that the full configuration graph of an EN system is completely determined by its sequential configuration graph. Finally we demonstrate how fundamental situations of concurrent systems can be naturally expressed in the model of EN systems.

Section 4 discusses a number of structural and/or behavioural normal forms for EN systems. This is done in the framework of several fundamental behavioural equivalence relations for EN systems, based on their state spaces. A first, basic, normal form is that for every EN system an equivalent EN system can be constructed that has no redundant transitions or places. For such "reduced" EN systems, a structural characterization is given for the behavioural notion of a sequential EN system, i.e., an EN system of which every global state consists of one local state only. After formalizing the notion of a sequential component of an EN system (i.e., a subsystem that, on its own, is a sequential EN system), a second, more involved, normal form is shown: for every EN system an equivalent EN system can be constructed that can be viewed as a concurrent set of communicating sequential components. Such systems are conceptually easier to understand.

In the last three sections we turn to the partial order view of concurrent behaviour, where the partial order represents the causal dependencies between the events in a run of the EN system. In Section 5 the notion of a concurrent run of an EN system is formalized; as usual in Petri net theory, it is called a "process". It is one of the nice features of Petri net theory that such a process is in fact itself an EN system, called a process net (or causal net); it consists moreover of a mapping that labels the places and transitions of the process net with those of the given EN system. Through this mapping, the firing sequences of the processes correspond to a partition of the set of firing sequences of the EN system. Since a process net is acyclic, the transitions of a process naturally represent a partial order that is labelled by the transitions of the EN system. There is a one-to-one correspondence between the processes of an EN system and the corresponding labelled partial orders. EN systems with the same set of labelled partial orders are said to be lpo-equivalent; this is a natural concurrent behavioural equivalence of EN systems.

In Section 6.1 it is demonstrated that two EN systems are lpo-equivalent if and only if they have the same firing sequences. This shows that all various notions of behaviour considered in this chapter actually amount to two equivalences: configuration equivalence (where two EN systems have the same sequential configuration graph) and firing sequence equivalence (where two EN systems have the same firing sequences). It is remarkable that these two equivalences formalize the *sequential* behaviour of EN systems. This means that the concurrent behaviour of an EN system (as, e.g., formalized by lpo-equivalence) can be derived from its sequential behaviour. In Section 6.2 it is demonstrated that the above partition of the set of firing sequences of an EN system corresponds to the so-called trace-equivalence relation on firing sequences: two firing sequences are trace-equivalent if the one can be obtained from the other by the interchange of causally independent transitions.

Section 7 discusses the notion of a branching process of an EN system, which formalizes the combination of a finite number of concurrent runs of the system, showing where these runs are in conflict and, thus, have to branch. The theory of branching processes (or unfoldings) of EN systems is very similar to the theory of processes, with the notion of partial order replaced by the notion of event structure, which is a partial order with a conflict relation. Finally, Section 8 contains a conclusion.

We note here that we only consider the *finite* behaviour of an EN system, i.e., we only formalize finite executions or runs of the system. Thus, infinite executions or runs of the systems should be understood through their finite initial parts.

This chapter is based on [Rei82, RozThi86, Thi87, Roz87] and on the lecture notes for the lecture "Theory of Concurrency I" given for several years already at the Department of Computer Science of Leiden University, The Netherlands.

2 Preliminaries

In this section we recall some well-known concepts and notation concerning sets and words.

The sets of non-negative and positive integers are denoted by $\mathrm{N} = \{0, 1, 2, \ldots\}$ and $\mathrm{N}_+ = \{1, 2, 3, \ldots\}$, respectively.

For a set A, $\mathcal{P}(A)$ is the set of all subsets of A, and $\#A$ denotes the number of elements of A. We consider total functions only, i.e., if $f : A \to A'$, then $f(a)$ is defined (and in A') for every $a \in A$. If $f : A \to A'$ and $B \subseteq A$, then $f \upharpoonright B$ denotes the restriction of f to B, i.e., the function $B \to A'$ defined by $(f \upharpoonright B)(b) = f(b)$. A function $f : A \to A'$ is injective if $f(a_1) \neq f(a_2)$ whenever $a_1 \neq a_2$, f is surjective if for every $a' \in A'$ there exists an $a \in A$ such that $f(a) = a'$, and f is a bijection (between A and A') if f is injective and surjective.

For a binary relation $R \subseteq A \times A$, the transitive (and reflexive) closure of R is denoted by R^+ (R^*, respectively). Hence, $(a, b) \in R^+$ iff there exist $a_1, \ldots, a_n \in A$, with $n \geq 1$, such that $a_1 = a, a_n = b$ and $(a_i, a_{i+1}) \in R$ for all $1 \leq i \leq n-1$, and $(a, b) \in R^*$ iff $(a, b) \in R^+$ or $a = b$.

For an alphabet Σ, Σ^* is the set of all words over Σ, and λ is the empty word. The length of a word $x \in \Sigma^*$ is denoted by $|x|$; thus, $|\lambda| = 0$. A word y is a prefix of a word x if there exists a word z such that $x = yz$. A language over Σ is a subset of Σ^*. We note that any finite set can be viewed as an alphabet.

Let Σ and Δ be alphabets and let $h : \Sigma \to \Delta$. We extend h to a function from Σ^* to Δ^* in the following way: $h(\lambda) = \lambda$ and, for a word $x = \sigma_1 \cdots \sigma_n$, with $n > 0$ and $\sigma_i \in \Sigma$ for all $1 \leq i \leq n$, $h(x) = h(\sigma_1) \cdots h(\sigma_n)$. Furthermore, for a language $L \subseteq \Sigma^*$ we define $h(L) = \{h(x) \mid x \in L\}$. In formal language theory, h is called a letter-to-letter homomorphism.

3 EN Systems

3.1 Informal Introduction

Elementary net systems form the most fundamental class of Petri nets. Like most of the models that fall under the generic name Petri nets, elementary net systems are a net-based model. The basic intuition behind net-based models is that such a model consists of a net, and of the rules of a token game played on the net. The net describes the static structure of a concurrent system, and the rules describe the dynamic behaviour of the concurrent system. Different classes of Petri nets differ by the sort of underlying nets and/or rules of the dynamic token game.

In this section we introduce the notion of a net for elementary net systems and the rules for playing the token game on such a net. These rules tell us how to get from one global state of the (elementary net) system to another global state. They give thus a potential state space of the system. In order to get the actual state space one has to specify the initial global state of the system: the actual state space is then obtained by starting from this initial global state and using the rules of the game. Hence a specific elementary net system is given by a net and an initial global state. The rules of the token game are the same for all elementary net systems.

A characteristic property of Petri nets is that the global state is a set (or multiset) of local states and that the transition from a global state to another global state is given by one or more local transitions, where local transitions act on local states. More precisely, a local transition replaces a subset of the local states of a global state by another such subset. Thus, Petri nets can be viewed as set transition systems or set replacement systems (as opposed to the string rewriting systems of formal language theory).

The local states of an elementary net system are also called places, and the local transitions are just called transitions. The net is a finite directed graph; its nodes are the potential places and transitions of the system, and its edges assign input and output places to each transition. Thus the net defines for each transition which local states are replaced by which other local states. The actual global state is indicated by putting tokens on the places that are the actual local states, and the token game is played by moving the tokens around, according to the transitions that are executed.

Thus, in elementary net systems there are only finitely many potential local states, and a global state is a finite set of local states. An important consequence of this setup is that the state space of an elementary net system is finite. This means that elementary net systems are a model of finite-state concurrent systems (just as finite automata model finite-state sequential systems).

This section discusses the basic notions of a net, the token game played on the net, and the state space of an elementary net system. However, before turning to the formal definitions, we wish to give the reader some intuitive feeling for the net-based approach to concurrent systems, through two easy and well-known examples.

Very often, a concurrent system consists of several communicating concurrent components, each of which is sequential (i.e., executes its actions one after the other). Concurrency of the components means that one component can perform (some of) its actions independently of (and thus simultaneously with) the actions of another component. From time to time the components communicate with each other, which means that they have to wait for each other and interact by synchronizing their actions. In modelling such a concurrent system by an elementary net system, a local state is a state of a component, and a local transition is either an action of one component that is independent of the other components, or consists of a synchronized action shared by several communicating components.

The producer/consumer problem Fig. 1 shows an elementary net system that models the well-known producer/consumer problem. Places (i.e., local states) are indicated by circles, and transitions by rectangles. For each transition, the local states with an edge to the transition (the input places) are replaced by the local states with an edge from the transition (the output places).

The system can be viewed as consisting of three components: the producer, the consumer, and the buffer. The producer puts "production units" in the buffer (which can contain only one such unit), and the consumer takes units from the buffer.

The producer is always in exactly one of its local states p_1 or p_2. If it is in state p_1, it can execute transition p (i.e., produce a unit) and go into state p_2 (which means that the token is moved from p_1 to p_2). Then it can synchronize with the buffer component through the execution of the shared transition f

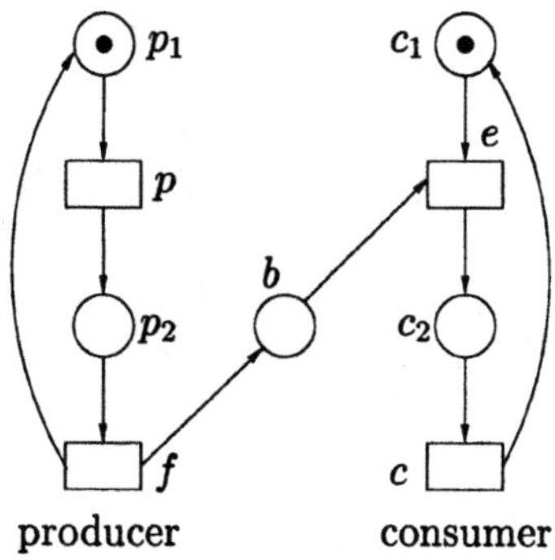

Fig. 1. The producer/consumer problem.

(i.e., fill the buffer). The producer then returns to state p_1, and the buffer place b now also contains a token, to indicate that the buffer is full. The buffer can be emptied again by the shared transition e (i.e., empty the buffer) in which the consumer and the buffer synchronize, and the consumer moves to its local state c_2. Then the consumer can execute transition c (i.e., consume the unit) and return to its local state c_1. In the meantime the producer may have executed transition p independently, and may have executed transition f as soon as the buffer was empty.

Note that the producer and the consumer never communicate directly, i.e., never perform shared actions. However, they cooperate asynchronously via the buffer.

The mutual exclusion problem Fig. 2 shows an elementary net system that models the well-known mutual exclusion problem. The system consists of three components: component 1, component 2, and the "permission" component. Components 1 and 2 compete for access to the same shared resource (such as a printer). At any moment of time at most one of these components can use the resource. It is the task of the permission component to schedule access to the resource. The availability of the resource is represented by a "permission" in place p; this permission is indicated by the presence of a token in place p.

A "critical section" is a part of a component which uses the shared resource and hence needs protection against a possible "disturbance" by the other component. The critical section of component 1 is represented by place c_1 and that of component 2 by place c_2. The noncritical part of component i (for $i = 1, 2$) is represented by places r_i (the remainder) and w_i (wait). Thus, component i has local states c_i, r_i, and w_i. Component i can perform the actions in_i, out_i, and d_i (entering the critical section, exiting the critical section, and an action outside the critical section). To enter or exit the critical section it has to synchronize with the permission component, which has local states p (the resource is not

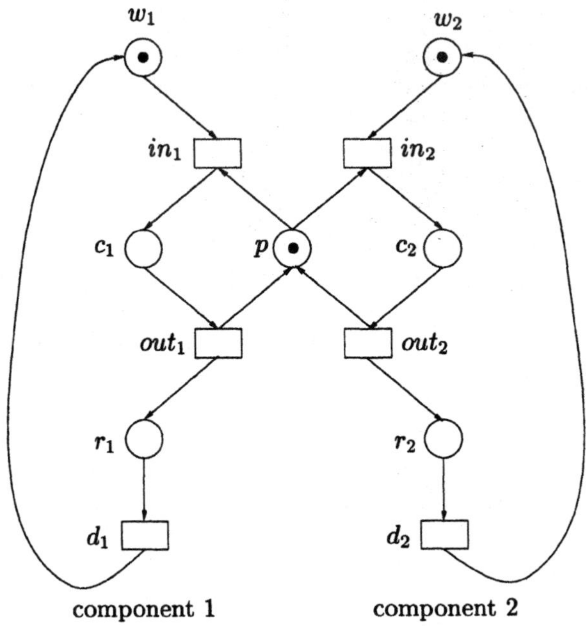

Fig. 2. The mutual exclusion problem.

used), c_1 (the resource is used by component 1), and c_2 (the resource is used by component 1), and can perform the actions in_i and out_i, $i = 1, 2$.

In the global state given in Fig. 2, components 1 and 2 compete for permission. Clearly, only one of the transitions in_1 or in_2 can be executed, permitting either component 1 or component 2 (respectively) to access the resource. When component i has obtained permission and has finished using the resource, it returns the permission to place p through transition out_i. Note that components 1 and 2 never communicate directly, i.e., never perform shared actions. However, they solve their conflict by communicating with the permission component. Note also that the permission component has places in common with the other two components. Thus, place c_1 represents both a local state of component 1 and a local state of the permission component.

3.2 Nets

The main part of an elementary net system is its net, which is defined as follows.

Definition 1. A *net* is a triple $N = (P, T, F)$, where:
(1) P and T are finite sets with $P \cap T = \varnothing$,
(2) $F \subseteq (P \times T) \cup (T \times P)$,
(3) for every $t \in T$ there exist $p, q \in P$ such that $(p, t), (t, q) \in F$, and
(4) for every $t \in T$ and $p, q \in P$, if $(p, t), (t, q) \in F$, then $p \neq q$.

Elements of P are called *places*, elements of T are called *transitions*, elements of $X = P \cup T$ are called *elements* (*of* N), and F is called the *flow relation* (*of* N). We will also use the notations P_N, T_N, X_N, and F_N for P, T, X, and F, respectively. Note that N can be the empty net, i.e., $N = (\varnothing, \varnothing, \varnothing)$.

For each $x \in X$, ${}^\bullet x = \{y \in X \mid (y, x) \in F\}$ is the *input-set* (or *pre-set*) *of* x and $x^\bullet = \{y \in X \mid (x, y) \in F\}$ is the *output-set* (or *post-set*) *of* x; the set ${}^\bullet x \cup x^\bullet$ is called the *neighbourhood* of x, denoted by $\mathbf{nbh}(x)$. Whenever we want to indicate the net under consideration, the notations $({}^\bullet x)_N$, $(x^\bullet)_N$, and $\mathbf{nbh}_N(x)$ will be used. For $Y \subseteq X$, we write ${}^\bullet Y = \bigcup_{x \in Y} {}^\bullet x$, $Y^\bullet = \bigcup_{x \in Y} x^\bullet$, and $\mathbf{nbh}(Y) = {}^\bullet Y \cup Y^\bullet$ and the terminology is carried over correspondingly. Note that the flow relation F is completely determined if ${}^\bullet t$ and $t^\bullet$ are known for every transition $t \in T$.

Conditions (3) and (4) of Definition 1 can now also be formulated as follows. For each transition t: ${}^\bullet t \neq \varnothing$, $t^\bullet \neq \varnothing$, and ${}^\bullet t \cap t^\bullet = \varnothing$. These requirements do not always appear in the literature, but we use them for two reasons. Firstly because they are quite natural, and secondly because they allow us to avoid many unnecessary technicalities. In the sequel, we will now and then indicate where these conditions are used.

It is clear that a net $N = (P, T, F)$ can in fact be seen as a directed graph G_N: the nodes of G_N are the elements of X and there is an edge from x to y iff $(x, y) \in F$. Thus, the reflexive and transitive closure F^* and the transitive closure F^+ indicate the paths in G_N: $(x, y) \in F^*$ iff there is a (possibly empty) directed path from x to y, and $(x, y) \in F^+$ iff there is a nonempty directed path from x to y.

Note that in fact G_N is a bipartite graph since $\{P, T\}$ is a partition of X and an edge leads either from P to T or from T to P. A difference between a net and an arbitrary bipartite graph is that the partition is explicitly given and that an explicit distinction is made between the two sets P and T of the partition, by the order in which they appear in the tuple (P, T, F).

Since nets are graphs, the standard conventions for drawing graphs can be applied to nets. In a drawing of a net, places are represented by circles and transitions by rectangles.

Example 1. Let $N = (P, T, F)$ be the net with $P = \{p_1, p_2, p_3\}$, $T = \{t_1, t_2, t_3\}$, and $F = \{(p_1, t_2), (p_1, t_3), (p_2, t_2), (p_2, t_3), (p_3, t_1), (t_1, p_1), (t_2, p_3), (t_3, p_3)\}$. N is drawn in Fig. 3.

Since every net N corresponds in a natural way to a graph G_N, one can classify nets by "structural", i.e., graph-theoretic properties. In particular, the following notions will be used in the sequel.

Definition 2. A net $N = (P, T, F)$
(1) is *acyclic* if, for every $x \in X$, $(x, x) \notin F^+$,
(2) is *P-simple* if, for all $p, q \in P$, (${}^\bullet p = {}^\bullet q$ and $p^\bullet = q^\bullet$) implies $p = q$,
(3) is *T-simple* if, for all $s, t \in T$, (${}^\bullet s = {}^\bullet t$ and $s^\bullet = t^\bullet$) implies $s = t$, and
(4) has *no isolated places* if, for all $p \in P$, $\mathbf{nbh}(p) \neq \varnothing$.

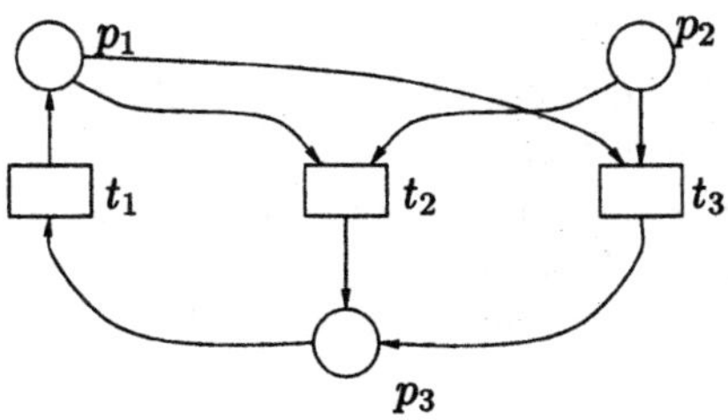

Fig. 3. A net.

Note that, due to condition (3) in Definition 1, transitions are never isolated.

Example 2. The net N from Example 1 (Fig. 3) is cyclic since, e.g., $(p_1, p_1) \in F^+$. N is P-simple, but not T-simple because ${}^\bullet t_2 = {}^\bullet t_3$ and $t_2{}^\bullet = t_3{}^\bullet$. N has no isolated places. The net in Fig. 4 is P-simple and T-simple. It has an isolated place and is cyclic.

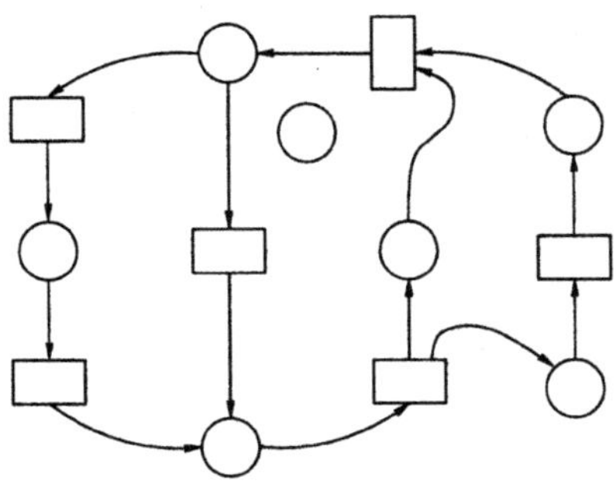

Fig. 4. A net that is P-simple and T-simple.

In considerations concerning the structure of a net, the concrete elements of the net are not important. To express this, we use the following notion of isomorphism. Note that, in the definition, the order of P and T in the tuple (P, T, F) is taken into account; this is the way in which places and transitions are distinguished formally (cf. Example 3(1)).

Definition 3. Two nets $N = (P, T, F)$ and $N' = (P', T', F')$ are *isomorphic*, denoted by $N \equiv N'$, if there exist two bijections $\alpha : P \rightarrow P'$ and $\beta : T \rightarrow T'$, such that for every $p \in P$ and $t \in T$,
$(p, t) \in F$ iff $(\alpha(p), \beta(t)) \in F'$ and
$(t, p) \in F$ iff $(\beta(t), \alpha(p)) \in F'$.

If we want to be more specific, we can say in the above situation that N and N' are (α, β)-*isomorphic*, denoted by $N \equiv_\beta^\alpha N'$. The ordered pair (α, β) is called an *isomorphism between N and N'*.

The conditions for α and β in Definition 3 can also be formulated as follows: for every $t \in T$, ${}^\bullet\beta(t) = \alpha({}^\bullet t)$ and $\beta(t)^\bullet = \alpha(t^\bullet)$, where, clearly, ${}^\bullet\beta(t)$ stands for $({}^\bullet\beta(t))_{N'}$, ${}^\bullet t$ stands for $({}^\bullet t)_N$, and likewise for the output-sets.

Example 3. (1) Let N and N' be the two nets in Figs. 5 and 6, respectively. It is clear that $N \equiv N'$ does not hold; in fact, in N the output-set of every transition is a singleton but in N' there are transitions with an output-set of cardinality 2. However, the graphs G_N and $G_{N'}$ are isomorphic graphs: there is a graph isomorphism γ between G_N and $G_{N'}$, viz. γ defined by $\gamma(t_1) = p_2$, $\gamma(t_2) = p_1$, $\gamma(t_3) = p_3$, $\gamma(p_1) = t_3$, $\gamma(p_2) = t_1$ and $\gamma(p_3) = t_2$). This isomorphism does not preserve the two sorts of the bipartition of the nets because it maps transitions into places, and places into transitions!

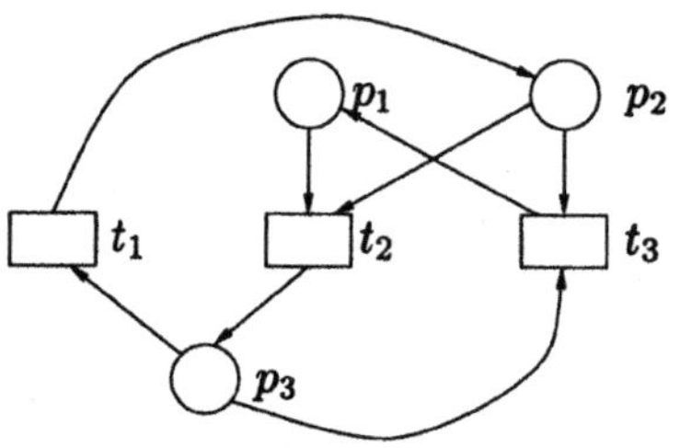

Fig. 5. A net N.

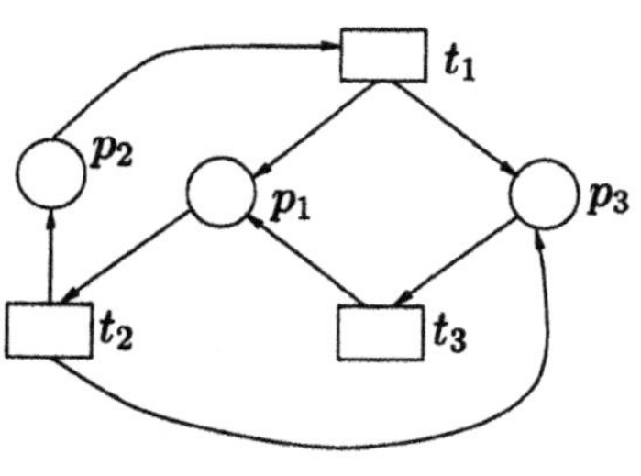

Fig. 6. A net N', not isomorphic with the net N of Fig. 5.

(2) Let N'' be the net in Fig. 7. Let α and β be the bijections from P_N to $P_{N''}$ and from T_N to $T_{N''}$ defined by $\alpha(p_1) = p_8$, $\alpha(p_2) = p_5$, $\alpha(p_3) = p_1$,

$\beta(t_1) = t_5$, $\beta(t_2) = t_3$, and $\beta(t_3) = t_4$. Now N and N'' are (α, β)-isomorphic, and so $N \equiv^{\alpha}_{\beta} N''$. The isomorphism (α, β) does preserve the sorts: it maps transitions into transitions, and places into places.

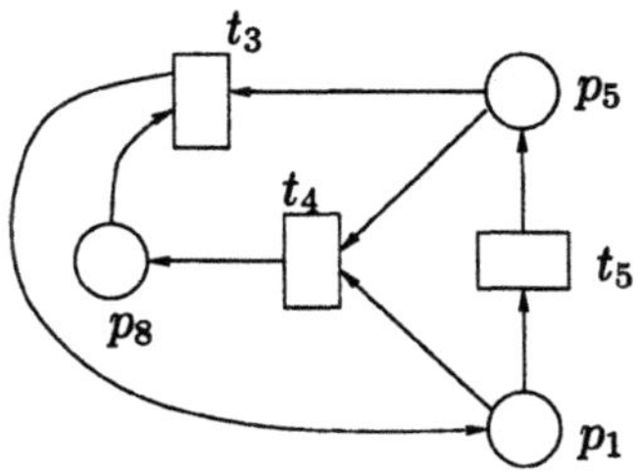

Fig. 7. A net N'', isomorphic with the net N of Fig. 5.

3.3 The Firing of Transitions

A global state of a concurrent system consists of (is a set of) local states, where a local state can often be viewed as the state of a component of the system. A global state transition consists of local transitions. During a local transition a "small" number of local states change, for instance as the result of a communication between some components of the system. Thus, the global states are "distributed", and so are the global state transitions.

When modelling a concurrent system by a net, the local states of the system are represented by the places of the net and its local transitions are represented by the transitions of the net, together with the flow relation. A global state of the system is thus represented by a set of places; such a set of places will be called a configuration.

Definition 4. A *configuration* of a net $N = (P, T, F)$ is a subset of P.

Graphically, a configuration $C \subseteq P$ is represented by placing a "token" (i.e., a fat dot, or a thumbtack when we use transparencies) in every circle corresponding to a place in C. Hence a single token represents a local state of the system, often corresponding to the state of a component of the system. Since we mark the net with tokens, a configuration is also called a *marking* of the net.

We are now ready to define the most fundamental Petri Net model, the "elementary net system". It is a net together with the initial configuration of the system.

Definition 5. An *elementary net system*, EN system for short, is a quadruple $M = (P, T, F, C_{in})$, where:

(1) (P, T, F) is a net and
(2) $C_{in} \subseteq P$ is the *initial configuration.*

For an EN system $M = (P, T, F, C_{in})$, we denote by $\mathbf{und}(M)$, P_M, T_M, F_M, and $(C_{in})_M$ the net (P, T, F), P, T, F, and C_{in}, respectively. Furthermore, we will call $\mathbf{und}(M)$ the *underlying net of M*, the places will also be called *conditions*, and the transitions will also be called *events* (in the literature, when places are called conditions, a configuration is often called a *case*). Finally, the terminology and notations concerning nets carry over to EN systems through their underlying nets. Note that M can be the empty EN system, i.e., $M = (\varnothing, \varnothing, \varnothing, \varnothing)$.

We represent an EN system M graphically by representing $\mathbf{und}(M)$ graphically and marking the initial configuration $(C_{in})_M$ by tokens.

Example 4. Let $N = (P, T, F)$ be the net in Fig. 8. Then $M = (P, T, F, C_{in})$, with $C_{in} = \{p_1, b, c_1\}$, is an EN system with underlying net N. The EN system M is drawn in Fig. 9. Note that M is the producer/consumer system of Fig. 1, with a full buffer.

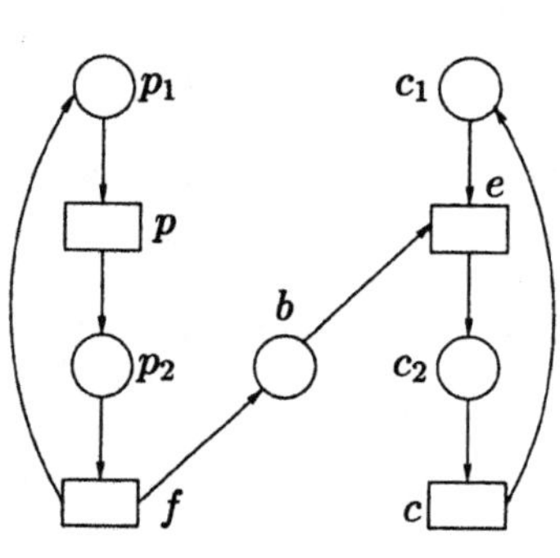

Fig. 8. The underlying net of the EN system of Fig. 9

Until now we have considered nets and EN systems only as static objects. We now turn to the definition of the dynamic behaviour of EN systems. This, however, is not as easy as it is for sequential systems (such as finite-state automata). In fact, it is hard to express in a mathematical, convincing way that certain events, such as the local state transitions of a concurrent system, occur independently, "at the same time". For this reason we will consider several definitions of the behaviour of an EN system, in which the concurrency of events is captured in different ways (without going into philosophical discussions about the concept of time). We start with the simplest case, where we consider how a configuration of an EN system is transformed by the execution of one transition

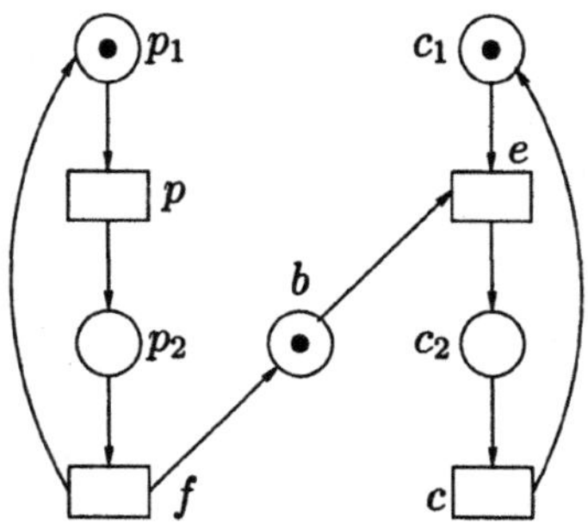

Fig. 9. An EN system.

of the system. Hence, in this case we do not yet consider concurrent events. In other words, we start by defining the sequential behaviour of the EN system.

A local transition of a concurrent system replaces a subset of the local states of a global state by another such subset. In the corresponding EN system, when the input-set of the transition is a subset of the current configuration, execution of the transition consists of replacing the input-set by the output-set of the transition. This is also called a firing of the transition.

Definition 6. Let $M = (P, T, F, C_{in})$ be an EN system and let $t \in T$.

(1) Let $C \subseteq P$ be a configuration. Then t *has concession in* C (or t *can be fired in* C, or t *is enabled in* C) if ${}^\bullet t \subseteq C$ and $t^\bullet \cap C = \varnothing$, written as t **con** C.

(2) Let $C, D \subseteq P$. Then t *fires from* C *to* D if t **con** C and $D = (C - {}^\bullet t) \cup t^\bullet$, written as $C[t\rangle D$; t is also called a *sequential step from* C *to* D.

When it is necessary to indicate the EN system M under consideration, the index M is used: we then write $t\ \mathbf{con}_M\ C$ and $C[t\rangle_M D$ instead of t **con** C and $C[t\rangle D$. In the literature $C[t\rangle$ is often written instead of t **con** C.

Note that a transition t can only be fired if it has both *input-concession* (i.e., ${}^\bullet t \subseteq C$) and *output-concession* (i.e., $t^\bullet \cap C = \varnothing$). Note also that, though we write $C[t\rangle D$, the new configuration D is uniquely determined by C and t, i.e., if $C[t\rangle D_1$ and $C[t\rangle D_2$, then $D_1 = D_2$.

The firing of a transition t is also called the *occurrence* of transition t, or the occurrence of event t. To represent the firing of a transition graphically, we play the so-called "token game" as follows. If, in a given configuration (token marking) C, there exists a transition (rectangle) t such that every circle that corresponds to an element of the input-set of t *contains a token*, and every circle that corresponds to an element of the output-set of t *does not contain a token*, then t can be fired (i.e., t has concession). The firing of t consists of removing a token from every circle that corresponds to an element of the input-set of t and placing a token in every circle that corresponds to an element of the output-set of t, cf. Fig. 10. In this way, the token marking that corresponds to the

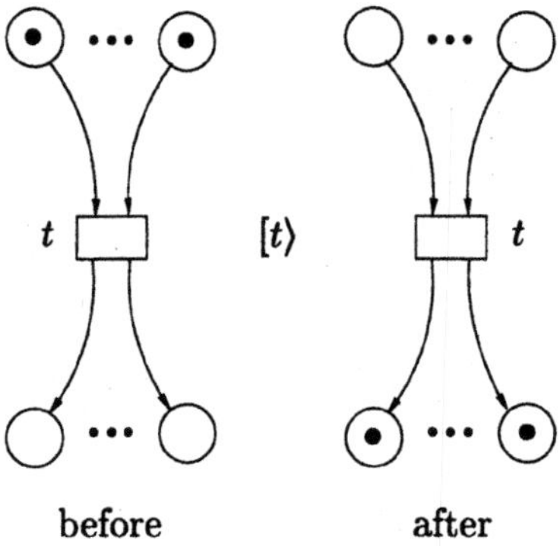

Fig. 10. Firing of transition t.

configuration $(C - {}^\bullet t) \cup t^\bullet$ is obtained. Note that this configuration can also be written as $(C \cup t^\bullet) - {}^\bullet t$, since for arbitrary sets A, B and C the following holds if A and B are disjoint: $(C - A) \cup B = (C \cup B) - A$ (and ${}^\bullet t \cap t^\bullet = \varnothing$ due to Definition 1(4)).

Example 5. Let M be the EN system from Example 4 (Fig. 9). Then transitions p and e have concession in C_{in}, i.e., p **con** C_{in} and e **con** C_{in}; transitions f and c do not have concession in C_{in}. Both $C_{in}[p\rangle\{p_2, b, c_1\}$ and $C_{in}[e\rangle\{p_1, c_2\}$ hold. The configuration $\{p_1, c_2\}$ is drawn in Fig. 11. Also, e.g., $\{p_2, b, c_1\}[e\rangle\{p_2, c_2\}$, $\{p_2, c_2\}[f\rangle\{p_1, b, c_2\}$, and $\{p_1, b, c_2\}[c\rangle C_{in}$ hold.

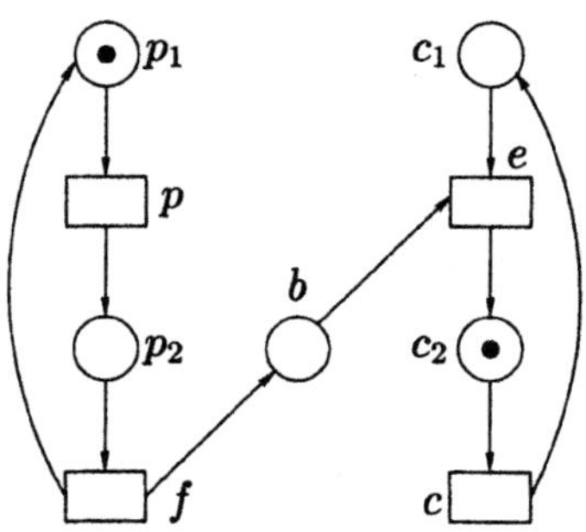

Fig. 11. The EN system of Fig. 9 after firing transition e.

We can also view the occurrence of event t as a *change of* the values of certain *conditions* (= places). The idea behind the definition of concession (Definition 6) is that the result of the occurrence of an event must be observable in each component of the system that is involved in this event. Hence the event t can only occur if all pre-conditions hold (i.e., ${}^\bullet t \subseteq C$) and all post-conditions do not hold (i.e., $t^\bullet \cap C = \varnothing$). After the event has occurred, the post-conditions hold and the pre-conditions no longer hold. Hence, viewing places as booleans, the occurrence of t can be intuitively described by the execution of the following conditional statement, in which all booleans in ${}^\bullet t = \{p_1, \ldots, p_m\}$ and $t^\bullet = \{q_1, \ldots, q_n\}$ change their value:

if p_1 **and** $\cdots$ **and** p_m **and** (**not** q_1) **and** $\cdots$ **and** (**not** q_n)
then begin $p_1 :=$ false ;...; $p_m :=$ false ;
$q_1 :=$ true ;...; $q_n :=$ true
end.

It must be observed that the execution of this statement is an atomic, indivisible, action and that the execution only takes place if the boolean condition (between the **if** and the **then**) is satisfied. Also note that the order of the assignments is irrelevant.

The next result shows that there is an alternative, symmetric, way of defining the firing of a transition.

Lemma 7. *Let $M = (P, T, F, C_{in})$ be an EN system. Let $t \in T$ and let $C, D \subseteq P$. Then $C[t\rangle D$ iff $C - D = {}^\bullet t$ and $D - C = t^\bullet$.*

Proof. For arbitrary sets A, B, C, D, the following two statements are equivalent: (1) $A \subseteq C$, $B \cap C = \varnothing$, and $D = (C - A) \cup B$, (2) $C - D = A$ and $D - C = B$. □

Let $M = (P, T, F, C_{in})$ be an EN system. Assume that an event t takes place in a configuration C and leads to configuration D ($C[t\rangle D$ in our notation). Then the amount of change of the configuration that is caused by t, can be given by the pair $(C - D, D - C)$: the conditions in $C - D$ stop being valid and the conditions in $D - C$ start being valid. From Lemma 7 now follows that the amount of change caused by an occurrence of an event is determined only by the event itself, i.e., it is independent of the configuration in which it takes place. Thus we have:

$(\forall t \in T)(\forall C_1, D_1, C_2, D_2 \subseteq P)$: if $C_1[t\rangle D_1$ and $C_2[t\rangle D_2$, then
$C_1 - D_1 = C_2 - D_2$ and $D_1 - C_1 = D_2 - C_2$.

This holds because, according to Lemma 7, $C_1 - D_1 = {}^\bullet t = C_2 - D_2$ and $D_1 - C_1 = t^\bullet = D_2 - C_2$. Thus, each event t determines unambiguously the amount of change it causes (when it occurs); this amount of change is given by its *characteristic pair* $\mathbf{cp}(t) = ({}^\bullet t, t^\bullet)$. However, there can be two distinct events t_1 and t_2 that cause the same amount of change, i.e., they have the same characteristic pair: ${}^\bullet t_1 = {}^\bullet t_2$ and ${t_1}^\bullet = {t_2}^\bullet$. This cannot happen in a T-simple EN system, because in such a system a transition t is completely determined by its characteristic pair. Consequently, a T-simple EN system satisfies the following *principle of extensionality*:

$(\forall t_1, t_2 \in T)(\forall C_1, D_1, C_2, D_2 \subseteq P):$ if $C_1[t_1\rangle D_1$ and $C_2[t_2\rangle D_2$, then $t_1 = t_2$ iff $C_1 - D_1 = C_2 - D_2$ and $D_1 - C_1 = D_2 - C_2$.

This means that, in a T-simple EN system, distinct transitions cause different amounts of change in all configurations.

For a given EN system we are only interested in those configurations that can be reached from the initial configuration by the repeated firing of transitions, and in the (sequences of) transitions that lead to these configurations. Knowing the effect of firing a single transition allows us to define the effect of several transitions that fire one after the other. Such a sequence of transitions can also be considered as an observation of the system by a sequential observer (i.e., an observer that can observe only one event at a time). Formally, a sequence of transitions is a word over the alphabet T, i.e., an element of T^*.

Definition 8. Let $M = (P, T, F, C_{in})$ be an EN system.

(1) Let $t_1 \cdots t_n \in T^*$, with $n \geq 0$ and $t_1, \ldots, t_n \in T$. Let $C, D \subseteq P$. Then $t_1 \cdots t_n$ *fires from* C *to* D if there exist configurations $C_0, C_1, \ldots, C_n \subseteq P$ with $C_0 = C$, $C_n = D$ and $C_{i-1}[t_i\rangle C_i$ for all $1 \leq i \leq n$, written as $C[t_1 \cdots t_n\rangle D$.

(2) Let $x \in T^*$ and $C \subseteq P$. Then x *has concession in* C (or x *can be fired in* C, or x *is enabled in* C) if there exists a $D \subseteq P$ such that $C[x\rangle D$, written as x **con** C.

(3) $x \in T^*$ is a *firing sequence of* M if x **con** C_{in}. The set of all firing sequences of M is denoted by $\mathrm{FS}(M)$.

(4) $C \subseteq P$ is a *reachable configuration of* M if there exists an $x \in \mathrm{FS}(M)$ with $C_{in}[x\rangle C$. The set of all reachable configurations of M is denoted by $\mathbb{C}_M$.

(5) $t \in T$ is a *useful transition of* M if there exists a reachable configuration C of M such that t **con** C. The set of useful transitions of M is denoted by $\mathbf{use}_M(T)$, or just $\mathbf{use}(T)$ when M is clear from the context.

(6) $t \in T$ is a *live transition of* M if for each $C \in \mathbb{C}_M$ there exists an $x \in T^*$ with xt **con** C.

Note that a configuration C is reachable (i.e., $C \in \mathbb{C}_M$) if there exist transitions $t_1, \ldots, t_n \in T$ (with $n \geq 0$) and configurations $C_1, \ldots, C_n$ such that $C_n = C$ and $C_{in}[t_1\rangle C_1[t_2\rangle C_2 \cdots [t_n\rangle C_n$. Hence, $\mathbb{C}_M$ can also be defined as the smallest set of configurations for which the following holds: (1) $C_{in} \in \mathbb{C}_M$ and (2) if $C \in \mathbb{C}_M$ and $C[t\rangle D$ for some $t \in T$, then $D \in \mathbb{C}_M$. Properties of reachable configurations can thus be proved by induction in the following way. Let $P(C)$ be a property of (reachable) configurations C of M. Assume that it has been proved that (1) $P(C_{in})$ and (2) if $C[t\rangle D$ and $P(C)$, then $P(D)$. Then $P(C)$ holds for all $C \in \mathbb{C}_M$. We call this a *proof by induction on* C of the statement $\forall C \in \mathbb{C}_M : P(C)$.

Intuitively, a transition is useful when it can eventually be fired starting from the initial configuration, and it is live when it can eventually be fired starting from any reachable configuration. Thus every live transition is useful.

Example 6. Let $M = (P, T, F, C_{in})$ be the EN system of Example 4 (Fig. 9). From the discussion in Example 5 it follows that the sequence of transitions

$pef \in T^*$ fires from $\{p_1, b, c_1\}$ to $\{p_1, b, c_2\}$, and so $\{p_1, b, c_1\}[pef\rangle\{p_1, b, c_2\}$. Consequently, pef has concession in $C_{in} = \{p_1, b, c_1\}$, written as pef **con** $\{p_1, b, c_1\}$, and pef is a firing sequence of M, written as $pef \in \text{FS}(M)$. The configurations $\{p_1, b, c_1\}$, $\{p_2, b, c_1\}$, $\{p_2, c_2\}$, and $\{p_1, b, c_2\}$ that are "encountered" during the firing of pef are thus reachable configurations of M, written as $\{p_1, b, c_1\}$, $\{p_2, b, c_1\}$, $\{p_2, c_2\}$, $\{p_1, b, c_2\} \in \mathbb{C}_M$. Since c **con** $\{p_1, b, c_2\}$, all transitions of M are useful. In the comments following Example 7 we observe that all transitions of M are even live. The transitions of the EN system of Fig. 12 are all useful, but none is live: from the (reachable) configuration $\{p_6\}$ no other configuration can be reached, and so no transition can be fired anymore.

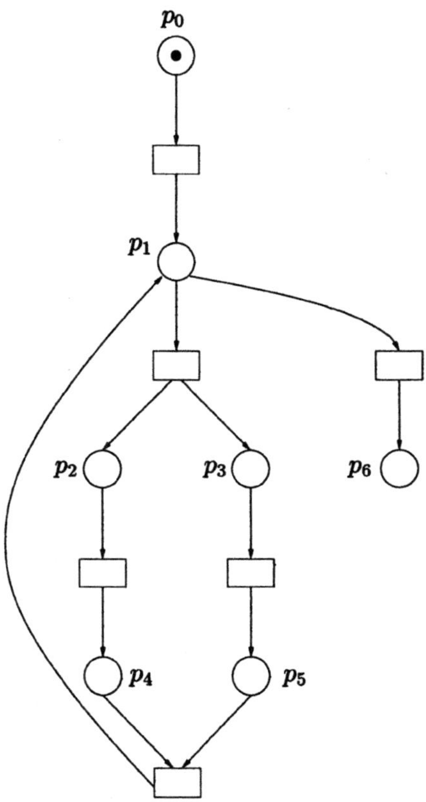

Fig. 12. An EN system of which all transitions are useful, but none is live.

When analyzing the (sequential) behaviour of an EN system M, its (sequential) configuration graph is often useful. It directly represents the way in which the set $\mathbb{C}_M$ is constructed from C_{in} by firing transitions.

Before presenting its formal definition, we first recall the notion of a (directed) edge-labelled graph. Here we want such a graph to be "initialized", i.e., to have an "initial node". Also, we allow an edge to have a set of labels rather than a single label, because this will be needed when we consider concurrent configuration graphs. In the literature, initialized edge-labelled graphs are also known as *transition systems.*

Definition 9. An (initialized) *edge-labelled graph* is a quadruple $(V, \Gamma, \Sigma, v_{in})$, where V is a finite set of *nodes*, v_{in} is the *initial node*, Σ is a finite set of *(edge-) labels*, and $\Gamma \subseteq V \times \mathcal{P}(\Sigma) \times V$ is a set of *(labelled) edges.*

For an edge $e = (v, U, w) \in \Gamma$ of such a graph $G = (V, \Gamma, \Sigma, v_{in})$, $U \subseteq \Sigma$ is the *set of labels of* e. If U is a singleton, $U = \{\sigma\}$, then we also write the edge e as (v, σ, w) rather than $(v, \{\sigma\}, w)$.

Isomorphism of edge-labelled graphs is defined in the following way. Note that this notion of isomorphism also allows the labels to change.

Definition 10. Let $G_1 = (V_1, \Gamma_1, \Sigma_1, v_1)$ and $G_2 = (V_2, \Gamma_2, \Sigma_2, v_2)$ be two edge-labelled graphs. Then G_1 and G_2 are *isomorphic*, denoted by $G_1 \equiv G_2$, if there exist two bijections $\alpha : V_1 \to V_2$ and $\beta : \Sigma_1 \to \Sigma_2$ such that $\alpha(v_1) = v_2$ and, for all $v, w \in V_1$ and all $U \subseteq \Sigma_1$, $(v, U, w) \in \Gamma_1$ iff $(\alpha(v), \beta(U), \alpha(w)) \in \Gamma_2$.

Here, as usual, $\beta(U) = \{\beta(\sigma) \mid \sigma \in U\}$. If we want to be more specific, then we say that G_1 and G_2 are *(α, β)-isomorphic*, written as $G_1 \equiv^{\alpha}_{\beta} G_2$.

The (sequential) configuration graph of an EN system is now formally defined as follows; here each edge has exactly one label, i.e., $\Gamma \subseteq V \times \Sigma \times V$. In the literature it is often called the sequential *case graph*, or the *transition system* of the EN system.

Definition 11. Let M be an EN system. The *sequential configuration graph of* M, denoted by SCG(M), is the edge-labelled graph $(V, \Gamma, \Sigma, v_{in})$, where $V = \mathbb{C}_M$, $v_{in} = (C_{in})_M$, $\Sigma = \mathbf{use}(T_M)$, and $\Gamma = \{(C, t, D) \mid C, D \in \mathbb{C}_M,\ t \in T_M,\ C[t\rangle_M D\}$.

Example 7. For the EN system M from Example 4 (Fig. 9), SCG(M) is given in Fig. 13. The "wriggly" arrow indicates its initial node (i.e., the initial configuration of M). To simplify notation, we use $a_1 a_2 \cdots a_n$ for the set $\{a_1, a_2, \ldots, a_n\}$.

Note that the useful transitions of an EN system M are precisely the labels that actually occur on the edges of the sequential configuration graph of M. Also the liveness of a transition t of an EN system M can easily be decided by analyzing the sequential configuration graph of M: from each node of SCG(M) there must exist a path to a node with an outgoing edge labelled by t. Thus, for the EN system M of Fig. 9, SCG(M) of Fig. 13 shows that all transitions of M are live.

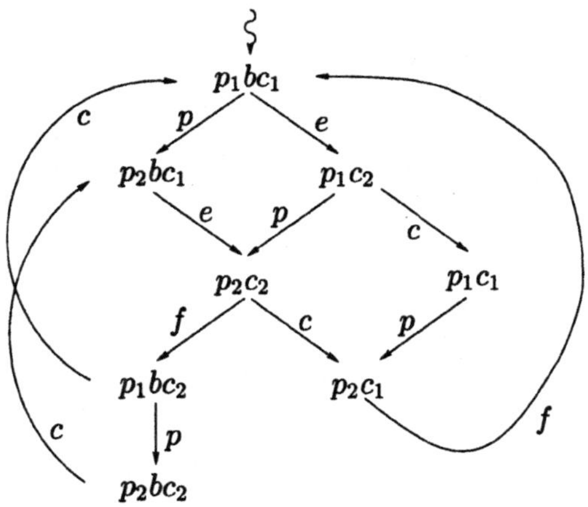

Fig. 13. A sequential configuration graph.

The class of transition sytems corresponding to EN systems (i.e., the class of all sequential configuration graphs SCG(M)) is characterized and studied in [EhrRoz90, NieRozThi92a, NieRozThi92b, NieRozThi95].

The sequential configuration graph SCG(M) can be considered as the (sequential) "state space" of the EN system M. Since M has only finitely many configurations, its state space is finite. Thus, an elementary net system M is a model of a finite-state concurrent system, and SCG(M) models its sequential behaviour. Since finite automata model finite-state sequential systems, there is a clear relationship between elementary net systems and finite automata. In fact, to the reader familiar with formal language theory (see, e.g., [HopUll79]) it should be clear that the sequential configuration graph SCG(M) can be seen as a finite automaton with initial state C_{in} (and all states being final states). A firing sequence of M is a word in T_M^* that forms the concatenation of the labels of a path in SCG(M) that starts in C_{in}. Hence, the firing sequences of M are precisely the words that are accepted by the finite automaton SCG(M). This gives the following result.

Theorem 12. *For every EN system M, FS(M) is a regular language.*

Proof. Let $M = (P, T, F, C_{in})$. Consider the (deterministic) finite automaton $\mathcal{A}$ with input alphabet $\mathbf{use}(T)$, set of states $\mathbb{C}_M$, initial state C_{in}, set of final states $\mathbb{C}_M$ (thus each state is a final state), and set of state transitions $\{(C, t, D) \mid C[t\rangle_M D\}$. Then FS($M$) is the language accepted by $\mathcal{A}$. □

Note that the language FS(M) is prefix-closed, i.e., if a word x is contained in FS(M), then also each prefix of x is contained in FS(M). This follows directly from the definition of a firing sequence (Definition 8) and explains why all states of the automaton $\mathcal{A}$ are taken as final states in the above proof. Hence not every regular language is the set of firing sequences of an EN system (e.g., the language $\{ab\}$, with $a, b \in \Sigma$, is regular but not prefix-closed). Also, not every prefix-closed regular language is the set of firing sequences of an EN system; e.g., $\{\lambda, t, tt\}$ with $t \in \Sigma$ is such a language, because no transition can fire twice consecutively (if $C[t\rangle D$, then ${}^\bullet t \cap D = \varnothing$, and hence t does not have concession in D).

3.4 Concurrency

Intuitively it is not so difficult to get an idea of a "run" of an EN system in which transitions can fire concurrently, i.e., independently of each other. In fact, in everyday life, people can be viewed as communicating concurrent components of a large system. However, to formalize this intuition is not so easy. We will now give a first attempt to capture the concurrent firing of transitions in a formal definition. The second, more sophisticated, attempt is presented in Section 5. In this first attempt, we still view the behaviour of an EN system in terms of global state transitions that are made in a step-wise fashion. However, as opposed to Definition 6(2), we now allow several transitions to fire in one such step.

Intuitively, two transitions such that each of them has concession in a given configuration, can be fired concurrently provided they are disjoint, i.e., have no common places. Also, it is intuitively clear how the configuration is transformed by firing both transitions. We will now formalize these intuitions; we do this right away for an arbitrary number of transitions (rather than two).

Definition 13. Let $M = (P, T, F, C_{in})$ be an EN system.

(1) Let $U \subseteq T$. U is a *disjoint set of transitions* if $U \neq \varnothing$ and for every two distinct transitions $t_1, t_2 \in U$: $\mathbf{nbh}(t_1) \cap \mathbf{nbh}(t_2) = \varnothing$; this is denoted by $\mathbf{disj}(U)$.

(2) Let $U \subseteq T$ and let $C \subseteq P$. Then U *has concession in* C (or U *can be fired in* C, or U *is enabled in* C) if $\mathbf{disj}(U)$, ${}^\bullet U \subseteq C$, and $U^\bullet \cap C = \varnothing$; this is denoted by U **con** C.

(3) Let $U \subseteq T$ and let $C, D \subseteq P$. Then U *fires from* C *to* D, written as $C[U\rangle D$, if U **con** C and $D = (C - {}^\bullet U) \cup U^\bullet$. If $\#U \geq 2$, then we also say that U is a *concurrent step from* C *to* D.

The firing of a disjoint set of transitions U can be seen as a global state transition of the system M, consisting of local state transitions (namely the transitions in U). Thus, states and state transitions are now treated in a similar way: global states (configurations) are sets of local states (places) and global state transitions (concurrent steps) are sets of local state transitions (sequential steps).

Analogous to Lemma 7 we obtain the following lemma.

Lemma 14. *Let* $M = (P, T, F, C_{in})$ *be an EN system. Let* $U \subseteq T$ *and let* $C, D \subseteq P$. *Then* $C[U\rangle D$ *holds iff* $\mathbf{disj}(U)$, $C - D = {}^\bullet U$, *and* $D - C = U^\bullet$.

The amount of change caused by a set of transitions is thus cumulative, i.e., it equals the sum of the changes caused by the separate transitions. As in the case of single transitions this means that the amount of change caused by the occurrence of several disjoint events is determined only by these events themselves, not by the configuration in which they occur. However, in a T-simple EN system the principle of extensionality does not necessarily hold for concurrent steps. For example, it is easy to construct an EN system with three transitions t_1, t_2, t_3 for which ${}^\bullet t_3 = {}^\bullet\{t_1, t_2\}$ and $t_3{}^\bullet = \{t_1, t_2\}^\bullet$.

Example 8. Let M be once again the EN system of Example 4 (Fig. 9). Then $\{p, e\}$ is a disjoint set of transitions (thus $\mathbf{disj}(\{p, e\})$), but $\{f, e\}$ is not. The only disjoint sets U with $\#U \geq 2$ are $\{p, e\}$, $\{p, c\}$ and $\{f, c\}$. The set $\{p, e\}$ has concession in C_{in} (thus $\{p, e\}$ **con** $\{p_1, b, c_1\}$), and $\{p_1, b, c_1\}[\{p, e\}\rangle\{p_2, c_2\}$.

The concession of a set U of transitions (in a given configuration) can be expressed through the concessions of the transitions in U, together with a simplified disjointness condition, as follows.

Lemma 15. *Let $M = (P, T, F, C_{in})$ be an EN system. Let $C \subseteq P$ and let $U \subseteq T$ with $U \neq \emptyset$. Then U* **con** *C iff*
(1) t **con** *C for all $t \in U$, and*
(2) for all $t_1, t_2 \in U$ with $t_1 \neq t_2$, ${}^\bullet t_1 \cap {}^\bullet t_2 = \emptyset$ and $t_1{}^\bullet \cap t_2{}^\bullet = \emptyset$.

Proof. Obviously, ${}^\bullet U \subseteq C$ and $U^\bullet \cap C = \emptyset$ if and only if t **con** C for all $t \in U$. Moreover, if t_1 **con** C and t_2 **con** C, then ${}^\bullet t_1 \cap t_2{}^\bullet = \emptyset$ and $t_1{}^\bullet \cap {}^\bullet t_2 = \emptyset$. □

For a concurrent step U from configuration C to configuration D, each splitting of U into nonempty sets U_1 and U_2 yields two (concurrent or sequential) steps U_1, U_2 which, when executed sequentially in arbitrary order (first U_1 then U_2, or first U_2 then U_1), lead to the same configuration, i.e., fire from C to D. This is formally expressed as follows.

Lemma 16. *Let $M = (P, T, F, C_{in})$ be an EN system, let $C, D \subseteq P$, and let $U \subseteq T$. Let $\{U_1, U_2\}$ be a partition of U, i.e., $U = U_1 \cup U_2$, $U_1 \cap U_2 = \emptyset$ and $U_1, U_2 \neq \emptyset$. If $C[U\rangle D$, then there exists a configuration $E \subseteq P$ such that $C[U_1\rangle E$ and $E[U_2\rangle D$.*

Proof. It is intuitively clear that E is the configuration that is obtained by firing U_1, i.e., $E = (C - {}^\bullet U_1) \cup U_1{}^\bullet$. The formal proof is as follows.

To begin with, $\mathbf{disj}(U_1)$ and $\mathbf{disj}(U_2)$ follow from $\mathbf{disj}(U)$. From Lemma 14 it follows that $C \cap D$, ${}^\bullet U$, and $U^\bullet$ are mutually disjoint sets with $C = (C \cap D) \cup {}^\bullet U$ and $D = (C \cap D) \cup U^\bullet$. From $\mathbf{disj}(U)$ and the fact that $\{U_1, U_2\}$ is a partition of U it then follows that $C \cap D$, ${}^\bullet U_1$, ${}^\bullet U_2$, $U_1{}^\bullet$, and $U_2{}^\bullet$ are mutually disjoint sets, with ${}^\bullet U = {}^\bullet U_1 \cup {}^\bullet U_2$ and $U^\bullet = U_1{}^\bullet \cup U_2{}^\bullet$. Hence $C = (C \cap D) \cup {}^\bullet U_1 \cup {}^\bullet U_2$ and $D = (C \cap D) \cup U_1{}^\bullet \cup U_2{}^\bullet$. Now consider the configuration $E = (C \cap D) \cup U_1{}^\bullet \cup {}^\bullet U_2$ (i.e., $E = (C - {}^\bullet U_1) \cup U_1{}^\bullet$). Then $C - E = {}^\bullet U_1$ and $E - C = U_1{}^\bullet$, and thus $C[U_1\rangle E$ according to Lemma 14. Likewise $E - D = {}^\bullet U_2$ and $D - E = U_2{}^\bullet$, and thus $E[U_2\rangle D$ according to Lemma 14. □

This lemma expresses a so-called *diamond property*: if $C[U\rangle D$ and $\{U_1, U_2\}$ is a partition of U, then there exist two configurations E_1 and E_2 such that $C[U_1\rangle E_1[U_2\rangle D$ and $C[U_2\rangle E_2[U_1\rangle D$. A drawing (in Fig. 14) of the four steps $C[U_1\rangle E_1$, $E_1[U_2\rangle D$, $C[U_2\rangle E_2$, and $E_2[U_1\rangle D$ gives a "diamond"; the step $C[U\rangle D$ is then a diagonal of this diamond. More general diamond properties are studied in [HooRoz91].

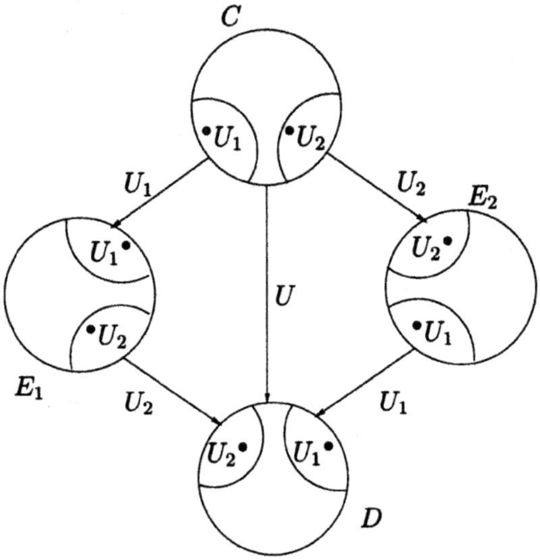

Fig. 14. A diamond.

It follows directly from Lemma 16 that a concurrent step can be realized by the firing of its elements in an arbitrary order. Such a realization of a concurrent step intuitively corresponds to a possible way in which a sequential observer sees the step as a sequence of sequential steps. This is expressed by the following lemma.

Lemma 17. *Let* $M = (P, T, F, C_{in})$ *be an EN system, let* $C, D \subseteq P$ *and let* $U \subseteq T$. *If* $C[U\rangle D$, *then* $C[t_1 \cdots t_n\rangle D$ *for each ordering* $(t_1, \ldots, t_n)$ *of the elements of* U.

Proof. By induction on $\#U$. The induction step follows directly from Lemma 16. □

This thus means that by allowing concurrent steps no new reachable configurations are obtained (and no new useful transitions). Adding all steps (C, U, D) with $\#U \geq 2$ to the sequential configuration graph leads only to new edges labelled by sets of transitions.

Definition 18. Let M be an EN system. The *configuration graph of* M, denoted by $\mathrm{CG}(M)$, is the edge-labelled graph $(V, \Gamma, \Sigma, v_{in})$, where $V = \mathbb{C}_M$, $v_{in} = (C_{in})_M$, $\Sigma = \mathbf{use}(T_M)$, and $\Gamma = \{(C, U, D) \mid C, D \in \mathbb{C}_M,\ U \subseteq T_M,\ C[U\rangle_M D\}$.

Note that if $C[U\rangle_M D$ and $C \in \mathbb{C}_M$, then $U \subseteq \mathbf{use}(T_M)$, according to Lemma 15.

Example 9. Let M be the EN system from Example 4 (Fig. 9). The configuration graph $\mathrm{CG}(M)$ is drawn in Fig. 15. Compare $\mathrm{CG}(M)$ with the sequential configuration graph $\mathrm{SCG}(M)$ of Fig. 13. Also note that $\mathrm{CG}(M)$ contains several "diamonds" (see Fig. 14) such as, e.g., the diamond at the concurrent step $\{p_1, b, c_1\}[\{p, e\}\rangle\{p_2, c_2\}$.

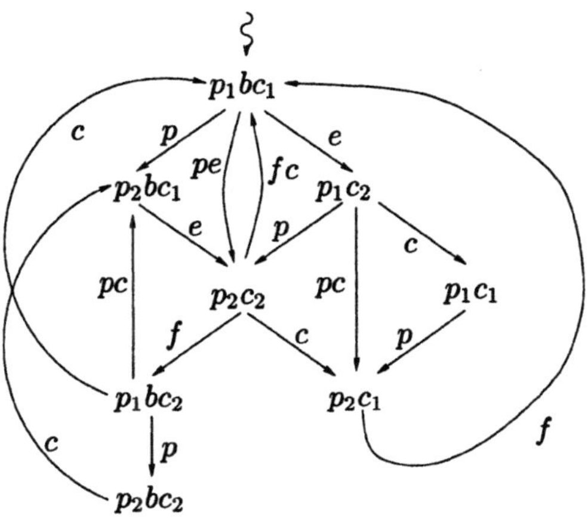

Fig. 15. A configuration graph.

Rather surprisingly, Lemma 17 also holds the other way around. To prove this (in Theorem 20) we use the following diamond property, which shows that a diamond exists whenever three of its sides are given, cf. [HooRoz91].

Lemma 19. *Let* $M = (P, T, F, C_{in})$ *be an EN system, let* $C \subseteq P$ *and let* $s, t \in T$. *If* st **con** C *and* t **con** C, *then* $\{s, t\}$ **con** C.

Proof. Since st **con** C, s **con** C certainly holds. Since t **con** C also holds, it suffices according to Lemma 15 to show that ${}^\bullet s \cap {}^\bullet t = \varnothing$ and $s^\bullet \cap t^\bullet = \varnothing$. Let $C[s\rangle D$. Then ${}^\bullet s \cap D = \varnothing$ and $s^\bullet \subseteq D$. Since t **con** D, ${}^\bullet t \subseteq D$ and $t^\bullet \cap D = \varnothing$. Hence ${}^\bullet s \cap {}^\bullet t = \varnothing$ and $s^\bullet \cap t^\bullet = \varnothing$. □

Now we prove the so-called *sequentialization property.*

Theorem 20. *Let $M = (P, T, F, C_{in})$ be an EN system, let $C, D \subseteq P$ and let $U \subseteq T$ with $U \neq \varnothing$. Then*

*(1) U **con** C iff $t_1 \cdots t_n$ **con** C for every ordering $(t_1, \ldots, t_n)$ of the elements of U, and*

(2) $C[U\rangle D$ iff $C[t_1 \cdots t_n\rangle D$ for every ordering $(t_1, \ldots, t_n)$ of the elements of U.

Proof. (Only-if) The only-if-part of (2) is Lemma 17. From this the only-if-part of (1) directly follows.

(If) We first prove the if-part of (1). For each $t \in U$ there exists an ordering $(t_1, \ldots, t_n)$ of U with $t_1 = t$. This implies that t **con** C for all $t \in U$. What remains to be proved (see Lemma 15), is $\mathbf{disj}(\{s, t\})$ for every two distinct elements s and t of U. To this aim, we use Lemma 19. We already know that t **con** C. If we now consider an ordering of U of the form $(s, t, t_3, \ldots, t_n)$, then, by assumption, $stt_3 \cdots t_n$ **con** C. Hence st **con** C, and so, by Lemma 19, $\{s, t\}$ **con** C. Thus $\mathbf{disj}(\{s, t\})$.

The if-part of (2) can now easily be deduced from the if-part of (1) and the only-if-part of (2). □

According to this theorem, the concession and the effect of a concurrent step are completely determined by the concession and the effect of sequences of sequential steps. Hence we can use the properties of firing sequences of transitions when reasoning about concurrent steps. In particular, it follows from Theorem 20 that we can construct the configuration graph CG(M) of an EN system M from its sequential configuration graph SCG(M), even if we do not know M itself (see Section 4 of [HooRoz91]). Thus, the sequential configuration graph contains already all the information about concurrency! This is formally expressed in the next theorem.

Theorem 21. *For EN systems M and M',*
SCG(M) $\equiv$ SCG(M') iff CG(M) $\equiv$ CG(M').

Proof. (If) It follows directly from the definitions that if CG(M) $\equiv^{\alpha}_{\beta}$ CG(M'), then SCG(M) $\equiv^{\alpha}_{\beta}$ SCG(M').

(Only-if) Assume that SCG(M) $\equiv^{\alpha}_{\beta}$ SCG(M'). This means that for all $C, D \in \mathbb{C}_M$ and $t \in \mathbf{use}(T_M)$, $C[t\rangle_M D$ iff $\alpha(C)[\beta(t)\rangle_{M'} \alpha(D)$. It is now easy to prove, by induction on $|x|$, that for all $C, D \in \mathbb{C}_M$ and $x \in \mathbf{use}(T_M)^*$, $C[x\rangle_M D$ iff $\alpha(C)[\beta(x)\rangle_{M'} \alpha(D)$; note that, for $x = t_1 \cdots t_n$, $\beta(x) = \beta(t_1) \cdots \beta(t_n)$ according to Section 2. From this and Theorem 20 it follows that for all $C, D \in \mathbb{C}_M$ and $U \subseteq \mathbf{use}(T_M)$, $C[U\rangle_M D$ iff $\alpha(C)[\beta(U)\rangle_{M'} \alpha(D)$. Thus CG($M$) $\equiv^{\alpha}_{\beta}$ CG(M'). □

It is shown in [HooRoz91] that this result (and in particular its If direction) is still true if the isomorphism between the configuration graphs disregards the fact that the edges are labelled by sets, viewing them as abstract symbols (i.e., if in Definition 10, β is taken as a partial injective function from $\mathcal{P}(\Sigma_1)$ to $\mathcal{P}(\Sigma_2)$).

3.5 Fundamental Situations

An attractive feature of Petri Net models is that important notions concerning concurrent systems can be formulated in terms of, e.g., EN systems in a very natural way. To illustrate this aspect of EN systems, we discuss several fundamental situations that may occur in the dynamic behaviour of (concurrent systems that can be modelled by) EN systems. In what follows, we assume that an EN system $M = (P, T, F, C_{in})$ is given.

There are three fundamental relationships that may hold between two events t_1 and t_2 in a given configuration C: causality, concurrency, and conflict.

(1) Causality (of events t_1 and t_2 in configuration C).

This notion is illustrated in Figs. 16 and 17: t_2 **con** C does not hold, but $t_1 t_2$ **con** C does hold. Thus, t_1 needs to occur to grant concession to t_2 (input-concession in Fig. 16, and output-concession in Fig. 17). In other words, t_1 is one of the causes of t_2. An equivalent formal definition is: $t_1 t_2$ **con** C holds, but $t_1{}^{\bullet} \cap {}^{\bullet}t_2 \neq \varnothing$ or ${}^{\bullet}t_1 \cap t_2{}^{\bullet} \neq \varnothing$.

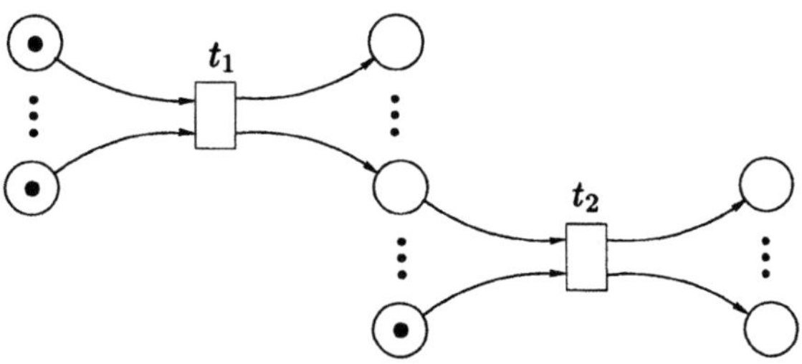

Fig. 16. Causality: $t_1{}^{\bullet} \cap {}^{\bullet}t_2 \neq \varnothing$.

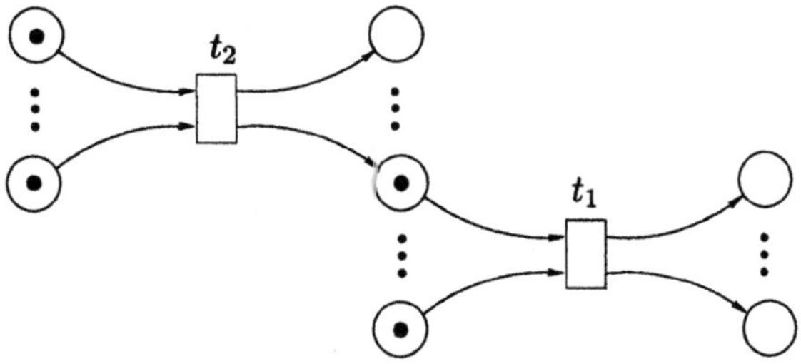

Fig. 17. Causality: ${}^{\bullet}t_1 \cap t_2{}^{\bullet} \neq \varnothing$.

(2) Concurrency (of events t_1 and t_2 in configuration C).

This notion is illustrated in Fig. 18: $\{t_1, t_2\}$ **con** C. Figure 19 shows a more complete picture of the situation: it gives a representation of C and $P - C$, and it shows how $^\bullet t_1, {}^\bullet t_2$ and $t_1{}^\bullet, t_2{}^\bullet$ fit into C and $P - C$, respectively.

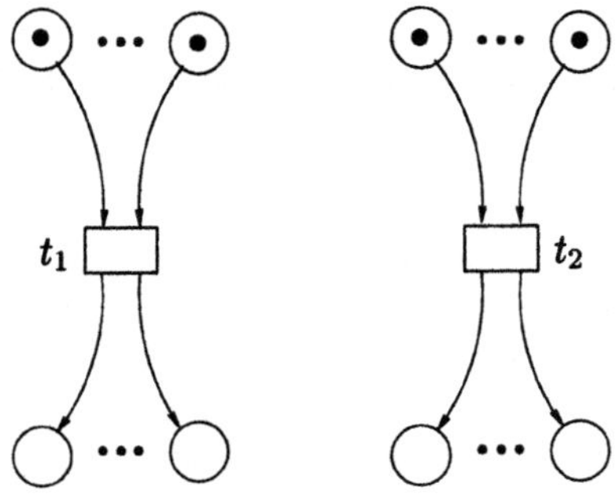

Fig. 18. Concurrency.

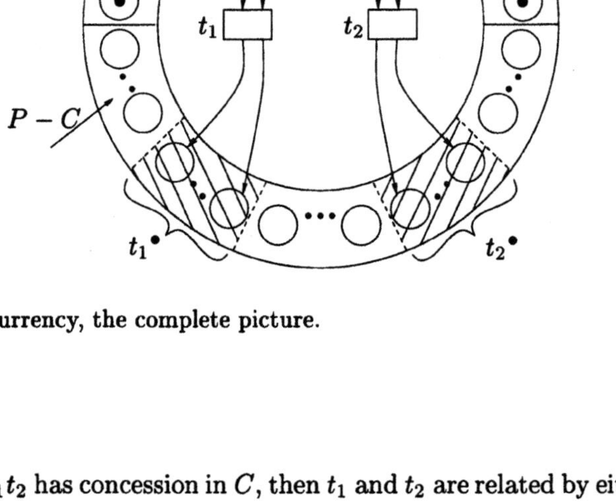

Fig. 19. Concurrency, the complete picture.

Hence if $t_1 t_2$ has concession in C, then t_1 and t_2 are related by either causality or concurrency: if t_2 **con** C does not hold, then there is causality, and if t_2 **con** C does hold, then (by Lemma 19) there is concurrency.

(3) Conflict (between events t_1 and t_2 in configuration C).

This notion is illustrated in Figs. 20 and 21: both t_1 **con** C and t_2 **con** C hold, but $\{t_1, t_2\}$ **con** C does not hold. Consequently, by Lemma 15, ${}^\bullet t_1 \cap {}^\bullet t_2 \neq \varnothing$ or $t_1{}^\bullet \cap t_2{}^\bullet \neq \varnothing$. If the former holds, then we have an input-conflict (represented in Fig. 20), and if the latter holds, then we have an output-conflict (represented in Fig. 21). Obviously, input-conflict and output-conflict can also both be present.

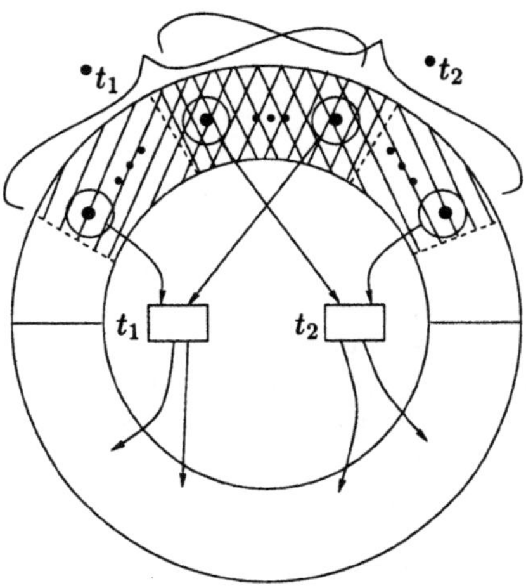

Fig. 20. Input-conflict.

Hence if t_1 and t_2 both have concession in configuration C, then there is either conflict or concurrency. In the case of concurrency t_1 and t_2 are independent, whereas in the case of a conflict they are not independent. That is why a conflict intuitively leads to a nondeterministic choice between the transitions (either t_1 occurs, or t_2 occurs). Clearly there is no need for choice if t_1 and t_2 are concurrent. EN systems in which no choice is made are particularly easy to understand; they are the concurrent equivalent of deterministic finite automata in the sequential case.

Definition 22. An EN system $M = (P, T, F, C_{in})$ is *conflict-free* if for every $C \in \mathbb{C}_M$ and all transitions $t_1, t_2 \in T$: if t_1 **con** C and t_2 **con** C, then $\{t_1, t_2\}$ **con** C.

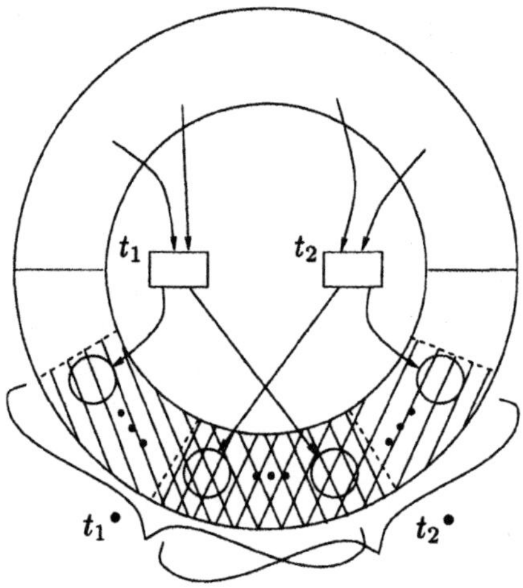

Fig. 21. Output-conflict.

Since in a conflict-free EN system choices are never made, it has only one run. The EN system of Fig. 1 is conflict-free, for "structural reasons": both $p^\bullet$ and ${}^\bullet p$ contain at most one transition, for every $p \in P$; such systems are clearly conflict-free: $\{t_1, t_2\}$ **con** C follows from ${}^\bullet t_1 \cap {}^\bullet t_2 = \varnothing$ and $t_1{}^\bullet \cap t_2{}^\bullet = \varnothing$, by Lemma 15. In the literature (see, e.g., [LanRob78]) conflict-free systems are usually called ***persistent***, in which case the term 'conflict-free' is reserved for a structural subclass such as the one above.

The interplay between concurrency and conflict may be quite intricate, and in particular it can lead to confusion - a phenomenon that seems to be fundamentally present in nature and appears in various disguises, depending on the chosen level and way of description of a concurrent system. Here we discuss confusion in the framework of EN systems.

(4) Confusion.
Consider the EN system in Fig. 22, hence $C_{in} = \{p_1, p_2, p_3\}$. Let $C = \{p_4, p_5\}$; thus $C_{in}[\{t_1, t_2\}\rangle C$. Different sequential realizations ($t_1 t_2$ and $t_2 t_1$) of this concurrent step have drastically different properties. Since sequential realizations of a concurrent step correspond to observations of the step by sequential observers, assume that we have two honest sequential observers O_1 (corresponding to $t_1 t_2$) and O_2 (corresponding to $t_2 t_1$). They will report their observations as follows:
O_1: "t_1 occurred first without having been in conflict with another event; then t_2 occurred", and
O_2: "t_2 occurred first; this resulted in a conflict between t_1 and t_3 which was resolved in favour of t_1, and so t_1 occurred".

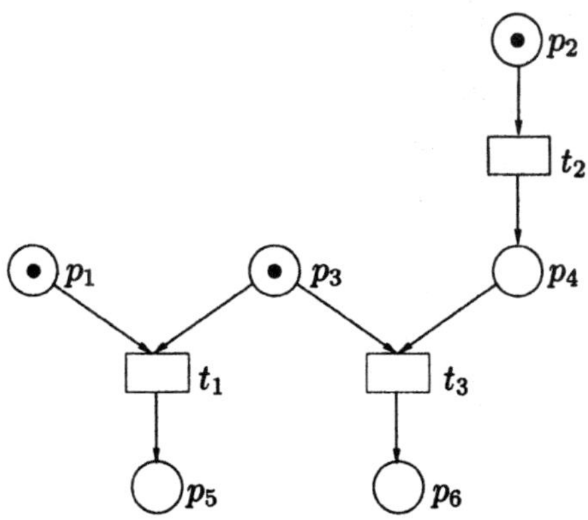

Fig. 22. A conflict-increasing confusion.

This is therefore a confusing situation, which resulted from the interplay between concurrency (between t_1 and t_2) and conflict (between t_1 and t_3). Systems where confusion occurs are in general difficult to analyze. This is due to the fact that the intermediate configurations determined by the different sequential realizations of a concurrent step can differ drastically from each other, as we have seen in the example above. Consequently, in general one may have to analyze all possible sequentializations of the step rather than just one. The theory of Petri Nets suggests that it is not the combination of concurrency and conflict as such that causes difficulties. Only those combinations of concurrency and conflict that result in confusion create problems. Unfortunately it is not always possible to avoid confusion. Even the rather simple "mutual exclusion problem" discussed in Section 3.1 contains confusion (in configuration $\{w_1, p, r_2\}$).

We now turn to a formal discussion of confusion.

Definition 23. Let $M = (P, T, F, C_{in})$ be an EN system, let $C \in \mathbb{C}_M$, and let $t \in T$ be such that t **con** C. The *conflict set of t in C*, denoted by $\mathbf{cfl}(t, C)$, is the set $\{t' \in T \mid t' \textbf{ con } C \text{ and } \neg \{t, t'\} \textbf{ con } C\}$.

Hence the conflict set of an event t in a configuration C is the set of all events that are in conflict with t in C. Note that t is in conflict in C only if t itself has concession in C.

Definition 24. Let $M = (P, T, F, C_{in})$ be an EN system, let $C \in \mathbb{C}_M$, and let $t_1, t_2 \in T$. The triple (C, t_1, t_2) is called a *confusion (in C)* if $t_1 \neq t_2$, $\{t_1, t_2\}$ **con** C, and $\mathbf{cfl}(t_1, C) \neq \mathbf{cfl}(t_1, D)$, where $C[t_2\rangle D$. Then M *is confused in C* if there is a confusion in C.

Hence a triple (C, t_1, t_2) is a confusion if $\{t_1, t_2\}$ is a step in C and the occurrence of t_2 in C changes the conflict set of t_1.

Example 10. Consider the EN system M depicted in Fig. 22. For the configuration $C = \{p_1, p_2, p_3\} = C_{in}$, $\mathbf{cfl}(t_1, C) = \varnothing$. Hence (C, t_1, t_2) is a confusion, because $\{t_1, t_2\}$ **con** C and $\mathbf{cfl}(t_1, C) = \varnothing \neq \{t_3\} = \mathbf{cfl}(t_1, D)$, where $D = \{p_1, p_3, p_4\}$.

It is natural to distinguish the following two types of confusion.

Definition 25. Let $M = (P, T, F, C_{in})$ be an EN system, let $C \in \mathbb{C}_M$, let $t_1, t_2 \in T$, let $\gamma = (C, t_1, t_2)$ be a confusion, and let $C[t_2\rangle D$.

(1) γ is a *conflict-increasing confusion, ci confusion* for short, if $\mathbf{cfl}(t_1, D) \supsetneq \mathbf{cfl}(t_1, C)$.

(2) γ is a *conflict-decreasing confusion, cd confusion* for short, if $\mathbf{cfl}(t_1, D) \subsetneq \mathbf{cfl}(t_1, C)$.

Example 11. (1) Consider the EN system M and the confusion (C, t_1, t_2) from Example 10 (Fig. 22). Since $\mathbf{cfl}(t_1, D) \supsetneq \mathbf{cfl}(t_1, C)$, (C, t_1, t_2) is a ci confusion.

(2) Consider the EN system in Fig. 23. For $C = C_{in} = \{p_1, p_2\}$, (C, t_1, t_2) is a confusion because $\{t_1, t_2\}$ **con** C and $\mathbf{cfl}(t_1, C) = \{t_3\} \neq \varnothing = \mathbf{cfl}(t_1, D)$, where $D = \{p_1, p_3\}$. Furthermore, since $\mathbf{cfl}(t_1, D) \subsetneq \mathbf{cfl}(t_1, C)$, (C, t_1, t_2) is a cd confusion.

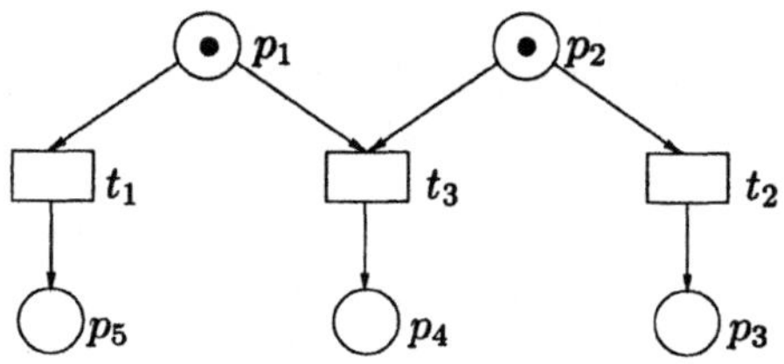

Fig. 23. A conflict-decreasing confusion.

(3) Consider the EN system in Fig. 24. For $C = C_{in} = \{p_1, p_2, p_4\}$, (C, t_1, t_2) is a confusion because $\mathbf{cfl}(t_1, C) = \{t_3\} \neq \{t_4\} = \mathbf{cfl}(t_1, D)$, where $D = \{p_1, p_3, p_4\}$. Note that (C, t_1, t_2) is neither a ci confusion nor a cd confusion.

As we have seen in the above example, the classification of confusions into ci confusions and cd confusions is not exhaustive: there exist confusions that are neither ci nor cd.

If (C, t_1, t_2) is a confusion, then the occurrence of t_2 in C has a rather strong impact on the occurrence of t_1 in C: it changes the conflict set of t_1. Hence the fact that (C, t_1, t_2) is a confusion shows an influence of t_2 on t_1 (in C). In order to better understand the mutual dependency between t_1 and t_2 it is important to know whether t_1 has the same sort of influence on t_2 (in C). This consideration leads to the notion of symmetric confusion.

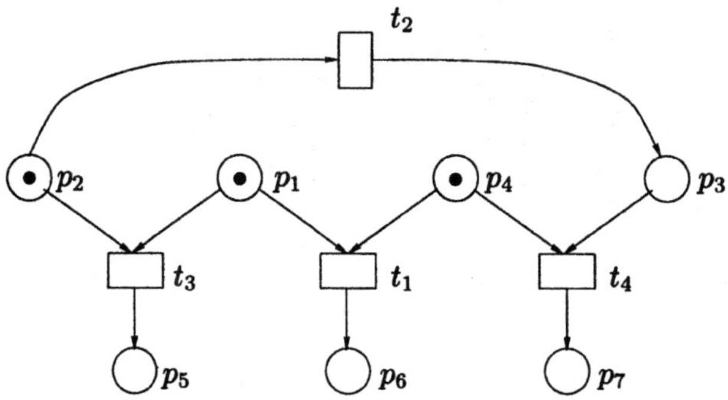

Fig. 24. A confusion that is neither conflict-increasing nor conflict-decreasing.

Definition 26. Let $M = (P, T, F, C_{in})$ be an EN system, let $C \in \mathbb{C}_M$, let $t_1, t_2 \in T$ and let $\gamma = (C, t_1, t_2)$ be a confusion. Then γ is ***symmetric*** if (C, t_2, t_1) is also a confusion, otherwise γ is ***asymmetric***.

Example 12. (1) Consider the EN system M and the confusion (C, t_1, t_2) from Example 10 (Fig. 22). Since (C, t_2, t_1) is not a confusion, (C, t_1, t_2) is a ci confusion that is asymmetric (see also Example 11(1)).

(2) Consider the EN system in Fig. 25. For $C = C_{in} = \{p_1, p_3\}$, (C, t_1, t_2) is a ci confusion and (C, t_2, t_1) is a ci confusion. Hence (C, t_1, t_2) is a ci confusion that is symmetric.

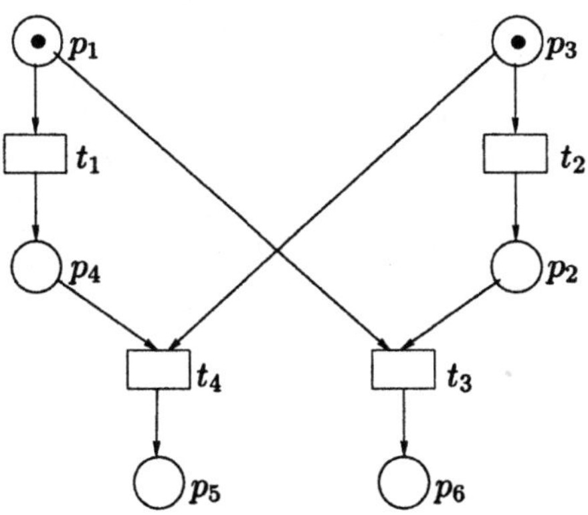

Fig. 25. A symmetric confusion.

(3) The confusion (C, t_1, t_2) from Example 11(2) (Fig. 23) is a symmetric cd confusion.

(4) The confusion (C, t_1, t_2) from Example 11(3) (Fig. 24) is a symmetric confusion that is neither a ci confusion nor a cd confusion.

The above example shows that the division into ci and cd confusions is rather independent from the division into symmetric and asymmetric confusions. We will not consider the topic of confusion in more detail here, however we would like to mention that the only nontrivial relation between these dividing lines is that cd confusions are always symmetric (in fact, if (C, t_1, t_2) is a cd confusion, then there is a transition t such that t **con** C and t is in conflict with both t_1 and t_2).

An EN system is said to be *free-choice* if, for all transitions t_1 and t_2, ${}^\bullet t_1 \cap {}^\bullet t_2 \neq \varnothing$ implies ${}^\bullet t_1 = {}^\bullet t_2$. This is a structural restriction on EN systems that guarantees the absence of confusion as far as input-conflicts are concerned. Requiring additionally that the system is contact-free (see Section 4.5), it is not difficult to show that there is no confusion. The class of free-choice EN (and P/T) systems is a large class with many interesting properties, see [Hac72, DesEsp95].

4 Equivalences and Normal Forms

When we wish to express in a formal, mathematical, way that two systems "are similar to each other" or that they "behave in the same way", then we have to define an equivalence relation on the class of all systems, such that the systems that "are similar to each other" or "behave in the same way" form an equivalence class. Such equivalence relations are particularly useful when one wants to transform or optimize a system without changing its "behaviour", i.e., transform a system into a "better", equivalent system. If R is a property of EN systems such that for each EN system there is an equivalent one satisfying R, then we say that R is a "normal form" for the class of EN systems. In this section we consider several notions of equivalence and normal forms for EN systems. In particular, we formalize the notion of a component of an EN system and show that every EN system is equivalent with one that can be viewed as consisting of communicating concurrent components (where each of the components is sequential). Such decompositions of Petri nets are studied, e.g., in [Hac72, DesEsp95]. A survey of various notions of equivalence for EN systems is presented in [PomRozSim92].

4.1 Equivalence

The simplest (and least interesting) definition of equivalence is isomorphism. Two EN systems are isomorphic if their underlying nets are isomorphic in such a way that the initial configurations correspond to each other.

Definition 27. Two EN systems $M = (P, T, F, C_{in})$ and $M' = (P', T', F', C'_{in})$ are *isomorphic*, denoted by $M \equiv M'$, if there exist two bijections $\alpha : P \to P'$ and $\beta : T \to T'$ such that $\mathbf{und}(M) \equiv^{\alpha}_{\beta} \mathbf{und}(M')$ and $\alpha(C_{in}) = C'_{in}$.

Thus, isomorphic EN systems have the same static structure. Now we will try to capture the dynamic "behaviour" of EN systems through an equivalence relation, as discussed above. For EN systems there are many possibilities to define such a notion of equivalence, some weaker than others. If two systems are isomorphic, then they will be equivalent for any of these notions. The underlying idea for all the notions of equivalence that we will consider, is that "behaviour" is mainly concerned with the actions (transitions) that occur during a run of the system and not so much with the distribution of the global state of the system (over the places).

Our first notion of dynamic equivalence defines two EN systems to be equivalent if there exist one-to-one correpondences between the transitions and the configurations (not the places!), such that the correspondence between configurations is preserved by firing corresponding transitions. This means that the systems "simulate each other's behaviour".

Definition 28. Let $M = (P,T,F,C_{in})$ and $M' = (P',T',F',C'_{in})$ be two EN systems. Then M and M' are *configuration equivalent*, denoted by $M \approx M'$, if there exist two bijections $\alpha : \mathbb{C}_M \to \mathbb{C}_{M'}$ and $\beta : \mathbf{use}_M(T) \to \mathbf{use}_{M'}(T')$ such that
(1) $\alpha(C_{in}) = C'_{in}$ and
(2) for all $C, D \in \mathbb{C}_M$ and $t \in \mathbf{use}_M(T)$, $C[t\rangle_M D$ iff $\alpha(C)[\beta(t)\rangle_{M'}\alpha(D)$.

If we want to be more specific, then we say that M and M' are (α,β)-*configuration equivalent*, denoted by $M \approx^{\alpha}_{\beta} M'$.

It can directly be seen that $\approx$ is indeed an equivalence relation on the class of EN systems. In fact, configuration equivalence is strongly related to the configuration graphs. It immediately follows from the definitions that two EN systems are configuration equivalent iff their sequential configuration graphs are isomorphic. Hence here the behaviour of an EN system is identified with (represented by) its sequential configuration graph, modulo isomorphism. According to Theorem 21 this notion of behaviour covers also the concurrent behaviour of the system.

Theorem 29. *Let M and M' be two EN systems.*
Then $M \approx M'$ iff $SCG(M) \equiv SCG(M')$ iff $CG(M) \equiv CG(M')$.

It is easy to see that $M \equiv M'$ implies $M \approx M'$, i.e., isomorphic EN systems are configuration equivalent. On the other hand there exist configuration equivalent EN systems that are not isomorphic (see the next example). Configuration equivalence is thus weaker than isomorphism of EN systems.

Example 13. Let M be the EN system of Fig. 26 and M' the EN system of Fig. 27. M and M' are not isomorphic. It is clear that $\mathbb{C}_M = \{\{p_1\}, \{p_3,p_4\}, \{p_2,p_4\}, \{p_5\}, \{p_6\}\}$ and $\mathbb{C}_{M'} = \{\{p_1\}, \{p_2\}, \{p_3\}, \{p_4\}, \{p_5\}\}$. All transitions are useful. Let $\alpha : \mathbb{C}_M \to \mathbb{C}_{M'}$ and $\beta : T_M \to T_{M'}$ be the bijections defined as follows: $\alpha(\{p_1\}) = \{p_1\}$, $\alpha(\{p_3,p_4\}) = \{p_2\}$, $\alpha(\{p_2,p_4\}) = \{p_3\}$, $\alpha(\{p_5\}) = \{p_4\}$, $\alpha(\{p_6\}) = \{p_5\}$, and $\beta(t_i) = t_i$ for all $1 \le i \le 6$. Now it should be clear that

$M \approx^{\alpha}_{\beta} M'$. The sequential configuration graphs SCG(M) and SCG(M') are given in Figs. 28 and 29. It is easy to see that SCG(M) $\equiv^{\alpha}_{\beta}$ SCG(M'). Note that for both M and M' the configuration graph is the same as the sequential configuration graph.

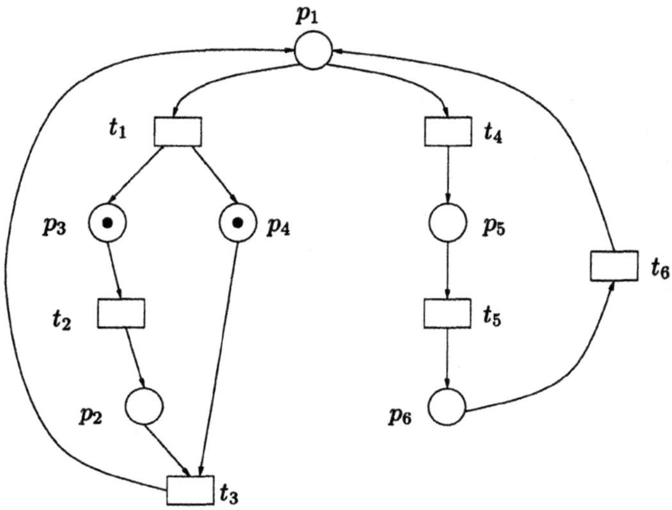

Fig. 26. An EN system M.

The following technical lemma is often useful in proofs of configuration equivalence.

Lemma 30. *Let $M = (P, T, F, C_{in})$ and $M' = (P', T', F', C'_{in})$ be two EN systems. If α is an injective function, $\alpha : \mathbb{C}_M \to \mathcal{P}(P')$, and β is a bijective function, $\beta : \mathbf{use}_M(T) \to T'$, such that*
(1) $\alpha(C_{in}) = C'_{in}$ and
(2) for all $C, D \in \mathbb{C}_M$ and $t \in \mathbf{use}_M(T)$,
$C[t\rangle_M D$ implies $\alpha(C)[\beta(t)\rangle_{M'}\alpha(D)$, and
$\beta(t)$ $\mathbf{con}_{M'}$ $\alpha(C)$ implies t $\mathbf{con}_M$ C,
then $M \approx^{\alpha}_{\beta} M'$.

Proof. We first prove the if-part of Definition 28(2). Let $\alpha(C)[\beta(t)\rangle_{M'}\alpha(D)$. Then $\beta(t)$ $\mathbf{con}_{M'}$ $\alpha(C)$ and thus t $\mathbf{con}_M$ C. Let $C[t\rangle_M E$. Then $\alpha(C)[\beta(t)\rangle_{M'}\alpha(E)$. Hence $\alpha(E) = \alpha(D)$. Since α is injective, $E = D$ and thus $C[t\rangle_M D$.

To prove that α is a bijection between $\mathbb{C}_M$ and $\mathbb{C}_{M'}$ we first show that $\alpha(C) \in \mathbb{C}_{M'}$ for every $C \in \mathbb{C}_M$. This is done by induction on C. For $C = C_{in}$, $\alpha(C_{in}) = C'_{in} \in \mathbb{C}_{M'}$ by (1). Now assume (as induction hypothesis) that

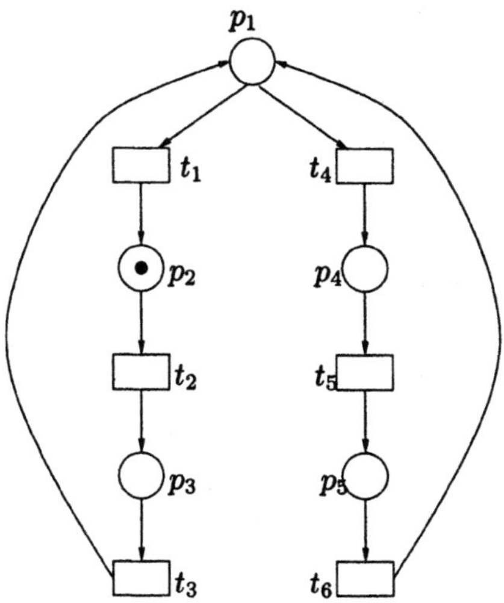

Fig. 27. An EN system M' configuration equivalent with the EN system M of Fig. 26.

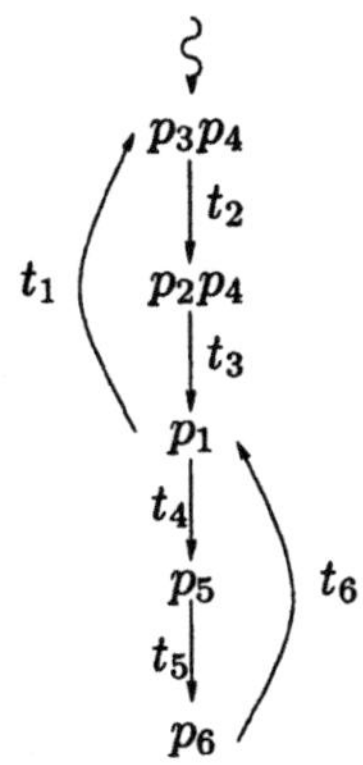

Fig. 28. The configuration graph of the EN system M of Fig. 26.

$\alpha(C) \in \mathbb{C}_{M'}$ and let $C[t\rangle_M D$. We have to prove that $\alpha(D) \in \mathbb{C}_{M'}$. From $C[t\rangle_M D$ and (2) it follows that $\alpha(C)[\beta(t)\rangle_{M'}\alpha(D)$. Since $\alpha(C) \in \mathbb{C}_{M'}$, we obtain that $\alpha(D) \in \mathbb{C}_{M'}$. Next we show that α is surjective, i.e., that for every $C' \in \mathbb{C}_{M'}$ there exists $C \in \mathbb{C}_M$ with $\alpha(C) = C'$. This is done in the same way, by induction on C'. For $C' = C'_{in}$, according to (1), we can take $C = C_{in}$. Now assume that $\alpha(C) = C'$ for $C \in \mathbb{C}_M$ and let $C'[t'\rangle_{M'} D'$. Since β is a bijection, there is a $t \in \mathbf{use}(T)$ with $\beta(t) = t'$. Hence $\beta(t)$ **con** $\alpha(C)$ and thus t **con** C according to (2). Let $C[t\rangle_M D$. Hence $D \in \mathbb{C}_M$. Then $\alpha(C)[\beta(t)\rangle_{M'}\alpha(D)$; in other words

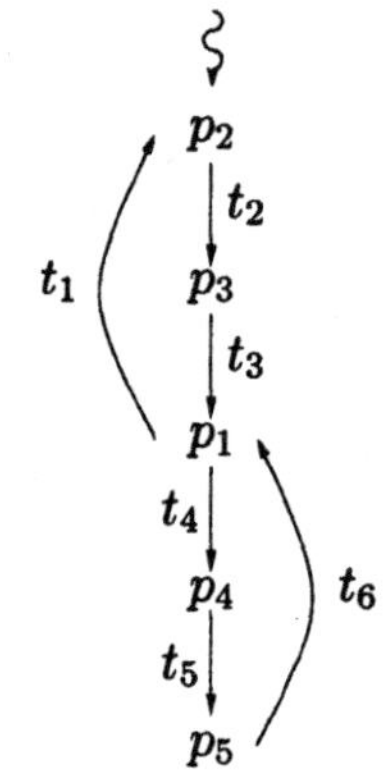

Fig. 29. The configuration graph of the EN system M' of Fig. 27.

$C'[t'\rangle_{M'}\alpha(D)$, and thus $\alpha(D) = D'$.

We still have to prove that β is a bijection from $\mathbf{use}_M(T)$ to $\mathbf{use}_{M'}(T')$, i.e., that $\beta(t) \in \mathbf{use}_{M'}(T')$ for every $t \in \mathbf{use}_M(T)$. If $t \in \mathbf{use}_M(T)$, then there exist $C, D \in \mathbb{C}_M$ such that $C[t\rangle_M D$. Hence $\alpha(C)[\beta(t)\rangle_{M'}\alpha(D)$. Since $\alpha(C) \in \mathbb{C}_{M'}$, $\beta(t) \in \mathbf{use}_{M'}(T')$. □

A similar, but less stringent, definition of equivalence is obtained by no longer requiring the correspondence α between configurations to be a bijection, but just a relation. In that way the systems can still simulate each other's behaviour "step by step", but the configuration graphs no longer need to be isomorphic. Again, the correspondence between the transitions is a bijection because we are mainly interested in the transitions (modulo their identity) of a system.

Definition 31. Let $M = (P, T, F, C_{in})$ and $M' = (P', T', F', C'_{in})$ be two EN systems. M and M' are *weakly configuration equivalent*, denoted by $M \approx_w M'$, if there exists a relation $\alpha \subseteq \mathbb{C}_M \times \mathbb{C}_{M'}$ and a bijection $\beta : \mathbf{use}(T) \to \mathbf{use}(T')$, such that
(1) $(C_{in}, C'_{in}) \in \alpha$,
(2) for all $C, D \in \mathbb{C}_M$, $C' \in \mathbb{C}_{M'}$, and $t \in \mathbf{use}(T)$: if $C[t\rangle_M D$ and $(C, C') \in \alpha$, then there is a $D' \in \mathbb{C}_{M'}$ such that $C'[\beta(t)\rangle_{M'} D'$ and $(D, D') \in \alpha$, and
(3) for all $C', D' \in \mathbb{C}_{M'}$, $C \in \mathbb{C}_M$, and $t' \in \mathbf{use}(T')$: if $C'[t'\rangle_{M'} D'$ and $(C, C') \in \alpha$, then there is a $D \in \mathbb{C}_M$ such that $C[\beta^{-1}(t')\rangle_M D$ and $(D, D') \in \alpha$.

In the literature the relation α is often called a *bisimulation*, and weak configuration equivalence is then called *bisimilarity* or *observation equivalence* (see, e.g., [Mil89]).

Condition (1) means that both systems start in corresponding configurations. Condition (2) means the following: if both systems are in corresponding configurations and M takes a step by firing transition t, then M' can simulate that step by firing the corresponding transition $\beta(t)$, after which M and M' are in corresponding configurations again. Condition (3) says the same as condition (2), with the roles of M and M' reversed.

As in the case of configuration equivalence, it can be shown also here (using Theorem 20) that conditions (2) and (3) also hold for concurrent steps $C[U\rangle_M D$ and $C'[\beta(U)\rangle_{M'} D'$.

It is clear that configuration equivalent EN systems are also weakly configuration equivalent. On the other hand there exist weakly configuration equivalent EN systems that are not configuration equivalent (see the next example). Hence weak configuration equivalence is weaker than configuration equivalence (as the name already suggests!).

Example 14. Let M and M' be the EN systems in Figs. 30 and 32, respectively; their (sequential) configuration graphs are given in Figs. 31 and 33, respectively. Since SCG(M) has less nodes than SCG(M'), SCG(M) and SCG(M') are not isomorphic, and so M and M' are not configuration equivalent. They are however weakly configuration equivalent, with the bisimulation α that consists of the following pairs: $(\{p_1\},\{p_1\})$, $(\{p_2\},\{p_2\})$, $(\{p_3\},\{p_3\})$, $(\{p_3\},\{p_3,q\})$, $(\{p_4\},\{p_4\})$, $(\{p_4\},\{p_4,q\})$, $(\{p_5\},\{p_5,q\})$, and $(\{p_6\},\{p_6\})$. For β we take the identity.

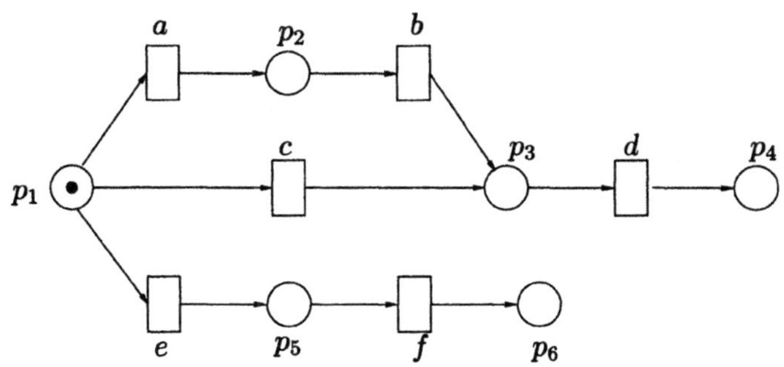

Fig. 30. An EN system M, weakly equivalent with the EN system M' of Fig. 32.

An even less stringent definition of equivalence is obtained by requiring only that the firing sequences of two equivalent EN systems (bijectively) correspond to each other, with no requirement at all on the configurations of the two EN systems. Thus here the behaviour of an EN system M is defined as (represented by) its set FS(M) of firing sequences, modulo the identity of the transitions. In this way we abstract completely from the global states of the system.

Note that if $\beta : \mathbf{use}(T_M) \to \mathbf{use}(T_{M'})$, where M and M' are EN systems, then $\beta(\mathrm{FS}(M)) \subseteq \mathbf{use}(T_{M'})^*$, see Section 2.

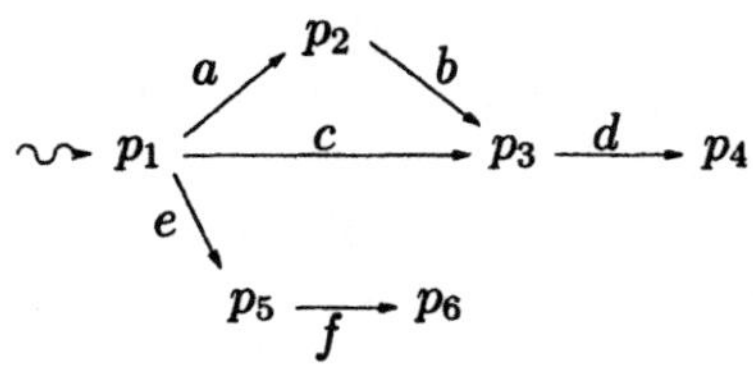

Fig. 31. The configuration graph of the EN system M of Fig. 30.

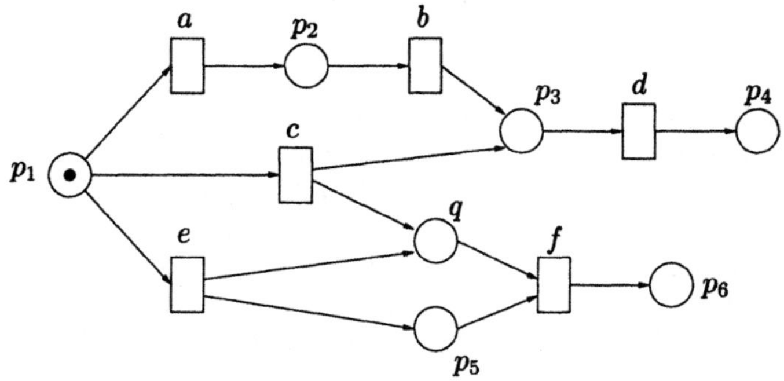

Fig. 32. An EN system M', weakly equivalent with the EN system M of Fig. 30.

Definition 32. Let M and M' be two EN systems. M and M' are *firing sequence equivalent*, denoted by $M \approx_{fs} M'$, if there exists a bijection $\beta : \mathbf{use}(T_M) \to \mathbf{use}(T_{M'})$ such that $\beta(\mathrm{FS}(M)) = \mathrm{FS}(M')$.

Example 15. The (weakly configuration equivalent) EN systems M and M' of Example 14 (Figs. 30 and 32) are firing sequence equivalent because $\mathrm{FS}(M) = \mathrm{FS}(M') = \{\lambda, a, c, e, ab, cd, ef, abd\}$.

Even though firing sequence equivalence abstracts completely from the configurations, it turns out to be the same as weak configuration equivalence! This

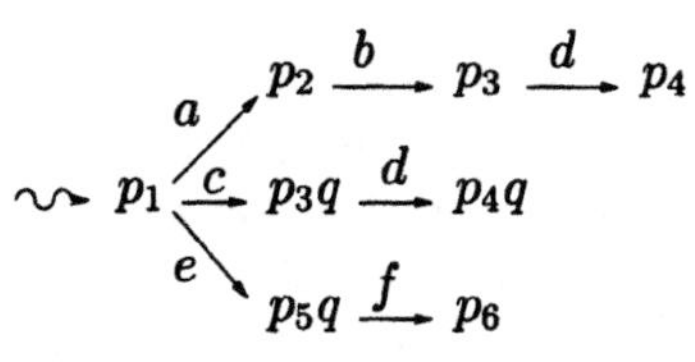

Fig. 33. The configuration graph of the EN system M' of Fig. 32.

is essentially due to the fact that firing a transition leads from a configuration to a unique next configuration (see [Eng85, Mil89, PomRozSim92]).

Theorem 33. *Two EN systems are firing sequence equivalent iff they are weakly configuration equivalent.*

Proof. Let M and M' be two EN systems.

First assume that $M \approx_w M'$, with bisimulation α and bijection β. It suffices to prove that $\beta(\mathrm{FS}(M)) \subseteq \mathrm{FS}(M')$: by symmetry this also proves that $\beta^{-1}(\mathrm{FS}(M')) \subseteq \mathrm{FS}(M)$.

It follows directly from the definition of weak configuration equivalence that each event t of M can be simulated by event $\beta(t)$ of M' in such a way that the relation α between the (old and new) configurations continues to hold. Hence each firing sequence x of M can be simulated in such a way by the sequence $\beta(x)$ of M'. This implies that $\beta(x)$ is a firing sequence of M'. Formally, we use the following extension of Definition 31(2), which can easily be proved by induction on $|x|$: for all $C, D \in \mathbb{C}_M$, $C' \in \mathbb{C}_{M'}$, and $x \in \mathbf{use}(T_M)^*$:

if $C[x\rangle_M D$ and $(C, C') \in \alpha$, then there exists $D' \in \mathbb{C}_{M'}$ such that $C'[\beta(x)\rangle_{M'} D'$ and $(D, D') \in \alpha$.

Taking $C = C_{in}$ and $C' = C'_{in}$, this shows, by Definition 31(1), that if $x \in \mathrm{FS}(M)$ then $\beta(x) \in \mathrm{FS}(M')$.

Now assume that $M = (P, T, F, C_{in})$ and $M' = (P', T', F', C'_{in})$ are firing sequence equivalent, with bijection β. Hence, if x is a firing sequence of M, then $\beta(x)$ is a firing sequence of M', and vice versa. This leads to the following definition of the relation $\alpha \subseteq \mathbb{C}_M \times \mathbb{C}_{M'}$:

$$(C, C') \in \alpha \text{ iff } \exists x \in T^* : C_{in}[x\rangle_M C \text{ and } C'_{in}[\beta(x)\rangle_{M'} C'.$$

It is straightforward to verify that α is a bisimulation, i.e., that α and β satisfy the three conditions in Definition 31. For condition (1) take $x = \lambda$. For condition (2), assume that $C_{in}[x\rangle_M C$ and $C'_{in}[\beta(x)\rangle_{M'} C'$, and that $C[t\rangle_M D$. Then $C_{in}[xt\rangle_M D$. Thus xt is a firing sequence of M, and so $\beta(xt)$ is a firing sequence of M'. Hence there exists $D' \in \mathbb{C}_{M'}$ such that $C'_{in}[\beta(xt)\rangle_{M'} D'$. Clearly $(D, D') \in \alpha$. It remains to prove that $C'[\beta(t)\rangle_{M'} D'$. Since $\beta(xt) = \beta(x)\beta(t)$, there exists $E' \in \mathbb{C}_{M'}$ such that $C'_{in}[\beta(x)\rangle_{M'} E'$ and $E'[\beta(t)\rangle_{M'} D'$. Since the firing of $\beta(x)$ leads from C'_{in} to a unique configuration, $E' = C'$ and thus indeed $C'[\beta(t)\rangle_{M'} D'$. Condition (3) can be proved analogously. □

Since configuration equivalence implies weak configuration equivalence, the following corollary is obtained from the above theorem.

Corollary 34. *If two EN systems are configuration equivalent, then they are also firing sequence equivalent.*

This corollary can also be seen directly: if two EN systems M and M' are configuration equivalent, then they have isomorphic configuration graphs and hence $\mathrm{FS}(M)$ and $\mathrm{FS}(M')$ are recognized by isomorphic finite automata (see Theorem 12).

To recapitulate, in this subsection we have introduced four equivalence relations for EN systems, of which one is static (isomorphism) and the other three are dynamic, depending on the intuitive notion of "behaviour". The following relationships hold between these four equivalences: isomorphism implies configuration equivalence, which in turn implies firing sequence equivalence, which equals weak configuration equivalence. Thus in formal notation we have, for EN systems M and M',

$$M \equiv M' \Rightarrow M \approx M' \Rightarrow M \approx_{fs} M' \Leftrightarrow M \approx_w M'.$$

4.2 Reduction

We now turn to normal forms for EN systems.

It is intuitively natural to assume that an EN system contains no "superfluous" transitions and places, hence in particular no useless transitions and no isolated places. Such transitions and places clearly play no part in the behaviour of the system. We will formalize this intuition by showing that for every EN system M there exists an equivalent EN system M' containing useful transitions and nonisolated places only. This holds for the strongest dynamic notion of equivalence from the previous subsection, viz. configuration equivalence.

Definition 35. An EN system M is *reduced* if all transitions of M are useful. M is *strongly reduced* if M is reduced and has no isolated places.

Our first normal form result is the following.

Theorem 36. *For every EN system M there exists a reduced EN system M' such that $M \approx M'$.*

Proof. Let $M = (P, T, F, C_{in})$. We construct M' by simply removing all useless transitions. Let M' be the EN system (P, T', F', C_{in}) with $T' = \mathbf{use}_M(T)$ and $F' = F \cap ((P \times T') \cup (T' \times P))$. It is clear that M and M' even have the same sequential configuration graph and thus are configuration equivalent. Hence M' contains only useful transitions and is thus reduced. □

This theorem can be strengthened as follows.

Theorem 37. *For every EN system M there exists a strongly reduced EN system M' such that $M \approx M'$.*

Proof. Let $M = (P, T, F, C_{in})$. According to Theorem 36 we may assume that M is reduced, i.e., $\mathbf{use}_M(T) = T$. We now simply construct M' by removing all isolated places. Define $\mathbf{use}(P) = \{p \in P \mid \mathbf{nbh}(p) \neq \varnothing\}$; thus $\mathbf{use}(P)$ is the set of nonisolated places of M. Let $M' = (P', T, F, C'_{in})$ with $P' = \mathbf{use}(P)$ and $C'_{in} = C_{in} \cap P'$. Obviously M' has no isolated places.

Define the function $\alpha : \mathbb{C}_M \to \mathcal{P}(P')$ with $\alpha(C) = C \cap P'$ for all $C \in \mathbb{C}_M$ and let β be the identity on T. We prove that $M \approx^{\alpha}_{\beta} M'$ using Lemma 30. It is easy to prove by induction on C that, for all $C \in \mathbb{C}_M$, $C - P' = C_{in} - P'$ (intuitively this holds because the marking of the isolated places never changes). This implies that α is injective. Also $\alpha(C_{in}) = C'_{in}$. By Lemma 30 it now suffices to prove that $C[t\rangle_M D$ iff $C \cap P'[t\rangle_{M'} D \cap P'$. This can easily be proved using Lemma 7, because $(C \cap P') - (D \cap P') = (C - D) \cap P'$ (for arbitrary sets) and ${}^\bullet t \cap P' = {}^\bullet t$, and analogously for $D - C$ and $t^\bullet$. Hence M' is (α, β)-configuration equivalent with M. Finally, by Definition 28, β is a bijection between $\mathbf{use}_M(T)$ and $\mathbf{use}_{M'}(T)$. This implies that $\mathbf{use}_{M'}(T) = T$. Hence M' is reduced and thus strongly reduced. □

Example 16. According to the constructions in the proofs of Theorems 36 and 37, the EN system M in Fig. 34 is transformed into the equivalent strongly reduced EN system M' of Fig. 35: first the useless transition t_3 is removed and then the isolated places p_7 and p_8 are removed.

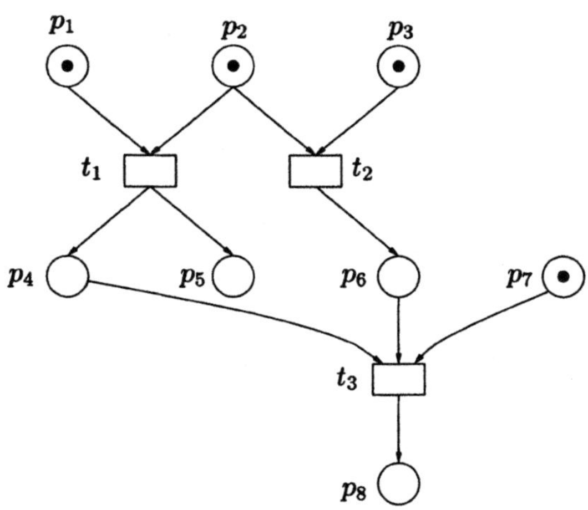

Fig. 34. An EN system with useless transition t_3.

We will now show that a strongly reduced EN system contains no "superfluous" places in the following sense: for every condition there exists a configuration in which this condition holds, and there exists a configuration in which this condition does not hold. Hence there are no void, static conditions that either always hold or never hold.

Theorem 38. *Let $M = (P, T, F, C_{in})$ be a strongly reduced EN system. For every $p \in P$ there exist configurations $C, D \in \mathbb{C}_M$ such that $p \in C$ and $p \notin D$.*

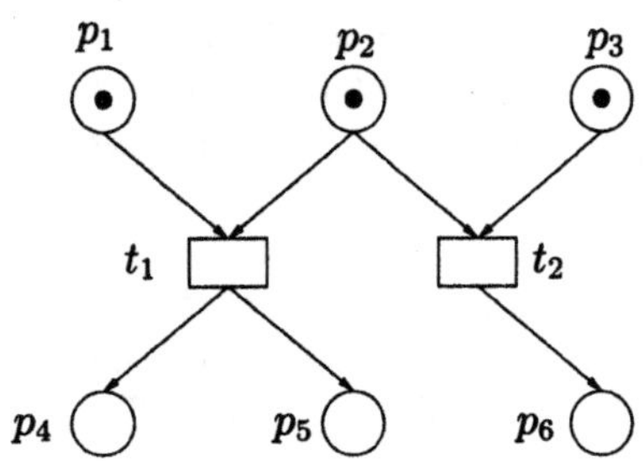

Fig. 35. A strongly reduced EN system, configuration equivalent with the EN system of Fig. 34.

Proof. Since p is not isolated, there is a transition t such that $t \in \mathbf{nbh}(p)$. Since t is useful, there is a reachable configuration E with t **con** E. Let $E[t\rangle E'$. If $t \in p^\bullet$, then $p \in E$ and $p \notin E'$. And if $t \in {}^\bullet p$, then $p \in E'$ and $p \notin E$. □

We have now proved that isolated places and useless transitions can be removed from an EN system M. But, in general, there are also other kinds of "superfluous" places and transitions, in particular when M is not simple (see Definition 2(2,3)). There are two types of simplicity, which we will now consider from this viewpoint.

First assume that M is not P-simple. Then there are two distinct places p and q such that ${}^\bullet p = {}^\bullet q$ and $p^\bullet = q^\bullet$. It is intuitively clear that one of these two places is superfluous. It can easily be shown that the removal of either p or q results in an EN system that is configuration equivalent with M. Hence for every EN system there exists a configuration equivalent strongly reduced EN system that is P-simple.

Now assume that M is not T-simple. Then there are two distinct transitions s and t such that ${}^\bullet s = {}^\bullet t$ and $s^\bullet = t^\bullet$. Intuitively one of these two transitions is again superfluous. However, if s and t are useful, then the removal of either s or t does not result in a configuration equivalent EN system, simply because the number of useful transitions of two configuration equivalent EN systems must be equal (β is a bijection). We will now show that T-simplicity is not a normal form for EN systems, not even with respect to the weakest kind of equivalence (viz. firing sequence equivalence).

Theorem 39. *There exists an EN system M such that for every EN system M': if $M' \approx_{fs} M$, then M' is not T-simple.*

Proof. The following technical property of an (arbitrary) EN system M' will be useful in our proof: if $s, u \in T_{M'}$ and there exists an $x \in T^*_{M'}$, such that $xsus \in \mathrm{FS}(M')$, then $s^\bullet \subseteq {}^\bullet u$ and ${}^\bullet s \subseteq u^\bullet$. This is proved as follows. Since $xsus \in \mathrm{FS}(M')$, there exist configurations C, D_1, D_2, and E of M' such that $C[s\rangle D_1$, $D_1[u\rangle D_2$, and $D_2[s\rangle E$. Since $D_1 - D_2 = {}^\bullet u$, $s^\bullet \subseteq D_1$, and $s^\bullet \cap D_2 = \varnothing$, we have $s^\bullet \subseteq {}^\bullet u$. Since $D_2 - D_1 = u^\bullet$, ${}^\bullet s \cap D_1 = \varnothing$, and ${}^\bullet s \subseteq D_2$, we have ${}^\bullet s \subseteq u^\bullet$.

Now consider the EN system $M = (P, T, F, C_{in})$ with $P = \{p, q\}$, $T = \{s, t, u\}$, $C_{in} = \{p\}$, ${}^\bullet s = {}^\bullet t = u^\bullet = \{p\}$ and $s^\bullet = t^\bullet = {}^\bullet u = \{q\}$. Suppose that M' is an EN system that is firing sequence equivalent with M, and assume for the sake of simplicity that $\mathrm{FS}(M') = \mathrm{FS}(M)$ (i.e., that β is the identity). Since $sus \in \mathrm{FS}(M) = \mathrm{FS}(M')$, the technical property given above implies that $s^\bullet \subseteq {}^\bullet u$ and ${}^\bullet s \subseteq u^\bullet$ hold in M'. Likewise it follows from $susu \in \mathrm{FS}(M')$ that $u^\bullet \subseteq {}^\bullet s$ and ${}^\bullet u \subseteq s^\bullet$ (take $x = s$ and interchange s and u in the statement of the above technical property). Hence $s^\bullet = {}^\bullet u$ and ${}^\bullet s = u^\bullet$ in M'. Analogously we have $t^\bullet = {}^\bullet u$ and ${}^\bullet t = u^\bullet$ in M'. Hence $s^\bullet = t^\bullet$ and ${}^\bullet s = {}^\bullet t$ in M', and M' is thus not T-simple. □

We would like to point out that the difference between P-simplicity and T-simplicity demonstrated above is a direct consequence of the fact that we are more interested in transitions than in places (which is formally expressed in our definitions of equivalence by requiring the existence of a bijection β between the useful transitions).

Note that it is also not possible to find for every EN system M a firing sequence equivalent EN system M' that has only live transitions. In fact, liveness of a transition t of M means (cf. Definition 8(6)) that for every $x \in \mathrm{FS}(M)$ there exists $y \in T_M^*$ such that $xyt \in \mathrm{FS}(M)$. This implies that if $\beta(\mathrm{FS}(M)) = \mathrm{FS}(M')$, then $t \in T_M$ is live iff $\beta(t) \in T_{M'}$ is live. Hence liveness of transitions is preserved by firing sequence equivalence.

4.3 Sequential EN Systems

An EN system can often be better understood when it can be seen as several communicating concurrent subsystems (or components), where each such subsystem is "simpler" (can be easier understood) than the whole system itself. It would be particularly desirable if the considered EN system could be decomposed into subsystems that no longer contain concurrency themselves, i.e., into sequential EN systems. Though this is not always directly possible, we will show in the next subsection that every EN system is equivalent with such a system. Hence "sequentially decomposable" systems are a normal form for EN systems.

In this subsection we define sequential EN systems and study several of their properties. We call an EN system sequential if its global states are not distributed (see the beginning of Section 3.3).

Definition 40. An EN system M is *sequential* if $\#C = 1$ for all $C \in \mathbb{C}_M$.

Example 17. The EN systems of Figs. 27 and 30 are sequential. This can be seen from their configuration graphs, which are given in Figs. 29 and 31, respectively.

Here is a sufficient (structural) condition for sequentiality; it is satisfied by the EN systems of the previous example.

Lemma 41. *If* $M = (P, T, F, C_{in})$ *is an EN system for which*
(1) $\#C_{in} = 1$, *and*
(2) $\#(^\bullet t) = \#(t^\bullet) = 1$ *for all* $t \in T$,
then M *is sequential.*

Proof. It is easy to prove by induction on C that $\#C = 1$ for all $C \in \mathbb{C}_M$. □

Systems that just satisfy the second property of this lemma are called *state machines* or *S-systems* in the literature, see, e.g., [Hac72, DesEsp95].

Finite automata, which are the usual model of finite-state sequential systems, are closely related to EN systems with the two properties of Lemma 41. The differences are that finite automata also have final states, that their transitions are labeled, and that transitions t with $^\bullet t = t^\bullet$ are allowed (cf. Definition 1(4)). These additional features have also been considered for Petri nets in the literature (see, e.g., [Pet81, Tau89, Och95]).

It should be clear that in the case of reduced EN systems the two properties of Lemma 41 characterize sequentiality. In fact, for a sequential EN system M, a transition with more than one place in its input- or output-set must be useless. If, moreover, M is strongly reduced, then every configuration $\{p\}$ is reachable, because every place p is the input- or output-set of at least one useful transition.

Lemma 42. *Let* $M = (P, T, F, C_{in})$ *be a reduced EN system.*
(1) M *is sequential iff*
(i) $\#C_{in} = 1$, *and*
(ii) $\#(^\bullet t) = \#(t^\bullet) = 1$ *for all* $t \in T$.
(2) If M *is strongly reduced and sequential, then* $\mathbb{C}_M = \{\{p\} \mid p \in P\}$.

From this lemma it easily follows that isomorphism and configuration equivalence coincide for strongly reduced sequential systems.

Theorem 43. *Let* M *and* M' *be two strongly reduced sequential EN systems. Then* $M \approx M'$ *iff* $M \equiv M'$.

Every strongly reduced sequential system is in fact isomorphic with the finite automaton constructed in the proof of Theorem 12.

The definition of a sequential EN system (Definition 40) is "place oriented": the global state of the system is not distributed. Another possibility is a "transition oriented" definition, where we require that the global state transitions are not distributed (see the beginning of Section 3.3), i.e., that concurrent steps do not occur (cf. the discussion of the notions of sequentiality and concurrency in Section 3.5). EN systems satisfying this property are "concurrency-free".

Definition 44. An EN system M is *concurrency-free* if there do not exist $C \in \mathbb{C}_M$ and $t_1, t_2 \in T_M$ such that $\{t_1, t_2\}$ **con** C.

It should be clear that every sequential EN system is concurrency-free. This does not hold the other way around: the EN systems of Figs. 26 and 32 are concurrency-free (as can be seen from their configuration graphs in Figs. 28 and 33, respectively), but not sequential. It is shown in Examples 13 and 14 that these EN systems are (weakly) configuration equivalent with sequential EN systems. However, there exist concurrency-free EN systems that are not even firing sequence equivalent with any sequential EN system. To show this we observe that for a sequential EN system $M = (P, T, F, C_{in})$ and for $x, y, z \in T^*$ and $t \in T$, if $xt, yt, xz \in \mathrm{FS}(M)$ then $yz \in \mathrm{FS}(M)$; this is because if $C_{in}[x\rangle C$ and $C_{in}[y\rangle D$, then $C = {}^\bullet t = D$. The concurrency-free EN system of Fig. 36 does not satisfy this property: take $x = t_1$, $y = t_2$, $t = t_3$, and $z = t_3 t_2$.

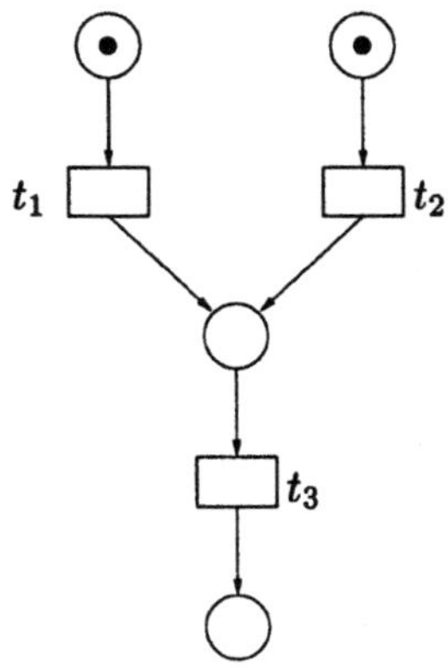

Fig. 36. A concurrency-free EN system that is not firing sequence equivalent with any sequential EN system.

Finally we observe that Theorem 43 does not hold for concurrency-free systems: the EN systems M (Fig. 26) and M' (Fig. 27) from Example 13 are configuration equivalent, but not isomorphic.

4.4 Subsystems and Sequential Components

In this subsection we will show that every EN system is configuration equivalent with an EN system that can be decomposed into sequential subsystems, i.e., in subsystems that, considered on their own, are sequential EN systems (see Definition 40). Sequential subsystems will also be called sequential components.

We first define the notions 'subsystem' and 'sequential component' and then study several of their characteristics. These notions, or variants of them, can be found in, e.g., [Hac72, DesEsp95].

Intuitively, a subsystem M' of M consists of a set of places of M (the local states of the subsystem) together with all transitions of M that can put tokens in these places and/or can remove tokens from these places, i.e., all transitions that belong to the neighbourhoods of these places. The flow relation F' of M' is completely determined by the flow relation F of M: a place and a transition of M' are connected by F' iff they are connected by F. The same holds for the inital configuration: a place of M' belongs to the initial configuration of M' iff it belongs to the initial configuration of M. Hence the subsystem is uniquely determined by its set of places.

Definition 45. Let $M = (P, T, F, C_{in})$ and $M' = (P', T', F', C'_{in})$ be EN systems. M' is a ***subsystem of*** M if:
(1) $P' \subseteq P$, $T' \subseteq T$, $F' \subseteq F$, $C'_{in} \subseteq C_{in}$,
(2) $\forall p \in P'$: $\mathbf{nbh}_M(p) \subseteq \mathbf{nbh}_{M'}(p)$, and
(3) $\forall p \in P'$: if $p \in C_{in}$, then $p \in C'_{in}$.
If, moreover, M' is a sequential EN system, then M' is a ***sequential component*** of M.

Note that the empty EN system $(\varnothing, \varnothing, \varnothing, \varnothing)$ and M itself are subsystems of M. They will be called the ***trivial*** subsystems of M.

In the next lemma we state some properties of subsystems that are easy to prove.

Lemma 46. *Let $M = (P, T, F, C_{in})$ and $M' = (P', T', F', C'_{in})$ be EN systems.*
(1) M' is a subsystem of M iff
$P' \subseteq P$, $T' = \mathbf{nbh}_M(P')$, $F' = F \cap ((P' \times T') \cup (T' \times P'))$, *and* $C'_{in} = C_{in} \cap P'$.
(2) If M' is a subsystem of M then:
for every $t \in T'$, $({}^\bullet t)_{M'} = ({}^\bullet t)_M \cap P'$ *and* $(t^\bullet)_{M'} = (t^\bullet)_M \cap P'$,
for every $t \in T - T'$, $\mathbf{nbh}_M(t) \cap P' = \varnothing$,
for every $p \in P'$, $({}^\bullet p)_{M'} = ({}^\bullet p)_M$ *and* $(p^\bullet)_{M'} = (p^\bullet)_M$.

Lemma 46(1) says that a subsystem is indeed completely determined by its set of places. However, not every set of places of an EN system determines a subsystem! Formally this means that, for a given EN system $M = (P, T, F, C_{in})$ and a given subset P' of P, the 4-tuple $M' = (P', T', F', C'_{in})$ with $T' = \mathbf{nbh}_M(P')$, $F' = F \cap ((P' \times T') \cup (T' \times P'))$, and $C'_{in} = C_{in} \cap P'$ does not have to be an EN system. For example, take $P = \{p, q\}$, $T = \{t\}$, $F = \{(p, t), (t, q)\}$, and $C_{in} = \{p\}$, and consider $P' = \{p\}$. Then $T' = \{t\}$, $F' = \{(p, t)\}$, and $C'_{in} = \{p\}$. But M' is not an EN system because (P', T', F') is not a net: $(t^\bullet)_{M'} = \varnothing$ (see Definition 1(3)).

If the 4-tuple M' determined by the set P' (as in Lemma 46(1)) is an EN system, then we call M' the ***subsystem*** of M ***determined by*** P'. We now characterize the sets of places that determine subsystems.

Lemma 47. *Let $M = (P, T, F, C_{in})$ be an EN system and let $S \subseteq P$. There exists a subsystem M' of M with $P_{M'} = S$ iff ${}^\bullet S = S^\bullet$.*

Proof. (Only-if) Assume that $M' = (S, T', F', C'_{in})$ is a subsystem of M. We first prove that ${}^\bullet S \subseteq S^\bullet$ (where the $^\bullet$ is of course the one of M). Take a $t \in {}^\bullet S$ and let $p \in S$ such that $t \in {}^\bullet p$. Then, according to Definition 45(2), $t \in \mathbf{nbh}_{M'}(p)$ and thus $t \in T'$. Definition 1(3) then implies that there exists $q \in S$ with $(q,t) \in F'$, and thus $(q,t) \in F$. This means that $t \in S^\bullet$. The inclusion $S^\bullet \subseteq {}^\bullet S$ can be proved analogously.

(If) Assume that ${}^\bullet S = S^\bullet$. Define $M' = (P', T', F', C'_{in})$ with $P' = S$ and T', F', and C'_{in} as in Lemma 46(1). It is easy to check that M' is an EN system, i.e., M' satisfies the conditions from Definitions 1 and 5. In particular, the equality ${}^\bullet S = S^\bullet$ guarantees that Definition 1(3) is satisfied. According to Lemma 46(1), M' is then a subsystem of M. □

Note that condition (3) from Definition 1 is essentially used in the above proof.

Since a subsystem is uniquely determined by its set of places, in the sequel we will *often make no distinction between a subsystem and its set of places.* Thus, according to Lemma 47: a set of places S is a subsystem iff ${}^\bullet S = S^\bullet$. This property of S has the following equivalent formulation: for all $t \in T$, ${}^\bullet t \cap S \neq \varnothing$ iff $t^\bullet \cap S \neq \varnothing$. Note also that, for a subsystem $M' = (S, T', F', C'_{in})$ of M, $T' = \mathbf{nbh}(S) = {}^\bullet S \cup S^\bullet$ according to Lemma 46(1), and hence $T' = {}^\bullet S = S^\bullet$. Finally note that, according to Lemma 47, the union of two subsystems is again a subsystem (or more precisely: the union of the sets of places of two subsystems determines a subsystem).

Example 18. (1) The subset $\{p_3, p_5\}$ of the set of places of the EN system M from Fig. 37 is a subsystem of M; this can easily be proved using Lemma 47. The subsystem determined by $\{p_3, p_5\}$ is $M' = (S, T, F, C_{in})$ with $S = \{p_3, p_5\}$, $T = \{t_3, t_4\}$, $F = \{(p_3, t_4), (t_4, p_5), (p_5, t_3), (t_3, p_3)\}$, and $C_{in} = \{p_3\}$. Since M' is sequential, M' is thus a sequential component of M. Another subsystem of M is $\{p_1, p_2, p_3, p_4\}$; for this subsystem $T = \{t_1, t_2, t_3, t_4\}$, $F = \{(t_1, p_1), (p_1, t_2), (t_2, p_2), (p_2, t_3), (t_3, p_3), (p_3, t_4), (t_4, p_4), (p_4, t_1)\}$, and $C_{in} = \{p_1, p_3\}$. Since $\#C_{in} > 1$ this is not a sequential component of M. Using Lemma 47 one can check that the only other subsystems of M are the trivial ones, i.e., the empty subsystem and the system M itself.

(2) Subsystems of the EN system M of Fig. 9 are $\{p_1, p_2\}$ (the consumer), $\{c_1, c_2\}$ (the producer), and $\{p_1, p_2, c_1, c_2\}$ (the union of the producer and the consumer). The producer and the consumer are both sequential components of M, but their union is not. M has no other nontrivial subsystems. In particular $\{b\}$ (the buffer) is not a subsystem of M. Thus, our notion of subsystem is more restricted than suggested in the discussion of the producer/consumer problem in Section 3.1, where the buffer was viewed as a component of the system. In Example 23 (Fig. 47) we will see that there is a slight variation of M in which the buffer is also a sequential component.

We will give (in Theorem 49) a characterization of the sets of places that determine sequential components. First we prove an important property of subsystems: when we restrict a reachable configuration of an EN system to a subsystem, then we obtain again a reachable configuration of the subsystem.

Lemma 48. *Let $M' = (S, T', F', C'_{in})$ be a subsystem of an EN system $M = (P, T, F, C_{in})$.*

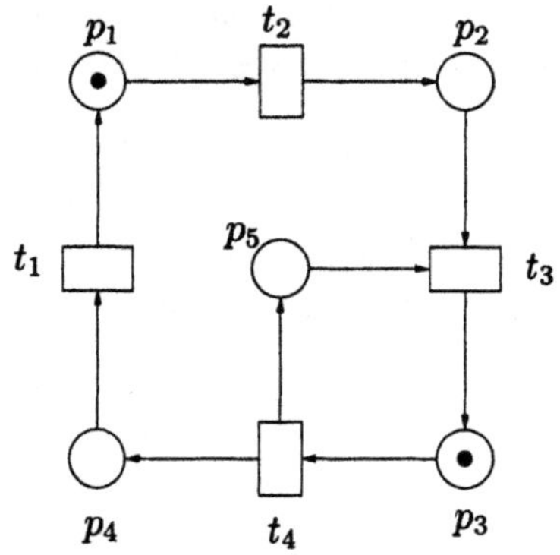

Fig. 37. An EN system with two nontrivial subsystems: $\{p_3, p_5\}$ and $\{p_1, p_2, p_3, p_4\}$.

(1) For all $C \subseteq P$, if $C \in \mathbb{C}_M$ then $C \cap S \in \mathbb{C}_{M'}$.
(2) For all $t \in T'$, if $t \in \mathbf{use}_M(T)$ then $t \in \mathbf{use}_{M'}(T')$.

Proof. Before proving (1) and (2), let us consider configurations C, D of M and a transition t of M such that $C \cap S \in \mathbb{C}_{M'}$ and $C[t\rangle_M D$. Then we claim the following: if $t \notin T'$ then $D \cap S = C \cap S$, and if $t \in T'$ then $(C \cap S)[t\rangle_{M'}(D \cap S)$. To prove this, note first that, by Lemma 7, $C - D = ({}^\bullet t)_M$ and $D - C = (t^\bullet)_M$, and hence $(C \cap S) - (D \cap S) = ({}^\bullet t)_M \cap S$ and $(D \cap S) - (C \cap S) = (t^\bullet)_M \cap S$. If $t \notin T'$, then, by Lemma 46(2), $({}^\bullet t)_M \cap S = \varnothing$ and $(t^\bullet)_M \cap S = \varnothing$, and so $D \cap S = C \cap S$. If $t \in T'$, then, by Lemma 46(2), $({}^\bullet t)_{M'} = ({}^\bullet t)_M \cap S$ and $(t^\bullet)_{M'} = (t^\bullet)_M \cap S$, and so, by Lemma 7, $(C \cap S)[t\rangle_{M'}(D \cap S)$. This proves our claim.

(1) We prove this by induction on C. By Lemma 46(1) the statement holds for $C = C_{in}$. Now assume that $C \cap S \in \mathbb{C}_{M'}$ and let $C[t\rangle_M D$ for a $t \in T$. By the above claim, either $t \notin T'$ and $D \cap S = C \cap S$, or $t \in T'$ and $(C \cap S)[t\rangle_{M'}(D \cap S)$. In both cases, $D \cap S \in \mathbb{C}_{M'}$.

(2) Assume that $t\ \mathbf{con}_M\ C$ for a $C \in \mathbb{C}_M$ and let $C[t\rangle_M D$. By (1), $C \cap S \in \mathbb{C}_{M'}$. Thus, by the above claim, $(C \cap S)[t\rangle_{M'}(D \cap S)$. Hence $t\ \mathbf{con}_{M'}\ C \cap S$. □

In the next example we will show that the mapping from $\mathbb{C}_M$ to $\mathbb{C}_{M'}$ that maps C to $C \cap S$ (see Lemma 48(1)) need not be surjective nor injective.

Example 19. Let M be the EN system of Fig. 38. Its configuration graph is drawn in Fig. 39.

(1) Figure 40 shows a subsystem M_1 of M, and Fig. 41 gives the configuration graph of M_1. For $\{p_1, p_5\} \in \mathbb{C}_{M_1}$ there does not exist $C \in \mathbb{C}_M$ such that $C \cap P_{M_1} = \{p_1, p_5\}$; this also holds for $\{p_4, p_5\} \in \mathbb{C}_{M_1}$. Furthermore, note that though M is a concurrency-free EN system, M_1 is not.

(2) Figure 42 shows a subsystem M_2 of M (and of M_1), and Fig. 43 gives the configuration graph of M_2. This subsystem is a sequential component of M. The configurations $\{p_1, p_2, p_3\}, \{p_2, p_3, p_4\}, \{p_2, p_6\} \in \mathbb{C}_M$ all give the configuration $\{p_2\} \in \mathbb{C}_{M_2}$ when intersected with P_{M_2}.

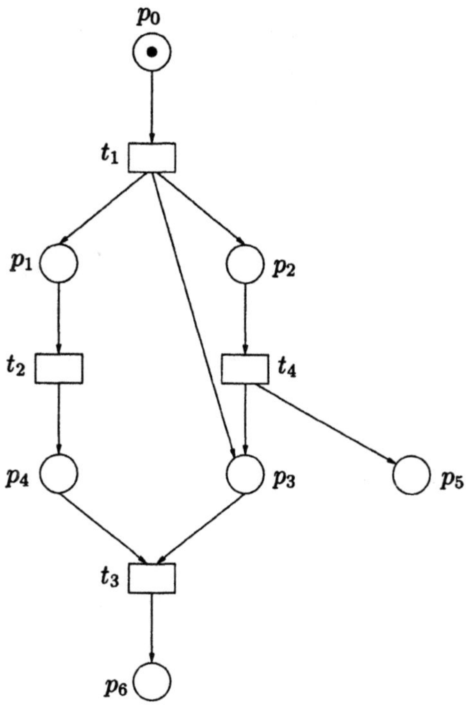

Fig. 38. An EN system M.

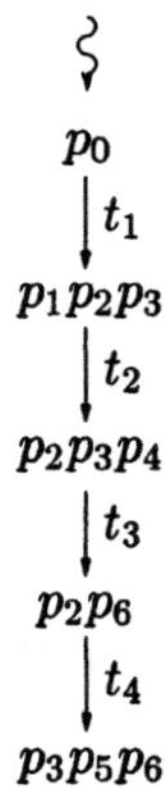

Fig. 39. The configuration graph of M.

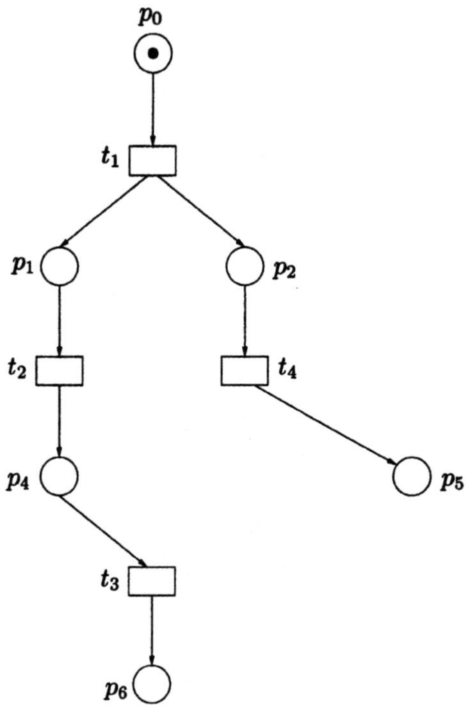

Fig. 40. A subsystem M_1 of M.

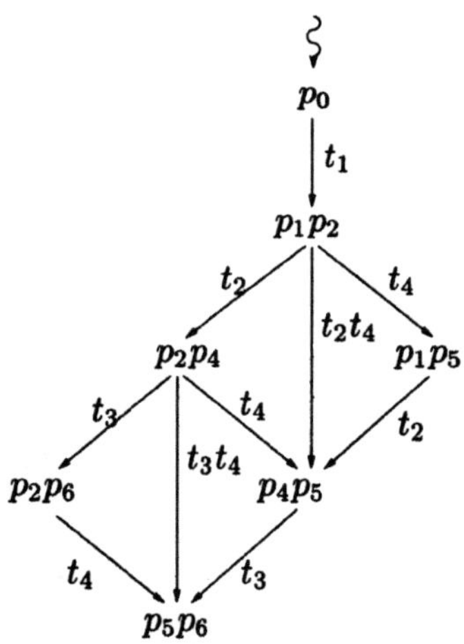

Fig. 41. The configuration graph of M_1.

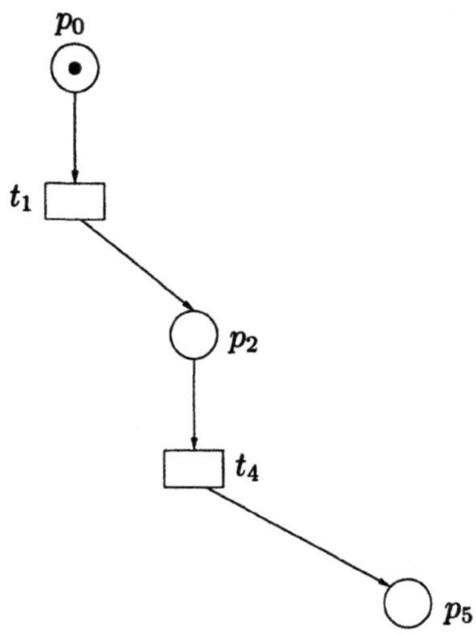

Fig. 42. A subsystem M_2 of M.

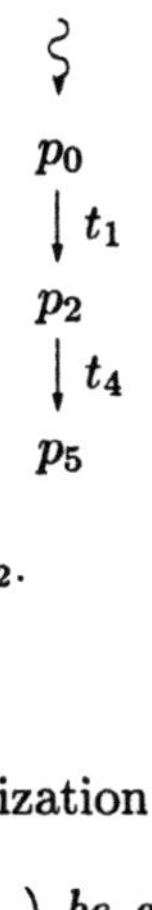

Fig. 43. The configuration graph of M_2.

Compare the following characterization with Lemma 42(1).

Theorem 49. *Let $M = (P, T, F, C_{in})$ be a reduced EN system and let $S \subseteq P$. Then the following statements are equivalent.*
(1) There is a sequential component M' of M with $P_{M'} = S$.
(2) $\#(C \cap S) = 1$ for all $C \in \mathbb{C}_M$.
(3) (i) $\#(C_{in} \cap S) = 1$, and
(ii) $\forall t \in T : \#({}^\bullet t \cap S) = \#(t^\bullet \cap S) = 1$ or $\#({}^\bullet t \cap S) = \#(t^\bullet \cap S) = 0$.

Proof. (1) implies (2): Follows directly from Lemma 48(1) and the definition of a sequential EN system (Definition 40).

(2) implies (3): To prove (ii), consider a $t \in T$. Since M is reduced, there is a $C \in \mathbb{C}_M$ with t **con** C. Since $\#(C \cap S) = 1$, $\#({}^\bullet t \cap S) \leq 1$. Now let $C[t\rangle D$. Since $\#(D \cap S) = 1$, it is easy to see that $\#({}^\bullet t \cap S) = \#(t^\bullet \cap S)$.

(3) implies (1): Condition (3)(ii) implies that ${}^\bullet S = S^\bullet$. Hence, according to Lemma 47, there exists a (unique) subsystem M' of M with $P_{M'} = S$. Then, by Lemma 46(1), $\#((C_{in})_{M'}) = 1$ and, by Lemma 46(2), $\#(({}^\bullet t)_{M'}) = \#((t^\bullet)_{M'}) = 1$ for all $t \in T_{M'}$. Lemma 41 now implies that M' is sequential. □

According to Theorem 49(2) a sequential component of an EN system is always in exactly one local state (there is always exactly one token in its set of places).

Subsystems S that just satisfy property 3(ii) of Theorem 49 and, moreover, are strongly connected (viewed as graphs), are called *state machine components* or *S-components* in the literature (see, e.g., [DesEsp95]). Such subsystems contain an arbitrary, but fixed number of tokens.

We will now show that, for strongly reduced EN systems, sequential components themselves do not have nontrivial subsystems and are thus not decomposable.

Lemma 50. *Let M be a strongly reduced sequential EN system and let M' be a subsystem of M. Then M' is trivial.*

Proof. Let $M = (P, T, F, C_{in})$ and $M' = (S, T', F', C'_{in})$. By Lemma 47, ${}^\bullet S = S^\bullet$. By Lemma 46 it suffices to prove that $S = \varnothing$ or $S = P$. Assume that $S \neq \varnothing$; then it remains to prove that $S = P$. Since M is strongly reduced, $\mathbb{C}_M = \{\{p\} \mid p \in P\}$ according to Lemma 42(2). Thus for all $t \in T$ and $p, q \in P$, if $\{p\}[t\rangle_M\{q\}$, then ($p \in S$ iff $q \in S$). This implies that for all $x \in T^*$ and $p, q \in P$, if $\{p\}[x\rangle_M\{q\}$, then ($p \in S$ iff $q \in S$). Now let $C_{in} = \{p_0\}$. Take a $q \in S$. Since $\{q\} \in \mathbb{C}_M$, $\{p_0\}[x\rangle_M\{q\}$ for some $x \in T^*$. Thus $p_0 \in S$. To prove that $P \subseteq S$, consider an arbitrary $p \in P$. Since $\{p\} \in \mathbb{C}_M$, $\{p_0\}[x\rangle_M\{p\}$ for some $x \in T^*$. Thus $p \in S$, and so $S = P$. □

Theorem 51. *Let M be a strongly reduced EN system, and let M' be a sequential component of M. Then M' has no nontrivial subsystems.*

Proof. Since M is reduced, Lemma 48(2) implies that M' is also reduced. Since M has no isolated places, neither does M' (by Definition 45(2)). Hence M' is strongly reduced, and so, by Lemma 50 has no nontrivial subsystems. □

The following example demonstrates that this theorem does not hold the other way around.

Example 20. Let M be the EN system from Example 19 (Fig. 38). Figure 44 shows a subsystem M_3 of M that is not sequential. It is straightforward to verify that M_3 has only trivial subsystems.

We now define the notion of a decomposition of an EN system into subsystems, and in particular into sequential components. It is called a "covering" of the EN system (see, e.g., [Hac72] or Chapter 5 of [DesEsp95]).

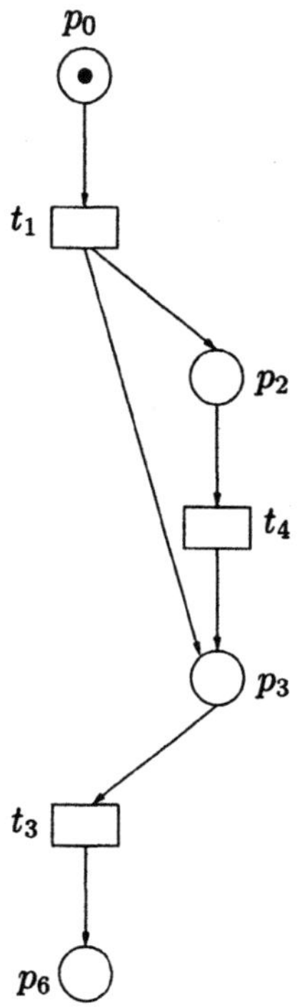

Fig. 44. A subsystem M_3 of M.

Definition 52. Let $M = (P, T, F, C_{in})$ be an EN system.
(1) A set $\{M_1, \ldots, M_n\}$ of subsystems of M, $n \geq 0$, with $M_i = (S_i, T_i, F_i, (C_{in})_i)$ for $1 \leq i \leq n$, is a *covering* of M if $P = \bigcup_{i=1}^n S_i$, $T = \bigcup_{i=1}^n T_i$, $F = \bigcup_{i=1}^n F_i$, and $C_{in} = \bigcup_{i=1}^n (C_{in})_i$.
(2) M is *covered by sequential components* if there exists a covering $\{M_1, \ldots, M_n\}$, $n \geq 0$, of M such that M_i is a sequential component of M for every $1 \leq i \leq n$.

Since a subsystem can be identified with its set of places, it should be clear that a set of subsystems of M is a covering of M if their union is the set of places of M.

Lemma 53. *Let $M = (P, T, F, C_{in})$ be an EN system, and let, for every $1 \leq i \leq n$ (with $n \geq 0$), $M_i = (S_i, T_i, F_i, (C_{in})_i)$ be a subsystem of M. Then $\{M_1, \ldots, M_n\}$ is a covering of M iff $P = \bigcup_{i=1}^n S_i$.*

An EN system that is covered by sequential components $M_1, \ldots, M_n$ can intuitively be viewed as a system consisting of the communicating concurrent subsystems $M_1, \ldots, M_n$, where each subsystem M_i is sequential. The communication between components takes place through synchronization on shared transitions. Components may also share places, meaning that the sets of places $S_1, \ldots, S_n$ of $M_1, \ldots, M_n$, respectively, need not be disjoint. If there is a token in place $p \in S_1 \cap S_2$, then this means that both components M_1 and M_2 are in the same local state p. Hence a token in p represents all components to which p belongs. Note that, in general, an EN system can have several coverings by sequential components, i.e., several interpretations as a set of communicating components.

Example 21. (1) The EN system M of Fig. 45 is covered by one sequential component, viz. M itself. The system itself is sequential. With $C_{in} = \{p_1, p_2\}$ the system has no covering by sequential components.

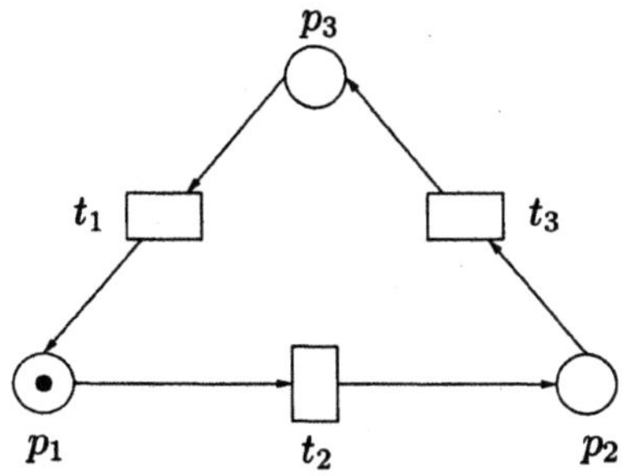

Fig. 45. A sequential EN system.

(2) The only sequential component of the EN system of Fig. 37 is $\{p_3, p_5\}$, see Example 18(1). We already saw in that example that Lemma 47 can be used to conclude that p_1, p_2, and p_4 do not belong to any sequential component. It is often quicker to use Theorem 49 for that purpose: for example, if S is the set of places of a sequential component and $p_2 \in S$, then $p_1 \in S$ and $p_3 \in S$ by Theorem 49(3)(ii), and hence $\#(C_{in} \cap S) = 2$, contradicting Theorem 49(2).

(3) The EN system of Fig. 12 has a covering by two sequential components, viz. $\{p_0, p_1, p_2, p_4, p_6\}$ and $\{p_0, p_1, p_3, p_5, p_6\}$. Intuitively, these two components work on the same job in places p_0, p_1, and p_6, but work on different jobs when they are in places p_2, p_4 and p_3, p_5, respectively.

(4) The EN system of Fig. 2 is covered by three sequential components. In terms of the sets of places, the components are: $\{w_1, c_1, r_1\}$ (component 1), $\{w_2, c_2, r_2\}$ (component 2), and $\{p, c_1, c_2\}$ (the permission component). If we give the complete specification, then component 1 is the EN system (S, T, F, C_{in}) with $S = \{w_1, c_1, r_1\}$, $T = \{in_1, out_1, d_1\}$, $F = \{(w_1, in_1), (in_1, c_1), (c_1, out_1), (out_1, r_1), (r_1, d_1), (d_1, w_1)\}$, and $C_{in} = \{w_1\}$; and analogously for component 2. The permission component is then the EN system (S, T, F, C_{in}) with $S = \{p, c_1, c_2\}$, $T = \{in_1, out_1, in_2, out_2\}$, $F = \{(p, in_1), (in_1, c_1), (c_1, out_1), (out_1, p), (p, in_2), (in_2, c_2), (c_2, out_2), (out_2, p)\}$, and $C_{in} = \{p\}$.

(5) The EN system of Fig. 9 is not covered by sequential components. Both the producer ($\{p_1, p_2\}$) and the consumer ($\{c_1, c_2\}$) are sequential components, but the buffer b belongs to no sequential component (see also Example 18(2)).

(6) See Examples 19 and 20. M is the EN system in Fig. 38, and M_1 and M_3 are the subsystems of M in Figs. 40 and 44, respectively. It should be clear that $\{M_1, M_3\}$ is a covering of M. However, there does not exist a covering of M by sequential components, since each subsystem M' that contains p_3 must also contain p_2 and t_1; then $\#(t_1{}^\bullet \cap P_{M'}) \geq 2$, and so, by Theorem 49, M' is not a sequential component. M_1 is covered by sequential components, but M_3 is not.

We will now show that for every EN system M there exists a configuration equivalent EN system M' that is covered by sequential components.

Theorem 54. *For every EN system M there exists a reduced EN system M' that is configuration equivalent with M and that is covered by at most $\#P_M$ sequential components.*

In the remainder of this subsection we will prove this theorem. Note that according to Theorem 36 we may assume that M is reduced. The proof technique is based on the so-called *complement construction*.

Definition 55. Let M be an EN system and let $p, q \in P_M$. Then p and q are *complementary*, denoted by p **com** q, if $p^\bullet = {}^\bullet q$ and ${}^\bullet p = q^\bullet$.

Example 22. Let M be the EN system of Fig. 37. It is clear that p_3 **com** p_5, and that there are no other complementary places.

In general a place can have several complementary places. If the EN system is P-simple, then each place has at most one complementary place.

The complement construction is based on the following property of two complementary places.

Lemma 56. *Let $M = (P, T, F, C_{in})$ be a reduced EN system. For all $p, q \in P$, $\{p, q\}$ is a sequential component of M iff $\#(C_{in} \cap \{p, q\}) = 1$ and p **com** q.*

Proof. We use Theorem 49(3). It is easy to check that, for arbitrary $p, q \in P$, p **com** q iff $S = \{p, q\}$ satisfies condition (ii) of Theorem 49(3) (where we use Definition 1(4): ${}^\bullet t \cap t^\bullet = \varnothing$ for all $t \in T$). □

If, moreover, M is strongly reduced, then the condition $\#(C_{in} \cap \{p, q\}) = 1$ can be omitted from the statement of Lemma 56. Thus, for strongly reduced EN systems, the complementary places are exactly the sequential components of size two.

By Lemma 56, in a reduced EN system two complementary places p and q (of which exactly one is in C_{in}) form a sequential component. That means that there is a token in q iff there is no token in p. Viewing p and q as booleans, q is the negation of p. The complement construction constructs a complement for those places that do not yet belong to a sequential component. We do this place by place, as follows.

Theorem 57. *Let M be a reduced EN system and let $p_0 \in P_M$. Then there exists a reduced EN system M' that is configuration equivalent with M, such that:*
(1) $P_{M'} = P_M \cup \{q_0\}$ with $q_0 \notin P_M$,
(2) $\{p_0, q_0\}$ is a sequential component of M', and
(3) for every $S \subseteq P_M$,
S is a sequential component of M iff S is a sequential component of M'.

Proof. Let $M = (P, T, F, C_{in})$. We define $M' = (P', T', F', C'_{in})$ by setting
$P' = P \cup \{q_0\}$, where q_0 is a new place (i.e., $q_0 \notin P \cup T$),
$T' = T$,
$F' = F \cup \{(q_0, t) \mid (t, p_0) \in F\} \cup \{(t, q_0) \mid (p_0, t) \in F\}$, and
$C'_{in} = C_{in}$ if $p_0 \in C_{in}$, and $C'_{in} = C_{in} \cup \{q_0\}$ if $p_0 \notin C_{in}$.
To prove that M and M' are configuration equivalent, we define the function $\alpha : \mathbb{C}_M \to \mathcal{P}(P')$ by $\alpha(C) = C$ if $p_0 \in C$, and $\alpha(C) = C \cup \{q_0\}$ if $p_0 \notin C$. Note that α is injective (because $\alpha(C) \cap P = C$) and that $\alpha(C_{in}) = C'_{in}$. According to Lemma 30 (with β the identity on T) it now suffices to prove that for all $C, D \in \mathbb{C}_M$ and $t \in T$,

$$t \ \mathbf{con}_{M'} \ \alpha(C) \text{ implies } t \ \mathbf{con}_M \ C \text{, and} \qquad (*)$$

$$C[t\rangle_M D \text{ implies } \alpha(C)[t\rangle_{M'} \alpha(D). \qquad (**)$$

Implication (*) follows from: $({}^\bullet t)_M = ({}^\bullet t)_{M'} \cap P$, $(t^\bullet)_M = (t^\bullet)_{M'} \cap P$, and $\alpha(C) \cap P = C$. To prove (**) we distinguish three cases.
Case 1: $p_0 \in (t^\bullet)_M$. Then $({}^\bullet t)_{M'} = ({}^\bullet t)_M \cup \{q_0\}$ and $(t^\bullet)_{M'} = (t^\bullet)_M$. Moreover $p_0 \notin C$ and $p_0 \in D$, and so $\alpha(C) = C \cup \{q_0\}$ and $\alpha(D) = D$. Hence $\alpha(C) - \alpha(D) = (C - D) \cup \{q_0\}$ and $\alpha(D) - \alpha(C) = D - C$. Thus $C - D = ({}^\bullet t)_M$ implies $\alpha(C) - \alpha(D) = ({}^\bullet t)_{M'}$, and $D - C = (t^\bullet)_M$ implies $\alpha(D) - \alpha(C) = (t^\bullet)_{M'}$. By Lemma 7, this proves implication (**) for this case.
Case 2 ($p_0 \in ({}^\bullet t)_M$) and Case 3 ($p_0 \notin ({}^\bullet t)_M \cup (t^\bullet)_M$) can be proved in a completely analogous way. Hence, by Lemma 30, M and M' are (α, β)-configuration equivalent. By Definition 28, $\mathbf{use}_{M'}(T') = T$, and so M' is reduced.

It is clear that (1) holds, (2) follows from Lemma 56, and (3) follows easily from Theorem 49(3). □

Now, Theorem 54 follows directly from the repeated (at most $\#P_M$ times) application of Theorem 57. In this way, each place that did not yet belong to a sequential component will be complemented and consequently, by Lemma 56, covered by a sequential component.

Example 23. (1) The EN system of Fig. 46 is obtained from the one of Fig. 37 by complementing places p_1, p_2, and p_4. (2) The EN system of Fig. 47 is obtained from the EN system of Fig. 9 by complementing the buffer place $b = b_f$. It is covered by three sequential components: the producer $\{p_1, p_2\}$, the consumer $\{c_1, c_2\}$, and the buffer $\{b_f, b_e\}$. It should be clear that this EN system still models the producer/consumer problem as discussed in Section 3.1. Rather than treating place b itself as the buffer (which is full iff it contains a token), we now represent the two possible states of the buffer by the two places b_f (full buffer) and b_e (empty buffer).

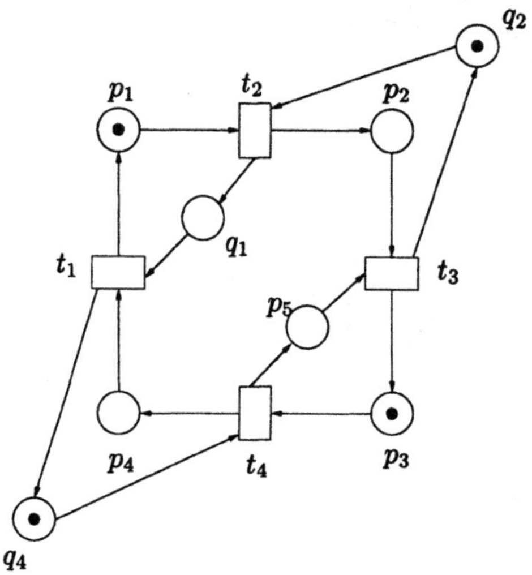

Fig. 46. The result of complementing places p_1, p_2, and p_4 of the EN system of Fig. 37.

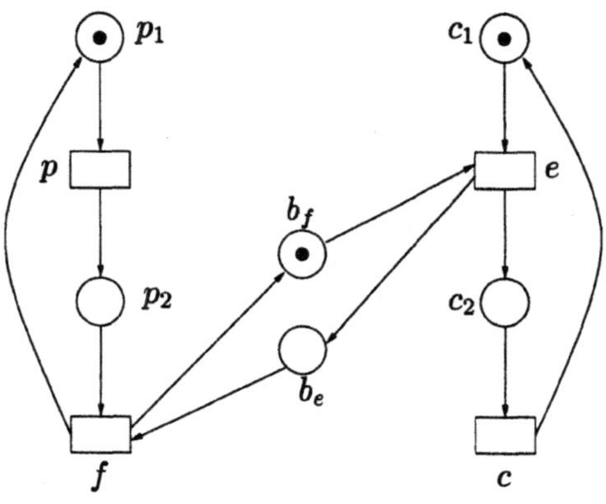

Fig. 47. The producer/consumer system with three sequential components.

4.5 Contact-freeness

In an EN system M, an event t has concession in a configuration C if it has both input-concession (i.e., ${}^\bullet t \subseteq C$) and output-concession (i.e., $t^\bullet \cap C = \varnothing$). A transition that has input-concession in C, need not have output-concession in C. This is called *contact*, and is illustrated in Fig. 48. As an example, transition f has contact in configuration $\{p_2, b, c_1\}$ of the producer/consumer system in Fig. 9; it cannot be fired because the buffer is full. Note that in the corresponding configuration $\{p_2, b_f, c_1\}$ of the producer/consumer system of Fig. 47 transition f does not have input-concession.

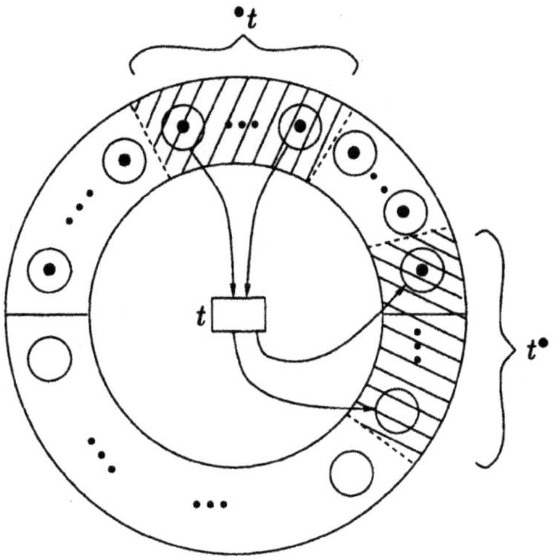

Fig. 48. Contact.

In general, to decide whether t has concession in C, one has to check both the pre-conditions and the post-conditions of t. However, if there is never contact in M, then t has concession in C iff all the pre-conditions of t in C are satisfied: thus one does not have to check the post-conditions of t!

For this reason EN systems without contact play an important part in the theory of the behaviour of EN systems. They are formally defined as follows.

Definition 58. Let $M = (P, T, F, C_{in})$ be an EN system.
M is *contact-free* if for all $t \in T$ and $C \in \mathbb{C}_M$, if ${}^\bullet t \subseteq C$ then $t^\bullet \cap C = \varnothing$.

Contact-free EN systems are also called *safe* EN systems. In fact, they are the same as the safe P/T systems (assuming that the nets of P/T systems are defined as in Definition 1).

Example 24. The EN system of Fig. 45 is contact-free (because it is sequential). With $C_{in} = \{p_1, p_2\}$ it is not contact-free.

We will now show that for every EN system there exists a configuration equivalent reduced EN system that is contact-free. Hence contact-free EN systems are a normal form for EN systems. This is a simple corollary of the normal form from the previous subsection.

Theorem 59. *If a reduced EN system M is covered by sequential components, then M is contact-free.*

Proof. Let $M = (P, T, F, C_{in})$ and let $C \in \mathbb{C}_M$. Let ${}^\bullet t \subseteq C$ with $t \in T$ and assume that $t^\bullet \cap C \neq \varnothing$. Let $p \in t^\bullet \cap C$ and let $S \subseteq P$ be a sequential component of M with $p \in S$. Then $p \in t^\bullet \cap S$ and so $\#(t^\bullet \cap S) = 1 = \#({}^\bullet t \cap S)$ by Theorem 49(3). Hence there is a place $q \in {}^\bullet t \cap S \subseteq C \cap S$. Thus $\#(C \cap S) \geq 2$, contradicting Theorem 49(2). □

Theorems 54 and 59 imply that contact-freeness is a normal form.

Theorem 60. *For every EN system there exists a configuration equivalent reduced contact-free EN system.*

Theorem 59 does not hold the other way around! The EN system of Fig. 49 has no sequential components, but it is contact-free. By (the proofs of) Theorems 3.15 and 3.18 of [DesEsp95] (see also Section 7.3 of [Rei82]), the converse of Theorem 59 *does* hold for the so-called *T-systems* (or *marked graphs*); in such systems both $p^\bullet$ and ${}^\bullet p$ contain exactly one transition, for every $p \in P$. In fact, assuming the EN system to be strongly reduced, the converse of Theorem 59 even holds for the (larger) class of EN systems that are free-choice (see the end of Section 3.5) and that have live transitions only (see Definition 8(6)); see Section 4.2 of [Hac72] or Theorem 5.6 of [DesEsp95] (the S-coverability Theorem). In these cases, the sequential components are even strongly connected.

Theorem 59 also provides a method to deduce contact-freeness, as demonstrated by the following example.

Example 25. (1) The EN system M in Fig. 45, with $C_{in} = \{p_1\}$, is sequential and hence contact-free according to Theorem 59. With $C_{in} = \{p_1, p_2\}$ the system clearly is not contact-free and hence, according to Theorem 59, it is not covered by sequential components (see Example 21(1)).

(2) The EN system M of Fig. 12 is covered by two sequential components (see Example 21(3)). Hence M is contact-free.

(3) The EN system M of Fig. 2 is covered by three sequential components (see Example 21(4)) and is thus also contact-free.

(4) The EN systems of Figs. 46 and 47 are covered by sequential components (see Example 23) and are thus contact-free.

The notion of 'contact-freeness' should not be confused with the notion of 'conflict-freeness' (see Definition 22). An example of an EN system that is both contact-free and conflict-free, is the producer/consumer system of Fig. 47. Such systems can be characterized as follows.

Theorem 61. *Let $M = (P, T, F, C_{in})$ be an EN system.*
M is contact-free and conflict-free iff
for all $C \in \mathbb{C}_M$ and all $U \subseteq T$ with $U \neq \varnothing$, if ${}^\bullet U \subseteq C$, then U **con** *C.*

If two transitions of a contact-free EN system have an output-conflict, in some configuration, then they also have an input-conflict in that configuration (cf. (3) of Section 3.5). Using this it can be shown that there is no confusion in contact-free free-choice systems (see the end of Section 3.5).

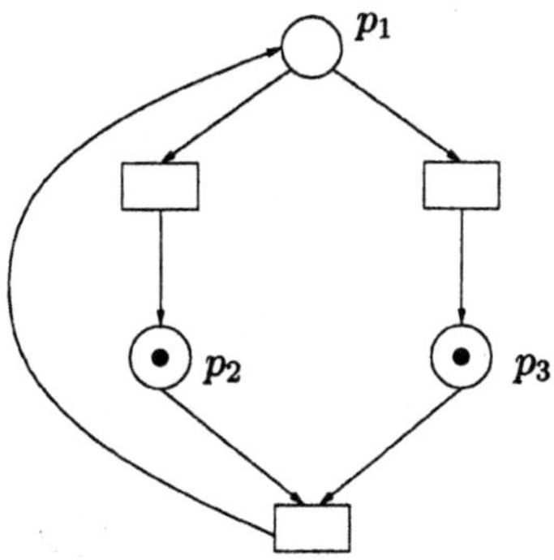

Fig. 49. A contact-free EN system without sequential components.

5 Processes

The way in which we have formalized the concurrent behaviour of an EN system in Section 3.4 still has a sequential flavour. The only difference with the sequential behaviour in Section 3.3 is that at each step, i.e., during each global state transition, several transitions can be fired simultaneously. This may be viewed as a formalization of simultaneity rather than concurrency. However, in general, actions do not occur simultaneously but they may overlap in time, in an arbitrary fashion. Thus, one component of a system can execute six actions while, independently, another component executes two actions and part of a third. In this section we will define the notion of a "process" of an EN system, which formalizes a concurrent run of the system, taking into account this feature of concurrency. In order to abstract from the notion of time, as we did before, we will only formalize that one action should be executed "before" another action, or, that one action is one of the "causes" of another action (cf. (1) in Section 3.5). Such a notion of causality between the events that occur during a run of the system, is, in general, a partial order. It will be represented by a special type of acyclic net, called "process net" (or causal net, or occurrence net).

In Section 5.1 we recall a number of notions concerning partial orders, and in Section 5.2 we consider process nets and some of their formal properties. In Sections 5.3 and 5.4 we introduce and study the processes of an EN system.

The theory of process nets and processes originated in [Pet76], and is presented in detail in [BesDev87, BesFer88] (for EN systems see in particular Section 4.4 of [BesFer88]).

5.1 Partial Orders

We start this subsection with the usual definition of a (strict) partial order.

Definition 62. Let A be a finite set. A binary relation $\rho \subseteq A \times A$ is a *partial order on A* if ρ is irreflexive and transitive; (A, ρ) is also called a *partially ordered set*. A subset B of A is *linearly ordered* if for all $a, b \in B$: $a \ \rho \ b$ or $b \ \rho \ a$ or $a = b$.

With every partial order ρ we can associate two important relations $\mathbf{li}_\rho$ and $\mathbf{co}_\rho$.

Definition 63. Let (A, ρ) be a partially ordered set. Then $\mathbf{li}_\rho \subseteq A \times A$ and $\mathbf{co}_\rho \subseteq A \times A$ are the binary relations such that, for every $a, b \in A$,
(1) $a \ \mathbf{li}_\rho \ b$ iff $a \ \rho \ b$ or $b \ \rho \ a$ or $a = b$, and
(2) $a \ \mathbf{co}_\rho \ b$ iff $\neg \ a \ \rho \ b$ and $\neg \ b \ \rho \ a$.

For a partially ordered set (A, ρ), $\mathbf{li}_\rho$ is called the *line relation of ρ* and $\mathbf{co}_\rho$ is called the *concurrency relation of ρ*.

Note that the irreflexivity of ρ implies that $a \ \mathbf{co}_\rho \ a$ for every $a \in A$. Two distinct elements of A are either comparable ($\mathbf{li}$) or incomparable ($\mathbf{co}$).

Lemma 64. *Let (A, ρ) be a partially ordered set. Then, for every $a, b \in A$,*
(1) $a \ \mathbf{li}_\rho \ b$ or $a \ \mathbf{co}_\rho \ b$, and
(2) ($a \ \mathbf{li}_\rho \ b$ and $a \ \mathbf{co}_\rho \ b$) iff $a = b$.

The maximal cliques of $\mathbf{li}_\rho$ and $\mathbf{co}_\rho$ play an important part in what follows. Maximal cliques are now defined for arbitrary reflexive symmetric relations (and note that $\mathbf{li}_\rho$ and $\mathbf{co}_\rho$ are reflexive and symmetric).

Definition 65. Let A be a finite set, let $\sigma \subseteq A \times A$ be a reflexive symmetric relation, and let $B \subseteq A$. B is a *σ-clique* if $a \ \sigma \ b$ for all $a, b \in B$, and B is a *maximal σ-clique* if B is a σ-clique and for every $a \in A - B$ there exists $b \in B$ such that $\neg \ a \ \sigma \ b$.

Lemma 66. *Let A be a finite set and let $\sigma \subseteq A \times A$ be a reflexive symmetric relation. For every σ-clique B there exists a maximal σ-clique C with $B \subseteq C$.*

Proof. If B is maximal then we are ready. Otherwise there exists $a_1 \notin B$ such that $a_1 \ \sigma \ b$ for all $b \in B$. Then $B_1 = B \cup \{a_1\}$ is a σ-clique. If B_1 is maximal then we are ready. Otherwise there exists $a_2 \notin B_1$ such that $B_2 = B_1 \cup \{a_2\}$ is a σ-clique. We iterate this procedure. Since A is finite it must terminate with a maximal σ-clique B_n. Clearly $B \subseteq B_n$. □

In particular, for every $a \in A$ there exists a maximal σ-clique C with $a \in C$ (because every singleton $\{a\}$ is a σ-clique, by the reflexivity of σ).

For a partially ordered set (A, ρ), a $\mathbf{li}_\rho$-clique is a linearly ordered subset of A, and a $\mathbf{co}_\rho$-clique is a set of mutually incomparable elements.

Definition 67. Let (A, ρ) be a partially ordered set. A maximal $\mathbf{li}_\rho$-clique is a *line of ρ* and a maximal $\mathbf{co}_\rho$-clique is a *cut of ρ*.

Note that if $A = \varnothing$ (and thus $\rho = \varnothing$), then the empty set is both a line and a cut of ρ. If $A \neq \varnothing$, then lines and cuts are nonempty sets.

It is clear from Lemma 64(2) that, for every line L and every cut C of ρ, $\#(L \cap C) \leq 1$. This leads to the following definition (see [Pet76] and Section 2.3 of [BesFer88], where it is called K-density).

Definition 68. Let (A, ρ) be a partially ordered set. The ordering ρ is *dense* if every line and every cut of ρ have a nonempty intersection.

The empty partially ordered set is clearly not dense. There also exist nonempty partially ordered sets that are not dense. Consider, e.g., the N-shaped partially ordered set (A, ρ) with $A = \{a, b, c, d\}$ and $\rho = \{(a, b), (c, b), (c, d)\}$. Then the line $\{c, b\}$ and the cut $\{a, d\}$ do not intersect.

A cut divides a partially ordered set into two parts: the part "before" (preceding) the cut, and the part "after" (following) the cut. This can be defined, for arbitrary subsets of A instead of cuts, as follows. We also define the sets of minimal and maximal elements of (a subset of) A.

Definition 69. Let (A, ρ) be a partially ordered set and let $B \subseteq A$. Then
$({}^{\rightarrow}B)_\rho = \{a \in A \mid \exists b \in B : a\ \rho\ b \text{ or } a = b\}$,
$(B^{\rightarrow})_\rho = \{a \in A \mid \exists b \in B : b\ \rho\ a \text{ or } b = a\}$,
$({}^{\circ}B)_\rho = \{b \in B \mid \neg\exists b' \in B : b'\ \rho\ b\}$, and
$(B^{\circ})_\rho = \{b \in B \mid \neg\exists b' \in B : b\ \rho\ b'\}$.

If ρ is clear from the context, then we will just write ${}^{\rightarrow}B$, $B^{\rightarrow}$, ${}^{\circ}B$, and B°. Intuitively, ${}^{\rightarrow}B$ is the part of A *before* B (including B), $B^{\rightarrow}$ is the part of A *after* B (including B), ${}^{\circ}B$ is the *initial* part of B, and B° is the *final* part of B. In the literature, ${}^{\rightarrow}B$ and $B^{\rightarrow}$ are often denoted $\downarrow B$ and $\uparrow B$, respectively.

The following technical lemma shows that every element of B is after its initial part and before its final part.

Lemma 70. *Let (A, ρ) be a partially ordered set and let $B \subseteq A$. Then $B \subseteq ({}^{\circ}B)^{\rightarrow}$ and $B \subseteq {}^{\rightarrow}(B^{\circ})$.*

Proof. To prove that $B \subseteq ({}^{\circ}B)^{\rightarrow}$, we have to show that for every $b \in B$ there exists $m \in {}^{\circ}B$ such that $m\ \rho\ b$ or $m = b$. Let $b \in B$. If $b \in {}^{\circ}B$ then we are ready. Otherwise there exists $b_1 \in B$ with $b_1\ \rho\ b$. If $b_1 \in {}^{\circ}B$ then we are ready. Otherwise there exists $b_2 \in B$ with $b_2\ \rho\ b_1$ and thus $b_2\ \rho\ b$. We iterate this procedure. Since B is finite it must terminate with a $b_n \in {}^{\circ}B$ such that $b_n\ \rho\ b$. Formally all this can be proved by induction on $\#\{b' \in B \mid b'\ \rho\ b\}$.

The proof of $B \subseteq {}^{\rightarrow}(B^{\circ})$ is "dual", i.e., follows from the above by considering the partially ordered set (A, ρ^{-1}), with $\rho^{-1} = \{(a, b) \mid (b, a) \in \rho\}$. □

Lemma 66 is a special case of (the second inclusion of) Lemma 70, for the partially ordered set $(\mathbb{C}, \subsetneq)$ where $\mathbb{C}$ is the set of all σ-cliques.

The following elementary properties of cuts can easily be proved from the definitions, using Lemma 70.

Theorem 71. *Let (A, ρ) be a partially ordered set and let B be a cut of ρ.*
(1) $^\circ A$ and A° are cuts of ρ,
(2) $(^\circ A)^\rightarrow = A$, $^\rightarrow(^\circ A) = {}^\circ A$, $(A^\circ)^\rightarrow = A^\circ$, and $^\rightarrow(A^\circ) = A$,
(3) $^\rightarrow B \cup B^\rightarrow = A$ and $^\rightarrow B \cap B^\rightarrow = B$,
(4) $^\circ(^\rightarrow B) = {}^\circ A$, $(^\rightarrow B)^\circ = B$, $^\circ(B^\rightarrow) = B$, and $(B^\rightarrow)^\circ = A^\circ$.

Proof. As an example we prove that $^\circ A$ is a cut. From Definition 69 it is obvious that $^\circ A$ is a **co**-clique. By Lemma 70, for every $a \in A - {}^\circ A$ there exists $m \in {}^\circ A$ such that $m \ \rho \ a$. This shows that $^\circ A$ is a maximal **co**-clique. □

We now present another result that uses Lemma 70: every line intersects the initial and the final cut (hence a "partial" density).

Lemma 72. *Let (A, ρ) be a partially ordered set with $A \neq \varnothing$ and let L be a line of ρ. Then $L \cap {}^\circ A \neq \varnothing$ and $L \cap A^\circ \neq \varnothing$.*

Proof. By duality it suffices to prove that $L \cap {}^\circ A \neq \varnothing$. Since A is nonempty, L is nonempty. Then, by Lemma 70 (with $B = L$), $^\circ L$ is nonempty. Since L is a **li**-clique, $^\circ L$ consists of precisely one element, say a. By Lemma 70 (with $B = A$) there exists $m \in {}^\circ A$ with $m \ \rho \ a$ or $m = a$. This implies that $L \cup \{m\}$ is a **li**-clique. Since L is maximal, $m \in L$. Hence $m \in L \cap {}^\circ A$ (and $m = a$). □

5.2 Process Nets

For the description of the concurrent runs of an EN system we will define so-called processes. In defining processes, nets of a special kind are used: process nets. These nets will be treated in this subsection.

Definition 73. A net $N = (P, T, F)$ is a *process net* if:
(1) N is acyclic, and
(2) $\#(^\bullet p) \leq 1$ and $\#(p^\bullet) \leq 1$ for all $p \in P$.

Hence, process nets are nets without cycles and with "unbranching" places only. They are also called *occurrence* nets or *causal* nets.

For every acyclic directed graph with edge relation F, the relation F^+ (which indicates the nonempty paths in the graphs) is a partial order on the set of nodes of the graph. Applying this to the directed graph G_N corresponding to the net N (see Section 3.2) gives the next result.

Lemma 74. *For every process net N, F_N^+ is a partial order on X_N.*

The above result allows us to consider a process net $N = (P, T, F)$ as a partially ordered set (X, F^+). In this way, all terminology and notations concerning partial orders introduced so far can be carried over to process nets. In particular, we write $\mathbf{li}_N$ and $\mathbf{co}_N$ instead of $\mathbf{li}_{F^+}$ and $\mathbf{co}_{F^+}$ and speak about lines and cuts of N instead of lines and cuts of F^+. Thus, for $x, y \in X$, $x \ \mathbf{co}_N \ y$ iff $\neg \ x \ F^+ \ y$ and $\neg \ y \ F^+ \ x$, and $x \ \mathbf{li}_N \ y$ iff $x \ F^+ \ y$ or $y \ F^+ \ x$ or $x = y$.

We are especially interested in cuts of a process net that consist of places only, i.e., that are configurations of the process net.

Definition 75. A *slice* of a process net N is a cut C of N such that $C \subseteq P_N$.

Lemma 76. *Let $N = (P,T,F)$ be a process net and let $C \subseteq P$.*
C is a slice of N iff
(1) for all $p,q \in C$, p $\mathbf{co}_N$ q, and
(2) for every $p \in P - C$ there exists $q \in C$ such that $\neg$ p $\mathbf{co}_N$ q.

Proof. If C is a slice of N, then (1) and (2) follow directly from the definition of a cut of N. Now assume that (1) and (2) hold. We have to prove that C is a cut of N. By (1), C is a $\mathbf{co}_N$-clique. To show that it is maximal, it suffices, by (2), to prove that for every $t \in T$ there exists $q \in C$ such that $t\ F^+\ q$ or $q\ F^+\ t$. Let $p \in t^\bullet$, i.e., $t\ F\ p$. If $p \in C$, then we are ready. Now assume $p \notin C$. According to (2) there exists $q \in C$, such that $p\ F^+\ q$ or $q\ F^+\ p$. If $p\ F^+\ q$, then $t\ F^+\ q$ (since $t\ F\ p$). If $q\ F^+\ p$, then $q\ F^+\ t$, because ${}^\bullet p = \{t\}$. This last fact is based on property (2) in the definition of a process net. □

This lemma says that a slice is the same as a maximal clique of the relation $\mathbf{co}_N$ restricted to the set P. The next result then follows immediately from Lemma 66.

Lemma 77. *Let $N = (P,T,F)$ be a process net. For every $\mathbf{co}_N$-clique $B \subseteq P$ there exists a slice C of N with $B \subseteq C$.*

From now on we write ${}^\circ N$ for ${}^\circ X_N$ (the minimal elements of the net N). Likewise we write N° for $X_N{}^\circ$ (the maximal elements of N). By Theorem 71(1), ${}^\circ N$ and N° are cuts. They are even slices of N, because, by Definition 1(3), ${}^\bullet t \neq \varnothing$ and $t^\bullet \neq \varnothing$, for every transition t. Note that ${}^\circ N = \{p \in P_N \mid {}^\bullet p = \varnothing\}$ and $N^\circ = \{p \in P_N \mid p^\bullet = \varnothing\}$.

In the sequel, we will view each process net $N = (P,T,F)$ as the EN system $(P,T,F,{}^\circ N)$, i.e., with the initial slice ${}^\circ N$ as initial configuration. Property (2) of a process net (Definition 73) guarantees that this EN system is conflict-free, see the discussion following Definition 22.

If two process nets N and N' are isomorphic via a bijection $\alpha : P_N \to P_{N'}$, then, obviously, $\alpha({}^\circ N) = {}^\circ N'$. In other words: then they are also isomorphic as EN systems (cf. Definition 27).

Example 26. Fig. 50 shows an example of a process net N. The places in ${}^\circ N$ are marked with tokens. The places marked with crosses show another slice C of N. Note that ${}^\to C$ is the part of N above (and including) the crossed places, whereas $C^\to$ is the part below (and including) them.

Here are some basic properties of **co**-cliques and slices of process nets.

Lemma 78. *Let $N = (P,T,F)$ be a process net.*

(1) For every $U \subseteq T$, if U is a **co**-*clique, then ${}^\bullet U$ and $U^\bullet$ are* **co**-*cliques. In particular, ${}^\bullet t$ and $t^\bullet$ are* **co**-*cliques for every $t \in T$.*

(2) For every **co**-*clique $U \subseteq T$ there exists a slice C such that ${}^\bullet U \subseteq C$.*

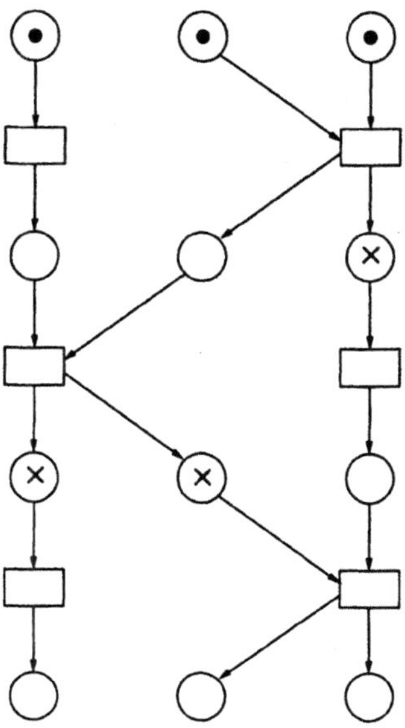

Fig. 50. A process net.

(3a) For every slice C and every $t \in T$, if ${}^\bullet t \subseteq C$, then $t^\bullet \cap C = \varnothing$ and $D = (C - {}^\bullet t) \cup t^\bullet$ is a slice such that ${}^{\rightarrow}D = {}^{\rightarrow}C \cup \{t\} \cup t^\bullet$.

(3b) For every slice C and every $t \in T$, if $t^\bullet \subseteq C$, then ${}^\bullet t \cap C = \varnothing$ and $D = (C - t^\bullet) \cup {}^\bullet t$ is a slice such that ${}^{\rightarrow}D = {}^{\rightarrow}C - t^\bullet - \{t\}$.

(4) For every slice C and every transition t, if $t \in {}^{\rightarrow}C$, then $\mathbf{nbh}(t) \subseteq {}^{\rightarrow}C$.

(5) For every slice $C \neq {}^\circ N$ there exists $t \in T$ such that $t^\bullet \subseteq C$.

Proof. (1) Let U be a co-clique. To show that ${}^\bullet U$ is a co-clique, let $p_1 \in {}^\bullet t_1$ and $p_2 \in {}^\bullet t_2$ with $t_1, t_2 \in U$, and suppose that $p_1 \; F^+ \; p_2$. Since ${p_1}^\bullet = \{t_1\}$ by Definition 73(2), $t_1 \; F^+ \; p_2$ and so $t_1 \; F^+ \; t_2$, contradicting the fact that U is a co-clique. In the same way it can be shown that $U^\bullet$ is a co-clique.

(2) follows directly from (1) and Lemma 77.

(3a) Let ${}^\bullet t \subseteq C$. Since C is a co-clique, $t^\bullet \cap C = \varnothing$. To prove that D is a slice, we first show that it is a co-clique. Since both C and $t^\bullet$ are co-cliques, it suffices to consider places $p \in C - {}^\bullet t$ and $q \in t^\bullet$ such that $p \; F^+ \; q$ or $q \; F^+ \; p$. If $p \; F^+ \; q$, then, since ${}^\bullet q = \{t\}$, there exists $q' \in {}^\bullet t$ such that $p \; F^+ \; q'$. If $q \; F^+ \; p$, then $q' \; F^+ \; p$ for any $q' \in {}^\bullet t$. In both cases, since $p, q' \in C$, this contradicts the fact that C is a co-clique. Using Lemma 76, the maximality of D easily follows from the maximality of C. Thus, D is a slice. The proof of the remaining property of D is left to the reader.

(3b) Here t "fires backwards" in C yielding D. The proof is similar to the proof of (3a).

(4) Let $t \in {}^{\rightarrow}C$. Thus, $t\ F^+\ q$ for some $q \in C$. Obviously ${}^\bullet t \subseteq {}^{\rightarrow}C$. Now consider $p \in t^\bullet$ and suppose that $p \notin {}^{\rightarrow}C$. Then, by Theorem 71(3), $p \in C^{\rightarrow} - C$ and so there exists $q' \in C$ such that $q'\ F^+\ p$. Hence $q'\ F^+\ t$, and so $q'\ F^+\ q$, contradicting the fact that C is a **co**-clique.

(5) Since $C \neq {}^\circ N$, $T \cap {}^{\rightarrow}C \neq \varnothing$. Lemma 70 then implies that $(T \cap {}^{\rightarrow}C)^\circ$ is also nonempty. Take a transition t that is a maximal element of $T \cap {}^{\rightarrow}C$. We claim that $t^\bullet \subseteq C$. In fact, suppose that there exists $p \in t^\bullet$ such that $p \notin C$. Since, by (4), $p \in {}^{\rightarrow}C$, there exists a place $q \in C$ such that $p\ F^+\ q$. Hence there is a transition t' such that $p\ F\ t'\ F^+\ q$. Consequently $t\ F^+\ t'$, which contradicts the maximality of t, because $t' \in T \cap {}^{\rightarrow}C$. □

We will now show that the reachable configurations of a process net N are exactly the slices of N.

Theorem 79. *Let $N = (P, T, F, {}^\circ N)$ be a process net and let $C \subseteq P$. $C \in \mathbb{C}_N$ iff C is a slice of N.*

Proof. (Only-if) By induction on C. The base of the induction holds because ${}^\circ N$ is a slice. The induction step follows directly from Lemma 78(3a).

(If) By induction on $\#({}^{\rightarrow}C \cap T)$, i.e., the number of events before the slice C. If ${}^{\rightarrow}C \cap T = \varnothing$, then $C = {}^\circ N$. If ${}^{\rightarrow}C \cap T \neq \varnothing$, then $C \neq {}^\circ N$ and so, by Lemma 78(5), there is a $t \in T$ such that $t^\bullet \subseteq C$. Then, by Lemma 78(3b), ${}^\bullet t \cap C = \varnothing$ and $D = (C - t^\bullet) \cup {}^\bullet t$ is a slice. Also ${}^{\rightarrow}D = {}^{\rightarrow}C - t^\bullet - \{t\}$ and hence $\#({}^{\rightarrow}D \cap T) = \#({}^{\rightarrow}C \cap T) - 1$. Thus, by the induction hypothesis, $D \in \mathbb{C}_N$. Obviously $D[t\rangle C$ and hence $C \in \mathbb{C}_N$. □

Theorem 79 and Lemma 78(3a) imply that every process net is contact-free. The conflict-freeness and contact-freeness of a process net can be expressed together as follows (see Theorem 61).

Theorem 80. *Let N be a process net, let $C \in \mathbb{C}_N$, and let $U \subseteq T_N$ with $U \neq \varnothing$. If ${}^\bullet U \subseteq C$, then U* **con** *C.*

In general a process net is not strongly reduced, because it may contain isolated places. However, as we will show now, a process net is always reduced. We also give a characterization of the concurrent steps of a process net (i.e., the sets of labels that appear in its configuration graph).

Theorem 81. *Let $N = (P, T, F, {}^\circ N)$ be a process net.*
(1) N is reduced.
(2) For every $U \subseteq T$, ($\exists C \in \mathbb{C}_N : U$ **con** *C) iff U is a* **co**-*clique.*

Proof. (1) follows from (2), because for every $t \in T$, $\{t\}$ is a **co**-clique.

(2) If U is a **co**-clique then, by Lemma 78(2) and by Theorems 79 and 80, U **con** C for a $C \in \mathbb{C}_N$. The other way around, if U **con** C, then ${}^\bullet U \subseteq C$ and hence ${}^\bullet U$ is a **co**-clique by Theorem 79. To show that U is a **co**-clique, suppose that $t_1\ F^+\ t_2$ for $t_1, t_2 \in U$. Then there exists $p_2 \in {}^\bullet t_2$ such that $t_1\ F^+\ p_2$. Hence, by Definition 1(3), there exists $p_1 \in {}^\bullet t_1$ such that $p_1\ F^+\ p_2$, contradicting the fact that ${}^\bullet U$ is a **co**-clique. □

Theorem 79 gives a characterization of the reachable configurations of a process net in terms of the partial order F^+. We will now do the same for sequential components. Whereas reachable configurations correspond to slices, sequential components correspond to lines.

Lemma 82. *Let $N = (P, T, F, {}^\circ N)$ be a process net.*
(1) If L is a $\mathbf{li}_N$-clique, then $L \cup \mathbf{nbh}(L \cap P)$ is a $\mathbf{li}_N$-clique.
(2) If L is a line of N, then for every $t \in T$:
$^\bullet t \cap L \neq \varnothing$ *iff* $t \in L$ *iff* $t^\bullet \cap L \neq \varnothing$.

Proof. (1) Let L be a **li**-clique. We have to prove that $L \cup {}^\bullet(L \cap P) \cup (L \cap P)^\bullet$ is a **li**-clique. Take $x \in L$ and $t \in (L \cap P)^\bullet$, hence $(p, t) \in F$ for a $p \in L$. If $x\ F^+\ p$ then $x\ F^+\ t$, and if $p\ F^+\ x$ then $t\ F^*\ x$ (by Definition 73(2)). A similar reasoning holds for all other cases.

(2) We will prove that $^\bullet t \cap L \neq \varnothing$ iff $t \in L$ (the proof that $t \in L$ iff $t^\bullet \cap L \neq \varnothing$ can be done in the same way). First let $p \in {}^\bullet t \cap L$. Then $t \in (L \cap P)^\bullet$ and hence $L \cup \{t\}$ is a **li**-clique according to (1). Then $t \in L$, because L is a line. Now, for the implication in the other direction, let us assume that $t \in L$ and let $Y = \{x \in L \mid x\ F^+\ t\}$. Lemma 72 implies that $Y \neq \varnothing$, and so, by Lemma 70, $Y^\circ \neq \varnothing$. Since L is a **li**-clique, Y° consists of one element, say x_m. Since $x_m\ F^+\ t$, there exists $p \in {}^\bullet t$ with $x_m\ F^*\ p$. This implies that $L \cup \{p\}$ is a **li**-clique and hence that $p \in L$. Thus $p \in {}^\bullet t \cap L$. □

Theorem 83. *Let $N = (P, T, F, {}^\circ N)$ be a process net with $P \neq \varnothing$.*
(1) If M is a sequential component of N, then $P_M \cup T_M$ is a line of N.
(2) If L is a line of N, then $(L \cap P, L \cap T, (L \times L) \cap F, L \cap {}^\circ N)$ is a sequential component of N.

Proof. (1) Let M be a sequential component of N and let $S = P_M$. Then $T_M = S^\bullet$ (see Lemmas 46 and 47). We have to show that $S \cup S^\bullet$ is a line of N. We first show that S is a **li**-clique. Let $p, q \in S$ and assume that $p \neq q$ and p **co** q. Then, by Lemma 77, there is a slice C with $p, q \in C$. Theorem 79 implies that $C \in \mathbb{C}_N$. But then $\#(C \cap S) \geq 2$, contradicting Theorem 49(2) (which is applicable because, by Theorem 81(1), N is reduced). Hence S is a **li**-clique. Then, by Lemma 82(1), $S \cup S^\bullet$ is a **li**-clique. Now it remains to prove that $S \cup S^\bullet$ is maximal. First consider $p \in P - S$. By Lemma 77 and Theorem 79 there is a slice $C \in \mathbb{C}_N$ with $p \in C$. Since $\#(C \cap S) = 1$, there exists $q \in S$ with $q \in C$. Then, for this $q \in S \cup S^\bullet$, p **li** q does not hold. Now consider $t \in T - S^\bullet$. Then $^\bullet t \cap S = \varnothing$. Again there is a slice $C \in \mathbb{C}_N$ with $^\bullet t \subseteq C$ (see also Lemma 78(1)) and again there exists $q \in C \cap S$. From the fact that p **co** q for every $p \in {}^\bullet t$ it follows that t **li** q does not hold for this $q \in S \cup S^\bullet$.

(2) Let L be a line and $S = L \cap P$. It is easy to check that $M = (S, L \cap T, (L \times L) \cap F, L \cap {}^\circ N)$ is a subsystem of N (where condition (2) of Definition 45 follows from Lemma 82(1) and the fact that L is a line). Now it remains to prove that M is sequential. By Theorem 49(2) and the fact that M is determined by S, it suffices to prove that $\#(C \cap S) = 1$ for every slice $C \in \mathbb{C}_N$. Since $\#(L \cap C) \leq 1$ and $C \subseteq P$, we only need to show that $L \cap C \neq \varnothing$. This is done by induction

on C. The base of the induction follows from Lemma 72 and the induction step from Lemma 82(2): if $C[t\rangle D$ and ${}^\bullet t \cap L \neq \varnothing$, then $t^\bullet \cap L \neq \varnothing$. □

From Lemma 53, Theorem 83(2), and Lemma 66 (for $\sigma = \mathbf{li}$) it follows that every process net is not only contact-free (as shown after Theorem 79), but is even covered by sequential components (cf. Theorem 59). As an example, the process net of Example 26 (Fig. 50) is covered by the three sequential components that correspond to the four places in each vertical row.

Also, Theorems 83(2), 79, and 49(2) imply that $L \cap C \neq \varnothing$ for every line L and every slice C. This is already close to density (see Definition 68).

Theorem 84. *Every process net $N = (P, T, F, {}^\circ N)$, with $P \neq \varnothing$, is dense.*

Proof. Let L be a line and C be a cut of N. It is not hard to prove (analogously to Lemma 78(3a)) that $C' = (C \cap P) \cup (C \cap T)^\bullet$ is a slice. By Theorems 83(2), 79, and 49(2), $L \cap C' \neq \varnothing$. Let $p \in L \cap C'$. If $p \in C \cap P$, then $p \in L \cap C$. If $p \in (C \cap T)^\bullet$, then there is a $t \in C$ such that $p \in t^\bullet \cap L$. Lemma 82(2) then implies that $t \in L$, and hence $t \in L \cap C$. □

Hence, density is an abstract version of the fact that in every reachable configuration every sequential component is in one particular state.

5.3 Processes

Process nets will be used in defining the notion of a process, which formalizes a concurrent run of a system. Informally speaking, a process of an EN system M describes a transformation of the initial configuration C_{in} of M to a configuration C of M; it is a record of all occurrences of events that lead from C_{in} to C, together with all conditions involved in these events. Two occurrences of events are (partially) ordered in this record if there is a condition that starts to hold as a result of the first occurrence of an event, and ceases to hold as a result of the second (or if the first is connected to the second by a chain of occurrences of events related in this way). This partial order represents the causal connection between the occurrences of the events, cf. Fig. 16. Note that the linear order of occurrences of events in a firing sequence also represents the fact that they are observed by a sequential observer.

Consider for example the (contact-free) EN system $M = (P, T, F, C_{in})$ of Fig. 51. We start by recording the initial configuration $\{p_1, p_2\}$ (see Fig. 52). Next we record the occurrence of event t_1 (see Fig. 53). What we obtain in this way, is a process (a record of a transformation) from the configuration $C_{in} = \{p_1, p_2\}$ to the configuration $C = \{p_3, p_4\}$. We can continue by also recording the occurrence of (for example) the concurrent step $\{t_2, t_3\}$ in configuration C, see Fig. 54. What we obtain now, is a process (a record of a transformation) leading from the configuration $C_{in} = \{p_1, p_2\}$ to itself. By continuing one more step, we again record the occurrence of t_1 and again obtain a process (a record of a transformation) from C_{in} to C (see Fig. 55). Finally we can (for example)

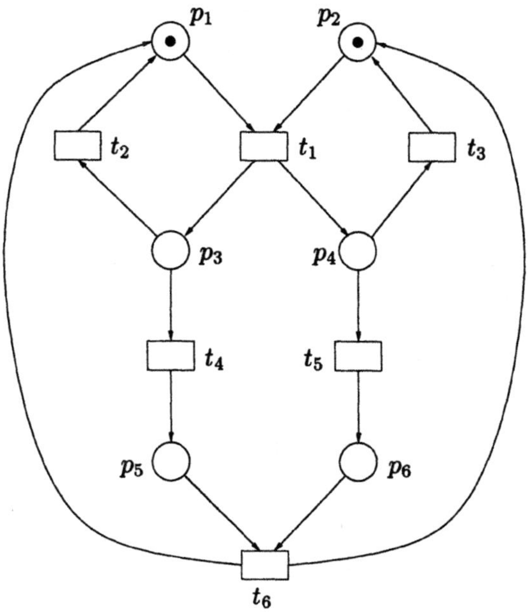

Fig. 51. An EN system M.

Fig. 52. A process of the EN system M of Fig. 51.

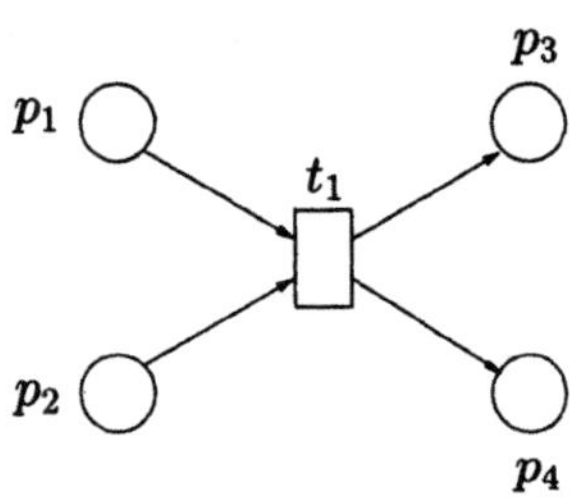

Fig. 53. A process that extends the process of Fig. 52.

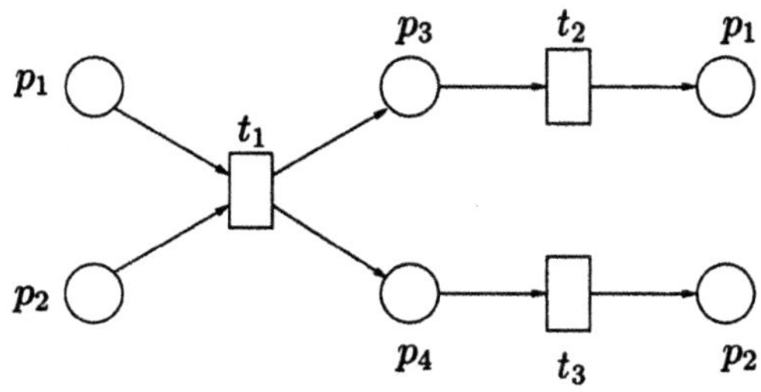

Fig. 54. A process that extends the process of Fig. 53.

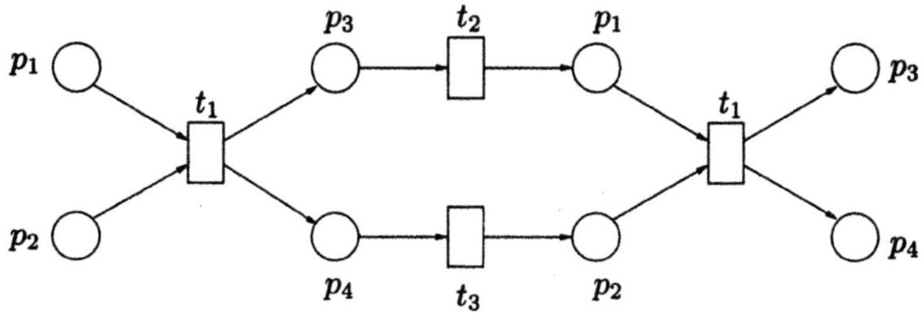

Fig. 55. A process that extends the process of Fig. 54.

record the occurrence of event t_4 and thus obtain a process leading from C_{in} to the configuration $\{p_4, p_5\}$ (see Fig. 56).

Thus, a process does not describe a complete run of the system (which can be infinite), but rather an initial finite part of it.

Note that our processes are themselves nets. These nets record the *occurrences of events* together with the *occurrences of the conditions that belong to these events.* Moreover, processes are nets of a special kind, viz. process nets: (1) if an event occurs, a possible conflict is resolved, and (2) different occurrences of the same condition and different occurrences of the same event are recorded by different copies of the corresponding condition and event, respectively. Therefore all places are unbranched and no cycles occur. The flow relation of the net represents the causal partial order between the events and conditions (see Lemma 74).

It is important to note that our processes only record that conditions hold, not that conditions do *not* hold. Hence a process faithfully describes a run of the system under consideration *only if* this system is contact-free, i.e., we need not record which conditions do *not* hold when we record the occurrence of an event. This is the reason for restricting our attention to processes of contact-free EN systems. Fortunately, Theorem 60 says that for every EN system there exists a configuration equivalent contact-free EN system. Hence, without loss

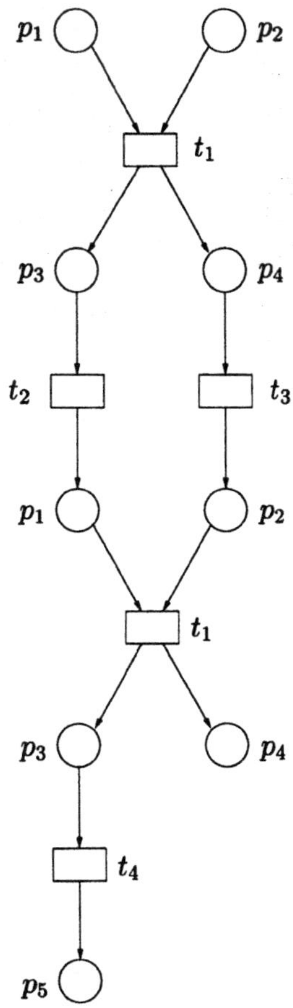

Fig. 56. A process of the EN system M of Fig. 51, leading from $\{p_1, p_2\}$ to $\{p_4, p_5\}$.

of generality, we can assume contact-freeness in our study of the behaviour of EN systems. More specifically, when dealing with an EN system M that is not contact-free, if we want to study its behaviour through its processes, then we will study the processes of a contact-free EN system that is configuration equivalent with M (e.g., the system that is obtained by the complement construction in the proof of Theorem 57).

The above considerations lead to the formal notion of a process in a contact-free EN system. However, before giving the formal definition, we need a number of auxiliary notions.

Definition 85. Let Σ_1 and Σ_2 be disjoint alphabets. A (Σ_1, Σ_2)-*labelled net* is a 5-tuple $N = (P, T, F, \phi_1, \phi_2)$, where
(P, T, F) is a net (the *underlying net of* N, denoted by $\mathbf{und}(N)$),

ϕ_1 is a function from P to Σ_1 (the ***place labelling of** N*), and
ϕ_2 is a function from T to Σ_2 (the ***transition labelling of** N*).

N is also called a ***labelled net***, and, if $\mathbf{und}(N)$ is a process net, then N is also called a ***(Σ_1, Σ_2)-labelled process net*** or simply a ***labelled process net***. All notations and terminology concerning (process) nets carry over, through the underlying nets, to labelled (process) nets. We will also use the notation ϕ_{1N}, ϕ_{2N} for ϕ_1, ϕ_2, respectively.

To compare labelled nets we need the following notion of isomorphism, which expresses the fact that the identity of the places, the transitions, and the labels is irrelevant.

Definition 86. Let $N = (P, T, F, \phi_1, \phi_2)$ and $N' = (P', T', F', \phi_1', \phi_2')$ be two (Σ_1, Σ_2)-, respectively (Σ_1', Σ_2')-labelled nets. Then N and N' are ***isomorphic***, denoted by $N \equiv N'$, if there exist bijections $\alpha : \Sigma_1 \to \Sigma_1'$, $\beta : \Sigma_2 \to \Sigma_2'$, $\gamma : P \to P'$, and $\delta : T \to T'$, such that:
(1) $\mathbf{und}(N) \equiv_\delta^\gamma \mathbf{und}(N')$,
(2) for all $p \in P$, $\phi_1'(\gamma(p)) = \alpha(\phi_1(p))$, and
(3) for all $t \in T$, $\phi_2'(\delta(t)) = \beta(\phi_2(t))$.

For isomorphic N and N' as above, we also say that N and N' are ***(α, β)-isomorphic***, denoted by $N \equiv_\beta^\alpha N'$.

Condition (1) above means that the underlying nets are isomorphic, condition (2) means that corresponding places (via γ) have corresponding labels (via α), and condition (3) means that corresponding transitions (via δ) also have corresponding labels (via β).

This notion of isomorphism between labelled nets can naturally be extended to isomorphism between sets of labelled nets in the following way.

Definition 87. Let $\mathcal{P}$ and $\mathcal{P}'$ be two sets of (Σ_1, Σ_2)-, respectively (Σ_1', Σ_2')-labelled nets. Then $\mathcal{P}$ and $\mathcal{P}'$ are ***isomorphic***, denoted by $\mathcal{P} \equiv \mathcal{P}'$, if there exist bijections $\alpha : \Sigma_1 \to \Sigma_1'$ and $\beta : \Sigma_2 \to \Sigma_2'$, such that
(1) for every $N \in \mathcal{P}$ there exists $N' \in \mathcal{P}'$ such that $N \equiv_\beta^\alpha N'$, and
(2) for every $N' \in \mathcal{P}'$ there exists $N \in \mathcal{P}$ such that $N \equiv_\beta^\alpha N'$.

For isomorphic $\mathcal{P}$ and $\mathcal{P}'$ as above, we also say that $\mathcal{P}$ and $\mathcal{P}'$ are ***(α, β)-isomorphic***, denoted by $\mathcal{P} \equiv_\beta^\alpha \mathcal{P}'$.

Condition (1) means that every net in $\mathcal{P}$ is isomorphic with a net in $\mathcal{P}'$ and condition (2) means that, the other way around, every net in $\mathcal{P}'$ is isomorphic with a net in $\mathcal{P}$. Note that the isomorphisms from (1) and (2) always use the same, fixed a priori, bijections α and β between the alphabets.

We now present the formal definition of the notion of a process. Recall from Section 2 that $f \restriction B$ denotes the restriction of function $f : A \to A'$ to the set $B \subseteq A$. The requirement that $f \restriction B$ is injective thus means that for all $b_1, b_2 \in B$, if $b_1 \neq b_2$ then $f(b_1) \neq f(b_2)$.

Definition 88. Let $N = (P_N, T_N, F_N, \phi_1, \phi_2)$ be a (Σ_1, Σ_2)-labelled process net and let $M = (P, T, F, C_{in})$ be a contact-free EN system.
Then N is a *process of* M if
(1) $\Sigma_1 = P$ and $\Sigma_2 = \mathbf{use}(T)$,
(2) $\phi_1 \restriction {}^\circ N$ is injective,
(3) $\phi_1({}^\circ N) = C_{in}$,
(4) for every $t \in T_N$, $\phi_1 \restriction {}^\bullet t$ is injective and $\phi_1 \restriction t^\bullet$ is injective, and
(5) for every $t \in T_N$, $\phi_1({}^\bullet t) = {}^\bullet(\phi_2(t))$ and $\phi_1(t^\bullet) = (\phi_2(t))^\bullet$.

From now on, for the sake of simplicity, we will also write ϕ instead of ϕ_1 and ϕ_2, when the subscript is clear from the context.

For a contact-free EN system M, PROC(M) denotes the set of all processes of M.

Requirement (1) above says that the places of the process net are labelled with the places of the system, and the transitions of the process net are labelled with the useful transitions of the system. Requirements (2) and (3) say that, via the labelling ϕ, the minimal places of the process net N are in one-to-one correspondence with the initial configuration of the system. The non-minimal places of N represent conditions that are set by the occurrence of some event. Thus, ${}^\circ N$ faithfully records the conditions that hold initially. Requirements (4) and (5) say that, via the labelling ϕ, the places in the input- and output-set of a transition s of the process net are in one-to-one correspondence with the places in the input- and output-set, respectively, of a transition t of the system. This means that s is indeed a faithful record of the occurrence of the event $t = \phi(s)$. Note that requirement (5) implies that for all $x, y \in X_N$, if $x \; F_N \; y$ then $\phi(x) \; F \; \phi(y)$.

Example 27. (1) Let M be the (contact-free) EN system of Fig. 47. A process N of M is drawn in Fig. 57 (the underlying process net is the one of Fig. 50). Note that the symbols next to the places and transitions of N are labels, i.e., places and transitions of M (the identities of the places and transitions of N are not given in the figure). The process N leads from the initial configuration $\{p_1, b_f, c_1\}$ of M to the configuration $\{p_2, b_e, c_2\}$ of M.

(2) A process of the EN system of Fig. 2 (mutual exclusion) is drawn in Fig. 58. It leads from the initial configuration $\{w_1, p, w_2\}$ to the configuration $\{c_1, w_2\}$.

We will now show that the firing of transitions in a process is mapped (by ϕ) to the firing of transitions in the system. Hence, playing the token game in a process corresponds to playing the token game in the system. To prove this, we need the following lemma.

Lemma 89. *Let $M = (P, T, F, C_{in})$ be a contact-free EN system and let $N = (P_N, T_N, F_N, \phi_1, \phi_2)$ be a process of M. Let $C, D \in \mathbb{C}_N$ and $t \in T_N$.*
If $\phi \restriction C$ is injective, $\phi(C) \in \mathbb{C}_M$, and $C[t\rangle_N D$, then $\phi \restriction D$ is injective and $\phi(C)[\phi(t)\rangle_M \phi(D)$.

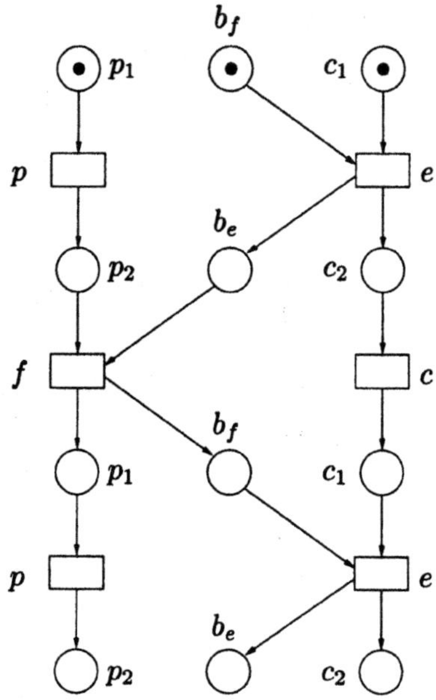

Fig. 57. A process of the producer/consumer system of Fig. 47.

Proof. From ${}^\bullet t \subseteq C$ it follows that $\phi({}^\bullet t) \subseteq \phi(C)$ and hence, by requirement (5) of Definition 88, that ${}^\bullet\phi(t) \subseteq \phi(C)$. Since M is contact-free, $\phi(t)^\bullet \cap \phi(C) = \varnothing$ and hence, again by requirement (5), $\phi(t^\bullet) \cap \phi(C) = \varnothing$. This, and the fact that $\phi \restriction t^\bullet$ is injective (requirement (4)), implies that $\phi \restriction (C \cup D)$ is injective. Consequently $\phi(C-D) = \phi(C) - \phi(D)$ and $\phi(D-C) = \phi(D) - \phi(C)$. By Lemma 7, ${}^\bullet t = C - D$ and $t^\bullet = D - C$. Hence ${}^\bullet\phi(t) = \phi({}^\bullet t) = \phi(C-D) = \phi(C) - \phi(D)$ and similarly $\phi(t)^\bullet = \phi(D) - \phi(C)$. Lemma 7 now implies that $\phi(C)[\phi(t)\rangle_M \phi(D)$. □

Theorem 90. *Let $M = (P, T, F, C_{in})$ be a contact-free EN system and let $N = (P_N, T_N, F_N, \phi_1, \phi_2)$ be a process of M.*
(1) For every $C \in \mathbb{C}_N$, $\phi \restriction C$ is injective and $\phi(C) \in \mathbb{C}_M$.
(2) For every $C, D \in \mathbb{C}_N$ and $t \in T_N$, if $C[t\rangle_N D$ then $\phi(C)[\phi(t)\rangle_M \phi(D)$.

Proof. (1) can easily be proved by induction on C. The base of the induction (i.e., $C = {}^\circ N$) directly follows from requirements (2) and (3) of Definition 88, and the induction step follows from Lemma 89.

(2) directly follows from (1) and Lemma 89. □

Theorem 91. *Let $M = (P, T, F, C_{in})$ be a contact-free EN system and let $N = (P_N, T_N, F_N, \phi_1, \phi_2)$ be a process of M.*
(1) For every **co**-*clique D of X_N, $\phi \restriction D$ is injective.*
(2) For every $C, D \in \mathbb{C}_N$ and $U \subseteq T_N$, if $C[U\rangle_N D$ then $\phi(C)[\phi(U)\rangle_M \phi(D)$.
(3) For every **co**-*clique $U \subseteq T_N$ there exist $C, D \in \mathbb{C}_M$ such that $C[\phi(U)\rangle_M D$.*

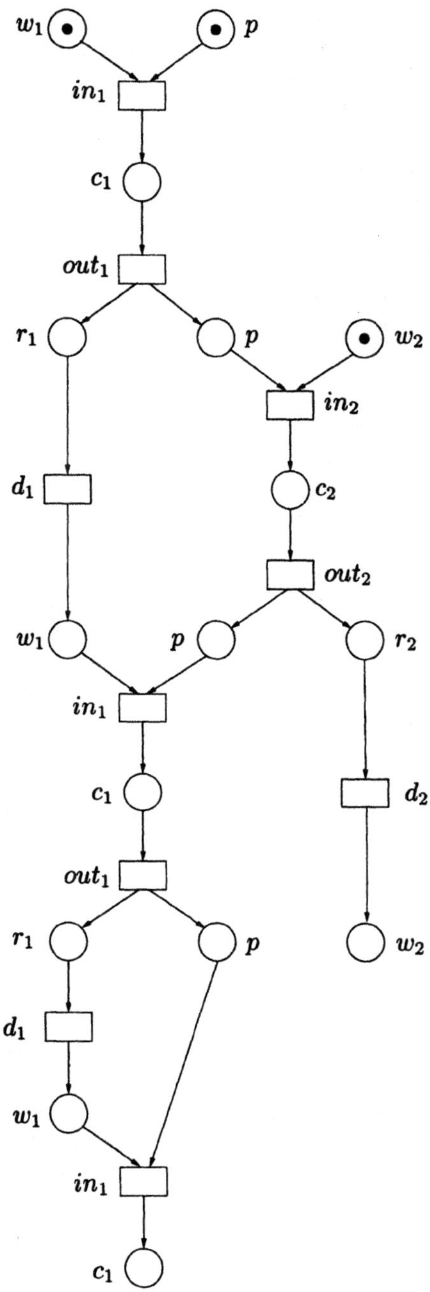

Fig. 58. A process of the mutual exclusion system of Fig. 2.

Proof. (1) We have to show that for every two distinct elements x and y of X_N, if x **co** y then $\phi(x) \neq \phi(y)$. This is obvious if one of the two is a place and the other a transition. If $\{x,y\} \subseteq P_N$, then by Lemma 77 there exists a slice $C \in \mathbb{C}_N$ with $\{x,y\} \subseteq C$, and Theorem 90(1) then implies that $\phi \restriction \{x,y\}$ is injective. If $\{x,y\} \subseteq T_N$, then Lemma 78(1) implies that $^\bullet x \cup {}^\bullet y$ is a **co**-clique. According to the previous case, $\phi \restriction ({}^\bullet x \cup {}^\bullet y)$ is injective. Definition 73(2) now implies that $\phi({}^\bullet x) \neq \phi({}^\bullet y)$. Hence $^\bullet\phi(x) \neq {}^\bullet\phi(y)$ and thus $\phi(x) \neq \phi(y)$.

(2) Assume that $C[U\rangle_N D$. By Theorem 81(2), U is a **co**-clique. Then (1) implies that $\phi \restriction U$ is injective. This means that the set of orderings of the elements of $\phi(U)$ equals $\{(\phi(t_1),\ldots,\phi(t_n)) \mid (t_1,\ldots,t_n)$ is an ordering of the elements of $U\}$. This, together with Theorems 20(2) and 90(2), implies that $\phi(C)[\phi(U)\rangle_M \phi(D)$.

(3) If $U \subseteq T_N$ is a **co**-clique, then there exists $C \in \mathbb{C}_N$ with U **con** C by Theorem 81(2). Theorem 90(1) implies that $\phi(C) \in \mathbb{C}_M$. Now apply (2). □

There is a clear connection between the firing sequences of an EN system and the firing sequences of its processes. Theorem 90 implies that every firing sequence of a process N (of an EN system M) is mapped to a firing sequence of M by ϕ. We will now show that this also holds the other way around: for every firing sequence of M there exists a firing sequence of a process N of M which is mapped to it by ϕ; moreover, we can guarantee that this is a "complete" firing sequence of N, i.e., a firing sequence from $^\circ N$ to N° (note that, by Theorem 79, every process net has such a complete firing sequence). This resembles the construction which we presented as an example at the beginning of this subsection (in Figs. 52 to 56).

Theorem 92. *Let $M = (P,T,F,C_{in})$ be a contact-free EN system, let $t_1,\ldots,t_n$ be transitions in T, and let $C \subseteq P$. Then $C_{in}[t_1 \cdots t_n\rangle_M C$ iff there exists a process $N = (P_N,T_N,F_N,\phi_1,\phi_2)$ of M and there exist transitions $s_1,\ldots,s_n$ in T_N such that*
(1) $\phi(s_i) = t_i$ for $1 \leq i \leq n$,
(2) $\phi(N^\circ) = C$, and
(3) $^\circ N[s_1 \cdots s_n\rangle_N N^\circ$.

Proof. (If) This follows from Definition 88(3) and Theorem 90(2).

(Only-if) The proof is by induction on n.

For $n = 0$ we have to show the existence of a process N with $^\circ N = N^\circ$. If $C_{in} = \{q_1,\ldots,q_m\}$, then such a process N is defined by: $P_N = \{p_1,\ldots,p_m\}$, $T_N = \varnothing$, $F_N = \varnothing$, $\phi_2 = \varnothing$, and $\phi_1(p_i) = q_i$ for $1 \leq i \leq m$.

Now assume that $C_{in}[t_1 \cdots t_n\rangle C[t\rangle D$ and assume (the induction hypothesis) that there exists a process N satisfying requirements (1-3). Let $^\bullet t = \{q_1,\ldots,q_k\}$ and $t^\bullet = \{q'_1,\ldots,q'_m\}$. Since $C = \phi(N^\circ)$ and $^\bullet t \subseteq C$, there are (unique) places $p_1,\ldots,p_k \in N^\circ$ such that $\phi(p_i) = q_i$ for $1 \leq i \leq k$. Now take an $s \notin T_N$ and $p'_1,\ldots,p'_m \notin P_N$. We extend N by adding transition s and places $p'_1,\ldots,p'_m$, in such a way that $^\bullet s = \{p_1,\ldots,p_k\}$, $s^\bullet = \{p'_1,\ldots,p'_m\}$, $\phi(s) = t$, and $\phi(p'_i) = q'_i$ for $1 \leq i \leq m$. It is easy to see that in this way a new process is obtained, and that it satisfies requirements (1-3) for $C_{in}[t_1 \cdots t_n t\rangle D$. □

It is also straightforward to prove that, for given firing sequence $t_1 \cdots t_n$, the process N is unique (modulo isomorphism, i.e., modulo $\equiv_\beta^\alpha$, where α is the identity on P_M and β the identity on $\mathbf{use}(T_M)$). Since every process has a complete firing sequence, this implies that *every* process of M is obtained in this way from a firing sequence of M (cf. [BesDev87], Theorem 3.5.3 of [BesFer88], and Section 3 of [NieRozThi90]). Note, however, that different firing sequences $t_1 \cdots t_n$ can lead to the same process N. The resulting equivalence relation between firing sequences is studied in Section 6.2. The uniqueness of N will also follow from later results (see the remark after Theorem 122).

Note that Theorem 92 implies that, for every contact-free EN system M, $\mathrm{PROC}(M) \neq \varnothing$ (since λ is always a firing sequence of M).

Example 28. Consider the EN system M of Fig. 47 and the process N of M of Fig. 57. Examples of firing sequences of M that correspond to process N in the way indicated in Theorem 92 are: *pefcep* and *ecpfpe*.

We now prove the converse of Theorem 90(2).

Theorem 93. *Let M be a contact-free EN system. Let $C, D \in \mathbb{C}_M$ and $t \in T_M$. If $C[t\rangle_M D$, then there exists a process N of M and there exist $C', D' \in \mathbb{C}_N$ and $s \in T_N$, such that $C'[s\rangle_N D'$, $\phi_N(C') = C$, $\phi_N(s) = t$, and $\phi_N(D') = D$.*

Proof. There is a firing sequence x such that $C_{in}[x\rangle_M C$. If we now apply Theorem 92 to $C_{in}[xt\rangle_M D$, then we obtain a process N of M and a firing sequence ys (with $y \in T_N^*$ and $s \in T_N$) such that $\phi(y) = x$, $\phi(s) = t$, $\phi(N^\circ) = D$, and ${}^\circ N[ys\rangle_N N^\circ$ (where $\phi = \phi_N$). Assume that ${}^\circ N[y\rangle_N C'[s\rangle_N N^\circ$ and let $D' = N^\circ$. By Theorem 90, $\phi({}^\circ N)[\phi(y)\rangle_M \phi(C')$ holds, i.e., $C_{in}[x\rangle_M \phi(C')$. Hence $\phi(C') = C$. □

Analogously the following converse of Theorem 91(2,3) can be proved. The details of the proof of (1) are left to the reader; (2) follows immediately from (1) and Theorem 81(2).

Theorem 94. *Let M be a contact-free EN system. Let $C, D \in \mathbb{C}_M$ and let $U \subseteq T_M$. If $C[U\rangle_M D$, then:*
(1) there exists a process N of M and there exist $C', D' \in \mathbb{C}_N$ and $V \subseteq T_N$, such that $C'[V\rangle_N D'$, $\phi_N(C') = C$, $\phi_N(V) = U$, and $\phi_N(D') = D$,
(2) there exist a process N of M and a **co**-*clique $V \subseteq T_N$ such that $\phi_N(V) = U$.*

It would be natural to say that two EN systems are equivalent if they have isomorphic sets of processes (see Definition 87). The next result shows that two (reduced) EN systems are equivalent in this sense iff they are isomorphic. Thus, as discussed in the next subsection, this equivalence relation does not capture equivalent behaviour of EN systems.

Theorem 95. *Let M and M' be two contact-free reduced EN systems. Then $PROC(M) \equiv PROC(M')$ iff $M \equiv M'$.*

Proof. Let $M = (P, T, F, C_{in})$ and $M' = (P', T', F', C'_{in})$. It is easy to see that if $M \equiv_\beta^\alpha M'$, then $\mathrm{PROC}(M) \equiv_\beta^\alpha \mathrm{PROC}(M')$. Now assume that $\mathrm{PROC}(M) \equiv_\beta^\alpha \mathrm{PROC}(M')$. Then α and β are bijections, $\alpha : P \to P'$ and $\beta : T \to T'$. It now suffices to show that (1) $\alpha(C_{in}) = C'_{in}$ and that (2) for every $t \in T$, $\alpha({}^\bullet t) = {}^\bullet\beta(t)$ and $\alpha(t^\bullet) = \beta(t)^\bullet$.

(1) Consider a process N of M (this is possible because $\mathrm{PROC}(M) \neq \varnothing$) and consider a process N' of M' that is (α, β)-isomorphic with N. Then ${}^\circ N$ corresponds to ${}^\circ N'$ and hence $\alpha(\phi_N({}^\circ N)) = \phi_{N'}({}^\circ N')$. Since $\phi_N({}^\circ N) = C_{in}$ and $\phi_{N'}({}^\circ N') = C'_{in}$, this means that $\alpha(C_{in}) = C'_{in}$.

(2) Since M is reduced, there exist $C, D \in \mathbb{C}_M$ with $C[t\rangle_M D$. Then, by Theorem 93, there exist a process N of M and an $s \in T_N$ such that $\phi_N(s) = t$. Let N' be a process of M' that is (α, β)-isomorphic with N and assume that $s' \in T_{N'}$ corresponds to s. Then $\phi_{N'}(s') = \beta(t)$. Furthermore, ${}^\bullet s$ corresponds to ${}^\bullet(s')$ and $s^\bullet$ to $(s')^\bullet$, and hence $\alpha(\phi_N({}^\bullet s)) = \phi_{N'}({}^\bullet(s'))$ and $\alpha(\phi_N(s^\bullet)) = \phi_{N'}((s')^\bullet)$. Now $\phi_N({}^\bullet s) = {}^\bullet\phi_N(s) = {}^\bullet t$, $\phi_{N'}({}^\bullet(s')) = {}^\bullet\phi_{N'}(s') = {}^\bullet\beta(t)$, $\phi_N(s^\bullet) = \phi_N(s)^\bullet = t^\bullet$, and $\phi_{N'}((s')^\bullet) = \phi_{N'}(s')^\bullet = \beta(t)^\bullet$, and hence $\alpha({}^\bullet t) = {}^\bullet\beta(t)$ and $\alpha(t^\bullet) = \beta(t)^\bullet$. □

5.4 Pruned Contracted Processes

We can interpret the set $\mathrm{PROC}(M)$ of processes of an EN system M as the behaviour of M: it is the set of all concurrent runs of M. However, by Theorem 95, this definition of the behaviour of an EN system is too strong, since the system M is uniquely determined by $\mathrm{PROC}(M)$, modulo isomorphism, which means that the behaviour of the EN system would be identified with its structure! In this way it would be impossible to transform systems while preserving their behaviour. The intuitive reason why Theorem 95 holds, is that in a process both the events and the conditions are recorded: in that way we are able to read the flow relation of the system from the flow relation of its processes and their labels (see Definition 88(4,5)). However, since we are more interested in the events than in the conditions of the system when defining behaviour (see Section 4.1), we will, in this subsection, remove the conditions from every process N. In this way the set T_N of recorded events remains, together with their causal order (the partial order F_N^+, restricted to T_N). This can be considered as a "labelled partially ordered set", with labels in T_M (or, more precisely, in $\mathbf{use}(T_M)$). Then we will define the behaviour of M as the set of all labelled partially ordered sets obtained in this way. With this definition of the behaviour of M, the system M is no longer uniquely determined, i.e., Theorem 95 no longer holds. Note that such partially ordered sets of occurrences of events are similar to firing sequences, which are linearly ordered sets of occurrences of events.

Modulo isomorphism, labelled partially ordered sets (with labels in an alphabet Σ) are also called partially ordered multisets (with elements in Σ) or ***pom-sets***, see, e.g., [Pra86]. Here, we model labelled partially ordered sets by node-labelled acyclic graphs, which are easier to compare to processes. There is a clear connection between partial orders and acyclic graphs. A partially ordered set (A, ρ) with $\rho \subseteq A \times A$ is a (special kind of) acyclic directed graph. The other way around it is clear that for every acyclic directed graph (V, Γ), (V, Γ^+) is a partially ordered set (cf. Lemma 74). We will say that the graph (V, Γ) represents the partially ordered set (V, Γ^+). In general a partially ordered set can thus be represented by several acyclic graphs.

To begin with, we recall several notions concerning node-labelled directed graphs which, for the sake of simplicity, we will simply call labelled graphs. Note that we have so far only considered (initialized) edge-labelled graphs (in particular configuration graphs). We have also considered labelled nets (in particular processes) in the previous section. When we view a net as a graph, labelled nets can be viewed as node-labelled graphs. The following definitions are thus analogous to those for labelled nets.

Definition 96. Let Σ be an alphabet. A (Σ-)*labelled graph* is a quadruple $G = (V, \Gamma, \Sigma, \phi)$, where V is a finite set of *nodes*, $\Gamma \subseteq V \times V$ is a set of *edges*, and ϕ is a function from V to Σ (the *labelling of* G). G is *acyclic* if Γ^+ is irreflexive.

The components of G are also indicated by V_G, Γ_G, Σ_G, and ϕ_G.

To compare graphs labelled over distinct alphabets we need the following notion of isomorphism.

Definition 97. Let Σ and Σ' be two alphabets and let $G = (V, \Gamma, \Sigma, \phi)$ and $G' = (V', \Gamma', \Sigma', \phi')$ be two Σ-, respectively Σ'-labelled graphs. Then G and G' are *isomorphic*, denoted by $G \equiv G'$, if there exist bijections $\beta : \Sigma \to \Sigma'$ and $\delta : V \to V'$, such that
(1) for all $v, w \in V$, $(v, w) \in \Gamma$ iff $(\delta(v), \delta(w)) \in \Gamma'$, and
(2) for all $v \in V$, $\phi'(\delta(v)) = \beta(\phi(v))$.

For isomorphic G and G' as above, we also say that G and G' are β-*isomorphic*, denoted by $G \equiv_\beta G'$.

The above notion of isomorphism between labelled graphs can naturally be extended to isomorphism between sets of labelled graphs in the following way.

Definition 98. Let $\mathcal{P}$ and $\mathcal{P}'$ be two sets of Σ-, respectively Σ'-labelled graphs. Then $\mathcal{P}$ and $\mathcal{P}'$ are *isomorphic*, denoted by $\mathcal{P} \equiv \mathcal{P}'$, if there exists a bijection $\beta : \Sigma \to \Sigma'$, such that
(1) for every $G \in \mathcal{P}$ there exists $G' \in \mathcal{P}'$ such that $G \equiv_\beta G'$, and
(2) for every $G' \in \mathcal{P}'$ there exists $G \in \mathcal{P}$ such that $G \equiv_\beta G'$.

For isomorphic $\mathcal{P}$ and $\mathcal{P}'$ as above, we also say that $\mathcal{P}$ and $\mathcal{P}'$ are β-*isomorphic*, denoted by $\mathcal{P} \equiv_\beta \mathcal{P}'$. Note that the isomorphisms from (1) and (2) always use the same, fixed a priori, bijection β between the alphabets (cf. Definition 87).

A labelled graph $G = (V, \Gamma, \Sigma, \phi)$ for which (V, Γ) is a partially ordered set is also called a *labelled partially ordered set* or, shorter, a *labelled partial order*. To every acyclic labelled graph such a labelled partial order is naturally associated.

Definition 99. Let $G = (V, \Gamma, \Sigma, \phi)$ be an acyclic labelled graph. The *transitive closure* of G, denoted by $\mathbf{tra}(G)$, is the labelled graph $(V, \Gamma^+, \Sigma, \phi)$. We also say that G *represents* $\mathbf{tra}(G)$.

Note that **tra**(G) is a labelled partial order. In general a labelled partial order is represented by several labelled graphs. One of these graphs is the "smallest" (and is also known as the "Hasse diagram" of the partial order).

Definition 100. Let $G = (V, \Gamma, \Sigma, \phi)$ be an acyclic labelled graph. The ***pruned version of*** G, denoted by **pru**(G), is the labelled graph $(V, \Gamma', \Sigma, \phi)$ with $\Gamma' = \{(v, w) \in \Gamma \mid \neg\, \exists u \in V : (v, u) \in \Gamma^+ \text{ and } (u, w) \in \Gamma^+\}$.

This means that **pru**(G) is the graph that is obtained from G by removing all edges (v, w) such that there exists a path of length ≥ 2 from v to w in G, i.e., by removing the so-called *transitive* edges.

Example 29. Figure 59 shows a graph with its transitive closure and its pruned version. The labels have been omitted.

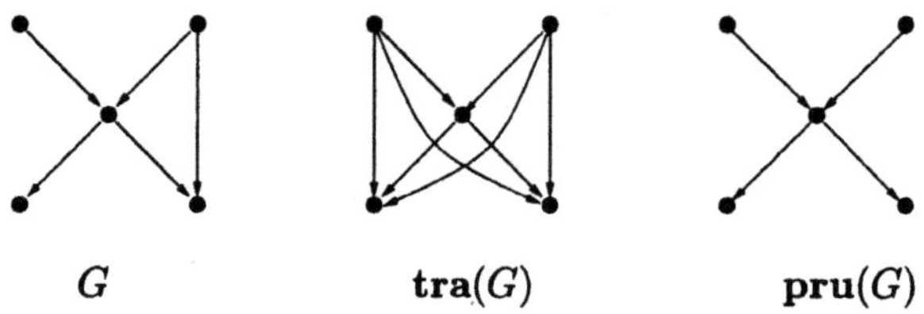

Fig. 59. A graph with its transitive closure and its pruned version.

We now recall the well-known fact that **tra**(G) is represented by **pru**(G), and that **pru**(G) is uniquely determined by **tra**(G).

Theorem 101. *For every acyclic labelled graph* G, **tra**(**pru**(G)) $=$ **tra**(G) *and* **pru**(**tra**(G)) $=$ **pru**(G).

Proof. Let $G = (V, \Gamma, \Sigma, \phi)$. We will prove the inclusion **tra**$(G) \subseteq$ **tra**(**pru**(G)); the other inclusions are immediate. We have to prove that, for all $v, w \in V$, if there is a path from v to w, then there is a path from v to w that uses only non-transitive edges. This is done by induction on the length $n(v, w)$ of the longest path from v to w. If $n(v, w) = 1$, then the longest path from v to w is the edge (v, w). Hence that edge is not transitive. Now take $n(v, w) \geq 2$ and assume that the claim holds for all lengths $< n(v, w)$. Consider the longest path from v to w and take a node u on that path that is distinct from v and w. Then the subpath from v to u is also the longest path from v to u, and hence $n(v, u) < n(v, w)$. This implies that there is a path from v to u with non-transitive edges only. Likewise there is a path from u to w with non-transitive edges only. □

This implies that two labelled partial orders are isomorphic iff their pruned versions are isomorphic.

Theorem 102. *Let Σ and Σ' be alphabets. Let G and G' be two acyclic Σ-, respectively Σ'-labelled graphs and let $\beta : \Sigma \to \Sigma'$ be a bijection.*
Then $\mathbf{tra}(G) \equiv_\beta \mathbf{tra}(G')$ *iff* $\mathbf{pru}(G) \equiv_\beta \mathbf{pru}(G')$.

Proof. It is clear that, for labelled graphs G_1 and G_2, if $G_1 \equiv_\beta G_2$, then $\mathbf{tra}(G_1) \equiv_\beta \mathbf{tra}(G_2)$ and $\mathbf{pru}(G_1) \equiv_\beta \mathbf{pru}(G_2)$. Hence, if $\mathbf{tra}(G) \equiv_\beta \mathbf{tra}(G')$, then $\mathbf{pru}(\mathbf{tra}(G)) \equiv_\beta \mathbf{pru}(\mathbf{tra}(G'))$ and thus $\mathbf{pru}(G) \equiv_\beta \mathbf{pru}(G')$. Likewise in the other direction. □

We now return to the processes of an EN system M. As observed before, we are interested in the labelled partial order $(T_N, F_N^+ \cap (T_N \times T_N), T_M, \phi_{2N})$ for a process N of M. There is an easy way to construct a labelled graph that represents this labelled partial order: we remove the places from N and replace them by edges. When we then remove the transitive edges from this graph we obtain a unique representation of the labelled partial order. The precise formulation of this is given in the following definition, cf. Fig. 16.

Definition 103. Let $N = (P, T, F, \phi_1, \phi_2)$ be an acyclic (Σ_1, Σ_2)-labelled net.

(1) The *contracted version of* N, denoted by $\mathbf{ctr}(N)$, is the labelled graph $(T, \Gamma, \Sigma_2, \phi_2)$ such that, for all $s, t \in T$, $(s, t) \in \Gamma$ iff $s^\bullet \cap {}^\bullet t \neq \varnothing$.

(2) The *pruned contracted version of* N is the labelled graph $\mathbf{pru}(\mathbf{ctr}(N))$.

It is easy to see that $\mathbf{ctr}(N)$ indeed represents the labelled partial order $(T_N, F_N^+ \cap (T_N \times T_N), \Sigma_2, \phi_{2N})$. Then, by Theorem 101, the same holds for $\mathbf{pru}(\mathbf{ctr}(N))$.

Lemma 104. *Let $N = (P, T, F, \phi_1, \phi_2)$ be an acyclic (Σ_1, Σ_2)-labelled net and let* $\mathbf{pru}(\mathbf{ctr}(N)) = (T, \Gamma, \Sigma_2, \phi_2)$. *Then,*
for all $s, t \in T$, $(s, t) \in \Gamma^+$ iff $(s, t) \in F^+$.

If N is a process of an EN system M, then $\mathbf{ctr}(N)$ is called a *contracted process of* M and $\mathbf{pru}(\mathbf{ctr}(N))$ a *pruned contracted process of* M. For a contact-free EN system M we denote by $\mathrm{LPO}(M)$ the set of all pruned contracted processes of M (where LPO stands for Labelled Partial Orders). Hence

$$\mathrm{LPO}(M) = \{\mathbf{pru}(\mathbf{ctr}(N)) \mid N \in \mathrm{PROC}(M)\}.$$

Example 30. (1) Let M be the EN system of Fig. 47 (the producer/consumer system) and let N be the process of M in Fig. 57. Then $\mathbf{ctr}(N) = \mathbf{pru}(\mathbf{ctr}(N))$ is given in Fig. 60. A larger pruned contracted process of M is drawn in Fig. 61.

(2) Let M be the EN system of Fig. 2 (the mutual exclusion system) and let N be the process of M in Fig. 58. Then $\mathbf{ctr}(N)$ and $\mathbf{pru}(\mathbf{ctr}(N))$ are given in Fig. 62.

We can now call two EN systems equivalent if their sets of pruned contracted processes are isomorphic (see Definition 98). The behaviour of an EN system M is thus defined as $\mathrm{LPO}(M)$, modulo isomorphism.

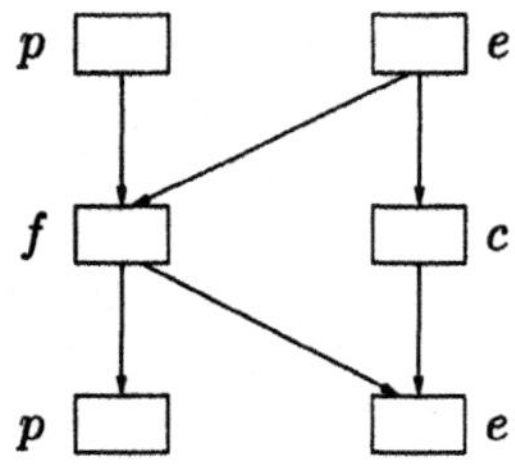

or

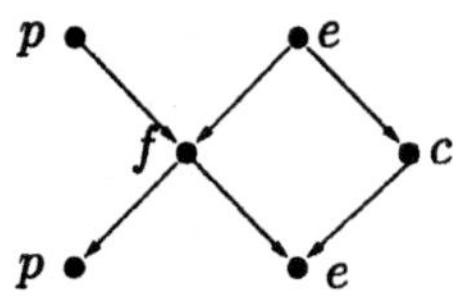

Fig. 60. The (pruned) contracted version of the process of Fig. 57.

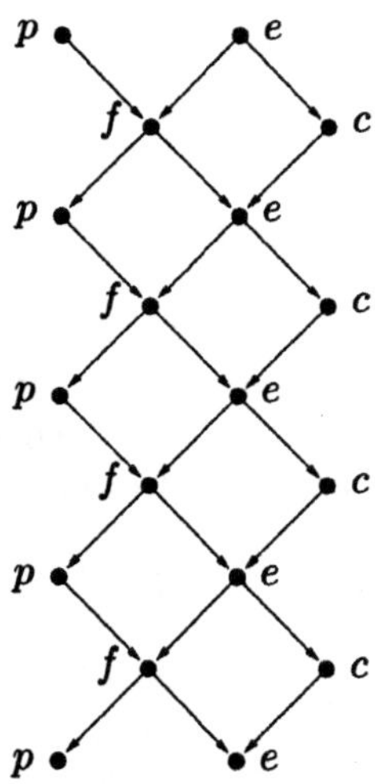

Fig. 61. Another pruned contracted process of the producer/consumer system of Fig. 47.

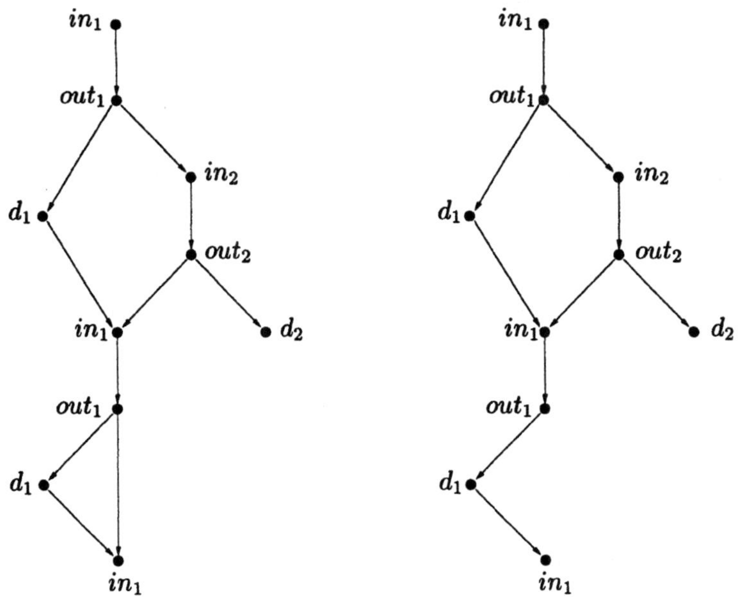

ctr(N) **pru**$($**ctr**$(N))$

Fig. 62. The contracted process and the pruned contracted process corresponding to the process of Fig. 58.

Definition 105. Two contact-free EN systems M and M' are *lpo-equivalent* if $\mathrm{LPO}(M) \equiv \mathrm{LPO}(M')$.

An intuitive way to view a concurrent system, is to see it as a system with its actions not linearly ordered but partially ordered, where a linear order can be interpreted as an ordering in time, whereas a partial order can be interpreted as showing the causal relationships between the actions. This means that the above definition of behaviour is intuitively attractive and perhaps the most natural one.

We now show (as mentioned already in the introduction of this subsection) that, in contrast with ordinary processes (see Theorem 95), the set of pruned contracted processes does not uniquely determine an EN system (modulo isomorphism).

Theorem 106. *There exist reduced contact-free EN systems M and M' such that $LPO(M) \equiv LPO(M')$ but $M \equiv M'$ does not hold.*

Proof. The EN systems M and M' of Figs. 30 and 32 have isomorphic sets of contracted processes, and those contracted processes are already pruned, i.e., contain no transitive edges. Hence LPO(M) $\equiv$ LPO(M'). Note that in this case the contracted processes correspond to FS(M) = FS(M') = $\{\lambda$, a, c, e, ab, cd, ef, $abd\}$, see Example 15.

Another example of two lpo-equivalent EN systems M and M' that are not isomorphic is depicted in Fig. 63. The sets of contracted processes of M and M' are not isomorphic, but the sets of pruned contracted processes are. □

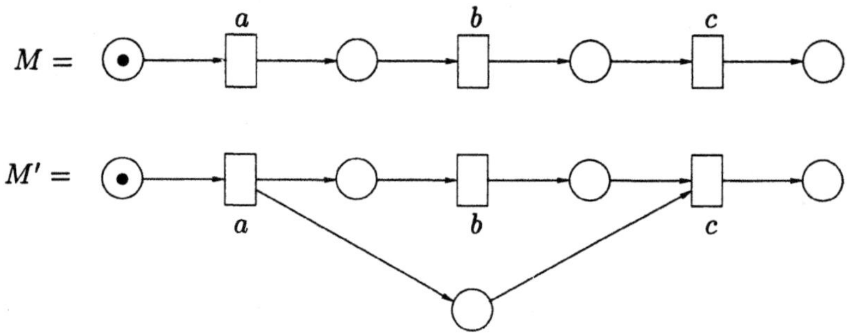

Fig. 63. Two non-isomorphic lpo-equivalent EN systems.

In the remainder of this subsection we will prove that the function **pructr** : PROC(M) $\rightarrow$ LPO(M), defined by **pructr**(N) = **pru**(**ctr**(N)), is a bijection (modulo isomorphism); i.e., though certain information is lost by the function **pructr**, it does not identify distinct processes. Hence LPO(M) is still a "faithful" modelling of the behaviour of M. This is formulated as follows.

Theorem 107. *Let M be a contact-free EN system and let N, N' be two processes of M. Let α be the identity on P_M and β the identity on* **use**(T_M). *Then* $N \equiv_\beta^\alpha N'$ *iff* **pru**(**ctr**(N)) $\equiv_\beta$ **pru**(**ctr**(N')).

This implies that if M and M' are lpo-equivalent EN systems, then there is a bijection f : PROC(M) $\rightarrow$ PROC(M') between their processes such that **pru**(**ctr**(N)) $\equiv_\beta$ **pru**(**ctr**($f(N)$)) for every process N of M.

We now turn to the proof of Theorem 107. The Only-if direction of the proof is clear (both **ctr** and **pru** preserve isomorphism). A simple proof of the implication in the other direction will be given after Theorem 115. The proof that follows is, however, more transparent, because it shows how to reconstruct the process N directly from the pruned contracted process **pru**(**ctr**(N)), i.e., from $(T_N, F_N^+ \cap (T_N \times T_N), T_M, \phi_{2N})$. This technique will be useful in Theorem 149. To explain the construction we need two lemmas.

Lemma 108. *Let $N = (P, T, F, \phi_1, \phi_2)$ be a process of a contact-free EN system M and let $q_1, q_2 \in P$.*

(1) For all $s_1, s_2 \in T$,
if $s_1 \in {}^\bullet q_1$, $s_2 \in q_2{}^\bullet$, $(s_1, s_2) \in F^+$, *and* $\phi(q_1) = \phi(q_2)$, *then* $(q_1, q_2) \in F^*$.
(2) If ${}^\bullet q_1 = \varnothing$ *and* $\phi(q_1) = \phi(q_2)$, *then* $(q_1, q_2) \in F^*$.

Proof. (1) If $q_1 \neq q_2$, then, by Theorem 91(1), q_1 **co** q_2 does not hold. Hence q_1 **li** q_2. Then $(q_1, q_2) \in F^+$, because $(q_2, q_1) \in F^+$ implies that $(s_2, s_1) \in F^*$ and hence that N is cyclic. The proof of (2) is analogous. □

In the next lemma we show how the position of the places of N can be determined from the labelled partial order $(T_N, F_N^+ \cap (T_N \times T_N), T_M, \phi_{2N})$.

Lemma 109. *Let* $N = (P, T, F, \phi_1, \phi_2)$ *be a process of a contact-free EN system* M *and let* $p \in P_M$.

(1) For all $s_1, s_2 \in T$:
$\exists q \in P$ *with* $s_1 \in {}^\bullet q$, $s_2 \in q^\bullet$, *and* $\phi(q) = p$
iff
$\phi(s_1) \in {}^\bullet p$, $\phi(s_2) \in p^\bullet$, $(s_1, s_2) \in F^+$, *and*
$\neg\exists s \in T : \phi(s) \in p^\bullet$ *and* $(s_1, s) \in F^+$ *and* $(s, s_2) \in F^+$.

(2) For all $s_2 \in T$:
$\exists q \in P$ *with* ${}^\bullet q = \varnothing$, $s_2 \in q^\bullet$, *and* $\phi(q) = p$
iff
$p \in (C_{in})_M$, $\phi(s_2) \in p^\bullet$, *and*
$\neg\exists s \in T : \phi(s) \in p^\bullet$ *and* $(s, s_2) \in F^+$.

Proof. First note that the right-hand side of the equivalence in (1) means that s_2 is minimal (with respect to the partial order F^+) in the set of all $s \in T$ with $\phi(s) \in p^\bullet$ and $(s_1, s) \in F^+$ (and, in fact, it is the minimum of that set). Similarly for (2), s_2 is minimal in the set of all $s \in T$ with $\phi(s) \in p^\bullet$.

(1) (Only-if) Obviously $(s_1, s_2) \in F^+$ and, by Definition 88(5), $\phi(s_1) \in {}^\bullet p$ and $\phi(s_2) \in p^\bullet$. Now assume that $\phi(s) \in p^\bullet$, $(s_1, s) \in F^+$, and $(s, s_2) \in F^+$. By Definition 88(5) there exists $q' \in P$ with $s \in (q')^\bullet$ and $\phi(q') = p$. Hence, by Lemma 108(1), $(q, q') \in F^*$. Then, since $q^\bullet = \{s_2\}$, also $(s_2, s) \in F^*$. This contradicts the fact that N is acyclic.

(If) Since $\phi(s_1) \in {}^\bullet p$, there exists $q_1 \in P$ with $s_1 \in {}^\bullet q_1$ and $\phi(q_1) = p$. Likewise there exists q_2 with $s_2 \in q_2{}^\bullet$ and $\phi(q_2) = p$. We prove that $q_1 = q_2$ (which is then the required q). By Lemma 108(1), $(q_1, q_2) \in F^*$. Suppose that $q_1 \neq q_2$. Then there is a transition s such that $q_1 \; F \; s \; F^+ \; q_2$. Clearly $\phi(s) \in p^\bullet$, $(s_1, s) \in F^+$, and $(s, s_2) \in F^+$, contradicting the assumption.

(2) can be proved analogously, using Lemma 108(2) and Definition 88(3). □

We now prove the If-part of Theorem 107. Let $M = (P_M, \mathbf{use}(T_M), F_M, C_{in})$. To show that the function **pructr** : PROC(M) → LPO(M) is a bijection modulo isomorphism, it suffices to define a function **proc** : LPO(M) → PROC(M) such that

(1) for all $G, G' \in$ LPO(M), if $G \equiv_\beta G'$, then $\mathbf{proc}(G) \equiv_\beta^\alpha \mathbf{proc}(G')$, and
(2) for all $N \in$ PROC(M), $\mathbf{proc}(\mathbf{pructr}(N)) \equiv_\beta^\alpha N$.

It is straightforward to define the function **proc**, on the basis of Lemma 109, as follows. Let $G = (T, \Gamma, \mathbf{use}(T_M), \phi)$ be an element of LPO(M). Then $\mathbf{proc}(G)$ is the $(P_M, \mathbf{use}(T_M))$-labelled net (P, T, F, ϕ_1, ϕ), with P, ϕ_1, and F defined as follows. P consists of newly created places of two types: all places q_p with $p \in (C_{in})_M$, and all places $q_{s,p}$ with $s \in T$ and $p \in \phi(s)^\bullet$. Their labels are defined by $\phi_1(q_p) = \phi_1(q_{s,p}) = p$. Intuitively, the places q_p form the initial slice of $\mathbf{proc}(G)$, and the places $q_{s,p}$ form the post-set of the transition s in $\mathbf{proc}(G)$. With this in mind we define $F \cap (T \times P) = \{(s, q_{s,p}) \mid s \in T, p \in \phi(s)^\bullet\}$. The remaining part of F is defined on the basis of Lemma 109 (and Lemma 104): $F \cap (P \times T)$ consists of all pairs $(q_{s_1,p}, s_2)$ such that

$\phi(s_2) \in p^\bullet$, $(s_1, s_2) \in \Gamma^+$, and

$\neg\exists s \in T : \phi(s) \in p^\bullet$ and $(s_1, s) \in \Gamma^+$ and $(s, s_2) \in \Gamma^+$,

and all pairs (q_p, s_2) such that

$\phi(s_2) \in p^\bullet$ and

$\neg\exists s \in T : \phi(s) \in p^\bullet$ and $(s, s_2) \in \Gamma^+$.

This ends the definition of the function **proc**. It should be clear that it satisfies property (1) above. To show property (2), let $N = (P, T, F, \phi_1, \phi_2)$. Then $\mathbf{proc}(\mathbf{pructr}(N))$ and N have the same transitions, with the same labels. Note now that, in N, the sets ${}^\circ N$ and all $s^\bullet$, $s \in T$, form a partition of P. Thus, by Definition 88, there is an obvious bijection between the places of $\mathbf{proc}(\mathbf{pructr}(N))$ and N: q_p corresponds to the unique place in ${}^\circ N$ with label p, and $q_{s,p}$ corresponds to the unique place in $s^\bullet$ with label p. It is straightforward to show from Lemmas 109 and 104 that this correspondence defines an isomorphism between $\mathbf{proc}(\mathbf{pructr}(N))$ and N. Note that this also shows that $\mathbf{proc}(G)$ is in PROC(M), for every $G \in$ LPO(M).

This ends the proof of Theorem 107.

6 Comparison of Partial and Linear Order

In this section we compare the partial order behaviour LPO(M) of an EN system M with its linear order behaviour FS(M). In the first subsection we show that lpo-equivalence and firing sequence equivalence are the same (cf. [Pom88, PomRozSim92]). The basic concepts used to prove this are the independency relation between the transitions of an EN system, and the dependency graph of a firing sequence. These concepts are at the basis of the so-called theory of traces (see, e.g., [AalRoz88] and [DieRoz95], in particular [Maz95] and [HooRoz95]). In the second subsection we show that LPO(M) can be viewed as the set of equivalence classes of a natural equivalence relation on FS(M) that models concurrency. This equivalence relation, called lpo-equivalence of firing sequences, is proved to be the same as the trace equivalence of firing sequences based on the independency relation (see [Maz95]).

This section is largely based on Section 5 of [AalRoz88].

6.1 LPO-Equivalence and Firing Sequence Equivalence

To recapitulate, we now have in total three definitions of the behaviour of a contact-free EN system M: SCG(M) by configuration equivalence, FS(M) by firing sequence equivalence, and LPO(M) by lpo-equivalence, where firing sequence equivalence is weaker than configuration equivalence (Corollary 34). Though firing sequence equivalence describes the sequential (linearly ordered) behaviour of M and lpo-equivalence the non-sequential (partially ordered) behaviour of M, we will prove in this subsection the rather surprising result that two contact-free EN systems are lpo-equivalent iff they are firing sequence equivalent. In one direction this result is as expected: if two EN systems are lpo-equivalent, then they are firing sequence equivalent. This intuitively holds because we can obtain the linear orders in FS(M) from the acyclic graphs in LPO(M) by ordering these graphs topologically. Topological order of acyclic graphs, as defined next, gives a fundamental connection between partial orders and linear orders.

Definition 110. A *topological order* of an acyclic labelled graph $G = (V, \Gamma, \Sigma, \phi)$ is a sequence $u_1 \cdots u_n \in V^*$, with $u_i \in V$ for $1 \leq i \leq n$, and
(1) all u_i are distinct,
(2) $V = \{u_1, \ldots, u_n\}$, and
(3) for all $1 \leq i, j \leq n$, if $(u_i, u_j) \in \Gamma$, then $i < j$.
A *word of* G is a word $\phi(u_1) \cdots \phi(u_n) \in \Sigma^*$, where $u_1 \cdots u_n$ is a topological order of G.

For an acyclic labelled graph $G = (V, \Gamma, \Sigma, \phi)$, the set of all topological orders of G is denoted by $\mathbf{top}(G)$. It is well known that every acyclic graph can be ordered topologically, i.e., $\mathbf{top}(G) \neq \varnothing$. Furthermore, $\mathbf{words}(G)$ denotes the set of all words of G, i.e., $\mathbf{words}(G) = \{\phi(u_1) \cdots \phi(u_n) \mid u_1 \cdots u_n \in \mathbf{top}(G)\}$.

Two graphs G and G' that represent the same partial order have the same topological orders (and hence the same words).

Lemma 111. *Let G and G' be acyclic labelled graphs.*
If $\mathbf{tra}(G) = \mathbf{tra}(G')$, *then* $\mathbf{top}(G) = \mathbf{top}(G')$ *and* $\mathbf{words}(G) = \mathbf{words}(G')$.

Example 31. Consider the acyclic labelled graph G shown in Fig. 64, with $V_G = \{v_1, \ldots, v_6\}$ and $\Sigma_G = \{p, f, e, c\}$. Then, e.g., $v_3v_1v_4v_2v_6v_5$ and $v_1v_2v_3v_4v_5v_6$ are topological orders of G, and *pefcep* and *ecpfpe* are the corresponding words of G. Sequences that are *not* topological orders of G are, e.g., $v_3v_4v_6v_4v_1$ and $v_3v_4v_1v_2v_6v_5$.

We want to show that for every contact-free EN system M,

$$\text{FS}(M) = \bigcup\{\mathbf{words}(G) \mid G \in \text{LPO}(M)\},$$

i.e., the firing sequences of M are the words of the pruned contracted processes of M. We know from Theorem 92 that there is a relationship between the firing sequences of M and the "complete" firing sequences of the processes of M.

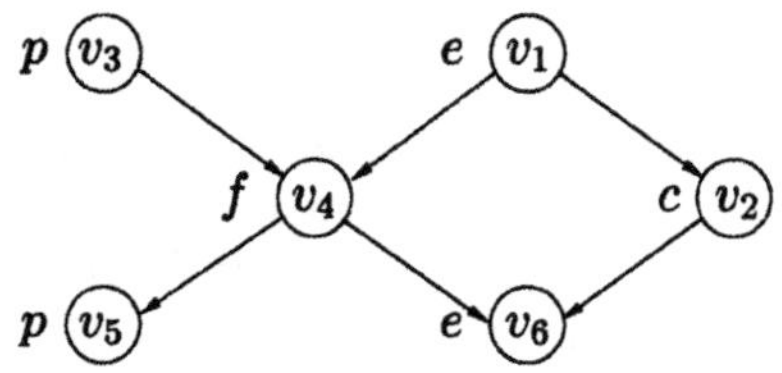

Fig. 64. An acyclic labelled graph.

More precisely, $\mathrm{FS}(M) = \{\phi_N(s_1)\cdots\phi_N(s_n) \mid {}^\circ N[s_1\cdots s_n\rangle N^\circ$ for a process N of $M\}$. Now it only remains to show that these complete firing sequences $s_1\cdots s_n$ of N are precisely the topological orders of the partially ordered set $(T_N, F_N^+ \cap (T_N \times T_N))$.

Theorem 112. *Let $N = (P, T, F)$ be a process net and let $t_1, \ldots, t_n \in T$. Then ${}^\circ N[t_1\cdots t_n\rangle N^\circ$ iff*
(1) all t_i are distinct,
(2) $T = \{t_1, \ldots, t_n\}$, and
(3) for all $1 \leq i, j \leq n$, if $(t_i, t_j) \in F^+$, then $i < j$.

This theorem is an immediate consequence (using Theorem 71(2)) of the following lemma, which is a simple extension of Theorem 79.

Lemma 113. *Let $N = (P, T, F, {}^\circ N)$ be a process net, $C \subseteq P$, and $t_1, \ldots, t_n \in T$. Then ${}^\circ N[t_1\cdots t_n\rangle C$ iff*
(1) all t_i are distinct,
(2) ${}^{\to}C \cap T = \{t_1, \ldots, t_n\}$,
(3) for all $1 \leq i, j \leq n$, if $(t_i, t_j) \in F^+$, then $i < j$, and
(4) C is a slice.

Proof. The proof is by induction on n, and is analogous to the proof of Theorem 79. The details are left to the reader. Note that if conditions (1-4) hold, then t_n is a maximal element of ${}^{\to}C \cap T$ (with respect to F^+) and hence ${t_n}^\bullet \subseteq C$. □

Theorem 114. *Let M be a contact-free EN system. Then $FS(M) = \bigcup\{\mathbf{words}(G) \mid G \in LPO(M)\}$.*

Proof. By Theorem 92, $\mathrm{FS}(M) = \{\phi_N(s_1)\cdots\phi_N(s_n) \mid \exists N \in \mathrm{PROC}(M) : {}^\circ N[s_1\cdots s_n\rangle N^\circ\}$. For a process $N = (P, T, F, \phi_1, \phi_2)$ of M, Theorem 112 now implies that ${}^\circ N[s_1\cdots s_n\rangle N^\circ$ iff $s_1\cdots s_n$ is a topological order of $\mathbf{ctr}(N)$, i.e., iff $s_1\cdots s_n$ is a topological order of $\mathbf{pru}(\mathbf{ctr}(N))$, see Theorem 101 and Lemma 111. Note that the labelling ϕ of $\mathbf{pru}(\mathbf{ctr}(N))$ is the restriction of ϕ_N to T_N. This implies that $\mathrm{FS}(M) = \{\phi_G(u_1)\cdots\phi_G(u_n) \mid u_1\cdots u_n$ is a topological order of $G \in \mathrm{LPO}(M)\} = \bigcup\{\mathbf{words}(G) \mid G \in \mathrm{LPO}(M)\}$. □

Example 32. Let M be the EN system of Fig. 47 and N the process of M in Fig. 57, with **pru(ctr(**N**))** shown in Fig. 60. The words of **pru(ctr(**N**))** that are given in Example 31 (see Fig. 64) are the firing sequences of M given in Example 28.

The next theorem now follows from Theorem 114 and the following simple fact: if G and G' are labelled graphs such that $G \equiv_\beta G'$, and δ is the corresponding bijection between V_G and $V_{G'}$ (see Definition 97), then:
(1) $u_1 \cdots u_n \in \mathbf{top}(G)$ iff $\delta(u_1) \cdots \delta(u_n) \in \mathbf{top}(G')$, and
(2) $\mathbf{words}(G') = \beta(\mathbf{words}(G))$.

Theorem 115. *Let M and M' be two contact-free EN systems and let β be a bijection from* **use**(T_M) *to* **use**$(T_{M'})$.
If $LPO(M) \equiv_\beta LPO(M')$ then $\beta(FS(M)) = FS(M')$.

Similar arguments can be used to give an alternative proof of Theorem 107, as follows. Let $\mathbf{pru}(\mathbf{ctr}(N)) \equiv_\beta \mathbf{pru}(\mathbf{ctr}(N'))$ where β is the identity on $\mathbf{use}(T_M)$. Then, by the simple fact above, $\mathbf{words}(\mathbf{pru}(\mathbf{ctr}(N))) = \mathbf{words}(\mathbf{pru}(\mathbf{ctr}(N')))$. By Theorem 112, $\mathbf{words}(\mathbf{pru}(\mathbf{ctr}(N))) = \{\phi_N(s_1) \cdots \phi_N(s_n) \mid {}^\circ N[s_1 \cdots s_n\rangle N^\circ\}$ and similarly for N'. From this and the uniqueness (modulo $\equiv_\beta^\alpha$) of the process in the statement of Theorem 92, it follows that $N \equiv_\beta^\alpha N'$. Note that, consequently, the function $\mathbf{proc} : \mathrm{LPO}(M) \to \mathrm{PROC}(M)$ mentioned in the proof of Theorem 107 can simply be defined as follows: for $G \in \mathrm{LPO}(M)$, let $t_1 \cdots t_n$ be any element of $\mathbf{words}(G)$; then $\mathbf{proc}(G)$ is the process N corresponding to $t_1 \cdots t_n$ as constructed in the proof of Theorem 92.

We are now going to prove the, more surprising, converse of Theorem 115 (see Theorem 125). More specifically, we will show that we can "break" the linear orders in FS(M) in such a way that partial orders in LPO(M) are obtained. To this purpose we use the "independency relation" between the transitions of an EN system M, defined as follows.

Definition 116. Let Σ be an alphabet. A relation $I \subseteq \Sigma \times \Sigma$ is an *independency relation* (*over* Σ) if I is irreflexive and symmetric.

Definition 117. Let $M = (P, T, F, C_{in})$ be an EN system.

(1) The *independency relation of* M is the independency relation **ind**(M) over **use**(T) defined by
$\mathbf{ind}(M) = \{(s,t) \in T \times T \mid s \neq t \text{ and } \exists C \in \mathbb{C}_M : \{s,t\} \ \mathbf{con}\ C\}$.

(2) The *dependency relation of* M is the relation **dep**(M) defined by
$\mathbf{dep}(M) = (\mathbf{use}(T) \times \mathbf{use}(T)) - \mathbf{ind}(M)$.

Example 33. Consider the EN system M of Fig. 47. Then $\mathbf{ind}(M) = \{(p,e), (e,p), (p,c), (c,p), (f,c), (c,f)\}$ and $\mathbf{dep}(M) = \{(p,f), (f,p), (f,e), (e,f), (e,c), (c,e)\} \cup \{(x,x) \mid x \in \{p,f,e,c\}\}$. See also Examples 8 and 9, and Fig. 15.

Usually (see [Maz95]), the independency relation of M is defined to be $\{(s,t) \in T_M \times T_M \mid s \neq t \text{ and } \mathbf{disj}(\{s,t\})\}$. The above, stronger definition serves the same purposes (cf. [Hoo94]), and moreover it satisfies the next lemma.

We will show that, using $\mathbf{ind}(M)$, we can construct for every firing sequence $t_1 \cdots t_n$ of M a pruned contracted process G of M such that $t_1 \cdots t_n$ is a word of G (without knowing the system M). First we show that we can determine $\mathbf{ind}(M)$ from $\mathrm{FS}(M)$ (without knowing M).

Lemma 118. *Let $M = (P, T, F, C_{in})$ be an EN system. Then* $\mathbf{ind}(M) = \{(s,t) \in T \times T \mid \exists x \in T^* : xst \in FS(M) \text{ and } xts \in FS(M)\}$.

Proof. If $(s,t) \in \mathbf{ind}(M)$, then there is a $C \in \mathbb{C}_M$ such that $\{s,t\}$ **con** C. Hence, by Lemma 17, st **con** C and ts **con** C. Let $x \in T^*$ with $C_{in}[x\rangle C$. Then xst and xts are firing sequences of M. The other way around, assume that $xst, xts \in \mathrm{FS}(M)$. Let $C \in \mathbb{C}_M$ with $C_{in}[x\rangle C$. Then st **con** C and ts **con** C. Lemma 19 then implies that $\{s,t\}$ **con** C, and hence $(s,t) \in \mathbf{ind}(M)$. □

We now "break" the linear order of every firing sequence of M, i.e., for every firing sequence we construct an acyclic graph, and then show that this graph is an element of $\mathrm{LPO}(M)$. In the next definition we use a predetermined countable set of nodes $\{v_1, v_2, v_3, \ldots\}$. This set is used to canonically construct graphs. For later usage we present the definition for an arbitrary independency relation, see [HooRoz95].

Definition 119. Let Σ be an alphabet and I an independency relation over Σ. Let $x = t_1 \cdots t_n \in \Sigma^*$, with $n \geq 0$ and $t_1, \ldots, t_n \in \Sigma$.

(1) The *dependency graph of x (over I)*, denoted by $\mathbf{dep}_I(x)$, is the labelled graph $(V, \Gamma, \Sigma, \phi)$, where $V = \{v_1, \ldots, v_n\}$, $\phi(v_i) = t_i$ for all $1 \leq i \leq n$, and, for all $1 \leq i, j \leq n$, $(v_i, v_j) \in \Gamma$ iff $i < j$ and $(t_i, t_j) \notin I$.

(2) The *pruned dependency graph of x (over I)* is $\mathbf{pru}(\mathbf{dep}_I(x))$.

Note that $v_1 \cdots v_n$ is a topological order of $\mathbf{dep}_I(x)$, and hence x is a word of $\mathbf{dep}_I(x)$.

For an EN system M we write $\mathbf{dep}_M(x)$ instead of $\mathbf{dep}_{\mathbf{ind}(M)}(x)$.

Example 34. Take $\Sigma = \{p, f, e, c\}$ and $I = \mathbf{ind}(M)$ as in Example 33. Then $\mathbf{dep}_I(ecpfpe)$ is given in Fig. 65 and $\mathbf{pru}(\mathbf{dep}_I(ecpfpe))$ in Fig. 66. Since the graphs in Figs. 64 and 66 are the same, $\mathbf{pru}(\mathbf{dep}_I(ecpfpe))$ can thus also be found in Fig. 64.

We now want to show that, for every contact-free EN system M, $\mathrm{LPO}(M) \equiv \{\mathbf{pru}(\mathbf{dep}_M(x)) \mid x \in \mathrm{FS}(M)\}$. First we show a simple connection between $\mathbf{ind}(M)$ and the processes of an EN system M.

Lemma 120. *Let $N = (P, T, F, \phi_1, \phi_2)$ be a process of a contact-free EN system M and let s, t be distinct elements of T. Then:*
(1) if s $\mathbf{co}_N$ t, then $(\phi(s), \phi(t)) \in \mathbf{ind}(M)$,
(2) if $s^\bullet \cap {}^\bullet t \neq \varnothing$, then $(\phi(s), \phi(t)) \in \mathbf{dep}(M)$.

Proof. (1) This follows directly from Theorem 91(3).
(2) If $s^\bullet \cap {}^\bullet t \neq \varnothing$, then also $\phi(s)^\bullet \cap {}^\bullet\phi(t) \neq \varnothing$. □

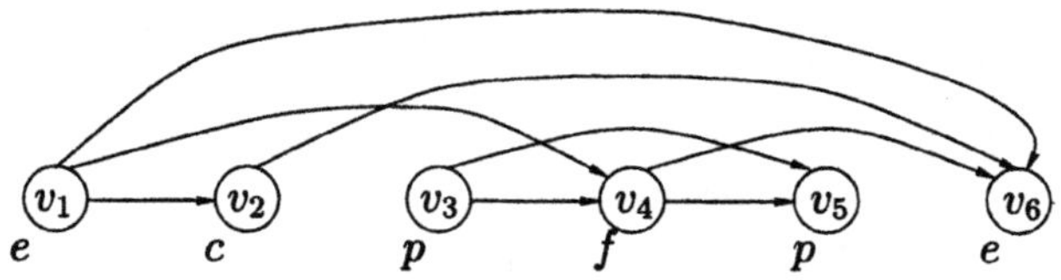

Fig. 65. A dependency graph.

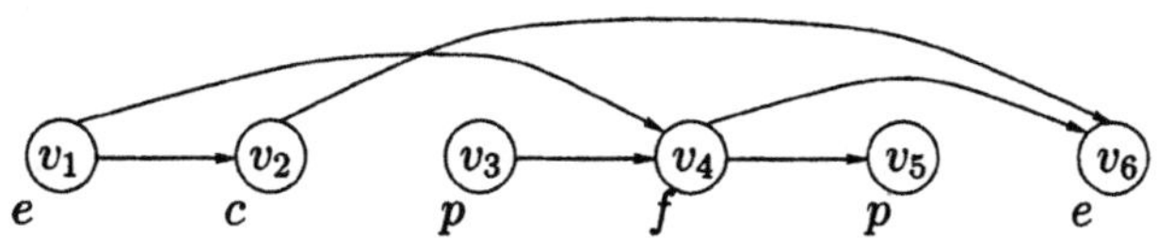

Fig. 66. A pruned dependency graph.

In the next theorem we show the connection between a complete firing sequence of a process N of M and the dependency graph of the corresponding firing sequence of M. To begin with, we prove the following lemma.

Lemma 121. *Let $N = (P, T, F, \phi_1, \phi_2)$ be a process of a contact-free EN system M. Let ${}^\circ N[s_1 \cdots s_n\rangle N^\circ$ with $T = \{s_1, \ldots, s_n\}$, and let $G = \mathbf{dep}_M(\phi(s_1) \cdots \phi(s_n))$. Then for all $1 \le i, j \le n$: $(s_i, s_j) \in F_N^+$ iff $(v_i, v_j) \in \Gamma_G^+$.*

Proof. (Only-if) If $s_i^\bullet \cap {}^\bullet s_j \neq \varnothing$, then, by Lemma 120(2), $(\phi(s_i), \phi(s_j)) \in \mathbf{dep}(M)$ and, by Theorem 112(3), $i < j$. Hence $(v_i, v_j) \in \Gamma_G$ according to Definition 119.

(If) If $(v_i, v_j) \in \Gamma_G$, then $i < j$ and $(\phi(s_i), \phi(s_j)) \in \mathbf{dep}(M)$. Then, by Lemma 120(1), $s_i\ \mathbf{co}_N\ s_j$ does not hold. Hence $(s_i, s_j) \in F_N^+$ or $(s_j, s_i) \in F_N^+$. Theorem 112(3) then implies that $(s_i, s_j) \in F_N^+$. □

Theorem 122. *Let $N = (P, T, F, \phi_1, \phi_2)$ be a process of a contact-free EN system M and let ${}^\circ N[s_1 \cdots s_n\rangle N^\circ$. Let β be the identity on $\mathbf{use}(T_M)$. Then $\mathbf{pru}(\mathbf{ctr}(N)) \equiv_\beta \mathbf{pru}(\mathbf{dep}_M(\phi(s_1) \cdots \phi(s_n)))$.*

Proof. Lemma 121 implies $\mathbf{tra}(\mathbf{ctr}(N)) \equiv_\beta \mathbf{tra}(\mathbf{dep}_M(\phi(s_1) \cdots \phi(s_n)))$ via the bijection δ with $\delta(s_i) = v_i$ between the nodes of these labelled graphs. Now, by Theorem 102, the result holds. □

Thus, for every firing sequence $t_1 \cdots t_n$ of the EN system M, its dependency graph $\mathbf{dep}_M(t_1 \cdots t_n)$ represents the same labelled partial order as (the contracted version of) a process N of M corresponding to $t_1 \cdots t_n$, as expressed in Theorem 92. Together with Theorem 107 this proves the uniqueness (modulo $\equiv_\beta^\alpha$) of this process N, cf. the remark following Theorem 92.

Example 35. Let M be the EN system of Fig. 47 and N the process of M in Fig. 57. Then there is a firing sequence $^\circ N[s_1 \cdots s_n\rangle N^\circ$ with $\phi(s_1)\cdots\phi(s_n) = ecpfpe$ (see Example 28). The graphs $\mathbf{pru}(\mathbf{ctr}(N))$ and $\mathbf{pru}(\mathbf{dep}_M(ecpfpe))$ are drawn in Figs. 60 and 64, respectively.

Theorem 123. *Let M be a contact-free EN system, and let β be the identity on $\mathbf{use}(T_M)$. Then $LPO(M) \equiv_\beta \{\mathbf{pru}(\mathbf{dep}_M(x)) \mid x \in FS(M)\}$.*

Proof. Directly by Theorems 92 and 122. □

We need one more lemma to show that firing sequence equivalence implies lpo-equivalence.

Lemma 124. *Let M and M' be EN systems and let $\beta : \mathbf{use}(T_M) \to \mathbf{use}(T_{M'})$ be a bijection. If $\beta(FS(M)) = FS(M')$ then*
$\{\mathbf{pru}(\mathbf{dep}_M(x)) \mid x \in FS(M)\} \equiv_\beta \{\mathbf{pru}(\mathbf{dep}_{M'}(x)) \mid x \in FS(M')\}$.

Proof. If $\beta(\mathrm{FS}(M)) = \mathrm{FS}(M')$ then, by Lemma 118, $\mathbf{ind}(M') = \{(\beta(s), \beta(t)) \mid (s,t) \in \mathbf{ind}(M)\}$. Then, for $x \in \mathrm{FS}(M)$, $\mathbf{pru}(\mathbf{dep}_M(x)) \equiv_\beta \mathbf{pru}(\mathbf{dep}_{M'}(\beta(x)))$. □

We are now ready to prove the main result of this section.

Theorem 125. *Let M and M' be two contact-free EN systems and let β be a bijection from $\mathbf{use}(T_M)$ to $\mathbf{use}(T_{M'})$.*
If $\beta(FS(M)) = FS(M')$ then $LPO(M) \equiv_\beta LPO(M')$.

Proof. By Theorem 123 and Lemma 124. □

Theorem 126. *Two contact-free EN systems are lpo-equivalent iff they are firing sequence equivalent.*

Proof. By Theorems 115 and 125. □

The following two corollaries of this characterization are interesting.

Corollary 127. *If two contact-free EN systems are configuration equivalent, then they are lpo-equivalent.*

Proof. Directly by Corollary 34 and Theorem 126. □

Corollary 128. *There is an algorithm that, for two arbitrary contact-free EN systems M and M', decides whether or not M and M' are lpo-equivalent.*

Proof. According to Theorem 126 we have to check whether there exists a bijection β from $\mathbf{use}(T_M)$ to $\mathbf{use}(T_{M'})$ such that $\beta(\mathrm{FS}(M)) = \mathrm{FS}(M')$. We first construct $\mathrm{SCG}(M)$ and $\mathrm{SCG}(M')$. From these we can obtain $\mathbf{use}(T_M)$ and $\mathbf{use}(T_{M'})$. Now we test all bijections β. The languages $\mathrm{FS}(M)$, $\beta(\mathrm{FS}(M))$, and $\mathrm{FS}(M')$ are regular, and finite automata can easily be constructed for them (see Theorem 12). Since there is an algorithm to decide whether two finite automata are equivalent (see, e.g., [HopUll79]), we can now apply it to the automata for $\beta(\mathrm{FS}(M))$ and $\mathrm{FS}(M')$. □

For an arbitrary (not necessarily contact-free) EN system M, let $\mathrm{PD}(M) = \{\mathbf{pru}(\mathbf{dep}_M(x)) \mid x \in \mathrm{FS}(M)\}$: the set of pruned dependency graphs of M. Then Theorem 123 says that for every contact-free EN system M: $\mathrm{LPO}(M) \equiv_\beta \mathrm{PD}(M)$, where β is the identity on $\mathbf{use}(T_M)$. Now note that $\mathrm{PD}(M)$ is also defined (and meaningful) for EN systems M that are not contact-free. Lemma 124 implies that, for arbitrary EN systems M and M', if $M \approx M'$ then $\mathrm{PD}(M) \equiv \mathrm{PD}(M')$. As we have observed already, for an EN system M that is not contact-free we consider the processes of a contact-free EN system M' that is configuration equivalent with M. Consequently, in such a case $\mathrm{PD}(M) \equiv \mathrm{PD}(M') \equiv \mathrm{LPO}(M')$; in other words $\mathrm{PD}(M)$ is isomorphic with $\mathrm{LPO}(M')$. Hence we can view $\mathrm{PD}(M)$ as the behaviour of M, and we can thus meaningfully define lpo-equivalence for arbitrary EN systems M and M': M and M' are lpo-equivalent if $\mathrm{PD}(M) \equiv \mathrm{PD}(M')$. The above results can then easily be extended to arbitrary EN systems (e.g., Lemma 124 would then be the generalization of Theorem 125).

6.2 Equivalence of Firing Sequences

Another way to understand the concurrent behaviour of an EN system is by calling two firing sequences of an EN system equivalent if they correspond to two distinct sequential observations of the same run of the system. This notion of equivalence can be formulated as follows.

Definition 129. Let M be a contact-free EN system and let $x, x' \in \mathrm{FS}(M)$. Then x and x' are *lpo-equivalent*, denoted by $x \approx_{lpo} x'$, if there exist a process N of M and two firing sequences $y, y' \in T_N^*$ such that ${}^\circ N[y\rangle N^\circ$, $\phi_N(y) = x$, ${}^\circ N[y'\rangle N^\circ$, and $\phi_N(y') = x'$.

Note that the process N is unique modulo $\equiv_\beta^\alpha$ (cf. the remarks following Theorems 92 and 122). Note also that, by Theorem 112, $x \approx_{lpo} x'$ iff $x, x' \in \mathbf{words}(G)$ for some $G \in \mathrm{LPO}(M)$.

Example 36. The two firing sequences *pefcep* and *ecpfpe* given in Example 28 are lpo-equivalent, i.e., *pefcep* $\approx_{lpo}$ *ecpfpe*.

Due to the relationship between pruned contracted processes and dependency graphs of firing sequences discussed in the previous subsection, lpo-equivalence can be characterized in terms of dependency graphs as follows (and for non-contact-free EN systems we can take this as the definition of lpo-equivalence). This characterization is a part of Theorem 3.12 of [NieRozThi90].

Theorem 130. *Let M be a contact-free EN system and let $x, x' \in FS(M)$. Then $x \approx_{lpo} x'$ iff $\mathbf{pru}(\mathbf{dep}_M(x)) \equiv_\beta \mathbf{pru}(\mathbf{dep}_M(x'))$, where β is the identity on $\mathbf{use}(T_M)$.*

Proof. (Only-if) If $x \approx_{lpo} x'$, then there exist a process N of M and two firing sequences y, y' of N with ${}^\circ N[y\rangle N^\circ$, $\phi_N(y) = x$, ${}^\circ N[y'\rangle N^\circ$, and $\phi_N(y') = x'$. Hence, by Theorem 122, $\mathbf{pru}(\mathbf{ctr}(N)) \equiv_\beta \mathbf{pru}(\mathbf{dep}_M(\phi_N(y)))$ and $\mathbf{pru}(\mathbf{ctr}(N)) \equiv_\beta \mathbf{pru}(\mathbf{dep}_M(\phi_N(y')))$.

(If) By Theorem 123 there exists a process N of M such that $\mathbf{pru}(\mathbf{ctr}(N)) \equiv_\beta \mathbf{pru}(\mathbf{dep}_M(x))$. Since $v_1 \cdots v_n$ is a topological order of $\mathbf{pru}(\mathbf{dep}_M(x))$ (where $n = |x|$), there is a topological order y of $\mathbf{pru}(\mathbf{ctr}(N))$ with the same labels, i.e., $\phi_N(y) = x$. Then, by Theorem 112, ${}^\circ N[y\rangle N^\circ$. Since $\mathbf{pru}(\mathbf{ctr}(N)) \equiv_\beta \mathbf{pru}(\mathbf{dep}_M(x'))$, analogously there exists y' with $\phi_N(y') = x'$ and ${}^\circ N[y'\rangle N^\circ$. □

We will now show a very simple characterization of isomorphism of dependency graphs, in terms of the independency relation $\mathbf{ind}(M)$ induced by the EN system M. This characterization can be proved very generally, without reference to EN systems; it is part of the theory of traces, see [HooRoz95].

For a given independency relation we define an associated equivalence relation on words as follows.

Definition 131. Let I be an independency relation over Σ.
The relation $\doteq_I \subseteq \Sigma^* \times \Sigma^*$ is defined as follows: for $x, y \in \Sigma^*$,
$x \doteq_I y$ iff there exist $a, b \in \Sigma$ and $x_1, x_2 \in \Sigma^*$, such that
$x = x_1abx_2$, $y = x_1bax_2$, and $(a, b) \in I$.
The relation $\approx_I \subseteq \Sigma^* \times \Sigma^*$ is then defined as the smallest equivalence relation that contains $\doteq_I$. If $x \approx_I y$ then x and y are *trace equivalent* (*over* I).

Thus, two words are trace equivalent if one can be obtained from the other by interchanging independent symbols, repeatedly.

Lemma 132. *Let I be an independency relation over Σ, and let $x, y \in \Sigma^*$.*

(1) $x \approx_I y$ iff there exist $n \geq 0$ and $x_0, \ldots, x_n \in \Sigma^*$ such that $x_0 = x$, $x_n = y$, and $x_{i-1} \doteq_I x_i$ for all $1 \leq i \leq n$.

(2) $x \approx_I y$ implies $|x| = |y|$.

Intuitively the definition of trace equivalence is based on the following property of firing sequences of an EN system $M = (P, T, F, C_{in})$: for $x, y \in T^*$ and $s, t \in T$, if $xsty \in \mathrm{FS}(M)$ and $(s, t) \in \mathbf{ind}(M)$, then $xtsy \in \mathrm{FS}(M)$. The proof of this property is easy: If $(s, t) \in \mathbf{ind}(M)$ then $\mathbf{disj}(\{s, t\})$. Let $C_{in}[x\rangle C[st\rangle D$. Then st **con** C and $\mathbf{disj}(\{s, t\})$. This implies that $\{s, t\}$ **con** C and thus (by Lemma 17) that $C[ts\rangle D$. Hence $xtsy \in \mathrm{FS}(M)$. In other words, transitions in $\mathbf{ind}(M)$ are interchangeable in a firing sequence, because they actually occur concurrently. Thus, $\mathrm{FS}(M)$ is closed under $\approx_{\mathbf{ind}(M)}$: if $x \in \mathrm{FS}(M)$ and $x \approx_{\mathbf{ind}(M)} y$, then $y \in \mathrm{FS}(M)$. A similar argument shows that, for $x, y \in \mathrm{FS}(M)$, $x \approx_{\mathbf{ind}(M)} y$ iff $x \approx_I y$, where $I = \{(s, t) \in T_M \times T_M \mid s \neq t, \mathbf{disj}(\{s, t\})\}$ (the "usual" independency relation, cf. the remark following Definition 117).

Example 37. Again, let $\Sigma = \{p, f, e, c\}$ and $I = \{(p, e), (e, p), (p, c), (c, p), (f, c), (c, f)\}$. Then I is an independency relation, $pefc \doteq_I epfc \doteq_I epcf$, $pefc \approx_I epcf$, and $[pefc]_I = \{pefc, pecf, epfc, epcf, ecpf\}$.

For an independency relation I over Σ and an $x \in \Sigma^*$, $[x]_I$ denotes the equivalence class of $\approx_I$ that contains x. An equivalence class of $\approx_I$ is called a *trace* over I; intuitively it is the set of all sequential observations of one run of some system. In Example 37 $[pefc]_I$ is a trace over I. We will prove that $[x]_I = \mathbf{words}(\mathbf{dep}_I(x))$. To this purpose we use the following well-known result from graph theory (where the labelling of the graph is irrelevant).

Lemma 133. *Let $G = (V, \Gamma, \Sigma, \phi)$ be an acyclic graph. Let $J \subseteq V \times V$ be the independency relation (of G) defined as follows: for all $u, w \in V$,*
$(u, w) \in J$ iff $u \neq w$, $(u, w) \notin \Gamma$, and $(w, u) \notin \Gamma$.
Then, for every topological order $u_1 \cdots u_n$ of G,
$\mathbf{top}(G) = \{w_1 \cdots w_n \in V^* \mid u_1 \cdots u_n \approx_J w_1 \cdots w_n\}$.

Proof. (1) It is clear that $\{w_1 \cdots w_n \in V^* \mid u_1 \cdots u_n \approx_J w_1 \cdots w_n\} \subseteq \mathbf{top}(G)$ for every $u_1 \cdots u_n \in \mathbf{top}(G)$.

(2) Now it remains to show that $\mathbf{top}(G) \subseteq \{w_1 \cdots w_n \in V^* \mid u_1 \cdots u_n \approx_J w_1 \cdots w_n\}$. Let $w_1 \cdots w_n \in \mathbf{top}(G)$. We prove by induction on k that for every k, $1 \leq k \leq n+1$, there exists a topological order $t_1 \cdots t_n$ of G such that $t_1 \cdots t_n \approx_J w_1 \cdots w_n$ and $t_i = u_i$ for all $1 \leq i \leq k-1$. For $k = 1$ we take $t_1 \cdots t_n = w_1 \cdots w_n$. For the induction step, assume that the statement holds for k, and consider the topological order $t_1 \cdots t_n$ which satisfies the requirements for k. There exists m, $k \leq m \leq n$, such that $u_k = t_m$. Then $t_k, \ldots, t_{m-1}$ succeed u_k in the topological order $u_1 \cdots u_n$. Hence there are no edges $(t_k, t_m), \ldots, (t_{m-1}, t_m)$ in Γ (and of course no edges in the other direction because $t_1 \cdots t_n$ is a topological order). Thus $(t_k, t_m), \ldots, (t_{m-1}, t_m) \in J$. This implies that $t_1 \cdots t_n \approx_J t_1 \cdots t_{k-1} t_m t_k \cdots t_{m-1} t_{m+1} \cdots t_n$. And this is a topological order (according to (1)) that satisfies the requirements for $k+1$. □

For $I = \mathbf{ind}(M)$, the next result is a refinement of Theorem 114, as can be seen from Theorem 123.

Theorem 134. *Let I be an independency relation over Σ, and $x \in \Sigma^*$. Then* $[x]_I = \mathbf{words}(\mathbf{dep}_I(x))$.

Proof. Let J be the independency relation of the graph $G = \mathbf{dep}_I(x)$, as defined in Lemma 133. Then it is easy to check that, for $u, w \in V_G$, $(u, w) \in J$ iff $(\phi(u), \phi(w)) \in I$. Hence, for all $y \in \Sigma^*$, $x \approx_I y$ iff there is a $w_1 \cdots w_n \in V_G^*$ with $v_1 \cdots v_n \approx_J w_1 \cdots w_n$ and $y = \phi(w_1) \cdots \phi(w_n)$. Now $v_1 \cdots v_n$ is a topological order of $\mathbf{dep}_I(x)$. Hence, according to Lemma 133, $v_1 \cdots v_n \approx_J w_1 \cdots w_n$ iff $w_1 \cdots w_n$ is a topological order. Thus, for all $y \in \Sigma^*$, $x \approx_I y$ iff there exists a topological order $w_1 \cdots w_n$ of $\mathbf{dep}_I(x)$ such that $y = \phi(w_1) \cdots \phi(w_n)$. □

This implies that two words are trace equivalent iff they have isomorphic dependency graphs.

Theorem 135. *Let I be an independency relation over Σ, and let $x, y \in \Sigma^*$. Then the following four statements are equivalent.*

(1) $x \approx_I y$,
(2) $[x]_I = [y]_I$,
(3) $\mathbf{dep}_I(x) \equiv_\beta \mathbf{dep}_I(y)$, *and*
(4) $\mathbf{pru}(\mathbf{dep}_I(x)) \equiv_\beta \mathbf{pru}(\mathbf{dep}_I(y))$,
where β *is the identity on* Σ.

Proof. (1) implies (3): Assume $x = x_1abx_2$ and $y = x_1bax_2$, with $x_1, x_2 \in \Sigma^*$ and $(a,b) \in I$. Then it is easy to see that $\mathbf{dep}_I(x)$ and $\mathbf{dep}_I(y)$ are isomorphic: if $|x_1| = i - 1$, then take the bijection $\delta : \{v_1, \ldots, v_n\} \to \{v_1, \ldots, v_n\}$ such that $\delta(v_i) = v_{i+1}$, $\delta(v_{i+1}) = v_i$, and $\delta(v_j) = v_j$ for $j \neq i, i+1$.

(3) implies (2): This follows from Theorem 134, because $\mathbf{dep}_I(x)$ and $\mathbf{dep}_I(y)$ have corresponding topological orders with the same labels. Likewise (4) implies (2), because $\mathbf{pru}(\mathbf{dep}_I(x))$ has the same topological orders as $\mathbf{dep}_I(x)$. □

Note that Theorems 134 and 135 do not refer to EN systems. The equality of lpo-equivalence and trace equivalence of firing sequences of EN systems is now a direct consequence of Theorems 130 and 135.

Theorem 136. *Let* M *be a contact-free EN system and* $x, x' \in FS(M)$. *Then* $x \approx_{lpo} x'$ *iff* $x \approx_{\mathbf{ind}(M)} x'$.

A *trace language* (*over* an independency relation I) is a set of equivalence classes of $\approx_I$. Now the behaviour of an EN system M can also be defined as the trace language $\mathrm{TR}(M) = \{[x]_{\mathbf{ind}(M)} \mid x \in \mathrm{FS}(M)\}$, i.e., the language $\mathrm{FS}(M)$ in which trace equivalent words are grouped together. According to Theorems 123 and 135 the function $\mathbf{prudep} : \mathrm{TR}(M) \to \mathrm{LPO}(M)$, defined by $\mathbf{prudep}([x]) = \mathbf{pru}(\mathbf{dep}_M(x))$, is a bijection between $\mathrm{TR}(M)$ and $\mathrm{LPO}(M)$, modulo isomorphism (and note that, by Theorem 134, **words** is its inverse). Thus, $\mathrm{TR}(M)$ can also be seen as a formalization of the set of runs of the system M (see Theorem 107). Theorems 130 and 136 (i.e., Theorem 3.12 of [NieRozThi90]) show that the mapping that assigns a process with each firing sequence, as defined in Theorem 92, is a bijection between $\mathrm{TR}(M)$ and $\mathrm{PROC}(M)$, modulo isomorphism. We can define two EN systems M and M' to be *trace equivalent* if there exists a bijection $\beta : \mathbf{use}(T_M) \to \mathbf{use}(T_{M'})$ such that $\beta(\mathrm{TR}(M)) = \mathrm{TR}(M')$, where $\beta(\mathrm{TR}(M))$ is defined in the obvious way. Then, clearly, trace equivalence is the same as lpo-equivalence.

7 Branching Processes

To obtain a more complete picture of the relationship between different runs of a system that is not conflict-free, we will consider "branching runs" (also called "unfoldings"). Intuitively, a branching run combines several conflicting runs of the system, with an indication of where the conflicts occur. At each point of conflict, one may view the system as splitting into several "parallel" copies of itself, one for each resolution of the conflict (just as a splitting universe in science-fiction). In this section we model these branching runs by "branching processes", which are a natural extension of the processes that we have considered in the previous sections.

The theory of branching processes or unfoldings was initiated in [NiePloWin81], and developed in, e.g., [Win87, NieRozThi90, RozThi91, Eng91, NieRozThi95, WinNie95a, WinNie95b]. In [McM93, McM95, Esp94, EspRömVog96] branching processes are used to develop an efficient model checker for contact-free EN systems, i.e., an algorithm that verifies logical properties of such systems (for a fragment of so-called branching time temporal logic in [Esp94]). Most papers consider the unique maximal unfolding of the EN system, which is a (usually infinite) branching run containing *all* runs of the system. Here we consider arbitrary (finite) branching runs, as in [Eng91].

Just as processes are based on process nets, branching processes are based on branching process nets. A branching process net is like a process net, except that its places may have arbitrary output-sets. A conflict is modelled by a place with more than one transition in its output-set. Whenever a conflict occurs, the conflicting parts of the branching run should be separated after the conflict. This is formalized by requiring the following "conflict relation" to be irreflexive.

Definition 137. Let $N = (P, T, F)$ be a net. The *conflict relation* of N is the binary relation $\otimes \subseteq X_N \times X_N$ defined as follows: for all $x_1, x_2 \in X_N$, $x_1 \otimes x_2$ if there exist distinct transitions $t_1, t_2 \in T$ such that ${}^\bullet t_1 \cap {}^\bullet t_2 \neq \varnothing$ and $t_i\ F^*\ x_i$ for $i = 1, 2$.

The conflict relation of a net N will also be denoted by $\otimes_N$. Note that it is a symmetric relation. We now use it to define branching process nets, introduced in [NiePloWin81] (where they are called occurrence nets).

Definition 138. A net $N = (P, T, F)$ is a *branching process net*, abbreviated *b-process net*, if:
(1) N is acyclic,
(2) $\#({}^\bullet p) \leq 1$ for all $p \in P$, and
(3) $\otimes_N$ is irreflexive.

Note that the irreflexivity of $\otimes_N$ can also be expressed as follows: for all distinct transitions $t_1, t_2 \in T$, if ${}^\bullet t_1 \cap {}^\bullet t_2 \neq \varnothing$, then $\{x \in X_N \mid t_1\ F^*\ x\} \cap \{x \in X_N \mid t_2\ F^*\ x\} = \varnothing$. Intuitively this means that conflicting transitions t_1 and t_2 have disjoint futures.

As for process nets, we will view a b-process net N as an EN system with initial configuration ${}^\circ N$. Every process net is a b-process net, because $\otimes_N$ is empty for a process net N.

Example 38. The EN systems in Figs. 22, 23, 24, and 25 are b-process nets. This shows that confusion can be present in b-process nets. The acyclic EN system in Fig. 30 is not a b-process net because $\#({}^\bullet p_3) = 2$. The EN system of Fig. 34 satisfies (1) and (2) above but it is not a b-process net, because $t_3 \otimes t_3$. In fact, ${}^\bullet t_1 \cap {}^\bullet t_2 \neq \varnothing$ and both $t_1\ F^*\ t_3$ and $t_2\ F^*\ t_3$. Thus, t_3 is in the conflict relation with itself and so $\otimes$ is not irreflexive.

In a b-process net N we define the notion of a slice in the same way as for a process net, but we additionally require that its places are not in the conflict relation $\otimes_N$. We will denote the complement of $\otimes_N$ by $\overline{\otimes}_N$; note that it is a reflexive symmetric relation (cf. Definition 65).

Definition 139. A *slice* of a b-process net N is a maximal ($\mathbf{co}_N \cap \overline{\otimes}_N$)-clique C of N such that $C \subseteq P_N$.

Example 39. Consider the b-process net N of Fig. 23. Its conflict relation is $\otimes_N = \{(x,y),(y,x) \mid x \in \{t_3,p_4\}, y \in \{t_1,p_5,t_2,p_3\}$. The slices of N are $\{p_1,p_2\}$, $\{p_4\}$, $\{p_5,p_2\}$, $\{p_1,p_3\}$, and $\{p_5,p_3\}$. Note that $\{p_4\}$ and $\{p_5,p_3\}$ are not cuts.

If one systematically considers the relation $\mathbf{co}_N \cap \overline{\otimes}_N$ instead of the relation $\mathbf{co}_N$, then Lemmas 76, 77, and 78, and Theorems 79 and 81, also hold for b-process nets. The details of the proofs are left to the reader. Thus, by Lemma 78 and Theorem 81, b-process nets are contact-free and reduced. In the next theorem we state the generalization of Theorem 79: the slices of a b-process net are exactly its reachable configurations.

Theorem 140. *Let $N = (P,T,F,{}^\circ N)$ be a b-process net and let $C \subseteq P$. $C \in \mathbb{C}_N$ iff C is a slice of N.*

Based on b-process nets, we now define branching processes of an EN system, see [Eng91]. They can be viewed as records of all events that occur during a branching run of the system.

Definition 141. Let $N = (P_N,T_N,F_N,\phi_1,\phi_2)$ be a (Σ_1,Σ_2)-labelled b-process net and let $M = (P,T,F,C_{in})$ be a contact-free EN system.
Then N is a *branching process of M*, abbreviated *b-process*, if
(1)–(5) of Definition 88 hold, and
(6) for all $s,t \in T_N$, if ${}^\bullet s = {}^\bullet t$ and $\phi_2(s) = \phi_2(t)$, then $s = t$.

For a contact-free EN system M, we denote the set of all b-processes of M by BPROC(M).

Condition (6) above says that a conflict is always between two distinct transitions of M. This is a natural requirement that prevents the same run to appear twice in the record of a branching run. However, many properties of b-processes also hold without the requirement.

Example 40. (1) Let M be the (contact-free) EN system of Fig. 51. A branching process N of M is drawn in Fig. 67. Note that the process of M that is given in Fig. 56 is "part" of N, i.e., it is one of the runs of M that is combined in the branching run corresponding to N.

(2) A b-process of the EN system of Fig. 2 (mutual exclusion) is drawn in Fig. 68. Intuitively, it is a combination of four possible runs of the system: component i gets permission to access its critical section, and then component j gets permission, for every combination of $i,j \in \{1,2\}$. This can be compared with the process in Fig. 58, corresponding to one run during which components 1,2,1, and 1 get permission, respectively.

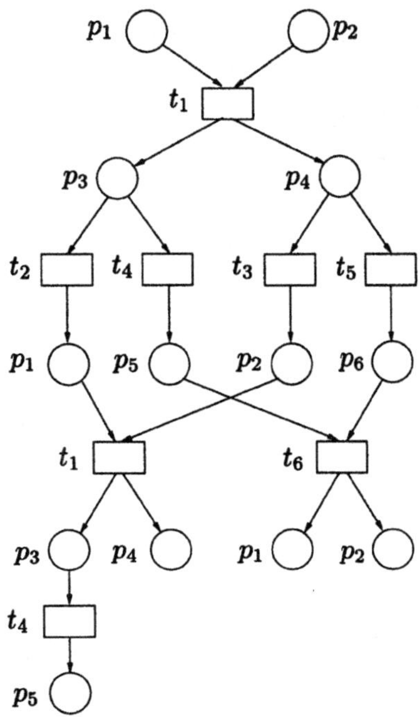

Fig. 67. A branching process of the EN system of Fig. 51.

It can be proved that Lemma 89 and Theorems 90 and 91 are also true for b-processes (again with $\mathbf{co}_N \cap \overline{\otimes}_N$ instead of $\mathbf{co}_N$). In fact, they are even true when condition (6) is dropped from the definition of b-process (Definition 141); this will be needed in the proof of Theorem 149.

As for processes, to compare the behaviour of two different EN systems we are mainly interested in the events, and their relationships, rather than in the conditions of a branching process. Thus, we will remove the conditions and consider pruned contracted b-processes (see Section 5.4). In the case of branching processes we are not only interested in the causal relationship between events but also in their conflict relation. However, since the conflicts in a b-process are modelled by places with more than one transition in their output-set, the conflict relation is lost when the places are removed. Consequently, in pruned contracted b-processes, we have to model the conflict relation explicitly.

Just as a pruned contracted process represents a labelled partial order, a pruned contracted b-process will represent a labelled partial order together with a conflict relation. Such a partial order with conflict relation is called an "event structure", introduced in [NiePloWin81] (see also, e.g., [Win87, WinNie95a]).

Definition 142. An *event structure* is a triple $(A, \rho, \otimes)$ where
(1) (A, ρ) is a partially ordered set,

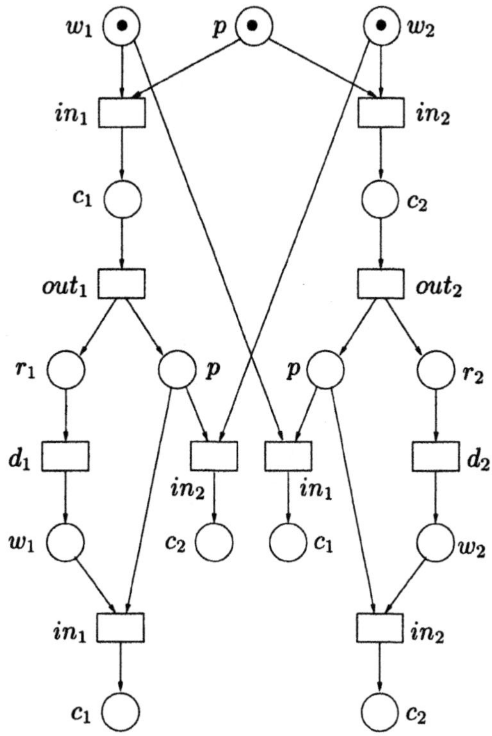

Fig. 68. A b-process of the mutual exclusion system of Fig. 2.

(2) $\otimes \subseteq A \times A$ is an irreflexive symmetric relation, and
(3) for all $a, b, a', b' \in A$, if $a \otimes b$, $a\ \rho\ a'$, and $b\ \rho\ b'$, then $a' \otimes b'$.

Condition (3) relates the causal relation ρ to the conflict relation $\otimes$. It expresses the fact that ρ inherits $\otimes$, in the sense that a conflict between a and b is inherited by all ρ-descendants of a and b (as in a vendetta).

It is easy to see that for every b-process net N, $(X_N, F_N^+, \otimes_N)$ is an event structure (cf. Lemma 74). For a b-process N of an EN system M, we will be interested in the "labelled event structure"

$$(T_N, F_N^+ \cap (T_N \times T_N), \otimes_N \cap (T_N \times T_N), T_M, \phi_{2N}).$$

As in the case of processes and labelled partial orders, we will represent such a labelled event structure by an acyclic labelled graph, with an additional relation that models conflict. We will call this a "labelled branching graph", defined as follows.

Definition 143. Let Σ be an alphabet. A (Σ-)*labelled b-graph* is a quintuple $G = (V, \Gamma, \otimes, \Sigma, \phi)$, where $(V, \Gamma, \Sigma, \phi)$ is an acyclic Σ-labelled graph, and

$\otimes \subseteq V \times V$ is the *conflict relation* of G, such that
$\neg\exists v_1, v_2, v \in V : v_1 \otimes v_2,\ v_1\ \Gamma^*\ v$, and $v_2\ \Gamma^*\ v$.

It is left to the reader to define the appropriate notion of isomorphism of labelled b-graphs and sets of labelled b-graphs (cf. Definitions 97 and 98). Note that a labelled b-graph G can be viewed as a graph with two types of edges: directed edges in Γ_G, and undirected edges in $\otimes_G$.

Example 41. Figure 69 shows an example of a labelled b-graph G. The conflict relation of G is indicated by undirected dashed lines. The nodes of G are labeled

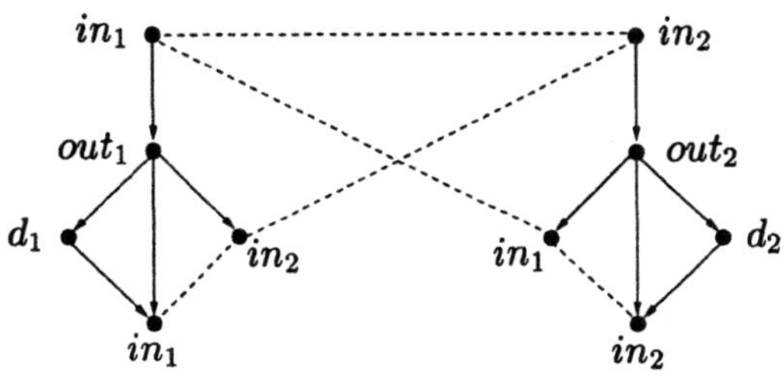

Fig. 69. A labelled b-graph.

by the transitions of the mutual exclusion system of Fig. 2. We will see later that G is in fact a contracted b-process of that EN system.

We now define the labelled event structure that is represented by a labelled b-graph, cf. Definition 99.

Definition 144. Let $G = (V, \Gamma, \otimes, \Sigma, \phi)$ be a labelled b-graph. The *transitive closure* of G, denoted by $\mathbf{tra}(G)$, is the labelled b-graph $(V, \Gamma^+, \mathbf{tra}(\otimes), \Sigma, \phi)$, where $\mathbf{tra}(\otimes) = \{(v_1, v_2) \mid \exists v_1', v_2' \in V : v_1' \otimes v_2' \text{ and } v_i'\ \Gamma^* v_i \text{ for } i = 1, 2\}$. We also say that G *represents* $\mathbf{tra}(G)$.

It should be clear that $(V, \Gamma^+, \mathbf{tra}(\otimes))$ is an event structure, and so $\mathbf{tra}(G)$ is a labelled event structure. The pruned version of a labelled b-graph is defined next, cf. Definition 100.

Definition 145. Let $G = (V, \Gamma, \otimes, \Sigma, \phi)$ be a labelled b-graph. The *pruned version of* G, denoted by $\mathbf{pru}(G)$, is the labelled b-graph $(V, \Gamma', \otimes', \Sigma, \phi)$ with $\Gamma' = \{(v, w) \in \Gamma \mid \neg\ \exists u \in V : (v, u) \in \Gamma^+ \text{ and } (u, w) \in \Gamma^+\}$ and $\otimes' = \{(v_1, v_2) \in \otimes \mid \neg\exists (v_1', v_2') \in \otimes : (v_1', v_2') \neq (v_1, v_2) \text{ and } v_i'\ \Gamma^*\ v_i \text{ for } i = 1, 2\}$.

It is not difficult to show that Theorems 101 and 102 still hold for labelled b-graphs.

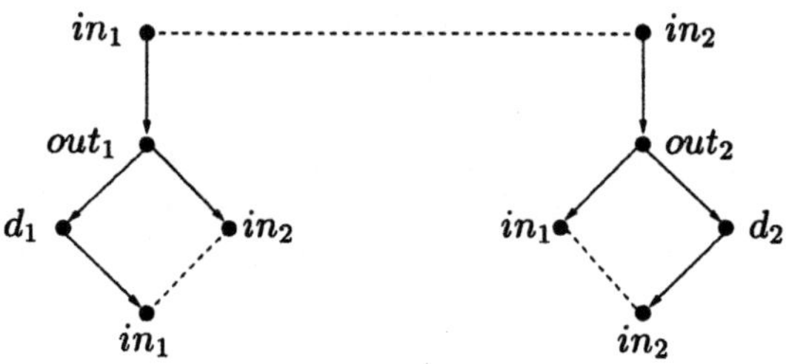

Fig. 70. The pruned version of the labelled b-graph of Fig. 69.

Example 42. Figure 70 shows the pruned version $\mathbf{pru}(G)$ of the labelled b-graph G of Fig. 69.

After discussing labelled b-graphs, we return to b-processes and show how to contract and prune them, cf. Definition 103.

Definition 146. Let $N = (P, T, F, \phi_1, \phi_2)$ be a b-process of an EN system M.

(1) The *contracted version of* N, denoted by $\mathbf{ctr}(N)$, is the labelled b-graph $(T, \Gamma, \otimes, T_M, \phi_2)$ such that, for all $s, t \in T$,
$(s,t) \in \Gamma$ iff $s^\bullet \cap {}^\bullet t \neq \varnothing$, and $(s,t) \in \otimes$ iff ${}^\bullet s \cap {}^\bullet t \neq \varnothing$.

(2) The *pruned contracted version of* N is the labelled b-graph $\mathbf{pru}(\mathbf{ctr}(N))$.

Example 43. Let N be the b-process given in Fig. 68. Its contracted version $\mathbf{ctr}(N)$ and pruned contracted version $\mathbf{pru}(\mathbf{ctr}(N))$ are shown in Figs. 69 and 70, respectively. They can be compared with the contracted and pruned contracted process of Fig. 62.

It is easy to see (using the analogue of Lemma 104 for $\mathbf{ctr}(N)$) that, for $\mathbf{ctr}(N) = (T_N, \Gamma, \otimes, \mathbf{use}(T_M), \phi_2)$, $\mathbf{tra}(\otimes) = \otimes_N \cap (T_N \times T_N)$. This implies that $\mathbf{ctr}(N)$ is indeed a labelled b-graph, and that both $\mathbf{ctr}(N)$ and $\mathbf{pru}(\mathbf{ctr}(N))$ represent the labelled event structure $(T_N, F_N^+ \cap (T_N \times T_N), \otimes_N \cap (T_N \times T_N), T_M, \phi_{2N})$. We note here that, by condition (6) of Definition 141, this labelled event structure is deterministic in the sense of [Vaa91].

For a contact-free EN system M we denote by $\mathrm{LES}(M)$ the set of all pruned contracted b-processes of M (where LES stands for Labelled Event Structures). Hence

$$\mathrm{LES}(M) = \{\mathbf{pru}(\mathbf{ctr}(N)) \mid N \in \mathrm{BPROC}(M)\}.$$

Definition 147. Two contact-free EN systems M and M' are *les-equivalent* if $\mathrm{LES}(M) \equiv \mathrm{LES}(M')$.

Obviously, for every contact-free EN system M, $\mathrm{LPO}(M) = \{G \in \mathrm{LES}(M) \mid \otimes_G = \varnothing\}$. This implies that if two EN systems are les-equivalent, then they are lpo-equivalent. To show that this also holds the other way around, we need the concept of a "configuration" of an event structure. We define it for labelled b-graphs. Intuitively, a configuration of a branching run is one of the conflict-free runs that the branching run consists of.

Definition 148. Let $G = (V, \Gamma, \otimes, \Sigma, \phi)$ be a labelled b-graph.
A *configuration* of G is a subset R of V such that
(1) $\forall v_1, v_2 \in R : \neg\, v_1 \otimes v_2$, and
(2) $\forall v_1, v_2 \in V$: if $v_2 \in R$ and $v_1\ \Gamma\ v_2$, then $v_1 \in R$.
For a configuration R of G, the (non-branching) labelled graph *induced by R*, denoted $G[R]$, is $(R, \Gamma \cap (R \times R), \Sigma, \phi \restriction R)$.

It is well known that configurations of unfoldings of an EN system correspond to processes of the system (cf. Theorem 4.6 of [NieRozThi90] and the discussion at the end of that paper). In the next theorem we show that, due to this correspondence, LES(M) can be recovered from LPO(M) without knowing M.

Theorem 149. *Let M be a contact-free EN system, and let $G = (T, \Gamma, \otimes, \Sigma, \phi)$ be a labelled b-graph with $\Sigma = \mathbf{use}(T_M)$.*
Then, $G \in LES(M)$ iff
(1) $\mathbf{pru}(G) = G$,
(2) $G[R] \in LPO(M)$ for every configuration R of G, and
(3) $\forall s_1, s_2 \in T$: if $s_1 \otimes s_2$, then $(\phi(s_1), \phi(s_2)) \in$ $\mathbf{dep}(M)$ *and $\phi(s_1) \neq \phi(s_2)$.*

Proof. (Only-if) This direction of the proof consists of verifying a number of rather obvious properties of pruned contracted b-processes. Let $G = \mathbf{pru}(\mathbf{ctr}(N))$ for a b-process $N = (P, T, F, \phi_1, \phi)$ of M. Property (1) is obvious. For a configuration $R \subseteq T$, let $N[R]$ be the labelled net $(P', T', F', \phi_1', \phi')$ with $T' = R$, $P' = {}^\circ N \cup \mathbf{nbh}(R)$, and F', ϕ_1', ϕ' are the restrictions of F, ϕ_1, ϕ to P' and T'. It is straightforward to show that $N[R]$ is a process of M with $\mathbf{pru}(\mathbf{ctr}(N[R])) = G[R]$. This proves property (2). To show property (3), let $s_1, s_2 \in T$ with $s_1 \otimes s_2$. Then $s_1 \neq s_2$ by Definition 143, and ${}^\bullet s_1 \cap {}^\bullet s_2 \neq \varnothing$ by Definition 146. This implies that ${}^\bullet\phi(s_1) \cap {}^\bullet\phi(s_2) \neq \varnothing$, and so $(\phi(s_1), \phi(s_2)) \in \mathbf{dep}(M)$. Now we assume that $\phi(s_1) = \phi(s_2)$ and derive a contradiction. Since $s_1 \neq s_2$, condition (6) of Definition 141 implies that ${}^\bullet s_1 \neq {}^\bullet s_2$. Thus, there exist distinct places $q_1 \in {}^\bullet s_1$ and $q_2 \in {}^\bullet s_2$ with the same label. Hence, by the b-analogue of Theorem 91(1), either $q_1\ \mathbf{li}_N\ q_2$ or $q_1 \otimes_N q_2$. In both cases it is easy to see that there exist transitions s_1', s_2' such that ${}^\bullet s_1' \cap {}^\bullet s_2' \neq \varnothing$, $s_i'\ F^*\ s_i$ for $i = 1, 2$, and $(s_1', s_2') \neq (s_1, s_2)$. Since G is pruned, this contradicts the fact that $s_1 \otimes s_2$.

(If) This direction of the proof is based on Lemma 109 and on the proof of Theorem 107. Let $G = (T, \Gamma, \otimes, \Sigma, \phi)$ satisfy properties (1–3). We have to show the existence of a b-process N of M such that $\mathbf{pru}(\mathbf{ctr}(N)) = G$. The construction of N is exactly the same as the construction of $\mathbf{proc}(G)$ (for $G \in$ LPO(M)) in the proof of the If-part of Theorem 107, at the end of Section 5. Thus, we define N to be the $(P_M, \mathbf{use}(T_M))$-labelled net (P, T, F, ϕ_1, ϕ), where P, ϕ_1, and F are defined in exactly the same way as for $\mathbf{proc}(G)$. In what follows we prove that N is a b-process of M such that $\mathbf{pru}(\mathbf{ctr}(N)) = G$. As usual, we will drop the subscript of ϕ_1.

Conditions (1)-(5) of Definition 141 (see Definition 88) are all obvious, except that we have to show that for every $s \in T$, $\phi({}^\bullet s) = {}^\bullet(\phi(s))$ and $\phi \restriction {}^\bullet s$ is injective. For a given s, let $R_s = \{s' \in T \mid s' \ \Gamma^* \ s\}$. It is easy to see that R_s is a configuration of G, and so $G[R_s] \in \mathrm{LPO}(M)$. From the definitions of N and $\mathbf{proc}(G[R_s])$, it should be clear that ${}^\bullet s$ contains the same places in N and in $\mathbf{proc}(G[R_s])$, with the same labels. Since $\mathbf{proc}(G[R_s])$ is a process of M, it satisfies the above two requirements.

Next we claim that

$$\text{for } s, s' \in T, \ \ s' \ F^+ \ s \text{ iff } s' \ \Gamma^+ \ s. \tag{1}$$

In the Only-if direction this follows directly from the definition of N. In the If direction it follows from an argument about $\mathbf{proc}(G[R_s])$ similar to the one above, using the fact that the partial order on the transitions of $\mathbf{proc}(G[R_s])$ is represented by $G[R_s]$, see Lemma 104. Equivalence (1) implies that N is acyclic, which is condition (1) of a b-process net, cf. Definition 138.

Similarly, for the conflict relations we claim that

$$\text{for } s_1, s_2 \in T, \ \ s_1 \otimes_N s_2 \text{ iff } (s_1, s_2) \in \mathbf{tra}(\otimes), \tag{2}$$

where $\otimes$ is the conflict relation of G.

We first show the Only-if direction of equivalence (2). By equivalence (1), it suffices to prove this for the case that ${}^\bullet s_1 \cap {}^\bullet s_2 \neq \varnothing$. Suppose that $(s_1, s_2) \notin \mathbf{tra}(\otimes)$. Then $R = \{s \in T \mid s \ \Gamma^* \ s_1 \ \text{ or } s \ \Gamma^* \ s_2\}$ is a configuration of G, and so $G[R]$ is in $\mathrm{LPO}(M)$. Again, this implies that also ${}^\bullet s_1 \cap {}^\bullet s_2 \neq \varnothing$ in $\mathbf{proc}(G[R])$, contradicting the fact that this is a process of M. From this Only-if direction, and the fact that $\mathbf{tra}(\otimes)$ is irreflexive, it follows that $\otimes_N$ is irreflexive (condition (3) of a b-process net, cf. Definition 138). Since condition (2) of a b-process net is immediate from the definition of N, this shows that N is a b-process net. Thus, we have almost proved that N is a b-process: only condition (6) of Definition 141 is still missing.

Next we show the If direction of equivalence (2). By equivalence (1), it suffices to prove this for the case that $s_1 \otimes s_2$. This implies, by property (3), that $(\phi(s_1), \phi(s_2)) \in \mathbf{dep}(M)$. Assume that $\neg \ s_1 \otimes_N s_2$. If $s_1 \ \mathbf{li}_N \ s_2$, then, by equivalence (1), $s_1 \ \Gamma^* \ s_2$ or $s_2 \ \Gamma^* \ s_1$, which contradicts the irreflexivity of $\mathbf{tra}(\otimes)$. If $s_1 \ \mathbf{co}_N \ s_2$, then (s_1, s_2) is in the relation $\mathbf{co}_N \cap \overline{\otimes}_N$. Thus, by the b-analogue of Theorem 91(3) (which is also valid without condition (6) of Definition 141), there exists $C \in \mathbb{C}_M$ such that $\{\phi(s_1), \phi(s_2)\} \ \mathbf{con} \ C$. This contradicts the fact that $(\phi(s_1), \phi(s_2)) \in \mathbf{dep}(M)$.

Equivalences (1) and (2) show that $\mathbf{pru}(\mathbf{ctr}(N))$ and G represent the same labelled event structure. Hence, because they are both pruned, they are the same (see Theorem 101). Note that we have applied $\mathbf{ctr}$ to N without knowing whether condition (6) of Definition 141 is satisfied; it should be clear that Definition 146 can also be used in this case.

It remains to show condition (6) of Definition 141. Consider distinct transitions $s_1, s_2 \in T$ with ${}^\bullet s_1 = {}^\bullet s_2$. It is straightforward to show that (s_1, s_2) is in the conflict relation of $\mathbf{pru}(\mathbf{ctr}(N))$. Thus, since $\mathbf{pru}(\mathbf{ctr}(N)) = G$, $s_1 \otimes s_2$. It now follows from property (3) that $\phi(s_1) \neq \phi(s_2)$. □

We note here that the condition $\phi(s_1) \neq \phi(s_2)$ in property (3) of Theorem 149 corresponds precisely to condition (6) of Definition 141. Theorem 149 is still true when both conditions are dropped.

Theorem 149 allows us to show that lpo-equivalence implies les-equivalence. This result can also be deduced from the results in [NieRozThi90, Eng91] together with Lemma 5.3 of [Vaa91].

Theorem 150. *Let M and M' be two contact-free EN systems and let β be a bijection from $\mathbf{use}(T_M)$ to $\mathbf{use}(T_{M'})$.*
If $LPO(M) \equiv_\beta LPO(M')$, then $LES(M) \equiv_\beta LES(M')$.

Proof. By Theorem 115, $\beta(\mathrm{FS}(M)) = \mathrm{FS}(M')$. Hence, by Lemma 118, $\mathbf{ind}(M') = \{(\beta(s), \beta(t)) \mid (s,t) \in \mathbf{ind}(M)\}$ (as in the proof of Lemma 124). It now follows from Theorem 149 that $\mathrm{LES}(M) \equiv_\beta \mathrm{LES}(M')$. □

Together with the remark after Definition 147 this shows that les-equivalence is the same as lpo-equivalence (and hence the same as firing sequence equivalence by Theorem 126).

Theorem 151. *Two contact-free EN systems are les-equivalent iff they are lpo-equivalent.*

The analogue of Theorem 107 also holds for b-processes, i.e., the function **pructr** is a bijection between BPROC(M) and LES(M); the proof is by Lemmas 108 and 109, which also hold for b-processes (even when the conflict relations are dropped from the pruned contracted b-processes, because in this case the system M is known).

It is shown in, e.g., [RozThi91, WinNie95a] that there is a general relationship between trace languages and event structures, similar to the relationship between traces and dependency graphs considered in Section 6.2. Possibly, this could be used to give an alternative proof of Theorem 151 that is similar to the one of Theorem 126 in Section 6.1.

8 Conclusion

In this chapter we have presented a comprehensive introduction to the theory of Elementary Net systems. We have discussed here both the structural and the behavioural aspects of the theory.

However, the behavioural aspects are prevalent in developing the theory of EN systems, because even such structural issues like reduction and decomposition make sense only modulo some behavioural equivalence of the systems. For example, if a normal form for EN systems is proved, then it is proved with respect to a specific equivalence. Thus typically we say: "For every EN system there exists an *equivalent* EN system that satisfies the conditions of the normal form". In this sense, the study of (behavioural) equivalences is central to the theory of EN systems.

We have studied a number of equivalence relations between contact-free EN systems and the relationships between them: isomorphism, which implies configuration equivalence, which in its turn implies firing sequence equivalence, which equals lpo-equivalence, weak configuration equivalence, trace equivalence, and les-equivalence. This leaves us with one notion of equivalent structure of EN systems (isomorphism), and essentially two notions of equivalent behaviour of EN systems, one stronger than the other: configuration equivalence, for which we require the two systems to have the same state space, and firing sequence equivalence (or lpo-equivalence), for which we require the two systems to have the same runs (i.e., "paths" in the state space). These are the two notions of behaviour that play an important role in system theory in general.

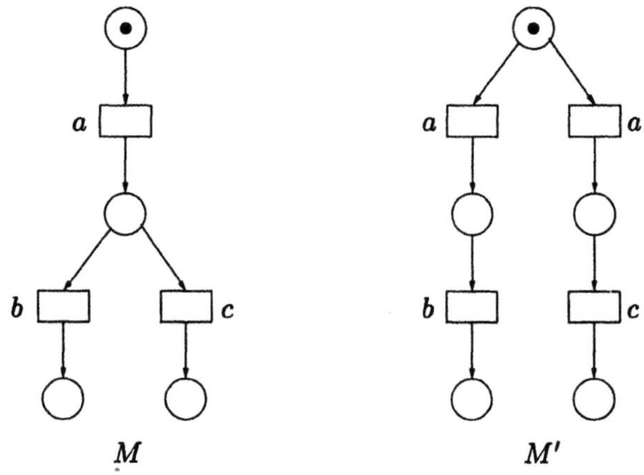

Fig. 71. Two labelled EN systems M and M' that are lpo-equivalent but not weakly configuration equivalent.

Finally we make a general remark concerning Theorem 126 which says that lpo-equivalence and firing sequence equivalence are the same. A technical reason for this surprising result is that all our considerations have been centered around the transitions: we compare (by means of equivalences) only EN systems with essentially the same transitions (where 'essentially' means: modulo bijections). In a more general approach, also EN systems with distinct transitions could be compared. To indicate that distinct transitions actually perform the same task we give them the same label (where the label thus in fact represents the task)

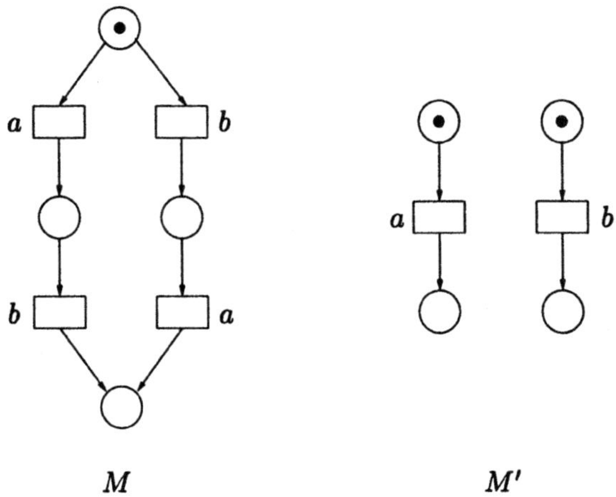

Fig. 72. Two labelled EN systems M and M' that are configuration equivalent but not lpo-equivalent.

from a given alphabet. Then, in all definitions of behaviour, the labels of the transitions are used instead of the transitions themselves. Hence, instead of firing sequences we take the sequences of labels corresponding with the firing sequences, and the labelled partial orders are not labelled with transitions of the system, but with their labels. Note that this may drastically change the properties of the behaviour of EN systems. For example, even such a fundamental property that a transition cannot fire twice consecutively (cf. the discussion following Theorem 12), does not hold for labelled EN systems: the same label can occur twice consecutively in a firing sequence (take any two transitions that fire one after the other, and give them the same label).

Under this approach Theorem 126 most certainly no longer holds, i.e., the new lpo-equivalence still implies the new firing sequence equivalence, but not the other way around. Similarly, Theorem 33 is no longer true: the new weak configuration equivalence still implies the new firing sequence equivalence, but not the other way around. More precisely, the new lpo-equivalence is incomparable with the new (weak) configuration equivalence; see Figs. 71 and 72 for two well-known counter-examples (where a, b, and c are the labels of the transitions rather than the transitions themselves). The relationships between various equivalences of labelled EN systems are presented in [PomRozSim92].

Note that each (old) equivalence relation implies the corresponding new one. Thus all normal forms discussed in Section 4 also hold for the new equivalences. As an example, Theorem 54 implies that for every EN system there is an lpo-equivalent (in the new sense) reduced EN system that is covered by sequential components; this is because configuration equivalence implies lpo-equivalence which, in its turn, implies lpo-equivalence in the new sense.

Acknowledgments

We wish to thank IJsbrand Jan Aalbersberg, Iko Keesmaat, George Leih, and Peter Hoogers for their contributions to these lecture notes. We thank Hendrik Jan Hoogeboom, Jetty Kleijn, and P.S.Thiagarajan for valuable comments throughout the years. We are grateful to Maurice ter Beek, Jurriaan Hage, and Nikè van Vugt for their assistance with the preparation of this paper.

References

[AalRoz88] IJ.J.Aalbersberg, G.Rozenberg; Theory of traces, Theor. Comput. Sci. 60 (1988), 1–82

[BesDev87] E.Best, R.Devillers; Sequential and concurrent behaviour in Petri net theory; Theor. Comput. Sci. 55 (1987), 87–136

[BesFer88] E.Best, C.Fernández C.; *Nonsequential Processes*, EATCS Monographs on Theoretical Computer Science, Vol.13, Springer-Verlag, Berlin, 1988

[BraReiRoz87] W.Brauer, W.Reisig, G.Rozenberg, eds.; *Petri Nets: Central Models and their Properties*, Lecture Notes in Computer Science 254, Springer-Verlag, Berlin, 1987

[DesEsp95] J.Desel, J.Esparza; *Free Choice Petri Nets*, Cambridge University Press, Cambridge, 1995

[DieRoz95] V.Diekert, G.Rozenberg, eds.; *The Book of Traces*, World Scientific, Singapore, 1995

[EhrRoz90] A.Ehrenfeucht, G.Rozenberg; Partial 2-structures; Part II: State spaces of concurrent systems, Acta Informatica 27 (1990), 348–368

[Eng85] J.Engelfriet; Determinacy $\rightarrow$ (observation equivalence = trace equivalence), Theor. Comput. Sci. 36 (1985), 21–25

[Eng91] J.Engelfriet; Branching processes of Petri nets, Acta Informatica 28 (1991), 575–591

[Esp94] J.Esparza; Model checking using net unfoldings, Science of Computer Programming 23 (1994), 151–195

[EspRömVog96] J.Esparza, S.Römer, W.Vogler; An improvement of McMillan's unfolding algorithm, in *Tools and Algorithms for the Construction and Analysis of Systems* (T.Margaria, B.Steffen, eds.), Proc. TACAS'96, Lecture Notes in Computer Science 1055, Springer-Verlag, Berlin, 1996, pp.87–106

[Gen87] H.J.Genrich; Predicate/Transition nets, in [BraReiRoz87], pp.207–247

[Hac72] M.H.T.Hack; *Analysis of production schemata by Petri nets*, M.S. Thesis, MIT, Project MAC, Cambridge, Mass., 1972

[Hoo94] P.W.Hoogers; *Behavioural aspects of Petri nets*, Ph.D.Thesis, Leiden University, 1994

[HooRoz91] H.J.Hoogeboom, G.Rozenberg; Diamond properties of elementary net systems, Fundamenta Informaticae 14 (1991), 287–300

[HooRoz95] H.J.Hoogeboom, G.Rozenberg; Dependence graphs, Chapter 2 of [DieRoz95]

[HopUll79] J.E.Hopcroft, J.D.Ullman; *Introduction to Automata Theory, Languages, and Computation*, Addison-Wesley, Reading, Mass., 1979

[Jen92] K.Jensen; *Coloured Petri Nets*, EATCS Monographs on Theoretical Computer Science, in three Volumes, Springer-Verlag, Berlin, 1992, 1994, and 1997

[LanRob78] L.H.Landweber, E.L.Robertson; Properties of conflict free and persistent Petri nets, J. of the ACM 25 (1978), 352–364

[Maz95] A.Mazurkiewicz; Introduction to trace theory, Chapter 1 of [DieRoz95]

[McM93] K.L.McMillan; Using unfoldings to avoid the state explosion problem in the verification of asynchronous circuits, in *Computer Aided Verification* (G.v.Bochmann, D.K.Probst, eds.), Proc. CAV'92, Lecture Notes in Computer Science 663, Springer-Verlag, Berlin, 1993, pp.164–177

[McM95] K.L.McMillan; A technique of a state space search based on unfolding, Formal Methods in System Design 6 (1995), 45–65; see also [McM93]

[Mil89] R.Milner; *Communication and Concurrency*, Prentice-Hall, Englewood Cliffs, NJ., 1989

[NiePloWin81] M.Nielsen, G.Plotkin, G.Winskel; Petri nets, event structures and domains, Part 1, Theor. Comput. Sci. 13 (1981), 85–108

[NieRozThi90] M.Nielsen, G.Rozenberg, P.S.Thiagarajan; Behavioural notions for elementary net systems, Distributed Computing 4 (1990), 45–59 (see also [Thi90])

[NieRozThi92a] M.Nielsen, G.Rozenberg, P.S.Thiagarajan; Elementary transition systems, Theor. Comput. Sci. 96 (1992), 3–33

[NieRozThi92b] M.Nielsen, G.Rozenberg, P.S.Thiagarajan; Elementary transition systems and refinement, Acta Informatica 29 (1992), 555–578

[NieRozThi95] M.Nielsen, G.Rozenberg, P.S.Thiagarajan; Transition systems, event structures, and unfoldings, Inform. and Comput. 118 (1995), 191–207

[Och95] E.Ochmański; Recognizable trace languages, Chapter 6 of [DieRoz95]

[Pet62] C.A.Petri; *Kommunikation mit Automaten*, Rheinisch-Westfälisches Institut für Instrumentelle Mathematik an der Universität Bonn, Schrift Nr.2, 1962

[Pet76] C.A.Petri; *Nicht-sequentielle Prozesse*, Universität Erlangen - Nürnberg, Arbeitsberichte des IMMD, Vol.9, Nr.8, 57–82, 1976

[Pet81] J.L.Peterson; *Petri Net Theory and the Modeling of Systems*; Prentice-Hall, Englewood Cliffs, 1981

[Pom88] L.Pomello; *Osservatore, reti di Petri, processi*, Ph.D.Thesis, Universities of Milano and Turino, 1988

[PomRozSim92] L.Pomello, G.Rozenberg, C.Simone; A survey of equivalence notions for net based systems, in *Advances in Petri Nets 1992* (G.Rozenberg, ed.), Lecture Notes in Computer Science 609, Springer-Verlag, Berlin, 1992, pp.410–472

[Pra86] V.R.Pratt; Modelling concurrency with partial orders, Int. J. of Parallel Programming 15 (1986), 33–71

[Rei82] W.Reisig; *Petri Nets*, EATCS Monographs on Theoretical Computer Science, Vol.4, Springer-Verlag, Berlin, 1982

[Roz87] G.Rozenberg; Behaviour of elementary net systems, in [BraReiRoz87], pp.60–94

[RozThi86] G.Rozenberg, P.S.Thiagarajan; Petri nets: basic notions, structure, behaviour, in *Current Trends in Concurrency* (J.W.de Bakker, W.-

P.de Roever, G.Rozenberg, eds.), Lecture Notes in Computer Science 224, Springer-Verlag, Berlin, 1986, pp.585–668

[RozThi91] B.Rozoy, P.S.Thiagarajan; Event structures and trace monoids, Theor. Comput. Sci. 91 (1991), 285–313

[Tau89] D.Taubner; *Finite Representations of CCS and TCSP Programs by Automata and Petri Nets*, Lecture Notes in Computer Science 369, Springer-Verlag, Berlin, 1989

[Thi87] P.S.Thiagarajan; Elementary net systems, in [BraReiRoz87], pp.26–59

[Thi90] P.S.Thiagarajan; Some behavioural aspects of net theory, Theor. Comput. Sci. 71 (1990), 133–153

[Vaa91] F.W.Vaandrager; Determinism $\rightarrow$ (event structure isomorphism $=$ step sequence equivalence), Theor. Comput. Sci. 79 (1991), 275–294

[Win87] G.Winskel; Event structures, in *Petri Nets: Applications and Relationships to other Models of Concurrency* (W.Brauer, W.Reisig, G.Rozenberg, eds.), Lecture Notes in Computer Science 255, Springer-Verlag, Berlin, 1987, pp.325–392

[WinNie95a] G.Winskel, M.Nielsen; Models for concurrency, in *Handbook of Logic in Computer Science* (S.Abramsky, D.M.Gabbay, T.S.E.Maibaum, eds.), Vol.4: Semantic Modelling, pp.1–148, Clarendon Press, Oxford, 1995

[WinNie95b] G.Winskel, M.Nielsen; Trace structures and other models for concurrency, Chapter 9 of [DieRoz95]

Place/Transition Petri Nets

Jörg Desel[1] and Wolfgang Reisig[2]

[1] Institut für Angewandte Informatik und Formale Beschreibungsverfahren
Universität Karlsruhe, Germany
desel@aifb.uni-karlsruhe.de
[2] Institut für Informatik
Humboldt-Universität zu Berlin, Germany
reisig@informatik.hu-berlin.de

Abstract. This contributions provides an introduction to the theory of place/transition Petri nets. Topics include the sequential and the concurrent behavior of place/transition Petri nets, marking graphs and coverability trees, and some analysis techniques that are based on the structure of place/transition Petri nets.

1 Introduction

Place/transition Petri nets (p/t-nets, for short) are the most prominent and best studied class of Petri nets. This class is sometimes just called *Petri nets*. In contrast to *elementary (Petri) net systems* [RoEn98], in p/t-nets a place can hold any number of tokens. In contrast to *high-level Petri nets* [Jens98], these tokens are indistinguishable. This contribution surveys some of the the most important concepts and analysis techniques for p/t-nets. The reader is assumed to have a rough idea about Petri nets, and in particular about elementary net systems. However, this knowledge is not a prerequisite for this paper.

A state of a p/t-net is determined by a distribution of tokens on its places. It is formally defined as a mapping that assigns each place a nonnegative integer. This mapping is usually called a *marking*. A p/t-net equipped with an initial marking constitutes a Petri net system[3], i.e. a system model capturing both static structural and dynamic behavioral aspects.

This paper mainly concentrates on a basic class of marked p/t-nets, which only consist of a net and an initial marking. These Petri net systems are called *marked nets*. We only roughly survey some additional features of p/t-nets. The basic concepts of arbitrary p/t-nets can be presented more lucidly for marked nets. Most of them can easily be generalized to more general classes of p/t-nets.

The elementary behavior of a marked net is defined by a local occurrence rule that determines the possible elementary marking changes caused by transition occurrences. A transition of a marked net is enabled at a marking if each arc to the transition originates at a place which carries at least one token. The

[3] sometimes called place/transition-system

occurrence of an enabled transition is described by means of the following picture, representing a transition and its vicinity before and after its occurrence:

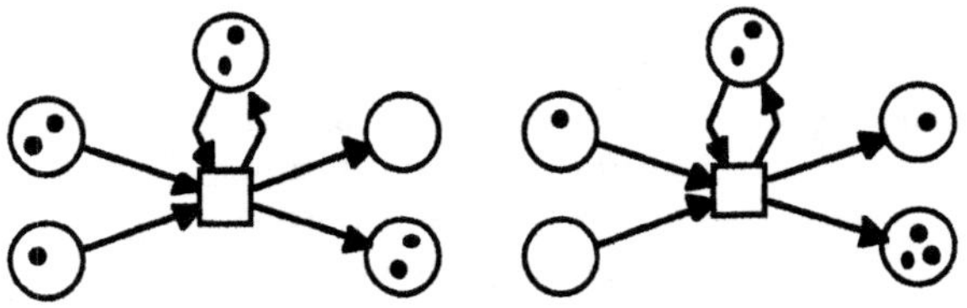

Starting with the initial marking, transition occurrences yield new markings, that in turn give rise to further transition occurrences. A run of a marked net consists of a set of transition occurrences that altogether might happen in some order and hence do not exclude each other. The global behavior of a marked net is determined by its runs.

There are various ways to formalize runs. We will emphasize a sequential view, given by sequences of transition occurrences (*occurrence sequences*), and a concurrent view, given by partially ordered *process nets*. Sequential runs are technically simple. However, concurrent occurrences of transitions have to be ordered in sequential runs whereas concurrency and dependency is explicitly represented in concurrent runs.

Occurrence sequences will be used to define behavioral properties of marked nets, such as *deadlock-freedom* (no total deadlock is reachable), *liveness* (no partial deadlock is reachable), *boundedness* (on each place, the token load does not grow unlimited), and *reversibility* (the initial marking can always be reached again). Analysis of a marked net w.r.t. these properties will be based on finite graphs that comprise the sequential behavior of a marked net, namely *marking graphs* and *coverability trees*. It will also be demonstrated that some of the properties can be characterized in terms of process nets, and that this view allows more insight to behavior and to more elegant proofs.

This contribution compiles fundamentals of theoretical results about p/t-nets. Some pointers to literature should provide a starting point for further work. For textbooks on Petri nets in general see [Pete81, Reis85]. Another good introductory text is [Mura89]. A collection of surveys is given in [BRR87]. There, the papers [Reis87, Laut87] are particularly related to p/t-nets.

Decidability and complexity results on p/t-nets are important topics, though missing in this contribution. The most important result in this field is the decidability of the problem whether a given marking is reachable in a marked p/t-net. A solution to this *reachability problem* was given in [Mayr84, Kosa82], see also [Reut88]. The paper [Jant87], also contained in [BRR87], gives in overview on some more results in this area. More recent papers surveying in particular results on *model checking* and p/t-nets are [EsNi94], [Espa98] and [Valm98], the latter two in this volume.

A marking of a p/t-net can be conceived as a tuple of numbers, each of which describing the number of tokens residing at some place. This view suggests a marking to be conceived as a vector of nonnegative integers. The change of a marking, caused by the occurrence of a transition corresponds to the addition of an integer-valued vector. Therefore, p/t-nets are isomorphic to a class of vector addition systems. In fact, a lot of Petri net research has originally focussed on vector addition systems. The survey paper [Dese98] in this volume presents application of linear algebraic techniques to the analysis and verification of marked nets.

A further research area focusses on structurally defined subclasses of Petri nets. The best known examples of such classes are *marked graphs*, *state machines* and *free choice nets.* In a marked graph, no place is branching. Therefore, marked graphs behave deterministically and do not allow to model *choice* between actions. A state machine has no branching transitions and hence does not allow for *synchronization* of actions. In free-choice nets, the combination of forward branched places and backward branched transitions is restricted, thus ruling out a too close interplay between choice and synchronization. Free-choice nets generalize the classes mentioned above. As pointed out in the present paper and in [Dese98], efficient analysis and verification techniques for arbitrary nets stick to either sufficient or necessary conditions for behavioral properties, whereas decision algorithms for behavioral properties suffer from the huge inherent complexity of most problems. For free-choice nets, the picture looks significantly better because many interesting behavioral properties of marked free-choice nets can be characterized in terms of the net structure. For a detailed treatment of the theory of free-choice Petri nets, the authors warmly recommend the textbook [DeEs95]. This book also contains many references to the original sources.

The following section provides an introductory example. In the third section, marked nets will be defined formally. We present the *Strongly-Connectedness Theorem* as an example for relations between structure and behavior of a marked net The fourth section is devoted to *marking graphs* and *coverability trees* and to the representation of behavioral properties of marked nets by means of structural properties of these graphs. In section five, more efficient analysis techniques based on the structure of nets are presented, namely *place invariants*, *transition invariants*, *siphons*, and *traps.* Advantages of process nets representing the concurrent runs of marked nets as well as their relation to occurrence sequences are presented in section six. Finally, the seventh section provides a quick survey on additional features of p/t-nets, including *arc weights*, *capacity restrictions* and *inhibitor arcs.*

2 An Introductory Example

Marked p/t-nets generalize *contact-free elementary net systems* [RoEn98]. In other words: every contact-free elementary net system is at the same time a marked p/t-net[4]. We start with an example of an elementary net system and subsequently add features of p/t-nets which are not available at the elementary level.

[4] Elementary net systems can nevertheless be considered *elementary* since all basic concepts of p/t-nets can be derived from respective concepts of elementary net systems, see [Golt90, DeMe90].

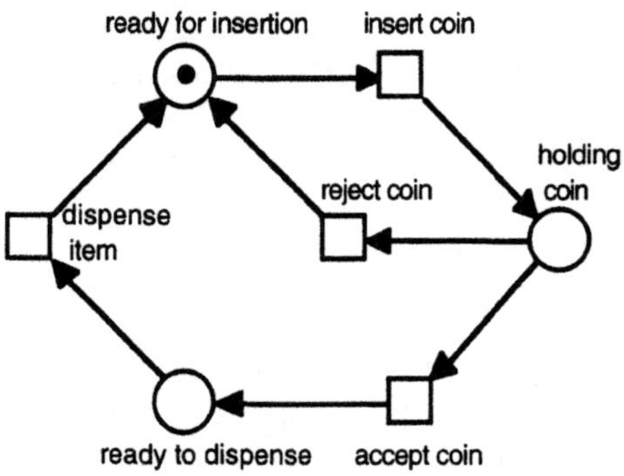

Fig. 1. A marked net representing the control structure of a vending machine

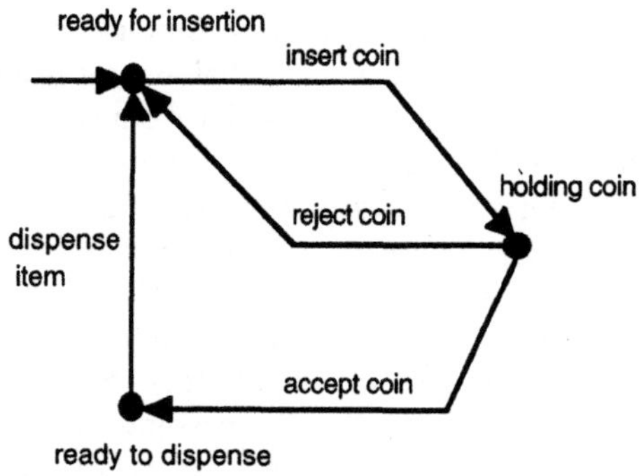

Fig. 2. The marking graph of the above marked net

Figure 1 shows an elementary net system modeling the control part of a vending machine. At the initial state shown in the figure, the machine is waiting for a coin to be inserted. This is modeled by the token on the place **`ready for insertion`**. An inserted coin is either rejected or accepted, depending on the result of a validity check which is not part of the system model. If the coin was rejected, the system returns to its initial state. Otherwise the system first dispenses an item and then returns to its initial state.

This system model is also a marked p/t-net. Terminology for p/t-nets usually differs from terminology for elementary net systems. Nonetheless we shall introduce and use the place/transition based terminology already here:

- every condition of an elementary net system is a *place* of the p/t-net,
- every event of an elementary net system is a *transition* of the p/t-net,
- every state or case of an elementary net system is a *marking* of the p/t-net; formally, a marking is represented as a mapping assigning each place, the number of tokens it carries – here 0 or 1.

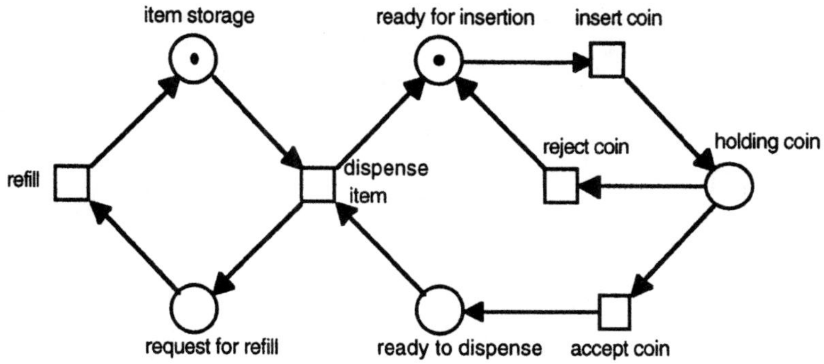

Fig. 3. The vending machine with capacity 1

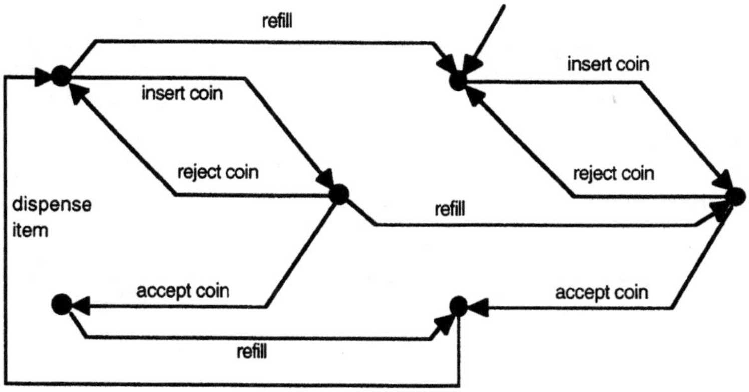

Fig. 4. The marking graph of the above marked net

The sequential behavior of a p/t-net can be represented by its *marking graph*. In case there is always only one token in the net, as in the above example, the marking graph is quite small, and it resembles the net itself (see Figure 2). In general, the marking graph is a compact representation of all possible sequences of subsequent transition occurrences. Starting with the distinguished *initial vertex* (`ready for insertion` in the example), each directed path of the marking graph represents a possible sequence of transition occurrences, and vice versa.

Now we add the system component delivering items to the vending machine. Assume the capacity of the machine is 1, i.e., after delivering one item, the machine has to be refilled (by another component not considered here), as shown in Figure 3. This system is still an elementary net system. Its marking graph is shown in Figure 4. For sake of readability, markings annotating vertices are dropped. The shape of the graph should help to compare it with the graph of Figure 2.

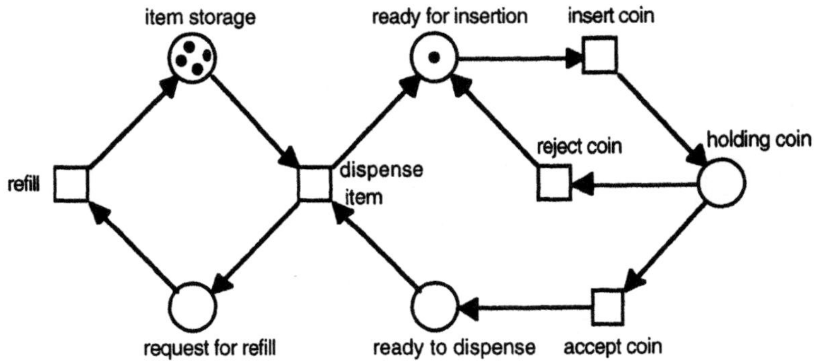

Fig. 5. The vending machine with capacity 4

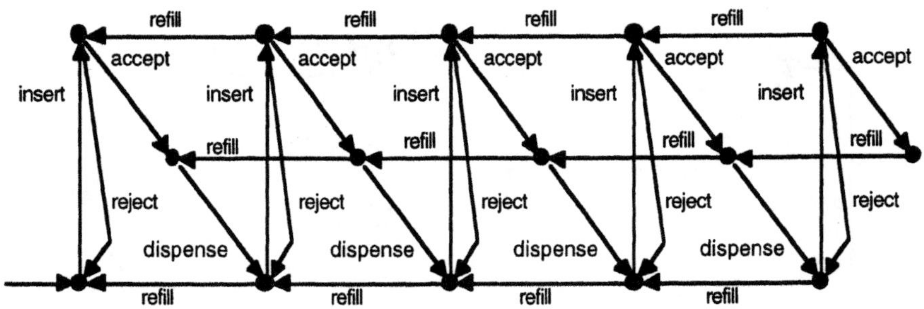

Fig. 6. The marking graph of the above marked net

The marking graph suggests that the system exhibits no concurrency. After the occurrences of transitions **insert coin**, **accept coin** and **dispense item**, both transitions **refill** and **accept coin** are enabled. The corresponding marking corresponds to the upper left hand vertex in the graph. Although the transitions **refill** and **accept coin** can occur concurrently, the sequential view enforces an order between these two transition occurrences. We will come back to this example and discuss its concurrent semantics in section six.

The storage capacity of real vending machines exceeds 1, of course. Figure 5 shows the same net as the previous example. However, the initial marking of the net is different: the place **item storage** carries four tokens initially. This systems is no elementary net system anymore: the place **item storage** cannot be interpreted as a condition with two states *true* and *false*. Instead, it has five possible states:

unmarked, marked with one token, ..., marked with four tokens.

The transition **dispense item** is only enabled if each of the two places **ready to dispense** and **item storage** contains at least one token. An occurrence of **dispense item** removes a token from **ready to dispense** and a token from

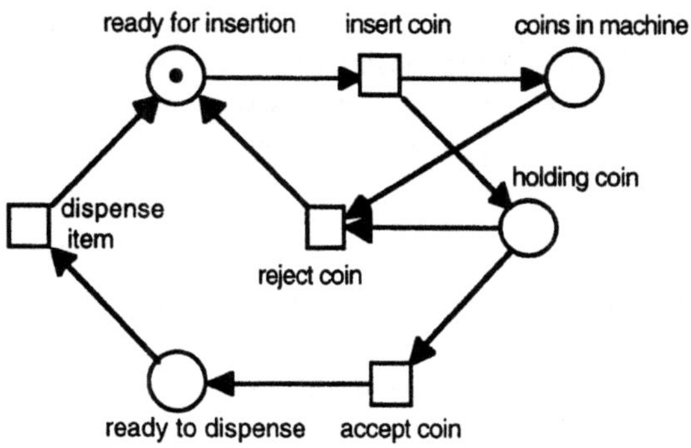

Fig. 7. The control structure of a vending machine with a counter

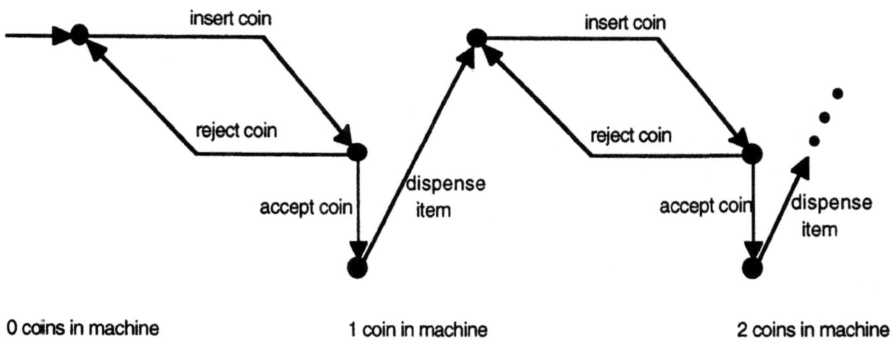

Fig. 8. The marking graph of the above marked net

item storage, and adds tokens to both places **request for refill** and **ready for insertion**.

All places of the marked net shown in Figure 5 are *bounded*: every place possesses an upper bound for the number of tokens it will carry at any reachable marking. This is not the case for the marked net shown in Figure 7. For simplicity, the left part of the previous net is omitted here, and an additional place represents the number of coins that are currently in the machine. Since this number can grow without limit, this place is *unbounded*. It can take an infinite number of different states, whence the marking graph of the marked net is infinite (see Figure 8). We will show in section 4 how finite *coverability trees* are used for analysis of unbounded marked nets.

As a further property of marked nets, the notion of *liveness* will intensely be studied in the sequel. A marked net is live if every transition can always occur again; more precisely, if from any reachable marking it is possible to reach some marking that enables the transition. All marked nets modeling the vending machine considered so far are live.

The property of *deadlock-freedom* is weaker than liveness: a marked net is deadlock-free if every reachable marking enables some transition.

A marked net is *reversible* if it can always reach its initial marking again. The marked nets of Figures 1, 3 and 5 are reversible, whereas the net of Figure 7 is not.

3 Basic Definitions and Elementary Results

This section formally defines marked nets and their behavior. As an example for the interplay between statical properties of nets and dynamic properties of marked nets, the Strongly-Connectedness-Theorem is proven.

3.1 Petri Nets

A *net* N is constituted by

- a set S of *places*,
- a set T of *transitions* such that $S \cap T = \emptyset$, and
- a set F of *directed arcs* (*flow relation*), $F \subseteq (S \cup T) \times (S \cup T)$, satisfying

$$F \cap (S \times S) = F \cap (T \times T) = \emptyset.$$

All places and transitions are said to be *elements* of N. A net is *finite* if its set of elements is finite.

In the graphical representation of a net, places are drawn as circles, transitions as squares and arcs as arrows. An arrow with two arrowheads between elements x and y indicates that (x, y) as well as (y, x) are directed arcs.

The sets of places, transitions and arcs of a net N are denoted by S_N, T_N and F_N, respectively. The set of all elements of the net N is sometimes also denoted by N, i.e., $N = S_N \cup T_N$.

For an element x of N, its *pre-set* ${}^\bullet x$ is defined by

$${}^\bullet x = \{y \in N \mid (y, x) \in F_N\}$$

and its *post-set* $x^\bullet$ is defined by

$$x^\bullet = \{y \in N \mid (x, y) \in F_N\}.$$

For $X \subseteq N$, define

$${}^\bullet X = \bigcup_{x \in X} {}^\bullet x \text{ and } X^\bullet = \bigcup_{x \in X} x^\bullet.$$

An element x is *isolated* if ${}^\bullet x = x^\bullet = \emptyset$.

A *directed path* (*path* for short) of a net is a nonempty sequence $x_0 \dots x_k$ of elements satisfying $x_i \in x_{i-1}^\bullet$ for each i $(1 \leq i \leq k)$. This path *leads from* x_0 *to* x_k. The net is *strongly connected* if for each two elements x and y there exists a directed path leading from x to y.

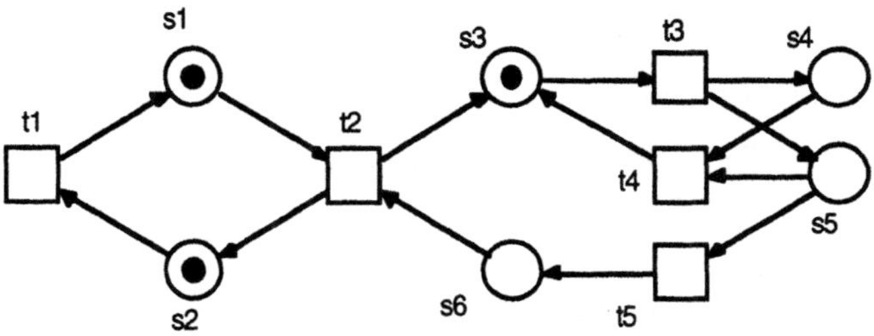

Fig. 9. A marked Petri net

An *undirected path* is a nonempty sequence $x_0 \ldots x_k$ of elements satisfying $x_i \in {}^\bullet x_{i-1} \cup x_{i-1}^\bullet$ for each i ($1 \leq i \leq k$). It *leads from x_0 to x_k*. The net is *weakly connected* if, for each two elements x and y, there exists an undirected path leading from x to y.

In Figure 9, a net is shown with places $\{\mathtt{s1}, \mathtt{s2}, \mathtt{s3}, \mathtt{s4}, \mathtt{s5}, \mathtt{s6}\}$ and transitions $\{\mathtt{t1}, \mathtt{t2}, \mathtt{t3}, \mathtt{t4}, \mathtt{t5}\}$. The pre-set of **t2** is $\{\mathtt{s1}, \mathtt{s6}\}$. The post-set of $\{\mathtt{s3}, \mathtt{s4}, \mathtt{s5}\}$ is $\{\mathtt{t3}, \mathtt{t4}, \mathtt{t5}\}$. The sequence **t2 s2 t1 s1 t2 s3** is a path. This net is strongly connected. The net obtained by removing **t1** and its adjacent arcs is connected but not strongly connected. The net obtained by removing **t2** and its adjacent arcs is not connected.

The following proposition follows immediately from the definitions.

Proposition 1. *A weakly connected net is strongly connected if and only if for each directed arc (x, y) there is a directed path leading from y to x.* □

3.2 Markings and Behavior

The states of a Petri net are defined by its markings, to be defined next. Elementary state changes are caused by the occurrences of transitions. The *occurrence rule* distinguishes the *enabling condition* and the *marking transformation*.

A *marking* of a net N is a mapping $m: S_N \to \mathbb{N}$ where $\mathbb{N} = \{0, 1, 2, \ldots\}$. A place s is *marked* by a marking m if $m(s) > 0$. The *null marking* is the marking which maps every place to 0.

A transition t is *enabled* by a marking m if m marks all places in ${}^\bullet t$. In this case t can *occur*. Its occurrence transforms m into the marking m', defined for each place s by

$$m'(s) = \begin{cases} m(s) - 1 & \text{if } s \in {}^\bullet t - t^\bullet, \\ m(s) + 1 & \text{if } s \in t^\bullet - {}^\bullet t, \\ m(s) & \text{otherwise.} \end{cases}$$

Notice that a place in ${}^\bullet t \cap t^\bullet$ is marked whenever t is enabled but does not change its token count by occurrence of t. Sometimes we will use the following equivalent equation for the marking m':

$$m'(s) = m(s) - |F_N \cap \{(s,t)\}| + |F_N \cap \{(t,s)\}|.$$

A marking is called *dead* if it enables no transition of N.

Consider the net shown in Figure 9 with the depicted marking, mapping **s1**, **s2**, and **s3** to 1 and all other places to 0. Transition **t1** is enabled, and the marking reached after its occurrence maps **s1** to 2 and **s2** to 0. Transition **t3** is enabled, too. The null marking is an example for a dead marking.

3.3 Occurrence sequences

Let m be a marking of a net. A finite sequence $t_1 \ldots t_k$ of transitions is called a *finite occurrence sequence, enabled at* m, if there are markings $m_1, \ldots, m_k$ such that

$$m \xrightarrow{t_1} m_1 \xrightarrow{t_2} \cdots \xrightarrow{t_k} m_k.$$

We write $m \xrightarrow{\sigma} m_k$ in this case, where $\sigma = t_1 \ldots t_k$. The empty sequence ε is enabled at any marking m and satisfies $m \xrightarrow{\varepsilon} m$. A marking m' is *reachable* from a marking m if there is a finite occurrence sequence $m \xrightarrow{\sigma} m'$.

An infinite sequence $t_1\, t_2\, t_3 \ldots$ is called *infinite occurrence sequence, enabled at* m, if there are markings $m_1, m_2, \ldots$ such that

$$m \xrightarrow{t_1} m_1 \xrightarrow{t_2} m_2 \xrightarrow{t_3} \cdots.$$

The following propositions follow immediately from the definitions.

Proposition 2. *If* $m \xrightarrow{\sigma} m'$ *is a finite occurrence sequence and* m' *enables a (finite or infinite) occurrence sequence* σ', *then the sequence* $\sigma\,\sigma'$ *is enabled at* m, *too.* □

Proposition 3. *An infinite sequence* σ *of transitions is enabled at a marking* m *if and only if every finite prefix of* σ *is enabled at* m. □

Proposition 4. *If* $m \xrightarrow{\sigma} m'$ *and* $l \xrightarrow{\sigma} l'$, *then* $m'(s) - m(s) = l'(s) - l(s)$ *for each place* s. □

Using the previous two propositions and the occurrence rule, it is easy to prove the following result.

Proposition 5. *If* m *and* l *are markings satisfying* $m(s) \geq l(s)$ *for each place* s *then every occurrence sequence enabled at* l *is also enabled at* m. □

Occurrence sequences do not provide full information about the causal relationships between transitions. Assume an occurrence sequence $v\,u$, enabled at a marking m. Transition u might be enabled only after the occurrence of v, or v and u might be *concurrently* enabled at m. In the latter case, $u\,v$ is an occurrence sequence, too. The following lemma identifies a sufficient condition for exchanging concurrent transitions.

Lemma 6. *Assume that u and v are transitions of a net satisfying $v^\bullet \cap {}^\bullet u = \emptyset$. If $m_1 \xrightarrow{v\,u} m_3$ then $m_1 \xrightarrow{u\,v} m_3$.*

Proof. Let $m_1 \xrightarrow{v} m_2$. We have $m_1(s) \geq m_2(s)$ for $s \in {}^\bullet u$ because $v^\bullet \cap {}^\bullet u = \emptyset$. Moreover, $m_2(s) \geq 1$ for $s \in {}^\bullet u$ because m_2 enables u. Therefore, m_1 enables u.

We have $m_1(s) \geq 1$ for $s \in {}^\bullet v$ because m_1 enables v. Moreover, $m_1(s) \geq 2$ for $s \in {}^\bullet u \cap {}^\bullet v$ because m_1 enables $v\,u$ and $v^\bullet \cap {}^\bullet u = \emptyset$. Therefore, m_1 enables $u\,v$.

By Proposition 4, the sequences $u\,v$ and $v\,u$ cause the same marking change. Therefore, the occurrence of $u\,v$ at m_1 leads to the marking m_3, too. □

3.4 Marked Nets and their Properties

A *marked net* is a net equipped with a marking, called *initial marking*. Throughout this paper we always denote the initial marking of a net by m_0. The initial marking of a marked net is graphically depicted by tokens in the places, as shown in Figure 9.

An *occurrence sequence of a marked net* is an occurrence sequence enabled by the initial marking. The *reachable markings* of a marked net are the markings reachable from the initial marking (including the initial marking itself).

Let $b \in \mathbb{N}$. A place s of a marked net is called *bounded* with bound b (*b-bounded*) if $m(s) \leq b$ for each reachable marking m. It is *bounded* if it is b-bounded for some b. A marked net is called *b-bounded* (*bounded*) if all its places are b-bounded (bounded, respectively).[5]

Proposition 7. *A finite marked net is bounded if and only if there is a bound b such that all its places are b-bounded.* □

Notice that this property fails for infinite nets because there might be infinitely many different bounds.

A transition t is called *dead* at a marking m if it is not enabled at any marking reachable from m. A marked net is said to be *live* if there is no dead transition at any reachable marking. It is *deadlock-free* if no reachable marking is dead.

Proposition 8. *Each live marked net with at least one transition is deadlock-free.* □

Notice that each marked net without transitions is live but not deadlock-free.

Often it is required that every place of a marked net eventually can be marked.

Proposition 9. *If no transition of a marked net is dead then each non-isolated place is marked at some reachable marking.* □

A marked net is called *reversible* if its initial marking can be reached from any other reachable marking.

[5] 1-bounded places are sometimes called 1-safe or just safe places. Accordingly, marked nets with only 1-bounded places are sometimes called 1-safe or safe marked nets.

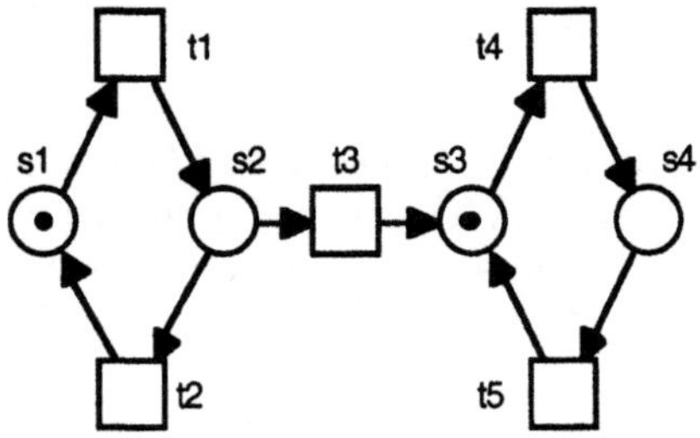

N1: bounded, not live

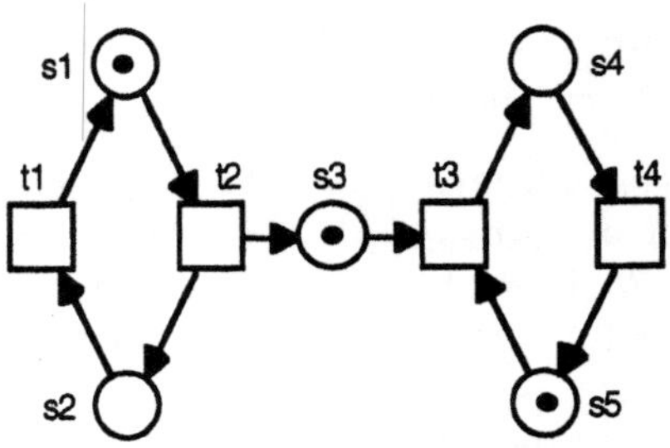

N2: live, not bounded

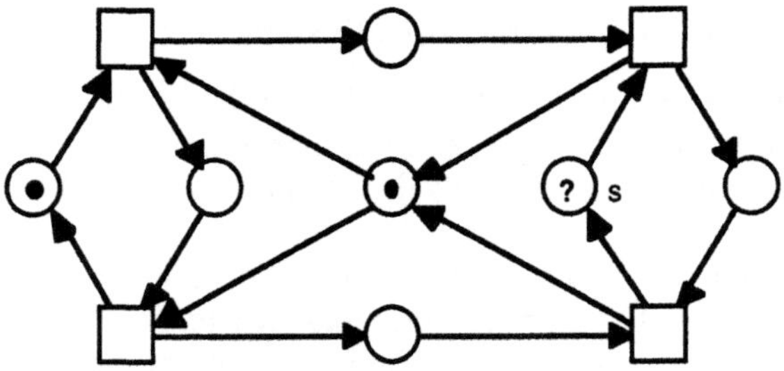

N3: bounded, live only if $m_0(s) = 1$

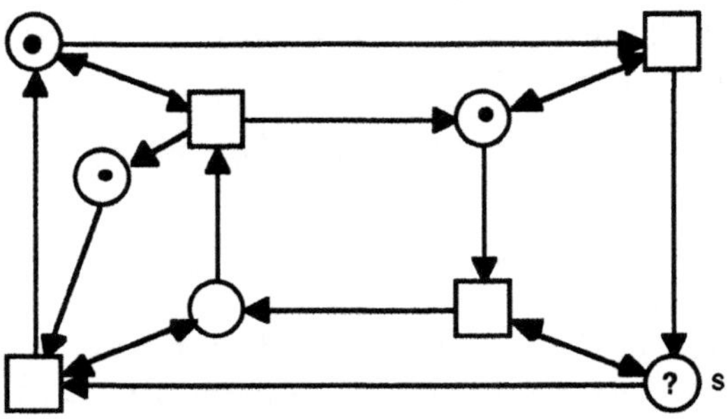

N4: live, bounded only if $m_0(s) = 0$

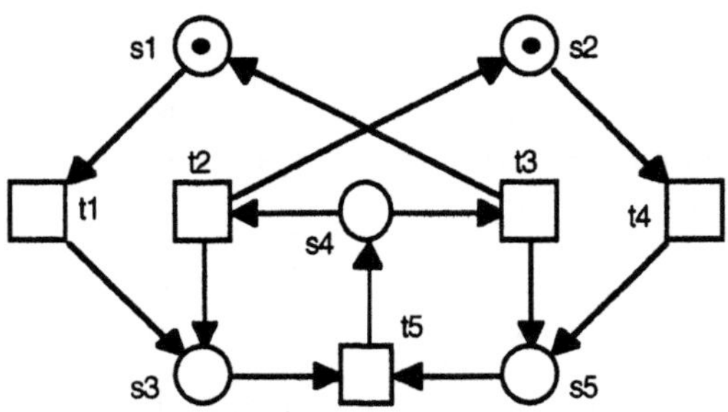

N5: live, 1-bounded, not reversible

Fig. 10. Examples for liveness, boundedness and reversibility of marked nets

Proposition 10. *A reversible marked net is deadlock-free if and only if the initial marking is not dead.* □

The marked net shown in Figure 9 is live, 2-bounded and reversible. Figure 10 presents further examples for the above notions. These examples also show that neither liveness of bounded marked nets nor boundedness of live marked nets is monotonous in the sense that the property is preserved when tokens are added to places.

3.5 The Strongly-Connectedness-Theorem

Live and bounded marked nets enjoy many interesting properties. One important example is the Strongly-Connectedness-Theorem which states that such a net is strongly connected, provided it is weakly connected. This result implies that each of the first two nets of Figure 10 either is not live or not bounded, for any initial marking:

Theorem 11. *Every weakly connected live and bounded marked net is strongly connected.*

Proof:[6] Assume a weakly connected live and bounded marked net. By Proposition 1 it suffices to prove for each arc (x, y) that there is a directed path leading from y to x.

Case 1: x is a place and y is a transition. Let V be the set of all transitions t for which there is a directed path from y to t. Let U be the set of all other transitions of the net. Then $u^\bullet \cap {}^\bullet v = \emptyset$ for every u in U and v in V.

Let b be the bound of place y. Liveness implies that the initial marking enables a finite occurrence sequence σ with $b+1$ occurrences of transition y. By Lemma 6, transitions of U and of V can repeatedly be swapped, resulting in an occurrence sequence $\sigma_1\, \sigma_2$, such that σ_1 contains all occurrences of transitions of U in σ and σ_2 contains all occurrences of transitions of V in σ.

Transition y is in the set V by definition of V. So y occurs $b+1$ times in σ_2. Since x is b-bounded and $y \in x^\bullet$, some transition of ${}^\bullet x$ occurs in σ_2. Since σ_2 contains only transitions of V, there is a transition in ${}^\bullet x \cap V$. By definition of V, there is a directed path from y to that transition. This path can be extended by x.

Case 2: x is a transition and y is a place. Let U be the set of all transitions t for which there is a directed path from t to x. Let V be the set of all other transitions of the net. Then $u^\bullet \cap {}^\bullet v = \emptyset$ for every u in U and v in V.

Let b be the bound of place y. Liveness implies that the initial marking enables a finite occurrence sequence σ with $b+1$ occurrences of transition x. We continue as in Case 1: By Lemma 6, transitions of U and of V can repeatedly be swapped, resulting in an occurrence sequence $\sigma_1\, \sigma_2$, such that σ_1 contains all

[6] This proof employs occurrence sequences and Lemma 6. For a more elegant proof based on partially ordered processes see Section 6.

occurrences of transitions of U in σ and σ_2 contains all occurrences of transitions of V in σ.

Transition x is in the set U by definition of U. So x occurs $b+1$ times in σ_1. Since y is b-bounded and $x \in {}^\bullet y$, some transition of $y^\bullet$ occurs in σ_1. Since σ_1 contains only transitions of U, there exists a transition in $y^\bullet \cap U$. By definition of U, there is a directed path from that transition to x. Concatenation of y and this path yields a directed path from y to x. □

4 Marking Graphs and Coverability Graphs

4.1 Marking Graphs

The sequential behavior of a marked net is given by its set of occurrence sequences. Each sequence of transition occurrences

$$m_0 \xrightarrow{t_1} m_1 \xrightarrow{t_2} m_2 \xrightarrow{t_3} \cdots,$$

can be considered a linear directed, acyclic graph

$$\overset{m_0}{\bullet} \xrightarrow{t_1} \overset{m_1}{\bullet} \xrightarrow{t_2} \overset{m_2}{\bullet} \xrightarrow{t_3} \cdots$$

with vertices labeled by reachable markings and edges labeled by transitions. In this way, the set of all occurrence sequences yield an acyclic graph consisting of many disconnected components: one component for each occurrence sequence. Equally labeled verices may be identified, yielding a connected graph, which is not acyclic in general. Its vertices correspond bijectively to the reachable markings of the marked net. Each transition enabled at a marking m yields one outgoing edge of the corresponding vertex. A distinguished vertex represents the initial marking. Each directed path starting at this vertex uniquely represents an occurrence sequence. Vice versa, each occurrence sequence is uniquely represented by a path.

Figures 11 shows the marking graphs of the marked nets **N1**, **N2** and **N5** given in Figure 10. In these graphs, each marking m is denoted by a vector such that the i-th entry of a vector is $m(\texttt{si})$. Initial markings are depicted by arrows without source.

Formal definition of marking graphs requires some notions about graphs:

An *arc-labeled directed graph* is given by

- a set V of *vertices* and
- a set of *labeled edges* (v, l, v') (representing source vertex, label, and target vertex, respectively) where $v, v' \in V$ and l is a label of some given set L.

If (v, l, v') is a labeled edge then v' is an *immediate successor* of v.

A *path* of an arc-labeled graph is a (finite or infinite) sequence of labeled edges

$$(v, l_1, v_1), (v_1, l_2, v_2), (v_2, l_3, v_3), \ldots.$$

All vertices in $\{v, v_1, v_2, v_3, \ldots\}$ are called *successors* of v.

The *marking graph* of a marked net is an arc-labeled directed graph with a distinguished initial vertex and edges labeled by transitions:

- the *vertices* are the reachable markings,
- the distinguished *initial vertex* is the initial marking,
- labeled edges are all triples (m, t, m') such that m and m' are reachable markings satisfying $m \xrightarrow{t} m'$.

4.2 Marking Graphs and Properties of Marked Nets

Proposition 12. *Given a marked net with initial marking* m_0,

$$m_0 \xrightarrow{t_1} m_1 \xrightarrow{t_2} m_2 \xrightarrow{t_3} \cdots$$

is a (finite or infinite) sequence of subsequent transition occurrences if and only if there is a path

$$(m_0, t_1, m_1), (m_1, t_2, m_2), (m_2, t_3, m_3), \ldots$$

of the marking graph. □

Behavioral properties of marked nets are closely related to structural properties of their marking graphs, as stated in the following results.

Lemma 13. *A finite marked net is bounded if and only if its marking graph has finitely many vertices.*

Proof. By definition, the set of vertices of the marking graph is finite if and only if the set of reachable markings is finite.
($\Longleftarrow$) If a marked net with initial marking m_0 has k reachable markings then any reachable marking can be reached by at most $k-1$ transition occurrences. Since a transition occurrence increases the number of tokens on a place by at most one, each place s is bounded by $m_0(s) + k - 1$.
($\Longrightarrow$) Assume that the net has n places. By Proposition 7 there is a common bound b for all places. Hence each place s has no more than $b+1$ different states, given by $m(s) = 0, \ldots, m(s) = b$. So there are no more than $(b+1)^n$ reachable markings. □

Lemma 14. *A marked net is deadlock-free if and only if its marking graph has no vertex without outgoing edge.*

Proof. Follows immediately from Proposition 12. □

Lemma 15. *A finite marked net is live if and only if, for each vertex m of its marking graph, there exists a path*

$$(m, t_1, m_1), (m_1, t_2, m_2), \ldots, (m_{k-1}, t_k, m_k)$$

such that the sequence $t_1\, t_2 \ldots t_k$ contains all transitions of the net.

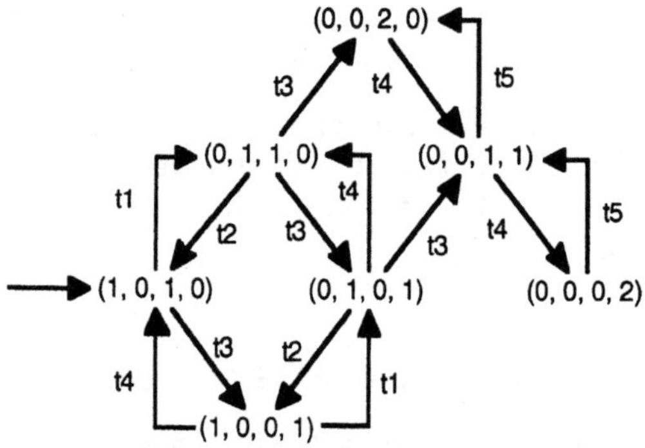

marking graph of **N1**

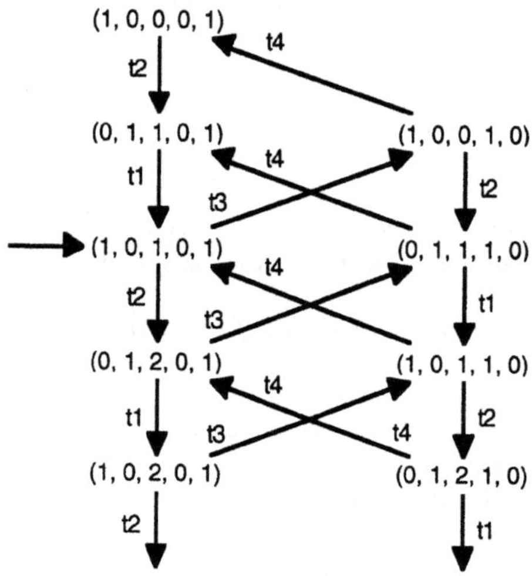

initial part of the marking graph of **N2**

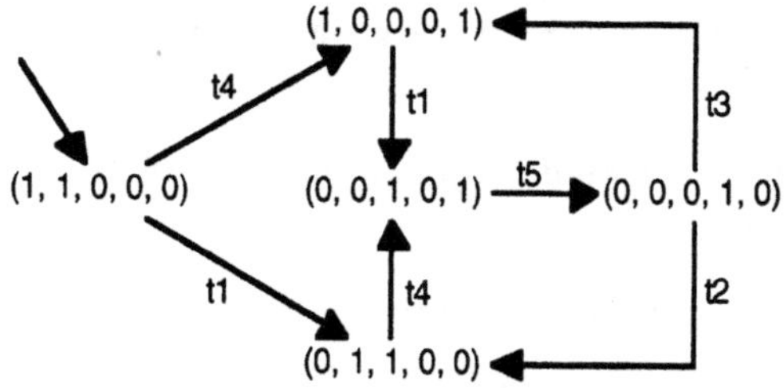

marking graph of **N5**

Fig. 11. Marking graphs of marked nets given in Figure 10

Proof.
($\Longleftarrow$) Each reachable marking m is represented by a vertex of the marking graph. By Proposition 12, the path $(m, t_1, m_1), (m_1, t_2, m_2), \ldots, (m_{k-1}, t_k, m_k)$ corresponds to an occurrence sequence $m \xrightarrow{t_1 \ldots t_k} m_k$. So each transition is enabled by some marking reachable from m.
($\Longrightarrow$) Let m be a reachable marking. Let $t_1, \ldots, t_m$ be a list of all transitions of the net. Liveness implies that some marking reachable from m enables t_1.
Let $m \xrightarrow{\sigma_1} m' \xrightarrow{t_1} m_1$.
Again by liveness, we find an occurrence sequence $m_1 \xrightarrow{\sigma_2} m_1' \xrightarrow{t_2} m_2$.
Repeated application of this construction eventually leads to an occurrence sequence that contains every transition at least once. The sequence of edge labels of the corresponding path of the marking graph contains all transitions. □

Lemma 16. *A marked net is reversible if and only if its marking graph is strongly connected.* [7]

Proof.
($\Leftarrow$) Since the marking graph is strongly connected, there are directed paths from any vertex to the initial vertex. Hence the initial marking can be reached from any reachable marking.
($\Rightarrow$) Let m_1 and m_2 be arbitrary reachable markings. We show that the marking graph has a directed path from m_1 to m_2. By reversibility of the marked net, there is an occurrence sequence from m_1 to the initial marking m_0. Since m_2 is reachable, there is an occurrence sequence from m_0 to m_2. The concatenation of these sequences leads from m_1 to m_2. The corresponding path of the marking graph leads from the vertex m_1 to the vertex m_2 □

The marking graphs of Figure 11 provides examples. The first one represents the behavior of a bounded deadlock-free marked net which is not live. The second one stems from of an unbounded live marked net. The marked net of the third marking graph is live and bounded but not reversible.

4.3 ω-Markings and ω-Occurrence Sequences

By Lemma 13, unbounded finite marked nets have infinitely many reachable markings and thus infinite marking graphs. Therefore, analysis techniques based on the marking graph are not applicable to them. A finite graph can be constructed instead, providing important information about the behavior of the marked net. In particular, this graph can be used to decide boundedness of places and deadness of transitions. The graph is actually a tree, called *coverability tree.*

The core idea of the following sections is the concept of *ω-markings* which generalize conventional markings.

Formally, an *ω-marking* of a net N is a mapping $\overline{m}: S_N \to \mathbb{N} \cup \{\omega\}$ where $\omega \notin \mathbb{N}$. Clearly, every (conventional) marking can be viewed as a particular ω-marking without ω-entries.

[7] For that reason, reversible marked nets are sometimes called *cyclic.*

ω-markings are interpreted as follows: If a marking m' is reachable from a marking m and satisfies $m'(s) \geq m(s)$ for each place s, the occurrence sequence leading from m to m' can be iterated arbitrarily often (Proposition 5). If moreover $m'(s_0) > m(s_0)$ for some place s_0 then the number of tokens on s_0 increases with each iteration of the occurrence sequence. This increasing sequence of markings is now replaced by one ω-marking $\overline{m'}$ with $\overline{m'}(s_0) = \omega$, denoting that, for each $b \in \mathbb{N}$, there is a reachable marking that coincides with m' for all places except s_0 and assigns at least b tokens to s_0. More generally, several places may map to ω, representing simultaneous growth of the token count on these places.

ω-markings are constructed by help of ω-occurrence sequences. The following definition of ω-occurrence sequences demonstrates that the occurrence rule can be applied for ω-markings in a similar way as for usual markings. An ω-entry of a place is not affected by any transition occurrence. Moreover, no ω-marking, except occasionally the last one, should occur twice in an ω-occurrence sequence. We will first define finite and infinite ω-occurrence sequences and then prove that every ω-occurrence sequence of a finite marked net is finite.

A (finite or infinite) sequence of transitions $t_1\, t_2\, t_3 \ldots$ is an *ω-occurrence sequence* of a marked net with initial marking m_0 if there exist ω-markings $\overline{m}_0, \overline{m}_1, \overline{m}_2, \ldots$ such that m_0 and $\overline{m}_0$ coincide for all places and, for each index i occurring in the sequence $t_1\, t_2\, t_3 \ldots$ the following conditions hold:

(1) For each place s in ${}^\bullet t_i$, either $\overline{m}_{i-1}(s) > 0$ or $\overline{m}_{i-1}(s) = \omega$ (the enabling condition).

(2) For each place s satisfying $\overline{m}_i(s) \neq \omega$,

$$\overline{m}_i(s) = \overline{m}_{i-1}(s) - |F_N \cap \{(s,t_i)\}| + |F_N \cap \{(t_i,s)\}|$$

(the conventional marking transformation).

(3) A place s satisfies $\overline{m}_i(s) = \omega$ if and only if
- either $\overline{m}_{i-1}(s) = \omega$ (places marked by ω remain marked by ω),
- or $\overline{m}_{i-1}(s) \neq \omega$ and there exists an index j, $j < i$, such that $\overline{m}_j(s) \neq \omega$ and $\overline{m}_j(s) < \overline{m}_{i-1}(s) - |F_N \cap \{(s,t_i)\}| + |F_N \cap \{(t_i,s)\}|$ and $\overline{m}_j(s') \leq \overline{m}_{i-1}(s') - |F_N \cap \{(s',t_i)\}| + |F_N \cap \{(t_i,s')\}|$ for each place s' satisfying $\overline{m}_j(s') \neq \omega$ and $\overline{m}_{i-1}(s') \neq \omega$ (places with increasing token count are marked by ω).

(4) If $i > 1$ then $\overline{m}_{i-1} \notin \{\overline{m}_0, \ldots, \overline{m}_{i-2}\}$ (after reaching an ω-marking the second time, the sequence stops).

We call an ω-marking $\overline{m}$ *reachable* in a marked net if some ω-occurrence sequence leads to $\overline{m}$.

Figure 12 shows an unbounded marked net. A part of its marking graph is given in Figure 13 (all markings with less than 5 tokens are shown). The ω-occurrence sequences of this marked net are:

t1 t1, t1 t2 t3 t3, t1 t2 t3 t4 t1, t1 t2 t4, t2 t4,

as well as all prefixes of these sequences.

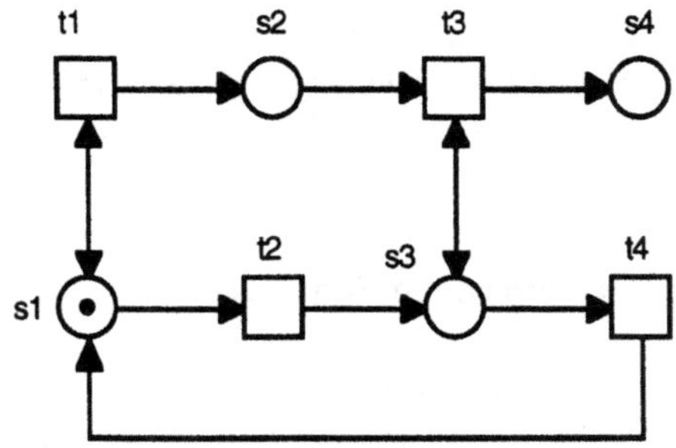

Fig. 12. An unbounded marked Petri net

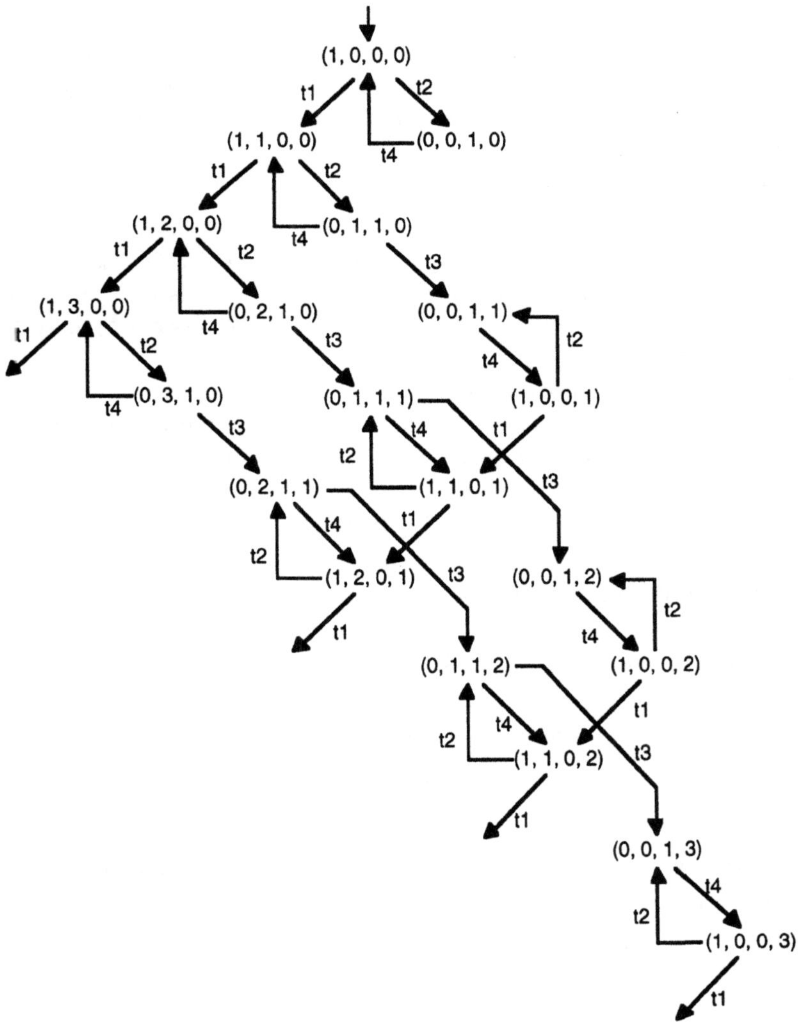

Fig. 13. A part of the marking graph of the marked net of Figure 12

Note that `t1 t2 t3 t3` is an ω-occurrence sequence but not an occurrence sequence. Conversely, `t1 t1 t1` is an example for an occurrence sequence which is not an ω-occurrence sequence, although no marking is reached twice during this sequence.

4.4 Finiteness of ω-Occurrence Sequences

The following theorem states that every ω-occurrence sequence of a finite marked net is finite. For technical convenience, we assume a linear order $\leq$ on the set $\mathbb{N} \cup \{\omega\}$, given by $x \leq y$ if and only if $y = \omega$ or $x \leq y$ in $\mathbb{N}$. The following lemma is known as *Dickson's Lemma.*

Lemma 17. *Let S be a finite set and let $\varphi_1\, \varphi_2\, \varphi_3 \ldots$ be an infinite sequence of mappings from S to $\mathbb{N} \cup \{\omega\}$. There exists an infinite sequence of indices $i_1\, i_2\, i_3\, \ldots$ which is strongly monotonic (i.e., $i_1 < i_2 < i_3 < \cdots$) such that, for each s in S,*

$$\varphi_{i_1}(s) \leq \varphi_{i_2}(s) \leq \varphi_{i_3}(s) \leq \cdots.$$

Proof. We prove the following stronger proposition: For each subset S' of S, there exists an infinite strongly monotonic sequence of indices $i_1, i_2, i_3, \ldots$ such that, for each s in S', $\varphi_{i_1}(s) \leq \varphi_{i_2}(s) \leq \varphi_{i_3}(s) \leq \cdots$. We proceed by induction on the number of elements in S'.
Base. If $S' = \emptyset$ then nothing has to be shown.
Step. Assume $S' \neq \emptyset$ and let $s \in S'$. By the induction hypothesis, there exists an infinite strongly monotonic sequence $i_1, i_2, i_3, \ldots$ such that, for each s' in $S' \setminus \{s\}$,

$$\varphi_{i_1}(s') \leq \varphi_{i_2}(s') \leq \varphi_{i_3}(s') \leq \cdots.$$

Now we restrict the sequence $i_1, i_2, i_3, \ldots$ to indices i_k satisfying

$$\varphi_{i_k}(s) \leq \varphi_{i_{k+1}}(s), \quad \varphi_{i_k}(s) \leq \varphi_{i_{k+2}}(s), \quad \varphi_{i_k}(s) \leq \varphi_{i_{k+3}}(s) \ldots$$

Clearly, the obtained sequence $i_{k_1}, i_{k_2}, i_{k_3}, \ldots$ satisfies the required property

$$\varphi_{i_{k_1}}(s) \leq \varphi_{i_{k_2}}(s) \leq \varphi_{i_{k_3}}(s) \leq \cdots$$

for each place s in S'. This sequence is infinite because, for each index i_k, every index i_l in $\{i_{k+1}, i_{k+2}, i_{k+3} \ldots\}$ satisfying

$$\varphi_{i_l}(s) \leq \varphi_{i_{k+1}}(s), \varphi_{i_{k+2}}(s), \varphi_{i_{k+3}}(s) \ldots$$

belongs to the sequence, too. Such an index i_l always exists because every nonempty subset of $\mathbb{N} \cup \{\omega\}$ has a minimal element. □

Theorem 18. *Every ω-occurrence sequence of a finite marked net is finite.*

Proof. By contraposition, assume a finite marked net that has an infinite ω-occurrence sequence $t_1\, t_2\, t_3 \ldots$,

$$\overline{m}_1 \xrightarrow{t_1} \overline{m}_2 \xrightarrow{t_2} \overline{m}_3 \xrightarrow{t_3} \cdots.$$

By Dickson's Lemma (Lemma 17), there exists an infinite strongly monotonic sequence of indices $i_1, i_2, i_3 \ldots$ such that, for each place s,

$$\overline{m}_{i_1}(s) \leq \overline{m}_{i_2}(s) \leq \overline{m}_{i_3}(s) \leq \cdots.$$

Let i and j be two subsequent indices of the sequence $i_1, i_2, i_3 \ldots$. By the definition of ω-occurrence sequences (4) no ω-marking appears twice in an infinite ω-occurrence sequence. Hence $\overline{m}_i(s) \neq \overline{m}_j(s)$ for at least one place s. By the definition of ω-occurrence sequences (3), $\overline{m}_i(s) \neq \omega$ and $\overline{m}_j(s) = \omega$. Again by (3), no place s satisfies $\overline{m}_i(s) = \omega$ and $\overline{m}_j(s) \neq \omega$. Hence $\overline{m}_j$ has more places with ω-entries than $\overline{m}_i$. Therefore, the set of places with ω-entries increases infinitely, contradicting the finiteness of the set of all places of the net. □

4.5 Coverability Trees

Next we show that a finite marked net has only finitely many ω-occurrence sequences and hence, together with the previous result, only finitely many reachable ω-markings. To this end, we introduce coverability trees.

As shown before for occurrence sequences, every ω-occurrence sequence can be conceived as a linear directed, acyclic graph with vertices labeled by ω-markings and edges labeled by transitions. Starting with a single disconnected graph representing all ω-occurrence sequences of a marked net, we identify common prefixes of ω-occurrence sequences. More precisely, we identify two vertices of the graph if and only if they are reached by the same ω-occurrence sequence. This construction yields the coverability tree of the marked net.

Formally, the *coverability tree* of a marked net is defined as a directed graph with a distinguished initial vertex and edges labeled by transitions:

- the vertices are the finite ω-occurrence sequences,
- a distinguished initial vertex is given by the empty sequence ε (which by definition is an ω-occurrence sequence),
- labeled edges are all triples $(\sigma,\ t,\ \sigma\, t)$ such that σ as well as $\sigma\, t$ are ω-occurrence sequences.

It is obvious that the coverability tree is a tree with the empty sequence ε being the root. In the graphical representation, any vertex σ is labeled by the ω-marking reached by σ.

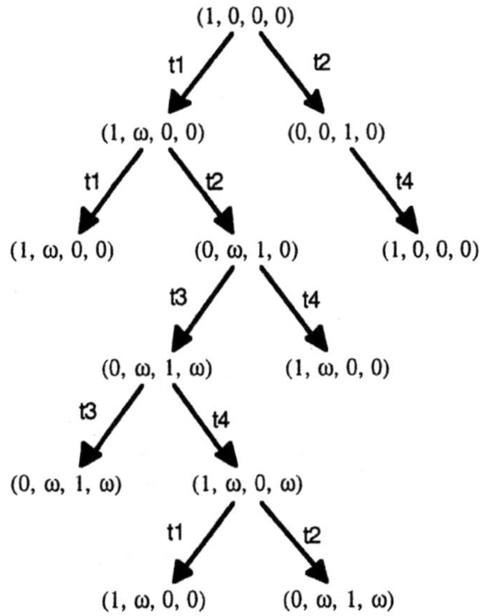

Fig. 14. The coverability tree of the marked net of Figure 12

Figure 14 shows the coverability tree of the marked net of Figure 12. Each vertex is the ω-occurrence sequence given by the arc labels of the path leading from the root to that vertex. The vertices are annotated by the respective reached ω-markings (in vector notation).

Theorem 19. *The coverability tree of a finite marked net is finite.*[8]

Proof. By contraposition, assume a finite marked net with an infinite coverability tree. Each vertex σ of the coverability tree has only finitely many immediate successors, one for each transition enabled by the ω-marking reached by σ. Hence every vertex σ with infinitely many successors has at least one immediate successor which also has infinitely many successors. By assumption, the initial vertex ε has infinitely many successors. Hence, starting with ε, we can construct an infinite directed path of the tree. The concatenation of the labels of the edges of this path yields an infinite ω-occurrence sequence — contradicting Theorem 18. □

Corollary 20. *A finite marked net has finitely many reachable ω-markings.* □

4.6 Reachable Markings and Reachable ω-Markings

The following two theorems formulate relations between reachable markings and reachable ω-markings of a marked net.

[8] This result is based on the well-known lemma of König

Theorem 21. *Let $\overline{m}$ be a reachable ω-marking of a finite marked net. For each b in $I\!N$, there is a reachable marking m such that every place s satisfies:*

- *if $\overline{m}(s) \neq \omega$ then $m(s) = \overline{m}(s)$,*
- *if $\overline{m}(s) = \omega$ then $m(s) \geq b$.*

Proof. Let N be a net with initial marking m_0. Let

$$\overline{m}_0 \xrightarrow{t_1} \overline{m}_1 \xrightarrow{t_2} \cdots \xrightarrow{t_k} \overline{m}_k$$

be an ω-occurrence sequence. By definition, $m_0(s) = \overline{m}_0(s)$ for each place s.

Let S_ω be the set of places satisfying $\overline{m}_k(s) = \omega$. Let $b \in I\!N$; we prove that there exists a reachable marking m which coincides with $\overline{m}_k$ on all places not in S_ω and assigns at least b tokens to each place in S_ω. To this end, we construct an occurrence sequence

$$m_0 \xrightarrow{t_1} m_1 \xrightarrow{\sigma_1} m_1' \xrightarrow{t_2} m_2 \xrightarrow{\sigma_2} m_2' \xrightarrow{t_3} \cdots \xrightarrow{\sigma_{k-1}} m_{k-1}' \xrightarrow{t_k} m_k \xrightarrow{\sigma_k} m_k'$$

such that m_k' satisfies the above property.

Let s be an arbitrary place in S_ω. By definition of ω-occurrence sequences, there is a unique index $i(s)$ such that

$$\overline{m}_0(s), \overline{m}_1(s), \ldots, \overline{m}_{i(s)-1}(s) \neq \omega,$$

$$\overline{m}_{i(s)}(s), \overline{m}_{i(s)+1}(s), \ldots, \overline{m}_k(s) = \omega$$

and an index $j(s)$, $j(s) < i(s)$, such that the transition sequence $\tau(s)$,

$$\tau(s) = t_{j(s)+1}\, t_{j(s)+2} \cdots t_{i(s)},$$

increases the token count of s and does not change the token count of any place s' satisfying $\overline{m}_{i(s)}(s') \neq \omega$. However, the sequence $\tau(s)$ might decrease the token count of places s' satisfying $\overline{m}_{i(s)-1}(s') = \omega$. It decreases the token count of a place s by at most $|\tau(s)|$, the length of $\tau(s)$. Any marking that coincides with $\overline{m}_{i(s)}$ on each place s' satisfying $\overline{m}_{i(s)}(s') \neq \omega$ and associates at least $|\tau(s)|$ tokens to each other place, enables $\tau(s)$.

For $n = 1, \ldots, k$, let $S_n = \{s \in S_\omega | i(s) = n\}$ and assume $S_n = \{s_n^1, \ldots, s_n^{|S_n|}\}$. In the sequel, $[\tau]^m$ stands for the sequence $\underbrace{\tau\,\tau \ldots \tau}_{m \text{ times}}$.

Define

$$\sigma_k = \left[\tau(s_k^1)\,\tau(s_k^2)\,\ldots\,\tau(s_k^{|S_k|})\right]^b$$

to ensure that each place s in S_k satisfies $m_k'(s) \geq b$. Since the token count of places in S_{k-1} can be decreased by t_k and by σ_k, define

$$\sigma_{k-1} = \left[\tau(s_{k-1}^1)\,\tau(s_{k-1}^2)\,\ldots\,\tau(s_{k-1}^{|S_k|})\right]^{(|\sigma_k|+1+b)}$$

to ensure that $t_k\,\sigma_k$ is enabled after the occurrence of σ_{k-1} and moreover leaves at least b tokens on each place in S_{k-1}.

In general, for $i = k, k-1, \ldots, 1$ the sequence σ_i is defined by

$$\sigma_i = \left[\tau(s_i^1)\,\tau(s_i^2)\,\ldots\,\tau(s_i^{|S_i|})\right]^{(|\sigma_{i+1}|+\cdots+|\sigma_k|+(k-i)+b)}.$$

Notice that some of the sequences σ_i might be empty. Since the marking $\overline{m}_0$ has no ω-entries, the sequence σ_1 does not decrease the token count of any place and is hence enabled at m_1.

Since the token count of a place not in S_ω is not changed by any occurrence sequence τ_i, it is not changed by any σ_i. Hence m'_k and $\overline{m}_k$ coincide on all these places. □

Theorem 22. *Let m be a reachable marking of a finite marked net. There exists a reachable ω-marking $\overline{m}$ such that $m(s) = \overline{m}(s)$ for each place s satisfying $\overline{m}(s) \neq \omega$.*

Proof. A sequence of transitions is called *generalized ω-occurrence sequence* of a marked net if it satisfies conditions (1) to (3) of the definition of ω-occurrence sequences, i.e., generalized ω-occurrence sequences are very much like ω-occurrence sequences except they do not necessarily stop after reaching an ω-marking twice.

We claim that, for each generalized ω-occurrence sequence σ leading to an ω-marking $\overline{m}$, there also exists an ω-occurrence sequence leading to $\overline{m}$. The claim is proven by induction on the length k of σ.
Base: $k = 0$. The empty sequence is an ω-occurrence sequence.
Step: $k > 0$. If σ is an ω-occurrence sequence then there is nothing to be shown. So assume that this is not the case. Let $\sigma = t_1\,t_2\,\ldots\,t_k$ and

$$\overline{m}_0 \xrightarrow{t_1} \overline{m}_1 \xrightarrow{t_2} \cdots \xrightarrow{t_k} \overline{m}_k.$$

Since σ is not an ω-occurrence sequence, there is a smallest index i $(1 \leq i \leq k)$ such that $\overline{m}_i$ coincides with some $\overline{m}_j$ $(0 \leq j < i)$. The sequence $t_1 \ldots t_j t_{i+1} \ldots t_k$ is a generalized ω-occurrence sequence, too. The ω-marking reached by this sequence is $\overline{m}_k$, too. Since the length of this sequence is smaller than k, the induction hypothesis can be applied, which finishes proof of the claim.

Let $m_0 \xrightarrow{t_1} m_1 \xrightarrow{t_2} \cdots \xrightarrow{t_k} m_k$ be an occurrence sequence such that $m_k = m$. Then, obviously $t_1\,t_2 \ldots t_k$ is a generalized ω-occurrence sequence. This generalized ω-occurrence sequence leads to an ω-marking $\overline{m}$ satisfying the required property. By the above claim, some ω-occurrence sequence leads to $\overline{m}$, too. □

In turn, the previous result implies a sufficient condition for non-reachability of a marking.

4.7 Analysis Through Coverability Trees

By construction of the coverability tree of a marked net, we obtain a finite representation of all reachable ω-markings. The following theorems show that by inspection of this finite set of ω-markings, boundedness of places and deadness of transitions can be decided.

Theorem 23. *A place s of a marked net is not bounded if and only if some reachable ω-marking $\overline{m}$ satisfies $\overline{m}(s) = \omega$ (i.e., some vertex of the coverability tree represents the ω-marking $\overline{m}$).*

Proof.
($\Longleftarrow$) follows immediately from Theorem 21.
($\Longrightarrow$) Since there are only finitely many reachable ω-markings by Theorem 19 there is a number $b \in \mathbb{N}$ such that each reachable ω-marking $\overline{m}$ satisfies either $\overline{m}(s) = \omega$ or $\overline{m}(s) < b$. Since s is not bounded, some reachable marking m satisfies $m(s) \geq b$. Since $m(s)$ does not coincide with $\overline{m}(s)$ for any reachable ω-marking $\overline{m}(s)$, there exists some reachable ω-marking $\overline{m}$ satisfying $\overline{m}(s) = \omega$ by Theorem 22. □

Together with Theorem 22, this result implies the following corollary.

Corollary 24. *A place s of a marked net is b-bounded if and only if each reachable ω-marking $\overline{m}$ satisfies $\overline{m}(s) \neq \omega$ and $\overline{m}(s) \leq b$.*

Theorem 25. *A transition t of a marked net is dead if and only if t does not occur in any ω-occurrence sequence (i.e., some arc of the coverability tree is labeled by t).*

Proof.
($\Longleftarrow$) Assume some reachable marking m enables t. By Theorem 22, a corresponding reachable ω-marking $\overline{m}$ satisfies $\overline{m}(s) \neq 0$ for each place s in ${}^\bullet t$. Hence, this ω-marking enables t, too.
($\Longrightarrow$) Assume some reachable ω-marking $\overline{m}$ enables t. By Theorem 21, there is a corresponding reachable marking m that marks all places satisfying $\overline{m} = \omega$ at least once. This marking m enables t, too. □

4.8 Coverability Trees and Marking Graphs

Theorem 23 immediately implies the following corollary.

Corollary 26. *A marked net is bounded if and only if no reachable ω-marking contains an ω-entry.*

In other words: reachable markings and reachable ω-markings coincide for bounded marked nets.

The *coverability graph* of a marked net is defined as an arc-labeled directed graph with a distinguished initial vertex and edges labeled by transitions:

- the *vertices* are the reachable ω-markings,
- the distinguished *initial vertex* is given by the ω-marking that coincides with the initial marking for each place,
- labeled edges are given by all triples $(\overline{m}, t, \overline{m}')$ such that $\overline{m}$ and $\overline{m}'$ are reachable ω-markings satisfying $\overline{m} \xrightarrow{t} \overline{m}'$.

The coverability graph is easily constructed from the coverability tree by identification of any two vertices representing ω-occurrence sequences that lead to identical markings. Note that, in general, a vertex of the coverability graph can have more than one outgoing edge labeled by the same transition, because different ω-occurrence sequences can lead to the same ω-marking. However, this is not the case for bounded marked nets:

Theorem 27. *The coverability graph and the marking graph of a bounded marked net are identical (up to different co-domains of markings and ω-markings).*

Proof. The result follows immediately from Corollary 26 and the definition of ω-occurrence sequences. □

5 Analysis Techniques

5.1 Behavioral Complexity of Marked Nets

As shown in the previous section, for bounded finite marked nets the marking graph can be constructed to analyze behavioral aspects. For unbounded finite marked nets, the coverability tree provides some information about the behavior. However, even if the marking graph of a marked net is finite, it is not always feasible to explicitly construct this graph because its size can explode with the size of the marked net[9].

Figure 15 shows that the set of reachable markings can grow exponentially with the size of a net: with $n+2$ transitions the marked net has 2^n+1 reachable markings. In the marked net shown in Figure 16, the smallest bound of the place **s** depends on the initial token count of the place **s0**: if **s0** carries n tokens initially then this bound is 2^n.

Analysis techniques that do not explicitly construct the set of all reachable markings (or ω-markings) have to stick to the structure of the net and its initial marking. Petri net theory provides many such techniques. In this section, we will introduce place and transition invariants, siphons and traps and discuss their relation to deadlock-freedom, liveness, boundedness and reversibility.

Often it is useful to prove that every reachable marking satisfies some given property. The following lemma gives a sufficient condition for this property.[10]

Lemma 28. *A set M of markings of a marked net with initial marking m_0 contains all reachable markings if*

(1) $m_0 \in M$ and

(2) for each marking m of N and each transition t of N, $m \in M$ and $m \xrightarrow{t} m'$ imply $m' \in M$.

[9] This is sometimes called *state explosion problem*, see [Valm98].

[10] This technique is not specific for Petri nets. In general, it is sometimes called *assertional reasoning*.

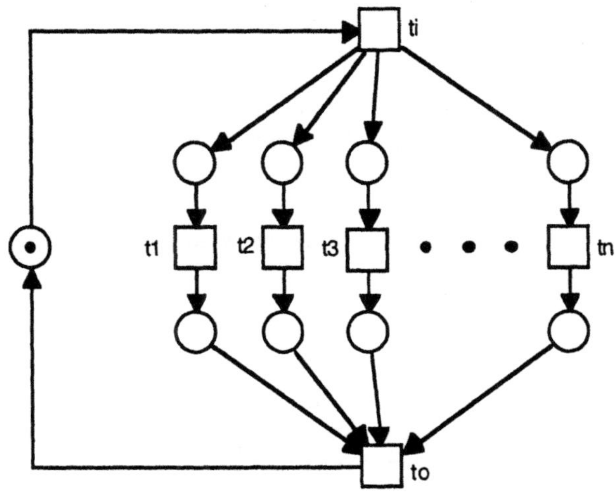

Fig. 15. The set of reachable markings explodes with the size of the net

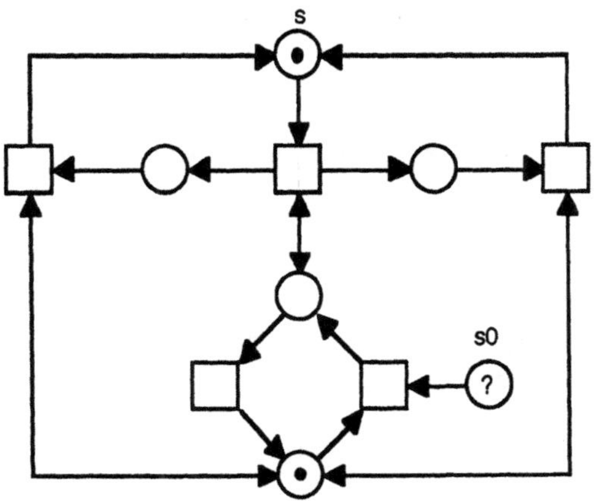

Fig. 16. The set of reachable markings explodes with the number of initial tokens

Proof. Assume that M satisfies conditions (1) and (2). Let m be a reachable marking and let $m_0 \xrightarrow{\sigma} m$, $\sigma = t_1\, t_2\, \ldots t_k$. Assume

$$m_0 \xrightarrow{t_1} m_1 \xrightarrow{t_2} m_2 \xrightarrow{t_3} \cdots \xrightarrow{t_k} m_k.$$

Then $m = m_k$.

We prove that all markings m_i, $0 \leq i \leq k$ are in M, by induction on i.
Base: $i = 0$. Then $m = m_0$. By (1), $m_0 \in M$.

Step: $i > 0$. By the induction hypothesis, $m_{i-1} \in M$. This implies by (2) that $m_i \in M$. □

Notice that conditions (1) and (2) of the above lemma are not necessary for a set M to include all reachable markings. This is because condition (2) applies to all markings m in M and not only to reachable markings in M. Obviously, it would be sufficient to require this condition for reachable markings only. However, the set of reachable markings is unknown in general and hence this modified condition would not be easy to verify.

5.2 Place Invariants

Occurrences of transitions transform the token distribution of a net, but often respect some global properties of markings. For example, the total token count of a set of places remains unchanged if the pre-set and the post-set of the transition contain the same number of places of this set. Place invariants formalize such invariant properties. They moreover allow to weight tokens on places by positive or negative values such that the sum of weighted tokens remains constant.

Given a net N, a *place invariant* is a mapping $i: S_N \to \mathbb{Z}$ satisfying $i(s) \neq 0$ for finitely many places and

$$\sum_{s \in {}^\bullet t} i(s) = \sum_{s \in t^\bullet} i(s)$$

for each transition t of N. A place invariant is *nonnegative* if it maps no place to a negative number.

Theorem 29. *If m is a reachable marking of a marked net with initial marking m_0 and i is a place invariant then*

$$\sum_{s \in S_N} i(s) \cdot m(s) = \sum_{s \in S_N} i(s) \cdot m_0(s).$$

Proof. Let M be the set of markings satisfying the above equation. We employ Lemma 28.

Clearly m_0 is in M. Assume a marking m in M and a transition occurrence $m \xrightarrow{t} m'$. Then, for each place s,

$$m'(s) = m(s) - |F \cap \{(s,t)\}| + |F \cap \{(t,s)\}|$$

by the occurrence rule. Hence

$$\sum_{s \in S_N} i(s) \cdot m'(s) = \sum_{s \in S_N} i(s) \cdot m(s) - \sum_{s \in S_N} i(s) \cdot |F \cap \{(s,t)\}| + \sum_{s \in S_N} i(s) \cdot |F \cap \{(t,s)\}|.$$

Observe that $|F \cap \{(s,t)\}| = 1$ if $s \in {}^\bullet t$, and $|F \cap \{(s,t)\}| = 0$ otherwise. Together with the corresponding property for $|F \cap \{(t,s)\}|$ and $t^\bullet$ we obtain

$$\sum_{s \in S_N} i(s) \cdot m'(s) = \sum_{s \in S_N} i(s) \cdot m(s) - \sum_{s \in {}^\bullet t} i(s) + \sum_{s \in t^\bullet} i(s).$$

By definition of place invariants, the last two sums are equal. Hence

$$\sum_{s \in S_N} i(s) \cdot m'(s) = \sum_{s \in S_N} i(s) \cdot m(s).$$

So $m' \in M$.

By Lemma 28, each reachable marking is in M, which implies the result.

□

The previous theorem constitutes the salient property of place invariants. For marked nets without dead transitions, the converse direction holds as well:

Theorem 30. *Assume a marked net N without dead transitions. Let m_0 be the initial marking. Let $i: S_N \to \mathbb{Z}$. If each reachable marking m satisfies*

$$\sum_{s \in S_N} i(s) \cdot m(s) = \sum_{s \in S_N} i(s) \cdot m_0(s)$$

then i is a place invariant.

Proof. Let t be a transition. Since t is not dead, there are reachable markings m and m' such that $m \xrightarrow{t} m'$. The assumption implies

$$\sum_{s \in S_N} i(s) \cdot m(s) = \sum_{s \in S_N} i(s) \cdot m'(s).$$

As shown in the proof of Theorem 29,

$$\sum_{s \in S_N} i(s) \cdot m'(s) = \sum_{s \in S_N} i(s) \cdot m(s) - \sum_{s \in {}^\bullet t} i(s) + \sum_{s \in t^\bullet} i(s).$$

Hence

$$\sum_{s \in {}^\bullet t} i(s) = \sum_{s \in t^\bullet} i(s)$$

which finishes the proof. □

Place invariants can be used to prove that a place is bounded:

Theorem 31. *Let s be a place of a marked net N with initial marking m_0. If there is a nonnegative place invariant i satisfying $i(s) \geq 1$ then s is bounded by*

$$\frac{1}{i(s)} \cdot \sum_{s' \in S_N} i(s') \cdot m_0(s').$$

Proof. Since i is nonnegative, $i(s') \geq 0$ for each place s'. Hence

$$i(s) \cdot m(s) \leq \sum_{s' \in S_N} i(s') \cdot m(s').$$

Hence, by Theorem 29, each reachable marking m satisfies

$$i(s) \cdot m(s) \leq \sum_{s' \in S_N} i(s') \cdot m_0(s').$$

Division by $i(s)$ yields the result. □

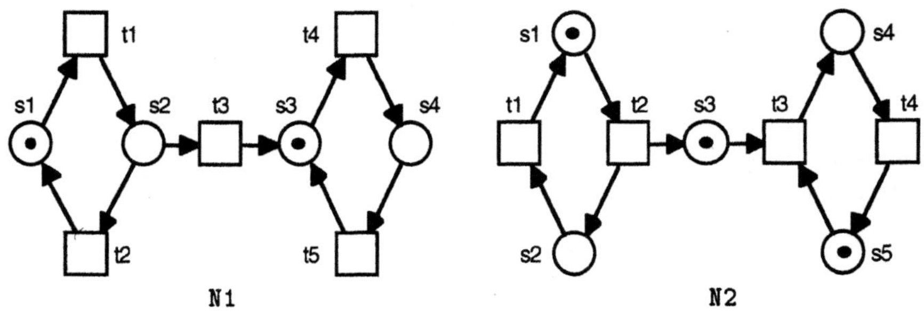

Fig. 17. Two marked nets (copied from Figure 10)

Corollary 32. *A finite marked net is bounded if it has a place invariant i that maps all places to positive numbers.*

Theorem 33. *Let N be a marked net with initial marking m_0 and let i be a nonnegative place invariant. Let S_i be the set of places s satisfying $i(s) > 0$. If $m_0(s) = 0$ for each place in S_i then every transition in ${}^\bullet S_i \cup S_i^\bullet$ is dead at the initial marking.*

Proof. Since i is a place invariant, no reachable marking marks any place in S_i. Hence no reachable marking enables a transition in $S_i^\bullet$. Each transition t in ${}^\bullet S_i$ is also dead because otherwise the occurrence of t would lead to a reachable marking that marks a place in S_i. □

Corollary 34. *If no transition of a marked net is dead at the initial marking then every nonnegative place invariant i either maps all non-isolated places to 0 or satisfies $i(s) \geq 1$ for at least one initially marked non-isolated place.*

This corollary applies in particular to live marked nets.

Figure 17 shows again the two marked nets **N1** and **N2** from Figure 10. In **N1**, the set of place invariants is given by all mappings $i: S_{\mathtt{N1}} \to \mathbb{Z}$ satisfying

$$i(\mathtt{s1}) = i(\mathtt{s2}) = i(\mathtt{s3}) = i(\mathtt{s4}).$$

An example is the place invariant that maps each places to 1. Corollary 32 applies and proves all places bounded. Theorem 33 and Corollary 34 do not apply because no transition is dead. Only if the initial marking was the null marking, these results would prove deadness of all transitions.

In **N2**, the set of place invariants is given by all mappings $i: S_{\mathtt{N2}} \to \mathbb{Z}$ satisfying

$$i(\mathtt{s1}) = i(\mathtt{s2}) \text{ and } i(\mathtt{s3}) = 0 \text{ and } i(\mathtt{s4}) = i(\mathtt{s5}).$$

An example is the place invariant that maps places `s1` and `s2` to 1 and the other places to 0. Using this invariant, Theorem 31 proves that `s1` and `s2` are bounded. If the token on `s1` was missing then Theorem 33 would prove deadness of `t1` and `t2`, using the same place invariant.

5.3 Transition Invariants

Given a net N, a *transition invariant* is a mapping $j: T_N \to \mathbb{Z}$ satisfying $j(t) \neq 0$ for finitely many transitions and

$$\sum_{t \in {}^\bullet s} j(t) = \sum_{t \in s^\bullet} j(t)$$

for each place s of N.

Let σ be a finite sequence of transitions of a net N. The *Parikh mapping* $p_\sigma: T_N \to \mathbb{Z}$ maps each transition t to the number of occurrences of t in σ.

Theorem 35. *If $m \xrightarrow{\sigma} m'$ is a finite occurrence sequence of a net then $m = m'$ if and only if the Parikh mapping p_σ is a transition invariant.*

Proof. Let s be a place. We prove that the equation

$$\sum_{t \in {}^\bullet s} p_\sigma(t) = \sum_{t \in s^\bullet} p_\sigma(t)$$

holds if and only if $m(s) = m'(s)$. The first equation is equivalent to

$$\sum_{t \in {}^\bullet s \setminus s^\bullet} p_\sigma(t) = \sum_{t \in s^\bullet \setminus {}^\bullet s} p_\sigma(t).$$

A token is added to s by a transition occurrence if and only if this transition is in ${}^\bullet s \setminus s^\bullet$. A token is removed if and only if the transition is in $s^\bullet \setminus {}^\bullet s$. Hence the last equation holds if and only if the number of transition occurrences in σ that increase the token count of s coincides with the number of transition occurrences in σ that decrease the token count of s. The latter is exactly the case if $m(s) = m'(s)$. □

Theorem 36. *If the initial marking of a finite bounded marked net enables an infinite occurrence sequence σ then the net has a transition invariant j satisfying*

- *$j(t) = 0$ for each transition t not occurring in σ,*
- *$j(t) = 0$ for each transition t occurring in σ finitely often,*
- *$j(t) \geq 1$ for each transition occurring in σ infinitely often.*

Proof. Let T_ω the set of transitions occurring infinitely often in the sequence σ. Assume $\sigma = t_1\, t_2\, t_3 \ldots$ and

$$m_0 \xrightarrow{t_1} m_1 \xrightarrow{t_2} m_2 \xrightarrow{t_3} \cdots.$$

Let k_1 be the first index $(1 \leq k_1)$ such that the sequence $t_{k_1+1}\, t_{k_1+2}\, t_{k_1+3} \cdots$ contains only transitions of T_ω.

Define an infinite sequence $k_1\, k_2\, k_3 \ldots$ of indices, $k_1 < k_2 < k_3 < \cdots$, such that, for $i = 1, 2, 3, \ldots$, the transition sequence

$$t_{k_i+1}\, t_{k_i+2}\, t_{k_i+3} \ldots t_{k_{i+1}}$$

contains each transition of T_ω at least once.

Since the marked net is finite and bounded, only a finite set of markings is reachable. Therefore, the markings $m_{k_1}, m_{k_2}, m_{k_3}, \ldots$ cannot be pairwise different. So there are indices i and j, $(1 \leq i < j)$ such that $m_{k_i} = m_{k_j}$. By Theorem 35, the Parikh vector of the transition sequence $t_{k_i+1}\, t_{k_i+2} \ldots t_{k_j}$ is a transition invariant. It is immediate that this transition invariant satisfies the conditions formulated above. □

Corollary 37. *If a finite marked net is live and bounded then it has a transition invariant that maps each transitions to a positive number.*

Proof. By liveness, every transition can always occur again. Therefore there exists an infinite occurrence sequence containing all transitions of the net infinitely often. The result follows by Theorem 36. □

In the example of Figure 17, `N1` is a finite and bounded marked net. It has an infinite occurrence sequence `t4 t5 t4 t5`.... In fact, the net has a transition invariant mapping transitions `t4` and `t5` to 1 and the other transitions to 0.

5.4 Siphons and Traps

Next we consider sets of places, S, that never gain a token once none of their places is marked. As shown in Theorem 33, if a nonnegative place invariant has positive entries only for places of S then no place of S will ever gain a token. Another sufficient condition is that each transition that adds one or more tokens to a place of this set also removes at least one token from some place of the set. This condition is formalized by the notion of a siphon, to be defined next.

A *siphon* is a set S of places satisfying ${}^\bullet S \subseteq S^\bullet$. A siphon is *marked* by a marking m if at least one place of it is marked at m.

Theorem 38. *Assume a marked net with a siphon S. If S is not marked at the initial marking then S is not marked at any reachable marking.*

Proof. We apply Lemma 28 to show that every reachable marking marks no place of S.

Let M be the set of markings that do not mark S. By assumption, the initial marking is in M. Assume a marking m in M and a transition occurrence $m \xrightarrow{t} m'$. Then $t \notin S^\bullet$ because m enables t and m marks no place in S. Since S is a siphon, this implies $t \notin {}^\bullet S$. Hence no place of S can gain a token by the occurrence of t and m' belongs to M, too.

So, by Lemma 28, M includes all reachable markings, which implies the result. □

Consider again the examples shown in Figure 17. The net **N1** has the siphons

$$\emptyset, \{\texttt{s1}, \texttt{s2}\}, \{\texttt{s1}, \texttt{s2}, \texttt{s3}, \texttt{s4}\}.$$

After the occurrence of **t1** and **t3**, the siphon $\{\texttt{s1}, \texttt{s2}\}$ is unmarked and can never become marked again.

The siphons of the net **N2** are

$$\emptyset, \{\texttt{s1}, \texttt{s2}\}, \{\texttt{s1}, \texttt{s2}, \texttt{s3}\}, \{\texttt{s1}, \texttt{s2}, \texttt{s3}, \texttt{s4}\}, \{\texttt{s1}, \texttt{s2}, \texttt{s3}, \texttt{s4}, \texttt{s5}\}, \{\texttt{s4}, \texttt{s5}\}.$$

Siphons can be used to prove that transitions of a marked net are dead:

Theorem 39. *Assume a marked net with a siphon S which contains at least one non-isolated place. If no transition is dead at the initial marking then some non-isolated place of S is initially marked.*

Proof. Assume that S contains no initially marked non-isolated place. Let s be a non-isolated place in S. By Theorem 38, s is not marked at any reachable marking. Therefore, by Proposition 9 all transition in the set ${}^\bullet s \cup s^\bullet$ are dead. Since s is not isolated, this set is not empty and hence there is an initially dead transition. □

Similarly to siphons, we consider sets of places that never loose all tokens once at least one of their places is marked. A sufficient condition is that each transition that removes at least one token from this set also adds a token. This condition is formalized by the notion of a trap:

A *trap* is a set S of places satisfying $S^\bullet \subseteq {}^\bullet S$. A trap is *marked* by a marking m if at least one place of it is marked at m.

Theorem 40. *Assume a marked net with a trap S. If S is marked at the initial marking then it is marked at every reachable marking.*

Proof. We apply Lemma 28.

Let M be the set of markings of the net that mark at least one place of S. By assumption, the initial marking is in M. Now assume a marking m in M and a transition occurrence $m \xrightarrow{t} m'$. If $t \notin S^\bullet$ then the place of S marked by m remains marked. If $t \in S^\bullet$ then $t \in {}^\bullet S$ because S is a trap. Hence, in this case at least one place in $t^\bullet \cap S$ is marked at m'.

So, by Lemma 28, M includes all reachable markings, which implies the result. □

The traps of the net **N1** of Figure 17 are

$$\emptyset, \{\texttt{s3}, \texttt{s4}\}, \{\texttt{s1}, \texttt{s2}, \texttt{s3}, \texttt{s4}\}$$

and the traps of the net **N2** are

$$\emptyset, \{\texttt{s1}, \texttt{s2}\}, \{\texttt{s1}, \texttt{s2}, \texttt{s3}, \texttt{s4}, \texttt{s5}\}, \{\texttt{s2}, \texttt{s3}, \texttt{s4}, \texttt{s5}\}, \{\texttt{s3}, \texttt{s4}, \texttt{s5}\}, \{\texttt{s4}, \texttt{s5}\}.$$

By contraposition of the following result, traps can be used to show that a marked net is not reversible:

Theorem 41. *Assume a reversible marked net with a nonempty trap S which contains no isolated place. If no transition is dead at the initial marking then S is initially marked.*

Proof. Let s be a place in S and let t be a transition in ${}^\bullet s \cup s^\bullet$. Since t is not initially dead, there exists a reachable markings m that marks s, by Proposition 9. So the trap S is marked at some reachable marking. Since the initial marking m_0 can be reached from m by reversibility, S is also marked at m_0. □

A first example of this result is given by **N2** of Figure 17; this marked net is reversible and has no dead transitions. Hence all nonempty traps are initially marked. As a negative example, consider the marked net **N5** of Figure 10. It has no dead transitions. The set $\{\mathtt{s3}, \mathtt{s4}, \mathtt{s5}\}$ is a trap which is not marked initially. Hence the marked net is not reversible.
Finally, siphons and traps provide a sufficient condition for deadlock-freedom:

Theorem 42. *Assume a marked net with at least one transition. If each nonempty siphon without isolated places includes a trap marked at the initial marking then the marked net is deadlock-free.*

Proof. Assume that the marked net is not deadlock-free and let m be a dead reachable marking. Let S be the set of non-isolated places that are not marked at m. We show that S is a non-empty siphon that includes no initially marked trap.

Each transition t is dead at m and hence has an unmarked input place. So $S^\bullet$ contains the set of all transitions. Therefore, ${}^\bullet S \subseteq S^\bullet$. S is not empty because the net has some transition by assumption and S contains a place in the pre-set of this transition. So S is a non-empty siphon without isolated places. By definition, S is not marked at m. Hence, S includes no trap marked at m. Since initially marked traps remain marked, S includes no trap marked at the initial marking. □

Again the nets **N1** and **N2** of Figure 17 serve as examples. In **N1**, the siphon $\{\mathtt{s1}, \mathtt{s2}\}$ does not contain any nonempty trap. In **N2**, the condition formulated in Theorem 42 holds, whence the marked net is deadlock-free.

5.5 Minimal Place Invariants and Minimal Siphons

Theorems 31 and 33 employ nonnegative place invariants. In general, a net possesses infinitely many nonnegative place invariants. We show next that there is a finite set of minimal place invariants such that every nonnegative place invariant can be expressed by a sum of minimal place invariants.

A nonnegative place invariant i is *minimal* if

- $i(s) \geq 1$ for some place s

and there are no two nonnegative place invariants i_1 and i_2 such that

- $i_1(s_1) \geq 1$ for some place s_1,
- $i_2(s_2) \geq 1$ for some place s_2,
- $i(s) = i_1(s) + i_2(s)$ for each place s.

Theorem 43. *The set of minimal place invariants of a finite net is finite.*

Proof. Assume a finite net with an infinite set of minimal place invariants. Let $i_1\, i_2\, i_3\, \ldots$ be an infinite sequence of different minimal place invariants. Dickson's Lemma (Lemma 17) implies that this sequence contains place invariants i_j and i_k such that $i_j \neq i_k$ and, for each place s, $i_j(s) \leq i_k(s)$. Consider the mapping $i: S_N \rightarrow \mathbb{Z}$, defined by $i(s) = i_k(s) - i_j(s)$. It is straightforward to verify that i is a nonnegative place invariant, too. Since $i_j \neq i_k$, i_k is not minimal. This contradicts the assumption that the sequence contains only minimal place invariants. □

In Theorems 31 and 33 it suffices to consider minimal place invariants because every place mapped to a positive number by any nonnegative place invariant is mapped to a positive number by a minimal one.

There exists a place invariant i satisfying $i(s) \geq 1$ for each place s if and only if the sum of minimal place invariants enjoys this property. So for Corollary 32 it also suffices to consider minimal place invariants.

The example net **N1** has only one minimal place invariant. This invariant maps all places to 1. The example net **N2** has two minimal place invariants; one of them only maps **s1** and **s2** to 1, the other one only maps **s4** and **s5** to 1 (the other places are mapped to 0).

Clearly, the sets of siphons and traps of a finite net are finite. However, often not every siphon has to be checked for the application of the previous results. In particular for Theorems 39 and 42 it suffices to consider siphons that do not properly include other siphons except the empty set.

A *minimal siphon* S is a nonempty siphon such that no proper subset of S is a nonempty siphon.

If a siphon S contains an isolated place s then $S \setminus \{s\}$ is a siphon, too. So a minimal siphon either contains no isolated places or consists of only one isolated place.

Since the union of traps is again a trap, the *maximal trap* in a given set of places S is the union of all traps included in S. It is the empty set if S includes no nonempty trap. So Theorem 42 can be reformulated as:

Corollary 44. *Assume a marked net with at least one transition and no isolated place. If all maximal traps included in minimal siphons are marked at the initial marking, the marked net is deadlock-free.* □

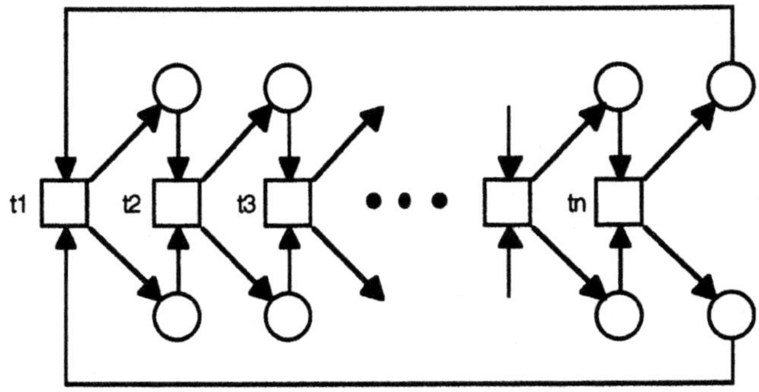

Fig. 18. The sets of minimal siphons, minimal traps and minimal place invariants grows exponentially in the size of the net

The net **N1** has only one minimal siphon, namely {s1, s2}. The maximal trap included in this siphon is empty. The net **N2** has minimal siphons {s1, s2} and {s4, s5}. Both sets are also traps; hence they are the maximal traps included in the siphons.

Minimal place invariants and siphons do not always help to significantly improve the efficiency of analysis techniques. Consider the nets shown in Figure 18. For a fixed number n of transitions, there are 2^n minimal place invariants, mapping one place of each vertical pair of places to 1 and the other place to 0, and there are 2^n minimal siphons, containing exactly one place of each vertical pair each.

6 Concurrent Semantics of Marked Nets

Occurrence sequences provide a *sequential semantics* of marked nets, because they represent each run by a sequence. In general, an occurrence sequence provides little information about dependencies between transition occurrences: if a transition name t' appears after a transition name t in an occurrence sequence then either t' can only occur after t has occurred, or both occurrences are concurrent. The aim of this section is to provide a *concurrent semantics* which explictly represents dependencies between transition occurrences. In particular, this semantics respects concurrency. Relations of this semantics to sequential semantics are studied in depth.

6.1 Causal Nets

Consider again the vending machine from the introduction. Figure 19 shows its representation as a marked net, as in Figure 9. In this example, t5 is only enabled after the occurrence of t3, whereas t3 and t1 can occur concurrently. A *concurrent run* of the transitions t1, t3 and t5 should explicitly represent these dependency relations .

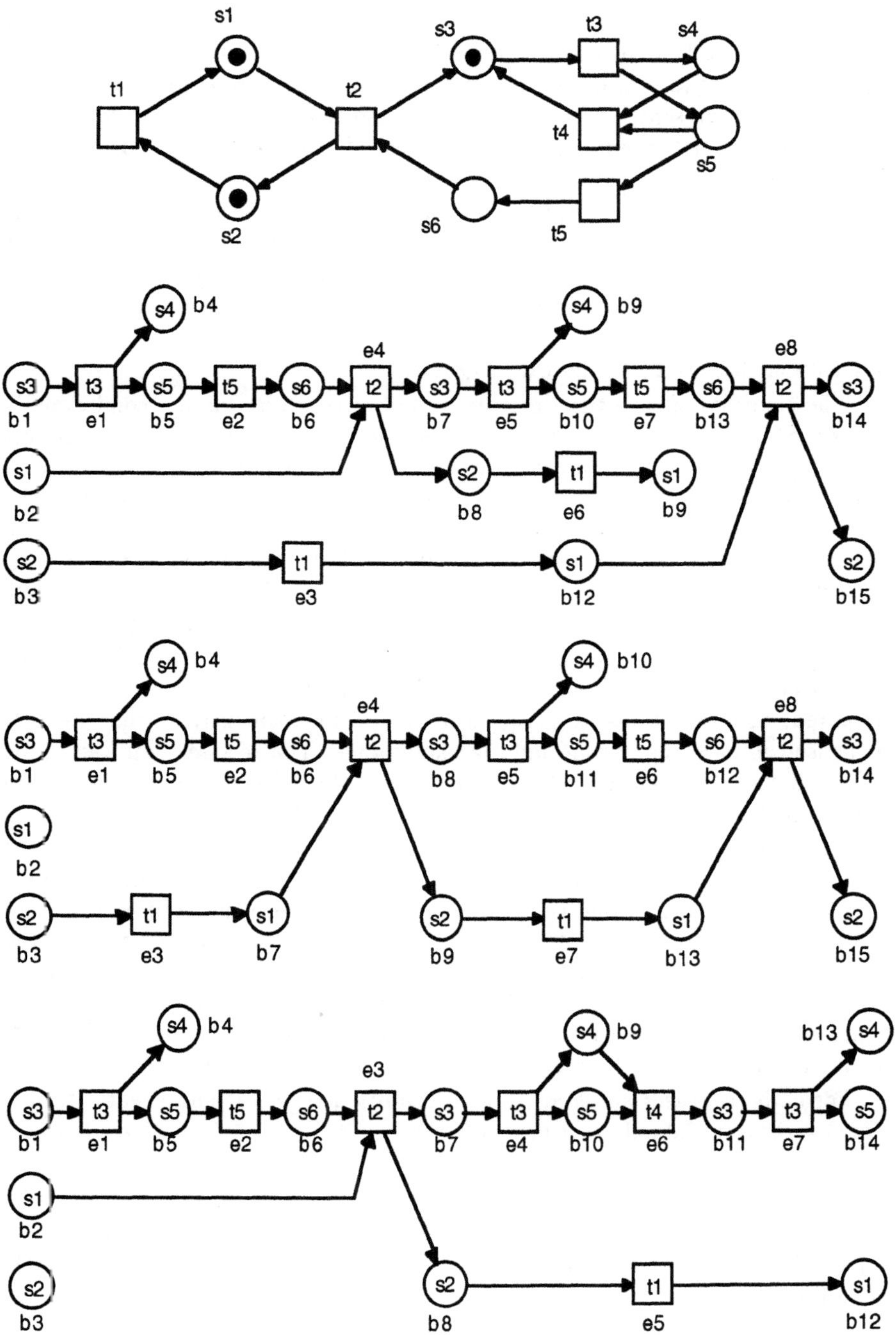

Fig. 19. A marked net with three concurrent runs

The occurrence of **t5** depends on the previous occurrence of **t3** because **t3** produces a token that is consumed by **t5**. Transition occurrences together with produced and consumed tokens and their mutual relationship are most conveniently represented by a special kind of labeled nets. Thus each concurrent run of a marked net is represented by a net, called causal net. To avoid confusion, the marked net will called *system net* in the sequel.

Figure 19 shows a marked system net and three causal nets representing concurrent runs. Each place of a causal net represents a single token occurrence in a system net. We call places of causal nets *conditions*, emphasizing their role as pre- and post-conditions of transition occurrences. Similarly, each transition of a causal net represents a single occurrence of a transition of a system net. Transitions of causal nets are called *events*.

The elements of the causal nets are usually written **b1**, **b2**, ... (conditions)[11] and **e1**, **e2**, ... (events). Moreover, each element is inscribed by the corresponding element of the system net: an event inscribed by t represents an occurrence of t, and a condition in the post-set of the event stands for a token produced by the transition occurrence. The condition inscription denotes the place of the system net, to which the token is added. The initial token distribution is reflected by according conditions of a causal net with empty pre-set. The causal nets of Figure 19 are finite, but causal nets may be infinite, thus representing infinite runs.

According to the vending machine interpretation of the example net, the first two causal nets represent concurrent runs where two coins are inserted and accepted, and two items are dispensed. In the first concurrent run, the first dispensed item is the one initially in the storage. The second item had to be refilled before it is dispensed. In the second run, the item initially in the storage is not touched. Instead both diepensed items depend on preceding occurrences of the refill transition **t1**. Concurrent semantics distinguishes these two runs because the mutual dependencies of transition occurrences are different. One could argue that the second run is more efficient because the occurrences of **t2** do not unnecessarily wait. Moreover, if items should not stay too long in the storage (such as warm beverages), the second run is not desirable. The third run is different. Here the second inserted coin is rejected. In contrast to the other runs, some condition inscribed by **s5** is followed by an event inscribed by **t4**. Thus, in this run a token an **s5** is removed by the occurrence of **t4**.

Formally, a *causal net* is a net with conditions (places) and events (transitions) satisfying the following properties:

(1) the flow relation is acyclic, i.e.
no path with at least two elements leads from an element to itself;
(2) conditions are not branched, i.e. each condition b satisfies $|{}^\bullet b| \leq 1$, $|b^\bullet| \leq 1$;
(3) only finitely many conditions b have an empty pre-set ${}^\bullet b$;
(4) for each event e, both sets ${}^\bullet e$ and $e^\bullet$ are finite and nonempty;
(5) for each element x, only finitely many different paths lead to x.

[11] from the German *Bedingungen*

For a given causal net K[12], the symbols B_K, E_K and F_K denote its sets of conditions, events and directed arcs, respectively. Since the flow relation of a causal net K is acyclic by definition, it induces a partial order $\preceq_K$ between the elements of K: $x \preceq_K y$ if and only if there is a path leading from x to y. We say that x *precedes* y in K if $x \neq y$ and $x \preceq_K y$. According to (4), each element is preceded by finitely many elements.

A *line* of a causal net K is a maximal set of elements that are pairwise ordered by $\preceq_K$ (maximality w.r.t. set inclusion).

In the first causal net of Figure 19, the set {b2, e4, b8, e6, b9} is a line.

Proposition 45. *For each path of a causal net, all elements appearing in the path belong to a single line.* □

A *cut* is a maximal set of elements such that every two distinct elements of the set are not ordered by $\preceq_K$.

In the first causal net of Figure 19, the set {b9, e7, b11, b12} is a cut.

The maximality of a cut implies the following proposition:

Proposition 46. *Given a cut X of a causal net K, each element y of K either belongs to X or precedes an element of X or is preceded by an element of X.* □

6.2 Causal Nets with Canonical Marking

Representing a single concurrent run of a system net, there is no need for markings of causal nets. Markings for causal nets are nevertheless useful for technical reasons. A *canonical* initial marking of a causal net K assigns to each minimal (w.r.t. $\preceq_K$) condition one token and no token to all other conditions. As will turn out, occurrence sequences of system nets and occurrence sequences of corresponding causal nets are closely related. This relation demonstrates that the concurrent semantics provided by causal net respects the sequential semantics provided by occurrence sequences.

For each causal net K, we call the marking l_0 defined by

$$l_0(b) = \begin{cases} 1 & \text{if } {}^\bullet b = \emptyset \\ 0 & \text{if } {}^\bullet b \neq \emptyset \end{cases}$$

the *canonical initial marking* of K.

In all causal nets of Figure 19, the canonical initial markings assign one token to the conditions b1, b2 and b3 and no token to any other condition.

[12] from the German *Kausalnetz*

Theorem 47. *Assume a causal net K with canonical initial marking l_0. A marking l of K is reachable from l_0 if and only if*

- *$l(b) \leq 1$ for each condition b, and*
- *the set $\{b \in B_K \mid l(b) = 1\}$ is a finite cut of K.*

Proof.
($\Longleftarrow$) Let B denote the finite cut $\{b \in B_K \mid l(b) = 1\}$. Let E be the set of events preceding conditions of B. By the definition of causal nets, E is finite because B is finite. We proceed by induction on $|E|$.
Base: $k = 0$. Then all conditions in B have an empty pre-set. Since B is a cut, B contains all conditions with empty pre-set. Therefore, $l = l_0$.
Step: $k > 0$. Assume a maximal event e in E (maximality w.r.t. $\preceq_K$).

We claim that $e^\bullet \subseteq B$. Let b in $e^\bullet$.

By the maximality of e, either $b^\bullet = \emptyset$ or $b^\bullet = \{e'\}$ for some event e' that does not belong to E. In both cases, b does not precede any condition of B. Since b is not branched, every condition preceding b also precedes e. Since e precedes some condition of B and conditions of B are not ordered, no condition preceding b belongs to B. Proposition 46 implies $b \in B$.

Define $B' = (B \setminus e^\bullet) \cup {}^\bullet e$. B' is finite because B as well as ${}^\bullet e$ are finite. We claim that B' is a cut. To this end, we first show that no condition in ${}^\bullet e$ is ordered with any other condition in B'. Then we prove that every element of K either belongs to B' or is ordered with at least one condition of B'.

Conditions in ${}^\bullet e$ are mutually not ordered because conditions are not branched and e does not precede a condition in ${}^\bullet e$ because K is acyclic. There is no path from a condition in ${}^\bullet e$ to a condition in $B \setminus e^\bullet$ because this path would pass through a condition in $e^\bullet$ but $e^\bullet \subseteq B$ and conditions in B are not ordered. There is no path from a condition in $B \setminus e^\bullet$ to a condition in ${}^\bullet e$ because this path could be extended by e and a condition in $e^\bullet$; since $e^\bullet \subseteq B$, this extended path would establish an order between distinct conditions of B.

Let x be an element of K. By Proposition 46, either x precedes some b in B, or some b in B precedes x, or x is equal to some b in B. If this condition b can be taken from $B \setminus e^\bullet$ then x is ordered with an element of B', too. Otherwise $b \in e^\bullet$. If $x \in e^\bullet$ or $x = e$ then any element of ${}^\bullet e$ precedes x. If $x \in {}^\bullet e$ then $x \in B'$. In all other cases, x precedes b if and only if x precedes a condition of ${}^\bullet e$ and b precedes x if and only if a condition of ${}^\bullet e$ precedes x. Since ${}^\bullet e \subseteq B'$, x is ordered with some condition of B'. This finishes the proof of the claim that B' is a cut.

Since the set of events preceding B' is $E \setminus \{e\}$, the induction hypothesis applies to B'. Hence some reachable marking l' marks exactly the conditions of B', by one token each. The event e is enabled at l'. Its occurrence leads to l.

($\Longrightarrow$) Let B be the set of conditions of some line X of K. We claim that, for any reachable marking, the number of all tokens on conditions of B is one. The claim is proven by help of Lemma 28. Let L be the set of markings satisfying the above property.

The line X contains at least one element with empty pre-set, because every element has a finite past. It contains at most one such element because every two elements of X are connected by a path. This unique minimal element of X is a condition, because events have non-empty pre-sets. By the definition of l_0, this condition is initially marked with one token, and all other conditions of X are initially unmarked. Hence l_0 belongs to L.

Assume a marking l in L and an event occurrence $l \xrightarrow{e} l'$. The token distribution on B is only changed if e belongs to the line X, because conditions are not branched. In this case, B contains exactly one condition in ${}^\bullet e$ and exactly one condition in $e^\bullet$ because e has nonempty pre- and post-sets. So the occurrence of e does not change the number of tokens on places in B. Therefore, l' also belongs to L, which finishes the proof of the claim.

Let l be a reachable marking. Since each condition is an element of a line, the claim implies that no condition carries more than one token at l, and that each condition is ordered with some condition marked at l. Proposition 45 implies that conditions marked at l are mutually not ordered. So the set of conditions marked at l constitutes a cut. This cut is finite because initially a finite set of conditions is marked and each event has a finite post-set. □

Corollary 48. *Each causal net marked with its canonical initial marking is a 1-bounded marked net.* □

The next result implies that occurence sequences of a causal net K respect the order $\preceq_K$.

Theorem 49. *Assume a causal net K with events e and e' satisfying $e \preceq_K e'$ (including the case $e = e'$). The canonical initial marking enables no occurrence sequence $\sigma_1\, e'\, \sigma_2\, e$.*

Proof. After the occurrence of e', the event e' precedes some marked condition. This situation is not changed by the occurrence of other events because all events have nonempty post-sets. Let l be a marking reached after an occurrence sequence $\sigma_1\, e' \sigma_2$. Since $e \preceq_K e'$, the event e precedes a condition marked at l. By Theorem 47, no condition in ${}^\bullet e$ is marked at l. Since ${}^\bullet e \neq \emptyset$, the event e is not enabled at l. □

Corollary 50. *No event of a causal net occurs more than once in an occurrence sequence enabled at the canonical initial marking.* □

Finally, the following theorem implies that each event of an occurrence sequence can eventually occur.

Theorem 51. *Let $e_1 e_2 e_3 \ldots$ be a sequence of events of a causal net K containing each event of E_K exactly once such that $e_i \preceq_K e_j$ implies $i \leq j$. Then this sequence is enabled at the canonical initial marking.*

Proof. By Proposition 3, it suffices to consider finite prefixes $e_1\, e_2 \ldots e_k$. The proof is by induction on k:
Base: $k = 0$. The empty sequence is enabled at any marking.
Step: $k > 0$. By the induction hypothesis, the canonical initial marking enables the occurrence sequence $e_1\, e_2 \ldots e_{k-1}$. Assume this sequence leads to l_{k-1}. Each event preceding e_k has occurred in this sequence. So each condition in $^\bullet e_k$ is marked at l_{k-1}. Hence l_{k-1} enables e_k. □

6.3 Process Nets

Since a causal net K represents a concurrent run of a marked net N, its conditions correspond to places of N and its events correspond to transitions of N. This relation is formalized by a *labeling function* $\pi\colon (B_K \cup E_K) \to (S_N \cup T_N)$.

A labeled causal net K is a *process net* of a net N with initial marking m_0 if the labeling function π enjoys the following properties:

(6) for each condition b in B_K, $\pi(b) \in S_N$;
(7) for each event e in E_K, $\pi(e) \in T_N$;
(8) for each event e in E_K, π generates bijections

$$\pi_e^- : {}^\bullet e \to {}^\bullet(\pi(e)) \quad \text{and} \quad \pi_e^+ : e^\bullet \to (\pi(e))^\bullet,$$

defined by $\pi_e^-(b) = \pi(b)$ for each b in $^\bullet e$ and $\pi_e^+(b) = \pi(b)$ for each b in $e^\bullet$;
(9) for each place s in S_N, $m_0(s) = |\{b \in B_K \mid {}^\bullet b = \emptyset \wedge \pi(b) = s\}|$.

Graphically, we denote the labels as inscriptions in the elements of a causal net.

The following two theorems tightly relate the occurrence sequences of a marked net and the occurrence sequences of its process nets.

Theorem 52. *Assume a marked net N with initial marking m_0 and a process net K with labeling function π. Let l_0 be the canonical initial marking of K and assume a finite occurrence sequence $e_1\, e_2 \ldots e_k$ of K, leading from l_0 to a marking l_k. Then $\pi(e_1)\, \pi(e_2) \ldots \pi(e_k)$ is an occurrence sequence of N, enabled at m_0. It leads to the marking m_k, satisfying for each place s the equation*

$$m_k(s) = |\{b \in B_K \mid l_k(b) = 1 \wedge \pi(b) = s\}|.$$

Proof. The proof is by induction on k.
Base: $k = 0$. The empty sequence is enabled by any marking. The equation for m_k coincides with the equation for m_0 in the definition of process nets.
Step: $k > 0$. Let $l_0 \xrightarrow{e_1 \ldots e_{k-1}} l_{k-1}$. By the induction hypothesis, m_0 enables the occurrence sequence $\pi(e_1)\pi(e_2) \ldots \pi(e_{k-1})$. Assume this sequence leads to m_{k-1}. Since e_k is enabled at l_{k-1}, all conditions in $^\bullet e_k$ are marked at l_{k-1}. By definition of a process net, π generates a bijection between these conditions and the places in $^\bullet(\pi(e_k))$.

Also by the induction hypothesis,

$$m_{k-1}(s) = |\{b \in B_K \mid l_{k-1}(b) = 1 \land \pi(b) = s\}|$$

for each place s in S_N. Hence each place in ${}^\bullet(\pi(e_k))$ is marked at m_{k-1}. So $\pi(e_k)$ is enabled at m_{k-1}. Since, for each place s in ${}^\bullet(\pi(e_k))$, e_k removes one token from a condition labeled by s, and, for each place s in $(\pi(e_k))^\bullet$, e_k adds one token to a condition labeled by s, we obtain

$$m_k(s) = |\{b \in B_K \mid l_k(b) = 1 \land \pi(b) = s\}|$$

for each place s in S_N. □

Proposition 2 implies:

Corollary 53. *Assume a marked net N with initial marking m_0 and a process net K with labeling function π. Let l_0 be the canonical initial marking of K and assume an infinite occurrence sequence $e_1\, e_2\, e_3 \ldots$ of K, enabled by l_0. Then $\pi(e_1)\, \pi(e_2)\, \pi(e_3) \ldots$ is an occurrence sequence of N, enabled at m_0.* □

For the converse direction we stick to marked nets satisfying the following property:

> **(N1)** *Both the pre-set and the post-set of each transition is nonempty and finite. The set of initially marked places is finite.*

By definition of a process net, occurrences of transitions with empty or infinite pre- or post-set cannot be represented in a causal net. If infinitely many places are initially marked then there is no corresponding finite initial cut of a process net.

Theorem 54. *Assume a marked net N with initial marking m_0 satisfying property* **(N1)**. *For each occurrence sequence $t_1\, t_2 \ldots t_k$, there exists a process net K with labeling function π such that its canonical initial marking l_0 enables an occurrence sequence $e_1\, e_2 \ldots e_k$ satisfying $\pi(e_i) = t_i$ for $1 \le i \le k$.*

Proof. The proof is by induction on k.
Base: $k = 0$. We only have to prove the existence of a process net K with appropriate labeling function π. To this end, a causal net without events and arcs is constructed: for each place s of N, the causal net contains $m_0(s)$ conditions labeled by s. It is easily verified that this net satisfies the defining properties of causal nets and process nets.
Step: $k > 0$. By the induction hypothesis there exists a process net K' with labeling function π', canonical initial marking l'_0, and an occurrence sequence

$$l'_0 \xrightarrow{e_1 \ldots e_{k-1}} l'_{k-1} \text{ such that } \pi(e_i) = t_i \text{ for } 1 \le i \le k-1.$$

Without loss of generality assume that every event of K occurs in this occurrence sequence. By Corollary 50, no event occurs more than once. So l'_{k-1} only marks conditions with empty post-set.

We construct the causal net K by adding to K' a new event e_k and, for each place s_i in $t_k^\bullet$, a new condition b_i. The labeling function π extends π' by $\pi(e_k) = t_k$ and $\pi(b_i) = s_i$ for each s_i in $t_k^\bullet$.

By Theorem 52,

$$m_{k-1}(s) = |\{b \in B_K \mid l'_{k-1}(b) = 1 \wedge \pi(b) = s\}|$$

for each place s. The marking m_{k-1} enables the transition t_k. Hence, for each place s_i in ${}^\bullet t_k$, at least one condition b_i labeled by s_i is marked at l'_{k-1} (there might be several suitable conditions). For each place s_i in ${}^\bullet t_k$, choose one of the conditions b_i labeled by s_i and add a directed arc from this condition to e_k. Finally, add directed arcs from e_k to all new conditions.

This construction obviously yields a causal net. The labeling function π satisfies the conditions formulated for process nets, because π' satisfies these conditions and the vicinity of e_k is respected by π. Since the new conditions are not marked initially, the occurrence sequence $e_1 \ldots e_{k-1}$ leads from the canonical initial marking of K to a marking l_{k-1} that coincides with l'_{k-1} on all conditions and maps the new conditions to 0. By construction, the event e_k is enabled at l_{k-1}. Thus, $e_1\, e_2 \ldots e_k$ is an occurrence sequence enabled at l_0. □

The construction of the causal net in the previous proof is not unique. However, if the marked net is 1-bounded then no reachable marking of the causal net marks two equally labeled conditions. Hence, for 1-bounded marked nets there is always exactly one way to connect marked conditions to a new event. Therefore, for 1-bounded marked nets, each occurrence sequence corresponds uniquely to a process net with a corresponding occurrence sequence. Figure 20 shows that the same does not hold for 2-bounded marked nets.

Consider occurrence sequences σ and σt of a marked net. Then the construction in the previous proof yields causal nets K and K' such that $B_K \subseteq B_{K'}$, $E_K \subseteq E_{K'}$ and $F_K \subseteq F_{K'}$. Moreover, the occurrence sequence of K is a prefix of the occurrence sequence of K'. Hence, for each infinite occurrence sequence σ of a marked net, the union of the respective sets of conditions, events, and arcs yields an infinite causal net. This infinite causal net has an infinite occurrence sequence such that the sequence of labels of its events is σ. So Theorem 54 implies:

Corollary 55. *Assume a marked net N with initial marking m_0 satisfying the conditions of* **(N1)**. *For each infinite occurrence sequence $t_1\, t_2\, t_3 \ldots$, there exists a process net K with labeling function π such that its canonical initial marking l_0 enables an occurrence sequence $e_1\, e_2\, e_3 \ldots$, satisfying $\pi(e_i) = t_i$ for $1 \leq i$.* □

Theorem 47 and Theorem 51, together with the previous results imply the following corollaries:

Corollary 56. *Assume a marked net N with initial marking m_0 satisfying the conditions of* **(N1)**. *A marking m is reachable from m_0 if and only if there is a process net K with labeling function π such that some finite cut B of K satisfies $B \subseteq B_K$ and, for each place s of N, $m(s) = |\{b \in B | \pi(b) = s\}|$.* □

The set of events E_K of a causal net K is partially ordered by $\preceq_K$. A (finite or infinite) sequence $e_1\, e_2\, e_3 \ldots$ of events is a *sequentialization* of E_K if

- every event of E_K appears exactly once in the sequence, and
- for each two events e and e' satisfying $e \preceq_K e'$, e appears before e' in the sequence.

Corollary 57. *Assume a marked net N with initial marking m_0 satisfying the conditions of* **(N1)**. *A sequence $e_1\, e_2\, e_3 \ldots$ is a sequentialization of the set of events of a process net with labeling function π if and only if $\pi(e_1)\, \pi(e_2)\, \pi(e_3) \ldots$ is an occurrence sequence of N, enabled at m_0.* □

6.4 Process Nets and Boundedness

Behavioral properties of a marked net based on reachable markings and transition occurrences can be reformulated in terms of process nets. For example, reversibility requires that for each finite process net leading to a cut X, there is another process net leading from X to a cut that corresponds to the initial marking. Boundedness can be characterized in a more elegant way:

Theorem 58. *A finite marked net N with initial marking m_0 satisfying the conditions of* **(N1)** *is unbounded if and only if some process net has an infinite cut.*

Proof.
($\Longleftarrow$) Assume a process net K with an infinite cut B. If B contains an event e then $(B \setminus \{e\}) \cup e^\bullet$ is also an infinite cut because events have nonempty post-sets and conditions are not branched. So we can assume without loss of generality that B contains only conditions.

We proceed indirectly and assume that N is bounded. Since N is finite, there is an upper bound k for the total number of tokens on all places. Consider an arbitrary finite subset B' of B with more than k conditions.

Let E be the set of all events preceding a condition of B'. Since B is a cut, no condition of B' precedes an event of E. By definition of a causal net, E is finite, because B' is finite. Since for each e in E, all events preceding E are contained in E as well, the canonical initial marking of K enables an occurrence sequence containing exactly the events of E. The marking reached by this sequence marks at least all conditions of B'. By Corollary 56, there is a corresponding reachable marking of the marked net such that the total token count of all places at this marking exceeds k, which contradicts the choice of k.

($\Longrightarrow$) Since the marked net is not bounded, some place is not bounded. By Theorem 23, there is a minimal ω-occurrence sequence σ leading to an ω-marking $\overline{m}$ such that $\overline{m}(s) = \omega$ for some place s. By definition of ω-occurrence sequences, $\sigma = \sigma_1\, \sigma_2$ such that, with

$$m_0 \xrightarrow{\sigma_1} m_1 \xrightarrow{\sigma_2} m_2,$$

each place s' satisfies $m_2(s') \geq m_1(s')$ and the place s satisfies $m_2(s) > m_1(s)$.

Now modify the net N as follows: Add a new transition $\overline{t}$, a new place $\overline{s}$, and new arcs $(s, \overline{t})$ and $(\overline{t}, \overline{s})$. This modified net will be called $\overline{N}$. Let $\overline{m}_0$ be the initial marking of $\overline{N}$ which coincides with m_0 on all places of N and assigns no token to $\overline{s}$.

Since the number of tokens on s increases by the occurrence of σ_2, there is an occurrence sequence

$$\overline{m}_0 \xrightarrow{\sigma_1} \overline{m}_1 \xrightarrow{\sigma_2\, \overline{t}} \overline{m}_2.$$

Then $\overline{m}_2(s') \geq \overline{m}_1(s')$ for each place s' of $\overline{N}$. So the sequence $\sigma_2 \overline{t}$ can be iterated any number of times:

$$\overline{m}_0 \xrightarrow{\sigma_1} \overline{m}_1 \xrightarrow{\sigma_2\, \overline{t}} \overline{m}_2 \xrightarrow{\sigma_2\, \overline{t}} \overline{m}_3 \xrightarrow{\sigma_2\, \overline{t}} \cdots .$$

By Theorem 54, there exists a process net K with labeling function π such that its canonical initial marking enables an infinite occurrence sequence of events where the sequence of event labels is

$$\sigma_1\, \sigma_2\, \overline{t}\, \sigma_2\, \overline{t}\, \sigma_2\, \overline{t} \ldots \; .$$

This causal net contains infinitely many events labeled by $\overline{t}$. Let B be the set of conditions in the pre-set of these events. Then B is also infinite.

Now consider the causal net gained from the previous one by deleting all events labeled by $\overline{t}$, all conditions labeled by $\overline{s}$, and all adjacent arcs. Then all conditions in B have an empty post-set. Hence they are mutually not ordered. It is easy to see that, taking the same labels as before, this causal net constitutes a process net of the original marked net. Clearly, the set B is again an infinite set of mutually not ordered conditions.

It remains to prove that B is included in an infinite cut. Let E be the set of all events preceding a condition of B. Let B' be the set of all minimal conditions (w.r.t. $\preceq_K$) not preceding any event in E. By minimality, conditions in B' are mutually not ordered. B' includes B because each condition in B either has an empty pre-set or the pre-set consists of an event in E. Finally, B' is maximal (w.r.t. set inclusion) because each minimal element not preceding a condition of B does not precede an event of E and hence belongs to B'. So B' is an infinite cut. □

Finally, the following proof of the Strongly-Connectedness-Theorem (Theorem 11) should help to demonstrate the advantages of causal semantics. Here we stick to nets satisfying the conditions of **(N1)**.

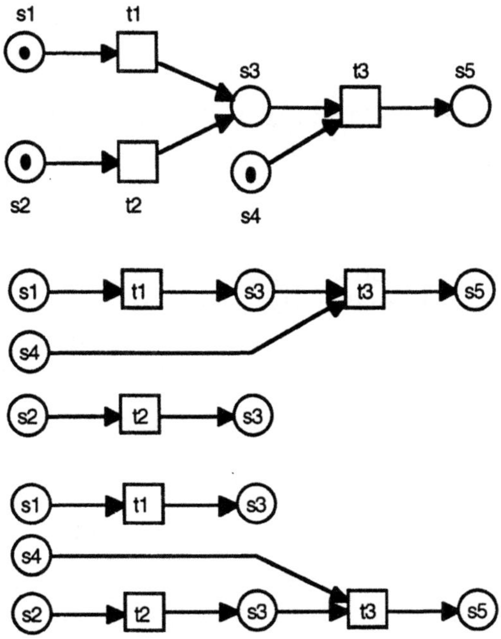

Fig. 20. A marked net and two of itsprocesses

Theorem 59. *Every weakly connected live and bounded marked net N satisfying the conditions of* **(N1)** *is strongly connected.*

Proof. By Proposition 1 it suffices to prove that, for each arc (x, y), there is a directed path leading from y to x.

Let t be the transition in $\{x, y\}$ and let s be the place in $\{x, y\}$. By liveness, there is an occurrence sequence with infinitely many occurrences of t. Hence, some process has infinitely many events labeled by t. Each of these events has a condition labeled by s in its pre-set or post-set. Let B be the infinite set of these conditions. By boundedness, the process net has no infinite cut. Hence there is a path connecting two distinct conditions of B. Since conditions are not branched, this path contains an event labeled by t.

By the definition of process nets, the sequence of labels of the elements of a path of a process net is a path of N. So some path of N leads from s via t back to s. In particular, some path leads from y to x. □

6.5 Expressive Power of Process Nets

The previous results have proven a close relationship between the sequential semantics of a marked net given by its occurrence sequences and the concurrent semantics given by process nets. Roughly speaking, an occurrence sequence determines a totally ordered set of transition occurrences. This order respects causal dependencies of transitions occurrences, but also concurrent transition occurrences are ordered.

In contrast, process nets reflect exactly the causal dependencies. The example of Figure 20 shows that causal dependencies of transition occurrences cannot be obtained from corresponding occurrence sequences. The figure shows a marked net and two of its process nets (labels are again depicted by inscriptions in conditions and events). In each process net, every transition of the marked net occurs exactly once. The first process net represents a concurrent run where the transition **t3** causally depends on the previous occurrence of **t1**. In the second process net, transition **t3** causally depends on the previous occurrence of **t2**. The first process net corresponds to the occurrence sequences

t1 t2 t3 , **t1 t3 t2** and **t2 t1 t3**.

The second process net corresponds to the occurrence sequences

t1 t2 t3 , **t2 t1 t3** and **t2 t3 t1**.

The occurrence sequences **t1t3t2** and **t2t3t1** belong to only one causal net each. The other two occurrence sequences, however, do not carry sufficient information to decide which concurrent run they respectively represent.

The reader is recommended to prove that the first two process nets of Figure 19 also have a common occurrence sequence.

7 Arc-Weights, Capacities, and Inhibitor Arcs

7.1 Arc-Weights

A *weighted arc* specifies that more than one token is removed from a place or added to a place by a single occurrence of a transition. As an example, consider the vending machine of the introduction, shown in Figure 5. If the arcs

(**item storage**, **dispense item**) and (**dispense item**, **request for refill**)

are both weighted by 2 then each occurrence of **dispense item** removes two tokens from **item storage** and adds two tokens to **request for refill**. Thus, this modified vending machine sells items in pairs. Similarly, if the arcs

(**refill**, **item storage**) and (**item storage**, **dispense item**)

are both weighted by 4, four items are sold by each occurrence of **dispense item**. In this case, a token on **request for refill** indicates that the storage is empty, whereas no token means that the storage carries four items.

The formal definition of a net N with arc-weights replaces the flow relation F_N by a weight function shaped

$$w_N: (S_N \times T_N) \cup (T_N \times S_N) \rightarrow \mathbb{N}.$$

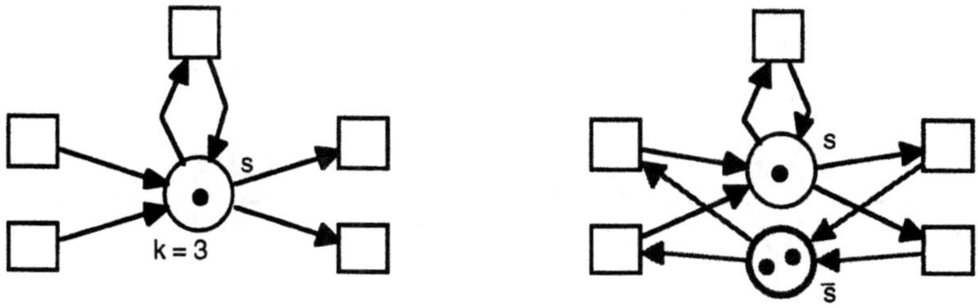

Fig. 22. A place with a weak capacity and a corresponding complement place

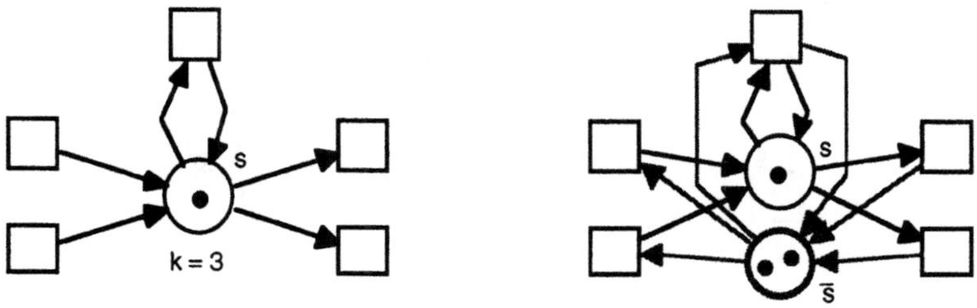

Fig. 23. A place with a strong capacity and a corresponding complement place

7.2 Capacities

A *capacity restriction* (*capacity*, for short) specifies that the token count of a place should never exceed a previously fixed value. Consider again the example of Figure 5. If the place **item storage** has a specified capacity 4, then no reachable marking should assign more than four tokens to the place. Since the place is 4-bounded, this capacity is respected.

Now assume the same marked net, but without the place **request for refill** and adjacent arcs. In this net, the transition **refill** always can occur, whence the place **item storage** is unbounded. If this place is given a capacity 4 then the behavior of the marked net with capacity restriction equals the behavior of the original marked net, i.e. both marked nets have the same occurrence sequences.

Formally, capacities are represented by assigning each place an element of $\{\omega, 1, 2, 3, \ldots\}$. The value ω means that there is no capacity restriction for this place. The initial marking has to respect all capacities. A modified occurrence rule prevents each transition occurrence that would violate any capacity restriction. Instead of presenting this modified model we will show how capacity restrictions are equivalently replaced by *complement places.* We distinguish *weak capacities* and *strong capacities.*

Figure 22 shows how to implement a weak capacity restriction of a place by means of a complement place. Consider a place s with a weak capacity k, $k \neq \omega$. The complement of s is a new place $\overline{s}$ (the shaded place in the figure) satisfying

$$^\bullet\overline{s} = s^\bullet \setminus {^\bullet s} \text{ and } \overline{s}^\bullet = {^\bullet s} \setminus s^\bullet.$$

This complement place $\overline{s}$ is initially marked by $k - m_0(s)$, where $m_0(s)$ is the

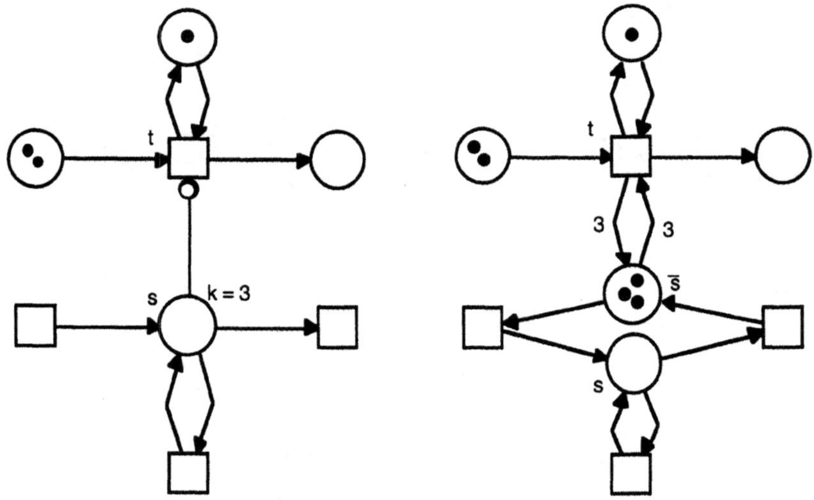

Fig. 24. A net with an inhibitor arc

initial marking of s. By construction, the total token count of s and $\overline{s}$ is equal to k for each reachable marking. Since markings are nonnegative, this construction guarantees that s is bounded by k.

A *strong* capacity differs from a weak capacity only in the presence of *self-loops*, i.e. of mutually connected elements. Given a strong capacity k, $k \neq \omega$ of a place s, not only the token count on s never exceeds k but moreover a transition t in ${}^{\bullet}s$ is only enabled at a marking m if $m(s) < k$. Figure 23 shows how to implement a strong capacity restriction of a place by means of a complement place. The complement of a place s with strong capacity k, $k \neq \omega$ is a new place $\overline{s}$ satisfying

$$^{\bullet}\overline{s} = s^{\bullet} \ \text{ and } \ \overline{s}^{\bullet} = {}^{\bullet}s.$$

Again, this place is initially marked by the value $k - m_0(s)$, and by construction s is k-bounded. Moreover, for each reachable marking m, $m(s) = k$ implies $m(\overline{s}) = 0$. Since a transition t in ${}^{\bullet}s$ is in $\overline{s}^{\bullet}$, t can only occur when s carries less than k tokens.

Strong capacities establish a close relationship between marked nets and elementary net systems: Each elementary net system can be considered a marked net with strong capacity 1 for all places, and vice versa.

7.3 Inhibitor Arcs

An *inhibitor arc* connecting a place and a transition specifies that the transition should only occur at a marking if the place is unmarked. Formally, inhibitor arcs of a net N constitute a subset of $S_N \times T_N$. The enabling condition of the occurrence rule is extended by the above constraint.

Graphically, an inhibitor arc is represented by a special arrow head. Figure 24 shows on the left hand side a marked net with inhibitor arc (**s**, **t**).

It is well known that inhibitor arcs can properly increase the expressive power of a net. The argument is that *two-counter machines* can be modeled by deterministic finite nets with (at least two) unbounded places and adjacent inhibitor arcs. Since two-counter machines can simulate Turing machines, it is not decidable if they ever halt. The two-counter machine halts if and only if the corresponding marked net is not deadlock-free, because this model has only one maximal occurrence sequence. However, deadlock-freedom for marked nets without inhibitor arcs easily reduces to the reachability problem [EsNi94], which is decidable [Mayr84].

Now consider inhibitor arcs connected to bounded places. Each b-bounded place s can be assigned the weak capacity b without changing its behavior. As shown above, a capacity restriction can equivalently be replaced by a complement place $\overline{s}$. Since, for each reachable marking, s is unmarked if and only if $\overline{s}$ carries b tokens, the inhibitor arc is equivalently replaced by conventional arcs from and to $\overline{s}$, weighted by b. Figure 24 shows an example.

References

[BeDe90] E. Best and J. Desel. Partial order behaviour and structure of Petri nets. Formal Aspects of Computing 2(2):123–138, 1990.

[BeFe88] E. Best and C. Fernandez C. Nonsequential Processes – a Petri Net View. EATCS Monographs on Theoretical Computer Science Vol. 13, Springer-Verlag, 1988.

[Brau80] W. Brauer (Ed.) Net Theory and Applications. Vol. 84 of Lecture Notes in Computer Science, Springer-Verlag, 1980.

[BRR87] W. Brauer, W. Reisig and G. Rozenberg (eds.). Petri nets: Central Models and Their Properties, Advances in Petri Nets 1986, part I, Vol. 254 of Lecture Notes in Computer Science, Springer-Verlag, 1987.

[DeEs95] J. Desel and J. Esparza. Free Choice Petri Nets, Vol. 40 of Cambridge Tracts in Theoretical Computer Science, Cambridge University Press, 1995.

[DeMe90] J. Desel and A. Merceron. P/T-systems as abstractions of C/E-systems. G. Rozenberg (ed.): Advances in Petri Nets 1989, Vol. 424 of Lecture Notes in Computer Science, Springer-Verlag, 1989.

[Dese98] J. Desel. Basic linear algebraic techniques for place/transition nets. Advanced Course on Petri Nets, 1996, in this volume.

[EsNi94] J. Esparza and M. Nielsen. Decidability issues for Petri nets. GI Petri Net Newsletter, 47:5–23, 1994.

[Espa98] J. Esparza. Decdability and complexity of Petri net problems – an introduction. Advanced Course on Petri Nets, 1996, in this volume.

[Golt90] U. Goltz. On condition/event representation of place/transition nets. K. Voss, H.J. Genrich and P.S. Thiagarajan (eds.): Concurrency and Nets, pp. 217–231, Springer-Verlag, 1990.

[GoRe83] U. Goltz and W. Reisig. The non-sequential behaviour of Petri nets. Information and Control, 3:125–147, 1983.

[Jant87] M. Jantzen. Complexity of place/transition nets. In [BRR87], pp. 413–435.

[JaVa80] M. Jantzen and R. Valk. Formal properties of place/transition nets. In [Brau80], pp. 165–21.

[Jens98] K. Jensen. An introduction to the practical use of coloured Petri nets. Advanced Course on Petri Nets, 1996, in this volume.

[Kosa82] S.R. Kosaraju. Decidability of reachablity in vector addition systems. 14th Annual ACM Symposium on Theory of Computing, San Francisco, pp. 267–281, 1982.

[Laut87] K. Lautenbach. Linear algebraic techniques for place/transition nets. In [BRR87], pp. 142-167.

[Mayr84] E.W. Mayr. An algorithm for the general Petri net reachability problem. SIAM Journal of Computing, 13(3):441-459, 1984.

[Mura89] T. Murata. Petri net: properties, analysis and application. Proceedings of the IEEE 77(4):541–580, 1989.

[Pete81] J. L. Peterson. Petri net Theory and the Modelling of Systems. Prentice–Hall, Englewood Cliffs, 1981.

[Reis85] W. Reisig. Petri Nets. EATCS Monographs on Theoretical Computer Science Vol. 4, Springer-Verlag, 1985.

[Reis87] W. Reisig. Place/transition systems. In [BRR87], pp. 117-141.

[Reut88] C. Reutenauer. The Mathematics of Petri Nets. Masson (1988).

[RoEn98] G. Rozenberg and J. Engelfriet. Elementary net systems. Advanced Course on Petri Nets, 1996, in this volume.

[Valm98] A, Valmari. The state explosion problem. Advanced Course on Petri Nets, 1996, in this volume.

Principles of High-Level Net Theory

Einar Smith

GMD, Sankt Augustin, Germany, esmith@gmd.de

Abstract. The paper gives an introduction to fundamentals and recent trends in the theory of high-level nets. High-level nets are first formally derived from low-level nets by means of a quotient construction. Based on a linear-algebraic representations, we develop an invariant calculus that essentially corresponds to the algebraic core of the well-known coloured nets. We demonstrate that the modelling power of high-level nets stems from the use of expressive symbolic annotation languages, where as a typical model we consider predicate-transition nets, both concrete models and net-schemes. As examples of specific high-level analysis-tools we discuss symbolic place-invariants and reachability-trees.

1 Introduction

Large scale modelling with nets would be impossible without the conceptual and notational convenience offered by high-level descriptions. The characteristic feature of high-level nets is that tokens represent individual data items (as opposed to boolean conditions in elementary nets or "dimensionless" multiplicities in place-transition nets). The use of structured tokens permits concise net representations of complex systems, ranging from distributed data bases to communication networks and flexible manufacturing systems. Throughout the years an ever increasing number of different high-level models has been proposed and put into use.

From a mathematical point of view, high-level models are essentially a concise representation of low-level nets, obtained by a suitable folding-operation. This approach has the advantage, that well-understood low-level concepts such as the semantics of non-sequential processes, but also analysis tools such as invariant-calculus can be lifted canonically from low-level to high-level models. We shall discuss the precise relationship in the first part of the paper.

However, the main interest in high-level nets for system modelling is due to the expressiveness of powerful description languages and specific analysis tools based on these. We shall develop and illustrate some of the central ideas, using *predicate-transition nets* as a paradigmatic example.

The paper is organized as follows. Section 2 contains standard mathematical prerequisites. Section 3 provides (or recalls) the basic concepts of *place-transition nets* and *elementary net systems*. In Section 4 we introduce the fundamental concepts leading from low- to high-level nets and the relationship between the levels. The key idea is folding of system representation along certain equivalence relations. In the following Section 5 we develop a linear-algebraic representation and a corresponding invariant-calculus. This leads to an approach similar to the original version of the well-known *coloured nets*. Section 6 gives a precise definition of *predicate-transition nets*, arguably *the* classical high-level model. Sections 7 and 8 are concerned with analysis methods. Among the topics covered are *invariants*, in particular *place-invariants*, and *reachability analysis*. In Section 9 we consider *abstract* and *algebraic predicate-transition nets*, where the logic symbols are not a priori bound to a fixed interpretation. Section 10 presents an overview over recent trends.

The paper is essentially self-contained and does in particular not require any formal knowledge of net theory. However for motivation purposes, some basic familiarity with modelling with nets will certainly be useful.

2 Prerequisites

We use standard notations for handling sets. In particular we write $A \subseteq B$ if A is a subset of B. The complement of B in A will be denoted by $A - B$. The cardinality of a set A will be denoted by $|A|$.

The symbol $\mathbb{N}$ denotes the set of natural numbers including 0. We use $\mathbb{Z}$ as symbol for the integers.

A *multiset* over a set A is a mapping from A to $\mathbb{N}$. The collection of all multisets over A is denoted by $\mathcal{M}(A)$.

For a binary relation R we usually write $x\,R\,y$ instead of $(x, y) \in R$.

Let A be a set, and $n, m \geqslant 1$. An *$n \times m$-matrix in A* is a matrix M with n rows and m columns, where each entry M_{ij} is an element from A. If $(A, +, \cdot)$ is a (non-necessarily commutative) ring over A, we assume the ring-operations to be continued to matrix operations as usual.

We shall often have use for expressions denoting *symbolic sums* over certain sets, formally defined as follows. For a set X the *free abelian group* over X is the pair $(\mathfrak{G}(X), +)$ where $\mathfrak{G}(X)$ the set of functions $f: X \to \mathbb{Z}$ such that $\{x \in X \mid f(x) \neq 0\}$ is finite. Addition between functions is defined componentwise as usual, as is scalar multiplication of the form $m \cdot f$ for $m \in \mathbb{Z}$. Identifying each element $x \in X$ with the characteristic function $\chi_x : x \mapsto 1$ and $\chi_x : y \mapsto 0$ for $y \neq x$, we may interpret any expression of the form $m_1 \cdot x_1 + \cdots + m_n \cdot x_n$, with $x_i \in X, m_i \in \mathbb{Z}$, as an element of the group $\mathfrak{G}(X)$.

3 Low-level net systems

This section gives a short introduction to the notions and concepts of low-level nets that will be needed in this paper. For a more detailed treatment we refer to [22,25] and the corresponding articles in this volume.

Example 3.1. As a running example we consider the notoriously known system of five philosophers. The philosophers are sitting around a table, and between any two neighbouring philosophers there is one fork. Each philosopher may assume one of two states, thinking or eating. For eating he needs both his left and the right fork. The situation for one philosopher is illustrated by the *net-system* in Fig. 1. (Formal definitions will be given below.)

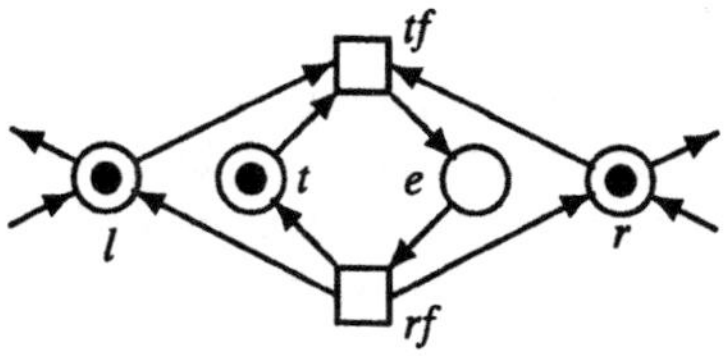

Fig. 1. Part of an elementary net model of the five-philosophers system

The philosopher is presently in the state of thinking (represented by the black token within the circle t). The tokens in the circles l and r indicate that the left and right forks are available. When the philosopher starts eating, he takes both forks, which in the figure would be represented by a removal of the tokens from t, l, r, and marking e with a token instead. When he has finished, he releases the forks (represented by the box *rf*) and returns to thinking. The additional arrows indicate that his neighbour philosophers share a common fork with him. If on of them picks up one of the shared forks, the transition from thinking to eating *tf* is disabled. If all of the possible actions and states of all philosophers are connected via the "interface-arrows" to the left of l and right of r, we arrive at a net-representation of the complete five-philosophers system, more precisely at an elementary-net representation.

3.1 Formal definitions

Definition 3.2. By a *net* we shall mean a triple $N := (P, T, F)$ satisfying the following properties:

(i) $P \cap T = \emptyset$,
(ii) $F \subseteq (P \times T) \cup (T \times P)$.

The elements of P and T are called *places* and *transitions*, respectively. The relation F is called *flow relation*.

As illustrated in Example 3.1, we follow the usual graphical conventions for representing nets: places are drawn as circles or ellipses, transitions as boxes and the flow relation F is indicated by appropriately directed arcs.

We use X to denote the *set of elements* $P \cup T$ of a net N. For an element x of X, the set ${}^\bullet x := \{y \in X | y\, F\, x\}$ is the *pre-set* of x, and $x^\bullet := \{y \in X | x\, F\, y\}$ is the *post-set*. The union of ${}^\bullet x$ and $x^\bullet$ is denoted by vic(x).

The net N is said to be *pure*, if ${}^\bullet x \cap x^\bullet = \emptyset$ for all $x \in X$.

We come to net *semantics*. For reasons of technical convenience we start with *place-transition nets*, PT-nets for short. Elementary net systems, though conceptually more fundamental, are defined as special PT-nets.

Definition 3.3. A *place-transition net* PT $= (N, W, K, M_{in})$ consists of a net N, a *weight function* $W: F \to \mathbb{N}^+$, a *partial* function K from P to $\mathbb{N}^+$, called *capacity constraint*, and a distinguished *initial marking* M_{in}, where in general a marking is a mapping from P to $\mathbb{N}$ such that $M(p) \leqslant K(p)$ whenever $K(p)$ is defined.

For technical convenience we usually assume the weight function to be trivially continued to all of $(P \times T) \cup (T \times P)$ by putting $W(x, y) = 0$ if $(x, y) \notin F$. Note that the flow-relation may be then be deducted from W by $x\ F\ y \Leftrightarrow W(x, y) \neq 0$.

A transition t is *enabled* in a marking M, in symbols $M \models t$,[1] iff for all $p \in {}^\bullet t$ we have $W(p, t) \leqslant M(p)$, and for all $p \in t^\bullet$ where K is defined, $M(p) + W(t, p) \leqslant K(p)$.

If t is enabled in M, *firing* of t yields the *follower marking* M', denoted $M[t\rangle M'$, given by $M' \colon p \mapsto M(p) - W(p, t) + W(t, p)$. If there is no need to mention the specific transition transforming M into M' we write $M \,\mathbf{r}\, M'$.

A marking M' is *reachable* from another marking M, in symbols $M' \in [M\rangle$, iff it can be reached from M by a finite sequence $M = M_0 \,\mathbf{r}\, M_1 \,\mathbf{r} \cdots \mathbf{r}\, M_n = M'$ of transition-firings.

Note that we use the letters PT as abbreviation for 'place-transition' and also as a symbol for actual PT-nets. This will usually not lead to any confusion. In fact, we implicitly adopt similar conventions also for other net classes.

We now define an elementary net system as a *strict* PT-net, i.e. a net where numeric values are limited to 0 or 1:

Definition 3.4. An *elementary net system*, EN-system or ENS for short, is a PT-net (N, W, K, M_{in}), where arc weights $W(x, y)$ are 1 for all arcs $(x, y) \in F$, the capacity $K(p)$ for each place is 1, and consequently all reachable markings are $\{0, 1\}$-valued. Due to their constant nature in EN-systems, the functions W and K are usually not mentioned. A $\{0, 1\}$-valued marking M is identified with the set $\{p \in P \mid M(p) = 1\}$, which is then again written as M.

Remark 3.5. Sometimes *isolated elements*, i.e. elements x such that $\text{vic}(x) = \emptyset$, are excluded from elementary net systems. This means in particular that every occurrence of a transition must have some visible effect.

Example 3.6. Disregarding the extra arrows connected to l and r, Fig. 1 in Example 3.1 shows an elementary net system.

3.2 Incidence matrix and invariants

In this paragraph, we assume nets to be *finite*, and the sets of places and transitions to be of the form $P = \{p_1, \ldots, p_n\}$ and $T = \{t_1, \ldots, t_m\}$ for some arbitrary but fixed order.

Definition 3.7. Let PT be a finite place-transition net as above. The *incidence matrix* of PT is the $n \times m$-matrix PT in $\mathbb{N}$ with entries

$$\mathsf{PT}_{ij} = W(t_j, p_i) - W(p_i, t_j).$$

The incidence matrix is the basis for *invariant-computations*, more precisely *place-invariants* and *transition-invariants*. In this paper we shall mainly be concerned with place-invariants, as these are by far the most relevant in theory and applications.

[1] This relation is often also written as $M[t\rangle$.

Definition 3.8. A *place-invariant* is an n-vector i in $\mathbb{N}$, written as a column-vector, such that

$$i^T \cdot \mathsf{PT} = \mathbf{0}, \tag{3.1}$$

where $\mathbf{0}$ denotes the 0-vector of length m.

The system-property characterized by such a place-invariant i is that

$$i^T \cdot M = i^T \cdot M_{\text{in}} \quad \text{for every } M \in [M_{\text{in}}\rangle, \tag{3.2}$$

where we assume markings to be written as column vectors with reference to the ordering $p_1, \ldots, p_n$ of P.

The proof of (3.2) is by induction using the following observation: Described in vector notation, the firing of a transition $t \in T$ in a marking M corresponds to the addition of the t-column from PT to M. Multiplication with i^T distributes over the sum, where $i^T \cdot t$ cancels out because of (3.1), and we are left with $i^T \cdot M$.

Place-invariants i can be used to show that a marking M is *not* reachable from the initial marking. A sufficient condition for this is that $i^T \cdot M \neq i^T \cdot M_{\text{in}}$.

A (trivial) example is the demonstration that not both of t and e in Example 3.1 can be marked at the same time. We leave this to the interested reader and return to place-invariants in more detail in the context of high-level nets below.

The notion of *transition-invariant* is defined dually to place-invariants: A transition-invariant is an m-vector $i = (i_1, \ldots, i_m)$ such that

$$\mathsf{PT} \cdot i = \mathbf{0},$$

where $\mathbf{0}$ now denotes the 0-vector of length n. The characteristic property of such a transition-invariant is that the effect of an i_k-fold occurrence of all transitions t_k for $k = 1, \ldots, m$ would reproduce the initial marking. Transition-invariants have attracted much less attention in the literature than place-invariants.

4 Quotient nets and basic high-level nets

After these preliminaries we now come to our main topic, high-level nets. The term is usually used rather informally for a whole class of nets, where the common idea is that a high-level net can be unfolded into a low-level net system, and thus conversely may be conceived as a condensed representation of the latter. This is also the point of view we shall adhere to in this paper.

In this section we show how high-level representations can in fact formally be derived from the structure and semantics of low-level nets. The central concept is *factorization* with respect to suitable *equivalence relations*, resulting in certain *net-quotients*. In the "condensed net", structure and behaviour of the original net is retained by the introduction of *individual* – instead of indistinguishable black – tokens and the distinction between different *firing-modes* for transitions. The development is a generalization of ideas from [29,28].

All definitions and constructions are formulated for place-transition nets. However, main emphasis will be put on elementary net systems *without multiplicities*. In particular, most examples will be of this type.

4.1 Quotient nets

We start with in informal example, which illustrates the main ideas.

Example 4.1. Consider the ENS defined by Fig. 2(a). The net (b) shows a quotient representation of (a). The "macro"-elements in the underlying net are sets of elements from the ENS. The black token on the place a_1 is translated into an individual token, again denoted a_1, on the macro-place $p_1 = \{a_1, a_2\}$. The macro-transitions may occur in various modes, corresponding to the original low-level transitions they contain. The specific effect of the original transitions does not appear in the figure, it will have to be represented by some additional annotation in the system description.

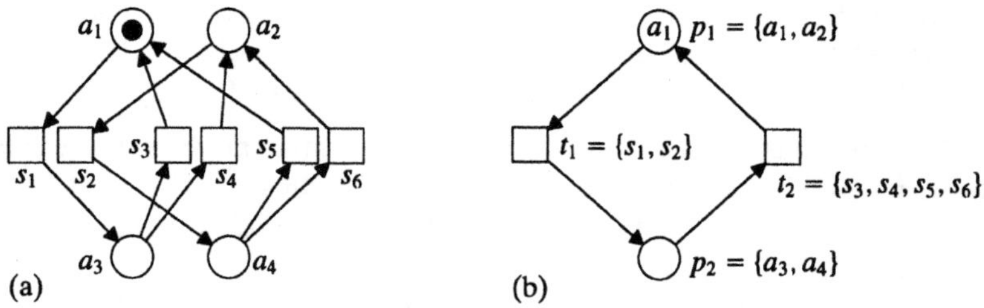

Fig. 2. An ENS (a) and a quotient representation (b).

Formalizing the idea behind Example 4.1, we shall now introduce high-level nets as quotient systems, derived from certain equivalence relations on low-level nets. Semantic notions will be represented in a condensed form as sets (or in general rather multisets) of low-level items.

The following definition collects the necessary notions concerning equivalences and quotient constructions. A more extensive treatment of quotient nets can be found in [29,27]. For the algebraic foundations of quotient constructions in general we refer to [7].

Definition 4.2. Let ρ be a fixed equivalence on the domain $X = P \cup T$ of a net N.

(i) For $x \in X$ let $\bar{x}$ denote the equivalence class $\{ y \in X \mid x \,\rho\, y \}$ of x.
(ii) For $Y \subseteq X$ let $\bar{Y} := \{ \bar{x} \mid x \in Y \}$.
(iii) ρ is *sort-respecting*, if $\rho \cap (P \times T) = \emptyset$.
(iv) The relation $\bar{F}$ on $\bar{X}$ is defined by

$$\bar{x} \,\bar{F}\, \bar{y} \;:\Leftrightarrow\; \exists x' \in \bar{x} \exists y' \in \bar{y} : x \,F\, y.$$

(v) We denote the triple $(\bar{P}, \bar{T}, \bar{F})$ by $\bar{N}$.

The definition of $\bar{F}$ ensures that arcs in $\bar{N}$ are inherited from N, no new arcs are introduced.

Clearly, *sort-respecting* means that each equivalence class contains *either* places or transitions, but not both. We are particularly interested in sort-respecting equivalences because of the following

Proposition 4.3. *Let ρ be an equivalence in a net N. The triple $\bar{N}$ is a net iff ρ is sort-respecting.*

Proof. It is sufficient to observe that $\bar{P} \cap \bar{T} = \emptyset \Leftrightarrow \rho \cap (P \times T) = \emptyset$. □

Definition 4.4. If ρ is sort-respecting, we call $\bar{N}$ a *quotient net*, more precisely the *quotient of N with respect to ρ.*

We shall now show how also the concept of marking and transition-occurrence may be suitably folded along with the net. We give the definition for general PT-nets, and then indicate more convenient representations for EN-systems.

Let N be a net, ρ a sort-respecting equivalence in N, and let $\bar{N}$ be the quotient of N with respect to ρ.

We first indicate how *markings* of N may be folded canonically to markings of $\bar{N}$. Recall that a marking of N is a map $M: P \to \mathbb{N}$, i.e. a multiset over P, symbolically $M \in \mathcal{M}(P)$. For such a marking we denote by $\bar{M}$ the map with domain $\bar{P}$ and as values multisets $\bar{M}(\bar{p}) \in \mathcal{M}(\bar{p})$, given by

$$\bar{M}(\bar{p})(q) = M(q). \tag{4.3}$$

By the same token, the capacity constraint K is translated into a mapping $\bar{K}$, the only difference being that K may be undefined for some values, which then also carries over to $\bar{K}$.

For the arc-weight function W we define $\bar{W}$ as the mapping with domain $\bar{F}$ and values $\bar{W}(\bar{x}, \bar{y}) \in \mathcal{M}(\bar{x} \times \bar{y})$ such that

$$\bar{W}(\bar{x}, \bar{y})(u, v) = W(u, v). \tag{4.4}$$

Transitions in $\bar{N}$ are *sets of transitions* from N. For a T-element $\bar{t}$ and $s \in \bar{t}$, we say that in a marking $\bar{M}$, $\bar{t}$ is *enabled in mode s*, if s is enabled in M, symbolically:

$$\bar{M} \models \bar{t}(s) :\Leftrightarrow M \models s. \tag{4.5}$$

In that case an *occurrence of $\bar{t}$ in mode s* yields the follower marking $\bar{B}$, where $B = A[s]$. Hence all behavioural notions are carried over canonically without the need for any specific new definition.

Putting all of the above together we arrive at the following definition of a quotient system:

Definition 4.5. Let $\mathrm{PT} = (N, W, K, M_{\mathrm{in}})$ be a PT-net, and ρ a sort-respecting equivalence in N. Then the system $\overline{\mathrm{PT}} := (\bar{N}, \bar{W}, \bar{K}, \bar{M}_{\mathrm{in}})$ is called the *quotient of* PT *with respect to ρ*. The *behaviour* of $\overline{\mathrm{PT}}$ is defined via the behaviour in PT according to (4.5).

Remark 4.6. Some of the behavioural notions may be simplified if we start out with an elementary net-system $\mathrm{ENS} = (N, M_{\mathrm{in}})$.

(i) Markings of ENS can be conceived as sets. Correspondingly, a marking $\bar{M}$ of the quotient can be conceived as a map with domain $\bar{P}$ and values $M(\bar{x}) \subseteq P$, such that

$$\bar{M}(\bar{x}) = \bar{x} \cap M. \tag{4.6}$$

(ii) The capacity constraint $\bar{K}$ does not have to be mentioned explicitly, since it always assumes the constant value 1 anyway; we usually write the quotient-representation in the form

$$\overline{\text{ENS}} = (\bar{N}, \bar{W}, \bar{M}_{\text{in}}). \tag{4.7}$$

(iii) The weight function $\bar{W}$ is related to the F-relation in N by

$$\bar{W}(\bar{x}, \bar{y})(u, v) = \begin{cases} 1, & \text{if } u \; F \; v \\ 0, & \text{otherwise.} \end{cases}$$

4.2 Basic high-level nets

From a mathematical point of view, Definition 4.5 already contains the quintessence of high-level nets. However, it is obviously not very convenient to work directly with net-quotients. It is usually sufficient to know that a high-level net *may be conceived* as such. To this end we introduce a notation of high-level nets, where all traces of quotient-origin have been discarded. We then show that it nonetheless does represent the same concept as the rigorous construction above.

The central idea is:

(i) places may be marked with *individual tokens*,

(ii) transitions fire in different *modes*, removing tokens from some places and adding tokens to some, where the only a-priori restriction is a *locality constraint*, namely that a transition in any mode may only remove tokens from its own pre-places and likewise add tokens only to its own post-places.

We formalize this as follows:

Definition 4.7. A *basic high-level system*, BHL-net or BHL for short, is a tuple

$$\text{BHL} = (N, D, \Phi, W, K, M_{\text{in}}),$$

where $N = (P, T, F)$ is a net and the remaining components are mappings explained in the following.

(i) D associates to each $p \in P$ a non-empty *domain* $D(p)$. i.e. a set of *individual tokens*.

(ii) A *marking* of BHL is a map M with domain P and values $M(p) \in \mathcal{M}(D(p))$.

(iii) The *capacity constraint* K is formally similar to a marking, except that $K(p)(a)$ may be undefined for some a in some $D(p)$. For all legal markings we require $M(p)(a) \leqslant K(p)(a)$, whenever the latter is defined.

(iv) M_{in} denotes a distinguished legal marking.

We now come to the meaning of Φ and W.

(v) Φ is a map which associates to each transition t a non-empty collection of *modes* $\Phi(t)$,

(vi) W is a map with domain F and values $W(p, t)$ in $\mathcal{M}\big(D(p) \times \Phi(t)\big)$ resp. $W(t, p)$ in $\mathcal{M}\big(\Phi(t) \times D(p)\big)$.

The *behaviour* of BHL is now defined as follows:

(vii) In a marking M a transition t is *enabled in a mode* $m \in \Phi(t)$ iff

$$\begin{aligned} &\forall p \in {}^\bullet t\ \forall a \in D(p): W(p,t)(a,m) \leqslant M(p)(a) \\ \text{and} \\ &\forall p \in t^\bullet\ \forall a \in D(p): M(p)(a) + W(t,p)(m,a) \leqslant K(p)(a), \end{aligned} \tag{4.8}$$

where again the capacity constraint is only relevant if $K(p)(a)$ is defined.

(viii) If the enabling condition is satisfied, a *firing* of t *in mode* m transforms the marking M into the follower marking M', where $M'(p) = M(p)$ for $p \notin \text{vic}(t)$ and for $p \in \text{vic}(t)$ and $a \in D(p)$ we have:

$$M'(p)(a) = M(p)(a) - W(p,t)(a,m) + W(t,p)(m,a). \tag{4.9}$$

Symbolically we use the notations $M \models t(m)$ and $M[t(m)\rangle M'$ to denote enabling, and transformation of markings through firing, respectively. Given these basic concepts, the definition of derived notions such as the *reachability class* $[M_{\text{in}}\rangle$ etc. is now straightforward.

The general definition may be simplified if item-multiplicities are excluded:

Definition 4.8. A basic high-level net BHL $= (N, D, \Phi, W, K, M_{\text{in}})$ is called *strict*, if all mappings involved are $\{0,1\}$-valued, i.e.

(i) for all markings M the values $M(p)(a)$ are in $\{0,1\}$, hence the $M(p)$ may be conceived as subsets of $D(p)$,

(ii) for every $p \in P$ and $a \in D(p)$, always $K(p)(a) = 1$, hence there is no need to mention the capacity constraint explicitly,

(iii) all $W(x,y)(u,v)$ belong to $\{0,1\}$, and may therefore be identified with the set $\{(u,v) \mid W(x,y)(u,v) = 1\}$.

We leave it to the reader to formulate simplifications for the general enabling rule and occurrence relation.

Remark 4.9. In Remark 3.5 we mentioned that in elementary net systems it is commonly required that each transition-occurrence must have some visible effect. The corresponding property for modes in $\Phi(t)$ is that for all $m \in \Phi(t)$ there is some $p \in {}^\bullet t$ or $p \in t^\bullet$, and some a in $D(p)$ such that $W(p,t)(a,m) \neq 0$ (resp. $W(t,p)(m,a) \neq 0$).

Example 4.10. Recall the elementary net model of the five-philosophers system in Example 3.1. We represent it as a (strict) basic high-level net BHL_{5Phil} as follows:

The underlying net consists of three places *think*, *eat*, and *avail*, denoting the currently thinking and eating philosophers, and available forks, respectively. We introduce two transitions *tf* "take forks" and *rf* "release forks" to describe the possible system dynamics when a philosopher decides to change his state between thinking and eating. The structure of the net (and the initial marking) is shown in Fig. 3.

We put $D(think) = D(eat) = \{ph_0, \ldots, ph_4\}$ and $D(avail) = \{fk_0, \ldots, fk_4\}$.

A marking of a place is a mapping which associates philosophers or forks to the respective places. Since we are dealing with strict nets, markings may be conceived as

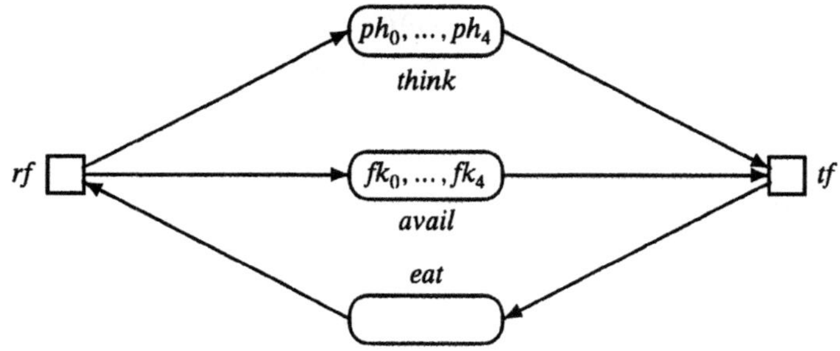

Fig. 3. The net N underlying a strict basic high-level model BHL_{5Phil} of the five-philosophers system, and the initial marking M_{in}, shown in set-representation.

subsets of the place-domains. The initial marking, where all philosophers are thinking and thus all forks are available, is given by $M_{\text{in}}(\mathit{think}) = \{ph_0, \ldots, ph_4\}$, $M_{\text{in}}(\mathit{eat}) = \emptyset$ and $M_{\text{in}}(\mathit{avail}) = \{fk_0, \ldots, fk_4\}$.

Both of the transitions may fire in five modes, one for each philosopher; we put $\Phi(\mathit{tf}) = \Phi(\mathit{rf}) = \{m_0, \ldots, m_4\}$. A priori, there is no relationship between names of individual tokens and names of transition-modes. However, in concrete models it is often convenient to indicate intended correspondence by suitable names for transition-modes. In the present example, a natural choice of mode-sets is $\Phi(\mathit{tf}) = \Phi(\mathit{rf}) = \mathit{Phil}$ instead of the anonymous m_i.

We come to the *weight function* W, which describes the effect of firing-modes. In mode ph_i, the transitions represent the change of state for philosopher i. We thus define $W(\mathit{think}, \mathit{tf})(ph_i ph_j) = 1$ if $i = j$, and 0 otherwise. Similarly, we define $W(\mathit{avail}, \mathit{tf})(fk_i, ph_j) = 1$ if $i = j$ or $i = j \oplus 1$, and 0 otherwise. (The operator $\oplus$ denotes addition modulo 5.) We leave the definition of the remaining W-values to the reader. Since W is $\{0, 1\}$-valued, it may be identified with the values mapped to 1, and we may for instance write $W(\mathit{think}, \mathit{tf}) = \{(ph_0, ph_0), \ldots, (ph_4, ph_4)\}$.

4.3 From quotients to BHL-nets

We shall show that basic high-level nets and quotient systems are essentially the same; the difference lies only in the *description* of tokens and firing-modes. To this end, we show that (1) quotients may be conceived as BHL-nets, and conversely, that (2) BHLs may be obtained as quotients from low-level nets, such that (3) both operations are inverse to each other.

Consider first a quotient $\overline{\text{PT}} = (\bar{N}, \bar{K}, \bar{W}, \bar{M}_{\text{in}})$, derived from a place-transition net $\text{PT} = (N, W, K, M_{\text{in}})$ via a sort-respecting equivalence. We wish to translate it into a BHL-net. But this is straightforward. We *forget* that places and transitions in quotient nets are *sets* containing the essential information about the systems's behaviour; in the BHL that information is then instead made explicit in the mappings D and Φ:

For $\bar{p} \in \bar{P}$ and $\bar{t} \in \bar{T}$ we set

$$D(\bar{p}) := \bar{p} \quad \text{and} \quad \Phi(t) := \bar{t}. \tag{4.10}$$

It is now easily verified that markings $\bar{M}$, the capacity constraint $\bar{K}$ as well as the weight-function $\bar{W}$ in the quotient $\overline{\text{PT}}$, defined according to (4.3) and (4.4) in Section 4.1, also satisfy the specifications for BHLs according to Definition 4.7 above. We show the claim for markings and leave the rest to the reader.

Now, a marking $\bar{M}$ of the quotient $\overline{\text{PT}}$ maps a $\bar{p}$ to a multiset over $\bar{p}$. On the other hand, in the BHL-interpretation a marking is a map of the place $\bar{p}$ to a multiset over $D(\bar{p})$. But, by definition, $D(p) = \bar{p}$, hence both characterizations coincide.

Similarly, it can easily be verified that transition-enabling (4.8) and -occurrence (4.9) according to the BHL-semantics coincide with the quotient semantics according to (4.5).

We summarize these observations in the following

Definition 4.11. Let $\overline{\text{PT}} = (\bar{N}, \bar{K}, \bar{W}, \bar{M}_{\text{in}})$ be a quotient system. The basic high level net $(\bar{N}, D, \Phi, \bar{W}, \bar{K}, \bar{M}_{\text{in}})$, where D and Φ are defined according to (4.10), is called the BHL *associated with* $\overline{\text{PT}}$.

For the special case of elementary net-systems, the BHL resulting from the quotient is obviously strict in the sense of Definition 4.8.

4.4 Unfolding BHL- to PT-nets

We turn to the converse and show that *any BHL can be conceived as a quotient.* To this end, we first canonically unfold a given BHL- into a PT-net. We then show that the original BHL may be regained from the unfolded net by a suitable factorization. Moreover, unfolding and factorization are inverse to each other, such that both transformations induce a correspondence between system behaviour.

Let BHL $= (N, D, \Phi, W, K, M_{\text{in}})$ be a basic high-level net. We wish to transform it into a PT-net with corresponding behaviour.

We start by unfolding the net $N = (P, T, F)$ to a net $\tilde{N} = (\tilde{P}, \tilde{T}, \tilde{F})$ as follows. First put

$$\tilde{P} := \{(p, a) \mid p \in P, a \in D(p)\}, \quad \tilde{T} := \{(t, m) \mid t \in T, m \in \Phi(t)\}. \tag{4.11}$$

$\tilde{P}$ is essentially the disjoint union of the domains $D(p)$. If these sets happen to be disjoint already, as for instance if they are derived directly from a quotient, there is no need for the extra tag p in (p, a), the a alone would then be sufficient.

$\tilde{T}$ consists of all firing-modes of BHL together with the transitions they belong to. Here we always have to be careful since two different modes of one transition can have the same effect, or conversely two transitions may share a common mode.

The flow-relation $\tilde{F}$ between elements $\tilde{p} := (p, a) \in \tilde{P}$ and $\tilde{t} := (t, m) \in \tilde{T}$ is then given by

$$\tilde{p}\,\tilde{F}\,\tilde{t} :\Leftrightarrow p\; F\; t \wedge W(p, t)(a, m) \neq 0,$$

and (4.12)

$$\tilde{t}\,\tilde{F}\,\tilde{p} :\Leftrightarrow t\; F\; p \wedge W(t, p)(m, a) \neq 0.$$

This takes care of the underlying net $\tilde{N}$.

We come to the transformation of markings, capacity constraint and the weight-function. A marking M of BHL is translated into the marking $\tilde{M}$ of $\tilde{N}$, such that

$$\tilde{M}(\tilde{p}) := M(p)(a), \qquad (4.13)$$

and similarly for the capacity constraint K. The weight function of $\tilde{N}$ is given by

$$\tilde{W}(\tilde{p},\tilde{t}) := W(p,t)(a,m) \quad \text{and} \quad \tilde{W}(\tilde{t},\tilde{p}) := W(t,p)(m,a). \qquad (4.14)$$

It is again straightforward to verify that transition-enabling and -occurrence are conserved in this transformation, i.e. we have:

Proposition 4.12. *Let m be a mode of $t \in T$, and let M be a marking of* BHL. *Then with $\tilde{t} = (t, m)$:*

$$M \models t(m) \Leftrightarrow \tilde{M} \models \tilde{t}$$

and

$$M[t(m)\rangle M' \Leftrightarrow \tilde{M}[\tilde{t}\rangle \tilde{M}'.$$

We summarize the construction in the following

Definition 4.13. The *unfolded system associated* with a basic high level net BHL is the PT-net $\widetilde{\text{BHL}} = (\tilde{P}, \tilde{T}, \tilde{F}, \tilde{W}, \tilde{K}, \tilde{M}_{\text{in}})$ as defined above.

It is immediately verified that a *strict* BHL gives rise to an *elementary net system.*

4.5 From BHL- to PT-nets and back

We finally show that folding of PT- into BHL-nets, and the unfolding of BHL into PT-nets are inverse to each other. In particular no behaviour-information is lost in either. More precisely we show that the transformation sequences

$$\text{(i)} \quad \text{PT} \rightarrow \overline{\text{PT}} \rightarrow \widetilde{\overline{\text{PT}}} \qquad \text{and} \qquad \text{(ii)} \quad \text{BHL} \rightarrow \widetilde{\text{BHL}} \rightarrow \overline{\widetilde{\text{BHL}}}$$

according to Definitions 4.5, 4.11, and 4.13 return the original PT- resp. BHL-net, when the quotient in (ii) is taken with respect to a canonic equivalence relation on $\widetilde{\text{BHL}}$.

Proposition 4.14.

(i) *Let $\overline{\text{PT}}$ be a quotient of a PT-net* PT. *Let $\widetilde{\overline{\text{PT}}}$ be the unfolding of $\overline{\text{PT}}$ according to Definition 4.13. Then, except for names of elements,* PT *and $\widetilde{\overline{\text{PT}}}$ are identical. In particular, transition-enabling and -occurrence in both systems coincide.*

(ii) *Let $\widetilde{\text{BHL}}$ be the PT-net obtained from a basic high-level net* BHL *according to Definition 4.13. Let $\overline{\widetilde{\text{BHL}}}$ be the quotient of $\widetilde{\text{BHL}}$ with respect to the equivalence ρ given by*

$$(x, y)\, \rho\, (x', y') :\Leftrightarrow x = x'.$$

Then the systems BHL *and $\overline{\widetilde{\text{BHL}}}$ are identical, except for names of elements. In particular, transition-enabling and -occurrence coincide.*

Proof. (i). It is easily checked that the map $\varphi: x \mapsto (\bar{x}, x)$ is a bijection between the elements of PT and $\widetilde{\overline{\mathrm{PT}}}$, such that moreover $x\ F\ z \Leftrightarrow \varphi(x)\,\bar{\bar{\tilde{F}}}\ \varphi(z)$. Hence the structure of the underlying nets coincide.

It remains to show that markings, capacity constraint and weight-function, as well as transition-enabling and -occurrence carry over. We compute markings and leave the rest to the reader:

For a place p we get, evaluating according to (4.13) and (4.3) in that order:

$$\bar{\bar{\tilde{M}}}(\varphi(p)) = \bar{\bar{\tilde{M}}}((\bar{p}, p)) = \bar{M}(\bar{p})(p) = M(p).$$

(ii). For every $x \in X$ chose y_x in $D(x)$ resp. $\Phi(x)$. (This is possible, because these sets are non-empty by definition.) We then put $\psi(x) := \overline{(x, y_x)}$. It is easily verified that ψ defines a bijection between the elements in BHL and $\widetilde{\overline{\mathrm{BHL}}}$, such that also $x\ F\ z \Leftrightarrow \psi(x)\ \bar{\bar{\tilde{F}}}\ \psi(z)$. This shows that the structure of the underlying nets coincide.

For $p \in P$, we have the bijection $\psi_p: a \mapsto (p, a)$ between the domains $D(p)$ and $D(\psi(p))$. For $t \in T$, the bijection ψ_t is defined analogously. We have to show that markings, capacity constraint and weight function, as well as transition-enabling and -occurrence are respected. This is again routine, as an example we establish the relationship for markings:

For $p \in P$ and $a \in D(p)$ we get, using (4.3) and (4.13):

$$\bar{\bar{\tilde{M}}}(\psi(p))(\psi_p(a)) = \tilde{M}(\psi_p(a)) = \tilde{M}((p, a)) = M(p)(a). \qquad \square$$

4.6 Lifting of low-level semantics: an example

Interdefinability between low- and high-level nets in particular also implies that behavioural properties carry over. We give an example which illustrates how the notion of *non-sequential process* may be lifted. Readers not acquainted with processes may skip the example, processes will not be mentioned any further in the sequel.

Example 4.15. We return to the philosopher system BHL_{5Phil} from Example 4.10. Fig. 4 shows a possible description of the process in which philosophers 0 and 2 start to eat concurrently, and once both have finished, philosopher 1 takes his turn.

What makes this description high-level specific is that each process-element carries *two* labels, the outer (e.g. *avail* or *tf*) indicates which system element is represented, and the inner (e.g. ph_0 or fk_1), which individual token (or transition-mode) is concerned.

Remark 4.16. Example 4.15 also illustrates that the concept of *concurrent behaviour* carries over to high-level nets. It should be noted, however, that concurrency in BHLs must be considered as a relation between transitions *plus* specific firing-modes, not between transitions per se.

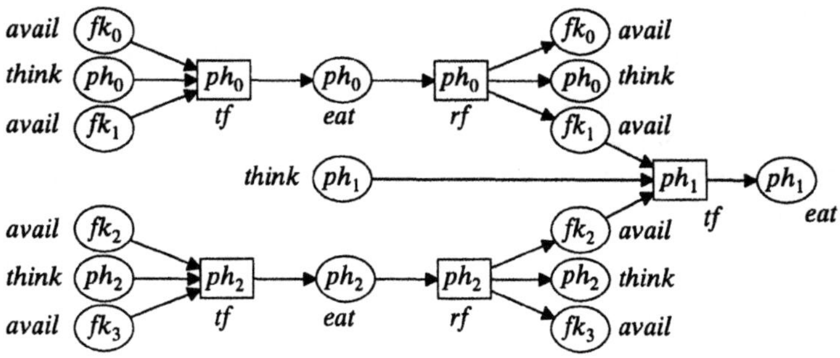

Fig. 4. A process of BHL_{5Phil}

5 Matrix-labelled nets

Until now we have not made any assumption on cardinality, all definitions so far apply to finite as well as to infinite nets. In this section we shall specifically consider BHLs, where the underlying net and the sets $D(p)$ and $\Phi(t)$ are all *finite*. This makes it possible to represent systems and their behaviour in form of finite matrices and corresponding linear-algebraic operations. In particular, it permits a precise invariant-calculus. The model we arrive at is essentially the original version of so-called *coloured Petri nets*, proposed by K. JENSEN around 1980 [13]. In order to avoid confusion with later definitions of coloured nets, we shall, however, prefer the term *matrix-labelled nets* from [19]. This section may be read independently from the technical development in Section 4.

We start with an informal example, which illustrates the main ideas:

Example 5.1. Recall the five-philosophers system BHL_{5Phil} from Example 4.10. All sets involved are finite, the underlying net as well as token-domains and mode-sets. In a matrix-labelled net we write place-markings $M(p)$ as (column)-vectors with reference to an arbitrary but fixed order on the domains $D(p)$. For instance, the marking $M(\mathit{think}) = \{ph_1, \cdots, ph_4\}$ is naturally represented by its characteristic $|D(p)|$-vector in $\{0, 1\}$, namely $(0, 1, 1, 1, 1)^T$.

Transition-firing may then be represented by vector addition. To illustrate, assume the initial situation M_{in} where all philosophers are thinking. Firing *tf* for ph_0 transfers ph_0 from *think* to *eat*, coincidently removing fk_0 and fk_1 from *avail*. In vector notation this corresponds to a

subtraction of $(1, 0, 0, 0, 0)^T$ from $M_{\text{in}}(\mathit{think}) = (1, 1, 1, 1, 1)^T$ yielding $(0, 1, 1, 1, 1)^T$,
subtraction of $(1, 1, 0, 0, 0)^T$ from $M_{\text{in}}(\mathit{avail}) = (1, 1, 1, 1, 1)^T$ yielding $(0, 0, 1, 1, 1)^T$,
and the addition of $(1, 0, 0, 0, 0)^T$ to $M_{\text{in}}(\mathit{eat})$ yielding $(1, 0, 0, 0, 0)^T$.

Generally speaking, for every arc (x, y) with $\{x, y\} = \{p, t\}$ between a place p and a transition t in a matrix-labelled net there is one such $|D(p)|$-vector $\mathbf{v}_m$ *for each* $m \in \Phi(t)$. The $\mathbf{v}_m$, $m \in \Phi(t)$ belonging to the same arc (x, y) may then be conceived as *column* vectors in a $|D(p)| \times |\Phi(t)|$-matrix $W(x, y)$. Firing t in mode m then amounts to the subtraction (resp. addition) of the m-column from each $W(p,t)$ (resp. $W(t,p)$) belonging to arcs surrounding t.

In the general definition, we consider also the possibility of *multiple* occurrences of identical tokens. To illustrate, assume for the moment that philosopher ph_0 has *two* right forks fk_1 and fk'_1, indistinguishable with respect to system behaviour. In vector notation the initial marking of the place *avail* may then be described by the vector $(1,2,1,1,1)$. Moreover, assume that ph_0 always uses both fk_1 and fk'_1 when eating. In matrix notation this may be represented by replacing the ph_0-column, i.e. column 1, by $(1,2,0,0,0)$ in the matrix $W(avail, tf)$.

5.1 Formal definitions

Note that the following definition is in accordance with the general Definition 4.7, it merely reformulates some basic concepts in terms of matrix language.

Definition 5.2. A *matrix-labelled net* $\mathrm{ML} = (N, D, \Phi, W, K, M_{\mathrm{in}})$ consists of a net N, where we assume the sets of places and transitions are finite of the form $P = \{p_1, \ldots, p_n\}$ and $T = \{t_1, \ldots, t_m\}$. The other components are explained in the following.

(i) The *domain-function* D is a mapping from P into finite non-empty sets, similarly the *mode-function* Φ is a map from T to non-empty sets. We assume each set $D(p)$ and $\Phi(t)$ to be enumerated in an arbitrary but fixed order.

(ii) W is a mapping with domain $(P \times T) \cup (T \times P)$, such that if $\{x, y\} = \{p, t\}$, then $W(x, y)$ is a $|D(p)| \times |\Phi(t)|$-matrix in $\mathbb{N}$. Moreover, for $(x, y) \notin F$ we let $W(x, y)$ be the 0-matrix of the appropriate dimension.

(iii) A *marking* of ML is a mapping M with domain P, such that $M(p)$ is a $|D(p)|$-vector in $\mathbb{N}$, written as a column-vector. M_{in} is the *initial marking*.

(iv) The capacity constraint K is again formally like a marking, except that it may be a partial function, and when defined always has values $\geqslant 1$.

(v) A transition t is *enabled in mode* $m \in \Phi(t)$ in a marking M iff for each $p \in {}^\bullet t$ and for each individual token $a_p \in D(p)$ the a_p-entry in $M(p)$ is $\geqslant$ the (a_p, m)-entry in $W(p, t)$, and whenever K is defined, the a_p-entry in $M(p)$ *plus* the (a_p, m)-entry in $W(t, p)$ does not exceed the a_p-entry in $K(p)$.

(vi) In that case, a *firing of t in mode m* leads to the follower marking M' obtained from M by subtracting the m-column of $W(p,t)$ from $M(p)$ for each $p \in {}^\bullet t$, and adding the corresponding m-column of $W(t, p)$ to $M(p)$ for each $p \in t^\bullet$.

(vii) We call the system *unbounded*, if K is nowhere defined.

(viii) It is *strict*, if all numeric values appearing in places and arc-matrices W are limited to 0 and 1, and the capacity constraint has the constant value 1.

Remark 5.3. If the function K is clear from the context, it is usually not mentioned explicitly. This may be the case for two diametrically opposite reasons: (1) either if the system is unbounded, (2) or if is strict.

5.2 The invariant-method

The ML-representation is the base for the following *incidence matrix*, which is particularly useful in system analysis:

Definition 5.4. The *incidence matrix* ML of a ML-net is the $|P| \times |T|$-matrix with entries

$$\mathsf{ML}_{ij} = W(t_j, p_i) - W(p_i, t_j).$$

The matrix representation usually gives a complete characterization, with one notable exception, namely when $W(t_j, p_i)_{kl} \neq 0 \neq W(p_i, t_j)_{kl}$ for some indices. In all examples we shall discuss here, the underlying net is pure, and we are therefore certainly on the safe side.

Example 5.5. We reformulate the system in Fig. 2 as an ML-net, where for simplicity we assume $D(p_1) = D(p_2) = \{a, b\}$. As transition-modes it appears natural to use the words aa, bb, ab, ba, such that $\Phi(t_1)$ becomes $\{aa, bb\}$ and $\Phi(t_2) = \{aa, bb, ab, ba\}$. Disregarding the columns M and $\mathbf{i}$ for a moment, the ML-net is then given by Fig. 5, where the four 'sub-matrices' separated by the dashed lines represent the arc-weights in the form $W(t, p)$ and $-W(p, t)$, respectively.

		t_1		t_2						
		aa	bb	aa	ab	ba	bb	M_{in}	M	$\mathbf{i}$
p_1	a	-1		1		1		1	0	1
	b		-1		1		1	0	1	1
p_2	a	1		-1	-1			0	0	1
	b		1			-1	-1	0	0	1

Fig. 5. The ML-net corresponding to Fig. 2(b)

We come to one of the main features of ML-nets, a powerful linear-algebraic invariant-calculus based on the incidence matrix ML.

In the following, we use the term *P-vector* for (column)-vectors $\mathbf{v}$ of the form

$$\mathbf{v}^T = (v_1, \ldots, v_n), \quad \text{where each entry } v_i \text{ is a } |D(p_i)|\text{-vector in } \mathbb{N}.$$

It is tedious but straightforward to verify, that for a P-vector $\mathbf{v}$, "dimension-compatibility" permits the definition of $\mathbf{v}^T \cdot \mathsf{ML}$ and $\mathbf{v}^T \cdot M$ for markings M, where the multiplication as usual in matrices reduces to a sum of products of the components. These in turn are easily defined, since the components themselves are matrices in $\mathbb{N}$.

We may now define a place-invariant as follows:

Definition 5.6. A *place-invariant* in an ML-net with incidence matrix **ML** is a *P*-vector $\mathbf{i}^T = (i_1, \dots, i_n)$, such that

$$\mathbf{i}^T \cdot \mathsf{ML} = \mathbf{0}, \tag{5.15}$$

where $\mathbf{0}$ denotes the appropriate 0-vector, i.e. the vector $(0_1, \dots, 0_m)$, $m = |T|$, such that each of the 0_j is the 0-vector in $\mathbb{N}$ of length $|C(t_j)|$.

Similar to equation (3.2) in Section 3 and Theorem 7.7 we have:

Theorem 5.7. *Let* ML *be a matrix-labelled net. If* $\mathbf{i}$ *is an invariant according to Definition* 5.6*, then*

$$\mathbf{i}^T \cdot M = \mathbf{i}^T \cdot M_{\text{in}}$$

for every reachable marking $M \in [M_{\text{in}}\rangle$.

Proof. If $M \in [M_{\text{in}}\rangle$, then by definition there is a sequence $t_1, \dots, t_k$ of transitions and transition-modes $m_i \in \Phi(t_i)$, $1 \leqslant i \leqslant k$, such that M results from M_{in} by addition of the m_i-columns in the $W(t_i, p)$ (resp. subtraction of the m_i-columns in $W(p, t_i)$). Multiplication with $\mathbf{i}$ distributes over that sum, and because of (5.15) the factors not involving M_{in} evaluate to 0. □

Example 5.8. We continue with Fig. 5 and verify that the vector

$$\mathbf{i}^T = (i_1, i_2) = \big((1,1), (1,1)\big)$$

is an invariant according to Definition 5.6:

We have to show that $\mathbf{i}^T \cdot t = 0$ for $t = t_1, t_2$. Now, for t_1 we get

$$\begin{aligned} \mathbf{i}^T \cdot t_1 &= i_1 \cdot W(p_1, t_1) + i_2 \cdot W(p_2, t_2) \\ &= (1,1) \cdot \begin{pmatrix} -1 & 0 \\ 0 & -1 \end{pmatrix} + (1,1) \cdot \begin{pmatrix} 1 & 0 \\ 0 & 1 \end{pmatrix} \\ &= (-1,-1) + (1,1) \\ &= (0,0). \end{aligned}$$

Similarly, it is easily verified that for t_2 we have $\mathbf{i}^T \cdot t_2 = (0,0,0,0)$.

To illustrate Theorem 5.7, we check that $\mathbf{i}^T \cdot M_{\text{in}} = \mathbf{i}^T \cdot M$ for the (obviously reachable) marking M in Fig. 5:

$$\begin{aligned} \mathbf{i}^T \cdot M_{\text{in}} &= i_1 \cdot M_{\text{in}}(p_1) + i_2 \cdot M_{\text{in}}(p_2) \\ &= (1,1) \cdot (1,0)^T + (1,1) \cdot (0,0)^T \\ &= 1, \end{aligned}$$

and an evaluation of $\mathbf{i}^T \cdot M_{\text{in}}$ clearly leads to the same result.

This last computation may also serve to illustrate the following: The product of an invariant with a marking is always an element of ℕ. The reason for this is quite simple: Essentially, what we do in the invariant-computation is to unfold the ML-net into a place-transition net PT, compute the invariants in PT by the usual method of solving a homogeneous linear equation in the integers, and finally rearrange the solutions into ML-representation.

For example, reconsider the incidence matrix in Fig. 5 above. If we disregard the dashed lines altogether, we are left with the incidence matrix PT of a place-transition net.

Likewise, the markings M_{in} and M are automatically translated into usual PT-markings. The same holds for the invariant **i**, which becomes a vector **i**′ in ℕ of length $\sum_{i=1}^{n} |D(p_i)|$, such that $\mathbf{i}'^T \cdot \mathsf{PT}$ evaluates to the 0-vector of length $\sum_{j=1}^{m} |\Phi(t_j)|$ in ℕ.

Conversely, we may define an ML-net as a place-transition net, which for convenience is partitioned into suitable "compartments". The walls can at liberty be withdrawn or inserted without any essential change to the corresponding mathematics.

In fact, an arbitrary place-transition net (where the sets of places and transitions are finite) can be turned into an ML-net at will, simply by a renumbering of the lines and columns of the incidence matrix, followed by an *arbitrary* insertion of border lines, and a suitable renaming of the compartments. Note that this transformation of a PT-net into an ML-net is again only a reformulation of the general quotient construction discussed in Section 4.

5.3 Lifting of low-level analysis: an example

The fact that ML-nets can be conceived as condensed PT-nets, has the positive consequence that all of the analysis methods developed for the latter may be lifted to ML-nets. A good example is the method of *stubborn sets* developed by A.Valmari, see e.g. [31]. Stubborn sets are used to construct a reduced reachability tree, which however still allows to determine dead states where no transition is enabled, or to find an infinite occurrence sequence if there is one. We illustrate the idea behind this approach.

For simplicity we give the definition for elementary net systems. In this case a set S of transitions is *stubborn in a marking* M, if the following three conditions hold:

(i) At least one transition $t \in S$ is enabled in M,
(ii) if $t \in S$ and t' is another transition with ${}^\bullet t \cap {}^\bullet t' \neq \emptyset$, i.e. such that t could be disabled by the firing of t', then also $t' \in S$,
(iii) if t is not enabled in M, then there is a $p \in {}^\bullet t$ such that $M(p) = 0$ and ${}^\bullet p \subseteq S$, i.e. t cannot be enabled without a direct predecessor that also belongs to S.

The reduced reachability tree is now inductively constructed as follows: For a marking M select a stubborn set S and include only follower-markings of M into the tree that result from occurrences of transitions in S. It is not hard to see that this construction in fact ensures that every dead marking will be reached, and conversely an infinite sequence will be generated if there is one.

Once the stubborn sets have been defined for low-level nets, it is then easy to lift the method. To this end it is essentially only necessary to replace "transition" with "pair of transition *plus* firing-mode". We refer to [31] for details.

6 Predicate-transition nets

The high-level models considered so far were derived directly by folding of low-level nets. This immediate approach does, however, not yet exploit the full possibilities of high-level nets. Their real advantage in modelling and analysis is invariably due to specific *languages* used in the characterization of individual tokens and firing modes. The rest of the paper is concerned with that aspect. It may be read independently from the technical development in the previous sections.

For definiteness, we shall discuss one particular model, the so-called *predicate-transition nets*, PrT-nets for short, introduced by H. GENRICH and K. LAUTENBACH in the late seventies [9]. It continues to be one of the central paradigms, and may moreover serve to illustrate ideas common to a wide variety of other approaches. We concentrate on strict nets without item-multiplicities, and only briefly indicate extensions to the non-strict case.

6.1 Introduction

PrT-nets combine dynamic system behaviour with the expressiveness of (first order) predicate logic. They are based on the observation that places in elementary nets are associated with changeable truth values, hence may be seen as a model in the realm of *dynamic propositional logic*. Similar to the extension from propositional logic to predicate logic, PrT-nets are built around *dynamically changing predicates* or *relations* over some universe. The process of change may be controlled by the evaluation of logic formulas associated with the transitions.

We start with an informal example:

Example 6.1. Fig. 6 shows a simple PrT net.

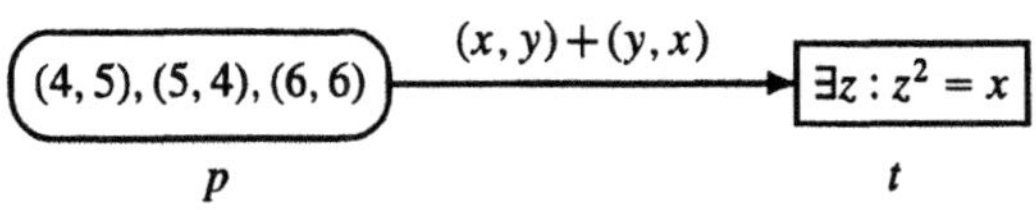

Fig. 6. A PrT-net

We assume that the place p may contain pairs of natural numbers, presently $(4,5)$, $(5,4)$, and $(6,6)$. The transition t is enabled for a binding of two number-pairs to the variable-pairs (x,y) and (y,x), for which additionally the condition $\exists z : z^2 = x$ holds. For example, if the number 4 is assigned to x and 5 to y, the pair $(4,5)$ is matched with (x,y) and $(5,4)$ with (y,x). The transition-constraint is then also satisfied. Now t may occur, withdrawing $(4,5)$ and $(5,4)$. This results in a new marking of p that contains only the pair $(6,6)$.

6.2 Predicate logic

We briefly recall the basic concepts from predicate logic needed for the definition of PrT-nets. For an exhaustive treatment the reader is referred to [6,30] or to any other standard text in logic.

Definition 6.2. We define a *language* to be a collection $\mathcal{L}$ of relation symbols, function symbols and (individual) constant symbols. Each relation symbol *Rel* is assumed to be an n-placed relation for some integer $n \geqslant 1$, depending on *Rel*. Similarly, each function symbol *func* of $\mathcal{L}$ denotes an m-placed function, where $m \geqslant 1$ and m depends on *func*.

We assume an infinite supply *Var* of individual variables plus the standard logical connectors and quantifiers.

Terms and *formulas* over $\mathcal{L}$ are defined as usual. We let *Term* denote the set of terms.

Definition 6.3. A *model* for a language $\mathcal{L}$ is a pair $\mathfrak{A} = (A, \mathcal{I})$ consisting of a *universe* A and an *interpretation* function $\mathcal{I}$ mapping the symbols of $\mathcal{L}$ to corresponding relations, functions and constant elements of A. We shall usually identify a symbol with its interpretation according to $\mathcal{I}$.

An *assignment* is a mapping $\alpha\colon Var \to A$.

The *value* $\theta[\alpha]$ of a term θ for an assignment α is defined inductively as usual. We extend this notion to *tuples* $\tau = (\theta_1, \dots, \theta_n)$ of terms, setting $\tau[\alpha] := (\theta_1[\alpha], \dots, \theta_n[\alpha])$. Finally, the definition $\mathfrak{A} \models \varphi[\alpha]$, meaning that the formula φ is satisfied in $\mathfrak{A}$ for α, is now straightforward.

6.3 The definition of predicate-transition nets

After these preliminaries, we come to the central definition of this section. We formulate it here for *strict* nets, and indicate extensions to the non-strict case later.

Definition 6.4. A *predicate-transition net* is a tuple PRT $:= (N, \mathcal{L}, \lambda, \mathfrak{A}, M_{\text{in}})$, where N is a net (P, T, F), such that

(i) P is a set of relation symbols, with given arity,
(ii) $\mathcal{L}$ is a language disjoint from P,
(iii) λ is a (partial) mapping with domain $T \cup F$, inscribing
   (a) some $t \in T$ (not necessarily all) with $\mathcal{L}$-formulas φ_t, called *transition-guards*,
   (b) each arc $(x, y) \in F$ with a finite set $\lambda(x, y) = \{\tau_1, \dots, \tau_m\}$ of n-tuples of $\mathcal{L}$-*terms*, where n is the arity of the place belonging to the arc,
(iv) $\mathfrak{A}$ is a model for $\mathcal{L}$,
(v) M_{in} is a distinguished marking called the *initial* marking, where in general a *marking* is a mapping M which interprets each $p \in P$ as a relation over A with the corresponding arity.

For technical reasons we always require $\mathcal{L}$ to contain at least one constant symbol (this is needed to ensure the existence of variable-free terms). Usually there is no need to mention the language $\mathcal{L}$, as it may be assumed to be implicitly given by $\mathfrak{A}$ and λ.

Note that each marking M extends the model $\mathfrak{A}$ to a model $\mathfrak{A}_M = (A, \mathcal{I} \cup M)$ for the extended language $\mathcal{L} \cup P$.

We usually denote arc-inscriptions $\lambda(x, y)$ as symbolic sums $\tau_1 + \cdots + \tau_m$, i.e. as particular elements of the free group $\mathfrak{G}(Term^k)$, where k is the arity of the place $p \in \{x, y\}$. We also use the notion $-\lambda(x, y)$ for the expression $-\tau_1 - \cdots - \tau_m$. We use 0 as a symbol for the 0-element in $\mathfrak{G}(Term^k)$. Occasionally, also markings will be denoted in the form of symbolic sums.

Example 6.5. We return to the five-philosophers system. Fig. 7 shows a PrT-model.

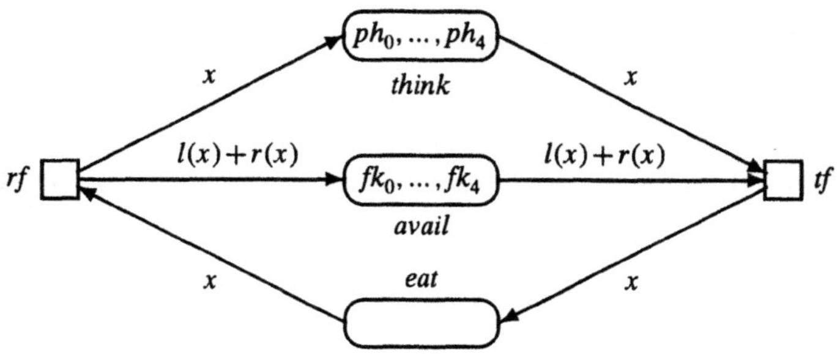

Fig. 7. PrT-model PRT_{5phil} of the five-philosophers system

The universe of discourse A is the union $Phil \cup Forks$, where $Phil = \{ph_0, \dots, ph_4\}$ is the set of philosophers, and $Forks = \{fk_0, \dots, fk_4\}$ denotes the set of forks.

The underlying net consists of three unary predicates *think*, *eat*, and *avail* with variable extension, denoting the currently thinking and eating philosophers, and available forks, respectively. The transitions *tf* "take forks" and *rf* "release forks" describe the possible system dynamics when a philosopher decides to change his state between thinking and eating. Fig. 7 shows the system in its initial state M_{in}, where all philosophers are thinking. So far the description corresponds directly to the basic model BHL_{5Phil} in Example 4.10.

What is different, is the use of arc-inscriptions to denote firing-modes and the weight-function: There are two functions l and r in $\mathcal{L}$, assigning a left fork $l(ph_i) = fk_i$ and right fork $r(ph_i) = fk_{i\oplus 1}$ to each philosopher (where again $\oplus$ denotes addition modulo 5). There are no relations or constants in $\mathcal{L}$.

We come to the *behaviour* of PrT-nets. It is based on the firing of transition-*instances* $t[\alpha]$, characterized by value-assignments α to arc-inscriptions. Undesired instances may be prevented from firing by means of transition-guards φ_t. Formally:

Definition 6.6. Let $\mathrm{PRT} := (N, \lambda, \mathfrak{A}, M_{\text{in}})$ be a PrT-net. An assignment $\alpha\colon Var \to A$ is called *feasible* for a transition t, if

(i) for any two different τ_1, τ_2 belonging to a common arc connected to t we have $\tau_1[\alpha] \neq \tau_2[\alpha]$,
(ii) $\mathfrak{A} \models \varphi_t[\alpha]$.

Note that *feasibility* expresses a *static*, i.e. marking-independent condition.

An *instance of a transition t* – also called a *firing-mode* – is a pair (t, α), denoted $t[\alpha]$, consisting of t *plus* a feasible assignment.

A transition-instance $t[\alpha]$ is *enabled* (equivalently we say that t is *α-enabled*) in a marking M, in symbols $M \models t[\alpha]$, iff

(iii) $\tau[\alpha] \in M(p)$ for all $p \in {}^\bullet t$ and $\tau \in \lambda(p, t)$,
(iv) $\tau[\alpha] \notin M(p)$ for all $p \in t^\bullet$ and $\tau \in \lambda(t, p)$.

If $M \models t[\alpha]$, a *firing* or *occurrence* of $t[\alpha]$, *α-occurrence of t* for short, leads to the follower marking

$$M' : p \mapsto \big(M(p) - \{\tau[\alpha] \| \tau \in \lambda(p, t)\}\big) \cup \{\tau[\alpha] \| \tau \in \lambda(t, p)\}. \qquad (6.16)$$

The abbreviation $M[t[\alpha]\rangle M'$ means that M' results from M according to (6.16). Again we write $M \,\mathbf{r}\, M'$ if reference to the particular $t[\alpha]$ is not of interest. The *reachability class* $[M\rangle$ of a marking M is defined as the set of all M' reachable by a sequence $M = M_0 \,\mathbf{r}\, M_1 \,\mathbf{r} \cdots \mathbf{r}\, M_n = M'$.

Example 6.7. Returning to the philosophers in Example 6.5, consider the assignment $\alpha : x \mapsto ph_0$. It is readily verified that $M_{\text{in}} \models tf[\alpha]$ and $M_{\text{in}}[tf[\alpha]\rangle M'$, where

$$M' : think \mapsto \{ph_1, \ldots, ph_4\},\ avail \mapsto \{fk_2, fk_3, fk_4\},\ eat \mapsto \{ph_0\}. \qquad (6.17)$$

In this new situation M', either of the following transition-instances may occur: $rf[\alpha]$, meaning that ph_0 puts his forks down and returns to thinking, or $tf[\beta]$ where β is one of the assignments $x \mapsto ph_2$ or $x \mapsto ph_3$.

Remark 6.8. Examples 6.5 and 6.7 should suffice to convince oneself that the behaviour of PrT-nets according to Definition 6.6 is compatible with the general basic high-level semantics in Section 4. A rigorous verification is straightforward but tedious. We leave the details to the interested reader.

Transition-guards

In the dynamics of the philosopher system in Example 6.5, we did not have to check condition (ii) in Definition 6.6, for the obvious reason that the system does not contain any transition-guards. To illustrate the meaning of transition-guards, assume for the moment that the system includes the formula $x \neq ph_0$ as a guard φ_{rf} for the transition "release forks". This guard would inhibit the return of ph_0 from eating to thinking.

Weak interpretation of PrT-nets

As mentioned, Definition 6.6 follows the *strict* interpretation of nets, where item-multiplicities are excluded. Sometimes, however, it is desirable to permit multiple appearances of tokens. This is possible in the *weak interpretation* of PrT-nets [10].

For the sake of argument, assume that there are *two* forks $fk_{i\oplus1}$ and $fk'_{i\oplus1}$ on the table between any two philosophers ph_i, $ph_{i\oplus1}$. Assume moreover, that every philosopher uses *one left* fork and *two right* forks for eating. To capture this in our (strict) model, we augment the set of forks by $Forks' := \{fk'_0, \dots, fk'_4\}$, such that the initial marking of $M_{\text{in}}(avail)$ now becomes $Forks \cup Forks'$. Additionally the expressions on the arcs connected to *avail* are replaced by, say, $l(x) + r(x) + r'(x)$.

Now, with respect to *system behaviour* the forks in each pair fk_i, fk'_i are actually indistinguishable from one another. It therefore appears as an overspecification to insist on their individuality within the model. Following the *weak interpretation* of PrT-nets, which permits token-multiplicities, we could model the situation by setting $M_{\text{in}}(avail) = \{2fk_0, \dots, 2fk_4\}$, and labelling the arcs with $l(x) + 2r(x)$. With regards to net behaviour this means that e.g. an occurrence of the *tf*-instance $tf[x \mapsto ph_0]$ in M_{in} would withdraw one left and both right forks, resulting in the follower marking $M'(avail) = \{fk_0, 2fk_2, 2fk_3 2fk_4\}$.

From a formal point of view there seems to be no need for multiplicities, as any n-placed predicate where we allow multiple appearances, may be turned into an $(n+1)$-placed predicate where distinguishing 'tags' are included in the additional component.

Moreover, sometimes it is not easy to get rid of the remainders of individuality. As a trivial but typical example, consider the arc function l above. It takes 'left fork' as value. But which one? If l is to remain a function, the value has to be unique. This means that we would have to make a choice among apparently indistinguishable forks after all.

6.4 Incidence matrix

As in PT- and ML-nets, it is often useful to represent a PrT-net by its *incidence matrix*. To this end we make use of the fact that arc inscriptions $\lambda(x, y)$ may be interpreted as elements of the free group $\mathfrak{G}(Term^k)$, where k denotes the arity of the place $p \in \{x, y\}$. Moreover, we trivially extend the domain of the mapping λ from F to all of $(P \times T) \cup (T \times P)$ by putting $\lambda(x, y) = 0$ for $(x, y) \notin F$.

Definition 6.9. Let PRT be a finite PrT-net, and assume $P = \{p_1, \dots, p_n\}$ and $T = \{t_1, \dots, t_m\}$. Then the *incidence matrix* of PRT is the $n \times m$-matrix PRT with entries

$$\mathsf{PRT}(p_i, t_j) := \lambda(t_j, p_i) - \lambda(p_i, t_j).$$

Example 6.10. Fig. 8 shows the dining philosophers system in matrix representation together with the marking (6.17) in Example 6.7.

Remark 6.11. Under certain conditions the structure of a PrT-net PRT is uniquely determined by its incidence matrix. This is the case if PRT has no transition-guards and, moreover, for all loops $(t, p), (p, t) \in F$ we have $\lambda(t, p) \cap \lambda(p, t) = \emptyset$. This last condition is necessary, since terms common to both label sets would be canceled out in the sum representation. In *pure* nets, for instance, we are on the safe side.

	tf	rf	M_{in}	M'
think	$-x$	x	$p_0+\cdots+ph_4$	$ph_1+\cdots+ph_4$
avail	$-r(x)-l(x)$	$r(x)+l(x)$	$fk_0+\cdots+fk_4$	$fk_3+fk_4+fk_5$
eat	x	$-x$		ph_0

Fig. 8. Incidence matrix and two markings of PRT_{5phil} in Fig. 7

7 Place-invariants

In this section we show how particular properties of the PrT-language may be used in system analysis. As a typical example we discuss the concept of symbolic invariants. We restrict the discussion to *place-invariants*, leaving aside the dual notion of *transition-invariants*.

The first approach to invariants in PrT-nets was proposed by GENRICH and LAUTENBACH in [9]. It is based on unification of terms by suitable variable-assignments. This method is not without problems, but as it continues to be in use – albeit in modified forms – we shall briefly discuss it. We then develop an approach proposed by J. VAUTHERIN and W. REISIG [24,23], based on symbolic term substitution.

All PrT-nets considered in this section are assumed to be finite.

7.1 The unification method

In the following let PRT be a finite PrT-net without transition-guards. For simplicity we assume the net to be pure. Assume that all places are unary predicates, and, moreover, that each arc-label consists of a *single variable*. Let **PRT** denote the incidence matrix of PRT according to Definition 6.9.

For the present discussion, a *P-vector* will be a (column) vector $\mathbf{v}^T = (v_1, \ldots, v_n)$ with components v_i from $\mathfrak{G}(Term)$.

Assume we have a commutative product "$\cdot$" between elements of $(\mathfrak{G}(Term), +)$, distributive over addition, such that for any $v_1, v_2, w_1, w_2 \in \mathfrak{G}(Term)$:

$$v_1 \cdot v_2 = w_1 \cdot w_2 \Rightarrow (v_1, v_2) = (w_1, w_2) \vee (v_1, v_2) = (w_2, w_1).$$

Formally, this product can be defined by extending the free group $\big(\mathfrak{G}(Term), +\big)$ to the ring $\big(\mathfrak{R}(Term), +, \cdot\big)$ of *formal polynomials* over *Term*. This is a standard construction, similar to the construction of the free group itself. We shall not bother with the details.

The ring-product may now as usual be extended canonically to cover also products $\mathbf{v}^T \cdot \mathbf{w}$ between P-vectors. Note in particular that markings M of PRT and the columns of **PRT** are P-vectors. Hence also the products $\mathbf{v}^T \cdot M$ and $\mathbf{v}^T \cdot \mathbf{PRT}$are defined.

Definition 7.1. As in PT-nets, a P-vector $\mathbf{i}$ is called a *place-invariant* iff

$$\mathbf{i}^T \cdot \mathbf{PRT} = \mathbf{0}, \tag{7.18}$$

where $\mathbf{0}$ is the zero-vector of length $|T|$ in $\mathfrak{R}(Term)$.

Recall that place-invariants in PT-nets were used to establish the invariance-equation (3.2). This naturally raises the question, whether a similar statement holds also for PrT-invariants $\mathbf{i}$ as in (7.18). In other words: can we prove that

$$\mathbf{i}^T \cdot M_{\text{in}} = \mathbf{i}^T \cdot M \tag{7.19}$$

for every reachable marking M? Unfortunately, this is not true in general. In order to make sense at all, (7.19) must be carefully interpreted. In the following three examples we shall discuss some of the possibilities and difficulties involved. In all examples we assume the underlying universe of discourse to be $A = \{a, b\}$.

Example 7.2. Consider the system Σ_1 given by the following table:

	t	M_{in}	M	$\mathbf{i}$
p_1	$-x$	a	0	y
p_2	y	0	b	x

(7.20)

In addition to the incidence matrix $\underline{\Sigma_1}$ (first column) and the initial marking we have included another marking M and a vector $\mathbf{i}$ in the table. Clearly, M is reachable from M_{in}, more precisely $M_{\text{in}}[t\,[x \mapsto a, y \mapsto b]\rangle M$. The vector $\mathbf{i}$ is a place-invariant according to (7.18), because

$$\mathbf{i}^T \cdot \underline{\Sigma_1} \;=\; \mathbf{i}^T \cdot t \;=\; -y \cdot x + y \cdot x \;=\; 0.$$

However, an immediate computation shows that (7.19) is *not* satisfied:

$$\mathbf{i}^T \cdot M_{\text{in}} \;=\; a \cdot y \;\neq\; b \cdot x \;=\; \mathbf{i}^T \cdot M. \tag{7.21}$$

It is, however, possible to *deduct* a "real" invariance-equality from (7.21). This is accomplished by a suitable *unification* of the expressions $a \cdot y$ and $b \cdot x$: Replacing the variable x by the constant a, and y by b, we obtain a P-vector $\mathbf{i}'$, which now satisfies (7.18) *as well as* (7.19).

The successful unification in (7.21) suggests the following re-interpretation of the invariance-equation (7.19):

For every reachable marking M there is a unifying assignment α of constants to the variables of $\mathbf{i}$, *such that* $\mathbf{i}[\alpha]^T \cdot M_{\text{in}} = \mathbf{i}[\alpha]^T \cdot M$. (7.22)

Indeed, (7.22) holds true whenever one and only one transition is involved. This is due to the fact that what the unification determines, is basically the variable-assignment needed for the transformation of M_{in} into the follower-marking M. The unification/assignment $x \mapsto a, y \mapsto b$ in Σ_1 illustrates this.

Unfortunately, (7.22) fails in general. Actually it may fail already in case of *two* transitions, as the following example shows.

Example 7.3. Consider the PrT-net Σ_2 given by the incidence matrix $\underline{\Sigma_2}$ in the following table:

	t_1	t_2	M_{in}	M	$\mathbf{i}$
p_1	$-x$	x	a	b	y
p_2	y	$-y$	0	0	x

(7.23)

It is again easily verified that $\mathbf{i}^T \cdot \underline{\Sigma_2} = \mathbf{0}$. Moreover, the marking M is obviously reachable from M_{in}. But a multiplication of $\mathbf{i}$ with the markings M_{in} and M yields

$$\mathbf{i}^T \cdot M_{\text{in}} = a \cdot y \neq b \cdot y = \mathbf{i}^T \cdot M. \tag{7.24}$$

Now, in contrast to (7.21), in (7.24) there cannot possibly exist any unification of the variables which would turn it into an equality, as such a unification would have to map y to a as well as to b!

The deeper reason for the failure of (7.22) in Example 7.3 is that *different* constants may be assigned to the variables in successive transition-firings, whereas such a dynamically changing assignment is not reflected in the "static" expression (7.19).

As an attempt to cope with dynamically changing assignment one could demand that every transition-environment (i.e. each column in the incidence matrix) should use a private set of variables. This remedy, however, would not help to deal with cases where the *same* transition successively occurs under two different assignments.

On the other hand, keeping variables local to transitions may lead to a converse problem since there will then usually not be sufficiently many identical variables to allow cancellation. In these cases, the unification approach may fail to find any invariant at all, even if there are obvious candidates. This is illustrated in the following

Example 7.4. Fig. 9 shows the PrT-version Σ_3 of the system in Figs. 2(b) and 5.

	t_1	t_2	M_{in}	M	$\mathbf{i}$
p_1	$-x$	z	a	b	v
p_2	x	$-y$	0	0	w

Fig. 9. A PrT-net Σ_3, an additional marking M, and a P-vector $\mathbf{i}$

In Example 5.8 we have already established, that there *is* an obvious invariant in the system, which *intuitively* may be described as follows. In every reachable marking of the system – as for instance in $M = (b, 0)$ in Fig. 9 – there is exactly one of a or b on exactly one of the places p_1 or p_2.

Unfortunately that invariant is not reflected in any place-invariant of Σ_3 according to (7.18): Now, if there were a vector $\mathbf{i} = (v, w)$ for some $v, w \in \mathfrak{G}(Term)$ satisfying (7.18), this would imply that both

$$-x \cdot v + x \cdot w = 0 \quad \text{and} \quad z \cdot v - y \cdot w = 0,$$

which would only be possible if $v = w$, and hence also $y = z$! But this is not the case.

7.2 The substitution method

We shall now develop an invariant-calculus, which does not depend on ex-post interpretations of the solutions. The method is based on the fact that transition-occurrence corresponds to function-application, mapping markings to follower markings. Consequently, firing-*sequences* may be conceived as function-*composition*. Since transitions operate via term-evaluation in PrT-nets, this suggests to use *term substitution* as a basis for invariant-computation. Moreover, as the 'occurrence-functions' are *additive* by nature, term-*substitution* may naturally be distributed over term-*addition*.

The method is applicable to PrT-nets in general, but for simplicity we shall here only consider *strict PrT-nets with unary places*. In particular this implies that all tuples appearing in arc-inscriptions consist of single terms. We also assume that system behaviour is not controlled by *transition-guards*. As before we concentrate on *place-invariants*, and only briefly mention the dual notion of *transition-invariants*.

Definition 7.5. Let PRT be a PrT-net as specified above. For terms θ, η in the language $\mathcal{L}$ of PRT, we define the *product* $\theta \cdot \eta$ to be the term $\theta[\eta]$ resulting from θ by a replacement of all variables of θ by η. Note that this multiplication is of course *not* commutative.

It is then straightforward to extend this operation to $\mathfrak{G}(Term)$ by distributing multiplication into sums of terms, putting

$$\sum_i \theta_i \cdot \sum_j \eta_j = \sum_{ij} \theta_i \cdot \eta_j . \tag{7.25}$$

In particular, for the 0-element of $\mathfrak{G}(Term)$ (the "empty sum") we have $\theta \cdot 0 = 0$ and $0[\eta] = 0$ for any terms θ, η.

We assume the operations to be continued to vector- and matrix-operations as usual.

By a *P-vector* we understand a (column) vector $\mathbf{v}^T = (v_1, \ldots, v_n)$ where each v_i is an element in $\mathfrak{G}(Term)$.

Based on this product we can define *place-invariants* as follows:

Definition 7.6. Let PRT denote the incidence matrix of of a PrT-net PRT according to Definition 6.9. A P-vector $\mathbf{i}$ is a *place-invariant* of PRT iff

$$\mathbf{i}^T \cdot \mathsf{PRT} = \mathbf{0}, \tag{7.26}$$

where $\mathbf{0}$ denotes the 0-vector in $\mathfrak{G}(Term)$ of appropriate length.

By essentially the same argument as in PT- and ML-nets we obtain:

Theorem 7.7. *Let* $\mathbf{i}$ *be a place-invariant (according to Definition* 7.6*) of a PrT-net* PRT *with initial marking* M_{in}*. Then*

$$\mathbf{i}^T \cdot M_{\text{in}} = \mathbf{i}^T \cdot M \tag{7.27}$$

for all reachable markings $M \in [M_{\text{in}}\rangle$.

Proof. By induction, using the following argument: Assume the marking M' results from another marking M by an α-occurrence of a transition t. In vector-notation this can be written as $M' = M + t[\alpha]$, where we identify t with the corresponding column in PRT. By multiplication with $\mathbf{i}^T$ we obtain $\mathbf{i}^T \cdot M + \mathbf{i}^T \cdot t[\alpha]$, where the second summand can be further evaluated to $(\mathbf{i}^T \cdot t)[\alpha]$ by associativity of substitution, and thus finally to $0[\alpha] = 0$, where again 0 is the 0-element in $\mathfrak{G}(\mathit{Term})$. □

Example 7.8. For illustration, we return to the dining philosophers. Consider the following P-vector $\mathbf{i}$ of the system PRT_{5phil} in Example 6.10:

$$\mathbf{i}^T = \big(r(y)+l(y), -y, 0\big). \tag{7.28}$$

We show that $\mathbf{i}$ is an invariant according to Definition 7.6, by verifying that both $\mathbf{i}^T \cdot \mathit{tf} = 0$ and $\mathbf{i}^T \cdot \mathit{rf} = 0$: Now,

$$\begin{aligned}\mathbf{i}^T \cdot \mathit{tf} &= \big(l(y)+r(y)\big)\cdot(-x)+(-y)\cdot\big(-l(x)-r(x)\big)\\ &= \big(-l(x)-r(x)\big)-\big(-l(x)-r(x)\big)\\ &= 0.\end{aligned}$$

The other equality can be treated similarly.

The system-invariant characterized by $\mathbf{i}$ is computed from the initial marking as follows:

$$\begin{aligned}\mathbf{i}^T \cdot M_{\mathrm{in}} &= (l(y)+r(y))\cdot \textstyle\sum ph_i - y\cdot\sum fk_i\\ &= \textstyle\sum l(ph_i)+\sum r(ph_i)-\sum fk_i\\ &= \textstyle\sum fk_i,\end{aligned}$$

where the last equality holds, because each of the two first summands evaluate to the expression $\sum fk_i$.

Application of invariants

As in PT- and ML-nets, place-invariants may be applied to establish that certain situations in a net-system are *not* reachable. As an example we use the invariant $\mathbf{i}$ in Example 7.8 to prove that not all philosophers may refrain from thinking at the same time. Suppose to the contrary, that M is a marking with $M(\mathit{think}) = \emptyset$. Then $\mathbf{i}^T \cdot M$ evaluates to a negative expression $-y \cdot M(\mathit{avail})$ and therefore certainly *not* to $\sum fk_i$.

Computation of invariants

At present the invariant-method is essentially limited to *verification* of invariants (as in Example 7.8). There is no systematic method known to *solve* the invariant-equation (7.26) in general.

In some cases, however, solutions can be found by straightforward reasoning. We give a simple example.

Example 7.9. Consider be the PrT-net Σ_3 defined by Fig. 9 above.

Let $\mathbf{i} = (v, w)$ be a vector, where v, w are sums of terms. If $\mathbf{i}$ is an invariant, we must have $\mathbf{i}^T \cdot t_1 = 0$ and $\mathbf{i}^T \cdot t_2 = 0$. From the first equality we deduce that both components of $\mathbf{i}$ are equal, such that $v = w$. From this we get $v \cdot z = v \cdot y$ by the second equality. Since z and y are different variables, this is true iff the value of v does not depend on the variables. Hence if for instance v is a constant c, we obtain the solution $\mathbf{i}^T = (c, c)$ for the invariant-equation (7.26).

Using the invariant $\mathbf{i}^T = (c, c)$, we can apply Theorem 7.7 to deduct the corresponding system-property: In every reachable marking – as for instance in $M = (b, 0)$ in Fig. 9 – there is exactly one of a or b on exactly one of the places p_1 or p_2.

Place-invariants in nets with transition-guards
In the discussion above we assumed the nets under consideration to be free from transition-guards. However, it is clear that Theorem 7.7 holds a fortiori if PRT *does* contain transition-guards, since these guards only *reduce* the class of reachable markings.

Transition-invariants
In Section 3 we briefly discussed the notion of transition-invariants in PT-nets. It is technically straightforward to define *transition-invariants* also for PrT-nets, namely as solutions to the equation $\mathsf{PRT} \cdot \mathbf{i} = \mathbf{0}$, where multiplication is again term-substitution according to Definition 7.5. At present, however, these transition-invariants do not play any significant role in system analysis.

Generalized place-invariants
The crucial property for the application of place-invariants is the invariance-equation (7.27). Closer analysis of the proof of Theorem 7.7 shows that (7.27) depends essentially only on the fact that the product is distributive over transition-occurrences; term-substitution is therefore only one possible approach (though probably the most natural one).

Another possibility to obtain more general invariants is to interpret the equality in (7.26) and (7.27) not in the restricted sense of "true" equality, but rather as a *congruence relation* compatible with addition. This could for instance be discussed in the context of term-equivalence in algebraic specifications.

8 Reachability analysis

Besides invariant-computation, reachability analysis is probably the most versatile analysis method in net theory. In its simplest form, reachability analysis is based on the generation of the complete state space, which of course tends to be intractably large in real modelling. This has lead to the investigation of various techniques of state space reduction. We shall take a closer look at the method of *equivalent markings* [12], which is probably the one most used. As an alternative approach we then briefly discuss the method of *parameterized markings* [20,21].

For simplicity we shall again only be concerned with strict interpretations of PrT-nets.

Reduction by equivalent markings

Definition 8.1. The reachability tree for a PrT-net PRT is a labelled tree $\mathcal{T}$ inductively constructed as follows:

(i) The root of $\mathcal{T}$ is labelled by the initial marking M_{in}.
(ii) If $\mathcal{T}$ contains a node n_M labelled by a marking M, then for each immediate follower marking M' of M a node $n_{M'}$ labelled by M' is included as successor node to n_M.

Sometimes it is also useful to assume the edges to be labelled by the transition-instances $t[\alpha]$ connecting the corresponding node labels.

The building of the reachability tree serves to determine all reachable markings. Hence if a node label appears repeatedly, only one of the occurrences has to be continued, as all subtrees with the same root label will be identical. The idea behind reduction by equivalent markings is an extension of cutting the tree at duplicate markings. It is based on the exploitation of certain symmetries in the system.

Definition 8.2. Let PRT be a PrT-net over the universe A. A bijection (or permutation) π on A is an *automorphism* iff

(i) $\pi(M_{\text{in}}) = M_{\text{in}}$, and
(ii) $M \mathbf{r} M' \Leftrightarrow \pi(M) \mathbf{r} \pi(M')$ for all markings M, M' of PRT.

Here the notion $\pi(M)$ stands for the marking obtained from M by renaming each constant a appearing in $M(p)$, $p \in P$, by $\pi(a)$. The relation $\mathbf{r}$ again denotes "reachable by *one* transition-occurrence".

Example 8.3. Recall that in the five-philosophers system PRT_{5phil} in Example 6.5, the universe of discourse is $A = \mathit{Phil} \cup \mathit{Forks}$. Let π be the permutation $ph_i \mapsto ph_{i\oplus 1}$, $fk_i \mapsto fk_{i\oplus 1}$, $i = 0, \dots, 4$. Both of the conditions (i) and (ii) in Definition 8.2 are easily verified.

To illustrate (i) we compute $\pi(M_{\text{in}})(\mathit{think})$:

$$\begin{aligned}\pi(M_{\text{in}})(\mathit{think}) &= \pi(M_{\text{in}}(\mathit{think}))\\ &= \{\pi(ph_0), \dots, \pi(ph_4)\}\\ &= M_{\text{in}}(\mathit{think}).\end{aligned}$$

To illustrate (ii), consider the assignment $\alpha{:}\, x \mapsto ph_0$. The transition tf is α-enabled in M_{in}, and an occurrence of $tf[\alpha]$ leads to a situation M where ph_0 is eating.

We show that also $\pi(M_{\text{in}})\ \mathbf{r}\ \pi(M)$: The difference between $\pi(M)$ and M is that ph_1 is eating in $\pi(M)$ instead of ph_0 With the assignment $\beta := \pi \circ \alpha$, which maps x to ph_1, and observing (i) we immediately get $\pi(M_{\text{in}}) = M_{\text{in}}[tf[\beta]\rangle\pi(M)$.

The following proposition is easily proved by induction:

Proposition 8.4. *Let π be an automorphism of* PRT.

(i) *Then for every marking M of* PRT, *M is reachable from M_{in} if and only if the marking $\pi(M)$ is.*

(ii) *Moreover, subtrees $\mathcal{T}_M$ and $\mathcal{T}_{\pi(M)}$ of $\mathcal{T}$ with respective root labels M and $\pi(M)$ are structurally isomorphic, such that each node label in $\mathcal{T}_{\pi(M)}$ is the π-image of the corresponding label in $\mathcal{T}_M$.*

Proposition 8.4 guarantees the correctness of the following algorithm (called "reduction by equivalent[2] markings") for the generation of the reachability class $[M_{\text{in}}\rangle$:

Algorithm 8.5 (Reduction by equivalent markings).

(i) Put $Reach := \{M_{\text{in}}\}$.
(ii) Then proceed to generate the tree $\mathcal{T}$. For each node label M encountered, check if it is already in *Reach* or if it is the π-image of a label in *Reach*. If either is the case, do not process that node any further. Otherwise include M into *Reach*.
(iii) Continuing in this way, we finally obtain $[M_{\text{in}}\rangle = Reach \cup \{\pi(M) | M \in Reach\}$.

Example 8.6. As an example of the effect of reduction by equivalent markings, we mention that in the five-philosophers system, reduction with respect to the rotation $ph_i \mapsto ph_{i\oplus 1}$ results in $|Reach| = 6$, whereas simple reduction by duplicate markings leaves 11 markings to be considered.

A crucial question for the application of the reduction algorithm is of course how to find suitable automorphisms. For an arbitrary PrT-net, this is probably a hopeless undertaking. Moreover, if automorphisms are to serve as a tool for simplified generation of the reachability tree, it appears paradox that condition (ii) in Definition 8.2 already depends on the prior knowledge of the whole reachability class.

In practical applications, however, most high-level models are usually already *developed* along some symmetry, which can then profitably be exploited.

In fact, this often results in a considerable number of usable automorphisms. In such cases, all of the available automorphisms may be applied to reduce the tree $\mathcal{T}$ even further. In the philosopher system, for instance, all rotations π_j mapping the ph_i to $ph_{i\oplus j}$ and rotating the forks correspondingly, may be used together in the algorithm above, reducing the number of non-equivalent markings further to $|Reach| = 4$.

In many cases the design behind a system implicitly relies on different *types* of individuals in the net's universe A, which are internally correlated by obvious bijections. Such knowledge often yields simplified definitions of usable automorphisms. For instance, in case of the philosophers, the set *Fork* is related to *Phil* via the bijection $l : ph_i \mapsto fk_i$. By means of l, every permutation π' of *Phil* has a unique extension to a permutation π of A, putting $\pi : fk_i \mapsto (l \circ \pi' \circ l^{-1})(fk_i)$. We refer to [12] for details.

Parameterized markings

An alternative method to reduce reachability trees is due to M. Lindqvist [20,21]. It is based on *symbolic* transition-occurrences, representing a whole class of concrete

[2] actually the relation between a marking M and its image $\pi(M)$ based on a single permutation is not an equivalence in the strict sense of the word.

firing-modes. Such a symbolic (or *parameterized*) occurrence results in a symbolic follower marking from which all corresponding "real" markings may be obtained by the assignment of appropriate values to the symbolic parameters. To give a (simplified) idea of the construction, in the philosopher system one would for instance fire the transition *tf* "for philosopher *i*" without binding *i* to a concrete value. In this way all markings and the whole reachability tree can be generated in parameterized form, which may lead to a considerable reduction in size. Unfortunately the necessary technical machinery turns out to be quite intricate. The method has recently been revised by K. SCHMIDT [26].

Partial analysis of reachability trees
Both methods discussed above, reduction by equivalent markings, and evaluation of parameterized markings, essentially construct the whole reachability tree, albeit in a condensed form. For special purposes it is often useful and sufficient to generate only a part of the reachability tree. For example, in the special case that the problem is whether a system will run into a deadlock where no transition is enabled, one may try the method of *stubborn sets*, mentioned at the end of Section 5.

9 Abstract PrT-nets and algebraic nets

Often it is convenient to deal with a whole class of structurally related systems within one generic framework instead of repeating similar arguments anew for each instance. For illustration, passing from 5 philosophers to an arbitrary number n, there are general statements which will still hold, for example it will still certainly be impossible for all philosophers to eat at the same time.

In PrT-nets, however, such a generalization within *one* representation is not immediately possible. According to Definition 6.4, every PrT-net is based on a *fixed* model $\mathfrak{A}$, hence for each n we would have to introduce a PrT-model of its own.

A promising approach to overcome this limitation is the introduction of *abstract PrT-nets* or net-*schemes* [8], where actual models $\mathfrak{A}$ are replaced by a set of *axioms* characterizing the whole class of intended models. Individual elements in the universe A are replaced by the notion of *variable-free* terms, as in the Herbrand theory of logic programming. We illustrate the basic concepts, where again we concentrate on strict nets.

9.1 Abstract PrT-nets

We use the notion $Ax \vdash \varphi$ to denote *derivability* of a formula φ from a set of (first order) formulas Ax. Again the details can be found in [6,30] or in any other standard text on logic.

Definition 9.1. By an *abstract predicate-transition net* we understand a tuple APRT $:= (N, \mathcal{L}, \lambda, Ax)$, where N, $\mathcal{L}$ and λ are as in Definition 6.4, and again without loss of generality we assume $\mathcal{L}$ to contain at least one constant symbol. Ax is a set of (first order) $\mathcal{L}$-formulas.

Note that in Definition 9.1 there is no reference to any initial marking. The reason is that it is not clear how to generically represent the class of all markings in question. For instance, in case of the philosophers, an initial marking of APRT would have to represent not a *fixed* number of philosophers but *all numbers n* within one and the same marking. (If desired, on the other hand, it is of course perfectly possible to include an initial marking into the definition.)

The general notions of *marking* and dynamic behaviour are defined as follows:

Definition 9.2. A *marking* of APRT is a mapping M which interprets each n-placed $p \in P$ as a set of n-tuples $\{\tau_1, \dots, \tau_k\}$ where the components $\theta_{i_1}, \dots, \theta_{i_n}$ of the τ_i, $1 \leqslant i \leqslant n$, are *variable-free terms* of $\mathcal{L}$. Moreover, as a syntactical counterpart to strictness we require all τ_i, τ_j to be mutually distinct in the sense that

$$Ax \nvdash \theta_{i_1} = \theta_{j_1} \wedge \cdots \wedge \theta_{i_n} = \theta_{j_n} . \tag{9.29}$$

We may now define behaviour in analogy with Definition 6.6. First note that in APRTs an assignment α can be conceived as a syntactical replacement with variable-free terms. We define *feasibility* as in Definition 6.6 (i), (ii), except that $\tau_1[\alpha] \neq \tau_2[\alpha]$ in (i) is re-interpreted according to (9.29) above, and (ii) is replaced by $Ax \vdash \varphi_t[\alpha]$.

With these modifications, the notions concerning *enabling*, *firing*, and *reachability* can now be copied literally from Definition 6.6.

A central question in APRTs is, to what extent properties carry over between abstract specifications and concrete net-models. We shall not go very much into detail here, only indicate some immediate possibilities and limitations.

Definition 9.3. A concrete PrT-net PRT $= (N, \mathcal{L}, \lambda, \mathfrak{A}, M_{\text{in}})$ is a *model* of an abstract PrT-net APRT $= (N, \mathcal{L}, \lambda, Ax)$ iff $\mathfrak{A} \models Ax$.

Example 9.4. A natural abstract representation for the dining philosophers is given by the abstract PrT-net APRT$_{phil} := (N, \mathcal{L}', \lambda, Ax)$, where the net N and the label-function λ are as in Example 6.5. The language $\mathcal{L}'$ is an extension of $\mathcal{L}$, containing an additional constant symbol a and a function symbol f with the intended meaning "right neighbour of philosopher x". Let the set Ax consist of the single equation $r(x) = l(f(x))$.

Consider the following markings of APRT$_{phil}$:

	M_1	M_2	M_3
think	$\sum_{i=0}^{4} f^i(a)$	$\sum_{i=1}^{4} f^i(a)$	$\sum_{i=0}^{3} f^i(a)$
avail	$\sum_{i=0}^{4} l(f^i(a))$	$\sum_{i=2}^{4} l(f^i(a))$	$\sum_{i=1}^{3} l(f^i(a))$
eat		a	$f^4(a)$

where, $f^i(a)$ is short for the term $\underbrace{f(\cdots f}_{i\text{-times}}(a)\cdots)$.

Let α be the assignment $x \mapsto a$. It is easily verified that $tf[\alpha]$ is enabled in M_1 according to Definition 9.2, and also that $M_1[tf[\alpha]\rangle M_2$.

A typical concrete model of APRT_{phil} is the five philosophers system PRT_{5phil}, where the additional symbols a and f are interpreted as respectively ph_0 and the function $ph_i \mapsto ph_{i\oplus 1}$, with $\oplus$ again denoting addition modulo 5.

Interpreted in PRT_{5phil}, the abstract marking M_1 becomes M_{in}, and M_2 becomes the marking M' in Fig. 8. Hence for these two markings reachability in the abstract and concrete nets coincide.

In contrast, it is a simple matter to check that the marking M_3 in the table above is *not* reachable from M_1 in APRT_{phil}, since a transfer of $f^4(a)$ from *think* to *eat* requires the presence of the term $l(f_5(a))$ on the place *avail*. In PRT_{5phil}, however, the interpretation corresponding to M_3 *is* reachable from M_{in}. The reason is of course that f^5 becomes the identity function in PRT_{5phil}, whereas in APRT_{phil} this is not the case, nor *could be*, because the abstract net must also permit models like PRT_{6phil} with 6 philosophers, where it is f^6 that is interpreted as identity rather than f^5.

Example 9.4 shows that reachability in actual models does not necessarily imply reachability in the APRT. We mention without formal demonstration that also the converse is not true. From reachability in an APRT we cannot deduce reachability in its models Σ: Suppose there is a marking M of APRT containing two terms θ and η. Then in particular these terms must be different with respect to Σ, in the sense that $Ax \not\vdash \theta = \eta$. However, it is still perfectly possible that the interpretations of θ and η in Σ might evaluate to one and the same element and thus violate the strictness-condition.

After these disappointments one might raise the question, whether abstract PrT-nets have any use at all. The answer is, yes, they do, and one keyword again is *invariants*:

Recall that invariants according to Definition 7.6 refer only to the incidence matrix. Now, by Definition 9.3 an abstract PrT-net shares its incidence matrix with all its realizations. Hence, if **i** is an invariant of an abstract PrT-net, then Proposition 7.7 will automatically hold for all its concrete models. For instance, the invariant **i** from Fig. 7.28 may now be used to show that actually in any system of arbitrary many philosophers, it is not possible that all of them eat at the same time.

9.2 Algebraic nets

Algebraic nets are a variant of abstract nets, which draw essential aspects of their semantics from the general framework of algebraic specifications and abstract data types. The notion is sometimes used quite informally, though there seems to be a general agreement on the following main features: Algebraic nets have no transition-inscriptions, the set of axioms Ax is empty or is concerned only with the specification of equalities between terms, strictness in the sense of (9.29) is not always required, and an initial marking usually forms part of the definition.

In this sense APRT_{phil} in Example 9.4 with a suitable initial marking is an algebraic net.

We refer to [23] for details.

10 Modern trends

Convergence
In the beginning there were essentially two different approaches to high-level nets, based *either* on expressive logic symbolism *or* a precise invariant-calculus. With the advent of unambiguous symbolic invariant-methods, there is a now a trend to define high-level nets mainly via *syntactic inscriptions*. Different models may be related to each another via a common "reference model", which relies on the purely semantic concepts of high-level nets, such as for instance the basic high-level nets proposed in Section 4.

Modeling and analysis requirements
The class of high-level nets is still under development. One reason for the introduction of new net classes is ease of modelling. As a typical example we mention hierarchical construction principles based on coloured nets [15,16].

Often new models are introduced as special-purpose nets, as for instance M-nets which allow composing operations similar to process-algebra [2].

Another impetus for the introduction of new nets is ease of analysis. As an example we mention *well formed coloured nets* where place-domains, firing-modes and arc-inscriptions are constructed from simple basic building blocks. These nets are fine-tuned to allow efficient use of various analysis tools [4,3,11].

Analysis of infinite nets
We have seen that for finite high-level nets with finite token domains, analysis methods easily carry over from low-level nets. Otherwise rather elaborate adaptations are necessary. As an example, we mention a symbolic variant of stubborn sets, which is discussed in [26].

Simulation
In many cases, where analysis-methods are not available, computer-aided simulation appears as an attractive alternative. There are various simulation tools available, among which Design/CPN, cf. [15,16] is probably the most widespread.

Extensions
Similar to the situation in low-level nets, there are many proposals to modify high-level net semantics by means of additional features such as *inhibitor arcs*, *test arcs*, *read only arcs*, *priorities*, etc. Examples can be found for instance in [5,18].

Of special interest in modelling are currently nets with semantics induced by *timing* or *stochastic* considerations. Also these classes of nets transcend high-level nets in the sense of this article. We refer to [1].

11 Assessment and concluding remarks

For application purposes, high-level nets in one form or another constitute the most versatile system model offered by net theory. This is primarily due to the fact that high-level nets permit concise representations of distributed systems while still retaining "locality aspects" on any desired level of detail. High-level nets support a great variety of data structures and permit the use of sophisticated behaviour control languages.

From a formal point of view it is possible to derive high-level nets and their semantic properties from elementary semantics of low-level nets by means of well understood translations. In particular this permits to lift mathematical methods from low- to high-level interpretations, as for instance the invariant-calculus of place-transition nets to matrix-labelled nets and related models such as coloured nets. Other analysis techniques are based on the high-level language proper such as the computation of symbolic invariants by term substitution in PrT-nets, or symbolic methods to simplify reachability analysis.

A recent approach in high-level nets is to proceed to even higher levels of abstraction, where nets are no more conceived as single concrete systems, but rather as a scheme for a whole class of related systems. Again it becomes possible to develop appropriate tools also on this level of discussion. As an example we have seen the generalization of symbolic invariants.

There is of course a lot more to high-level nets than we could cover here. As a starting point for further reading we recommend related papers in this volume and the collection [17], which among others contains various of the original papers mentioned here.

References

1. M. Ajmone Marsan et al. *Modelling with Generalized Stochastic Petri Nets*. Wiley, 1995.
2. E. Best, H. Fleischhack, W. Fraczak, R. Hopkins, H. Klaudel, E. Pelz. A class of composable high level Petri nets. In *Application and Theory of Petri Nets 1995*, vol. 935 of *LNCS*, pp. 103–120. Springer-Verlag, 1995.
3. R. Brgan, D. Poitrenaud. An efficient algorithm for the computation of stubborn sets of well formed Petri nets. In *Application and Theory of Petri Nets 1995*, vol. 935 of *LNCS*, pp. 121–140. Springer-Verlag, 1995.
4. G. Chiola, C. Dutheillet, G. Franceschinis, S. Haddad. On well formed colored Petri nets and their symbolic reachability graph. In [17], pp. 373–296.
5. S. A. Christensen, N. Hansen. Coloured Petri nets extended with place capacities, test arcs and inhibitor arcs. In *Application and Theory of Petri Nets 1993*, vol. 691 of *LNCS*, pp. 186–205. Springer-Verlag, 1993.
6. C. Chang and H. Keisler. *Model Theory*. North Holland, 1973.
7. P. Cohn. *Universal Algebra*. D. Reidel Publishing Company, 1985.
8. H. Genrich. Equivalence transformations of PrT-nets. In *Advances in Petri Nets 1989*, vol. 424 of *LNCS*, pp. 179–208. Springer-Verlag, 1990.
9. H. Genrich, K. Lautenbach. System modelling with high-level Petri nets. *Theoretical Comput. Sci.*, 13:109–136, 1981.
10. H. J. Genrich. Predicate/transition nets. In *Advances in Petri Nets 1986, part I*, vol. 254 of *LNCS*, pp. 207–247. Springer-Verlag, 1987.
11. S. Haddad, J. Ilié, M. Taghelit, B. Zouari. Symbolic reachability graph and partial symmetries. In *Application and Theory of Petri Nets 1995*, vol. 935 of *LNCS*, pp. 238–257. Springer-Verlag, 1995.

12. P. Huber, A. Jensen, L. Jepsen, K. Jensen. Reachability trees for high-level Petri nets. *Theoretical Comput. Sci.*, 45:261–292, 1986.
13. K. Jensen. Coloured Petri nets and the invariant-method. *Theoretical Comput. Sci.*, 14:317–336, 1981.
14. K. Jensen. Coloured Petri nets: A high level language for system design and analysis. In *Advances in Petri Nets 1990*, vol. 483 of *LNCS*, pp. 342–416. Springer-Verlag, 1990.
15. K. Jensen. *Coloured Petri Nets, vol. 1*. Springer-Verlag, 1992.
16. K. Jensen. *Coloured Petri Nets, vol. 2*. Springer-Verlag, 1995.
17. K. Jensen, G. Rozenberg (eds.) *High-level Petri Nets*. Springer-Verlag, 1991.
18. G. Lakos, S. Christensen. A general systematic approach to arc extensions for coloured Petri nets. In *Application and Theory of Petri Nets 1994*, vol. 815 of *LNCS*, pp. 338–357. Springer-Verlag, 1994.
19. K. Lautenbach, A. Pagnoni. *Invariance and duality in predicate/transition nets and in coloured nets*, Arbeitspapiere der GMD 132, 1985
20. M. Lindqvist. *Parameterized Reachability for Predicate/Transition Nets*, vol. 54 of *Acta Polytechnica Scandinavica, Mathematics and Computer Science*. Helsinki, 1989.
21. M. Lindqvist. Parameterized reachability for predicate/transition nets. In *Advances in Petri Nets*, vol. 674 of *LNCS*, pp. 321–324. Springer-Verlag, 1993.
22. W. Reisig. *Petri-Nets*. Springer-Verlag, 1985.
23. W. Reisig. Petri nets and algebraic specifications. *Theoretical Comput. Sci.*, 80:1–34, 1991.
24. W. Reisig, J. Vautherin. An algebraic approach to high level Petri nets. In *Proc. 8th Workshop on Applications and Theory of Petri Nets*, pp. 51–72, Zaragoza (Spain), 1987.
25. G. Rozenberg, P. S. Thiagarajan. Petri nets: Basic notions, structure, behaviour. In *Current Trends in Concurrency*, vol. 224 of *LNCS*, pp. 585–668. Springer-Verlag, 1986.
26. K. Schmidt. Parameterized reachability trees for algebraic Petri nets. In *Application and Theory of Petri Nets 1995*, vol. 935 of *LNCS*, pp. 392–411. Springer-Verlag, 1995.
27. E. Smith. On net systems generated by process foldings. In *Advances in Petri Nets 1991*, vol. 524 of *LNCS*, pp. 253–276. Springer-Verlag, 1991.
28. E. Smith. A primer on high-level Petri-net theory. In C. Fernandez et. al., editors, *Advanced Course on Petri Nets*, pp. 114–140. Editorial de la Universidad de Santiago de Chile, 1996.
29. E. Smith, W. Reisig. The semantics of a net is a net – an exercise in general net theory. In K. Voss, H. Genrich, and G. Rozenberg, editors, *Concurrency and Nets*, Advances in Petri Nets, pp. 461–480. Springer-Verlag, 1987.
30. V. Sperschneider, G. Antoniou. *Logic: A Foundation for Computer Science*. Addison-Wesley, 1991.
31. A. Valmari. Stubborn sets for coloured Petri nets. In *Proc. 12th Int. Conf. on Appl. and Theory of Petri Nets 1991*, pp. 102–121.

Petri Nets in Performance Analysis: An Introduction

M. Ajmone Marsan[1], A. Bobbio[2], and S. Donatelli[2]

[1] Dipartimento di Elettronica
Politecnico di Torino – Italy
ajmone@polito.it

[2] Dipartimento di Informatica
Università di Torino – Italy
bobbio,susi@di.unito.it

Abstract. *In this tutorial paper, the authors discuss the motivations that led to the adoption of Petri nets for performance evaluation, define the class of Petri nets that is most frequently used for performance analysis, and present the subclasses that allow a simpler derivation of performance metrics. Definitions and discussions are paralleled with examples, thus visualizing the strong and weak points of the different alternatives.*

KEY WORDS - Stochastic Petri nets, Performance evaluation, Markov chains, Queues.

1 Introduction

Petri nets (PN) [86, 87, 91] were originally developed and used for the study of the *qualitative* properties of systems exhibiting concurrency and synchronization characteristics.

The use of PN-based techniques for the *quantitative* analysis of systems requires the introduction of temporal specifications within the basic, untimed models.

This fact was recognized about twenty years ago, and led to several different proposals for the introduction of temporal specifications in PN. The main alternatives that characterize the different proposals concern

- the PN elements associated with timing (normally either places or transitions, but some authors also looked into the possibility of defining timed arcs or tokens),
- the firing semantics in the case of timed transitions (either atomic firing or firing in three phases),

- the nature of the temporal specification (either deterministic or probabilistic),
- the conflict resolution policy.

In this tutorial paper we consider PN models that are augmented with a temporal specification by associating a (possibly null) firing delay with transitions. The transition firing operation is assumed to be atomic, i.e., tokens are removed from input places and put into output places with a single, indivisible operation, after the transition firing delay has elapsed. The specification of the firing delay of timed transitions is of probabilistic nature, so that either the probability density function (pdf) or the cumulative distribution function (Cdf) of the delay associated with a transition needs to be specified. Such functions may be general, or even degenerate, thus allowing the definition of constant (possibly null) delays. We refer to this type of timed Petri nets as Generally Distributed Timed Transitions Stochastic Petri Nets (GDTT_SPN).

Such timed Petri nets are those most frequently used in the field of performance evaluation. Evidence of this fact can be gathered by the observation of the papers presented at the series of International Workshops on Petri Nets and Performance Models, the leading forum for the presentation of novel results in this field.

In this work we review the different classes of GDTT_SPN that have been proposed in the literature, and we shall consider in some details two special subclasses of GDTT_SPN: i) Stochastic Petri Nets (SPN), where all transition firing delays are non-null and have negative exponential pdf, ii) generalized SPN (GSPN), where immediate (null-delay) transitions are freely mixed with timed transitions associated with exponentially distributed non-null random firing delays.

The goal of this paper is to provide a tutorial introduction to GDTT_SPN and their subclasses, discussing several aspects related to their use for performance analysis of complex systems, specially emphasizing their strong and weak points, and the feasibility of their solution.

This tutorial paper is addressed to PN experts who are not familiar with the stochastic performance modelling field. For this reason, a brief overview of the classical approach to the performance evaluation of systems in a probabilistic framework is included in Section 2, where some elementary notions are summarized. Basic tools are discussed first, with special attention to stochastic processes with Markovian characteristics, and their generalizations. The most common model specification techniques are then presented, starting from queues and queueing networks, and arriving at PN-based approaches; along the path from one modelling paradigm to the other we shall note an increase in modelling power, that is however paid with a reduction in the ease of the model solution.

Section 3 contains the general definition of GDTT_SPN models, and a discussion of their possible use for performance analysis.

Sections 4 and 5 are specifically addressed to the two subclasses of GDTT_SPN that we discuss in more details: SPN, and GSPN, respectively. The presentation of the different modelling paradigms is paralleled with examples of their application, thus visualizing the strong and weak points of the different alternatives.

Finally, Section 6 provides some concluding remarks, and comments on the present and future trends of research in this field.

2 Performance Evaluation

The performance evaluation area can be initially subdivided into two subareas. The first one relates to *measuring*, and comprises three distinct fields that can be called

- measurements,
- benchmarks,
- prototypes.

Measurements are performed on a real system under real operating conditions. They provide the *actual* system performance in the particular condition in which the system is observed. However, measurement results have very little generality, since they are heavily dependent upon the detailed characteristics of the measured system, and on the particular workload imposed on the system during the measurement.

When the performances of two systems, say two supercomputers, have to be compared, it is not sufficient to rely upon measurements, since nothing guarantees that the operating conditions under which measurements are performed are equivalent. The comparison would thus be unfair. In order to overcome this problem, benchmarks were developed. They provide an artificial workload for the system, such that observations can be performed in equivalent operating conditions, and meaningful comparisons can be made.

Both measurements and benchmarks require the availability of the system to be studied, so that it can be observed. In the (many) cases in which the performance study concerns a system that is not available (maybe because it is not yet operational), it is necessary to develop a representative approximation of it, either in hardware or in software. Such approximations, which need to be fairly detailed, are normally called prototypes (the term emulator is also often used when the approximation is implemented in software). Observations are then made on such prototypes, possibly using benchmarks as artificial workloads.

In all three cases, the system performance is obtained by observing the behaviour of the system, or its approximations, in operation, i.e., when loaded by either the actual user requests, or the benchmark.

The study of the performance of a system, however, is not only an important task during and after the system implementation, but also during the early design stages, in order to compare possible alternate architectural choices. This is true in particular when the development of new systems is mainly motivated with the request for ever-increasing performance, like in the computer and telecommunications fields.

During the design process, measurements on real systems are obviously not possible, and also prototype implementations present insurmountable difficulties due to the necessity of specifying many details that are far from being decided.

The second subarea of performance evaluation thus comes into play: *modelling*. It can be partitioned into two fields:

– simulation models,
– analytical models.

In both cases the performance study is carried out using a description that includes only some "important" characteristics of the system. In the case of simulation models, the description is embedded into the computer program that simulates the system, whereas in the case of analytical models the description is given in mathematical terms. Goal of the analysis is to evaluate a set of "performance indices," like, for example, the percentage of time the system is idle, or the average amount of useful work performed by the system in a fixed period of operation, or the variability of the quality of the service offered to the final (human) user of the system.

Models (both simulative and analytical) can be either deterministic or probabilistic. While it is clear that most systems of interest exhibit a deterministic behaviour (we tend to like the fact that by running twice the same program with the same input data we obtain the same results, or that two transfers of the same file produce identical copies), it may be simpler to describe a very large number of complex, detailed deterministic phenomena by means of macroscopic probabilistic assumptions. This is often done because details are not known, and even when they are, their inclusion may lead to very complex models. Furthermore, the probabilistic approach may be advantageous because it may provide sufficient accuracy while yielding more general results, and it may permit the study of sensitivity to parameter variations.

It should be noted that a key element in the development of a model is the selection of the *level of abstraction* (also called level of detail). This amounts to selecting the system features to be included in the model. No precise rule exists for this selection, that rests mainly on the experience and ingenuity of the performance analyst. On the other hand, the level of abstraction is the element that differentiates a model from a prototype or an emulator. Simulation lends itself better to the development of more detailed models, whereas analytical models are normally more abstract.

An important characteristic of models concerns the representation of the system behaviour along the time scale. While it is obvious that any instrument for the measurement of time operates according to a discrete time scale, due to its finite precision, and that most interesting modern systems, being digital in nature, intrinsically use a discrete time scale, models often use a *continuous* time scale. The reason for this discrepancy lies in the greater simplicity of continuous-time models. Indeed, if the time axis is discrete, the model has to consider the fact that multiple events may occur between two consecutive time marks, and explore the effect of all possible combinations and orderings of these events. In the continuous time scale, instead, using appropriate probabilistic assumptions, it is possible to univocally order events, so that it is always possible to take into consideration only one event at a time.

In this paper we deal with models of a probabilistic nature operating on a continuous time scale.

The mathematical framework underlying this class of models, be they simulative or analytical, is the theory of stochastic processes.

2.1 Stochastic Processes

Random phenomena are close to our everyday experience, at least due to our familiarity with unpredictable weather changes, equipment failures, and games of chance based on dices or cards (excluding the tricks played by magicians, that, when successful, leave no space for casuality).

A stochastic process is a mathematical model useful for the description of phenomena of a probabilistic nature as a function of a parameter that usually has the meaning of time. Many text books on stochastic processes are available, see for example [70].

Since the definition of a stochastic process is based on the notion of a *random variable*, it is necessary to recall some elementary concepts of probability theory first.

A *random experiment* is an experiment which may have several different outcomes. The set of all possible elementary outcomes is the *sample space* of the experiment. A simple example of a random experiment is provided by the toss of a fair dice. The sample space is, in this case, comprised of six elementary outcomes. By associating a probability measure to all possible (elementary and complex) outcomes of a random experiment we construct a *probability space*. Continuing with our example, we can associate probability 1/6 with each elementary result of the dice toss, and appropriate probabilities to complex results such as "more than one and less or equal to five, but not equal to three".

A random variable is a real function defined over a probability space; for example, a random variable could associate the value $2\pi i$ (where $\pi = 3.1415\ldots$) to the elementary result $i, i = 1, 2, \ldots, 6$. The set of possible values of the function is the *state space* of the random variable.

The probabilistic characterization of a random variable X is given in terms of its Cdf

$$F_X(x) = P\{X \leq x\}$$

which is a real, nonnegative, nondecreasing function of x for which

$$\lim_{x \to -\infty} F_X(x) = 0$$

and

$$\lim_{x \to \infty} F_X(x) = 1$$

Alternatively, the random variable X can be described by its pdf

$$f_X(x) = \frac{d}{dx} F_X(x)$$

which is a nonnegative function for which

$$\int_{-\infty}^{\infty} f_X(x)dx = 1$$

In the case of random variables assuming values in a discrete set, instead of using their pdf, which is a generalized function, it may be more convenient to use their probability mass function (pmf)

$$\boldsymbol{p}_X = (p_1, p_2, p_3, \ldots)$$

which is a vector whose entries

$$p_i = P\{X = x_i\} \qquad i = 1, 2, \ldots$$

are the probabilities that the random variable equals one of the admissible values.

The probabilistic characterization of a random vector $\boldsymbol{X}$ comprising n random variables X_i, $i = 1, 2, \ldots, n$ is given either by the joint Cdf of the n random variables:

$$F_{\boldsymbol{X}}(\boldsymbol{x}) = P\{X_1 \leq x_1, X_2 \leq x_2, \ldots, X_n \leq x_n\}$$

or by their joint pdf

$$f_{\boldsymbol{X}}(\boldsymbol{x}) = \frac{\partial^n}{\partial x_1 \partial x_2 \ldots \partial x_n} F_{\boldsymbol{X}}(\boldsymbol{x})$$

The importance of the probabilistic characterization of a random variable or a random vector is in the fact that it provides the tool for the mathematical formulation of any problem involving the random quantity itself.

We are now ready to give the definition of a stochastic process. A stochastic process $\{X(t), t \in T\}$ is a family of random variables defined over the same probability space, taking values in the state space S, and indexed by the parameter t, which assumes values in the set T; normally $T = [0, \infty)$, and t is usually interpreted as "time".

A stochastic process can be visualized as a family of functions of time, called *sample paths* of the process. Each sample path defines a particular trajectory over the state space, and corresponds to a possible observed behaviour of the process. Consider for example the stochastic process modelling the state of a door (either closed or open). Each sample path is a function of time made of steps corresponding to the durations of the open and closed times. The observation of the process eliminates the uncertainty on its evolution, and thus yields (at least for the observed time period) a sample path (the same difference is found before and after the toss of a dice). Figure 1 depicts the sample path of a continuous-time stochastic process: each step represents the sojourn time in a state (black dots denote right-continuity). The set of all possible sample paths, together with a probability measure, may provide an alternate description of a stochastic process, which is however normally impractical.

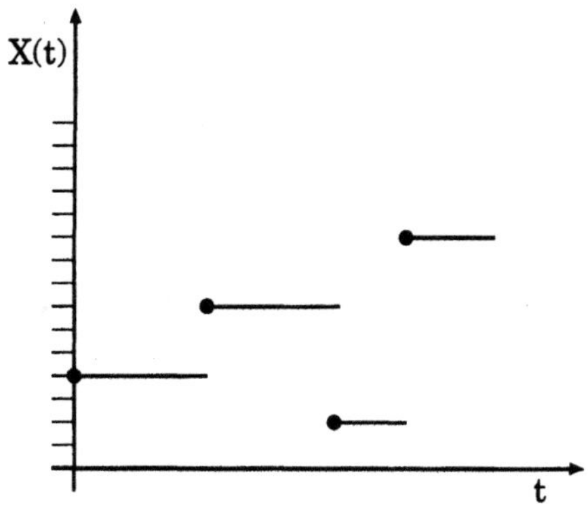

Fig. 1. Sample path of a continuous-time stochastic process.

2.2 Markov Processes and Their Generalizations

The complete probabilistic characterization of a random process requires the characterization of any random vector $\boldsymbol{X}$ comprising an arbitrary number k of random variables $X(t_i)$ extracted from the process at any set of time instants $\{t_i\}_{i=1}^k$.

In the general case, the complete probabilistic characterization of a stochastic process is a formidable task. Special classes of stochastic processes for which the probabilistic characterization is simpler are of particular interest.

One such class is comprised of *Markov processes*. A Markov process is a stochastic process that satisfies the *Markovian property*

$$P\{X(\tau) \leq x | X(t), t \in [0, \theta]\} = P\{X(\tau) \leq x | X(\theta) = y\}$$

for any $\tau > \theta$.

Note that the Markovian property defines a stochastic process for which the behaviour in the future (at some time τ) depends only on the present situation (at time θ), not on the past history. In other words, a Markov process has no memory of the trajectory followed to reach the present state. This condition is not met by many real life systems; nevertheless, Markovian processes are widely used for the construction of stochastic models of discrete event systems. Their main merit lies in their low analysis complexity, and in the possibility of coping with the main sources of memory in the system behaviour with an accurate definition of the process state.

Markov processes with a discrete state space are called *Markov chains*. If the parameter t is discrete, the process is a discrete-time Markov chain (DTMC). If the parameter t is continuous, the process is a continuous-time Markov chain (CTMC). The time spent in states of a CTMC is a random variable with negative exponential pdf, as we shall see.

Another class of stochastic processes with discrete state space that allow a simple description and analysis is that of *Semi-Markov Processes* (SMP). Calling

δ_i the instant of time in which the stochastic process $X(t)$ changes state for the *ith* time, and Y_i the random variable that describes the state of the process between δ_i and δ_{i+1}, then the stochastic process $X(t)$ is Semi-Markov if and only if:

$$P\{Y_{n+1} = j, \delta_{n+1} - \delta_n \leq \tau | (Y_k, \delta_k), k \in [0, n]\} = P\{Y_{n+1} = j, \delta_{n+1} - \delta_n \leq \tau | Y_n = i\}$$

the latter probability is often written as $H_{ij}(\tau)$, to emphasize the dependency of the process future behaviour on the current state and the elapsed sojourn time in that state.

The stochastic sequence $\{Y_n, n \geq 0\}$ is a DTMC, called the Embedded Markov Chain (EMC) of the SMP. Therefore, also in this case, at the instant in which the state of $X(t)$ changes, the future of the stochastic process only depends on the present state, but now the pdf of the time spent in states is no longer exponential.

A further generalization of both CTMC and SMP is obtained with Markov Regenerative Processes (MRP), that allow the identification of a number of instants when the process changes state and "regenerates": at these instant the future evolution of the process only depends on the current state.

A stochastic process $X(t)$ is a MRP if and only if it comprises a renewal sequence of random variables $\{(Y_n, t_n), n \geq 0\}$ such that

$$P\{X(t_n + \tau) = j | X(t), t \in [0, t_n], Y_n = i\} = P\{X(\tau) = j | Y_0 = i\} \qquad (1)$$

The sequence of states at regeneration instants forms a DTMC, but now between two consecutive regeneration points it is possible to have many (possibly infinitely many) changes of states, and the distribution of the time between two consecutive regeneration points needs not be exponential.

2.3 Continuous-time Markov chains

The Markovian property requires that sojourn times in states be exponentially distributed random variables. Indeed, the negative exponential pdf

$$f_X(x) = \mu e^{-\mu x} u(x)$$

where $u(x)$ is the unit step function[1], and μ is the parameter (or rate) of the pdf, is the only continuous pdf for which the *memoryless property*

$$P\{X \geq x + \alpha | X \geq \alpha\} = P\{X \geq x\}$$

holds. Hence, at any time instant, the residual sojourn time in a state does not depend on the time already spent in the state (i.e., on the past history), but only on the present state, as required by the Markovian property.

Note that the negative exponential pdf is characterized by just one parameter, μ, whose inverse μ^{-1} identifies the average value of the random variable.

[1] The unit step function $u(t)$ is such that $u(t) = 0$ for $t < 0$, and $u(t) = 1$ for $t \geq 0$.

These considerations imply that for the complete probabilistic description of a CTMC it is sufficient to give the pmf over the state space S at the initial time (typically 0), as well as the averages of the negative exponential pdf describing the sojourn times in all states in S, and the probabilities of moving from one state to another.

In practice, a CTMC is described through either a ***state transition rate diagram*** or a transition rate matrix, also called ***infinitesimal generator*** and denoted by $\boldsymbol{Q}$. The state transition rate diagram is a labelled directed graph whose vertices are labelled with the CTMC states, and whose arcs are labelled with the rate of the exponential distribution associated with the transition from a state to another. The infinitesimal generator is a matrix whose elements outside the main diagonal are the rates of the exponential distributions associated with the transitions from state to state, while the elements on the main diagonal make the sum of the elements of each row equal to zero.

In Figure 2 we show the state transition rate diagram for a CTMC with two states (closed and open door in our previous example), for which the average sojourn time in state 1 is λ^{-1}, and the average sojourn time in state 2 is μ^{-1}. The infinitesimal generator for such a CTMC is

$$\boldsymbol{Q} = \begin{bmatrix} -\lambda & \lambda \\ \mu & -\mu \end{bmatrix}$$

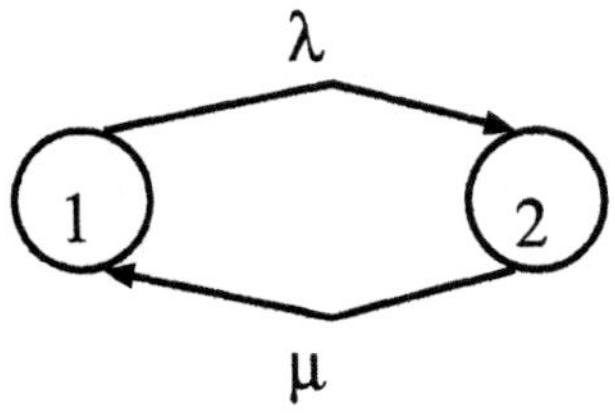

Fig. 2. State transition rate diagram for a CTMC with two states.

The solution of a CTMC model consists of the computation of the pmf over the state space S either at any arbitrary time instant t or in equilibrium conditions. When an equilibrium or ***steady-state*** pmf exists, and is independent of the initial state, the CTMC is said to be *ergodic* [42, 53, 64].

Denoting by

$$\pi_i(t) = P\{X(t) = i\}$$

the probability that the CTMC is in state i at time t, the pmf at time t

$$\boldsymbol{\pi}(t) = (\pi_1(t), \pi_2(t), \pi_3(t), \ldots)$$

is defined by the differential equation

$$\frac{d\boldsymbol{\pi}(t)}{dt} = \boldsymbol{\pi}(t)\,\boldsymbol{Q}$$

with initial condition $\boldsymbol{\pi}(0)$, whose solution can be expressed as

$$\boldsymbol{\pi}(t) = \boldsymbol{\pi}(0)e^{\boldsymbol{Q}t}$$

where $e^{\boldsymbol{Q}t}$ is the matrix exponential defined by

$$e^{\boldsymbol{Q}t} = \sum_{k=0}^{\infty} \frac{(\boldsymbol{Q}t)^k}{k!}$$

Letting

$$\pi_i = \lim_{t \to \infty} P\{X(t) = i\}$$

in the case of ergodic CTMC, the steady-state pmf

$$\boldsymbol{\pi} = (\pi_1, \pi_2, \pi_3, \ldots)$$

can be obtained as the solution of the system of linear equations

$$\boldsymbol{\pi Q} = \boldsymbol{0}$$

augmented with the normalization condition

$$\sum_i \pi_i = 1$$

The interested reader can find in [96, 99] an in-depth treatment of the problem posed by the solution of the above equations.

A modelling interpretation of the steady-state probabilities is the following: π_i is the probability according to which a random observer finds the system in equilibrium at state i, or equivalently the percentage of time that the system spends in state i when in equilibrium conditions.

The cost of solving the linear system $\boldsymbol{\pi Q} = \boldsymbol{0}$ is polynomial in the number of states. Iterative techniques are often applied, whose cost per iteration is of the order of the number of nonzero elements in $\boldsymbol{Q}$, i.e., of the order of the number of arcs in the state transition diagram.

From the steady-state pmf it is possible to derive many parameters of interest to quantify the system performance.

As an example, consider a lamp equipped with one lightbulb. The lamp may be turned on and off, and the lightbulb can fail while the lamp is on. Failed bulbs are replaced with new ones, and before the replacement operation is performed, the lamp switch is set in the off position.

We can easily identify three states in our system: 1) off, 2) on, and 3) failed.

The transitions from state to state obey the following rules:

- when the lamp is off, it may be turned on,
- when the lamp is on, either it can be turned off, or the bulb can fail,
- when the lightbulb fails, it is replaced by a new one, after switching off the lamp.

In order to obtain a CTMC model, we need to introduce temporal specifications such that the evolution in the future depends only on the present state, not on the past history. To this purpose we assume that:

- the time periods during which the lamp is off are exponentially distributed with parameter β,
- the time periods during which the lamp is on are exponentially distributed with parameter α,
- the lightbulb lifetime (sum of the durations of the on periods before a fault) is exponentially distributed with parameter μ,
- the lamp repair time is exponentially distributed with parameter λ.

The state transition rate diagram of the resulting CTMC is depicted in Figure 3, and the infinitesimal generator is

$$\boldsymbol{Q} = \begin{bmatrix} -\beta & \beta & 0 \\ \alpha & -(\alpha+\mu) & \mu \\ \lambda & 0 & -\lambda \end{bmatrix}$$

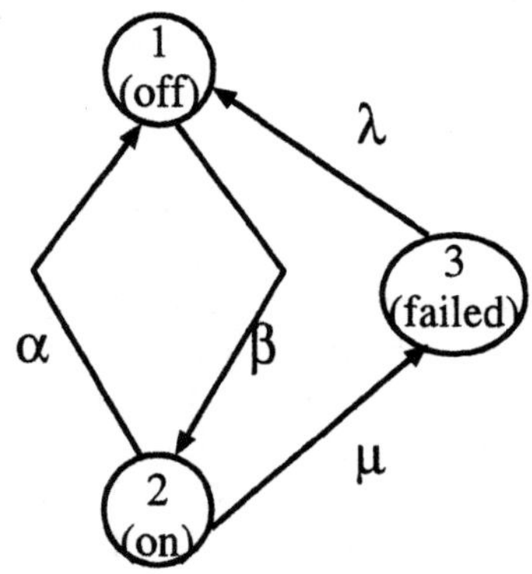

Fig. 3. State transition rate diagram for the CTMC describing the behaviour of a lamp.

The CTMC is ergodic, and the steady-state distribution is easily computed by solving the system of linear equations:

$$\begin{aligned} \beta\pi_1 &= \alpha\pi_2 + \lambda\pi_3 \\ (\alpha+\mu)\pi_2 &= \beta\pi_1 \\ \lambda\pi_3 &= \mu\pi_2 \\ \pi_1 + \pi_2 + \pi_3 &= 1 \end{aligned}$$

obtaining

$$\pi = \frac{1}{\lambda(\alpha+\beta)+\mu(\lambda+\beta)}(\lambda(\alpha+\mu),\ \lambda\beta,\ \beta\mu)$$

Note that the first three equations of the linear system above can be interpreted as equalities of the flow into and out of a given state, where the probability flow over an arc is the product of the steady-state probability of the state from which the arc originates times the arc label. Thus for example in the case of state 1 we get

$$\text{flow out} = \beta\pi_1$$

$$\text{flow in} = \alpha\pi_2 + \lambda\pi_3$$

This also implies the linear dependency of the equations; however, the solution is unique when considering the fact that probabilities must sum to 1.

From $\boldsymbol{\pi}$ it is possible to compute several steady-state performance indices:

- π_2 is the fraction of time in which the lamp is on,
- π_3 is the fraction of time in which the bulb is failed,
- $[\lambda\pi_3] = [\mu\pi_2]$ is the mean number of failures in unit time (the failing throughput),
- $[\lambda\pi_3]^{-1} = [\mu\pi_2]^{-1}$ is the average time between two consecutive failures,
- $[(\alpha + \mu)\pi_2]^{-1} = [\beta\pi_1]^{-1}$ is the average time between two consecutive instants at which the lamp is turned on.

From the transient solution $\boldsymbol{\pi}(t)$ we can derive other interesting performance indices:

- $\pi_2(t)$ probability of the lamp being on at time t,
- $\pi_3(t)$ probability of the bulb being failed at time t,

and, if no repair is possible in the system (i.e., the arc between states 3 and 1 in the CTMC model has rate 0, or is removed), we can compute the probability of the bulb not having yet failed at time t as $1 - \pi_3(t)$.

Another simple example of a CTMC is provided by the *Poisson process*. In this case the state space comprises all nonnegative integers, and transitions are possible only from state i to state $i + 1$ for all $i \geq 0$. Sojourn times in states are independent random variables with negative exponential pdf, and mean independent of the state. The parameter of such exponential pdf is the *rate* of the Poisson process. The Poisson process obviously never reaches an equilibrium condition and hence is not ergodic. The pmf at time t of a Poisson process with rate λ comprises probabilities

$$\pi_i(t) = \frac{(\lambda t)^i}{i!} e^{-\lambda t}\ u(t) \qquad i \geq 0$$

assuming that

$$\pi_0(0) = 1$$

These probabilities form a Poisson pmf.

Constructing models of complex systems directly at the CTMC level is generally *difficult*, mainly due to the need of choosing an appropriate state definition, and enumerating all states in the evaluation of transition rates. For this reason, *more abstract* probabilistic modelling tools were proposed. The main such tools are based on queueing theory or Petri nets.

2.4 Queues

A queue [7, 43, 45, 47, 55, 69, 71] is a system to which *customers* arrive to receive service by a *service station*. The service station may comprise one or more *servers*. When all servers are busy, customers are forced to wait in a *waiting room*. At the end of service, customers leave the queue. A pictorial representation of a queue is given in Figure 4.

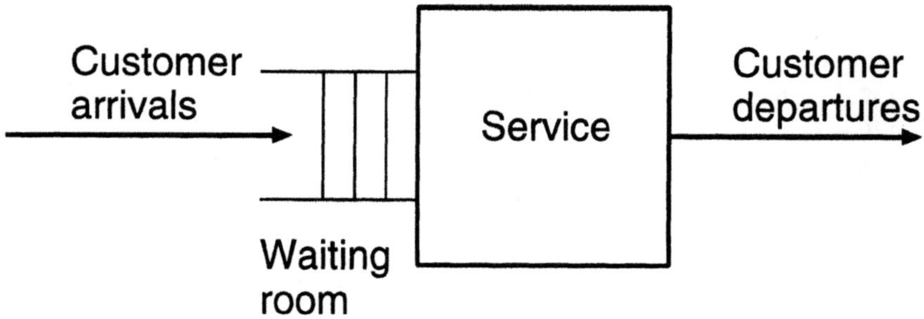

Fig. 4. Pictorial representation of a queue.

A queue is a compact description of a probabilistic (not necessarily Markovian) model in which users (customers) share resources (servers). The probabilistic characterization of the model is comprised of the stochastic process describing the arrival of customers, and the random variables describing the customer service times. Other parameters of a queueing model are:

- the number of servers in the service station,
- the size of the waiting room,
- the size of the customer population,
- the queueing discipline.

The simplest queue is known with the acronym M/M/1. The first symbol M identifies the arrival process as Markovian, and precisely as a Poisson process with a fixed rate, say λ. The second symbol M identifies the service time as Markovian (hence with negative exponential pdf); an average service time μ^{-1} is considered. The symbol 1 refers to the presence of only one server in the service station. Furthermore, the size of the waiting room and the customer population are taken to be unlimited, and the first-come-first-served discipline is used for the selection of the next customer to be served among those in the waiting room.

The CTMC corresponding to the M/M/1 queue has the state transition rate diagram depicted in Figure 5, which corresponds to the infinitesimal generator

$$Q = \begin{bmatrix} -\lambda & \lambda & 0 & 0 & 0 & 0 & \cdots \\ \mu & -(\lambda+\mu) & \lambda & 0 & 0 & 0 & \cdots \\ 0 & \mu & -(\lambda+\mu) & \lambda & 0 & 0 & \cdots \\ 0 & 0 & \mu & -(\lambda+\mu) & \lambda & 0 & \cdots \\ \cdot & \cdot & \cdot & \cdot & \cdot & \cdot & \cdot \\ \cdot & \cdot & \cdot & \cdot & \cdot & \cdot & \cdot \\ \cdot & \cdot & \cdot & \cdot & \cdot & \cdot & \cdot \end{bmatrix}$$

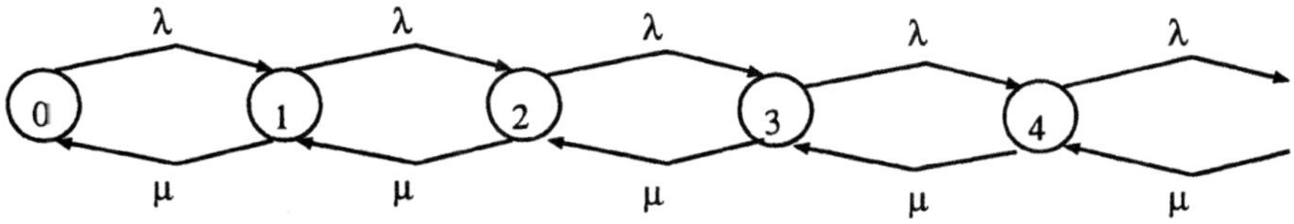

Fig. 5. State transition rate diagram of the CTMC generated by the M/M/1 queue.

The solution of the M/M/1 queue at time t, leads to a rather complicate expression, that can be found, for example, in [69]. When $\lambda < \mu$ the CTMC is ergodic, and the steady-state probabilities can be expressed as:

$$\pi_i = \left(1 - \frac{\lambda}{\mu}\right)\left(\frac{\lambda}{\mu}\right)^i \quad i \geq 0$$

Many queueing models exist with more elaborate characteristics for what concerns the customer arrival process, the pdf of the customer service time, the number of servers, the size of the waiting room, the size of the customer population, the queueing discipline.

Nevertheless, a single queue may not be adequate to describe complex system behaviours, where customers may require the service of many different servers, in different orders. For this reason, a more flexible formalism was introduced, and it is still one of the most popular in performance evaluation: queueing networks.

2.5 Queuing networks

A queueing network [7, 55, 71] is a system of interconnected queues in which customers circulate, and possibly arrive from, and leave to, the outside world. When no arrivals from, and departures to, the external world are possible, the queueing network is said to be closed; otherwise it is said to be open.

The path followed by customers in the network is determined by routing probabilities.

As an example, an open queueing network comprising three queues is depicted in Figure 6.

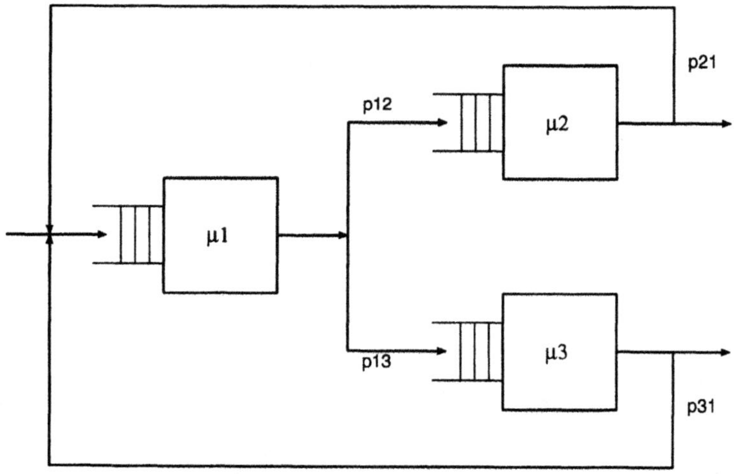

Fig. 6. An open queueing network comprising three queues.

With queueing networks it is possible to construct models of systems where the sharing of individual resources is represented in more detail than it would be possible if the system model had to be constructed using only one queue.

Queuing networks have become extremely popular in the applied stochastic modelling field for a wide gamut of different application areas, such as telecommunications, computers, manufacturing, and transportation. For example, Figure 6 can be considered as a description of a very simple manufacturing process: servers at queues represent machines, and customers are products that are being manufactured. Each product is first worked at machine 1, and it is then moved to either machine 2 or 3. After being worked at machine 2 or 3 the product is either considered finished, and it leaves the system, or it may require to repeat the sequence of operations on machine 1 followed by machine 2 or 3.

The main reason for which queueing networks have become so popular is due to the *product form* solution property that holds for a fairly wide class of these models. This property implies that the steady-state solution of the queueing network can be *factored* in the product of the steady-state solutions of the individual queues, and hence obtained with very limited complexity (typically polynomial in the number of queues and customers).

It should be stressed that queueing theory is fairly advanced in the case of continuous-time models. The analysis of discrete-time models is much more complex because of the reasons mentioned in Section 2, and, for example, the product-form characteristic of queueing networks is not retained in the case of discrete-time models, except for some cases of limited practical interest.

The shortcomings of queueing-based models are mainly due to their lack of descriptive power in presence of phenomena such as *synchronization, blocking, splitting of customers,* and to the fact that most of these features, quite common in distributed systems, generally destroy the product-form characteristic, so that even a simple queueing model must be translated into its corresponding CTMC for the solution phase.

To cope with the lack of modelling power of queueing networks, many authors have introduced special "queues", that allow the description of the phenomena mentioned above, but there is no commonly agreed language for extended queueing networks, and this special queues usually do not have a formally defined semantics.

2.6 Timed Petri nets

When blocking and synchronization phenomena are important characteristics of the system to be modelled, their description with queueing networks is not natural, and the model solution must be obtained (with few exceptions) from the CTMC translation of the queueing model.

The use of Petri nets (PN) for performance analysis comes into play in this environment, where they are basically equivalent to extended queueing models from the point of view of the model solution, since both require a translation into the underlying CTMC, but PN offer a language in which synchronization, blocking, and splitting are native in the formalism, and in its semantics. Moreover, PN models benefit from the availability of a wide range of qualitative results derived in a number of years of lively research; these qualitative results allow for example to check for deadlocks, livelocks, etc.

The use of PN-based techniques for performance analysis requires the introduction of temporal specifications within the basic, untimed models, thus generating the modelling paradigms that are usually named Timed Petri Nets (TPN).

Of the numerous TPN proposals that appeared in the literature, we consider only the case in which timing is associated with transitions that keep the atomic firing semantics typical of the "untimed" PN world. We thus neglect the approaches based on timed places, timed arcs, timed tokens, as well as timed transitions that operate in three phases, removing tokens from input places as soon as they become enabled, then letting the transition delay elapse, and finally generating tokens into output places. Although preferences about modelling paradigms are very personal, it may be fair to say that the class of TPN that we consider in this paper is the one that gained widespread acceptance among researchers in the field.

Consistently with the stochastic modelling technique that characterizes queueing approaches, delays associated with TPN transitions will be assumed to be of probabilistic nature.

In summary, we focus on TPN models where a random delay is associated with transitions whose firing is assumed to be atomic, i.e., tokens are removed from input places and put into output places with a single indivisible operation, after the transition firing delay has elapsed. The specification of the firing delay of timed transitions is of probabilistic nature, so that either the probability density function (pdf) or the cumulative distribution function (Cdf) of the delay associated with a transition needs to be specified. Such functions may be general, or even degenerate, thus allowing the definition of constant (possibly null) delays. We refer to this type of timed Petri nets as Generally Distributed Timed Transitions Stochastic Petri Nets (GDTT_SPN).

The class of TPN that we consider is however too wide to allow a simple solution of any GDTT_SPN model; for this reason we shall pay special attention to two special subclasses of GDTT_SPN, that have the nice property of permitting a reasonably simple evaluation of performance metrics:

- Stochastic Petri Nets (SPN), where all transition firing delays are non-null and have negative exponential pdf,
- generalized SPN (GSPN), where immediate (null-delay) transitions are freely mixed with timed transitions associated with exponentially distributed non-null random firing delays.

When a GDTT_SPN model of a system has been developed, this is normally translated into its underlying stochastic process, that is analyzed either in steady-state or in transient conditions. The computation of the corresponding probabilities can be translated into performance metrics that have a net-level semantics; for example, a net-level result may be the distribution of the number of tokens into a particular place at steady-state, or the average number of times a transition fires during a specified time interval. The mapping between the system features and their description within the GDTT_SPN model allows the translation of such net-level performance parameters into the system-level performance metrics of interest.

3 Generally Distributed Timed Transitions Stochastic Petri Nets

For the translation of a GDTT_SPN model into its underlying stochastic process, it is necessary to associate with the model an *execution policy*, comprising two specifications: a rule to choose the next transition to fire in any marking (the *firing policy*), and a criterion to account for the past history of the model whenever a transition fires (the *memory policy*)[2].

As regards the firing policy, two alternatives are basically possible: either use the delays associated with transitions to decide which one will fire next, or add a specific metrics for this purpose. The GDTT_SPN formalisms that we consider in this paper adopt the first option, that corresponds to the *race policy*: the transition with the minimum remaining delay is the one that fires first.

As regards the memory policy, again two basic alternatives are possible at every change of marking:

[2] In (untimed) Petri nets, the next transition to fire is chosen non deterministically, and there is no need to record the "past history," which is captured by the current state (an intrinsically Markovian assumption!)

continue: the timers[3] associated with transitions hold their present values and will continue being decremented later on;

restart: the timers associated with transitions are restarted, i.e., their present values are discarded, and new values will be generated when needed.

The memory policy is implemented whenever a transition fires. The memory policy thus affects transitions that fire as well as transitions that lose their enabling due to the change of marking, and transitions that keep their enabling in the new marking. The memory of transitions that fire is irrelevant, since in this case a new delay instance must always be generated. The memory of transitions that do not fire is often assumed to be of the following types:

Resampling - The timer of the transition is reset to a new value at any change of marking. The new value is sampled from the pdf of the delay associated with the transition.

Enabling memory - If in the new marking the transition is still enabled, the value of the timer is kept; it is instead reset to a new value if the transition is not enabled.

Age memory - The timer value is kept, even if the transition is not enabled in the new marking.

Table 3 summarizes the possible memory combinations, depending on the transition enabling in the new marking.

	transition remains enabled	transition loses enabling
Resampling	restart	restart
Enabling memory	continue	restart
Age memory	continue	continue

Table 1. Summary of the memory mechanisms.

A GDTT_SPN is a seven-tuple

$$GDTT\_SPN = (P, T, I, O, H, M_0, \mathcal{W}, \mathcal{E})$$

where (P, T, I, O, H, M_0) is the underlying PN system, which as usual comprises

- a set of places $P = (p_1, p_2, \ldots, p_m)$,
- a set of transitions $T = (t_1, t_2, \ldots, t_n)$,
- the input, output, and inhibitor functions $I, O, H : T \to N$,
- an initial marking $M_0 = (m_{01}, m_{02}, \ldots, m_{0m})$,

[3] We can describe the evolution of a GDTT_SPN by associating a timer with each transition: timers are decremented at constant speed while transitions are enabled, and when a timer runs down to zero the corresponding transition fires.

to which it is necessary to add

- a distribution function $\mathcal{W} : T \rightarrow \{\text{pdf}\}$, that assigns to each transition a random variable with a specified pdf,
- an execution policy function $\mathcal{E} : T \rightarrow \{\text{resampling, enabling, age}\}$ that assigns an execution policy to each transition.

The stochastic process that corresponds to the evolution of the GDTT_SPN over its state space (or reachability set) is called the *marking process.*

It should be noted that if $\mathcal{W}$ identifies only continuous functions, then the probability that two transitions are scheduled to fire at exactly the same instant is zero. In such case, a GDTT_SPN model evolves by firing transitions *one by one.*

A GDTT_SPN model can correspond to quite complex stochastic processes, whose definition, not to mention their solution, is not at all a trivial task. Researchers and practitioners have defined subclasses of GDTT_SPN corresponding to simpler classes of stochastic processes, that will be discussed in the sequel of this paper.

Stochastic Petri Nets (SPN)

When all the firing delays associated with transitions are exponentially distributed random variables, GDTT_SPN are called Stochastic Petri Nets (SPN).

The marking process generated by a SPN is a CTMC, with state space isomorphic to the reachability set.

This type of GDTT_SPN models is the most popular in the literature [1, 54, 80, 81, 83, 84], and a number of software tools are available for them [18, 36, 41, 46, 56, 73, 76].

We keep, in this context, the name that was initially assigned to this class of GDTT_SPN, although a more appropriate (less ambiguous) name could be Exponential Petri Nets. We discuss SPN at length in section 4.

Generalized Stochastic Petri Nets (GSPN)

Generalized SPN (GSPN) were originally proposed in [6], with the aim of allowing the simple modelling of complex state changes induced by the firing of a transition, as well as the representation within one model of activities that consume a significant amount of time and activities that require a negligible amount of time. GSPN models comprise therefore two types of transitions:

timed transitions, which are associated with random, exponentially distributed firing delays, as in SPN, and

immediate transitions, which fire in zero time, *with priority* over timed transitions.

We discuss GSPN at length in section 5.

Semi-Markov SPN

When all transitions in a GDTT_SPN are assigned a resampling policy, the marking process becomes a semi-Markov process, independently of the adopted $\mathcal{W}$ function.

This case was studied in [19, 84], but is of little interest in applications, since it is difficult to find systems where the firing of any transition of the corresponding GDTT_SPN has the effect of forcing the reset of the timers associated with all other transitions, even of those that are concurrently enabled.

A more interesting semi-Markov SPN model was defined in [52]. In this case, transitions are partitioned into three classes: exclusive, competitive and concurrent. Provided that the firing delays associated with all concurrent transitions are exponentially distributed, and that non-exponential competitive transitions are resampled whenever they become enabled, the associated marking process becomes a semi-Markov process.

Phase Type SPN (PHSPN)

A PHSPN is a GDTT_SPN in which:

- the function $\mathcal{W}$ associates with transitions PH (phase type) distributions [85] with a single initial stage and a single final stage,
- any timed transition is assigned a memory policy among the three defined alternatives: resampling, enabling or age memory.

The distinguishing feature of PHSPN is that it is possible to design a completely automated tool for their solution. The non-Markovian process generated by a PHSPN over the reachability set $\mathcal{R}(M_0)$ is converted into a CTMC defined over an expanded state space. The measures pertinent to the original process are defined at the PN level and can be evaluated by solving the expanded CTMC.

A program package that can solve this type of nets is ESP [48].

Deterministic SPN (DSPN)

Deterministic and Stochastic PN were defined in [10], with the aim of providing a technique for considering GDTT_SPN models in which not all transition firing delays are forced to be exponentially distributed.

In [10] only the steady-state solution of DSPN was studied. An improved algorithm for the evaluation of the steady-state probabilities was successively presented in [75, 76], and some structural extensions were proposed in [40].

A DSPN [10] is a GDTT_SPN in which:

- the function $\mathcal{W}(t)$ associates either exponentially distributed or deterministic firing times with timed transitions,
- at most one transition with deterministic delay is enabled in each marking,
- the execution policy for all deterministic transitions is *enabling memory.*

As a consequence of this definition, during the firing of a transition with a deterministic delay, the marking process can undergo state changes only due to exponentially timed transitions, thus describing a CTMC called the subordinated process.

The steady-state solution technique originally proposed in [10] is based on the evaluation of the subordinated CTMC at the firing time of the deterministic transition.

Tools currently supporting the steady-state analysis of DSPN models are DSPNexpress [76], UltraSAN [46] and TimeNET [56], this last one also supports transient analysis.

Choi et al. [37] proved that the marking process associated with a DSPN is a Markov regenerative process (MRP), for which steady-state and transient solution equations are available [42].

Markov Regenerative SPN (MRSPN)

A natural extension of DSPN was proposed in [38], where the deterministic distribution is substituted by a general one; this model is referred to by the authors as MRSPN, since the underlying model is again an MRP.

Similar restrictions as for DSPN apply also to MRSPN, that is to say:

- $\mathcal{W}(t)$ is either exponential or general,
- at most one transition with general pdf can be enabled in each marking,
- the only allowed execution policy for generally distributed transitions is *enabling memory*.

During the firing delay of a transition with general distribution, only exponential transitions can concurrently fire: the process subordinated to a generally distributed transition is thus a CTMC.

With the aim of extending the modelling power of MRSPN by including generally distributed transitions with age memory policy, Bobbio and Telek [20, 21] investigated a class of models characterized by the fact that the subordinated process between two consecutive regeneration epochs is a Semi Markov Reward Process [90].

GDTT_SPN Taxonomy

The relationships among the various subclasses of GDTT_SPN are depicted in Figure 7.

It should be noted that some of the definitions of the GDTT_SPN subclasses do not provide a structural characterization of the class. Indeed, the net structure is sufficient to decide whether a GDTT_SPN belongs to the SPN, PHSPN or GSPN classes (it is sufficient to check the $\mathcal{W}$ function). Instead, it may not be possible to determine from the GDTT_SPN structure, without building the reachability set, whether a GDTT_SPN is a DSPN or a MRSPN, since it is necessary to determine the set of effectively conflicting transitions. As a consequence, from the structure of the GDTT_SPN it is only possible to obtain

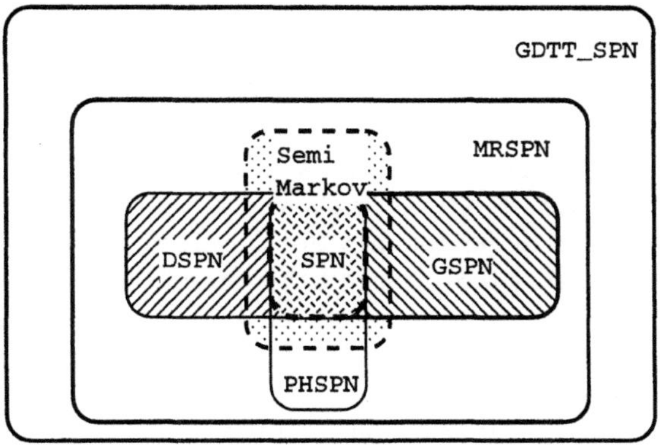

Fig. 7. Subclasses of GDTT_SPN.

sufficient conditions for the GDTT_SPN to be a DSPN or a MRSPN. Actually, even the state space construction may not be sufficient in the more general case of MRSPN, since it may be the case that additional informations, computable from the reachability graph, may be needed.

Another possible classification of GDTT_SPN could be based on the relationships between the behaviour of the timed models and the behaviour of the underlying P/T net. Indeed, in those cases in which the reachability graph of the timed net is identical to the one of the P/T net, the standard qualitative analysis techniques can be applied; in particular, all properties proved on the P/T net structure and/or on the reachability graph of the P/T net are valid also for the timed model, typically P- and T-invariants, boundedness, liveness, etc. This is the case for SPN.

In the case in which general firing delay pdf are allowed, it may happen that the probabilistic characteristics of the model have an impact on the qualitative behaviour, in the sense that, although two transitions may be simultaneously enabled, one of them cannot fire due to timing constraints, or not all interleavings of concurrent transitions can actually take place in any timed sequence.

A sufficient condition for the two reachability graphs to be identical is that $\mathcal{W}$ is taken from the set of pdf that have unlimited support [2, 3]. This condition in general is not met by GDTT_SPN with deterministic distributions, or uniform distributions over a finite interval.

In the case of GSPN, instead, the reachability set is identical to the one of the underlying PN system *with priorities*. Not too many qualitative analysis results are available for this class of nets, but it can be observed that the presence of inhibitor arcs and priorities only restricts the reachability set with respect to the one of the basic underlying net.

The restriction of the reachability set guarantees that all P-invariants found with the study of the basic underlying PN still hold for the GSPN; in principle, there might exist other place invariants which hold for the GSPN, but are not valid for the basic underlying PN. For what concerns T-invariants, the presence of priorities and inhibitor arcs may make non fireable a fireable invariant.

4 Stochastic Petri nets

SPN models were proposed by researchers active in the applied stochastic modelling field, with the goal of developing a tool which allowed the integration of formal description, proof of correctness, and performance evaluation. For what concerns the last aspect, the proposals aimed at an equivalence between SPN and CTMC models, while for the first two, it was chosen to introduce time in such a way as not to alter the untimed behaviour of the system.

In order to obtain an equivalence between a PN and a CTMC, it is necessary that the determination of the next reachable states depends only on the current one (true by definition in Petri nets), and that sojourn times in markings are random variables with negative exponential pdf.

This idea formed the basis of the doctoral dissertations of S.Natkin [84] at the Conservatoire National des Arts et Métiers in Paris, France, and of M.K.Molloy [80] at the Computer Science Department of the University of California at Los Angeles in the United States. These works were performed independently and approximately at the same time, in the late seventies. They led to the definition of almost identical models which even bore the same name: *Stochastic Petri Nets.* It should be mentioned, however, that the idea of associating an exponentially distributed random delay with PN transitions was already present in the doctoral dissertation of F.J.Symons within the definition of Numerical Petri nets [98, 97].

4.1 The Basic Model

An SPN is a GDTT_SPN in which the $\mathcal{W}$ function assigns to each transition an exponential pdf. Since the exponential distribution is fully characterized by its mean value (or by its inverse, the *rate*), and its memoryless characteristics makes inessential, as we shall explain later, the definition of an execution policy for transitions, then the definition of an SPN is simpler than the one for GDTT_SPN, namely:

$$SPN = (P, T, I, O, H, M_0, W)$$

where:

- (P, T, I, O, H, M_0) is the underlying PN system[4], as for GDTT_SPN,
- $W : T \rightarrow \Re$ is a *weight function*; $w(t)$ is the rate of the exponential distribution associated with transition t.

$w(t)$ is also called the *firing rate* of transition t.

According to some definitions of SPN, the weight function can depend on the current marking: in this case we write $w(t, M)$ to express the weight of transition

[4] Inhibitor arcs were not present in the first definitions of SPN, but they have become very popular in the field, so that we prefer to include them in the basic definition.

t in marking M. Common types of marking dependency are the ones borrowed from queueing theory, namely *single, multiple and infinite server*, but while in queueing systems the dependence of the service rate normally is on the number of customers in the queue, in SPN the dependence is on the transition enabling degree [8].

The average firing delay of transition t in marking M is therefore $[w(t, M)]^{-1}$.

Under the race policy that we are considering, the transition with the minimum delay is the one that fires first. The firing delay computed for each transition at the ingress in a new marking is either a new value sampled from the exponential distribution associated with the transition (for transitions that are "restarted"), or the residual firing time of the timer (for transitions that "continue"). However, due to the memoryless property of the exponential distribution, the distribution of the whole firing time is identical to the one of the residual firing time, hence the race is always among exponentially distributed variables, thus making not necessary the definition of the execution policy in SPN models. Since the minimum of two random variables with negative exponential pdf and parameters μ_1 and μ_2 is a random variable which still is exponentially distributed, with parameter $(\mu_1 + \mu_2)$, the sojourn time in marking M is a random variable with negative exponential pdf, with mean

$$\left[\sum_{t \in E(M)} w(t, M)\right]^{-1}$$

where $E(M)$ is the set of all enabled transitions in M.

The fact that all firing delays have exponential pdf permits a simple expression to be written for the probability that a given transition, say t, is the one for which the minimum delay has been sampled, and hence determines the change of marking by firing:

$$P\{t|M\} = \frac{w(t, M)}{\sum_{t' \in E(M)} w(t', M)} \qquad t \in E(M)$$

The reachability set of an SPN is identical to the one of the underlying untimed PN (with interleaving semantics) due to the unlimited support and to the continuity of the exponential distribution.

The state transition rate diagram of the CTMC corresponding to the SPN is obtained by constructing the reachability graph, and by labelling arcs with the firing rate of the transition whose firing produces the marking change. Indeed, if in a marking M two transitions t_1 and t_2 are enabled, either concurrently or in conflict, with $M[t_1\rangle M_1$ and $M[t_2\rangle M_2$, the sojourn time in M is a random variable distributed as the minimum of the two exponential distributions of the random variables associated with t_1 and t_2, that is to say, it is exponentially distributed with rate $w(t_1) + w(t_2)$. Since the probability of firing t_1 is $\frac{w(t_1)}{w(t_1)+w(t_2)}$, then the rate at which the system moves from M to M_1 is $[w(t_1) + w(t_2)]\frac{w(t_1)}{w(t_1)+w(t_2)} = w(t_1)$. A similar computation leads to $w(t_2)$ for t_2.

The steady-state solution of the model is then obtained by solving the system of linear equations

$$\begin{aligned} \boldsymbol{\pi Q} &= \mathbf{0} \\ \textstyle\sum_{M \in RS} \pi[M] &= 1 \end{aligned}$$

$\boldsymbol{\pi}$ is the equilibrium pmf over the reachable markings, and we write $\pi[M]$ for the steady-state probability of a given marking M.

The transient solution of the model is instead obtained solving the set of differential equations

$$\frac{d\boldsymbol{\pi}(t)}{dt} = \boldsymbol{Q\pi}(t)$$

where $\pi(t)[M]$ is the probability of the system being in state M at time t.

We may now go back to the lamp example discussed in Section 2.3. The SPN model describing the system considered in the example is depicted in Figure 8. The reader is adviced to compare the SPN model and the CTMC state transition rate diagrams, noting the similarity in the topology. This is due to the fact that the PN underlying our SPN model is a state machine [91].

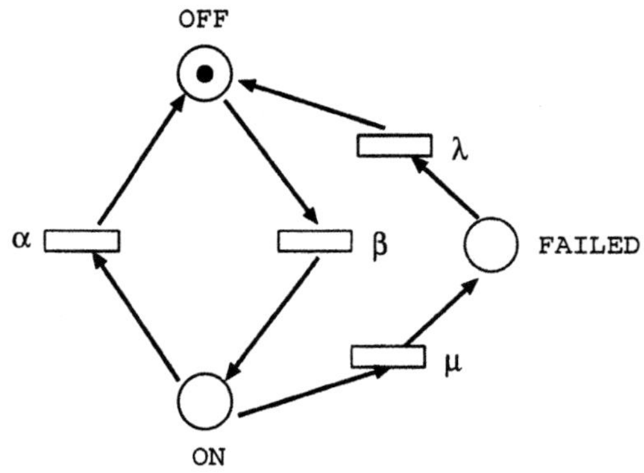

Fig. 8. SPN model of the lamp example.

In many cases this similarity in the topology does not exist.

Consider as a second example the SPN depicted in Figure 9. This is the SPN representation of the M/M/1 queue described in Section 2.4. Hence, the state transition rate diagram it generates is the one depicted in Figure 5, which comprises a denumerable infinity of states, in spite of the extremely simple SPN topology.

As an additional example, consider a system that exhibits choice, concurrency, as well as splitting and joining of customers, whose model is given in Figure 10. A process executes a choice between a fork activity (modelled by transition $T1$), and a computation (modelled by transition $T2$). The computation is followed by an additional activity modelled by transition $T4$. The execution of the two forked processes is modelled by the two independent transitions $T5$

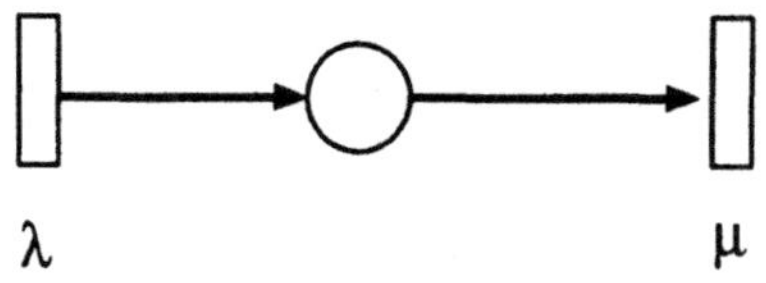

Fig. 9. SPN representation of the M/M/1 queue.

and $T6$, while transition $T3$ represents the join, and indeed its enabling requires that both subprocesses have finished their execution. Transition $T7$ represents a subprocess common to the two branches, that takes the process back to its initial state.

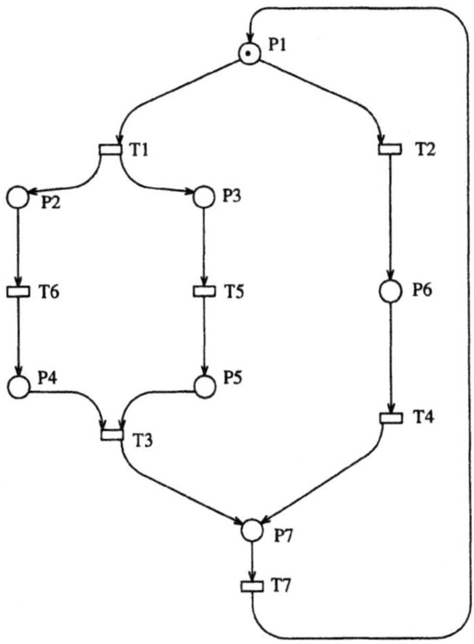

Fig. 10. SPN model of a fork and join system.

The state space of this SPN comprises seven states, shown in Table 2, while the infinitesimal generator is given in Table 3, and its state transition diagram is depicted in Figure 11.

It is important to remark that, in case of conflict, the rate of conflicting transitions accounts for both the *duration* of the activity and the *probability* of that activity. For example, if in the system we are modelling the average time

M_1	P1
M_2	P2 + P3
M_3	P6
M_4	P2 + P5
M_5	P3 + P4
M_6	P7
M_7	P4 + P5

Table 2. State space of the Fork and Join SPN model.

	M_1	M_2	M_3	M_4	M_5	M_6	M_7
M_1		$w(T1)$	$w(T2)$				
M_2				$w(T5)$	$w(T6)$		
M_3						$w(T4)$	
M_4							$w(T6)$
M_5							$w(T5)$
M_6	$w(T7)$						
M_7						$w(T3)$	

Table 3. Representation of the infinitesimal generator of the Fork and Join SPN model.

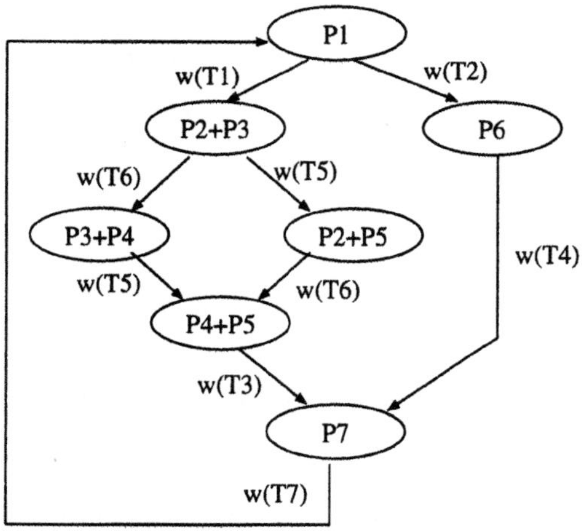

Fig. 11. State transition rate diagram of the Fork and Join SPN model.

to decide whether to perform a fork and join ($T1$) or a simple computation ($T2$) is equal to 0.0001 time units, and the probability of a fork and join is 99% against the 1% of the normal computation, we obtain a rate for $T1$ and $T2$ of: $w(T1) = 9,900$ and $w(T2) = 100$: in SPN there is no way to split the duration from the probabilistic choice.

4.2 Performance Indices

Several kinds of aggregate results are easily obtained from the steady-state or transient distributions over reachable markings. In this section we quote some of the most commonly and easily computed aggregate performance parameters [9].

- The *expected fraction time spent in a subset of markings* $\mathcal{M}$, in the interval $[0 \ldots t]$, can be computed as

$$\psi\{\mathcal{M}, t\} = \frac{1}{t} \sum_{M \in \mathcal{M}} \int_0^t \pi(z)[M] dz$$

  where $\pi(t)[M]$ is the probability of being in state M at time t. From the theory of Markov chains, it is well known that as t approaches infinity, $\psi\{\mathcal{M}, t\}$ becomes equal to the steady-state probability

$$\pi(\mathcal{M}) = \sum_{M \in \mathcal{M}} \pi[M]$$

- The *probability of an event* defined through place markings (e.g., no token in a subset of places, or at least one token in a place while another one is empty, etc.), can be computed by adding the probabilities of all markings in which the condition corresponding to the event definition holds true. Thus, for example, the steady-state probability of the event $\mathcal{A}$ defined through a condition that holds true for the markings $M \in \mathcal{M}$ is obtained as:

$$P\{\mathcal{A}\} = \sum_{M \in \mathcal{M}} \pi[M]$$

  while the probability of event A at time t is

$$P\{\mathcal{A}, t\} = \sum_{M \in \mathcal{M}} \pi(t)[M]$$

  This formula can also be used to compute the probability that a given condition is satisfied *for the first time* at time t, provided that the SPN is such that all and only the states satisfying the conditions that define event $\mathcal{A}$ are deadlock states (if this is not the case, the SPN should be modified accordingly). The same result can be obtained by changing the marking process rather than the SPN: it is only necessary to make the states in $\mathcal{M}$ absorbing.

- The *pmf of the number of tokens at steady-state in a place*, say p, can be obtained by computing the individual probabilities in the pmf as probabilities of the event "place p contains k tokens". The *pmf of the number of tokens in a place, at time* t, can be obtained similarly, using the event "place p contains k tokens at time t"
- The *average number of tokens in a place (at time t)* can be computed from the pmf of tokens in that place (at time t).
- The *expected number of firings of transition* t_k in the interval $[0,t]$, $f_{t_k}(t)$, can be computed integrating over the given interval the firing rate of transition t_k, expressed as the sum over all states M enabling t_k of the firing rate of t_k in M, weighted by the probability of being in M at time t

$$f_{t_k}(t) = \int_0^t \sum_{M:t_k \in E(M)} w(t_k, M)\; \pi(z)[M]\; dz$$

where $E(M)$ is the set of transitions enabled in M, and $w(t_k, M)$ is the firing rate of t_k in M. The sum and the integral can be exchanged to get:

$$f_{t_k}(t) = \sum_{M:t_k \in E(M)} \int_0^t w(t_k, M)\; \pi(z)[M]\; dz$$

- The *frequency of firing of a transition (throughput)*, i.e., the average number of times a transition t_k fires in unit time, can be computed as the weighted sum of the transition firing rate:

$$f_{t_k} = \sum_{M:t_k \in E(M)} w(t_k, M)\; \pi[M]$$

All the above performance indices can be defined using a unified approach based on rewards: a reward is a function $r(t) : M \rightarrow \Re$ for transient analysis and $r : M \rightarrow \Re$ for steady state. An *average reward* can be computed as the weighted sum

$$\begin{aligned} R(t) &= \sum_{M \in RS} r(M)\pi(t)[M] \\ R &= \sum_{M \in RS} r(M)\pi[M] \end{aligned}$$

for transient and steady state analysis, respectively.

For example, the mean number of tokens in a place p can be computed by defining $r(M) = M(p)$, and the throughput of a transition can be computed by the reward function

$$r(M) = \begin{cases} w(t, M) & \text{if } t \in E(M) \\ 0 & \text{otherwise} \end{cases}$$

From the above indices it is possible to compute the *average delay of a token* in traversing a subnet in steady-state conditions by using Little's formula [69, 77]

$$E[T] = \frac{E[N]}{E[\gamma]}$$

where $E[T]$ is the average delay, $E[N]$ is the average number of tokens in the process of traversing the subnet, and $E[\gamma]$ is the average input (or output) rate of tokens into (or out of) the subnet. This procedure can be applied whenever the interesting tokens can be identified inside the subnet (which can also comprise other tokens defining its internal condition, but these must be distinguishable from those whose delay is studied), so that their average number can be computed, and a relation can be established between input and output tokens (e.g., one output token for each input token).

As an example of a performance parameter which in the general case is difficult to compute, we may quote the distribution of the delay incurred by a token in traversing a subnet, or in completing a cycle through a net.

5 Generalized SPN

The key factor that limits the applicability of SPN models is the complexity of their analysis. This is due to many factors. The possibly very large number of reachable markings is by far the most critical one. Other aspects may however add to the model solution complexity. One of these is due to the presence in one model of activities that take place on a much faster (or slower) time scale than the one relating to the events that play a critical role on the overall performance. This results in systems of linear equations which are *stiff*, i.e., difficult to solve with an acceptable degree of accuracy by means of the usual numerical techniques. On the other hand, neglecting the "fast" (or "slow") activities may result in models which are logically incorrect. It may also happen that in the construction of the topology of an SPN model, the analyst inserts transitions that correspond to purely logical aspects of the system behaviour to ease the description of complex marking changes, so that no timing can be reasonably associated with them.

Generalized SPN (GSPN) were originally proposed in order to tackle these problems [6]. The definition was later improved in order to better exploit the structural properties of the modelling tool [4, 5]. The book [8] presents in detail the GSPN formalisms together with a number of application examples.

GSPN models comprise two types of transitions:

timed transitions, which are associated with random, exponentially distributed firing delays, as in SPN, and

immediate transitions, which fire in zero time, *with priority* over timed transitions.

Furthermore, different priority levels of immediate transitions can be used, and weights are associated with immediate transitions.

A GSPN is thus an eight-tuple

$$GSPN = (P, T, \Pi, I, O, H, M_0, W)$$

where $(P, T, \mathcal{P}, I, O, H, M_0)$ is the underlying untimed PN with priorities, that comprises

- the set P of places,
- the set T of transitions,
- the input, output and inhibitor functions $I, O, H : T \to N$,
- the initial marking M_0.

Additionally, the GSPN definition comprises the priority function $\Pi : \Pi : T \to N$ which associates lowest priority (0) with timed transitions and higher priorities (≥ 1) with immediate transitions:

$$\Pi(t) = \begin{cases} 0 & \text{if } t \text{ is timed} \\ \geq 1 & \text{if } t \text{ is immediate} \end{cases}$$

Finally, the last item of the GSPN definition is the function $W : T \to \Re$, that associates a real value with transitions. $w(t)$ is:

- the parameter of the negative exponential pdf of the transition firing delay, if t is a timed transition,
- a weight used for the computation of firing probabilities of immediate transitions, if t is an immediate transition.

In the graphical representation of GSPN, immediate transitions are drawn as segments, and exponential transitions as white rectangular boxes.

The untimed underlying model of a GSPN is a P/T net with inhibitor arcs and global priorities: the addition of priorities to a P/T system can reduce the number of reachable states, and it may destroy eventuality properties like liveness and home states, while all safety properties are maintained.

The stochastic interpretation of a GSPN model is very similar to that of an SPN model, with the changes necessary to account for immediate transitions.

When a marking is entered, it is first necessary to ascertain whether it enables timed transitions only, or at least one immediate transition. Markings of the former type are called *tangible*, whereas markings of the latter type are called *vanishing*.

In the case of a tangible marking, the timers of the enabled timed transitions either resume their decrement, or are re-initialized and then decremented, until one timed transition fires, exactly as in the case of SPN.

In the case of a vanishing marking, the selection of which transition to fire cannot be based on the temporal description, since all immediate transitions fire exactly in zero time. The choice is thus based on priorities and weights. The set of transitions with concession at the highest priority level is first found, and if it comprises more than one transition, the further selection, of probabilistic nature, is based on the transition weights according to the expression

$$P\{t\} = \frac{w(t)}{\sum_{t' \in E(M)} w(t')}$$

where $E(M)$ is the set of enabled immediate transitions in marking M, i.e., of the transitions with concession at the highest priority level.

Observe that in the above formula the probabilities are normalized over all enabled transitions, so that the normalization takes place also among nonconflicting transitions. From a modeller point of view it may be difficult to specify transitions weights if they are then normalized over the whole net. However, it was proved in [3] that if no confusion is present in the net (that is to say if no interplay exists between conflict and concurrency), it is possible to determine at a structural level the sets of possibly conflicting transitions, called "extended conflict sets (ECS)," and the normalization of weights can be done only among transitions that belong to the same ECS.

Note that the semantics of a GSPN model always assumes that transitions are fired one by one, even in a vanishing marking comprising nonconflicting enabled immediate transitions. The equivalence of this behaviour with the one resulting from the simultaneous firing of some immediate transitions in the model can be exploited to reduce the complexity of the solution algorithms [16].

The analysis of a GSPN model requires the solution of a system of linear equations comprising as many equations as the number of reachable *tangible* markings. The infinitesimal generator of the CTMC associated with a GSPN model is derived with a contraction of the reachability graph labelled with the rates or weights of the transitions causing the change of marking.

A different approach to the analysis of GSPN models, which also implies a different semantics, is presented in [12].

An example of construction of the tangible reachability graph is presented in Figure 12, that depicts a very simple GSPN in which a conflict exists among immediate transitions. Its reachability graph is shown in the upper right portion. Dotted lines for state $p2$ indicate that the state is vanishing: indeed, when transition $T1$ fires the system enters marking $p2$ in which two immediate transitions are enabled, and the marking changes in zero time to either $p3$ or $p4$, with probability $\frac{\alpha}{\alpha+\beta}$ and $\frac{\beta}{\alpha+\beta}$ respectively. The tangible reachability graph in the lower right portion is obtained by eliminating the vanishing marking $p2$. The rate at which the system moves from $p1$ to $p3$ ($p4$) is obtained by multiplying the rate μ of the state transition from $p1$ to $p2$ with the probability of going from state $p2$ to $p3$ ($p4$).

Figure 13 shows a GSPN model of a fork and join behaviour similar to the one described with the SPN in Figure 10. The GSPN model is obtained from the SPN model by making immediate those transitions that describe only a logical behaviour, or an activity of negligible duration; in particular the choice is modelled as a conflict between two immediate transitions and the join is also considered as an immediate action. Markings $P1$ and $P2+P3$ are now vanishing.

The application of SPN and GSPN modelling techniques has been very productive in several areas. The factor that has however limited their acceptance as a modelling tool lies in the (graphical and computational) *complexity* of the models of realistic systems. Different efforts for the problem solution are summarized in Section 6.

It must also be stressed that the use of SPN and GSPN heavily relies on the availability of adequate software *tools*, without which the model construction

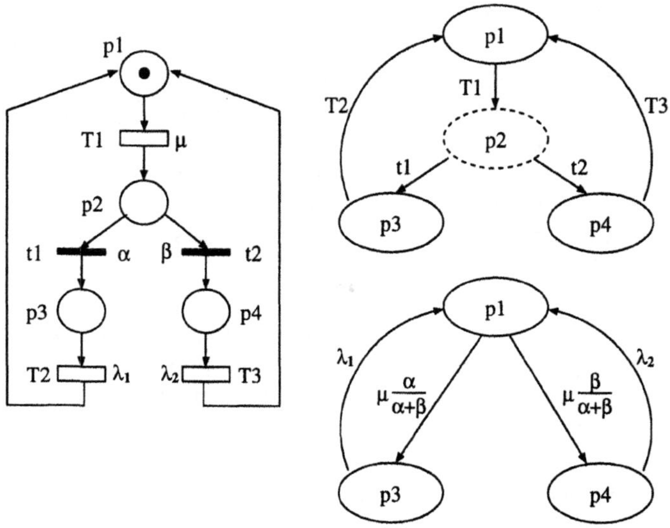

Fig. 12. Tangible reachability graph construction.

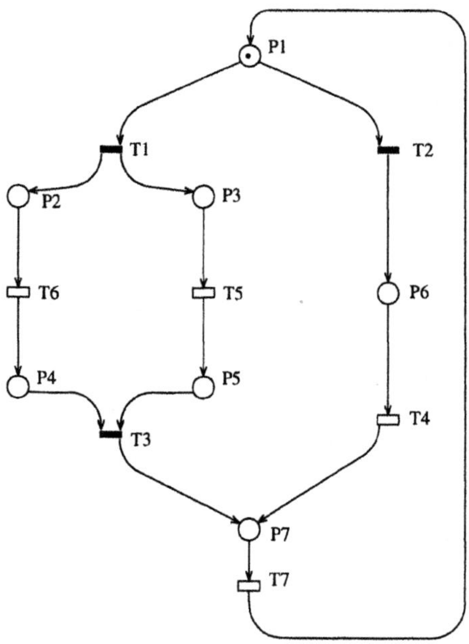

Fig. 13. GSPN representation of a fork and join system.

and solution is possible only for the smallest toy examples[5].

As a complete example of the use of GSPN both from a modelling and from an analysis point of view, we consider a multiserver multiqueue system, also known in the literature as a *multiserver cyclic polling system.*

5.1 GSPN Model of a Multiserver Cyclic Polling System

A single-server cyclic polling system comprises a set of waiting lines that receive arrivals from the external world, and one server that cyclically visits the queues, providing service to customers, if any is waiting. The GSPN description of such a polling system is provided in Figure 14.

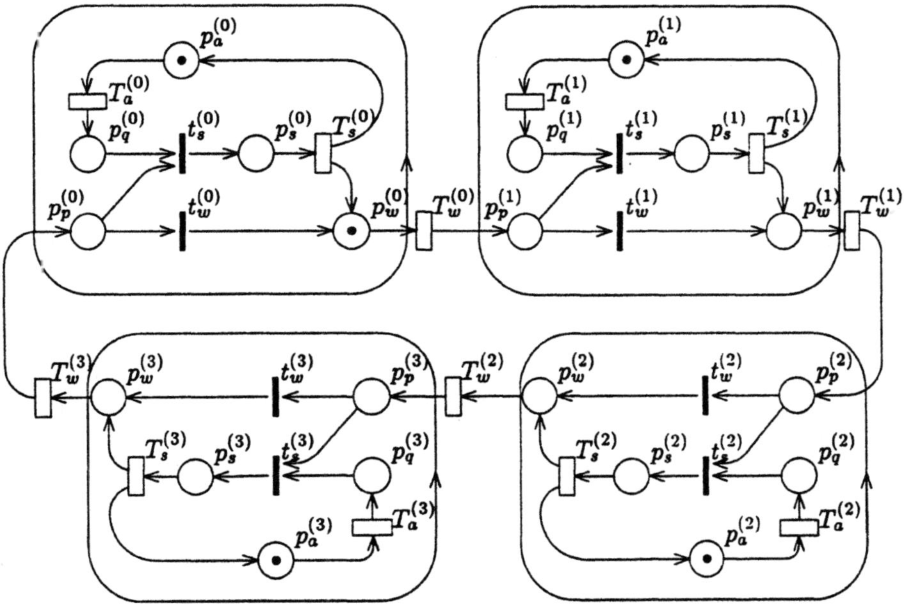

Fig. 14. GSPN representation of a single-server cyclic polling system.

The GSPN model in Figure 14 comprises four replicas of the subnet that describes the internal organization of each individual queue (enclosed within ovals), interconnected by four replicas of the subnet that describes the movement of the server from a queue to the next one.

[5] Readers interested in untimed and timed Petri nets tools can find many useful informations at the Petri net Web site "www.daimi.aau.dk/PetriNets"

Transition $T_a^{(q)}$ models the customer arrival process at queue[6] q ($q = 0, 1, 2, 3$). Customers waiting for a server are queued in place $p_q^{(q)}$, while a token in place $p_p^{(q)}$ represents the server when polling queue q. The two immediate transitions $t_s^{(q)}$, and $t_w^{(q)}$ have priorities 2, and 1, respectively ($\Pi(t_s^{(q)}) = 2, \Pi(t_w^{(q)}) = 1$). Transition $t_s^{(q)}$ fires if the server finds a waiting customer when it polls the queue, so that service can be provided; if $t_s^{(q)}$ cannot fire when the queue is polled by a server, i.e., if the server finds no waiting customers, then $t_w^{(q)}$ fires, and the server walks to the next queue.

One token in place $p_s^{(q)}$ represents a customer of queue q being served, as well as the server when providing service at queue q. $T_s^{(q)}$ is timed with a delay equal to the customer service time.

The server moves to place $p_w^{(q)}$ at the end of the visit at queue q (after the service completion represented by the firing of $T_s^{(q)}$, if a waiting customer was found; after the firing of $t_w^{(q)}$ if no waiting customer was found). From $p_w^{(q)}$ the server walks to the next queue. Transition $T_w^{(q)}$ models the server walk time from queue q to the next queue in the cyclic schedule.

The GSPN model in Figure 14 represents the case of a single customer in each queue ($M(p_a^{(q)}) = 1$), and of a single server initially placed at the exit of queue 0 ($M(p_w^{(0)}) = 1$). GSPN models for K customers in each queue, and S circulating servers can be obtained by assigning an initial marking with $M(p_a^{(q)}) = K$, and $M(p_w^{(0)}) = S$

The characteristics of the timed and immediate transitions in the GSPN model in Figure 14 are summarized in Tables 4 and 5, respectively: note that transitions that represent activities of servers ($T_s^{(q)}$ and $T_w^{(q)}$) are of infinite-server type, since we want each server to be able to either walk or serve in parallel with any other server; vice-versa, the transitions that represent an arrival of a customer to a queue ($T_a^{(q)}$) are single-server, to emulate a finite Poisson source of customers.

transition	rate	semantics
$T_a^{(q)}$	$\lambda^{(q)}$	single-server
$T_s^{(q)}$	$\mu^{(q)}$	infinite-server
$T_w^{(q)}$	$\omega^{(q)}$	infinite-server

Table 4. Characteristics of the timed transitions in the GSPN model of a cyclic single-server polling system ($q = 0, 1, 2, 3$).

[6] The superscript (q) indicates that the place (or transition) belongs to the model of queue q.

transition	weight	priority
$t_s^{(q)}$	1	2
$t_w^{(q)}$	1	1

Table 5. Characteristics of the immediate transitions in the GSPN model of a cyclic single-server polling system ($q = 0, 1, 2, 3$).

Typical aggregate performance figures of interest in the analysis of polling systems are the average customer delay, the probability of customers having to wait for a server, as well as the throughputs of the whole system, and of individual queues.

The structural analysis of the GSPN model detects five P-semiflows that cover all the places of the GSPN. Four of them cover the triplets of places $p_a^{(q)}$, $p_q^{(q)}$, $p_s^{(q)}$, with $q = 0, 1, 2, 3$. The resulting P-invariants are $M(p_a^{(q)}) + M(p_q^{(q)}) + M(p_s^{(q)}) = N^{(q)}$. This guarantees that places $p_a^{(q)}$, $p_q^{(q)}$, $p_s^{(q)}$, with $q = 0, 1, 2, 3$ are bounded, as was expected, since the number of customers either waiting or being served at any queue cannot exceed the number of customer in the closed arrival generation process. The fifth P-semiflow covers places $p_p^{(q)}$, $p_s^{(q)}$, $p_w^{(q)}$, with $q = 0, 1, 2, 3$. The token count of the resulting P-invariant is S, since the P-invariant refers to the conservation of the number of servers in the system. As a result, the GSPN model is bounded, and thus the number of reachable states is finite.

The GSPN model in Figure 14 is covered by 16 T-semiflows; each of them represents a possible path of a server in the cycle of queues. The simplest cycle (no customer is available at any queue) corresponds to the T-semiflow $t_w^{(q)}$, $T_w^{(q)}$, with $q = 0, 1, 2, 3$. The 16 T-semiflows correspond to the possible combinations of the conditions "customer ready for service" and "customer not ready for service" in the 4 queues. For example, the T-semiflow that represents a cycle in which the server provides service only at queue 0, is: $T_a^{(0)}$, $t_s^{(0)}$, $T_s^{(0)}$, $T_w^{(0)}$, and $t_w^{(q)}$, $T_w^{(q)}$, with $q = 1, 2, 3$.

The covering of the transitions of the net is a necessary condition for the system to be live and reversible, but not sufficient. A reachability graph analysis shows that the GSPN model indeed is live and reversible.

The size of the reachability graph depends on the number of queues, on the number of servers, and on the number K of customers at each queue. The second column of Table 6 shows the size of the reachability graph for 4 queues and 2 servers, for different values of K. Bigger values for the number of queues can be considered, although the reachability set grows quite fast (the GSPN model with $K = 1$, 2 servers, and 8 queues produces 287,328 tangible markings, the file that stores the reachability graph occupies about 14 megabytes, and the reachability set takes about 2.5 megabytes).

The third and fourth columns of Table 6 show the throughputs of transitions $T_a^{(q)}$ and $T_w^{(q)}$, respectively, when the weights of transitions $T_a^{(q)}$ and $T_s^{(q)}$ are taken equal to 1.0 s^{-1}, and the weight of $T_w^{(q)}$ is set to 5.0 s^{-1} (which implies a delay among queues of 0.2 s), independently of the queue index q. Due to the symmetry of the system, all transitions of equal name, but different queue index, have the same throughput. The last two columns report the probability that at any queue the source of arrival is empty ($P\{p_a^{(q)} = 0\}$), and that no customer is waiting ($P\{p_q^{(q)} = 0\}$).

The analysis was performed only for values of K up to 6, because we can observe that the system reaches the limit maximum throughput for transitions that describe servers; indeed, when the system is fully loaded (a customer is always waiting for service) a server will provide service at each queue, which implies a mean cycle time per server, equal to 4.8 s (4.0 s for service and 0.8 s to walk), and, considering that servers are independent in their services and walks, due to the infinite-server semantics, then the maximum throughput that can be observed for server transitions is $1/4.8 * 2 = 0.41666$ s^{-1}: this value is reached only when the number of customers is so high that the probability that a server finds no customer when polling the queue is negligible. As it can be observed from the last column, the system is very close to this limit condition already for $K = 4$, since the probability that no customer is waiting in the queue is close to zero.

K	$\|RS\|$	$f_{T_a^{(q)}}$	$f_{T_w^{(q)}}$	$P\{p_a^{(q)} = 0\}$	$P\{p_q^{(q)} = 0\}$
1	312	0.34500	0.77497	0.65499	0.69001
2	1,998	0.40964	0.45176	0.59035	0.22832
3	7,008	0.41563	0.42185	0.58437	0.03081
4	18,150	0.41649	0.42185	0.58371	0.00594
5	39,096	0.41651	0.41682	0.58349	0.00107
6	74,382	0.41661	0.41669	0.58334	0.00107

Table 6. Reachability set size, throughputs and probabilities for the polling model of Figure 14.

As an example of transient analysis for this system, consider the event E defined as "all servers are simultaneously providing service at queue 0 for the first time ". We can then compute $P\{E, t\}$, using the trick of making the first encountered state that satisfies E an absorbing state. At the net level this can be obtained by adding an immediate transition with an input arc of weight equal to S from place $p_s^{(0)}$, and an output arc of weight 1 to a new place p_{abs}. The probability of event E at time t can then be computed as the probability of a token in place p_{abs} at time t. Please observe that, for this probability to be non-zero, we need $K \geq S$. Table 7 shows this probability for a polling system with 4 stations, with the same rate parameters as used for steady-state analysis, and $S = K = 2$.

t	$P\{E,t\}$
1	0.0288
2	0.1124
3	0.1924
4	0.2674
5	0.3376
10	0.6048
50	0.9942

Table 7. $P\{E,t\}$ for the polling model of Figure 14.

6 Current Research on GDTT_SPN

Several groups of researchers are presently active in the field of GDTT_SPN. We mention here some of the current research efforts, aiming at a unitary view of the research field rather than at a comprehensive list of isolated activities.

Much of the current research work is devoted to the application of GDTT_SPN to performance and reliability studies of a very diverse gamut of systems, including distributed computing systems architectures, distributed software, object-oriented systems, data base, real-time systems, communication protocols, VLSI, manufacturing systems, inventory and logistics.

In some of these application studies the performance analysis aspect is integrated with the formal proof of correctness of the system under investigation, exploiting the formal system description obtained with the PN formalism (see for example [15]).

As we already mentioned, the main problem in the use of SPN techniques for the analysis of real-life systems originates from the complexity in the model solution. It is often the case that models comprise such monstrously large state spaces that the generation of the reachability set is too costly (both in time and in space) to be performed. Several research efforts are thus devoted to attempts to reduce the solution cost.

A possible approach is to resort to simulation techniques, rather than trying to numerically solve the model. By so doing, the problems originating from the space complexity are removed, since the generation of the reachability set is not necessary any more, but the burdens inherent to the time complexity remain, as always in the case of simulation. On the other hand, Haas and Shedler have shown [57, 58] that the modelling power of GDTT_SPN in the simulation framework is remarkable, since they are equivalent to generalized semi-Markov processes (GSMP).

A number of research teams are instead attacking the theoretical problems inherent in the management of largeness in GDTT_SPN models. Some of the main lines of research are concisely mentioned below.

Distributed solution algorithms - Numerical distributed algorithms have been specifically developed for both the generation of the reachability graph in a PN and for the solution of the underlying CTMC [11, 39, 33, 79].

Structured representation - An approach to increase the size of analyzable CTMC is to represent the generator matrix in a compact form as a combination of smaller component matrices, and to exploit this representation in the solution algorithm. A compositional technique based on Kronecker operators proposed in [88] was initially transferred to the SPN framework in [49, 50]. Efficient techniques were developed for reachability analysis [66, 67] and for numerical analysis [26, 27, 28, 32, 68], exploiting the structure of the generator matrix. In order to use the structured analysis technique, the PN model normally has to be described by means of submodels interacting via synchronizing transitions. However, structured schemes for asynchronously interacting submodels have also been presented in [25].

SPN and queueing networks - Results obtained in other stochastic modelling fields can be cast into PN. A very interesting line of research is aimed to combining the techniques available for the analysis of queueing networks into the language of PN. In some cases, it is possible to solve subsystems in the form of queueing networks, and to compose the results in a *higher level* GSPN representation specifying the interaction among submodels. The inverse procedure was also followed, first solving independent GSPN submodels that are then connected in a queueing network structure [13, 14]. A package supporting the replacement of GSPN places by queueing systems is presented in [17].

Particular classes of SPN, like those generating queueing models with matrix-geometric structure [85], were also considered, and a tool was built for their analysis [61, 62]

Product form SPN - Several proposals were recently documented to import the product form concept into the PN arena.

In [72], a class of SPN is identified for which a product form equilibrium equation can be written from the knowledge of partial balance equations. The generation of the reachability graph is needed to recognize this class of PNs. An extension of the same line is presented in [74].

Henderson et al. [63] developed a product form criterion based only on the structure of the PN, with no need to generate the reachability graph.

A comparative analysis of these two types of approaches was presented in [51], where the possibility of recognizing whether a PN admits a product form solution using results from the structural analysis was proved for the first time.

A complete characterization of product form SPN is provided in [22], and a necessary and sufficient condition for the existence of a positive solution for the traffic equation can be found in [23]. Specific algorithms for the computation of product form solutions were presented in [44, 94].

Mean value analysis for non-product form SPN was explored in [93].

PN-driven techniques - These techniques deal with the reduction of both memory requirements and time complexity of solution algorithms by using information about the structure of the untimed PN models.

High-Level Stochastic Petri Nets, such as for example Stochastic Well-formed Nets (SWN), often exhibit behavioural symmetries that can be exploited to reduce the size of the state space, and of the corresponding CTMC, by grouping states into *equivalence classes.* The desirable properties of a good analysis method based on this idea are the possibility of automatically discovering the symmetries using only the information contained in the model description at the PN level, and the possibility of directly generating the reduced state space (and the lumped Markov chain) with no need to build the complete reachability graph.

A completely automatic method for the construction of a lumped CTMC of a SWN model was presented in [35, 37]. In [59, 60] it was shown that in some cases it is possible to integrate this approach with decomposition methods based on Kronecker Algebra. In [92], the same idea was developed for Stochastic Activity Networks (SAN), and an automatic method was described for the construction of the lumped CTMC starting from the SAN description.

Deterministically Synchronized Sequential Processes (DSSP) are a class of SPN that can be obtained by resorting to simple modular design principles; for this class of models, a well-established theory exists for qualitative analysis [89, 95]. Net-driven techniques, developed for DSSPs, can recognize and extract from the original model a set of simpler auxiliary submodels that can then be analyzed through approximate iterative techniques [31] as well as exact manipulation [32].

Performance Bounds - A complementary approach to the development of efficient solution techniques for the computation of performance measures, is the search for bounds. Bounds require less computational effort with respect to exact analytical solutions, since they are estimated from PN-level equations, and do not require the knowledge of the reachability graph. Moreover, the evaluation of the bounds is, usually, not restricted to the Markovian assumption.

Results in this direction were derived since the beginning of the research on SPN [24, 82].

Subsequently, an extensive amount of work tried to exploit the structure of the PN to obtain efficient computation techniques. The evaluation of bounds for the subclass of marked graphs was presented in [29, 31], and in [30] for PN with a unique consistent firing count vector.

A general approach for the computation of bounds was formulated in [34],

based on operational analysis techniques applied at the PN level, with very weak assumptions on their timing semantics. The bounds can be obtained in polynomial time by solving suitable linear programming problems; they depend only on the mean values of firing times and are insensitive to distributions. In the case of Markovian SPN, an improved solution technique, based on the randomization algorithm, was presented in [78].
A time scale decomposition approach was proposed in [65]. This approach requires that transitions be classified into two classes: fast and slow.

Is summary, it can be observed that research in the GDTT_SPN field has been and still is lively and challenging, possibly quite rewarding for bright young researchers wishing to contribute to the field.

References

1. M. Ajmone Marsan. Stochastic Petri nets: an elementary introduction. In G. Rozenberg, editor, *Advances in Petri Nets 1989, Lecture Notes in Computer Science.* Springer Verlag, 1990.
2. M. Ajmone Marsan, G. Balbo, A. Bobbio, G. Chiola, G. Conte, and A. Cumani. On Petri nets with stochastic timing. In *Proceedings International Workshop on Timed Petri Nets*, pages 80–87, Torino (Italy), 1985. IEEE Computer Society Press no. 674.
3. M. Ajmone Marsan, G. Balbo, A. Bobbio, G. Chiola, G. Conte, and A. Cumani. The effect of execution policies on the semantics and analysis of stochastic Petri nets. *IEEE Transactions on Software Engineering*, SE-15:832–846, 1989.
4. M. Ajmone Marsan, G. Balbo, G. Chiola, and G. Conte. Generalized stochastic Petri nets revisited: random switches and priorities. In *Proceedings International Workshop on Petri Nets and Performance Models - PNPM87*, pages 44–53. IEEE Computer Society, 1987.
5. M. Ajmone Marsan, G. Balbo, G. Chiola, and G. Conte. Generalized stochastic Petri nets: a definition at the net level and its implications. *IEEE Transactions on Software Engineering*, 19:89–107, 1993.
6. M. Ajmone Marsan, G. Balbo, and G. Conte. A class of generalized stochastic Petri nets for the performance evaluation of multiprocessor systems. *ACM Transactions on Computer Systems*, 2:93–122, 1984.
7. M. Ajmone Marsan, G. Balbo, and G. Conte. *Performance Models of Multiprocessor Systems.* The MIT Press, 1986.
8. M. Ajmone Marsan, G. Balbo, G. Conte, S. Donatelli, and G. Franceschinis. *Modelling with Generalized Stochastic Petri Nets.* Wiley Series in Parallel Computing, 1995.
9. M. Ajmone Marsan, A. Bobbio, G. Conte, and A. Cumani. Performance analysis of degradable multiprocessor systems using generalized stochastic Petri nets. *IEEE Computer Society Newsletters*, 6, SI-1:47–54, 1984.
10. M. Ajmone Marsan and G. Chiola. On Petri nets with deterministic and exponentially distributed firing times. In *Lecture Notes in Computer Science*, volume 266, pages 132–145. Springer Verlag, 1987.

11. S.C. Allmaier, M. Kowarschik, and G. Horton. State space construction and steady-state solution of GSPNs on a shared-memory multiprocessor. In *7-th International Conference on Petri Nets and Performance Models - PNPM97*, pages 112–121. IEEE Computer Society, 1997.
12. H. H. Ammar and R. W. Liu. Analysis of the generalized stochastic Petri nets by state aggregation. In *Proceedings International Workshop on Timed Petri Nets*, pages 88–95, Torino (Italy), 1985. IEEE Computer Society Press no. 674.
13. G. Balbo, S.C. Bruell, and S. Ghanta. Combining queueing network and generalized stochastic Petri net models for the analysis of some software blocking phenomena. *IEEE Transactions on Software Engineering*, 12:561–576, 1986.
14. G. Balbo, S.C. Bruell, and S. Ghanta. Combining queueing network and generalized stochastic Petri nets for the solution of complex models of system behavior. *IEEE Transactions on Computers*, 37:1251–1268, 1988.
15. G. Balbo, G. Chiola, S.C. Bruell, and P. Chen. An example of modelling and evaluation of a concurrent program using coloured stochastic Petri nets: Lamport's fast mutual exclusion algorithm. *IEEE Transactions on Parallel and Distributed Systems*, 3(1), January 1992.
16. G. Balbo, G. Chiola, G. Franceschinis, and G. Molinar Roet. On the efficient construction of the tangible reachability graph of Generalized Stochastic Petri Nets. In *Proceedings International Workshop on Petri Nets and Performance Models - PNPM87*. IEEE Computer Society, 1987.
17. F. Bause. Queueing Petri nets: a formalism for the combined qualitative and quantitative analysis of systems. In *Proceedings International Workshop on Petri Nets and Performance Models - PNPM93*, pages 14–23. IEEE Computer Society, 1993.
18. C. Beounes and et al. SURF-2: a program for dependability evaluation of complex hardware and software systems. In *Proceedings 23-th International Symposium on Fault-Tolerant Computing*, TOULOUSE, June 1993. IEEE.
19. A. Bertoni and M. Torelli. Probabilistic Petri nets and semi Markov processes. In *Proceedings 2-nd European Workshop on Petri Nets*, 1981.
20. A. Bobbio and M. Telek. Transient analysis of a preemptive resume M/D/1/2/2 through Petri nets. Technical report, Department of Telecommunications - Technical University of Budapest, April 1994.
21. A. Bobbio and M. Telek. Markov regenerative SPN with non-overlapping activity cycles. In *International Computer Performance and Dependability Symposium - IPDS95*, pages 124–133. IEEE Computer Society Press, 1995.
22. R.J. Boucherie. A characterization of independence for competing Markov chain with applications to stochastic Petri nets. In *Proceedings 5th International Workshop on Petri Nets and Performance Models - PNPM93*, pages 117–126. IEEE Computer Society, 1993.
23. R.J. Boucherie and M. Sereno. On the traffic equations for batch routing queueing networks and stochastic Petri nets. Technical Report ERCIM-04/94-R032, European Research Consortium for Informatics and Mathematics, 1994. To appear in the journal *A. on Appl. Prob.*
24. S.C. Bruell and S. Ghanta. Throughput bounds for generalized stochastic Petri ne modelst. In *International Workshop Timed Petri Nets*, pages 250–261, Torino (Italy), 1985. IEEE Computer Society Press No. 674.
25. F. Buchholz. A hierarchical view of GCSPNs and its impact on qualitative and quantitative analysis. *Journal of Parallel and Distributed Computation*, 15:207–224, 1992.

26. P. Buchholz. Hierarchical structuring of superposed GSPNs. In *7-th International Conference on Petri Nets and Performance Models - PNPM97*, pages 81–90. IEEE Computer Society, 1997.
27. P. Buchholz and P. Kemper. Numerical analyisis of stochastic marked graphs. In *Proc. 6th Intern. Workshop on Petri Nets and Performance Models*, pages 32–41, Durham, NC, USA, October 1995. IEEE-CS Press.
28. P. Bucholz. Aggregation and reduction techniques for hierarchical GCSPNs. In *Proceedings International Workshop on Petri Nets and Performance Models - PNPM93*, pages 216–225. IEEE Computer Society, 1993.
29. J. Campos, G. Chiola, J.M. Colom, and M. Silva. Properties and performance bounds for timed marked graphs. *IEEE Transactions on Circuits and Systems - I: Fundamental Theory and Applications*, 39:386–401, 1992.
30. J. Campos, G. Chiola, and M. Silva. Ergodicity and throughput bounds of Petri nets with unique consistent firing count vector. *IEEE Transactions on Software Engineering*, 17:117–125, 1991.
31. J. Campos, J.M. Colom, H. Jugnitz, and M. Silva. Approximate throughput computation of stochastic marked graphs. *IEEE Transactions on Software Engineering*, 20:526–535, 1994.
32. J. Campos, M. Silva, and S. Donatelli. Structures solution of stochastic DSSP systems. In *7-th International Conference on Petri Nets and Performance Models - PNPM97*, pages 91–100. IEEE Computer Society, 1997.
33. S. Caselli, G. Conte, and P. Marenzoni. Massively parallel analysis GSPN models. In *Proceedings International Conference on Application and Theory of Petri Nets*, pages 181–200. Springer Verlag - LNCS, Vol 935, 1995.
34. G. Chiola, C. Anglano, J. Campos, J.M. Colom, and M. Silva. Operational analysis of timed Petri nets and application to the computation of performance bounds. In *Proceedings 5th International Workshop on Petri Nets and Performance Models - PNPM93*, pages 128–137. IEEE Computer Society, 1993.
35. G. Chiola, C. Dutheillet, G. Franceschinis, and S. Haddad. Stochastic Well-Formed coloured nets and multiprocessor modelling applications. In K. Jensen and G. Rozenberg, editors, *High-Level Petri Nets. Theory and Application.* Springer Verlag, 1991.
36. G. Chiola, G. Franceschinis, R. Gaeta, and M. Ribaudo. GreatSPN1.7: GRaphical Editor and Analyzer for Timed and Stochastic Petri Nets. *Performance Evaluation*, 24, 1995. Special Issues on Performance Modelling Tools.
37. Hoon Choi, V.G. Kulkarni, and K. Trivedi. Transient analysis of deterministic and stochastic Petri nets. In *Proceedings of the 14-th International Conference on Application and Theory of Petri Nets*, Chicago, June 1993.
38. Hoon Choi, V.G. Kulkarni, and K. Trivedi. Markov regenerative stochastic Petri nets. *Performance Evaluation*, 20:337–357, 1994.
39. G. Ciardo, J. Gluckman, and D. Nicol. Distributed state-space generation of discrete state stochastic models. *ORSA J. Comp. - To appear*, 1997.
40. G. Ciardo and C. Lindemann. Analysis of deterministic and stochastic Petri nets. In *Proceedings International Workshop on Petri Nets and Performance Models - PNPM93*, pages 160–169. IEEE Computer Society, 1993.
41. G. Ciardo, J. Muppala, and K.S. Trivedi. SPNP: stochastic Petri net package. In *Proceedings International Workshop on Petri Nets and Performance Models - PNPM89*, pages 142–151. IEEE Computer Society, 1989.
42. E. Cinlar. *Introduction to Stochastic Processes.* Prentice-Hall, Englewood Cliffs, 1975.

43. J.W. Cohen. *The Single Server Queue.* Elsevier, 1969.
44. J.L. Coleman, W. Henderson, and P.G. Taylor. Product form equilibrium distributions and an algorithm for classes of batch movement queueing networks and stochastic Petri nets. *Performance Evaluation*, 26:159–180, 1996.
45. R.B. Cooper. *Introduction to Queueing Theory.* MacMillan, New York, 1972.
46. J.A. Couvillon, R. Freire, R. Johnson, W.D. Obal, M.A. Qureshi, M. Rai, W. Sanders, and J.E. Tvedt. Performability modeling with UltrasSAN. *IEEE Software*, 8:69–80, September 1991.
47. D.R. Cox and W.L. Smith. *Queues.* Wiley, 1961.
48. A. Cumani. Esp - A package for the evaluation of stochastic Petri nets with phase-type distributed transition times. In *Proceedings International Workshop Timed Petri Nets*, pages 144–151, Torino (Italy), 1985. IEEE Computer Society Press no. 674.
49. S. Donatelli. Superposed stochastic automata: a class of stochastic Petri nets amenable to parallel solution. *Performance Evaluation*, 18:21–36, 1993.
50. S. Donatelli. Superposed generalized stochastic Petri nets: definition and efficient solution. In *Proceedings International Conference on Application and Theory of Petri Nets*, pages 258–277. Springer Verlag - LNCS, Vol 815, 1994.
51. S. Donatelli and M. Sereno. On the product form solution for stochastic Petri nets. In *Proceedings International Conference on Application and Theory of Petri Nets*, pages 154–172. Springer Verlag - LNCS, Vol 616, 1993.
52. J. Bechta Dugan, K. Trivedi, R. Geist, and V.F. Nicola. Extended stochastic Petri nets: applications and analysis. In *Proceedings PERFORMANCE '84*, Paris, 1984.
53. W. Feller. *An Introduction to Probability Theory and its Applications - Vol. I and II.* Wiley, New York, 1968.
54. G. Florin and S. Natkin. Les reseaux de Petri stochastiques. *Technique et Science Informatique*, 4:143–160, 1985.
55. E. Gelenbe and I. Mitrani. *Analysis and Synthesis of Computer Systems.* Academic Press, New York, 1980.
56. R. German, C. Kelling, A. Zimmermann, and G. Hommel. *TimeNET - A toolkit for evaluating non-markovian stochastic Petri nets.* Report No. 19 - Technische Universität Berlin, 1994.
57. P.J. Haas and G.S. Shedler. Regenerative stochastic Petri nets. *Performance Evaluation*, 6:189–204, 1986.
58. P.J. Haas and G.S. Shedler. Stochastic Petri nets with timed and immediate transitions. *Stochastic Models*, 5:563–600, 1989.
59. S. Haddad and P. Moreaux. Evaluation of high level Petri nets by means of aggregation and decomposition. In *6-th International Conference on Petri Nets and Performance Models - PNPM95*, pages 11–20. IEEE Computer Society, 1995.
60. S. Haddad and P. Moreaux. Asynchronous composition of high level petri nets: a quantitative approach. In *Proceedings of the 17-th International Conference on Application and Theory of Petri Nets, ICATPN '96*, pages 192–211, Osaka, Japan, June 1996. Springer Verlag - LNCS, Vol 1091.
61. B. Haverkort. Matrix-geometric solution of infinite stochastic Petri nets. In *International Computer Performance and Dependability Symposium - IPDS95*, pages 72–81. IEEE Computer Society Press, 1995.
62. B. Haverkort and A. Ost. Steady-state analysis of infinite stochastic Petri nets: a comparing between the spectral expansion and the matrix-geometric method. In *7-th International Conference on Petri Nets and Performance Models - PNPM97*, pages 36–45. IEEE Computer Society, 1997.

63. W. Henderson, D. Lucic, and P.G. Taylor. A net level performance analysis of stochastic Petri nets. *Journal of Australian Mathematical Soc. Ser. B*, 31:176–187, 1989.
64. R.A. Howard. *Dynamic Probabilistic Systems, Volume II: Semi-Markov and Decision Processes*. John Wiley and Sons, New York, 1971.
65. S.M.R Islam and H.H. Ammar. On bounds for token probabilities in a class of generalized stochastic Petri nets. In *Proceedings 3rd International Workshop on Petri Nets and Performance Models - PNPM89*, pages 221–227. IEEE Computer Society, 1989.
66. P. Kemper. Numerical analysis of superposed GSPNs. *IEEE Transactions on Software Engineering*, 22, 1996.
67. P. Kemper. Reachability analysis based on structured representation. In *Proceedings International Conference on Application and Theory of Petri Nets*, pages 269–288. Springer Verlag - LNCS, Vol 1091, 1996.
68. P. Kemper. Transient analysis of superposed GSPNs. In *7-th International Conference on Petri Nets and Performance Models - PNPM97*, pages 101–110. IEEE Computer Society, 1997.
69. L. Kleinrock. *Queuing systems, Volume 1: Theory*. Wiley Interscience, New York, 1975.
70. V.G. Kulkarni. *Modeling and Analysis of Stochastic Systems*. Chapman Hall, 1995.
71. S.S. Lavenberg. *Computer Performance Modeling Handbook*. Academic Press, New York, 1983.
72. A.A. Lazar and T.G. Robertazzi. Markovian Petri net protocols with product form solution. *Performance Evaluation*, 12:67–77, 1991.
73. R. Lepold. PEPNET: A new approach to performability modelling using stochastic Petri nets. In *Proceedings 1st International Workshop on Performability Modelling of Computer and Communication Systems*, pages 3–17, University of Twente - Enschede (NL), 1991.
74. M. Li and N.D. Georganas. Parametric analysis of stochastic Petri nets. In G. Balbo and G. Serazzi, editors, *Computer Performance Evaluation*. Elsevier Science Publishers, 1992.
75. C. Lindemann. An improved numerical algorithm for calculating steady-state solutions of deterministic and stochastic Petri net models. *Performance Evaluation*, 18:75–95, 1993.
76. C. Lindemann. DSPNexpress: a software package for the efficient solution of deterministic and stochastic Petri nets. *Performance Evaluation*, 22:3–21, 1995.
77. J.D.C. Little. A proof of the queueing formula $L = \lambda W$. *Operations Research*, 9:383–387, 1961.
78. Z. Liu. Performance bounds for stochastic timed Petri nets. In G. De Michelis and M. Diaz, editors, *Application and Theory of Petri Nets (16-th International Conference), Lecture Notes in Computer Science*, volume 935, pages 316–334. Springer Verlag, 1995.
79. P. Marenzoni, S. Caselli, and G. Conte. Analysis of large GSPN models: a distributed solution tool. In *7-th International Conference on Petri Nets and Performance Models - PNPM97*, pages 122–131. IEEE Computer Society, 1997.
80. M.K. Molloy. On the integration of delay and throughput measures in distributed processing models. Technical report, Phd Thesis, UCLA, 1981.
81. M.K. Molloy. Performance analysis using stochastic Petri nets. *IEEE Transactions on Computers*, C-31:913–917, 1982.

82. M.K. Molloy. Fast bounds for stochastic Petri nets. In *International Workshop Timed Petri Nets*, pages 244–249, Torino (Italy), 1985. IEEE Computer Society Press No. 674.
83. T. Murata. Petri nets: properties, analysis and applications. *Proceedings of the IEEE*, 77:541–580, 1989.
84. S. Natkin. Les reseaux de Petri stochastiques et leur application a l'evaluation des systemes informatiques. Technical report, These de Docteur Ingegneur, CNAM, Paris, 1980.
85. M.F. Neuts. *Matrix Geometric Solutions in Stochastic Models.* Johns Hopkins University Press, Baltimore, 1981.
86. J.L. Peterson. *Petri net theory and the modeling of systems.* Prentice Hall, Englewood Cliffs, 1981.
87. C.A. Petri. Kommunikation mit automaten. Technical report, Doctoral Thesis, University of Bonn, 1962. (Available in English as: *Communication with automata*, Technical Report RADC-TR-65-377, Rome Air Development Center, Griffiss NY, 1966).
88. B.D. Plateau. On the stochastic structure of parallelism and synchronization models for distributed algorithms. In *Proceedings ACM/SIGMETRICS Conference*, pages 147–154, Austin, 1985.
89. L. Recalde, E. Teruel, and M. Silva. On well-formedness analysis: the case of Deterministic Systems of Sequential Processes. In J. Desel, editor, *Structures in Concurrency Theory*, pages 279–293. Springer Verlag, 1995.
90. A. Reibman, R. Smith, and K.S. Trivedi. Markov and Markov reward model transient analysis: an overview of numerical approaches. *European Journal of Operational Research*, 40:257–267, 1989.
91. W. Reisig. *Petri nets - An introduction.* Springer-Verlag, 1982.
92. W.H. Sanders and J.F. Meyer. Reduced base model construction methods for stochastic activity networks. *IEEE Journal on Selected Areas in Communications*, 9(1):25–36, January 1991.
93. M. Sereno. Approximate mean value analysis for stochastic marked graphs. *IEEE Transactions on Software Engineering*, 22:654–664, 1996.
94. M. Sereno and G. Balbo. Computational algorithms for product form solution stochastic Petri nets. In *Proceedings 5th International Workshop on Petri Nets and Performance Models - PNPM93.* IEEE Computer Society, 1993.
95. M. Silva, L. Recalde, and E. Teruel. Linear algebraic techniques for the analysis of liveness of Petri nets. *Journal of Circuits, Systems and Computers*, 1998.
96. G.W. Stewart. *Introduction to the Numerical Solution of Markov Chains.* Princeton University Press, New Jersey, 1984.
97. F. J. W. Symons. The description and definition of queueing systems by numerical Petri nets. *Australian Telecommunications Research*, 13:20–31, 1980.
98. F. J. W. Symons. Introduction to numerical Petri nets, a general graphical model of concurrent processing systems. *Australian Telecommunications Research*, 14:28–33, 1980.
99. K. Trivedi. *Probability & Statistics with Reliability, Queueing & Computer Science applications.* Prentice-Hall, 1982.

Basic Linear Algebraic Techniques for Place/Transition Nets

Jörg Desel

Institut für Angewandte Informatik und Formale Beschreibungsverfahren
Universität Karlsruhe, Germany
desel@aifb.uni-karlsruhe.de

Abstract. Linear algebraic techniques for place/transition nets are surveyed. In particular, place and transition invariant vectors and their application to verification, proof and analysis of behavioral properties of marked Petri nets are presented. The considered properties are the non-reachability of a marking and conditions that hold true for all reachable markings. In addition, it is shown how the rank of the incidence matrix implies sufficient criteria and necessary criteria for liveness of bounded marked Petri nets.

1 Introduction

This contribution surveys the state of the art in linear algebraic techniques for Petri nets. More precisely, we show how the matrix representation of a Petri net together with appropriate equation systems or systems of inequalities can be exploited for gaining or proving properties of the net's behavior.

The main motivation for the linear algebraic approach is the behavioral complexity of Petri nets. In general, the set of reachable markings of an initially marked net can explode both with respect to the size of the net and the number of initial tokens, and it even can be infinite. Hence, in practice, the explicit construction of all reachable markings is not feasible. This implies that efficient analysis of behavioral properties can not be based on models of behavior such as marking graphs or sets of all occurrence sequences or even process nets. We rather have to stick to the structure of the considered net, together with its initial token distribution. Linear algebraic techniques exploit only those structural data. They yield at least sufficient *or* necessary conditions for properties of the behavior of a marked net. These conditions can be verified by help of efficient algorithms for linear algebraic techniques. The complexity of such algorithms heavily depend on the domain under consideration. For example, solutions of an equation system in the rational numbers are simpler to achieve than solutions in the integers or in the natural numbers. A trade-off frequently occurs between the complexity of an algorithm and the expressive power of its result.

Combining linear algebraic techniques and Petri nets is nearly as old as Petri net theory. The first relevant contributions appeared already in the mid 70ies. We collect both the old concepts and new results in a uniform and, hopefully, readable way. We do *not* consider arc weights, capacity restrictions, time annotations or other extensions. Most presented results can canonically be extended to these generalizations. We also do not consider specific properties of restricted classes of Petri nets such as free-choice Petri nets.

The applicability of equation systems to Petri nets is based on the duality of states and state changes that was emphasized by C. A. Petri [Petr73, Petr82]. In Petri nets, places, being local units of states, and transitions, being atomic units of state changes, are represented on the same level. This duality is reflected by linear methods: In the matrix representation of a net, places and transitions are represented by rows and columns, respectively. Linear algebra relates properties of a matrix to properties of its transposed matrix. This applies in particular to solubility criteria for the respective equation systems. Translated to Petri nets, these criteria yield relations between the places and the transitions of a net that are used in techniques described in this contribution.

All linear algebraic techniques for Petri nets are based on the observation that the occurrence of a transition always causes the same relative change of the marking of places: it decreases the number of tokens on each place which is in its pre-set but not in its post-set by one, and it increases the number of tokens on each place which is in its post-set but not in its pre-set by one. The markings of all other places remain unchanged. There exist modified occurrence rules for Petri nets that allow varying relative changes, too. For example, *preemptive arcs* allow to empty a place, no matter what its number of tokens was before. The more general *self-modifying nets* allow variable effects of a transition to a place, depending on the actual marking of some other place. These nets allow to copy the marking of some place to another place, similar to an assignment statement in a programming language. We will not consider such extensions in this paper.

As a general prerequisite for matrix and vector notations, we assume a finite set of places $S = \{s_1, \ldots, s_n\}$ and a finite set of transitions $T = \{t_1, \ldots, t_l\}$, that are indexed by the numbers $1, \ldots, n$ ($1, \ldots, l$, respectively). A marking associates to each place its actual number of tokens. Hence, every marking is a mapping $m: S \to \mathbb{N}$, where $\mathbb{N} = \{0, 1, 2, \ldots\}$ is the set of natural numbers (non-negative integers). A marking m can be represented by a vector $\mathbf{m} \in \mathbb{N}^n$ such that the i-th component of the vector $\mathbf{m}$ is the value $m(s_i)$. The constant relative change caused by a transition can be described by a vector, too. Its i-th component represents the effect of the transition's occurrence on the place s_i.

Figure 1 shows a vending-machine, as introduced in [DeRe98]. Its places are **s1**, **s2**, ..., **s6** and its transitions are **t1**, **t2**, ..., **t5**. The vector representation $\mathbf{m}_0$ of the initial marking and the vector representation **t2** of the transition **t2** read as

$$\mathbf{m}_0 = \begin{pmatrix} 1 \\ 1 \\ 1 \\ 0 \\ 0 \\ 0 \end{pmatrix} \quad \text{and} \quad \mathbf{t2} = \begin{pmatrix} -1 \\ 1 \\ 1 \\ 0 \\ 0 \\ -1 \end{pmatrix} .$$

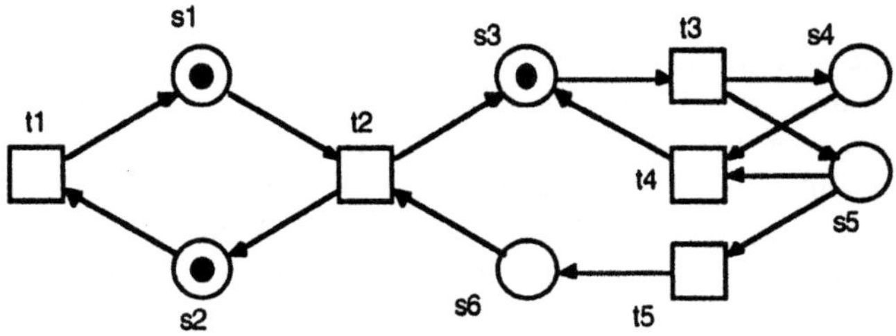

Fig. 1. A marked Petri net

A transition occurrence changes the number of tokens on a place at most by one. Hence, the vectors associated to the transitions have components of the set $\{-1, 0, 1\}$. Generalizations to arbitrary integers are not difficult and they are studied in the literature in depth (when nets with *weighted arcs* are considered).

If the occurrence of a transition t transfers a marking m to a marking m', then m' is uniquely determined by the vector equation

$$\mathbf{m}' = \mathbf{m} + \mathbf{t} \; .$$

The sequential behavior of a Petri net with initial marking m_0 is given by its *occurrence sequences*. An occurrence sequence is a finite or infinite sequence of transitions such that m_0 enables the first transition, the reached marking enables the second transition, and so on. The marking m reached by a finite occurrence sequence can be computed using the respective numbers of occurrences of transitions in the sequence: If k_t stands for the number of occurrences of a transition t in the occurrence sequence then the marking m can be computed by

$$\mathbf{m} = \mathbf{m}_0 + \sum_{t \in T} k_t \, \mathbf{t} \; .$$

Usually, this equation is given in its matrix form. The matrix with columns $\mathbf{t}_1 \ldots \mathbf{t}_l$ is called *incidence matrix* of the net N. We will denote the incidence matrix of N by $\mathbf{N}$, i.e., $\mathbf{N} = [\mathbf{t}_1 \cdots \mathbf{t}_l]$. Let $\mathbf{k}$ be the column vector with components $k_{t_1}, \ldots, k_{t_l}$. Then, the matrix form of the above equation reads

$$\mathbf{m} = \mathbf{m}_0 + \mathbf{N} \cdot \mathbf{k} \; .$$

This formula for the marking m is the basis of most concepts mentioned in this contribution. It is a compact way to express the interrelation between markings and numbers of transition occurrences in occurrence sequences. Its linear algebraic background turns out particularly appropriate because it allows to transfer and interpret concepts and results from linear algebra to Petri nets.

We provide one example already at this place: A slight reformulation of the equation given above yields a necessary condition for the a marking m to be reachable from the initial marking m_0 in a net N: the *marking equation* of the marking m,

$$\mathbf{N} \cdot \mathbf{x} = \mathbf{m} - \mathbf{m}_0 \; ,$$

must have a vector solution for $\mathbf{x}$ with components in $\mathbb{N}$. Hence, in particular, there must exist a solution with rational components. A suitable solution vector $\mathbf{k}$ for the variable $\mathbf{x}$ proves that the marking equation is soluble. Conversely, the non-reachability of a marking m can often be shown by proving that the marking equation is not soluble. There are efficient ways to prove non-solubility in the rationals, e.g. Gauss-elimination, but these techniques do not provide short and understandable proofs showing that no solution exists.

Another well-known necessary condition for the reachability of a marking is provided by *place invariants*. A place invariant can be viewed as an integral vector $\mathbf{i}$ satisfying $\mathbf{i} \cdot \mathbf{t} = 0$ for each transition t. Hence, a place invariant vector is a solution $\mathbf{i}$ for $\mathbf{y}$ to the equation system

$$\mathbf{y} \cdot \mathbf{N} = (0, \ldots, 0) .$$

The vectors

$$(1,1,0,0,0,0) \text{ and } (0,0,1,0,1,1)$$

are two place invariant vectors of our example net of Figure 1. It is not difficult to see that, for each marking m reachable from m_0, the products $\mathbf{i} \cdot \mathbf{m}_0$ and $\mathbf{i} \cdot \mathbf{m}$ coincide for any place invariant vector $\mathbf{i}$. Thus, place invariants yield very elegant proofs for the non-reachability of a marking m: every place invariant vector $\mathbf{i}$ satisfying $\mathbf{i} \cdot \mathbf{m}_0 \neq \mathbf{i} \cdot \mathbf{m}$ does the job.

A central theorem of linear algebra is the alternatives' theorem of Fredholm [Schr86]. It implies that the marking equation of a marking m possesses a rational solution if and only if $\mathbf{i} \cdot \mathbf{m}_0 = \mathbf{i} \cdot \mathbf{m}$ for each place invariant vector $\mathbf{i}$. Hence, solubility of the marking equation is shown by a suitable solution vector whereas its non-solubility can be proven by a suitable place invariant vector $\mathbf{i}$ satisfying $\mathbf{i} \cdot \mathbf{m} \neq \mathbf{i} \cdot \mathbf{m}_0$.

Decisive application oriented properties of Petri nets are mostly hard to decide, i.e., there exist no efficient algorithm to decide whether or not the property holds. However, there are necessary or sufficient conditions for many properties, that can efficiently be tested. We distinguish three kinds of techniques in this context:

- a verification technique is a deterministic algorithm to prove a desired property,
- a proof technique is intended to formulate a short proof of a property (formally, it is a non-deterministic algorithm that guesses the proof arguments and then verifies the proof),
- an analysis technique provides a variety of properties of a given marked net, as a basis for further arguments.

For the above discussed property – non-reachability of some marking m – a verification technique finds out that the marking equation has no solution. A proof technique for the same property employs a place invariant. The result of an analysis technique could be a collection of place invariants.

In section 2 we study basic concepts such as equation systems, systems of inequalities, Petri nets and their matrix representations, place and transition invariant vectors. Section 3 concentrates on the marking equation and related necessary conditions for the reachability of a marking. Linear algebraic characterizations of traps and siphons as well as consequences for reachability analysis are the topics of section 4. In section 5, we study state based properties of marked nets. In particular, we identify properties of markings of a net that are satisfied by all reachable markings. We provide a verification technique for such properties, based on the results of the previous sections. Finally, the 6th section is concerned with the so-called rank conditions. These conditions provide necessary criteria and sufficient criteria for liveness of marked nets.

Each section contains bibliographic notes including sources and related concepts.

Bibliographic Notes

The books [Pete81, Reis85, DeEs95] as well as the survey articles [Pete77, JaVa80, Mura89] contain sections on linear algebraic techniques for Petri nets. The papers [MeRo80, Laut87a] are surveys devoted to this topic. References to particular concepts such as place invariant vectors or the marking equation can be found at the end of the respective sections where they are introduced formally.

A more detailed presentation of the results of this contribution and of further applications of linear algebra and linear programming to verification, proofs and analysis of Petri net behavior is provided in the book [Dese98] (in German).

2 Definitions and Elementary Results

2.1 Systems of Linear Inequalities

The set of rational numbers is denoted by $\mathbb{Q}$, the set of integers by $\mathbb{Z}$ and the set of natural numbers (i.e., nonnegative integers) by $\mathbb{N}$. A *positive* integer is greater than 0.

We use bold capitals $\mathbf{A}, \mathbf{B}, \mathbf{C}$ as symbols for matrices, bold $\mathbf{a}, \mathbf{b}, \mathbf{c}$ for vectors and a, b, c for numbers. The transposed of a matrix $\mathbf{A}$ is denoted $\mathbf{A}^{\top}$, and similarly for vectors. Variable vectors are denoted by $\mathbf{x}$ and $\mathbf{y}$ and variable numbers by x and y. Products of a matrix with some other matrix or vector always assumes proper arities. The product of vectors or matrices with numbers is defined componentwise and is denoted by simple concatenation (without "·").

If $\mathbf{a} = (a_1, \ldots, a_n)$ and $\mathbf{b} = (b_1, \ldots, b_n)$ are vectors, then $\mathbf{a} \leq \mathbf{b}$ if $a_i \leq b_i$ for all i $(1 \leq i \leq n)$ and $\mathbf{a} < \mathbf{b}$ if $a_i < b_i$ for all i $(1 \leq i \leq n)$. We write $\mathbf{a} \neq \mathbf{b}$ if not $\mathbf{a} = \mathbf{b}$. Notice that $\mathbf{a} \leq \mathbf{b}, \mathbf{a} \neq \mathbf{b}$ does not imply $\mathbf{a} < \mathbf{b}$. We write $\mathbf{0}$ for vectors with the entry 0 in each component. Each vector $\mathbf{a} > \mathbf{0}$ is *positive*. Each vector $\mathbf{a} \geq \mathbf{0}$ is *nonnegative*. For a set A and a nonnegative integer k, A^k denotes the set of all vectors with k components and all components in A. We write A^* for the set of all vectors over the set A, i.e., vectors with all components in A.

A *linear inequality* is given by an integral vector $\mathbf{a}$ and an integer b. It is represented in the form

$$\mathbf{a} \cdot \mathbf{x} \leq b \, .$$

It is *soluble over a set* A if there exists some $\mathbf{k}$ in A^* satisfying $\mathbf{a} \cdot \mathbf{k} \leq b$.

A *system of linear inequalities* is a set of linear inequalities. It is soluble if there exists a vector that solves all inequalities of the set. If it is finite then it has a matrix based representation

$$\mathbf{A} \cdot \mathbf{x} \leq \mathbf{b} \, ,$$

where the vectors $\mathbf{a}$ of the linear inequalities are the rows of the matrix $\mathbf{A}$ and the numbers b are the components of the vector $\mathbf{b}$. Other representations are:

$\mathbf{A} \cdot \mathbf{x} \geq \mathbf{b}$	for $(-1)\, \mathbf{A} \cdot \mathbf{x} \leq (-1)\, \mathbf{b}$,
$\mathbf{A} \cdot \mathbf{x} \leq \mathbf{b}, \mathbf{C} \cdot \mathbf{x} \leq \mathbf{d}$	for the union of the two sets of inequalities,
$\mathbf{A} \cdot \mathbf{x} = \mathbf{b}$	for $\mathbf{A} \cdot \mathbf{x} \leq \mathbf{b}$, $\mathbf{A} \cdot \mathbf{x} \geq \mathbf{b}$.

Thus, each *linear equation system* can be interpreted as a particular system of linear inequalities that contains only equations.

We will frequently employ analysis techniques that are based on the solubility of systems of linear inequalities. The complexity of these techniques depend on the domain under consideration. The following well-known results can e.g. be found in [Schr86].

Proposition 1.

(1) Each system of linear inequalities over $\mathbb{Q}$ can be solved in polynomial time (linear programming).

(2) The solubility of systems of linear inequalities over $\mathbb{Z}$ is NP-complete (integer linear programming, variant 1.)

(3) The solubility of linear equation systems over $\mathbb{N}$ is NP-complete (integer linear programming, variant 2.)

2.2 Petri Nets and Linear Algebraic Representations

For a formal definition of Petri nets and their behavior as well as for usual notations we refer to [DeRe98], in this volume. Here, we only consider nets with finite and nonempty sets of places and transitions. All places and transitions of a net N are said to be *elements* of N. The sets of places, transitions and arcs of a net N are denoted by S_N, T_N and F_N, respectively. The net N is denoted by the triplet (S_N, T_N, F_N). The set of all elements of N is frequently also denoted by N, i.e., $N = S_N \cup T_N$.

For the application of linear algebraic techniques, we assume

$$S = \{s_1, \ldots, s_n\} \quad \text{and} \quad T = \{t_1, \ldots, t_l\} \, .$$

Then, a mapping $k\colon S \to A$ can be represented by a row or a column vector $\mathbf{k}$ with n components such that the i-th component of $\mathbf{k}$ is $k(s_i)$ $(1 \leq i \leq n)$.

Such vectors are called *place vectors.* Abusing terminology, we will frequently define place vectors via their associated mappings and write $k(s)$ instead of $\mathbf{k}(s)$. *Transition vectors* and related notions are defined correspondingly.

For example, a marking $m: S \rightarrow \mathbb{N}$ of N is represented by the place vector

$$\mathbf{m} = \begin{pmatrix} m(s_1) \\ \vdots \\ m(s_n) \end{pmatrix} \in \mathbb{N}^n .$$

If a transition t occurs at a marking m then the follower marking m' is given by

$$\mathbf{m}' = \mathbf{m} + \mathbf{t}$$

where $\mathbf{t}: S \rightarrow \mathbb{N}$ is a place vector defined by

$$\mathbf{t}(s) = \begin{cases} -1 & \text{if } s \in {}^\bullet t \setminus t^\bullet, \\ +1 & \text{if } s \in t^\bullet \setminus {}^\bullet t, \\ 0 & \text{otherwise} . \end{cases}$$

For each place s, the vector $\mathbf{e}_s: S \rightarrow \{0, 1\}$ is defined by $\mathbf{e}_s(s) = 1$ and $\mathbf{e}_s(s') = 0$ for $s \neq s'$. For a transition t, the transition vector $\mathbf{e}_t$ is defined correspondingly.

For a set A of places, $\chi(A)$ denotes the *characteristic place vector*, defined as the sum of all vectors $\mathbf{e}_s$ for places $s \in A$. *Characteristic transition vectors* of sets of transitions are defined correspondingly.

2.3 Incidence Matrix and Marking Equation

For a net N with n places and l transitions, the $n \times l$-matrix $\mathbf{N} = [\mathbf{t}_1 \cdots \mathbf{t}_l]$ is called *incidence matrix* of N. We will always denote the incidence matrix of a net N by the bold letter $\mathbf{N}$.

For a finite sequence σ of transitions of N, the transition vector $\mathbf{p}_\sigma: T \rightarrow \mathbb{N}$ denotes the *Parikh vector* of σ, where $\mathbf{p}_\sigma(t)$ is the number of occurrences of t in the sequence σ.

Each finite occurrence sequence terminates in a marking that can be computed by help of the incidence matrix and the Parikh vector of the sequence:

Theorem 2. *Let $m_0 \xrightarrow{\sigma} m$ be a finite occurrence sequence of a net N. Then*

$$\mathbf{m}_0 + \mathbf{N} \cdot \mathbf{p}_\sigma = \mathbf{m}. \qquad \square$$

The following necessary condition for the reachability of a marking is an immediate consequence:

Corollary 3. *Let N be a net. If a marking m of N is reachable from a marking m_0, then there exists a solution for $\mathbf{x}$ over $\mathbb{N}$ to the equation*

$$\mathbf{N} \cdot \mathbf{x} = \mathbf{m} - \mathbf{m}_0 . \qquad \square$$

For a given net N with initial marking m_0, this equation is called *the marking equation for m.*

Each finite sequence σ of transitions is enabled at some marking (if σ has length k, provide k tokens for each place). However, given a fixed initial marking, no reachable marking may enable σ. Every transition vector over $\mathbb{N}$ can be interpreted as a Parikh vector of a sequence of transitions; hence the following lemma:

Lemma 4. *Let* $\mathbf{k}$ *in* $\mathbb{N}^*$ *be a transition vector of a net* N. *Then there exist markings* m, m' *and an occurrence sequence* $m \xrightarrow{\sigma} m'$ *of* N *such that* $\mathbf{k}$ *is the Parikh vector of* σ. □

2.4 Place Invariants and Transition Invariants

Place invariants have already been mentioned in the introduction. A formal definition was provided in [DeRe98]. An equivalent definition based on linear algebraic concepts is given by the characterization of the following proposition.

Proposition 5. *Let* N *be a net. A mapping* $i: S_N \to \mathbb{Z}$ *is a place invariant if and only if its corresponding row vector* $\mathbf{i}$ *is a solution for* $\mathbf{y}$ *to the homogeneous equation system*

$$\mathbf{y} \cdot \mathbf{N} = \mathbf{0} .$$

□

The vector representation of a place invariant will be called *place invariant vector*.

The fundamental property of place invariants can be rephrased in terms of place invariant vectors. It follows immediately from the marking equation:

Theorem 6. *Let* N *be a net with a place invariant* i. *If a marking* m *is reachable from a marking* m_0, *then*

$$\mathbf{i} \cdot \mathbf{m} = \mathbf{i} \cdot \mathbf{m}_0 .$$

Proof. If m is reachable from m_0, the marking equation for m has some solution. Multiplication of both sides of the equation by $\mathbf{i}$ yields the result. □

Transition invariants and place invariants are defined correspondingly in [DeRe98]. The following proposition provides a linear algebraic characterization of transition invariants:

Proposition 7. *Let* N *be a net. A mapping* $j: T_N \to \mathbb{Z}$ *is a transition invariant if and only if its corresponding column vector* $\mathbf{j}$ *is a solution for* $\mathbf{x}$ *to*

$$\mathbf{N} \cdot \mathbf{x} = \mathbf{0} .$$

□

These transition vectors are called *transition invariant vectors*.

The fundamental property of transition invariants follows immediately from the marking equation:

Theorem 8. *Let N be a net with an occurrence sequence $m \xrightarrow{\sigma} m'$. The Parikh vector $\mathbf{p}_\sigma$ of σ is a transition invariant vector if and only if $m = m'$.*

Proof. $\mathbf{p}_\sigma$ is a transition invariant vector if and only if $\mathbf{N} \cdot \mathbf{p}_\sigma = \mathbf{0}$. This implies the result, because $\mathbf{m}_0 + \mathbf{N} \cdot \mathbf{p}_\sigma = \mathbf{m}$. □

A connectedness theorem was given in [DeRe98]: Each connected net with a live and bounded marking is strongly connected. We finish this section with a similar result:

Theorem 9. *Each connected net with a positive place invariant and a positive transition invariant is strongly connected.*

Proof. Let N be a connected net. Let i be a place invariant of N such that $\mathbf{i} > \mathbf{0}$ and let j be a transition invariant of N such that $\mathbf{j} > \mathbf{0}$. We only prove that, for each arc (u, v) of N, there exists a path from v to u. The results follows by the definition of weak and strong connectedness.

Case 1: $u \in S_N$ and $v \in T_N$.
Let the mapping $j' : T \to \mathbb{N}$ be given by

$$j'(t) = \begin{cases} j(t) & \text{if there is a path from } v \text{ to } t, \\ 0 & \text{otherwise.} \end{cases}$$

Consider the place vector $\mathbf{N} \cdot \mathbf{j}'$. Its component for a place s is given by

$$\sum_{u \in {}^\bullet s} j'(u) - \sum_{u \in s^\bullet} j'(u) .$$

We show $\mathbf{N} \cdot \mathbf{j}'(s) \leq 0$ for each place s. Let s be a place.

Assume first $j'(t) = 0$ for each transition t in ${}^\bullet s$. Since $j'(t) \geq 0$ for each transition t by definition,

$$0 = \sum_{t \in {}^\bullet s} j'(t) \leq \sum_{t \in s^\bullet} j'(t) .$$

If $j'(t) = j(t) > 0$ for some transition t in ${}^\bullet s$ then there is a path from v to s. In this case there exists a path from v to any t in $s^\bullet$. Therefore, $j'(t) = j(t) > 0$ for each transition t in $s^\bullet$. So, in this case,

$$0 < \sum_{t \in {}^\bullet s} j'(t) \leq \sum_{t \in {}^\bullet s} j(t) = \sum_{t \in s^\bullet} j(t) = \sum_{t \in s^\bullet} j'(t) .$$

We have shown for both cases:

$$\sum_{t \in {}^\bullet s} j'(u) \leq \sum_{t \in s^\bullet} j'(t) .$$

This finishes the proof of $\mathbf{N} \cdot \mathbf{j}' \leq 0$.

We have $\mathbf{i} \cdot \mathbf{N} = \mathbf{0}$, because i is a place invariant. Therefore, $\mathbf{i} \cdot \mathbf{N} \cdot \mathbf{j}' = 0$. Since $\mathbf{i} > \mathbf{0}$ and $\mathbf{N} \cdot \mathbf{j}' \leq \mathbf{0}$, the vector $\mathbf{N} \cdot \mathbf{j}'$ has no negative component, which implies $\mathbf{N} \cdot \mathbf{j}' = \mathbf{0}$. Hence, j' is a transition invariant.

So we obtain the following inequalities:

$$\begin{aligned} \sum_{t \in {}^\bullet u} j'(t) = \sum_{t \in u^\bullet} j'(t) & \quad (j' \text{ is a transition invariant}) \\ \geq j'(v) & \quad (v \in u^\bullet) \\ = j(v) & \quad (\text{definition of } j') \\ > 0 & \quad (j \text{ is a positive transition invariant vector})\ . \end{aligned}$$

Hence, there exists a transition $t \in {}^\bullet u$ satisfying $j'(t) > 0$. By the definition of j', there is a path from v to t. This path can be extended by the place u.

Case 2: $u \in T$ and $v \in S$.
Consider the net $N' = (T_N, S_N, F_N)$, where places and transitions of N are swapped. The incidence matrix $\mathbf{N}'$ of N' is given by $(-1)\,\mathbf{N}^{\mathsf{T}}$. So, by Propositions 5 and 7, i is a positive transition invariant of N' and j is a positive place invariant of N'. The arc (v, u) leads from a place of N' to a transition of N'. As shown above (case 1), the net N' contains a path from transition v to place u. In the net N, this path leads from place v to transition u. □

Bibliographic Remarks

Place invariants were first introduced in [LaSc74, Laut75] in a linear algebraic framework. The paper [Ramc74] considers transition invariants (using different notations). Other relevant early references are [Lien76] and [Mura77], where the latter paper concentrates particularly on the marking equation. [Sifa79] gives a proof of the connectedness theorem (Theorem 9) in a more general setting.

3 The Marking Equation

The problem whether a marking m is reachable from a given initial marking m_0 is known to be decidable, but its enormous complexity prevents effective algorithms (see [DeRe98] for references). In this section, we study the inverse problem: is a given marking m not reachable from m_0? As discussed in the introduction, verification and proof techniques based on linear algebra lead to more efficient algorithms that work for many, though not all, instances. Such techniques for verifying or proving non-reachability of a marking are studied in this section.

It was shown in the introduction that there is a solution of the marking equation over the natural numbers for each reachable marking. Thus, the marking equation provides a necessary condition for reachability. This condition can be employed to verify the non-reachability of the marking. Any algorithm can be used that decides integral solubility of the marking equation: The marking under consideration is not reachable in case there is no solution. Nevertheless, there might exist non-reachable markings of a net which do possess an integral solution to the marking equation. Therefore, this verification technique does not work for all non-reachable markings.

Moreover, unfortunately no efficient algorithm is known that decides integral solubility of linear equation systems (Proposition 1). We will consider weaker criteria for reachability, namely solubility of the marking equation over the rational numbers, over the non-negative rational numbers and over the integers. These conditions will lead to more efficient verification algorithms with, admittedly, less expressive power.

As any verification technique, the verification techniques mentioned above can be viewed as proof techniques, too. However, we can do better. As an example, a place invariant can be used to prove the non-reachability of some given marking. It is easy to check that a given place vector is in fact a place invariant vector and that the initial marking and the given marking disagree w.r.t. this invariant. In the introduction, we have shown that a place invariant can be used to prove non-reachability of a marking if and only if the marking equation has no rational solution. Here, we show that more powerful proof techniques correspond to the solubility of the marking equation over the positive rationals and over the integers.

3.1 Solubility of the Marking Equation over the Natural Numbers

The marking equation provides a necessary criterion for the reachability of a marking m: if m is reachable from the initial marking m_0 of a net N then

$$\mathbf{N} \cdot \mathbf{x} = \mathbf{m} - \mathbf{m}_0$$

has a solution for $\mathbf{x}$ in $\mathbb{N}^*$. Hence the following reachability condition for m:

(M1) *The marking equation for m has a solution for* $\mathbf{x}$ *in* $\mathbb{N}^*$.

This condition cannot be decided efficiently in general because (M1) is an instance of "Integer Linear Programming", hance an NP-complete problem (Proposition 1(3)).

The example given in Figure 2 shows that the marking equation provides no sufficient condition for reachability: The marking

$$\mathbf{m}_1 = (0,1,0,0,1,0,0)^{\mathsf{T}}$$

is not reachable from the initial marking

$$\mathbf{m}_0 = (0,0,1,1,0,0,0)^{\mathsf{T}} .$$

However, there is the following solution to the marking equation:

$$\mathbf{N} \cdot (1,1,0,2,2,0,2)^{\mathsf{T}} = \mathbf{m}_1 - \mathbf{m}_0 .$$

All other markings of this net with solvable marking equation are reachable. So the technique only fails for the marking $\mathbf{m}_1$.

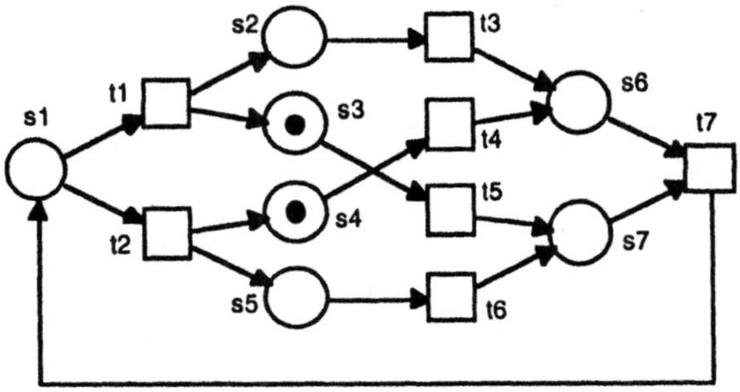

Fig. 2. Limits of the marking equation

The marking equation does not characterize the set of reachable markings because the token count on a place never turns negative. If tokens could be "borrowed" temporarily to avoid negative markings, each marking satisfying the marking equation was reachable. This observation can formally be expressed by addition of another marking at both sides of the marking equation:

Proposition 10. *Let N be a net with markings m_0 and m. If*

$$\mathbf{m}_0 + \mathbf{N} \cdot \mathbf{k} = \mathbf{m}$$

for some $\mathbf{k}$ *in* $\mathbb{N}^*$ *then there exists a marking* $\overline{m}$ *and a transition sequence* σ *such that* $\mathbf{k}$ *is the Parikh vector of* σ *and*

$$\overline{\mathbf{m}} + \mathbf{m}_0 \xrightarrow{\sigma} \overline{\mathbf{m}} + \mathbf{m} \, .$$

Proof. Choose a marking $\overline{m}$ which associates to each place more tokens than the sum of all components of $\mathbf{k}$. □

3.2 Solubility of the Marking Equation over the Rational Numbers

Integral solutions of the marking equation with negative components correspond to "backward occurrences" of transitions. Of course, the marking transformation caused by a transition occurrence is not reversible in general. Another important aspect of the occurrence rule is atomicity of tokens. If tokens could be "broken into pieces", each token could be replaced by suitable "token pieces". If the marking equation has a non-negative rational solution $\mathbf{k}$ with a common denominator d then

$$\mathbf{N} \cdot \mathbf{x}' = d\,\mathbf{m} - d\,\mathbf{m}_0$$

has the non-negative integral solution $d\,\mathbf{k}$ for $\mathbf{x}'$. The vectors $d\,\mathbf{m}_0$ and $d\,\mathbf{m}$ are the initial and the final marking after replacing each token by d "token pieces".

For these reasons, the following condition (M2) properly weakens (M1), because it allows negative and rational numbers.

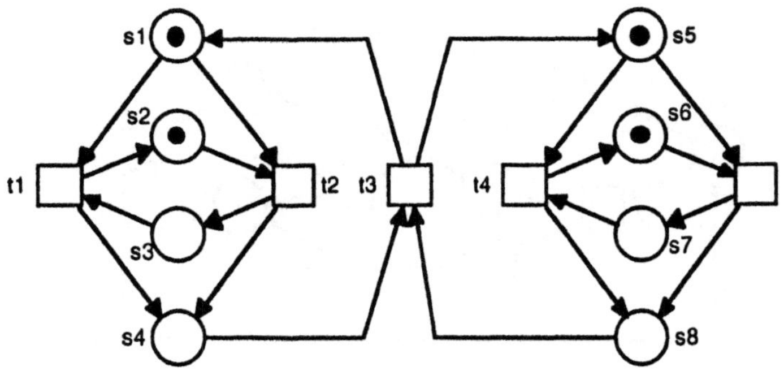

Fig. 3. Limits of rational solutions of the marking equation.

(M2) *The marking equation for m has a solution for* $\mathbf{x}$ *in* $\mathbb{Q}^*$.

The example shown in Figure 3 shows that (M2) has less expressive power than (M1)[1]. The marking

$$\mathbf{m}_1 = (1,0,1,0,1,1,0,0)^{\mathsf{T}}$$

is not reachable from the depicted initial marking m_0. There is no solution of the marking equation over $\mathbb{Z}$; therefore, there is no solution over $\mathbb{N}$. However, there is a solution over $\mathbb{Q}$, namely

$$\mathbf{m}_0 + \mathbf{N} \cdot \left(0,1,1,\frac{1}{2},\frac{1}{2}\right)^{\mathsf{T}} = \mathbf{m}_1 \ .$$

It is easy to see that the marking m_1 would be reachable from m_0 if transitions could occur for "half tokens".

(M2) was discussed in the introduction already. There are efficient algorithms to decide if a given equation system has a rational solution; e.g. the Gauss elimination works in in cubic time. For a related proof technique, there exists a more efficient way using place invariants. The equivalent expressive power of these techniques is stated in the following proposition:

Proposition 11. *Let N be a net with markings* m_0 *and* m. *There exists no solution of the marking equation for* m *over* $\mathbb{Q}$ *if and only if some place invariant* i *satisfies*

$$\mathbf{i} \cdot \mathbf{m}_0 \neq \mathbf{i} \cdot \mathbf{m} \ .$$

Proof. The proposition follows immediately from Fredholm's theorem of the alternatives (which was first observed by Gauss in the year 1809), see [Schr86, Corollary 3.1b, page 28]. □

[1] This can be shown using simpler examples but we shall need this example later in other contexts, too.

3.3 Solubility of the Marking Equation over the Non-Negative Rational Numbers

Let $\mathbb{Q}_+$ denote the set of non-negative rationals. Then, $\mathbb{Q}_+^*$ contains all rational vectors $\mathbf{x} \geq \mathbf{0}$.

If the marking equation for a marking m has no solution over $\mathbb{Q}_+$, there is in particular no solution over $\mathbb{N}$. So (M1) can be weakened to (M3):

(M3) *The marking equation for m has a solution for* $\mathbf{x}$ *in* $\mathbb{Q}_+^*$.

Clearly, this condition is stronger than (M2). It can be expressed by the solubility of the following system of linear inequalities:

$$\mathbf{N} \cdot \mathbf{x} = \mathbf{m} - \mathbf{m}_0 \,,$$
$$\mathbf{x} \geq 0 \,,$$

where $\mathbf{x}$ varies over $\mathbb{Q}^*$. This system can be solved in polynomial time, because it is an instance of *Linear Programming* (Proposition 1 (1)). So (M3) provides an efficient verification technique.

For a related *proof technique*, we use the following variant of Farkas' Lemma (see [Schr86, Corollary 7.1c, page 89]):

Proposition 12. *Exactly one of the following systems of inequalities has a solution over* $\mathbb{Q}$*:*

$$\mathbf{A} \cdot \mathbf{x} = \mathbf{b},\ \mathbf{x} \geq 0 \,,$$
$$\mathbf{A}^\mathsf{T} \cdot \mathbf{y} \geq 0,\ \mathbf{b} \cdot \mathbf{y} < 0 \,.$$

□

With $\mathbf{N} = \mathbf{A}$ and $\mathbf{b} = \mathbf{m} - \mathbf{m}_0$, (M3) corresponds to the first line of the system given in this proposition. Hence, the complement of (M3) is equivalent to the rational solubility of the following system of inequalities:

$$\mathbf{N}^\mathsf{T} \cdot \mathbf{y} \geq 0 \,,$$
$$(\mathbf{m} - \mathbf{m}_0) \cdot \mathbf{y} < 0 \,.$$

Similar to place invariant vectors, any solution of this system of inequalities provides an efficient proof showing that m is not reachable from m_0. Notice that for every rational solution there also is an integral solution, obtained by multiplication with the common denominator.

The example of Figure 3 demonstrates that (M3) has less expressive power than (M1), because the previously given rational solution has no negative components.

If a net has a positive transition invariant, solubility of the marking equation over $\mathbb{Q}$ and solubility of the marking equation over $\mathbb{Q}_+$ coincide, because repeated addition of a positive transition invariant vector to any solution will eventually yield a non-negative solution.

3.4 Solubility of the Marking Equation over the Integers

Another weakening of (M1) is given by

(M4) *The marking equation of the marking m has a solution for* $\mathbf{x}$ *in* $\mathbb{Z}^*$.

The conditions (M4) and (M3) are incomparable. Notice that (M3) and (M4) together do not imply (M1) because the existence of a non-negative solution and the existence of an integral solution does not imply that there is a solution that is both integral and non-negative.

The verification technique based on (M4) corresponds to the integral solubility of the marking equation. The theory of linear diophantic equations implies that the incidence matrix $\mathbf{N}$ can be transformed to its *Smith normal form* (see [Schr86], page 50):

Proposition 13. *Any integral matrix* $\mathbf{N}$ *can be transformed into its Smith normal form* $\mathbf{S}$. *This transformation is given by the matrix product* $\mathbf{S} = \mathbf{Q} \cdot \mathbf{N} \cdot \mathbf{P}$, *where* $\mathbf{Q}$ *and* $\mathbf{P}$ *are unimodular matrices of suitable size.* $\mathbf{S}$ *has the following form:*

$$\mathbf{S} = \begin{pmatrix} s_{1,1} & & & & & \\ & \ddots & & & 0 & \\ & & s_{a,a} & & & \\ & & & 0 & & \\ & 0 & & & \ddots & \\ & & & & & 0 \end{pmatrix},$$

where $s_{1,1}, \ldots, s_{a,a}$ *are positive integers and, for* $i = 1, \ldots, a-1$, *the diagonal element* $s_{i,i}$ *divides* $s_{i+1,i+1}$. *The Smith normal form is uniquely determined. Its diagonal elements are called elementary divisors of the matrix* $\mathbf{N}$.

$\mathbf{Q}$ *and* $\mathbf{P}$ *are invertible quadratic integral matrices such that their inverses* $\mathbf{Q}^{-1}$ *and* $\mathbf{P}^{-1}$ *are also integral (unimodularity of the matrices).* □

The next theorem shows that integral solubility of the marking equation can be decided by constructing the Smith normal form of the incidence matrix.

Theorem 14. *Let* N *be a net and* m_0, m *be markings of* N. *Let* $\mathbf{S} = \mathbf{Q} \cdot \mathbf{N} \cdot \mathbf{P}$ *be the Smith normal form of* $\mathbf{N}$. *Define*

$$\mathbf{b} = (b_1, b_2, \ldots)^{\mathsf{T}} = \mathbf{Q} \cdot (\mathbf{m} - \mathbf{m}_0) .$$

There exists an integral solution for $\mathbf{x}$ *to the marking equation*

$$\mathbf{N} \cdot \mathbf{x} = \mathbf{m} - \mathbf{m}_0$$

if and only if

(α) $s_{i,i}$ *divides* b_i $(1 \leq i \leq a)$, *and*
(β) $b_i = 0$ $(i > a)$.

Proof. Since $\mathbf{S} = \mathbf{Q} \cdot \mathbf{N} \cdot \mathbf{P}$, the marking equation can be transformed into

$$\mathbf{Q}^{-1} \cdot \mathbf{S} \cdot \mathbf{P}^{-1} \cdot \mathbf{x} = \mathbf{m} - \mathbf{m}_0 \ .$$

Define $\mathbf{y} = \mathbf{P}^{-1} \cdot \mathbf{x}$. Multiplication by $\mathbf{Q}$ yields

$$\mathbf{S} \cdot \mathbf{y} = \mathbf{Q} \cdot (\mathbf{m} - \mathbf{m}_0) = \mathbf{b} \ .$$

Since $\mathbf{P}$ as well as $\mathbf{P}^{-1}$ are integral matrices, this equation system has an integral solution for $\mathbf{y}$ if and only if the marking equation has an integral solution for $\mathbf{x}$. It is easy to see that, by the form of $\mathbf{S}$, there is an integral solution for $\mathbf{y}$ if and only if (α) and (β) hold true. □

The transformation matrix $\mathbf{Q}$ can be constructed in polynomial time [KaBa79]. Given the matrix $\mathbf{Q}$, the conditions (α) and (β) of Theorem 14 can obviously be decided in polynomial time. Hence, by constructing the Smith normal form (including the matrix $\mathbf{Q}$), it can be decided in polynomial time whether the marking equation possesses an integral solution. Thus we obtained an efficient verification technique based on condition (M4).

Place invariants do not suffice for a proof technique based on (M4). They can only prove the non-existence of rational solutions of the marking equation. As mentioned above, in Figure 3, the marking

$$\mathbf{m}_1 = (1, 0, 1, 0, 1, 1, 0, 0)^{\mathsf{T}}$$

agrees with the initial marking on all place invariants but its marking equation has no integral solution.

For this example, it is not difficult to verify that

$$m(\mathtt{s1}) + m(\mathtt{s2}) + m(\mathtt{s5}) + m(\mathtt{s6}) \quad \textit{is an even number}$$

holds for all reachable markings m: If a marking m satisfies the property and the occurrence of a transition leads to m' then m' satisfies the property, too. The same holds for the complementary property (odd token count on the places in $\{\mathtt{s1}, \mathtt{s2}, \mathtt{s5}, \mathtt{s6}\}$). For the markings m_0 and m_1 we obtain the following sums:

$$m_0(\mathtt{s1}) + m_0(\mathtt{s2}) + m_0(\mathtt{s5}) + m_0(\mathtt{s6}) = 4 \ ,$$

$$m_1(\mathtt{s1}) + m_1(\mathtt{s2}) + m_1(\mathtt{s5}) + m_1(\mathtt{s6}) = 3 \ .$$

Hence, since this sum is initially even and remains even when transitions occur, m_1 is not reachable from m_0.

The same argument can be stated as follows: For the vector

$$\mathbf{i} = (1, 1, 0, 0, 1, 1, 0, 0)$$

and each column vector $\mathbf{t}$ of the incidence matrix $\mathbf{N}$,

$$\mathbf{i} \cdot \mathbf{t} \equiv 0 \ (\mathrm{mod}\ 2) \ .$$

Each marking transformation $m \xrightarrow{t} m'$ satisfies

$$\mathbf{m}' = \mathbf{m} + \mathbf{t} \ ,$$

where $\mathbf{t}$ is the column of the incidence matrix associated to the transition t. Multiplication by $\mathbf{i}$ yields

$$\mathbf{i} \cdot \mathbf{m}' = \mathbf{i} \cdot \mathbf{m} + \mathbf{i} \cdot \mathbf{t} \ .$$

Since $\mathbf{i} \cdot \mathbf{t} \equiv 0 \pmod 2$,

$$\mathbf{i} \cdot \mathbf{m}' \equiv \mathbf{i} \cdot \mathbf{m} \pmod 2 \ .$$

Vectors with this property will be called modulo place invariant vectors:

Let $k \geq 2$ be a natural number, and let N be a net. A vector $\mathbf{i} \in \mathbb{Z}^*$ is called *modulo-k place invariant vector* of N if, for each column vector $\mathbf{t}$ of $\mathbf{N}$,

$$\mathbf{i} \cdot \mathbf{t} \equiv 0 \pmod k.$$

A vector $\mathbf{i} \in \mathbb{Z}^*$ is called *modulo place invariant vector* if it is a modulo-k place invariant vector for some $k \geq 2$.

Clearly, each place invariant vector is a modulo-k place invariant vector for any k. The inverse does not hold in general. In the above example, the vector $(1, 1, 0, 0, 1, 1, 0, 0)$ is a modulo-2 place invariant vector but no (classical) place invariant vector.

Each modulo place invariant vector induces a token conservation law:

Theorem 15. *Let m_0 be an initial marking of a net N and let m be reachable from m_0. Then each modulo-k place invariant vector $\mathbf{i}$ of N satisfies*

$$\mathbf{i} \cdot \mathbf{m}_0 \equiv \mathbf{i} \cdot \mathbf{m} \pmod k \ .$$

Proof. Since m is reachable from m_0, there is a solution for $\mathbf{x}$ in $\mathbb{N}^*$ of the marking equation

$$\mathbf{m}_0 + \mathbf{N} \cdot \mathbf{x} = \mathbf{m} \ .$$

Multiplication by $\mathbf{i}$ yields

$$\mathbf{i} \cdot \mathbf{m}_0 + \mathbf{i} \cdot \mathbf{N} \cdot \mathbf{x} = \mathbf{i} \cdot \mathbf{m} \ .$$

An immediate consequence is

$$\mathbf{i} \cdot \mathbf{m}_0 + \mathbf{i} \cdot \mathbf{N} \cdot \mathbf{x} \equiv \mathbf{i} \cdot \mathbf{m} \pmod k \ .$$

Since $\mathbf{i}$ is a modulo-k place invariant vector, all components of $\mathbf{i} \cdot \mathbf{N}$ are multiples of k. Since $\mathbf{x}$ is integral, multiplication by $\mathbf{x}$ yields a multiple of k. Hence,

$$\mathbf{i} \cdot \mathbf{N} \cdot \mathbf{x} \equiv 0 \pmod k \ .$$

So this product can be removed in the above congruence, which completes the proof. □

The inverse of Theorem 15 does not hold true in general. Figure 2 shows a counter example. However, modulo place invariants allow to characterize the integral solubility of the marking equation:

Theorem 16. *Let m_0 and m be markings of a net N. The following propositions are equivalent:*

- *For each integer $k \geq 2$ and each modulo-k place invariant vector* $\mathbf{i}$,

$$\mathbf{i} \cdot \mathbf{m} \equiv \mathbf{i} \cdot \mathbf{m}_0 \pmod{k} .$$

- *The marking equation*

$$\mathbf{m}_0 + \mathbf{N} \cdot \mathbf{x} = \mathbf{m}$$

  has a solution for $\mathbf{x}$ *over* $\mathbb{Z}$.

Proof. The direction ($\Longleftarrow$) was already part of the proof of Theorem 15. So only ($\Longrightarrow$) remains to be proven. Let $\mathbf{S} = \mathbf{Q} \cdot \mathbf{N} \cdot \mathbf{P}$ be the Smith normal form of $\mathbf{N}$ with diagonal elements $s_{1,1}, \ldots, s_{a,a}$. Define $\mathbf{b} = (\mathbf{b_1}, \mathbf{b_2}, \ldots)^{\mathsf{T}} = \mathbf{Q} \cdot (\mathbf{m} - \mathbf{m_0})$. By Theorem 14, it suffices to show:

(α) $s_{i,i}$ divides b_i ($1 \leq i \leq a$), and
(β) $b_i = 0$ ($i > a$).

Let k be the least multiple of $s_{a,a}$ which is greater than all values $|b_i|$, the components of $\mathbf{b}$. By assumption, every modulo-k place invariant vector $\mathbf{i}$ satisfies the congruence

$$\mathbf{i} \cdot (\mathbf{m} - \mathbf{m}_0) \equiv 0 \pmod{k} .$$

(α) Let $1 \leq i \leq a$. Since $s_{i,i}$ divides $s_{a,a}$ and since k is a multiple of $s_{a,a}$, the number $s_{i,i}$ divides k. So we can define the integral vector

$$\mathbf{y}_i = \left(0, \ldots, 0, \frac{k}{s_{i,i}}, 0, \ldots, 0\right)$$

with a positive i-th component. Multiplication by the matrix $\mathbf{S}$ yields

$$\mathbf{y}_i \cdot \mathbf{S} = (0, \ldots, 0, k, 0, \ldots, 0) .$$

Substituting $\mathbf{S}$ by $\mathbf{Q} \cdot \mathbf{N} \cdot \mathbf{P}$ and multiplying both sides of the equation by $\mathbf{P}^{-1}$ proves

$$\mathbf{y}_i \cdot \mathbf{Q} \cdot \mathbf{N} = (0, \ldots, 0, k, 0, \ldots, 0) \cdot \mathbf{P}^{-1} .$$

Since $\mathbf{P}^{-1}$ is an integral matrix, each component on the right hand side of this equation is a multiple of k. Therefore, $\mathbf{y}_i \cdot \mathbf{Q}$ is a modulo-k place invariant vector.

The assumption implies

$$\mathbf{y}_i \cdot \mathbf{Q} \cdot (\mathbf{m} - \mathbf{m}_0) = \mathbf{y}_i \cdot \mathbf{b} \equiv 0 \pmod{k} .$$

By the definition of $\mathbf{y}_i$ follows

$$\frac{k}{s_{i,i}} b_i \equiv 0 \pmod{k} .$$

This congruence holds true if and only if $s_{i,i}$ divides b_i.

(β) Let $i > a$. Then

$$\mathbf{e}_i \cdot \mathbf{S} = \mathbf{0} .$$

As in the previous case, substituting $\mathbf{S}$ by $\mathbf{Q} \cdot \mathbf{N} \cdot \mathbf{P}$ and multiplying both sides of the equation by $\mathbf{P}^{-1}$ proves

$$\mathbf{e}_i \cdot \mathbf{Q} \cdot \mathbf{N} = \mathbf{0} .$$

Hence, $\mathbf{e}_i \cdot \mathbf{Q}$ is a place invariant vector. So it is also a modulo-k place invariant vector. The assumption implies

$$\mathbf{e}_i \cdot \mathbf{Q} \cdot (\mathbf{m} - \mathbf{m}_0) = \mathbf{e}_i \cdot \mathbf{b} = b_i \equiv 0 \pmod{k} .$$

Since k was chosen greater than $|b_i|$, we finally obtain $b_i = 0$. □

This theorem states that, if there is no integral solution of the marking equation, then there exists a modulo place invariant proving the non-reachability of the marking under consideration. So the proof technique based on modulo place invariants is exactly as powerful as the verification technique based on (M4).

3.5 Calculating Modulo Place Invariants

Theorem 16 referred to all modulo place invariants of a net. In its proof we showed that it suffices to consider a fixed modulo number k which depends on both the net and the difference between the considered markings. For a given net, no number k suffices for all markings: Every marking $\mathbf{m}_0 + k\,\mathbf{e}_i$ agrees with $\mathbf{m}_0$ on all modulo-k place invariant vectors. However, in general there are no integral solutions for arbitrary vectors $\mathbf{e}_i$ of the corresponding equation $\mathbf{N} \cdot \mathbf{x} = k\,\mathbf{e}_i$.

A finite set of modulo place invariants with different numbers k is not sufficient, too. We could equivalently choose a common multiple of these numbers as a common modulo number and run into the same problem as above. Therefore, no finite set of modulo invariants identifies all markings without integral solution to the marking equation. The situation is different for classical place invariants: Every integral base of the vector space of solutions of $\mathbf{N} \cdot \mathbf{x} = \mathbf{0}$ suffices to prove the non-solubility of the marking equation in the rational numbers.

The following theorem shows that, combining classical place invariants and modulo place invariants, a finite set of vectors comprises the expressive power of all classical place invariants and modulo place invariants.

The following notations will be useful: Each place invariant vector $\mathbf{i}$ of a net N is called a *modulo-0 place invariant vector*. Each place vector $\mathbf{i} \in \mathbb{Z}^*$ is called *modulo-1 place invariant vector*. For arbitrary $u \in \mathbb{Z}$, we write $u \equiv u \pmod{0}$ and for arbitrary $u, v \in \mathbb{Z}$, we write $u \equiv v \pmod{1}$.

Proposition 17. *Let u and v be integers, and let $k \in \mathbb{N}$. Then $u \equiv v \pmod{k}$ if and only if $|u - v|$ is a multiple of k.* □

Proposition 18. *Let N be a net, and let $k \in \mathbb{N}$. An integral place vector $\mathbf{i}$ is a modulo-k place invariant vector if and only if $\mathbf{i} \cdot \mathbf{N} = k\,\mathbf{j}$ for some integral transition vector $\mathbf{j}$.*

Proof. For $k = 0$, the equation coincides with the definition of place invariants. For $k = 1$, choose $\mathbf{j} = \mathbf{i} \cdot \mathbf{N}$. For $k \geq 2$, we have $\mathbf{i} \cdot \mathbf{N} = k\,\mathbf{j}$ for some integral transition vector $\mathbf{j}$ if and only if each component of the vector $\mathbf{i} \cdot \mathbf{N}$ is a multiple of k. Since these components are the products $\mathbf{i} \cdot \mathbf{t}$ for column vectors $\mathbf{t}$ of $\mathbf{N}$, the equation holds exactly for modulo-k place invariant vectors. □

We will show that every row of the transformation matrix $\mathbf{Q}$ of the Smith normal form is a generalized modulo place invariant vector. The so defined set of modulo place invariant vectors is complete in the sense that no larger set allows more implications for the non-reachability of a marking.

Lemma 19. *Let N be a net with n places. Assume that $\mathbf{S} = \mathbf{Q} \cdot \mathbf{N} \cdot \mathbf{P}$ is the Smith normal form of $\mathbf{N}$ with diagonal elements $s_{1,1}, \ldots, s_{a,a}$. For $a < i \leq n$, define $s_{i,i} = 0$. Then, for $1 \leq i \leq n$, the i-th row vector of the matrix $\mathbf{Q}$ is a modulo-$s_{i,i}$ place invariant vector.*

Proof. For $1 \leq i \leq n$, let $\mathbf{q}_i$ denote the i-th row vector of the matrix Q.

The equation $\mathbf{S} = \mathbf{Q} \cdot \mathbf{N} \cdot \mathbf{P}$ implies

$$\mathbf{Q} \cdot \mathbf{N} = \mathbf{S} \cdot \mathbf{P}^{-1}.$$

Since $\mathbf{S}$ is a diagonal matrix with diagonal entries $s_{1,1}, s_{2,2}, \ldots$, each entry $s_{i,i}$ satisfies

$$\mathbf{q}_i \cdot \mathbf{N} = s_{i,i} \cdot \overline{\mathbf{p}}_i,$$

where $\overline{\mathbf{p}}_i$ is the i-th row of $\mathbf{P}^{-1}$. Since $\mathbf{P}^{-1}$ is an integral matrix, $\overline{\mathbf{p}}_i$ is integral, too. So Proposition 18 can be applied; it yields: $\mathbf{q}_i$ is a modulo-$s_{i,i}$ place invariant vector. □

Theorem 20. *Let m_0 and m be markings of a net N with n places. Assume that $\mathbf{S} = \mathbf{Q} \cdot \mathbf{N} \cdot \mathbf{P}$ is the Smith normal form of $\mathbf{N}$ with diagonal elements $s_{1,1}, \ldots, s_{a,a}$. For $a < i \leq n$, define $s_{i,i} = 0$. Then the following propositions are equivalent:*

- *For $1 \leq i \leq n$, the i-th row $\mathbf{q}_i$ of the matrix $\mathbf{Q}$ satisfies*

$$\mathbf{q}_i \cdot \mathbf{m} \equiv \mathbf{q}_i \cdot \mathbf{m}_0 \pmod{s_{i,i}}.$$

- *The marking equation*

$$\mathbf{N} \cdot \mathbf{x} = \mathbf{m} - \mathbf{m}_0$$

has an integral solution for $\mathbf{x}$.

Proof.
($\Longleftarrow$) For the rows $\mathbf{q}_i$ satisfying $s_{i,i} = 0$, the vector $\mathbf{q}_i$ is a classical place invariant vector. Since the marking equation has an integral solution, it has in particular a rational solution. So

$$\mathbf{q}_i \cdot \mathbf{m} = \mathbf{q}_i \cdot \mathbf{m}_0 \ .$$

This is equivalent to

$$\mathbf{q}_i \cdot \mathbf{m} \equiv \mathbf{q}_i \cdot \mathbf{m}_0 \ (\mathrm{mod}\ 0) \ .$$

For the rows $\mathbf{q}_i$ satisfying $s_{i,i} = 1$, nothing has to be shown. For the rows $\mathbf{q}_i$ satisfying $s_{i,i} \geq 2$, the proposition follows immediately from Theorem 16.

($\Longrightarrow$) Let

$$\mathbf{b} = (b_1, \ldots, b_n) = \mathbf{Q} \cdot (\mathbf{m} - \mathbf{m}_0) \ .$$

By Theorem 14, it suffices to show:

(α) $s_{i,i}$ divides b_i ($1 \leq i \leq a$), and
(β) $b_i = 0$ ($i > a$).

The assumption implies that every row $\mathbf{q}_i$ of $\mathbf{Q}$ satisfies

$$\mathbf{q}_i \cdot (\mathbf{m} - \mathbf{m}_0) \equiv 0 \ (\mathrm{mod}\ s_{i,i}) \ .$$

By the definition of $\mathbf{b}$, b_i is a multiple of $s_{i,i}$ for each i. This implies (α) for $1 \leq i \leq a$, and it implies (β) for $i > a$. □

Bibliographic Remarks

The material of this section is essentially based on [DeNR96], This applies in particular to modulo-invariants and their relation to the marking equation.

A survey on the reachability problem and related topics with emphasis on decidability and complexity issues can be found in [EsNi94]. [Jant87] studies the complexity of solutions of the marking equation in various domains.

[CoSi91] studies non-reachable markings that nevertheless possess solutions to the marking equation.

The reachability problem is simpler for certain subclasses of Petri nets. An example are live and bounded marked free-choice Petri nets where the initial marking can always be reached again. [DeEs93, DeEs95] show that in this class a marking is reachable if and only if the marking equation has a rational solution. So, for this class, place invariants can disprove reachability for all non-reachable markings.

4 Traps and Siphons

The marking equation does not always suffice to identify all non-reachable markings of a marked net. Figure 2 provides an example; the marking

$$\mathbf{m}_1 = (0,1,0,0,1,0,0)^\mathsf{T}$$

of the net shown in the figure is not reachable from the depicted initial marking although the marking equation has a nonnegative integral solution.

Another technique for proving non-reachability is given by traps, discussed in detail in [DeRe98]. A trap is a set A of places satisfying $A^\bullet \subseteq {}^\bullet A$. The following proposition formulates the salient property of traps:

Proposition 21. *Let m_0 and m be markings of a net N such that m is reachable from m_0. Then each trap A of N containing a place marked at m_0 contains some place marked at m.* □

In the net shown in Figure 2, the set

$$A = \{\mathtt{s1}, \mathtt{s3}, \mathtt{s4}, \mathtt{s6}, \mathtt{s7}\}$$

is an initially marked trap. So, for all reachable markings, at least one place of the trap A is marked. In particular, this trap proves that the marking $\mathbf{m}_1$ defined above is not reachable, because it marks no place of A. Thus, traps provide a proof technique for non-reachability that adds expressive power to the techniques based on the marking equation of the previous section.

4.1 Traps and Reachability

Traps generate a necessary condition for a marking m to be reachable from an initial marking m_0 of a net:

> **(T1)** *If some trap A contains a place marked at m_0 then it contains a place marked at m.*

The proof technique for non-reachability based on (T1) formulates a suitable set of places, checks the trap property, and verifies that the initial marking marks at least one place of this trap whereas the marking under consideration does not. Both the defining property of a trap and the conditions for the two markings are easily verified.

It is more difficult to use (T1) for verification. We present a linear algebraic approach in the sequel. A transition might have more input places that belong to some trap than output places. Thus, the token count of a trap can decrease by the occurrence of transitions. The trap condition only requires that, whenever the transition consumes at least one token from a place of a trap then it also adds at least one token to a place of this trap. Hence, by an appropriate weight of output places, we obtain the inequality expressed in the following lemma.

Lemma 22. *A set A of places is a trap if and only if, for each transition t,*

$$|{}^\bullet t| \cdot |A \cap t^\bullet| \geq |A \cap {}^\bullet t| .$$

Proof.
($\Longrightarrow$) Let t be a transition. If $t \notin A^\bullet$ then $|A \cap {}^\bullet t| = 0$ and we are finished. So assume $t \in A^\bullet$. Then $t \in {}^\bullet A$, because A is a trap. Therefore, $|A \cap t^\bullet| \geq 1$. So we obtain

$$|{}^\bullet t| \cdot |A \cap t^\bullet| \geq |{}^\bullet t| \geq |A \cap {}^\bullet t| .$$

($\Longleftarrow$) Let t be a transition in $A^\bullet$. Then $|A \cap {}^\bullet t| > 0$. The hypothesis yields

$$|{}^\bullet t| \cdot |A \cap t^\bullet| \geq |A \cap {}^\bullet t| > 0$$

and therefore $|A \cap t^\bullet| > 0$. This implies $t \in {}^\bullet A$. □

Recall that, for a transition t of a net, $\chi({}^\bullet t)$ is the characteristic vector of its pre-set and $\chi(t^\bullet)$ is the characteristic vector of its post-set. Both vectors are place vectors. Their sum is the column $\mathbf{t}$ of the incidence matrix.

Lemma 23. *A set A of places of a net is a trap if and only if there exists a nonnegative place vector* $\mathbf{x}$ *in* $\mathbb{Q}^*$ *such that*

- *a component of* $\mathbf{x}$ *is 0 if and only if the corresponding place belongs to A (i.e., A is the carrier set of* $\mathbf{x}$*),*
- *for each transition t of N,*

$$|{}^\bullet t|\, \chi(t^\bullet) \cdot \mathbf{x} \geq \chi({}^\bullet t) \cdot \mathbf{x} .$$

Proof.
($\Longrightarrow$) The characteristic vector of A enjoys the property required for $\mathbf{x}$. We have

$$\chi(t^\bullet) \cdot \chi(A) = |A \cap t^\bullet| \quad \text{and} \quad \chi({}^\bullet t) \cdot \chi(A) = |A \cap {}^\bullet t| .$$

The inequality follows by Lemma 22.

($\Longleftarrow$) Let $t \in A^\bullet$ be a transition. The assumption on $\mathbf{x}$ implies $\chi({}^\bullet t) \cdot \mathbf{x} > 0$. Hence,

$$|{}^\bullet t|\, \chi(t^\bullet) \cdot \mathbf{x} \geq \chi({}^\bullet t) \cdot \mathbf{x} > 0$$

by the second assumption. Therefore, $\chi(t^\bullet) \cdot \mathbf{x} > 0$. Again using the assumption on $\mathbf{x}$, this inequality implies $t \in {}^\bullet A$. Hence A is a trap. □

Deciding (T1) can be reduced by Lemma 23 to the rational solubility of the following system of inequalities:

Theorem 24. *Let N be a net and m_0, m be markings of N. There is a trap*

- *containing a place s satisfying $m_0(s) \geq 1$, and*
- *containing no place s satisfying $m(s) \geq 1$*

if and only if the following system of inequalities has a solution:

$$\begin{aligned} \mathbf{m}_0^\mathsf{T} \cdot \mathbf{x} &> 0\,, \\ \mathbf{m}^\mathsf{T} \cdot \mathbf{x} &\leq 0\,, \\ (|{}^\bullet t| \cdot \chi(t^\bullet) - \chi({}^\bullet t)) \cdot \mathbf{x} &\geq 0 \quad \textit{for each transition } t\,, \\ \mathbf{x} &\geq 0\,. \end{aligned}$$

Proof. The third and fourth inequality ensure that the carrier set of the solution vector is a trap, by Lemma 23. The first inequality states that initially at least one place of the trap is marked. The second inequality states that no place of the trap is marked at m. □

If the system of inequalities has any solution, then, by Lemma 22, the carrier set of this vector is a trap and the characteristic vector of this trap is a solution, too. So, if soluble at all, the system of inequalities has a solution $\mathbf{x}$ in $\{0,1\}^*$ which moreover satisfies $\mathbf{m}^\mathsf{T} \cdot \mathbf{x} = 0$.

(T1) can be decided efficiently, because the system of inequalities of Theorem 24 can be solved in polynomial time, by Proposition 1 (1).

4.2 Siphons and Reachability

Siphons and traps are very similar concepts. Whereas a trap A satisfies $A^\bullet \subseteq {}^\bullet A$, a siphon is a set of places satisfying ${}^\bullet A \subseteq A^\bullet$. Siphons containing no initially marked place will never gain a token (see [DeRe98]):

Proposition 25. *Let m_0 and m be markings of a net N such that m is reachable from m_0. Then each siphon A of N containing no place marked at m_0 contains no place marked at m.* □

So also siphons generate a necessary condition for a marking m to be reachable from the initial marking m_0:

> **(S1)** *If some siphon A contains no place marked at m_0 then it contains no place marked at m.*

As for traps, a proof technique for non-reachability based on (S1) formulates a suitable set of places, checks the siphon property, and verifies that the initial marking marks no place of this siphon whereas the marking under consideration marks at least one place of this siphon.

Using (S1) for verification of non-reachability follows closely the above lines for traps. The following lemma is an analogue to Lemma 23:

Lemma 26. *A set A of places of a net is a siphon if and only if there exists a nonnegative place vector* $\mathbf{x}$ *in* $\mathbb{Q}^*$ *such that*

- *a component of* $\mathbf{x}$ *is 0 if and only if the corresponding place belongs to* A,
- *for each transition t of* N,

$$|t^\bullet|\, \chi({}^\bullet t) \cdot \mathbf{x} \geq \chi(t^\bullet) \cdot \mathbf{x}\,.$$

Proof. By definition, a siphon of a net (S,T,F) is a trap of the net (S,T,F^{-1}) and vice versa, where $F^{-1} = \{(u,v)|(v,u) \in F\}$. The pre-set ${}^\bullet t$ of a transition t in (S,T,F) is equal to the post-set $t^\bullet$ of t in (S,T,F^{-1}) and vice versa. The result follows by Lemma 23. □

By Lemma 26, (S1) reduces to the rational solubility of the following system of inequalities:

Theorem 27. *Let* N *be a net and* m_0, m *be markings of* N. *There is a siphon*

- *containing no place s satisfying* $m_0(s) \geq 1$, *and*
- *containing some place s satisfying* $m(s) \geq 1$

if and only if the following system of inequalities has a solution:

$$\begin{aligned} \mathbf{m}_0 \cdot \mathbf{x} &\leq 0\,, \\ \mathbf{m} \cdot \mathbf{x} &> 0\,, \\ (|t^\bullet| \cdot \chi({}^\bullet t) - \chi(t^\bullet)) \cdot \mathbf{x} &\geq 0 \quad \textit{for each transition } t\,, \\ \mathbf{x} &\geq 0\,. \end{aligned}$$

Proof. Analogously to the proof of Theorem 24. □

There is a polynomial time algorithm to decide (S1) by Theorem 27 and Proposition 1 (1).

Bibliographic Remarks

Calculating traps and siphons using equation systems was suggested in [Laut87b]. The characterizations of traps and siphons used in this section are based on results from [EzCS93].

5 Verification of Facts

In contrast to properties like *liveness* and *boundedness*, considered in [DeRe98], properties of Petri nets that are specific for a given net are studied in this section. We concentrate on state based properties, i.e., properties formulated in terms of markings of a net. Figure 4 shows a typical example. This net models an algorithm that guarantees mutual exclusion of critical sections of two concurrent processes. The critical sections are modeled by the places **s2** and **s4**.
For proving mutual exclusion, it has to be shown:

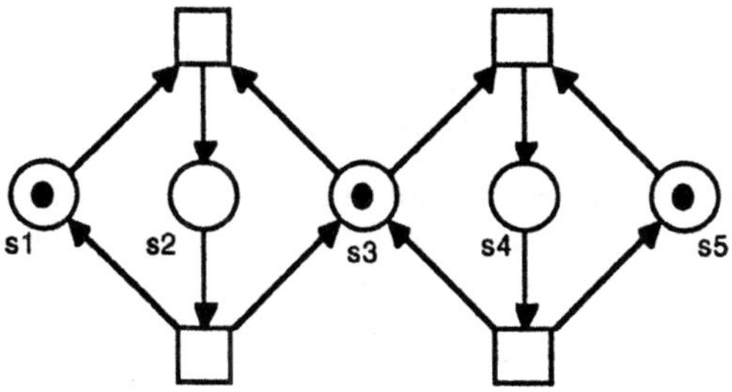

Fig. 4. Mutual exclusion

(M1) *At no reachable marking, both* **s2** *and* **s4** *are marked.*

Conversely, this specification states a property for all reachable markings, namely that either **s2** or **s4** is not marked. It is not difficult to verify that the example net satisfies this specification, by constructing the set of all reachable markings. However, we aim at proving such properties by linear algebraic techniques, because the explicit construction of all reachable markings is not feasible in general.

We restrict to state based properties shaped:

Every reachable marking satisfies φ,

where φ is a predicate over minimal or maximal token counts on places. The property (M1) belongs to this class, because it is equivalently formulated as

(M1′) *Every reachable marking* m *satisfies* $(m(s_2) = 0 \vee m(s_4) = 0)$.

5.1 Linear Predicates

A state of a system corresponds to a marking of its Petri net model. So a *predicate* of a net is a relation on the token count of places. Here, we consider any marking of a net and do not stick to markings reachable from some initial marking. A predicate is interpreted on markings; for each marking, it is either satisfied or not. Hence, a predicate is uniquely determined by the set of markings for which it holds true.

Predicates will be formulated by linear inequalities with variables varying over markings. Every such inequality denotes the predicate given by the set of markings satisfying the inequality. More formally, a predicate φ of a net N is called *linear* if it is satisfied by (and only by) the nonnegative integral solutions for the variable $\mathbf{m}$ of a linear inequality

$$\mathbf{u} \cdot \mathbf{m} \leq v \ ,$$

where $\mathbf{u}$ is an integral place vector and v is an integer. The predicate φ is said to be *denoted by* this inequality.

We will additionally employ the following notations:

- $\mathbf{u} \cdot \mathbf{m} \geq v$ stands for $(-1)\,\mathbf{u} \cdot \mathbf{m} \leq (-1)\,v$, and
- $\mathbf{u} \cdot \mathbf{m} = v$ represents the two linear predicates $\mathbf{u} \cdot \mathbf{m} \leq v$ and $\mathbf{u} \cdot \mathbf{m} \geq v$.

Notice that variables in these inequalities are markings whereas in previous sections we considered inequalities with transition vectors as variables.

Many important predicates are linear:

(1) If a place s of a net models a logical condition, the corresponding predicate is given by "$m(s) \geq 1$". A corresponding inequality in the above form is

$$\mathbf{e}_s \cdot \mathbf{m} \geq 1 .$$

(2) The set of markings that mark at least one place of a set A of places is denoted by the predicate

$$\chi(A) \cdot \mathbf{m} \geq 1 .$$

(3) Each place invariant vector $\mathbf{i}$ defines two linear predicates that are satisfied for all markings reachable from the initial marking m_0. They are denoted by

$$\mathbf{i} \cdot \mathbf{m} = (\mathbf{i} \cdot \mathbf{m}_0) .$$

(4) An upper bound k of a place s is denoted by

$$\mathbf{e}_s \cdot \mathbf{m} \leq k .$$

(5) Two places s and r model mutually exclusive sections if they are never marked together. If moreover both places are 1-bounded, this predicate can be denoted by

$$(\mathbf{e}_s + \mathbf{e}_r) \cdot \mathbf{m} \leq 1 .$$

Linear predicates are not closed under conjunction or disjunction in general. The complement of a linear predicate is linear, too:

Proposition 28. *If φ is a linear predicate, denoted by $\mathbf{u} \cdot \mathbf{m} \leq v$ then its complement $\neg\varphi$ is linear; it is denoted by $\mathbf{u} \cdot \mathbf{m} \geq v + 1$.* □

5.2 Implication of Linear Predicates

Implications between linear predicates can be characterized by means of systems of inequalities:

Proposition 29. *Assume linear predicates $\varphi_1, \ldots, \varphi_n$, denoted by inequalities*

$$\mathbf{u}_i \cdot \mathbf{m} \leq v_i \ (1 \leq i \leq n) .$$

Let φ be the linear predicate denoted by $\mathbf{u} \cdot \mathbf{m} \leq v$. The set $\{\varphi_1, \ldots, \varphi_n\}$ implies φ (i.e., every marking satisfying all predicates of this set also satisfies φ) if and only if the following inequality system has no solution for $\mathbf{m}$ over $\mathbb{N}$.

$$\begin{aligned} \mathbf{u}_1 \cdot \mathbf{m} &\leq v_1 , \\ \mathbf{u}_2 \cdot \mathbf{m} &\leq v_2 , \\ &\vdots \\ \mathbf{u}_n \cdot \mathbf{m} &\leq v_n , \\ \mathbf{u} \cdot \mathbf{m} &\geq v + 1 . \end{aligned}$$

□

In the marked net shown in Figure 2, all reachable markings satisfy the following inequalities:

$$\begin{array}{l} (-1, 0, -1, -1, 0, -1, -1) \cdot \mathbf{m} \leq -1 , \\ (\ \ 2, 1, \ \ 1, \ \ 1, 1, \ \ 1, \ \ 1) \cdot \mathbf{m} \leq \ \ 2 . \end{array}$$

Now we show that these predicates imply the mutual exclusion of s_2 and s_5. According to Proposition 29, we add the complement of the inequality

$$(0, 1, 0, 0, 1, 0, 0) \cdot \mathbf{m} \leq 1$$

to the above inequalities:

$$\begin{array}{l} (-1, \ \ 0, -1, -1, \ \ 0, -1, -1) \cdot \mathbf{m} \leq -1, \\ (\ \ 2, \ \ 1, \ \ 1, \ \ 1, \ \ 1, \ \ 1, \ \ 1) \cdot \mathbf{m} \leq \ \ 2, \\ (\ \ 0, -1, \ \ 0, \ \ 0, -1, \ \ 0, \ \ 0) \cdot \mathbf{m} \leq -2. \end{array}$$

All inequalities are formulated in terms of the relation "$\leq$". Hence, summing up the respective left hand sides and right hand sides yields one more inequality which is implied by the other inequalities. In the example, this sum is

$$(1, 0, 0, 0, 0, 0, 0) \cdot \mathbf{m} \leq -1 .$$

Clearly, this inequality does not have any nonnegative solution for $\mathbf{m}$. Therefore, the three inequalities above have no common nonnegative solution for $\mathbf{m}$, and we are finished.

The same result is shown easier if the property of the marking m being nonnegative is added in terms of inequalities. Then, it suffices to prove that there is no integral solution for $\mathbf{m}$. In the above example, the inequality

$$(-1, 0, 0, 0, 0, 0, 0) \cdot \mathbf{m} \leq 0$$

can be added, and we gain the contradiction "$0 \leq -1$".

Proposition 29 also states that a linear predicate φ does not follow from other linear predicates if the inequality system given in the proposition has an integral solution. In fact, every solution constitutes a marking that satisfies all premises but does not satisfy φ.

We mentioned already that neither the conjunction nor the disjunction of linear predicates is linear in general. However, the above proof technique can be extended to such predicates. The following results, recalled from propositional logic, will frequently be used:

Proposition 30. *Let $\varphi_1, \ldots, \varphi_n$ be predicates. The set $\{\varphi_1, \ldots, \varphi_n\}$ implies the conjunction of linear predicates $\psi_1, \ldots, \psi_k$ if and only if it implies every ψ_i $(1 \leq i \leq k)$. It implies the disjunction of linear predicates $\psi_1, \ldots, \psi_k$ if and only if the set $\{\varphi_1, \ldots, \varphi_n, \neg\psi_1, \ldots, \neg\psi_{k-1}\}$ implies the predicate ψ_k.* □

As an example, consider the predicate that holds for all markings of a net except one particular marking m_1 (non-reachability of m_1). This predicate is expressed by a disjunction of linear predicates, denoted by the inequalities

$$\mathbf{e}_s \cdot \mathbf{m} \le m_1(s) - 1 \ ,$$
$$\mathbf{e}_s \cdot \mathbf{m} \ge m_1(s) + 1 \ ,$$

for each place s. For proving that this predicate is implied by a set $\{\varphi_1, \ldots, \varphi_k\}$ of linear predicates, it suffices to show the non-solubility of the following system of inequalities:

$$\begin{aligned} \mathbf{u}_1 \cdot \mathbf{m} &\le v_1 \ , \\ &\vdots \\ \mathbf{u}_k \cdot \mathbf{m} &\le v_k \ , \\ \mathbf{e}_{s_1} \cdot \mathbf{m} &= m_1(s_1) \ , \\ &\vdots \\ \mathbf{e}_{s_n} \cdot \mathbf{m} &= m_1(s_n) \ , \end{aligned}$$

where the first k inequalities denote the predicates $\varphi_1, \ldots, \varphi_k$ and $s_1, \ldots, s_n$ are the places of the net.

Thus, the technique of Proposition 29 can be generalized from linear predicates to propositional formulas constructed from linear predicates. To this end, w.l.o.g. assume a formula in conjunctive normal form. Then, all clauses can be proven as shown above.

5.3 Facts and Place Invariants

We are particularly interested in *facts* of initially marked nets, i.e. predicates that hold for all markings reachable from the initial marking. Facts can be used to establish other facts:

Proposition 31. *Given an initially marked net, each predicate implied by a set of facts is a fact, too.* □

In particular, each weakening of a fact is also a fact.

A given predicate ψ is a fact if and only if the *strongest fact*

$$\psi_0 = \text{\textit{“is reachable from the initial marking”}}$$

implies ψ. In most cases, the proof of this implication is not trivial because ψ_0 characterizes the set of all reachable markings, a set difficult do deal with. Neither the predicate ψ_0 nor the predicate ψ is linear in general. Implications of linear predicates can nevertheless be used, when the following steps are taken.

(1) *The predicate ψ_0 implies linear predicates $\varphi_1, \ldots, \varphi_n$.*
(2) *The set of predicates $\{\varphi_1, \ldots, \varphi_n\}$ implies further linear predicates $\varphi'_1, \ldots, \varphi'_n$.*
(3) *The set of linear predicates $\{\varphi'_1, \ldots, \varphi'_n\}$ implies ψ.*

Step (2) was discussed in the previous subsection. Step (3) consists of logical transformations. As an example, ψ could be the conjunction or the disjunction of linear predicates. In our examples, ψ is always a linear predicate itself. Proof of step (1) requires inequalities from the structure of the net and from the initial marking such that the linear predicates $\varphi_1, \ldots, \varphi_n$ hold true for at least all reachable markings.

Place invariants provide an example, as shown above. In the example of Figure 2, the vector $\mathbf{i} = (2,1,1,1,1,1,1)$ is a place invariant vector. Together with the initial marking, it generates the second of the above inequalities:

$$(2,1,1,1,1,1,1) \cdot \mathbf{m} \leq 2 .$$

Other linear predicates are derived from traps; by Proposition 21, every initially marked trap A induces the linear predicate

$$\chi(A) \cdot \mathbf{m} \geq 1 .$$

In the above example, the first inequality

$$(-1,0,-1,-1,0,-1,-1) \cdot \mathbf{m} \leq -1$$

states that the set of places $\{s_1, s_3, s_4, s_6, s_7\}$ contains at least one marked place. This predicate is a fact because the set constitutes an initially marked trap. Finally, each initially marked siphon A generates the fact

$$\chi(A) \cdot \mathbf{m} \leq 0 .$$

It was previously shown that the marking equation has no solution if and only if some place invariant proves the considered marking non-reachable. Now this result is used for proving facts using place invariants. To this end, let the linear predicates $\varphi_1, \ldots, \varphi_n$ be denoted by the inequalities

$$\mathbf{u}_i \cdot \mathbf{m} \leq v_i \quad (1 \leq i \leq n) .$$

In each inequality, the variable marking $\mathbf{m}$ can be replaced by $\mathbf{m}_0 + \mathbf{N} \cdot \mathbf{x}$ as, for every reachable marking m, there exists some nonnegative integral solution to the marking equation $\mathbf{m}_0 + \mathbf{N} \cdot \mathbf{x} = \mathbf{m}$. Thus, for $1 \leq i \leq n$ we obtain the inequalities

$$\mathbf{u}_i \cdot (\mathbf{m}_0 + \mathbf{N} \cdot \mathbf{x}) \leq v_i .$$

Transformation yields

$$(\mathbf{u}_i \cdot \mathbf{N}) \cdot \mathbf{x} \leq v_i - \mathbf{u}_i \cdot \mathbf{m}_0 .$$

Each place invariant vector $\mathbf{i}$ induces two linear predicates, given by the equation

$$\mathbf{i} \cdot \mathbf{m} = (\mathbf{i} \cdot \mathbf{m}_0) .$$

Substitution into the above inequality leads to

$$(\mathbf{i} \cdot \mathbf{N}) \cdot \mathbf{x} = (\mathbf{i} \cdot \mathbf{m}_0) - \mathbf{i} \cdot \mathbf{m}_0 .$$

Since $\mathbf{i} \cdot \mathbf{N} = \mathbf{0}$, both sides evaluate to 0, independently from the value of $\mathbf{x}$. Hence, these inequalities generated by place invariants carry no information and can be dropped. In particular, if a linear predicate follows only from place invariants then it suffices to consider a single inequality:

Theorem 32. *Let N be a net with initial marking m_0. A linear predicate, denoted by $\mathbf{u} \cdot \mathbf{m} \leq v$, follows from linear predicates generated by place invariants if and only if the following system of inequalities has no rational solution for $\mathbf{x}$:*

$$(\mathbf{u} \cdot \mathbf{N}) \cdot \mathbf{x} \geq v + 1 - \mathbf{u} \cdot \mathbf{m}_0 ,$$
$$\mathbf{x} \geq \mathbf{0} .$$

□

5.4 Facts, Traps and Siphons

Each initially marked trap A yields a very simple linear predicate that can be used for proving facts of a marked net:

$$\chi(A) \cdot \mathbf{m} \geq 1,$$

i.e., the trap is marked at each reachable marking m (Proposition 21).

Now we are interested in a corresponding verification technique: Given a linear predicate, is there a trap that proves that this predicate is a fact? In general, the construction and investigation of all traps is not efficient because the number of traps can grow exponentially w.r.t. the size of the net.

The expressive power of all place invariants is characterized by solubility of the marking equation. Hence the verification of a fact using place invariants reduces to the solubility of a single system of inequalities, as shown in the previous theorem. Similarly, we will develop a system of inequalities that has a solution if and only if some trap proves that a given linear predicate is a fact. More formally, we aim at constructing a linear predicate characterizing the set of markings with at least one marked place in each initially marked trap. The inequality for this predicate does not only have variables for markings but also additional variables. The number of these additional variables exceeds the number of transitions of the net by one. Therefore, the solubility of corresponding systems of inequalities is still polynomial in the size of the net.

Recall (T1) from the previous section:

> **(T1)** *If some trap A contains a place marked at the initial marking m_0 then it contains a place marked at m.*

By Lemma 23, a set A is a trap if and only if there exists a nonnegative place vector $\mathbf{x}$ in $\mathbb{Q}^*$ such that

- a component of $\mathbf{x}$ is 0 if and only if the corresponding place belongs to A (i.e., A is the carrier set of $\mathbf{x}$), and
- for each transition t of N we have $|^\bullet t| \, \chi(t^\bullet) \cdot \mathbf{x} \geq \chi(^\bullet t) \cdot \mathbf{x}$.

The vector $|{}^\bullet t|\,\chi(t^\bullet) - \chi({}^\bullet t)$ is a place vector for each transition t. We employ these vectors as columns of a matrix $\mathbf{B}$. Then a set A is a trap if and only if it is the carrier set of a non-negative solution for $\mathbf{x}$ of the inequality

$$\mathbf{B}^\mathsf{T} \cdot \mathbf{x} \geq \mathbf{0} .$$

Hence, given a net with an initial marking m_0 and some marking m, there is a trap A containing at least one place marked at m_0 and containing no place marked at m if and only if there is a solution to the following system of inequalities:

$$\begin{aligned} \mathbf{m}_0 \cdot \mathbf{x} &> 0 , \\ \mathbf{m} \cdot \mathbf{x} &\leq 0 , \\ \mathbf{B}^\mathsf{T} \cdot \mathbf{x} &\geq \mathbf{0} , \\ \mathbf{x} &\geq \mathbf{0} . \end{aligned}$$

A variant of Farkas' Lemma was employed in the second section of this paper (Proposition 12). Here, another variant will be useful, which can e.g. be found in [Schr86, Corollary 7.1f, page 90].

Proposition 33. *Exactly one of the following systems of inequalities has a solution over* $\mathbb{Q}$*:*

$$\begin{gathered} \mathbf{A} \cdot \mathbf{y} \leq \mathbf{b},\ \mathbf{y} \geq \mathbf{0} , \\ \mathbf{A}^\mathsf{T} \cdot \mathbf{x} \geq \mathbf{0},\ \mathbf{b} \cdot \mathbf{x} < 0,\ \mathbf{x} \geq \mathbf{0} . \end{gathered}$$

□

Consider an additional column to the previously defined matrix $\mathbf{B}$ that contains the components of the vector $(-1)\,\mathbf{m}$. Substituting this extended matrix $[\mathbf{B}\,|\,(-1)\,\mathbf{m}]$ for $\mathbf{A}$ and the vector $(-1)\,\mathbf{m}_0$ for $\mathbf{b}$, the above system of inequalities exactly matches the second row of Proposition 33. Therefore, Farkas' Lemma proves that this system has no solution if and only if the following system is soluble:

$$\begin{aligned} [\mathbf{B} \mid (-1)\,\mathbf{m}] \cdot \mathbf{y} &\leq (-1)\,\mathbf{m}_0 , \\ \mathbf{y} &\geq \mathbf{0} . \end{aligned}$$

In other words, this system of inequalities has a solution for $\mathbf{y}$ if and only if the marking m marks all initially marked traps.

Let y be the last component of the variable vector $\mathbf{y}$, and decompose $\mathbf{y}$ into $\tilde{\mathbf{y}}\ y$. Then the above system is equivalently represented as:

$$\begin{aligned} y\,\mathbf{m} &\geq \mathbf{B} \cdot \tilde{\mathbf{y}} + \mathbf{m}_0 , \\ \tilde{\mathbf{y}} &\geq \mathbf{0} , \\ y &\geq 0 . \end{aligned}$$

If there is a solution satisfying $y = 0$, the same solution vector $\tilde{\mathbf{y}}$ completed by any positive value for y is also a solution, because $\mathbf{m}$ is a marking with no negative components. So the last line of the above system can be replaced by $y > 0$, without changing the solubility of the system. Division by y yields

$$\begin{aligned} \mathbf{m} &\geq \tfrac{1}{y}\,\mathbf{B} \cdot \tilde{\mathbf{y}} + \tfrac{1}{y}\,\mathbf{m}_0 , \\ \tilde{\mathbf{y}} &\geq \mathbf{0} , \\ y &> 0 . \end{aligned}$$

Setting $\tilde{\mathbf{z}} = \frac{1}{y} \cdot \tilde{\mathbf{y}}$ and $z = \frac{1}{y}$ yields

$$\begin{aligned} \mathbf{m} &\geq \mathbf{B} \cdot \tilde{\mathbf{z}} + z\,\mathbf{m}_0 , \\ \tilde{\mathbf{z}} &\geq \mathbf{0} , \\ z &> 0 . \end{aligned}$$

This form is linear in the variables $\mathbf{m}$, $\tilde{\mathbf{z}}$ and z. The following theorem recaptures the result of these transformations.

Theorem 34. *Let N be a net with initial marking m_0. Each trap containing an initially marked place contains a place marked at a given marking m if and only if the following system of inequalities has a rational solution for $\tilde{\mathbf{z}}$ and z:*

$$[\mathbf{B} \mid \mathbf{m}_0 \mid (-1)\,\mathbf{I}] \cdot \begin{bmatrix} \tilde{\mathbf{z}} \\ z \\ \mathbf{m} \end{bmatrix} \geq 0 , \; \tilde{\mathbf{z}} \geq 0 , \; z > 0 ,$$

where the matrix $\mathbf{B}$ is the matrix defined above and $\mathbf{I}$ is the identity matrix.

□

Since, for every reachable marking, there is a solution to this system of inequalities, every implied linear predicate is a fact. Its expressive power comprises the expressive power of all initially marked traps:

Corollary 35. *Given an initially marked trap A, its generated fact $\chi(A) \cdot \mathbf{m} > 0$ is implied by the system of inequalities given in Theorem 34.* □

Siphons can be used for the verification of facts in a similar manner. Recall the reachability criterion for a marking m generated by siphons from the previous section:

> **(S1)** *If some siphon A contains no place marked at the initial marking m_0 then it contains no place marked at m.*

By Lemma 26, a set of places A is a trap if and only if there exists a non-negative place vector $\mathbf{x}$ in $\mathbb{Q}^*$ such that

- a component of $\mathbf{x}$ is 0 if and only if the corresponding place belongs to A,
- for each transition t of N, we have $|t^\bullet|\, \chi(^\bullet t) \cdot \mathbf{x} \geq \chi(t^\bullet) \cdot \mathbf{x}$.

Similarly as above, we employ the vectors $|t^\bullet|\chi(^\bullet t) - \chi(t^\bullet)$ as columns of a matrix $\mathbf{C}$. Then a set A is a siphon if and only if it is the carrier set of a non-negative solution for $\mathbf{x}$ of the inequality

$$\mathbf{C}^{\mathsf{T}} \cdot \mathbf{x} \geq \mathbf{0} .$$

Hence, given a net with an initial marking m_0 and some marking m, there is a siphon A containing at least one place marked at m and containing no place marked at m_0 if and only if there is a solution to the following system of inequalities:

$$\begin{aligned} \mathbf{m} \cdot \mathbf{x} &> 0 \,, \\ \mathbf{m}_0 \cdot \mathbf{x} &\leq 0 \,, \\ \mathbf{C}^{\mathsf{T}} \cdot \mathbf{x} &\geq \mathbf{0} \,, \\ \mathbf{x} &\geq \mathbf{0} \,. \end{aligned}$$

We apply again Farkas' Lemma (Proposition 33), taking $[\mathbf{C} \,|\, (-1)\,\mathbf{m}_0]$ for $\mathbf{A}$ and $(-1)\,\mathbf{m}$ for $\mathbf{b}$. Thus, there is a solution of the above system if and only if the following system of inequalities has no solution:

$$\begin{aligned} [\mathbf{C} \,|\, (-1)\,\mathbf{m}_0] \cdot \mathbf{y} &\leq (-1)\,\mathbf{m} \,, \\ \mathbf{y} &\geq \mathbf{0} \,. \end{aligned}$$

The following theorem follows immediately:

Theorem 36. *Let N be a net with initial marking m_0. Each trap containing no initially marked place contains no place marked at a given marking m if and only if the following system of inequalities has a rational solution for $\tilde{\mathbf{y}}$ and y:*

$$[\mathbf{C} \,|\, (-1)\,\mathbf{m}_0 \,|\, \mathbf{I}] \cdot \begin{bmatrix} \tilde{\mathbf{y}} \\ y \\ \mathbf{m} \end{bmatrix} \leq 0 \,, \; \tilde{\mathbf{y}} \geq \mathbf{0} \,, \; y \geq 0 \,,$$

where $\mathbf{C}$ is the matrix defined above and $\mathbf{I}$ is the identity matrix. □

This system of inequalities has a solution for every reachable marking because unmarked siphons remain unmarked. Its expressive power comprises the expressive power of all initially unmarked siphons:

Corollary 37. *Given an initially unmarked siphon A, its generated fact $\chi(A) \cdot \mathbf{m} \leq 0$ is implied by the system of inequalities given in Theorem 36.* □

5.5 Deadlock-freeness and Facts

The enabling condition of a transition can be formulated by means of a conjunction of linear predicates: For each input place of the transition, a linear predicate demands a token. For transitions with only one input place, enabledness is thus described by exactly one linear predicate, which is called *enabling predicate* of the transition.

If a transition has more than one input place but all its input places are bounded by 1, then one enabling predicate is sufficient, too:

Proposition 38. *Let t be a transition of a marked net and assume that every place in ${}^\bullet t$ is 1-bounded. Then t is enabled at a reachable marking m if and only if*

$$\chi({}^\bullet t) \cdot \mathbf{m} \geq |{}^\bullet t| \,.$$

Proof. The value $\chi(^\bullet t) \cdot \mathbf{m}$ denotes the number of tokens on input places of t at a marking m. If none of these places carry more than one token, $\chi(^\bullet t) \cdot \mathbf{m} \geq |^\bullet t|$ implies that every input place is marked at m. Conversely, $\chi(^\bullet t) \cdot \mathbf{m} < |^\bullet t|$ implies that at least one place in $^\bullet t$ is unmarked at m. □

By 1-boundedness, the value $\chi(^\bullet t) \cdot \mathbf{m}$ never exceeds $|^\bullet t|$. Hence, the relation symbol "$\geq$" can be replaced by "$=$".

If every transition of a net has a single enabling predicate, a marking is dead if and only if the disjunction of these enabling predicates holds true for this marking. Hence, for showing deadlock-freeness, this disjunction has to be proven a fact. The following theorem states this result for 1-bounded marked nets:

Theorem 39. *A net N with a 1-bounded initial marking m_0 is deadlock-free if and only if every reachable marking m satisfies the linear predicate*

$$\chi(^\bullet t) \cdot \mathbf{m} \geq |^\bullet t|$$

for at least one transition t. □

In all previous examples, it is possible to prove deadlock-freeness with this technique, employing only place invariants. In other words, in all these marked nets, the disjunction of the enabling predicates is implied by linear predicates generated by place invariants.

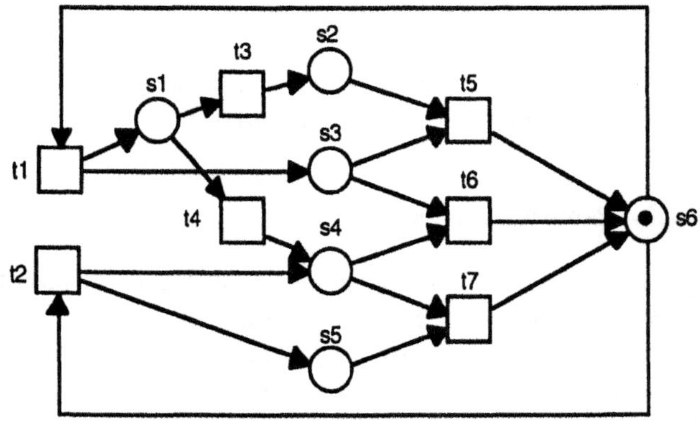

Fig. 5. A deadlock-free marked net

The marked net shown in Figure 5 is also deadlock-free but this property cannot be proven by help of place invariants only. The marking

$$\mathbf{m} = (0, 1, 0, 0, 1, 0)^{\mathsf{T}}$$

is a (non-reachable) dead marking. However, we have

$$\mathbf{m}_0 + \mathbf{N} \cdot (2,2,2,0,1,1,1)^{\mathsf{T}} = \mathbf{m} .$$

Hence, by Proposition 11, the marking m agrees with the initial marking m_0 on all place invariants. However, non-reachability of m can be proven by help of a trap; the set

$$A = \{\mathbf{s3}, \mathbf{s4}, \mathbf{s6}\}$$

constitutes an initially marked trap which is not marked at m.

Not only the dead marking m is non-reachable, but any dead marking, as will be shown next. Consider the place invariant vectors

$$\mathbf{i}_1 = (1,1,0,1,0,1) \text{ and } \mathbf{i}_2 = (0,0,1,0,1,1) .$$

The place invariant vector $\mathbf{i}_1$ proves that the places **s1**, **s2**, **s4** and **s6** are 1-bounded, whereas $\mathbf{i}_2$ moreover proves 1-boundedness of the places **s3** and **s5**. We show that these place invariants and the trap A suffice to prove that the disjunction of the enabling predicates is a fact. Equivalently, we show that the following system of inequalities has no integral solution for $\mathbf{m}$:

[1]	$(1,1,0,1,0,1) \cdot \mathbf{m} = 1$	place invariant $\mathbf{i}_1$,
[2]	$(0,0,1,0,1,1) \cdot \mathbf{m} = 1$	place invariant $\mathbf{i}_2$,
[3]	$(0,0,1,1,0,1) \cdot \mathbf{m} \geq 1$	trap A ,
[4]	$(0,0,0,0,0,1) \cdot \mathbf{m} = 0$	**t1** and **t2** are not enabled,
[5]	$(1,0,0,0,0,0) \cdot \mathbf{m} = 0$	**t3** and **t4** are not enabled,
[6]	$(0,1,1,0,0,0) \cdot \mathbf{m} \leq 1$	**t5** is not enabled,
[7]	$(0,0,1,1,0,0) \cdot \mathbf{m} \leq 1$	**t6** is not enabled,
[8]	$(0,0,0,1,1,0) \cdot \mathbf{m} \leq 1$	**t7** is not enabled,

Assume that some marking m satisfies all these inequalities; we aim at deriving a contradiction.

By [5], $m(\mathbf{s1}) = 0$. By [4], $m(\mathbf{s6}) = 0$.

Assume that $m(\mathbf{s3}) \geq 1$. Then, by [6], $m(\mathbf{s2}) = 0$ and, by [7], $m(\mathbf{s4}) = 0$, contradicting [1].

Now assume that $m(\mathbf{s3}) = 0$. Then, by [3], $m(\mathbf{s4}) \geq 0$ and hence, by [8], $m(\mathbf{s5}) = 0$, contradicting [2].

So we obtain a contradiction in both cases, which finishes the proof.

Notice that, for this example, we have to consider solubility of the system of inequalities over $\mathbb{Z}$ because there exists a rational solution; the marking

$$\mathbf{m} = \left(0, \frac{1}{2}, \frac{1}{2}, \frac{1}{2}, \frac{1}{2}, 0\right)^{\mathsf{T}}$$

satisfies all above inequalities.

Bibliographic Remarks

Place invariants and other structural techniques for proving facts are suggested in many publications. See e.g. [Reis85]. The main results on traps (Subsection 5.4) are taken from [EsMe96]. The representation of the enabling condition by a linear inequality was introduced in [Dese85]. The example of a live and safe marked net for which deadlock-freeness can not be proved by place invariants can be found in [Dese88]. The paper [TeCS93] contains a deeper discussion of this analysis technique. In particular, techniques for reducing the number of enabling predicates are presented.

6 The Rank Conditions

The rank conditions provide a sufficient criterion and a necessary criterion for liveness of bounded marked Petri nets. In particular, they establish relations between a net's behavioral property and the rank of its incidence matrix. This connection might be surprising because the rank of the incidence matrix has no obvious interpretation in terms of behavior. Moreover, it is invariant against addition of rows or columns, multiplication by numbers and even transposition. All the corresponding transformations of a net structure cause significant changes of behavior that do not preserve properties such as liveness.

As shown in [DeRe98], every live and bounded marked net has a positive transition invariant. The column rank of the incidence matrix of a live and bounded marked net is therefore not maximal; it never exceeds $|T| - 1$, where T is the set of transitions of the net. Transition invariants are closely related to occurrence sequences that lead from a marking back to itself. If a net has only one transition invariant (and its multiples) then the relative number of transition occurrences in an occurrence sequence that leads from a marking to itself is very restricted. In particular, no conflict can occur again and again. If there is a conflict between transitions t_1 and t_2 then t_1 could be chosen more often than t_2 or vice versa. In this case, there exist transition invariant vectors $\mathbf{j}_1 \geq \mathbf{0}$ and $\mathbf{j}_2 \geq \mathbf{0}$ such that

$$\mathbf{j}_1(t_1) > \mathbf{j}_1(t_2) \text{ and } \mathbf{j}_2(t_1) < \mathbf{j}_2(t_2).$$

Then the rank of the incidence matrix does not exceed $|T| - 2$ because $\mathbf{j}_1$ and $\mathbf{j}_2$ are linearly independent.

These considerations suggest that the rank of the incidence matrix of a net is related to the number of its possible conflict situations. The rank conditions formulate such relations.

Two distinct transitions are said to be in *potential conflict* if their pre-sets are not disjoint. The conflict appears when both transitions are enabled and a common input place carries only one token. The ***conflict area*** of a transition t is the minimal set of transitions that contains t and is closed under potentially conflicting tansitions. Formulated differently, it is an equivalence class of the

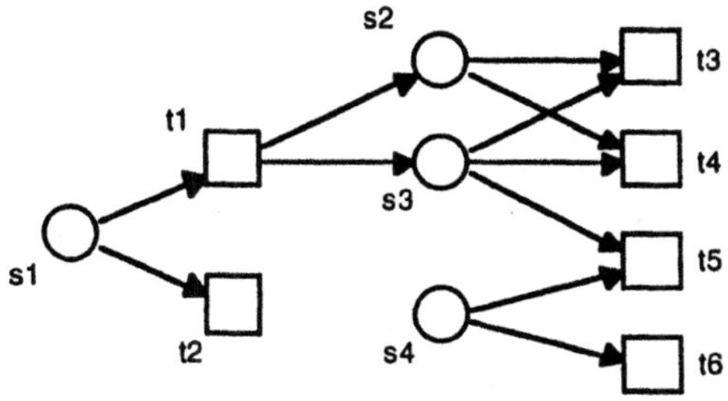

Fig. 6. A net with two conflict areas and four different pre-sets of transitions

transitive closure of potential conflict. In the example of Figure 6, the conflict areas are {t1, t2} and {t3, t4, t5, t6}.

The *sufficient rank condition* requires the rank of the incidence matrix to be smaller than the number of conflict areas of the net.

In the sequel, we consider only nets with each transition having a nonempty pre-set. Two transitions are in *independent conflict* if they have identical pre-sets. Every marking enabling one of the transitions enables the other one, too. This relation is an equivalence relation. The number of equivalence classes equals the number of different pre-sets of transitions. In the example of Figure 6, these different pre-sets are {s1}, {s2, s3}, {s3, s4} and {s4}.

The *necessary rank condition* requires the rank of the incidence matrix to be smaller than the number of different pre-sets of transitions.

In the first subsection, *strong occurrence sequences* will be introduced. This technical notation will be used to prove the sufficient rank condition. Strong occurrence sequences obey a more restrictive enabling rule in the occurrence condition. *Strongly live markings* are based on strong occurrence sequences. The second subsection considers the sufficient rank equation. This condition is sufficient for the liveness of a marking. It is sufficient and necessary for strong liveness of a marking, as will be shown in the third subsection. In the fourth subsection, the necessary rank condition will be proven. We do not only provide the upper bound for the rank of this condition but a lower bound, too.

We stick to nets with positive place and transition invariants. The existence of a positive place invariant implies that every marking is bounded. If a net with a positive place invariant can be marked lively, then it even has a live and bounded marking. In this case, there also exists a positive transition invariant (see [DeRe98]). Moreover, each net with both a positive place invariant and a positive transition invariant is strongly connected (Theorem 9).

We provide a sufficient condition for liveness of a marked net with positive place and transition invariants. First we give a condition for *potential liveness* of a net:

> *If the rank of the incidence matrix is smaller than the number of conflict areas then the net can be marked lively.*

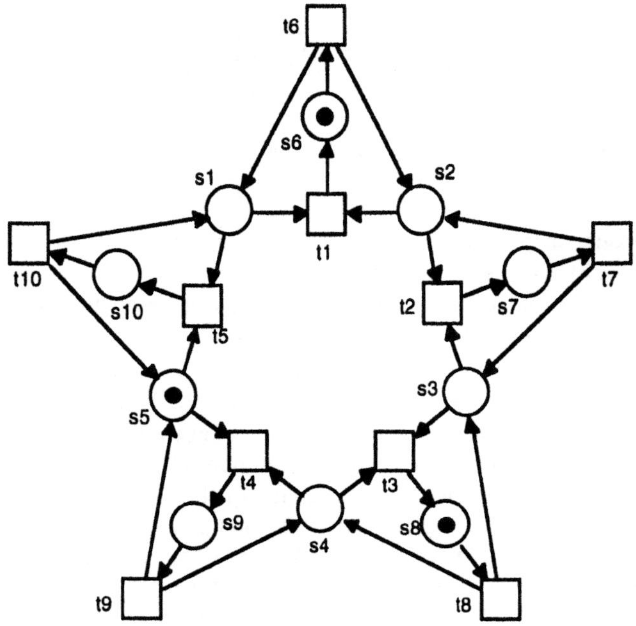

	t1	t2	t3	t4	t5	t6	t7	t8	t9	t10
s1	−1	0	0	0	−1	1	0	0	0	1
s2	−1	−1	0	0	0	1	1	0	0	0
s3	0	−1	−1	0	0	0	1	1	0	0
s4	0	0	−1	−1	0	0	0	1	1	0
s5	0	0	0	−1	−1	0	0	0	1	1
s6	1	0	0	0	0	−1	0	0	0	0
s7	0	1	0	0	0	0	−1	0	0	0
s8	0	0	1	0	0	0	0	−1	0	0
s9	0	0	0	1	0	0	0	0	−1	0
s10	0	0	0	0	1	0	0	0	0	−1

Fig. 7. An example net and its incidence matrix

Before proving this result, we demonstrate it by help of an example. Consider the net shown in Figure 7. It has the positive place invariant vector

$$\mathbf{i} = (1,1,1,1,1,2,2,2,2,2)$$

and the positive transition invariant vector

$$\mathbf{j} = (1,1,1,1,1,1,1,1,1,1).$$

The rank of the incidence matrix is 5. The set of vectors $\{\mathtt{t1},\mathtt{t2},\mathtt{t3},\mathtt{t4},\mathtt{t5}\}$ is a basis of its columns.

The net has 6 conflict areas,

$$\{\mathtt{t1},\mathtt{t2},\mathtt{t3},\mathtt{t4},\mathtt{t5}\},\{\mathtt{t6}\},\{\mathtt{t7}\},\{\mathtt{t8}\},\{\mathtt{t9}\},\{\mathtt{t10}\}.$$

So the premise of the above condition is fulfilled. Hence, the result states that the net can be marked lively. In fact, there are live markings; an example is the marking depicted in the figure.

6.1 Strong Occurrence Sequences and Strongly Live Markings

Two transitions are said to be in a *conflict situation* at a marking if both transitions are enabled and share a common input place. Conversely, a common input place does not guarantee that both transitions are only enabled in conflict situations. For example, if one of the transition becomes enabled before the other transition then it might occur before the other transition is enabled. Such a *confused* situation can sometimes be avoided by the following restricted enabling condition:

> *A transition can occur only in case all transitions of its conflict area are enabled, i.e., all places in the pre-set of transitions of its conflict area are marked.*

In this case, the transition is called *strongly enabled.*

Not every transition enabled at a marking is strongly enabled; a marking might enable only a subset of a conflict area. In the example, transition **t6** is strongly enabled, because {**t6**} is a conflict area. The marking obtained after the occurrence of **t6** marks the places **s1**, **s2**, **s5** and **s8**. It enables, among other transitions, **t1**. However, it does not strongly enable **t1** because **t2** belongs to the same conflict area and is not enabled.

A finite occurrence sequence $m_0 \xrightarrow{\sigma} m_n$ is called a *strong occurrence sequence* if, with $\sigma = t_1 \dots t_n$ and

$$m_0 \xrightarrow{t_1} m_1 \xrightarrow{t_2} \cdots \xrightarrow{t_n} m_n \;,$$

each marking m_{i-1} strongly enables the transition t_i $(1 \leq i \leq n)$. In the example, the sequence

$$\sigma = \mathtt{t6}\ \mathtt{t8}\ \mathtt{t1}$$

is a strong occurrence sequence, enabled at the initial marking.

A marking m_0 is *strongly live* if, for each reachable marking m and every transition t, there exists a strong occurrence sequence $m \xrightarrow{\sigma} m'$ such that t occurs in σ. The initial marking of the above example net is strongly live.

Clearly, each strongly enabled transition is enabled and each strong occurrence sequence is an occurrence sequence. Each strongly live marking is live, because, in the definition of strong liveness, the marking m is any reachable marking.

6.2 A Sufficient Condition for the Existence of a Live Marking

A net is potentially live if there exists a live initial marking. In this subsection we show that a net with positive place and transition invariants is potentially live if the rank of its incidence matrix is smaller than the number of its conflict areas. Actually, we show the contraposition: if the net has no live initial marking then the rank of its incidence matrix is not smaller than the number of conflict areas. The proof consists of the following steps:

(1) *A given net N with positive place and transition invariants has no live initial marking.*

The definition of strong liveness and (1) imply:

(2) *The net N has no strongly live initial marking.*

The following Lemma 40 will show that (2) implies:

(3) *For each initial marking of N, there is a reachable marking which does not strongly enable any transition.*

This result will be applied to the marking that associates one token to each place. Then the following Lemma 41 will prove:

(4) *There exists a set of transitions $\{t_1, \ldots, t_n\}$ of N, containing exactly one transition of each conflict area, and a place vector $\mathbf{x}$ satisfying $\mathbf{x} \cdot \mathbf{t}_i < 0$ $(1 \leq i \leq n)$.*

Consider the matrix $[\mathbf{t}_1 \ldots \mathbf{t}_n]$. By (4), the vector $\mathbf{x} \cdot [\mathbf{t}_1 \ldots \mathbf{t}_n]$ has only negative components. Assume a solution $\mathbf{j}$ of $[\mathbf{t}_1 \ldots \mathbf{t}_n] \cdot \mathbf{y} = 0$. Then either $\mathbf{j} = 0$ or $\mathbf{j}$ has both positive and negative components. In particular, we obtain:

(5) *Every solution $\mathbf{j}$ of $[\mathbf{t}_1 \ldots \mathbf{t}_n] \cdot \mathbf{y} = 0$, $\mathbf{y} \geq 0$ satisfies $\mathbf{j} = 0$.*

Together with (5), Lemma 42 will prove:

(6) *The set $\{\mathbf{t}_1, \ldots, \mathbf{t}_n\}$ is linearly independent.*

Finally, (6) immediately implies the result we are after:

(7) *The rank of* $\mathbf{N}$ *is not smaller than the number of conflict areas of N.*

Lemma 40. *Let N be a net with positive place and transition invariants and let m_0 be an initial marking of N which is not strongly live. Then there is a reachable marking m such that no transition is strongly enabled at m.*

Proof. Since m_0 is not strongly live, there exist a reachable marking m and a transition t such that m does not enable any strong occurrence sequence containing t. In this situation, t is said to be *excluded* at m. Every transition excluded at a marking m remains excluded at any marking reachable from m by a strong occurrence sequence. Now w.l.o.g. assume that m excludes a maximal set of transitions. Then all markings reachable from m by strong occurrence sequences exclude the same set of transitions.

If a transition is excluded at a marking m then so are all transitions of its conflict area. Let K be a conflict area such that all its transitions are excluded at m. Define $R = {}^\bullet K$. The net N is strongly connected because it has positive place and transition invariants (Theorem 9). Hence, R is not the empty set. The definition of conflict areas implies $K = R^\bullet$. Therefore, no strong occurrence sequence enabled at m contains transitions in $R^\bullet$.

The marking m is bounded because N has a positive place invariant. Hence, the number of tokens on places in R cannot grow unlimited and transitions in ${}^\bullet R$ can occur only finitely often in strong occurrence sequences enabled at m. Consider a strong occurrence sequence which is enabled at m and leads to some marking that also excludes all transitions in ${}^\bullet R$. We have chosen the marking m such that the number of excluded transitions is maximal at m. Hence, all transitions in ${}^\bullet R$ are already excluded at m.

Now assume that u is a transition excluded at m. Then u belongs to some conflict area K. We have shown that every transition in K and every transition in ${}^\bullet({}^\bullet K)$ is excluded at m. So every transition connected to u by a directed path ending at u is excluded at m. There exists at least one transition excluded at m, vic. transition t. Therefore, since N is strongly connected, every transition of N is excluded at m. In other words: no transition is strongly enabled at m. □

Lemma 41. *Let N be a net and let m_0 be the marking of N that associates one token to each place. Assume an occurrence sequence $m_0 \xrightarrow{\sigma} m$ such that no transition of N is strongly enabled at m. Then, for every set $\{K_1, \ldots, K_k\}$ of conflict areas of N there are transitions $t_1 \in K_1, \ldots, t_k \in K_k$ and a place vector $\mathbf{x}$ in $\mathbb{N}^*$ satisfying $\mathbf{x} \cdot \mathbf{t}_i < 0$ $(1 \leq i \leq k)$.*

Proof. We proceed by induction on k, the number of considered conflict areas.

Base. If $k = 0$, nothing has to be shown.

Step. Assume $k \geq 1$ and define $K = K_1 \cup \cdots \cup K_k$. Let R denote the set of places in ${}^\bullet K$ that are not marked at m. The definition of a conflict area implies $R^\bullet \subseteq K$.

No transition is strongly enabled at the marking m. Hence, every conflict area in $\{K_1, \ldots, K_k\}$ has at least one input place in R. Since m_0 marks all places, the occurrence sequence σ contains transitions in $R^\bullet$ (at least one for each conflict area). Assume that t is the last transition occurring in σ that belongs to $R^\bullet$. Then $t \notin {}^\bullet R$ because places in R are not marked at m.

W.l.o.g. assume that t belongs to the conflict area K_k. We apply the induction hypothesis to $\{K_1, \ldots, K_{k-1}\}$ and obtain transitions $t_1 \in K_1, \ldots, t_{k-1} \in K_{k-1}$ and a place vector $\mathbf{x}'$ in $\mathbb{N}^*$ such that $\mathbf{x}' \cdot \mathbf{t}_i < 0$ $(1 \leq i \leq k-1)$. Now choose the transition t for the conflict area K_k, i.e., set $t_k := t$. The place vector $\mathbf{x}$ is defined by

$$\mathbf{x}(s) = \begin{cases} (|R|+1)\,\mathbf{x}'(s) & \text{if } s \in {}^\bullet(K_1 \cup \cdots \cup K_{k-1}) \cap R, \\ 1 & \text{if } s \in {}^\bullet K_k \cap R, \\ 0 & \text{otherwise.} \end{cases}$$

Clearly all components of $\mathbf{x}$ are nonnegative integers. Moreover, every positive component belongs to a place in R. We have to show that $\mathbf{x} \cdot \mathbf{t}_i < 0$ $(1 \leq i \leq k)$.

Case 1. $i = k$. Then $t_i = t_k = t$. Recall that the vector $\mathbf{t}(s)$ equals 1 for $s \in t^\bullet \setminus {}^\bullet t$, -1 for $s \in {}^\bullet t \setminus t^\bullet$ and 0 otherwise. So

$$\mathbf{x} \cdot \mathbf{t} = \sum_{s \in t^\bullet} \mathbf{x}(s) - \sum_{s \in {}^\bullet t} \mathbf{x}(s) \ .$$

We have $t^\bullet \cap R = \emptyset$ because $t \notin {}^\bullet R$. So the first sum equals 0. Since $t \in K_k$ and $t \in R^\bullet$, there exists a place s in ${}^\bullet t$ satisfying $s \in {}^\bullet K_k \cap R$. Then $\mathbf{x}(s) = 1$. So the second sum evaluates to at least 1, and we obtain $\mathbf{x} \cdot \mathbf{t} \leq -1$.

Case 2. For $1 \leq i \leq k-1$, the definition of $\mathbf{x}$ implies:

$$\mathbf{x} \cdot \mathbf{t}_i \ \leq \ (|R|+1)\,\mathbf{x}' \cdot \mathbf{t}_i + |t_i^\bullet \cap ({}^\bullet K_k \cap R)| \ \leq \ (|R|+1)\,\mathbf{x}' \cdot \mathbf{t}_i + |R| \ .$$

Since $\mathbf{x}' \cdot \mathbf{t}_i$ is negative and integral, we have $\mathbf{x}' \cdot \mathbf{t}_i \leq -1$, and hence

$$(|R|+1)\,\mathbf{x}' \cdot \mathbf{t}_i + |R| \leq (-|R|-1) + |R| < 0 \ .$$

□

Lemma 42. *Let N be a net with a positive place invariant. Let $\{t_1, \ldots, t_k\}$ be a set of transitions of N, containing at most one transition of each conflict area. Then either the set $\{\mathbf{t}_1, \ldots, \mathbf{t}_k\}$ of associated columns of the incidence matrix is linearly independent or there are nonnegative coefficients $\lambda_1, \ldots, \lambda_k \geq 0$ such that $(\lambda_1, \ldots, \lambda_k) \neq 0$ and $\lambda_1 \mathbf{t}_1 + \cdots + \lambda_k \mathbf{t}_k = 0$.*

Proof. Assume the set $\{\mathbf{t}_1, \ldots, \mathbf{t}_k\}$ is not linearly independent. Then there exist coefficients $\mu_1, \ldots, \mu_k$ such that at least one of the μ_i is positive and

$$\mu_1 \mathbf{t}_1 + \cdots + \mu_k \mathbf{t}_k = 0 \ .$$

For each i, $1 \leq i \leq k$, define the coefficient λ_i by

$$\lambda_i := \begin{cases} \mu_i & \text{if } \mu_i \geq 0, \\ 0 & \text{otherwise.} \end{cases}$$

No λ_i is negative. Since at least one of the μ_i is positive this likewise holds for some λ_i. Now define the transition vectors $\mathbf{j}$ and $\mathbf{j}'$ as follows:

$$\mathbf{j} = \begin{pmatrix} \mu_1 \\ \vdots \\ \mu_k \\ 0 \\ \vdots \\ 0 \end{pmatrix} \quad \text{and} \quad \mathbf{j}' = \begin{pmatrix} \lambda_1 \\ \vdots \\ \lambda_k \\ 0 \\ \vdots \\ 0 \end{pmatrix}.$$

The vector $\mathbf{j}$ is a transition invariant vector, i.e., $\mathbf{N} \cdot \mathbf{j} = \mathbf{0}$. We will prove $\mathbf{N} \cdot \mathbf{j}' \geq \mathbf{0}$.

Let s_i be any place and $\mathbf{s}_i$ its associated row vector of the incidence matrix. The negative components in $\mathbf{s}_i$ belong to the transitions in $s_i^\bullet$. In particular, they all belong to a single conflict area. Hence, by assumption, $s_i^\bullet$ contains at most one transition t in $\{t_1, \ldots, t_k\}$. For any other negative component of $\mathbf{s}_i$, the component in $\mathbf{j}$ and the component $\mathbf{j}'$ equals 0. We show $\mathbf{s}_i \cdot \mathbf{j}' \geq 0$.

If $s_i^\bullet \cap \{t_1, \ldots, t_k\} = \emptyset$ then nothing has to be shown. Now assume that $s_i^\bullet \cap \{t_1, \ldots, t_k\} = \{t\}$.

Case 1. If $\mathbf{j}(t) \leq 0$ then $\mathbf{j}'(t) = 0$ and we are finished.

Case 2. If $\mathbf{j}(t) > 0$ then $\mathbf{j}'(t) = \mathbf{j}(t)$. The vector $(\mathbf{j}' - \mathbf{j})$ has no negative components. Furthermore, $(\mathbf{j}' - \mathbf{j})(t) = 0$. So, for each negative component of $\mathbf{s}_i$, the component in the vector $(\mathbf{j}' - \mathbf{j})$ has the value 0. Therefore, $\mathbf{s}_i \cdot (\mathbf{j}' - \mathbf{j}) \geq 0$. Finally, $\mathbf{s}_i \cdot \mathbf{j} = 0$ implies $\mathbf{s}_i \cdot \mathbf{j}' \geq 0$.

So we obtain

$$\mathbf{N} \cdot \mathbf{j}' = \begin{pmatrix} \mathbf{s}_1 \\ \vdots \\ \mathbf{s}_n \end{pmatrix} \cdot \mathbf{j}' \geq \mathbf{0}.$$

The net N has a positive place invariant i, i.e. $\mathbf{i} > \mathbf{0}$ and $\mathbf{i} \cdot \mathbf{N} = \mathbf{0}$. Hence, $\mathbf{i} \cdot \mathbf{N} \cdot \mathbf{j}' = 0$. Now $\mathbf{i} > \mathbf{0}$ and $\mathbf{N} \cdot \mathbf{j}' \geq \mathbf{0}$ implies $\mathbf{N} \cdot \mathbf{j}' = \mathbf{0}$.

By definition of $\mathbf{j}'$, we finally obtain

$$\lambda_1 \mathbf{t}_1 + \cdots + \lambda_k \mathbf{t}_k = \mathbf{0}$$

which completes the proof. □

Theorem 43. *Let N be a net with positive place and transition invariants. If the rank of* $\mathbf{N}$ *is smaller than the number of conflict areas of N then there exists a live initial marking of N.*

Proof. The contraposition of this theorem is given by the implication (1) ⇒ (7) of the previously defined propositions (1) and (7). With the above lemmas, the implications (1) ⇒ (2) ⇒ ⋯ ⇒ (7) are shown above. □

This sufficient condition for potential liveness can be checked efficiently, i.e., in polynomial time: the existence of positive invariants can be decided by solving the respective inequality systems and the rank inequality is checked by calculating the rank and counting the conflict areas.

6.3 Characterizing Strongly Live Markings

In this subsection we give a sufficient condition for liveness of a marking. This condition is also necessary for strong liveness.

Lemma 44. *Let N be a net with positive place and transition invariants. Let $\{s_1, \ldots, s_k\}$ be a set of places of N containing at most one place of each conflict area. Then either the set of row vectors $\{\mathbf{s}_1, \ldots, \mathbf{s}_k\}$ of the incidence matrix is linearly independent or there exist nonnegative coefficients $\lambda_1, \ldots, \lambda_k \geq 0$ such that $(\lambda_1, \ldots, \lambda_k) \neq \mathbf{0}$ and $\lambda_1 \mathbf{s}_1 + \cdots + \lambda_k \mathbf{s}_k = \mathbf{0}$.*

Proof. Assume $N = (S, T, F)$. Let $N^d = (T, S, F^{-1})$ be the *dual net* of N, where $F^{-1} = \{(y, x) | (x, y) \in F\}$. Then $\mathbf{N}^d$ is the transposed matrix of $\mathbf{N}$. Each positive transition invariant of N is a positive place invariant of N^d, and vice versa. The net N is strongly connected because it has positive place and transition invariants (Theorem 9). In particular, no conflict area consists of a single transition with an empty pre-set. By definition, every conflict area of N^d is the pre-set (in N) of a conflict area of N, and vice versa. Therefore, the set $\{s_1, \ldots, s_k\}$ is a set of transitions of N^d, containing at most one transition of each conflict area. Now Lemma 42, applied to N^d and the set $\{s_1, \ldots, s_k\}$, proves the result. □

Theorem 45. *Let N be a net with positive place and transition invariants. Assume that the rank of $\mathbf{N}$ is smaller than the number of conflict areas of N. Then a marking m_0 of N is strongly live if and only if every non-negative place invariant vector $\mathbf{i} \geq \mathbf{0}$ satisfies $\mathbf{i} \cdot \mathbf{m}_0 > 0$.*

Proof.
($\Rightarrow$) Each strongly live marking is live. As shown in [DeRe98], a live marking m_0 satisfies $\mathbf{i} \cdot \mathbf{m}_0 > 0$ for each place invariant $\mathbf{i} \geq \mathbf{0}, \mathbf{i} \neq \mathbf{0}$.
($\Leftarrow$) We prove the contraposition. Let m_0 be a marking of N which is not strongly live. By Lemma 40, some marking m is reachable from m_0 such that no transition is strongly enabled at m. Hence, for each conflict area K, there exists at least one place in ${}^\bullet K$ that is not marked at m.

W.l.o.g. assume that $\{s_1, \ldots, s_k\}$ is a set of places, all unmarked at m, containing for each conflict area K exactly one place of ${}^\bullet K$. The number of places in this set is equal to the number of conflict areas. By assumption, the rank of $\mathbf{N}$ is smaller than the number of conflict areas. Hence the set of vectors $\{\mathbf{s}_1, \ldots, \mathbf{s}_k\}$ is not linearly independent. By Lemma 44, there are coefficients $\lambda_1, \ldots, \lambda_k \geq 0$ such that $(\lambda_1, \ldots, \lambda_k) \neq \mathbf{0}$ and $\lambda_1 \mathbf{s}_1 + \cdots + \lambda_k \mathbf{s}_k = \mathbf{0}$.

Let $\mathbf{i}$ be the place vector of N, defined by

$$\mathbf{i} = (\lambda_1, \ldots, \lambda_k, 0 \ldots, 0) .$$

Then $\mathbf{i} \cdot \mathbf{N} = \mathbf{0}$, i.e., $\mathbf{i}$ is a place invariant vector satisfying $\mathbf{i} \geq \mathbf{0}, \mathbf{i} \neq \mathbf{0}$. Since every place in $\{s_1, \ldots, s_k\}$ is unmarked at m, we have $\mathbf{i} \cdot \mathbf{m} = 0$. Since m is reachable from m_0 and since $\mathbf{i}$ is a place invariant, finally follows $\mathbf{i} \cdot \mathbf{m}_0 = 0$. □

The following corollary formulates the sufficient condition for liveness of a marked net given by Theorems 43 and 45.

Corollary 46. *Let N be a net with initial marking m_0. Assume that N has positive place and transition invariants and that the rank of* $\mathbf{N}$ *is smaller than the number of conflict areas of N. If every place invariant* $\mathbf{i}$ *with* $\mathbf{i} \geq \mathbf{0}, \mathbf{i} \neq \mathbf{0}$ *satisfies* $\mathbf{i} \cdot \mathbf{m}_0 > 0$ *then m_0 is a live marking.*

The condition of the previous theorem and corollary can be checked efficiently by help of the following system of inequalities. The condition is satisfied if and only if there exists no solution

$$\begin{aligned}
\mathbf{N}^{\mathsf{T}} \cdot \mathbf{x} &= \mathbf{0} \quad (\mathbf{x}^{\mathsf{T}} \text{ is a place invariant vector})\\
(1,0,\ldots,0) \cdot \mathbf{x} &\geq 0 \quad (\mathbf{x} \geq 0)\\
&\vdots\\
(0,\ldots,0,1) \cdot \mathbf{x} &\geq 0\\
(1\ldots1) \cdot \mathbf{x} &\geq 1 \quad (\mathbf{x} \neq 0)\\
\mathbf{m}^{\mathsf{T}} \cdot x &\leq 0\,.
\end{aligned}$$

6.4 A Necessary Condition for the Liveness of a Marking

Now we show the counterpart to the previous results, i.e. we provide a necessary condition for liveness which is based on the rank of the incidence matrix, too.

As motivated at the beginning of this section, two transitions which are always enabled in conflict lead to different transition invariants. This observation holds in particular for transitions with identical pre-sets. In the sequel, we add *regulation circles* to nets, that ensure that sets of transitions with identical pre-sets occur cyclically in a fixed order.

Let $U = \{t_1, \ldots, t_k\}$ be a nonempty set of transitions of a net. For each transition $t_i \in U$, let s_i' be a new place. (s_i' is not an element of N). Then the net

$$N^U = (\{s_1', \ldots, s_k'\}, U, \{(t_1, s_1'), (s_1', t_2), \ldots, (t_k, s_k'), (s_k', t_1)\})$$

is called a *regulation circle* of U. The net obtained by the componentwise union of the nets N and N^U is called the *composition* of N and N^U.

Lemma 47. *Let N be a net, let N^U be the regulation circle of some set U of transitions of N, and let N' be the composition of N and N^U. If N has a positive place invariant then N' also has a positive place invariant.*

Proof. Let $\mathbf{i}$ be a positive place invariant vector of N. Let $\mathbf{i}'$ be a place vector of N' that coincides with $\mathbf{i}$ for all places of N and that has the value 1 for all places of the regulation circle. By construction, this vector is positive.

Let t be an arbitrary transition of N. If $t \notin U$ then ${}^\bullet t$ in N and ${}^\bullet t$ in N' are identical, and the same holds for $t^\bullet$. If $t \in U$, both ${}^\bullet t$ and $t^\bullet$ contain in N' exactly one additional place of the regulation circle. Both corresponding components of $\mathbf{i}'$ are 1. So, in both cases, $\mathbf{i}' \cdot \mathbf{t} = \mathbf{i} \cdot \mathbf{t} = 0$. □

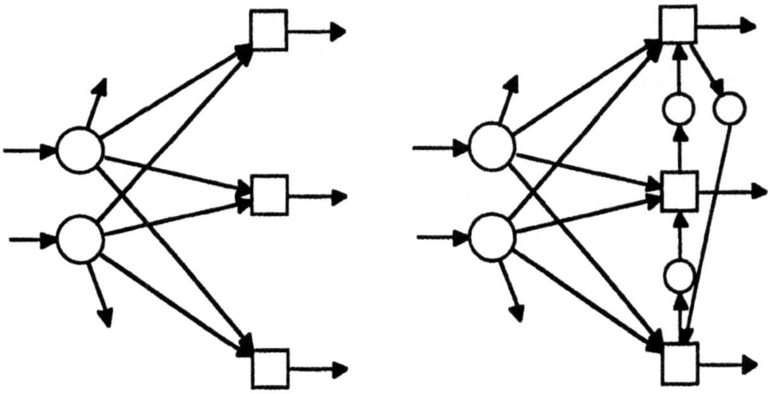

Fig. 8. A piece of a net with a regulation circle

Lemma 48. *Let N be a net with a positive place invariant. Let U be a nonempty set of transitions of N such that all transitions in U have identical pre-sets. Let N^U be a regulation circle of U, and let N' be the composition of N and N^U. If there exists a live marking of N then there also exists a live marking of N'.*

Proof. Let m_0 be a live marking of N. Since N has a positive place invariant , m_0 is bounded. Therefore, the set of markings reachable from m_0 is finite; let k be its cardinality. Define a marking m'_0 of the net N' that coincides with m_0 for all places of N and associates k tokens to every place of the regulation circle. We will show that m'_0 is a live marking of N'.

Let $m'_0 \xrightarrow{\sigma} m'_1$ be any occurrence sequence of N', and t be any transition. We will construct an occurrence sequence τ, enabled at m'_1, that includes an occurrence of t. The sequence τ is the concatenation of two sequences τ_1 and τ_2.

Since N' is constructed from N and N^U by identification of common transitions only, every occurrence sequence σ of N' induces occurrence sequences of N and N^U. More formally, for each marking m' of N' we denote its restriction to the places of N by m and its restriction to the places of N^U by m^U. The projection σ^U of σ to the transitions of N^U is obtained from σ by cancellation of all transitions that are not in U. Now it is easy to verify that there is an occurrence sequence

$$m'_0 \xrightarrow{\sigma} m'_1 \text{ of } N'$$

if and only if there are occurrence sequences

$$m_0 \xrightarrow{\sigma} m_1 \text{ of } N \text{ and}$$

$$m_0^U \xrightarrow{\sigma^U} m_1^U \text{ of } N^U \ .$$

(Notice that N and N' have identical sets of transitions.)

The net N^U is a simple circle. Therefore, its initial marking m_0^U (associating k tokens to every place) can be reached again from m_1^U. Let

$$m_1^U \xrightarrow{u_1 \ldots u_n} m_0^U$$

be an according occurrence sequence of the net N^U.

We show that the marking m_1 of the net N enables an occurrence sequence τ_1 such that its projection ${\tau_1}^U$ is $u_1 \ldots u_n$. To this end, we choose a minimal occurrence sequence enabled at m_1 such that the reached marking enables a transition in U (this might be the empty sequence). Since m_0 is live, m_1 is live, too. So, such an occurrence sequence exists. Since all transitions in U have identical pre-sets in N, they all are enabled at m_1. Now extend the occurrence sequence by the transition u_1. Starting with the marking reached after u_1, choose again a minimal occurrence sequence enabling the transitions of U, and then continue with u_2. By the minimality of the occurrence sequences, these sequences do not contain any transition of U. After n repetitions of this procedure, an occurrence sequence $m_1 \xrightarrow{\tau_1} m_2$ is gained such that the projection of τ_1 to the set U is the sequence $u_1 \ldots u_n$.

The sequence τ_1 is enabled at m_1 in N and leads to m_2. Its restriction $\tau_1^U = u_1 \ldots u_n$ is an occurrence sequence of N_U leading from m_1^U to m_0^U. Hence the occurrence sequence τ_1 is also enabled at m_1' in the net N'. It leads to the marking m_2' which coincides with m_2 on the places of N and with m_0' on the places of N^U.

The marking m_2 is again a live marking of N. So, m_2 enables a minimal occurrence sequence τ_2 that contains the transition t. By minimality, no marking is reached more than once during this occurrence sequence. So the length of τ_2 is smaller than k, the number of all reachable markings. Since, in N^U, the marking m_0^U associates k tokens to each place, it enables all sequences with length up to k and in particular the sequence τ_2^U. Therefore, the sequence τ_2 is enabled at m_2' in N'. Finally, the composition $\tau_1\,\tau_2$ is enabled at m_0' and contains t. □

For a net N, the set $P_N = \{{}^\bullet t \mid t \in T_N\}$ denotes all pre-sets of transitions of N (T_N is the set of transitions of N). If two transitions have identical pre-sets, they only contribute to the set P_N once.

Theorem 49. *Let N be a net with positive place and transition invariants. If N has a live marking then the rank of* **N** *is smaller than the value* $|P_N|$.

Proof. Assume that N has a live marking m_0. We proceed by induction on $k = |T_N| - |P_N|$.

Base. Assume $k = 0$. The matrix **N** has $|T_N|$ columns. The rank of **N** is at most $|T_N| - 1$ because N has a positive transition invariant. Since $k = 0$ we have $|T_N| = |P_N|$. So the rank of **N** is smaller than $|P_N|$

Step. Assume $k > 0$. Let U be a maximal set of transitions with identical pre-sets. We have $|U| \geq 2$ because $k = |T_N| - |P_N| > 0$.

Let N^U be the regulation circle of U and let N' be the composition of N and N^U. By Lemmas 47 and 48, N' has a positive place invariant and can be marked lively. So it has a live and bounded marking. As shown in [DeRe98], it has a positive transition invariant.

The set of transitions of N' is also T_N. Each transition in $T_N \setminus U$ has the same pre-set in N and in N'. In N', two distinct transitions of U have different pre-sets, since this is the case in the regulation circle. So we get

$$|P_{N'}| = |P_N| + |U| - 1 \ .$$

This implies that the number $|T_N| - |P_{N'}|$ is smaller than $|T_N| - |P_N|$. Hence, with the induction hypotheses applied to the net N', the rank of $\mathbf{N}'$ does not exceed $|P_{N'}| - 1$. So we have

$$\text{rank}(\mathbf{N}) + |U| - 1 \leq \text{rank}(\mathbf{N}') \leq |P_{N'}| - 1 = |P_N| + |U| - 2 \ .$$

It remains to show

$$\text{rank}(\mathbf{N}) + |U| - 1 \leq \text{rank}(\mathbf{N}'),$$

i.e. the rank of $\mathbf{N}'$ exceeds the rank of $\mathbf{N}$ by at least $|U| - 1$. W.l.o.g. assume $U = \{t_1, \ldots, t_n\}$. The n rows in $\mathbf{N}'$ of the places of the regulation circle are shaped:

$$\begin{array}{l}
(\,-1 \quad 0 \cdots \quad 0 \quad 1\,0 \cdots 0\,) \\
(\quad 1 \, -1 \quad 0 \cdots \quad 0\,0 \cdots 0\,) \\
\vdots \qquad \ddots \qquad\quad \vdots\;\vdots \quad\; \vdots\;\vdots \\
(\quad 0 \cdots \quad 1 \, -1 \quad 0\,0 \cdots 0\,) \\
(\quad 0 \cdots \quad 0 \quad 1 \, -1\,0 \cdots 0\,)
\end{array}$$

Clearly, every subset of this set containing $n-1$ vectors is linearly independent. We show that none of this rows is a linear combination of other rows of $\mathbf{N}'$, i.e. of rows of $\mathbf{N}$. We proceed indirectly and assume that the row $\mathbf{s}_i'$ is a linear combination of the rows of $\mathbf{N}$. Let $\mathbf{s}_i' = \mathbf{k} \cdot \mathbf{N}$, let t_i be the unique transition in the pre-set of s_i', and let t_j be the unique transition in the post-set of s_i'. By assumption, there exists a live marking m_0 of N. Since N has a positive place invariant, m_0 is a bounded marking and only finitely many markings can be reached. Let m be reachable from m_0 such that $\mathbf{k} \cdot \mathbf{m}$ has a maximal value. Define

$$\lambda := \mathbf{k} \cdot \mathbf{m} - \mathbf{k} \cdot \mathbf{m}_0 \ .$$

Since, in N, the transitions t_i and t_j have identical pre-sets, there exists an occurrence sequence $m_0 \xrightarrow{\sigma} m_1$ such that the transition t_i occurs $\lambda + 1$ times in σ and t_j does not occur in σ. The corresponding Parikh vector $\mathbf{p}_\sigma$ satisfies $\mathbf{s}_i' \cdot \mathbf{p}_\sigma > \lambda$. By $\mathbf{m}_0 + \mathbf{N} \cdot \mathbf{p}_\sigma = \mathbf{m}_1$ and by the choice of m follows

$$\mathbf{k} \cdot \mathbf{N} \cdot \mathbf{p}_\sigma = \mathbf{k} \cdot \mathbf{m}_1 - \mathbf{k} \cdot \mathbf{m}_0 \leq \lambda,$$

in contradiction to

$$\mathbf{k} \cdot \mathbf{N} \cdot \mathbf{p}_\sigma = \mathbf{s}_i' \cdot \mathbf{p}_\sigma > \lambda.$$

□

Bibliographic Remarks

The rank conditions were first introduced for *free-choice nets*. There the sufficient and the necessary condition coincide.

The rank conditions for free-choice nets were developed in [Espa90], [CaCS91] and [Dese92] (see also [DeEs95]). They were generalized to free-choice nets with arc weights (*equal-conflict nets*) in [ReTS95, TeSi96].

The necessary rank condition for general nets is from [CoCS90]; see also [DeEs95] and [TeSi96]. The sufficient condition is from [Dese94].

In [Reis79], conflict areas were introduced.

The paper [Mura77] provides further bounds for the rank of the incidence matrix: If each marking of a net is reachable from some initial marking then the row rank is maximal and equals the number of places. A net with this property has no place invariant except the vector 0.

References

[CaCS91] G. Campos, G. Chiola and M. Silva. Properties and performance bounds for closed free choice synchronized monoclass queueing networks. IEEE Transactions on Automation and Control, 36(12):1368–1382, 1991.

[CoCS90] J. M. Colom, J. Campos and M. Silva. On Liveness Analysis Through Linear Algebraic Techniques. Dpto. Ing. Electrica e Informatica, Universidad de Zaragoza RR 90-11, 1990.

[CoSi91] J. M. Colom and M. Silva. Improving the linearly based characterization of P/T nets. In G. Rozenberg (ed.), Advances in Petri nets 1990, Vol. 483 of Lecture Notes in Computer Science, pp. 113–145, Springer-Verlag, 1991.

[DeEs93] J. Desel and J. Esparza. Reachability in cyclic extended free choice systems. Theoretical Computer Science, 114:93–118, 1993.

[DeEs95] J. Desel and J. Esparza. Free Choice Petri Nets, Vol. 40 of Cambridge Tracts in Theoretical Computer Science. Cambridge University Press, 1995.

[DeNR96] J. Desel, K.-P. Neuendorf and M.-D. Radola. Proving non-reachability by modulo-invariants. Theoretical Computer Science, 153:49–64, 1996.

[Dese85] J. Desel. Another boatman story. GI Petri Net Newsletter, 22:3–6, 1985.

[Dese88] J. Desel. Wanted: Dead or alive? GI Petri Net Newsletter, 29:0–2, 1988.

[Dese92] J. Desel. A proof of the rank theorem for extended free choice nets. In K. Jensen (ed.), Application and Theory of Petri nets, Vol. 616 of Lecture Notes in Computer Science, pp. 134–153, Springer-Verlag, 1992.

[Dese94] J. Desel. Regular marked Petri nets. In J. van Leeuwen (ed.), Graph-Theoretic Concepts in Computer Science, Vol. 790 of Lecture Notes in Computer Science, pp. 264–275, Springer-Verlag, 1994.

[DeRe98] J. Desel and Wolfgang Reisig. Place/transition Petri nets. Advanved Course on Petri Nets, 1996, in this volume.

[Dese98] J. Desel. Petrinetze, Lineare Algebra und Lineare Programmierung. Teubner-Texte zur Informatik Vol. 26, Teubner Verlag, 1998.

[EsMe96] J. Esparza and S. Melzer. Checking system properties via Integer Programming. In H.R. Nielson (ed.), ESOP'96, Vol. 1058 of Lecture Notes in Computer Science, pp. 250-264, Springer-Verlag, 1996.

[EsNi94] J. Esparza and M. Nielsen. Decidability issues for Petri nets. GI Petri Net Newsletter, 47:5–23, 1994.

[Espa90] J. Esparza. Synthesis rules for Petri nets, and how they lead to new results. In J.C.M. Baeten and J.W. Klop (eds.), CONCUR 90, Vol. 458 of Lecture Notes in Computer Science, pp. 182–198. Springer-Verlag, 1990.

[EzCS93] J. Ezpeleta, J. M. Couvreur and M. Silva. A new technique for finding a generating family of siphons, traps and ST-components. Application to colored Petri nets. In G. Rozenberg (ed.), Advances in Petri nets 1993, Vol. 674 of Lecture Notes in Computer Science, pp. 126–147, Springer-Verlag, 1993.

[Jant87] M. Jantzen. Complexity of place/transition nets. In W. Brauer, W. Reisig and G. Rozenberg (eds.), Petri nets: Central Models and Their Properties, Advances in Petri Nets 1986, part I, Vol. 254 of Lecture Notes in Computer Science, pp. 413-435, Springer-Verlag, 1987.

[JaVa80] M. Jantzen and R. Valk. Formal properties of place/transition nets. In W. Brauer (ed.), Net Theory and Applications, Vol. 84 of Lecture Notes in Computer Science, pp. 165–212, Springer-Verlag, 1980.

[KaBa79] R. Kannan and A. Bachem. Polynomial algorithms for computing the Smith and Hermite normal forms of an integer matrix. SIAM Journal on Computing, 8:499–507, 1979.

[Laut75] K. Lautenbach. Liveness in Petri nets. GMD-ISF 75-02.1, Gesellschaft für Mathematik und Datenverarbeitung, Bonn, 1975.

[Laut87a] K. Lautenbach. Linear algebraic techniques for place/transition nets. In W. Brauer, W. Reisig and G. Rozenberg (eds.), Petri nets: Central Models and their Properties, Advances in Petri nets 1986, Part I, Vol. 254 of Lecture Notes in Computer Science, pp. 142–167. Springer-Verlag, 1987.

[Laut87b] K. Lautenbach. Linear algebraic calculation of deadlocks and traps. In H.J. Genrich, K. Voss and G. Rozenberg (Ed.), Concurrency and Nets, pp. 315–336, Springer-Verlag, 1987.

[LaSc74] K. Lautenbach and H. A. Schmidt. Use of Petri nets for proving correctness of concurrent process systems. In IFIP Congress 1974, pp. 187–191. North–Holland, 1974.

[Lien76] Y. E. Lien. Termination properties of generalized Petri nets. SIAM Journal on Computing, 5(2):251–265, 1976.

[MeRo80] G. Memmi and G. Roucairol. Linear algebra in net theory. In W. Brauer (ed.), Net Theory and Applications, Vol. 84 of Lecture Notes in Computer Science, pp. 213–223. Springer-Verlag, 1980.

[Mura77] T. Murata. State equation, controllability, and maximal matchings of Petri nets. IEEE Transactions on Automation and Control, 22(3):412–416, 1977.

[Mura89] T. Murata. Petri nets: Properties, analysis and applications. Proc. of the IEEE, 77(4):541–580, 1989.

[Pasc86] K.-H. Pascoletti. Diophantische Systeme und Lösungsmethoden zur Bestimmung aller Invarianten in Petri-Netzen. GMD–Berichte Nr. 160, Oldenbourg, 1986.

[Pete77] J. L. Peterson. Petri nets. ACM Computing Surveys, 9(3):223–252, 1977.

[Pete81] J. L. Peterson. Petri Net Theory and the Modelling of Systems. Prentice–Hall, Englewood Cliffs, 1981.

[Petr73] C. A. Petri. Concepts of net theory. In: Mathematical Foundations of Computer Science, Proceedings of a symposium and summer school, High

Tatras, September 3-8, 1973. Math. Inst. of the Slovak Acad. of Science, pp. 137–146, 1973.

[Petr82] C. A. Petri. State-transition structure in physics and computation. International Journal of Theoretical Physics, 21(10/11):979–992, 1982.

[Ramc74] C. Ramchandani. Analysis of asynchronous concurrent systems by Petri nets. PhD thesis, MIT, Dept. Electrical Engineering, Cambridge, Mass., 1974.

[Reis79] W. Reisig. On solving conflicts in Petri nets. In: U. Pape (ed.),Discrete Structures and Algorithms, Hanser-Verlag, pp. 241–254, 1979.

[Reis85] W. Reisig. Petri Nets. EATCS Monographs on Theoretical Computer Science Vol. 4, Springer-Verlag, 1985.

[ReTS95] L. Recalde, E. Teruel and M. Silva. On well-formed analysis: The case of deterministic systems of sequential processes. In J. Desel (ed.), Structures in Concurrency Theory, Workshops in Computing, pp. 279–293, Springer-Verlag, 1995.

[Schr86] A. Schrijver. Theory of Linear and Integer Programming. Wiley, 1986.

[Sifa79] J. Sifakis. Structural properties of Petri nets. In J. Winkowski (ed.), Mathematical Foundations of Computer Science, Vol. 64 of Lecture Notes in Computer Science, pp. 474–483, Springer-Verlag, 1979.

[TeCS93] E. Teruel, M. Colom and M. Silva. Linear analysis of deadlock-freeness of Petri net models. 2nd European Control Conference, Vol. 2, pp. 513–518. North–Holland, 1993.

[TeSi96] E. Teruel and M. Silva. Structure theory of equal conflict systems. Theoretical Computer Science, 153:271–300, 1996.

Linear Algebraic and Linear Programming Techniques for the Analysis of Place/Transition Net Systems

Manuel Silva, Enrique Teruel, and José Manuel Colom *

Dep. Informática e Ingeniería de Sistemas, CPS, Universidad de Zaragoza,
Maria de Luna 3, E-50015 Zaragoza, Spain

Abstract. The structure theory of Place/Transition net systems is surveyed — incorporating new contributions — in a tutorial style, mainly from a linear algebraic perspective. Topics included are: state equation based analysis of safety properties (e.g., boundedness, mutual exclusion, deadlock-freeness, etc.), linear invariants, siphons and traps, implicit places and their application to improve the accuracy of the state equation, and rank theorems (structural conditions for liveness and boundedness based on the rank of the incidence matrix).

Contents

* This work was partially supported by Project TIC-94-0242 of the Spanish CICYT and Contract CHRX-CT94-0452 (MATCH) within the HCM Programme of the EU.

1 Introduction

A *Petri net (PN)* model of a dynamic system, i.e., a *net system*, consists of two parts: A *net structure*, comprising the state variables (places) and their transformers (transitions) and a *marking*, that represents a distributed overall state on the structure. The system dynamics or behaviour is given by the evolution rules for the marking. This separation allows one to reason on net based models at two different levels: *structural* and *behavioural*. From the former we may derive some "fast" conclusions on the possible behaviours of the modelled system. Purely behavioural reasonings can be more conclusive, but they may require costly computations, or even they may not be feasible. The structural reasoning can be regarded as an *abstraction* of the behavioural one: for instance, instead of studying whether a given system, i.e., a net structure with an initial marking, has a finite state space, we might investigate whether the state space is finite *for every* possible initial marking; or we could study whether *there exists* an initial marking that guarantees infinite activity rather than deciding this for a given one, etc.

Two intimately related families of techniques have extensively been used for structural reasoning: *graph theory* and *linear algebra*. In this work we deal with both kinds of techniques from a linear algebraic viewpoint, often expressing or interpreting in linear algebraic terms some classical graph theory based notions (such as those related to siphons and traps). We are interested in giving a general framework, conceptually simple and reasonably efficient, rather than presenting the most efficient algorithm for each particular property (as an example, liveness and boundedness of a free choice system can be decided in polynomial time after Corollary 50, but a graph based algorithm [46] performs more efficiently).

The material is presented mainly in a tutorial style, covering the main developments in the field from the seventies and also introducing some new perspectives or revisiting previous works. In order to improve readability, most proofs are given as previous explanations of the results, concepts and results are illustrated by several examples, additional information is contained in separated remarks, and bibliographical remarks have been collected in a final section.

The basic notions are recalled in Section 2, where also the main notations are introduced. (A brief recall to notions and results in linear programming and duality theory that are used throughout the paper is included as an appendix.)

The starting point for structure theory is the description of the behaviour of the system in structural terms, based on the net *state equation* and other structural objects. The presentation and comparison of these concepts forms Section 3.

Section 4 covers the analysis of important safety properties of net systems (e.g., boundedness, mutual exclusion, deadlock-freeness, etc.) through the state equation. This method is at the same time more efficient and accurate than the classical *invariant method*, which on the other hand has salient merits for the understanding. Special emphasis is given to show the bridge between results in the fields of structure theory and linear algebra/convex geometry. For instance, classical results from the invariant method are derived applying duality theory to the linear programming formulations based on the state equation.

A major limitation of the state equation method to analyse net systems is the fact that structural descriptions of the behaviour, particularly those derived from the net state equation, are, in general, *relaxations*. Owing to this, the analysis allows only to semidecide the corresponding properties, i.e., find only necessary or sufficient conditions. Section 5 presents several techniques to improve the accuracy of structural descriptions, hence the resolution of structural methods.

Another limitation of the state equation method is that it is best suited to analyse safety properties, i.e., existence or non-existence of markings (and firing vectors). Other properties, particularly transition's liveness, cannot be dealt with directly. Nevertheless, some structural results based on the incidence matrix and the conflict structure of the net are helpful for this analysis, specially in the case of some net subclasses, as it is shown in Section 6.

The paper is concluded with bibliographical remarks in Section 7, to give an impression on the development of the field and to point at related topics that have not been exhaustively covered.

2 Nets and Net Systems: Basic Concepts and Notation

2.1 Vector and Matrix Notations

We denote vectors as $\mathbf{v} = [v_i]$; v_i is the i-th component of $\mathbf{v}$, alternatively written $\mathbf{v}[i]$. For matrices, we have $\mathbf{C} = [c_{ij}]$ and $\mathbf{C}[i,j] = c_{ij}$. Most often in net theory, vectors and matrices are indexed by the (arbitrarily) ordered sets of places and transitions. For instance, if $p \in P$ and $t \in T$, $\mathbf{C}[p,t]$ denotes the entry of $\mathbf{C}$ corresponding to row p and column t; if $P' \subseteq P$ and $T' \subseteq T$ we can write $\mathbf{C}[P',T']$ to refer to the submatrix of $\mathbf{C}$ corresponding to rows from P' and columns from T'; the column of $\mathbf{C}$ corresponding to transition $t \in T$ would be $\mathbf{C}[P,t]$. We often describe markings and other rather sparse vectors using a bag/formal sum notation. For instance, a marking that puts two tokens in p_1 and one in p_3 is denoted $2p_1 + p_3$ instead of $[2\,0\,1\,0 \cdots 0]$.

The transpose of a matrix is denoted by $\mathbf{C}^{\perp}$. (The transpose of a vector is not defined, i.e., vectors are not considered one-row or one-column matrices.) Operations are denoted as usual. For instance, $k \cdot \mathbf{C} = [k \cdot c_{ij}]$, $\mathbf{v} \cdot \mathbf{v}' = \sum_i v_i \cdot v_i'$, $\mathbf{v} \cdot \mathbf{C} = [\mathbf{v} \cdot \mathbf{C}[\cdot,i]]$, and $\mathbf{C} \cdot \mathbf{v} = [\mathbf{C}[i,\cdot] \cdot \mathbf{v}]$. Relational operators applied on vectors or matrices are interpreted componentwise. For instance, $\mathbf{v} > \mathbf{v}'$ means that $v_i > v_i'$ for every i. The following will be used too: $\mathbf{v} \gneq \mathbf{v}'$ means that $v_i \geq v_i'$ for every i and some i exists such that $v_i > v_i'$ (not to be confused with $\mathbf{v} \not\geq \mathbf{v}'$ meaning that $\mathbf{v} \geq \mathbf{v}'$ is false). The support of a vector $\mathbf{v}$ — the set of indices of non-null elements — is denoted by $\|\mathbf{v}\|$.

We denote $\mathbf{0}$ and $\mathbf{1}$ the vectors/matrices with every entry equal to zero and one, respectively, $\mathbf{I}$ the identity matrix, $\mathbf{1}_i$ the vector whose only non null entry is i, which takes value one (a characteristic vector), and $\mathbf{1}_S = \sum_{i \in S} \mathbf{1}_i$ (the characteristic vector of S).

2.2 Place/Transition Net Systems

We concentrate here on the formalism of *Place/Transition (P/T) net systems.* We denote a *P/T net* as $\mathcal{N} = \langle P, T, \mathbf{Pre}, \mathbf{Post} \rangle$, where P and T are the sets of *places* and *transitions*, and **Pre** and **Post** are the $|P| \times |T|$ sized, natural valued, *incidence matrices.* $\mathbf{Post}[p,t] = w$ means that there is an *arc* from t to p with *weight* (or *multiplicity*) w, and $\mathbf{Pre}[p,t] = 0$ indicates no arc from p to t. (We assume without loss of generality that nets are *connected.*)

A *marking* is a $|P|$ sized, natural valued, vector. A *P/T system* is a pair $\mathcal{S} = \langle \mathcal{N}, \mathbf{m_0} \rangle$, where $\mathbf{m_0}$ is the *initial marking.* A transition t is *enabled* at $\mathbf{m}$ iff $\mathbf{m} \geq \mathbf{Pre}[P,t]$; its *occurrence* or *firing*, denoted by $\mathbf{m} \xrightarrow{t} \mathbf{m}'$, yields a new marking $\mathbf{m}' = \mathbf{m} + \mathbf{C}[P,t]$, where $\mathbf{C} = \mathbf{Post} - \mathbf{Pre}$ is called the *token flow matrix.* (In *pure nets*, i.e., without self-loops, positive and negative entries in **C** completely represent the post- and pre- incidence functions, and then **C** can be properly called the *incidence matrix* of the pure net.)

An *occurrence sequence* from $\mathbf{m}$ is a sequence of transitions $\sigma = t_1 \cdots t_k \cdots$ such that $\mathbf{m} \xrightarrow{t_1} \mathbf{m}_1 \cdots \mathbf{m}_{k-1} \xrightarrow{t_k} \cdots$. The set of all the occurrence sequences, or *language*, from $\mathbf{m_0}$ is denoted by $\mathrm{L}(\mathcal{N}, \mathbf{m_0})$, and the set of all the reachable markings, or *reachability set*, from $\mathbf{m_0}$, is denoted by $\mathrm{RS}(\mathcal{N}, \mathbf{m_0})$. The reachability relation is conventionally represented by a *reachability graph* $\mathrm{RG}(\mathcal{N}, \mathbf{m_0})$ where the nodes are the reachable markings and there is an arc labeled t from node $\mathbf{m}$ to $\mathbf{m}'$ iff $\mathbf{m} \xrightarrow{t} \mathbf{m}'$.

For pre- and postsets we use the conventional dot notation, e.g., ${}^\bullet t = \{p \in P \mid \mathbf{Pre}[p,t] \neq 0\}$. A transition t such that $|t^\bullet| > 1$ (resp. $|{}^\bullet t| > 1$) is called a *fork* (resp. a *join*). A place p such that $|{}^\bullet p| > 1$ (resp. $|p^\bullet| > 1$) is called an *collector* (resp. a *distributor*). Distributor places are required to model conflicts. The output transitions of a distributor place are said to be in *structural conflict* relation. The *coupled conflict* relation is defined as the transitive closure of the structural conflict relation. The equivalence class (or *coupled conflict set*) of transition t is denoted by $\mathrm{CCS}(t)$ and the quotient set is SCCS. When $\mathbf{Pre}[P,t] = \mathbf{Pre}[P,t'] \neq \mathbf{0}$, t and t' are in *equal conflict* (EQ) relation, meaning that they are both enabled whenever one is. This is also an equivalence relation on the set of transitions. The equivalence class (or *equal conflict set*) of transition t is denoted by $\mathrm{EQS}(t)$ and the quotient set is SEQS.

Weighted (or multiple) arcs permit the abstract modelling of bulk services and arrivals. For instance, they appear naturally when the presence of symmetries allows to "decolour" a high level model. If $\mathbf{Pre}[p, p^\bullet] = w\mathbf{1}$ we say that the weighting is *homogeneous* on p (e.g., the weighting of the input places of an equal conflict set). If this holds for every place, the weighting of the net is homogeneous. A historically and conceptually interesting subclass of P/T nets with homogeneous weighting are *ordinary* nets, where every arc weight is one, which lead to a straightforward but important generalisation of automata models. Although it is possible to simulate/implement weighted P/T systems by ordinary ones preserving the (projected) language (with transformations like the one shown in Figure 12 later in the paper), several reasons justify dealing with weighted P/T systems directly rather than with their ordinary implementations: the models are more concise, the transformations do not preserve concurrent semantics appropriately in general, and the ordinary implementations fall typically out of the subclasses which enjoy strong analytical results, even in the simplest cases.

A subset of places $\Theta \subseteq P$ such that $\Theta^\bullet \subseteq {}^\bullet\Theta$ is called a *trap* because once it becomes marked it remains marked (tokens are "trapped"). A subset of places $\Sigma \subseteq P$ such that ${}^\bullet\Sigma \subseteq \Sigma^\bullet$ is called a ***siphon*** because once it becomes unmarked it remains unmarked (it cannot be "refilled" with tokens).

By reversing arcs or interchanging places and transitions we get the *reverse net*, $\mathcal{N}^r$, or the *dual net*, $\mathcal{N}^d$, of $\mathcal{N}$. Both transformations together lead to the *reverse-dual net*, $\mathcal{N}^{rd}$. Sometimes in net theory relations are established between a net and its reverse, dual, or reverse-dual, e.g., siphons and traps are reverse objects: a siphon of $\mathcal{N}$ is a trap of $\mathcal{N}^r$, etc.

$\mathcal{N}$	$\langle P, T, \mathbf{Pre}, \mathbf{Post}\rangle$	$\mathbf{C}$
$\mathcal{N}^r$	$\langle P, T, \mathbf{Post}, \mathbf{Pre}\rangle$	$-\mathbf{C}$
$\mathcal{N}^d$	$\langle T, P, \mathbf{Post}^\perp, \mathbf{Pre}^\perp\rangle$	$-\mathbf{C}^\perp$
$\mathcal{N}^{rd}$	$\langle T, P, \mathbf{Pre}^\perp, \mathbf{Post}^\perp\rangle$	$\mathbf{C}^\perp$

A net $\mathcal{N}'$ is *subnet* of $\mathcal{N}$ (written $\mathcal{N}' \subseteq \mathcal{N}$) when $P' \subseteq P$, $T' \subseteq T$ and its pre- and post-incidence matrices are $\mathbf{Pre}' = \mathbf{Pre}[P', T']$ and $\mathbf{Post}' = \mathbf{Post}[P', T']$. (In what follows, for the sake of readability, whenever a net or system is defined it "inherits" the definition of all the characteristic sets, functions, parameters,... with names conveniently marked.) Subnets are generated by subsets of places *and* transitions. When a subnet is generated by a subset V of nodes of a single kind, it is assumed that it is generated by $V \cup {}^\bullet V \cup V^\bullet$. Subnets generated by a subset of places (transitions) are called P- (T-) subnets.

2.3 Analysis of Logical Properties

A major goal of the mathematical modelling of systems is to allow their automatic analysis. In general, verification consists in checking that a system model satisfies its logic specification (e.g., some temporal logic formulae). Here we are interested in the verification of some selected properties of "good behaviour" that are often part of the specification of systems (specially reactive ones), or appear as precondition for the temporal analysis or performance evaluation of the interpreted model.

A P/T system is *bounded* when every place is bounded, i.e., its token content is less than some bound at every reachable marking (when the bound is one, then it is said to be *safe*). It is *live* when every transition is live, i.e., it can ultimately occur from every reachable marking, and it is *deadlock-free* when every reachable marking enables some transition. A marking is a *home state* when it is reachable from every reachable marking, and a net system is *reversible* when the initial marking (hence every marking) is a home state. Two places are in *mutual exclusion* when they are never marked simultaneously. Boundedness precludes overflows, liveness ensures that no single action in the system can become unattainable, existence of home states informs on the possibility to return to certain states, and mutual exclusion is required between places that represent the use of a common resource or the presence in a critical section.

Conventionally, analysis methods of PN models are classified as follows:

- Enumeration Techniques: If the system is bounded, the reachability graph can be used as the computational model for a proof system or for decision procedures and tools for automatic verification. Two major problems of this approach are the size of the state space of a concurrent system, that can be palliated in some cases where not every state needs to be computed, and the necessity to repeat the analysis for each initial marking of interest. Unbounded systems can be partially analysed using a similar approach.
- Transformation Techniques: To facilitate the analysis of a large and complex system it can be transformed (typically reduced) preserving the properties to be analysed. Transformation rules somehow preserve the behaviour while they are often supported by structural arguments as simple, and efficient, sufficient conditions.
- Structural Techniques: The basic idea is to obtain useful information about the behaviour reasoning on the structure of the net and the initial marking. Two crucial advantages of this approach are the *deep understanding* of the system behaviour that is gained, and the possible *efficiency* of the algorithms. Two intimately related families of techniques have extensively been used: graph theory and linear algebra/convex geometry. The rest of the paper is devoted to describe these techniques in some detail from a linear algebraic viewpoint (over the non-negative integers or reals).

The above groups of techniques are not to be understood as mutually exclusive, but they should be effectively combined for the analysis in practice.

General net systems are difficult to analyse. As in all theories, it is a common trend in net theory to consider particular *subclasses* of models by introducing appropriate restrictions, either on the behaviour or the structure (or syntax) of the model. A possible way of obtaining syntactical subclasses is restricting the inscriptions (e.g., nets with every weight equal to one are ordinary) or the topology, usually aiming at limiting the interplay between conflicts and synchronisations. The latter can be achieved either by giving a general restriction, typically on distributor places and/or join transitions (e.g., there are no distributors), or by giving rules to construct models (e.g., sequential functional entities are synchronised by some restricted message passing). These restrictions are intended to facilitate the analysis (we shall give some examples through the paper) at the price of losing some modelling capabilities. The designer must find a compromise between modelling power and availability of powerful analysis tools, while one of the theoretician's goals is obtaining better results for increasingly larger — and more practical — subclasses.

3 Linear Descriptions and Structural Objects

The starting point for the structural analysis of P/T systems by linear algebraic techniques is the description of the state space by some system of linear equations. In fact, as we shall see, these descriptions are often *relaxations* with a different degree of accuracy. In principle, there is a trade-off between the accuracy of the description and the efficiency of verification algorithms, but this is not necessarily the case: in some net subclasses, relaxations that — in general — are less accurate than others describe exactly the state space; it is also worth noticing that some linear descriptions that are extensively used are less accurate *and* less efficient than others (although they may have salient merits for the understanding).

3.1 The State Equation

Recall that when a transition t is enabled at $\mathbf{m}$ ($\mathbf{m} \geq \mathbf{Pre}[P,t]$), the new marking reached by its firing ($\mathbf{m} \xrightarrow{t} \mathbf{m}'$) is $\mathbf{m}' = \mathbf{m} + \mathbf{C}[P,t]$. Analogously, a *step* $\mathbf{s}$ (where $\mathbf{s}[t]$ is the number of times t occurs in that step) is enabled when $\mathbf{m} \geq \mathbf{Pre} \cdot \mathbf{s}$, and its firing leads to:

$$\mathbf{m}' = \mathbf{m} + \mathbf{C} \cdot \mathbf{s} \tag{1}$$

This equation resembles the state equation of a *discrete-time linear system*, where $\mathbf{m}$ is the *current state (vector)*, $\mathbf{m}'$ is the *next state*, and s is the *inputs vector* (there is one input per transition). Differently from general linear systems, here the *dynamic matrix* is identity (all the eigenvalues are one, what corresponds to integrators or counters: the state is memorised in the absence of inputs) and not every action is possible in a given state because inputs and state variables are defined to be non-negative integers.

Assume that it is known that every marking $\mathbf{m}$ is reachable in (at most) k steps from $\mathbf{m_0}$ (this is true for bounded systems, even it is possible to structurally compute a finite such k when the net is *structurally bounded*). Then we could describe *exactly* the set of reachable markings by the following system of linear inequalities ($\mathbf{m}_i \in \mathbb{N}^{|P|}, \mathbf{s}_i \in \mathbb{N}^{|T|}$):

$$\begin{aligned} \mathbf{m_0} &\geq \mathbf{Pre} \cdot \mathbf{s}_0 \\ \mathbf{m}_1 &= \mathbf{m_0} + \mathbf{C} \cdot \mathbf{s}_0 \\ &\dots \\ \mathbf{m}_{k-1} &\geq \mathbf{Pre} \cdot \mathbf{s}_{k-1} \\ \mathbf{m} &= \mathbf{m}_{k-1} + \mathbf{C} \cdot \mathbf{s}_{k-1} \end{aligned} \tag{2}$$

Although we can easily eliminate the $\mathbf{m}_i$ variables by substitution, this linear description is of course highly impractical due to the size of k, that moreover depends on the initial marking. In what follows, we look for more concise linear descriptions — although we expect that they are not exact descriptions but only more or less accurate approximations.

If we integrate Equation (1) over a sequence σ of inputs (transitions or steps) from the initial state $\mathbf{m_0}$ and yielding $\mathbf{m}$, denoting by $\boldsymbol{\sigma}$ the *firing count vector* of sequence σ ($\boldsymbol{\sigma}[t] = \#(t,\sigma)$ is the number of times t occurs in the sequence) we obtain:

$$\mathbf{m} = \mathbf{m_0} + \mathbf{C} \cdot \boldsymbol{\sigma} \tag{3}$$

Since every reachable marking is obtained by the occurrence of some sequence from the initial marking, it is clear that, for every reachable marking (state) there exists some $\boldsymbol{\sigma} \in \mathbb{N}^{|T|}$ such that Equation (3) holds. This is why (3) is referred to as the ***net state equation***.

It is straightforward to derive the following linear description of the set of reachable markings (notice that integrality of $\boldsymbol{\sigma}$ ensures integrality of $\mathbf{m}$):

Definition 1. Let $\mathcal{S}$ be a P/T system. Its *linearised reachability set (using the state equation)* is defined as:

$$\mathrm{LRS}^{\mathrm{SE}}(\mathcal{S}) = \{\mathbf{m} \in \mathbb{N}^{|P|} \mid \exists \boldsymbol{\sigma} \in \mathbb{N}^{|T|} \text{ such that } \mathbf{m} = \mathbf{m}_0 + \mathbf{C} \cdot \boldsymbol{\sigma}\}$$

This description is suitable for the incorporation of the state equation into a set of linear constraints, e.g., in the restrictions of an integer programming problem, for analysis purposes (see Section 4).

Remark 2. Although integer programming is NP-complete, some analysis problems can be solved efficiently using the state equation over the integers. For instance, a sufficient condition for non reachability of a given marking $\mathbf{m}$ in $\mathcal{S}$ is non existence of $\boldsymbol{\sigma}' \in \mathbb{Z}^{|T|}$ such that $\mathbf{C} \cdot \boldsymbol{\sigma}' = \mathbf{m} - \mathbf{m}_0$, which is polynomial time [45,76]. Actually, if $\mathcal{N}$ is *consistent*, i.e., an $\mathbf{x} > \mathbf{0}$ exists such that $\mathbf{C} \cdot \mathbf{x} = \mathbf{0}$, this is equivalent to $\mathbf{m} \notin \mathrm{LRS}^{\mathrm{SE}}(\mathcal{S})$, because from a $\boldsymbol{\sigma}' \in \mathbb{Z}^{|T|}$, a $\boldsymbol{\sigma} \in \mathbb{N}^{|T|}$ can be obtained as $\boldsymbol{\sigma} = \boldsymbol{\sigma}' + k\mathbf{x}$. □

The inclusion $\mathrm{RS}(\mathcal{S}) \subseteq \mathrm{LRS}^{\mathrm{SE}}(\mathcal{S})$ may well be proper, since Equation (3) does not check whether there is a sequence of intermediate markings such that some $\sigma \in \mathrm{L}(\mathcal{S})$ with firing count vector $\boldsymbol{\sigma}$ is actually fireable. (In other words, we have removed from (2) the inequalities requiring the fireability of the steps, i.e., the non-negativity of the intermediate $\mathbf{m}_i$ variables, and then we have eliminated these variables.) The markings in $\mathrm{LRS}^{\mathrm{SE}}(\mathcal{S}) - \mathrm{RS}(\mathcal{S})$ will be called *spurious (with respect to the state equation)*.

Similarly to the reachability graph, we can represent a *linearised reachability graph (using the state equation)*, $\mathrm{LRG}^{\mathrm{SE}}(\mathcal{S})$, where the nodes are the markings in $\mathrm{LRS}^{\mathrm{SE}}(\mathcal{S})$ and there is na arc labeled t from node $\mathbf{m}$ to $\mathbf{m}'$ iff $\mathbf{m} \xrightarrow{t} \mathbf{m}'$. Figure 1 shows a P/T system together with its $\mathrm{LRG}^{\mathrm{SE}}$, where the spurious markings are shaded. As another example, the marking p_2 is spurious in the system shown in Figure 2 top-left; it is reached by the "occurrence" of $t_1 + t_2 + t_3$ (notice that in every possible "sequence" with this firing count vector some intermediate marking variable becomes negative).

We can further relax the description by dropping integrality constraints, as it is typical in the mathematical modelling of systems with large state spaces (e.g., population models). This further relaxation introduces more spurious solutions. For instance, the marking $2p_2$ is spurious in the system shown in Figure 2 bottom-left; it is reached by the "occurrence" of $0.5t_1 + t_2 + t_3$. On the other hand, this relaxation allows to use *linear* programming instead of *integer* programming in the verification, leading to polynomial time algorithms. Sometimes

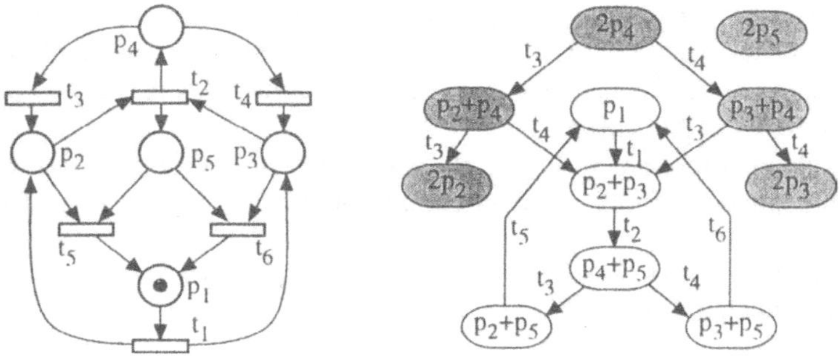

Fig. 1. A P/T system and its LRG^{SE}

only the integrality of firing variables is disregarded, either because other restrictions guarantee the integrality of the marking or because mixed integer linear programming is used. As an example, the following linear programming problem can be used to analyse reachability of a given marking $\mathbf{m}$ (if the problem is infeasible, then $\mathbf{m}$ is proven unreachable):

$$\max\{\mathbf{0} \cdot \boldsymbol{\sigma} \mid \mathbf{C} \cdot \boldsymbol{\sigma} = \mathbf{m} - \mathbf{m}_0 \wedge \boldsymbol{\sigma} \geq \mathbf{0}\} \tag{4}$$

In summary, we define a second linear description of the set of reachable markings:

Definition 3. Let $\mathcal{S}$ be a P/T system. Its *linearised reachability set (using the state equation over the reals)* is defined as:

$$\text{LRS}^{\text{SE}\mathbb{R}}(\mathcal{S}) = \{\mathbf{m} \in \mathbb{N}^{|P|} \mid \exists \boldsymbol{\sigma} \geq \mathbf{0} \text{ such that } \mathbf{m} = \mathbf{m}_0 + \mathbf{C} \cdot \boldsymbol{\sigma}\}$$

Remark 4. We have relaxed the description of the state space applying two principles: ***path integration*** and ***continuisation*** (or ***fluidisation***). These principles can also be applied in the reverse order, first continuisation and then path integration, leading again to $\text{LRS}^{\text{SE}\mathbb{R}}$. By disregarding first the integrality of variables, we get *continuous P/T net systems* [27]. In these models, "fluid tokens" are contained in "deposits" (the places), the "level" of which (the marking) captures the state of the system (as in *Forrester diagrams* [26]). Transitions are regarded as "mixing valves" whose firing (opening) consumes fluid from the input places and produces fluid onto the output places in a given proportion, according to the following firing rule: t is enabled in some amount $\lambda > 0$ at marking $\mathbf{m}$ when $\mathbf{m} \geq \lambda \mathbf{Pre}[P, t]$, and its occurrence leads to the marking $\mathbf{m}' = \mathbf{m} + \lambda \mathbf{C}[P, t]$. The set of reachable markings of system $\mathcal{S}$ is denoted by $\text{CRS}(\mathcal{S})$, standing for *continuous reachability set.* These nets are interesting in the modelling of certain continuous systems, and also as an approximation of systems where there are large amounts of (discrete) tokens. When used as an approximation, they naturally suffer from the presence of spurious solutions. For instance, the marking $2p_2$ is spurious in the system shown in Figure 2 top-right; it is reached by the "occurrence" of $0.5t_2$. □

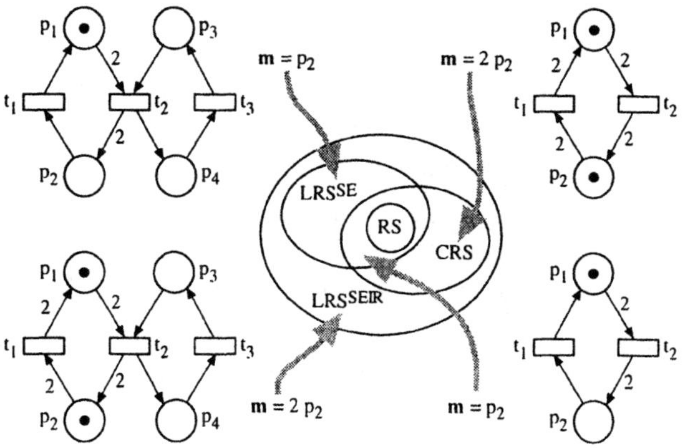

Fig. 2. Diverse relaxations of the state space applying path integration and continuisation.

Figure 2 illustrates the relaxations we have discussed so far. Notice that the spurious markings induced by the use of LRS^{SE} or CRS are not necessarily disjoint: the marking p_2 is spurious in the system shown in Figure 2 bottom-right; it is reached by the "occurrence" of either $t_1 + t_2$ or $0.5t_2$.

Clearly, the presence of spurious solutions may prevent reaching conclusions using analysis techniques based on a relaxed description of the state space. We shall give later techniques to remove spurious solutions by adding more information to the state equation after carefully considering the net structure. It is also worth noticing that some net subclasses enjoy special properties on their spurious solutions that palliate the problem, as we shall illustrate (e.g., see Subsection 3.4)

3.2 Linear Invariants from the State Equation

Since every reachable marking must satisfy the state equation, it can be regarded as a set of *linear invariant laws*, one per place, containing marking and firing variables. The firing variables can be eliminated, by multiplying the equation by a suitable vector, in order to obtain linear invariant laws involving marking variables only. For instance, if $\mathbf{y}$ is such that $\mathbf{y} \cdot \mathbf{C} = \mathbf{0}$ — vector $\mathbf{y}$ is called a *P-flow* — then, *for every initial marking* $\mathbf{m}_0$, every reachable marking $\mathbf{m}$ satisfies:

$$\mathbf{y} \cdot \mathbf{m} = \mathbf{y} \cdot \mathbf{m}_0 + \mathbf{y} \cdot \mathbf{C} \cdot \boldsymbol{\sigma} = \mathbf{y} \cdot \mathbf{m}_0 = k$$

This provides a "token balance law": if the positive and negative parts of $\mathbf{y}$ are separated: $\mathbf{y} = \mathbf{y}_+ - \mathbf{y}_-$, where $\mathbf{y}_+, \mathbf{y}_- \geq \mathbf{0}$, then for every reachable marking $\mathbf{y}_+ \cdot \mathbf{m} = \mathbf{y}_- \cdot \mathbf{m} + k$, that is, the tokens in $\|\mathbf{y}_+\|$ and $\|\mathbf{y}_-\|$ are somehow "balanced". Conversely, in net systems where all transitions can fire at least once *every* linear token conservation law is associated with a P-flow. This can be deduced as follows: Assume $\mathbf{y} \cdot \mathbf{m} = \mathbf{y} \cdot \mathbf{m_0}$ holds for every reachable marking. It must be shown that $\mathbf{y} \cdot \mathbf{C} = \mathbf{0}$. Consider an arbitrary t and $\mathbf{m}_t \xrightarrow{t} \mathbf{m}'_t$. Then $\mathbf{m}'_t = \mathbf{m}_t + \mathbf{C}[P,t]$, hence $\mathbf{y} \cdot \mathbf{m}'_t = \mathbf{y} \cdot \mathbf{m}_t + \mathbf{y} \cdot \mathbf{C}[P,t]$. Since $\mathbf{y} \cdot \mathbf{m}'_t = \mathbf{y} \cdot \mathbf{m}_t$, it follows that $\mathbf{y} \cdot \mathbf{C}[P,t] = 0$. In summary:

Theorem 5. *Let $\mathcal{N}$ be a P/T net.*

1. *If $\mathbf{y} \cdot \mathbf{C} = \mathbf{0}$ then for every $\mathbf{m_0}$: $\mathbf{y} \cdot \mathbf{m} = \mathbf{y} \cdot \mathbf{m_0}$ for every $\mathbf{m} \in \mathrm{RS}(\mathcal{N}, \mathbf{m_0})$.*
2. *Let $\mathcal{S} = \langle \mathcal{N}, \mathbf{m_0} \rangle$. Assume that for every $t \in T$ some $\mathbf{m}_t \in \mathrm{RS}(\mathcal{S})$ exists such that $\mathbf{m}_t \geq \mathbf{Pre}[P,t]$. If $\mathbf{y} \cdot \mathbf{m} = \mathbf{y} \cdot \mathbf{m_0}$ for every $\mathbf{m} \in \mathrm{RS}(\mathcal{S})$, then $\mathbf{y} \cdot \mathbf{C} = \mathbf{0}$.*

The P-flows of a net $\mathcal{N}$ form a vector space. Using $\mathbf{B}$, a matrix whose rows form a basis of P-flows, we obtain a new linear description of the set of reachable markings where only marking variables appear:

Definition 6. Let $\mathcal{S}$ be a P/T system. Its *linearised reachability set (using a basis* $\mathbf{B}$ *of P-flows)* is defined as:

$$\mathrm{LRS}^{\mathrm{Pf}}(\mathcal{S}) = \{\mathbf{m} \in \mathbb{N}^{|P|} \mid \mathbf{B} \cdot \mathbf{m} = \mathbf{B} \cdot \mathbf{m_0}\}$$

Token balance laws become specially useful when $\mathbf{y} \geq \mathbf{0}$ — in such case, $\mathbf{y}$ is called a *P-semiflow* — because, taking into account that $\mathbf{m} \geq \mathbf{0}$, from $\mathbf{y} \cdot \mathbf{m} = k$ we can deduce, for instance, that all the places in $\|\mathbf{y}\|$ are bounded. The invariants that we obtain are "token conservation laws": for every reachable marking the weighted sum of tokens in $\|\mathbf{y}\|$ remains constant.

Besides the actual invariant law, a major interest of P-semiflows is the *decomposed view* of the model that they provide. The P-subnet generated by the support of a P-semiflow is called a *conservative component* of the net, meaning that it is a part of the net that conserves its weighted token content. In the case that $\mathbf{y} > \mathbf{0}$ such that $\mathbf{y} \cdot \mathbf{C} = \mathbf{0}$ exists, the whole net is a conservative component, and it is said that the net is *conservative*, what obviously implies that it is bounded for every (finite) initial marking.

The token conservation laws induced by P-semiflows are by far the most popular invariant laws, to the point that the classical invariant method considers P-semiflows only, which, historically, are very often called *P-invariants* in the literature. Anyhow, it is important to realise that there are three notions that should be differentiated:

- The P-semiflow (a vector).
- The token conservation law or marking invariant (an equation).
- The conservative component (a net).

Actually, apart from those derived from P-semiflows, there are other invariant laws and components, as it shall be shown.

A P-semiflow $\mathbf{y}$ is said to be *minimal* when the positive y_i are relatively prime and no P-semiflow $\mathbf{y}'$ exists such that $\|\mathbf{y}'\| \subset \|\mathbf{y}\|$. In order to prove properties, only minimal P-semiflows need to be considered because every P-semiflow can be obtained as a non-negative linear combination — possibly with rational coefficients — of minimal P-semiflows. In a net $\mathcal{N}$, the set of all the minimal P-semiflows, called the *fundamental set of P-semiflows*, is unique.

Remark 7. The supports of two minimal P-semiflows are non comparable, since otherwise we would be able to obtain by difference another P-semiflow whose support would be contained in one of the former. Therefore, a bound for the cardinality of the fundamental set of P-semiflows — which is reached in some cases although it is generally quite high — is the number of possible combinations of $\lceil |P|/2 \rceil$ out of $|P|$ elements:

$$\binom{|P|}{\lceil |P|/2 \rceil}$$

□

Algorithm 8 gives a simple procedure to compute the fundamental set of P-semiflows from the incidence matrix of the net. A row $\mathbf{\Phi}[i]$ memorises the coefficients of the positive linear combination of rows of matrix $\mathbf{C}$ which generate $\mathbf{A}[i]$. In Step 3 of the algorithm, all the rows of $\mathbf{A}$ have been made null, so each row $\mathbf{\Phi}[i]$ is a P-semiflow: $\mathbf{\Phi}[i] \cdot \mathbf{C} = \mathbf{0}$.

Algorithm 8 (Computation of P-semiflows)

Input - The incidence matrix $\mathbf{C}$.
Output - A matrix $\mathbf{\Phi}$ whose rows are the fundamental set of P-semiflow.

1. Let $\mathbf{A} = \mathbf{C}$ and $\mathbf{\Phi} = \mathbf{I}$ { $\mathbf{I}$ is the identity matrix of dimension $|P|$ }
2. **for** $i = 1$ **to** $|T|$ **do**
   2.1 Add to the matrix $[\mathbf{\Phi}|\mathbf{A}]$ all rows which are natural linear combinations of pairs of rows of $[\mathbf{\Phi}|\mathbf{A}]$ and which annul the i-th column of $\mathbf{A}$
   2.2 Eliminate from $[\mathbf{\Phi}|\mathbf{A}]$ the rows in which the i-th column of $\mathbf{A}$ is non-null
3. Remove from $\mathbf{\Phi}$ all rows whose support is not minimal, and divide each other by the g.c.d. of its non-null elements

During execution of Algorithm 8, the number of rows in $[\mathbf{\Phi}|\mathbf{A}]$ typically grows beyond the cardinality of the fundamental set of P-semiflows. To improve the efficiency of the algorithm, instead of annulling the columns of $\mathbf{A}$ in their order, some heuristics for the selection of the column to annul next drastically reduce the growth of $[\mathbf{\Phi}|\mathbf{A}]$. Moreover, introducing some tests of non minimality during the execution of the algorithm — including the application of certain rank properties, combining net theory with linear algebraic techniques — it is also possible to discard many rows that cannot lead to a minimal P-semiflow before completing the computations (see [57,22]).

The fundamental set of P-semiflows provides another linear description of the set of reachable markings where only marking variables appear. Let $\mathbf{\Phi}$ be a matrix whose rows are the fundamental set of P-semiflows of $\mathcal{N}$:

Definition 9. Let $\mathcal{S}$ be a P/T system. Its *linearised reachability set (using the fundamental set of P-semiflows)* is defined as:

$$\mathrm{LRS}^{\mathrm{Psf}}(\mathcal{S}) = \{\mathbf{m} \in \mathbb{N}^{|P|} \mid \boldsymbol{\Phi} \cdot \mathbf{m} = \boldsymbol{\Phi} \cdot \mathbf{m}_0\}$$

The dual notion of P-flows are *T-flows* (in the sense that the P-flows of $\mathcal{N}^r$ or $\mathcal{N}^{rd}$ are the T-flows of $\mathcal{N}$). If $\mathbf{x}$ is such that $\mathbf{C} \cdot \mathbf{x} = \mathbf{0}$:

$$\mathbf{m} = \mathbf{m}_0 + \mathbf{C} \cdot \mathbf{x} = \mathbf{m}_0$$

T-flows become specially useful when $\mathbf{x} \geq \mathbf{0}$ — in such case, $\mathbf{x}$ is called a *T-semiflow* — because, in that case they correspond to cyclic sequences. (Note that the firing count vector of a cyclic sequence is a T-semiflow, but possibly for a given initial marking it is *not* possible to fire a sequence whose firing count vector is a given T-semiflow.) The fundamental set of T-semiflows can be readily computed applying Algorithm 8 to $\mathbf{C}^{\perp}$. Similarly to P-semiflows, T-semiflows provide an interesting decomposed view of the model. The T-subnet generated by the support of a T-semiflow is called a *consistent component* of the net. With an appropriate initial marking, a consistent component is able to exhibit a cyclic or repetitive behaviour. In the case that $\mathbf{x} > \mathbf{0}$ such that $\mathbf{C} \cdot \mathbf{x} = \mathbf{0}$ exists, the whole net is a consistent component, and it is said that the net is *consistent.* When a net is not consistent it cannot be lively and boundedly marked (see Proposition 10 and Theorem 45). In fact, historically the name consistent is due to the fact that in a live and bounded system, the equation system $\mathbf{C} \cdot \mathbf{x} = \mathbf{0}$ must be (algebraically) consistent.

Generalising flows and semiflows, other multipliers of $\mathbf{C}$ may provide useful information. For instance, a vector $\mathbf{y} \geq \mathbf{0}$ such that $\mathbf{y} \cdot \mathbf{C} \leq \mathbf{0}$ indicates that the weighted token content of the places in $\|\mathbf{y}\|$ cannot be increased (it can be decreased if $\mathbf{y} \cdot \mathbf{C} \lneq \mathbf{0}$). If $\mathbf{y} > \mathbf{0}$ it follows that the net is bounded for whichever initial marking, or *structurally bounded.* Similarly, a vector $\mathbf{x} \geq \mathbf{0}$ such that $\mathbf{C} \cdot \mathbf{x} \geq \mathbf{0}$ indicates that the occurrence of a sequence with firing count vector $\mathbf{x}$ (if some such sequence was fireable) would not decrease the marking, so it could be repeated once and again (the marking would be increased at each execution if $\mathbf{C} \cdot \mathbf{x} \gneq \mathbf{0}$). If $\mathbf{x} > \mathbf{0}$ the net is said to be *structurally repetitive.* A net that can be lively marked is structurally repetitive, because there must be sequences leading from a marking to a greater or equal one (equal in case of boundedness) involving every transition, the firing count vector of which proves structural repetitiveness:

Proposition 10. *If $\mathcal{S}$ is a live P/T system then $\mathcal{N}$ is structurally repetitive. If $\mathcal{S}$ is also bounded then $\mathcal{N}$ is consistent.*

3.3 Proving Properties Through Linear Invariants

Let us illustrate with an example the usability of (minimal) P-semiflows to prove properties, and the decomposed view that they provide.

A production cell and a P/T description of its local controller are shown in Figure 3 (taken from [82]). The places "wait_raw", "load", "op$_1$", "wait_dep.",

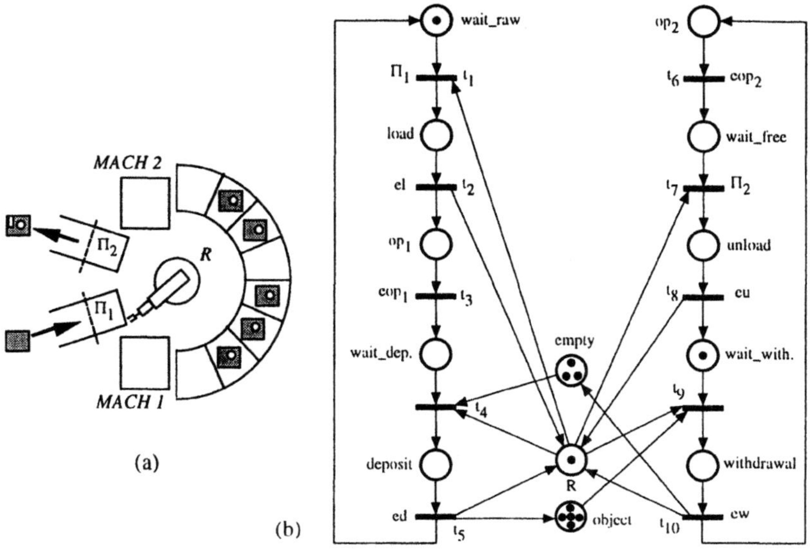

Fig. 3. A production cell with two machines, one robot, and a store, and a P/T description of its behaviour.

and "deposit" represent the possible states of *MACH 1*. The place "R" is marked when the robot is available. The places "empty" and "object" contain as many tokens as empty slots or parts are available in the temporary buffer, etc. In this model actions are associated with places, and transitions represent atomic instantaneous changes of state, e.g., *MACH 2* performs its operations while place "op_2" is marked, and the event of finishing is modelled by the firing of "eop_2".

The marking linear invariants induced by the minimal P-semiflows of the net in Figure 3 are the following:

$$\mathbf{m}[\text{wait\_raw}] + \mathbf{m}[\text{load}] + \mathbf{m}[\text{op}_1] + \mathbf{m}[\text{wait\_dep.}] + \mathbf{m}[\text{deposit}] = 1 \quad (5)$$

$$\mathbf{m}[\text{op}_2] + \mathbf{m}[\text{wait\_free}] + \mathbf{m}[\text{unload}] + \mathbf{m}[\text{wait\_with.}] + \mathbf{m}[\text{withdrawal}] = 1 \quad (6)$$

$$\mathbf{m}[\text{empty}] + \mathbf{m}[\text{deposit}] + \mathbf{m}[\text{object}] + \mathbf{m}[\text{withdrawal}] = 8 \quad (7)$$

$$\mathbf{m}[\text{R}] + \mathbf{m}[\text{load}] + \mathbf{m}[\text{unload}] + \mathbf{m}[\text{deposit}] + \mathbf{m}[\text{withdrawal}] = 1 \quad (8)$$

Since markings are non-negative, the following can be easily stated from the previous equations:

- The marking bound of every place is one, except for "empty" and "object", that is seven.
- The places in each of the following sets are in (pairwise) marking mutual exclusion:
  - {wait_raw, load, op_1, wait_dep., deposit}
  - {op_2, wait_free, unload, wait_with., withdrawal}
  - {R, load, unload, deposit, withdrawal}

Using the invariants in (5–8), it is also possible to prove that the net system in Figure 3 is deadlock-free. We proceed by contradiction, more precisely we try to construct a marking $\mathbf{m}$ that satisfies (5–8) and in which no transition is fireable. In such a marking, the places "load", "op$_1$", "deposit", "op$_2$", "unload", and "withdrawal" should be unmarked, because these are the only input places of their corresponding transitions, so the token conservation laws in (5–8) reduce to:

$$\mathbf{m}[\text{wait\_raw}] + \mathbf{m}[\text{wait\_dep.}] = 1 \tag{9}$$
$$\mathbf{m}[\text{wait\_free}] + \mathbf{m}[\text{wait\_with.}] = 1 \tag{10}$$
$$\mathbf{m}[\text{empty}] + \mathbf{m}[\text{object}] = 8 \tag{11}$$
$$\mathbf{m}[\text{R}] = 1 \tag{12}$$

Since the above implies that "R" should be marked, to prevent the firing of t_1 and t_7, the places "wait_raw" and "wait_free" should be unmarked too. The token conservation laws are reduced once more, leading to:

$$\mathbf{m}[\text{wait\_dep.}] = 1 \tag{13}$$
$$\mathbf{m}[\text{wait\_with.}] = 1 \tag{14}$$
$$\mathbf{m}[\text{empty}] + \mathbf{m}[\text{object}] = 8 \tag{15}$$
$$\mathbf{m}[\text{R}] = 1 \tag{16}$$

Since the above implies that "wait dep." and "wait with." should be marked, to prevent the firing of t_4 and t_9, both "empty" and "object" should be unmarked, against (15), so the net system is proven deadlock-free. The above "ad hoc" proof is generalised and fully automatised in Subsection 4.6.

As an example of the loss of information when non minimal P-semiflows are used instead of minimal ones, observe that summing up (5–8) we obtain a P-invariant involving all the places which does not allow to prove any of the properties we have deduced from the minimal ones (it allows to prove 10-boundedness of the net, though).

In the example of Figure 3, the only minimal T-semiflow is $\mathbf{1}$, meaning that every cyclic sequence fires all the transitions in the same proportion, so, due to boundedness, in the "long run" all the transitions occur the same number of times per time unit. Therefore, under boundedness, the existence of a unique minimal T-semiflow ensures that deadlock-freeness implies liveness, because every infinite behaviour must contain all the transitions, so from our previous proof of deadlock-freeness we deduce liveness.

Finally, in Figure 4, the decomposed view induced by the minimal P-semiflows is graphically presented. This view is even useful to derive an *implementation*. For instance, it shows that the net system in Figure 3 could be made up with two sequential processes (for *MACH 1* and *MACH 2*) and three semaphores: "object", "empty", and "R" — where "R" is a mutual exclusion semaphore. (Having a unique minimal T-semiflow, no T-decomposition exists in this case.)

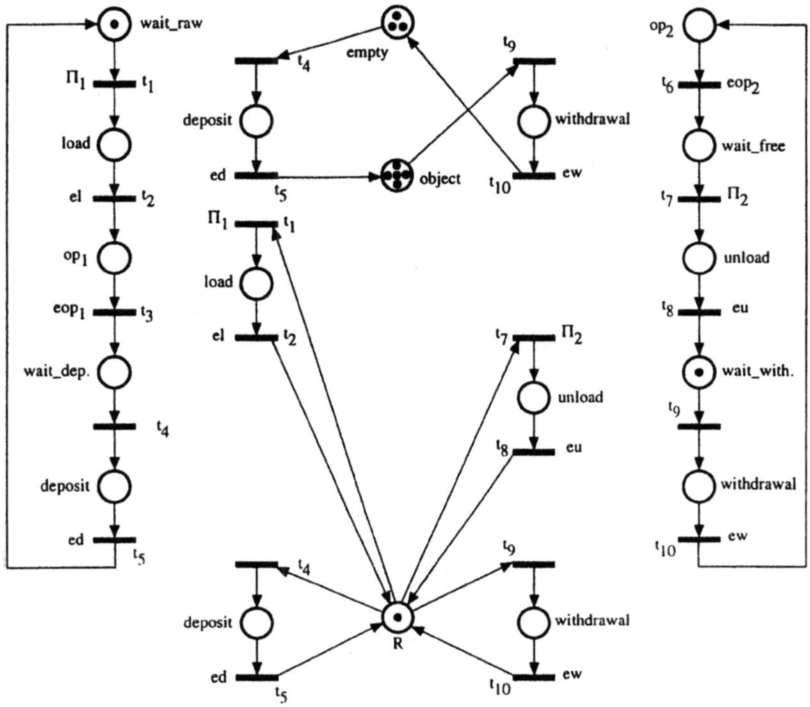

Fig. 4. A decomposed view of the net system in Figure 3.

3.4 Comparison of Linear Descriptions

We announced that the diverse linear descriptions have different degrees of accuracy. With respect to those presented so far, it can easily be established that:

Theorem 11. *Let $\mathcal{S}$ be a P/T system.*

$$\mathrm{RS}(\mathcal{S}) \subseteq \mathrm{LRS}^{\mathrm{SE}}(\mathcal{S}) \subseteq \mathrm{LRS}^{\mathrm{SE}\mathbb{R}}(\mathcal{S}) \subseteq \mathrm{LRS}^{\mathrm{Pf}}(\mathcal{S}) \subseteq \mathrm{LRS}^{\mathrm{Psf}}(\mathcal{S})$$

All the above inclusions may well be proper. We have already shown some examples for the first and second in Subsection 3.1. The P/T systems in Figure 5 give examples for the others.

The vector $p_1 + p_2 + p_5$ forms a basis of P-flows of the net of Figure 5 (a), so every marking $\alpha p_3 + p_5$, with $\alpha \geq 0$, is in $\mathrm{LRS}^{\mathrm{Pf}}(\mathcal{S})$. On the other hand, since $\mathbf{C}[p_4, T] > \mathbf{C}[p_5, T]$ and $\mathbf{m_0}[p_4] = \mathbf{m_0}[p_5]$, $\mathbf{m}[p_4] \geq \mathbf{m}[p_5]$ in every $\mathbf{m} \in \mathrm{LRS}^{\mathrm{SE}\mathbb{R}}(\mathcal{S})$, what is false in the case of $\alpha p_3 + p_5$.

For the net of Figure 5 (b), $p_1 + p_2$ is the only minimal P-semiflow, so every marking $p_1 + \alpha p_3 + \beta p_4$, with $\alpha, \beta \geq 0$, is in $\mathrm{LRS}^{\mathrm{Psf}}(\mathcal{S})$. Nevertheless, the dimension of the space of P-flows is two; the P-flow $p_3 - p_4$, that together with the minimal P-semiflow forms a basis, shows that markings $p_1 + \alpha p_3 + \beta p_4$ with $\alpha \neq \beta$ are not in $\mathrm{LRS}^{\mathrm{Pf}}(\mathcal{S})$.

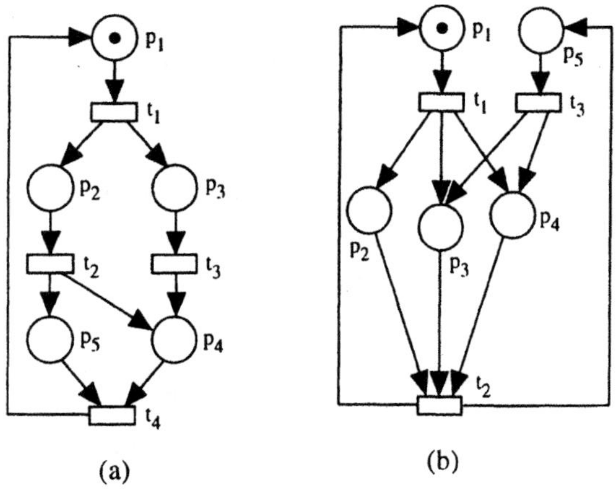

Fig. 5. Two P/T systems showing that $\mathrm{LRS}^{\mathrm{SE}\mathbb{R}} \overset{(a)}{\subsetneq} \mathrm{LRS}^{\mathrm{Pf}} \overset{(b)}{\subsetneq} \mathrm{LRS}^{\mathrm{Psf}}$.

In some cases, the above inclusions reduce to equalities. The interest of the following result is that it allows to use only the "more convenient" P-semiflows, instead of general P-flows, in the case of conservativeness, and that it guarantees that P-flows are as accurate as the state equation over the reals in the case of consistency:

Proposition 12. *Let $\mathcal{S}$ be a P/T system.*

1. *If $\mathcal{N}$ is conservative, then* $\mathrm{LRS}^{\mathrm{Pf}}(\mathcal{S}) = \mathrm{LRS}^{\mathrm{Psf}}(\mathcal{S})$.
2. *If $\mathcal{N}$ is consistent, then* $\mathrm{LRS}^{\mathrm{SE}\mathbb{R}}(\mathcal{S}) = \mathrm{LRS}^{\mathrm{Pf}}(\mathcal{S})$.

Proof. For Part 1, consider a $\mathbf{y} > \mathbf{0}$ such that $\mathbf{y} \cdot \mathbf{C} = \mathbf{0}$. A basis formed by P-semiflows only can easily be obtained from a basis of P-flows by adding a multiple of $\mathbf{y}$ to each P-flow.

For Part 2, if $\mathbf{m} \in \mathrm{LRS}^{\mathrm{Pf}}(\mathcal{S})$, then $\boldsymbol{\sigma}'$ (possibly $\boldsymbol{\sigma}' \not\geq \mathbf{0}$) exists such that $\mathbf{C} \cdot \boldsymbol{\sigma}' = \mathbf{m} - \mathbf{m_0}$. Using a $\mathbf{x} > \mathbf{0}$ such that $\mathbf{C} \cdot \mathbf{x} = \mathbf{0}$, a $\boldsymbol{\sigma} = \boldsymbol{\sigma}' + k\mathbf{x} \geq \mathbf{0}$ can be obtained so that the state equation is satisfied. □

In some net subclasses, stronger relations have been found. For instance, in live *state machines* (ordinary nets where every transition has one input and one output place; they are always conservative, and the only minimal P-semiflow is $\mathbf{1}$; liveness is equivalent to strong connectedness and non empty marking) $\mathrm{RS} = \mathrm{LRS}^{\mathrm{Psf}}$. In live *marked graphs* (ordinary nets where every place has one input and one output place; they are always consistent, and the only minimal T-semiflow is $\mathbf{1}$; liveness is equivalent to every directed circuit being marked) $\mathrm{RS} = \mathrm{LRS}^{\mathrm{SE}\mathbb{R}}$ [25,36,62]. In live consistent *weighted T-systems* (every place has one input and one output transition: marked graphs with weights; when consistent, they have only one minimal T-semiflow, $\mathbf{x}$, and then liveness is equivalent to $\mathbf{x}$ being fireable) $\mathrm{RS} = \mathrm{LRS}^{\mathrm{SE}}$ [88], although the inclusion $\mathrm{LRS}^{\mathrm{SE}} \subseteq \mathrm{LRS}^{\mathrm{SE}\mathbb{R}}$

can be proper because $\mathbf{C}$ is not *unimodular* as in marked graphs. Even in the case of weighted T-systems, these relations allow to prove properties more easily. For example, the reachability problem (i.e., is $\mathbf{m}$ reachable?) in a live consistent weighted T-system can be solved in polynomial time proceeding as indicated in Remark 2.

Other subclasses enjoy weaker but still useful properties. We give some examples. *Structurally persistent systems* are those with no distributor places, i.e., $|p^\bullet \leq 1$ for every place p. (These systems, when strongly connected and consistent, are conservative and have only one minimal T-semiflow, $\mathbf{x}$. In such case, liveness and reversibility is equivalent to $\mathbf{x}$ being fireable.) In live, bounded, and reversible structurally persistent systems the reachable markings are the vectors in LRS^{SE} from which $\mathbf{x}$ is fireable [90]. In live, bounded, and reversible (extended) *free choice systems* (ordinary nets where $\text{CCS}(t) = \text{EQS}(t)$ for every t; in the sequel, by free choice we always mean extended free choice) the reachable markings are the vectors in LRS^{Psf} that mark every trap [30]. In live *equal conflict systems* (for every t, $\text{CCS}(t) = \text{EQS}(t)$: free choice with weights) every two solutions of the state equation have a common successor [92]. This implies existence of home states in live and bounded equal conflict systems, and the property that no spurious solution with respect to the state equation is a deadlock marking, what allows to analyse liveness using the state equation (see Subsection 4.6).

To conclude, an important consequence of the comparison is that the state equation, even over the reals, provides *more* information than the P-semiflows, even though the state equation leads to *more* efficient verification techniques (remember that the cardinality of the fundamental set of P-semiflows may grow exponentially with $|P|$). Therefore, in what follows, the state equation is used as the basic linear description of the state space.

3.5 Traps and Siphons

Besides the marking invariants that can be obtained from the net state equation, that were described in Subsection 3.2, other marking invariants can be formulated for net systems. In particular, traps and siphons, which are also structural objects, lead to new kinds of invariants. Differently from those associated with flows, possibly the invariant laws associated with traps and siphons do not hold in every marking, but once they become true they remain true for whichever future evolution (i.e., they are *stable predicates*): traps remain marked once they become marked, and siphons remain unmarked once they become unmarked. The same as in the case of semiflows, invariants, and components, we can distinguish three notions (idem for siphons):

- The trap (a set of places).
- A trap invariant (a stable predicate).
- The trap subnet (the P-subnet generated by the trap).

When a set of places is both a siphon and a trap, e.g., the support of a P-semiflow, the P-subnet that it generates is called a *siphon-trap component*. Notice, though, that in the case of P-semiflows the P-invariant (token conservation law) is more informative than the siphon and trap invariants that are deduced from the corresponding siphon-trap component.

Traps and siphons have been extensively used for the structural analysis of (mainly ordinary) net systems (see Bibliographical Remarks). As a mere example, we state the following properties:

- In an ordinary deadlocked system, the subset of unmarked places is a siphon, because otherwise one of its input transitions would be enabled. Owing to this property of siphons, they are often called deadlocks, what is somehow misleading.
- Taking into account that traps remain marked, if every siphon of an ordinary net contains an initially marked trap, then the system is deadlock-free. In the case of *asymmetric choice* or *simple* systems (at most one of the input places to a join transition is a distributor place) the condition that every siphon contains a marked trap is sufficient for liveness, and in the case of free choice nets it is also necessary [41].
- If $\mathbf{m}$ is a home state of a live system, then every trap must be marked, because otherwise once the trap becomes marked — and it will eventually do by liveness — $\mathbf{m}$ cannot be reached any more. In the case of live and bounded free choice systems the converse is also true [7].

To our discussion, it is specially relevant to point out that some traps and siphons may contain information about the reachable markings that is *not* contained in the state equation. Therefore, they are potentially useful to improve the linear description of the state space provided by the state equation, as we shall discuss in Section 5. Consider the net in Figure 1. Clearly, $\Theta = \{p_1, p_2, p_5\}$ is an initially marked trap, so it must remain marked ($\mathbf{m}[p_1] + \mathbf{m}[p_2] + \mathbf{m}[p_5] \geq 1$), what allows to conclude that $2p_3$, $2p_4$, and $p_3 + p_4$ were spurious solutions to the state equation. Regarding siphons, for instance, in every spurious solution of the system of Figure 6 the (initially unmarked) siphon $\Sigma = \{p_1, p_2, p_3, p_6, p_7\}$ is marked, so taking the siphon invariant into account all the spurious solutions are proven non reachable. (Pragmatically, siphon invariants are less useful, since the existence of unmarked siphons in the initial marking is often considered undesirable: it implies that all its output transitions are dead.)

Next we describe a linear algebraic method to obtain a *generating family* of traps, i.e., a set such that every trap of the net is a union of traps in this set (observe that the union of two traps is a trap, they are *stable under union*). Notice that a generating family *includes* all the *minimal traps*, i.e., those which do not contain any other, but the set of minimal traps may not be generating. (The same approach can be applied to compute siphons, using that a siphon is a trap of the reverse net, or siphon-trap components.)

The method is based on the following property, which characterises traps in terms of non-negative solutions to a linear system of inequalities:

Theorem 13. *Let $\mathcal{N} = \langle P, T, \mathbf{Pre}, \mathbf{Post} \rangle$ be a P/T net. Define*

$$\mathcal{N}_\Theta = \langle P, T, \mathbf{Pre}, \mathbf{Post}_\Theta \rangle$$

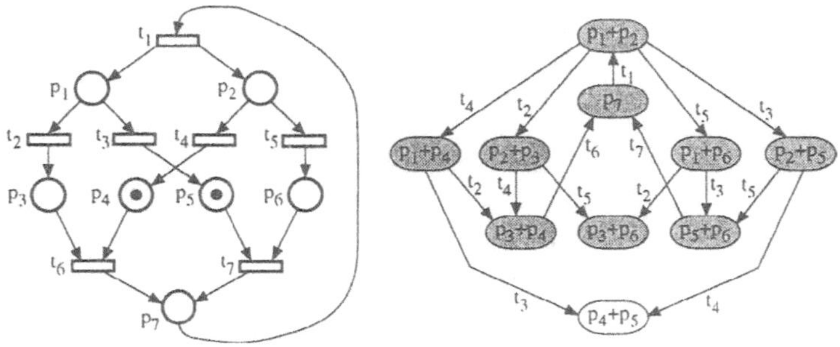

Fig. 6. A non live net system and its LRG^{SE}, where every spurious solution marks an initially unmarked siphon.

such that $\mathbf{Post}_\Theta[p,t] = 0$ *iff* $\mathbf{Post}[p,t] = 0$, *and* $\mathbf{Post}_\Theta[p,t] \geq \sum_{p' \in {}^\bullet t} \mathbf{Pre}[p',t]$ *otherwise.*

A set $\Theta \subseteq P$ *is a trap of* $\mathcal{N}$ *iff* $\mathbf{y} \geq \mathbf{0}$ *exists such that* $\|\mathbf{y}\| = \Theta$ *and* $\mathbf{y} \cdot \mathbf{C}_\Theta \geq 0$.

Proof. The inequality $\mathbf{y} \cdot \mathbf{C}_\Theta \geq \mathbf{0}$ means that in $\langle \mathcal{N}_\Theta, \mathbf{m_0} \rangle$ the weighted (according to $\mathbf{y}$) token content would never be decreased by the firing of transitions. If $\|\mathbf{y}\|$ was not a trap, then the firing of a transition in $\|\mathbf{y}\|^\bullet$ but not in ${}^\bullet\|\mathbf{y}\|$ would decrease the token content. On the other hand, by the definition of $\mathcal{N}_\Theta$, the firing of any transition puts at least as many tokens in each output place as it removes from the input places, so it is clear that when Θ is a trap the vector $\mathbf{1}_\Theta$ is a suitable $\mathbf{y}$. □

Steps 1 and 2 of Algorithm 8 can be used to compute the $\mathbf{y}$ vectors, hence the sought traps, by simply changing Step 2.2 so that only rows in which the i-th column of $\mathbf{A}$ is negative are eliminated. In order to provide a link to the well-known notion of P-semiflows, observe that the non-negative solutions to $\mathbf{y} \cdot \mathbf{C}_\Theta \geq \mathbf{0}$ are the non-negative solutions to

$$[\mathbf{y}\ \mathbf{z}] \cdot \begin{bmatrix} \mathbf{C}_\Theta \\ -\mathbf{I} \end{bmatrix} = \mathbf{0} \tag{17}$$

where $\mathbf{z}$ are *slack* variables. Therefore, the vectors $[\mathbf{y}\ \mathbf{z}]$ are P-semiflows of a net $\widetilde{\mathcal{N}_\Theta}$ that is obtained from $\mathcal{N}_\Theta$ by adding a source input place to every transition.

As an example, consider the net in Figure 1. A corresponding $\widetilde{\mathcal{N}_\Theta}$ is shown in Figure 7. The support of the P-semiflow of $\widetilde{\mathcal{N}_\Theta}$ that corresponds to the trap $\{p_1, p_2, p_5\}$ is shaded.

The above characterisation of traps allows to express some trap properties in linear algebraic terms. For instance:

- Initially marked traps remain marked: If $\mathbf{m}$ is reachable then the following system must be infeasible:

$$\mathbf{y} \cdot \mathbf{C}_\Theta \geq \mathbf{0} \wedge \mathbf{y} \geq \mathbf{0} \wedge \mathbf{y} \cdot \mathbf{m_0} > 0 \wedge \mathbf{y} \cdot \mathbf{m} = 0 \tag{18}$$

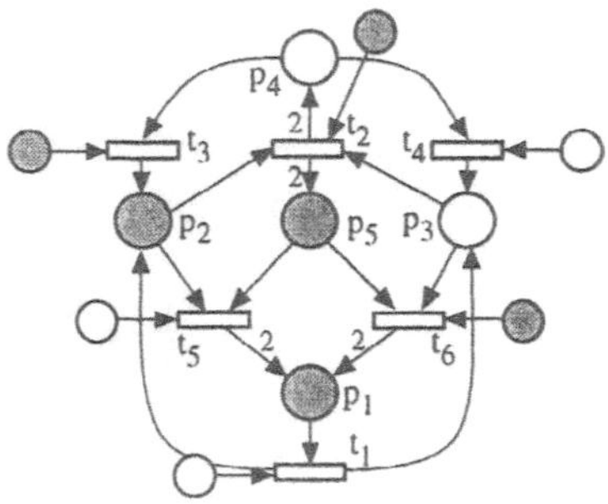

Fig. 7. A $\widetilde{\mathcal{N}_\Theta}$ corresponding to the net in Figure 1. The support of the P-semiflow associated with the trap $\{p_1, p_2, p_5\}$ is shaded.

where $\|\mathbf{y}\|$ is a trap (according to Theorem 13), that is initially marked ($\mathbf{y} \cdot \mathbf{m_0} > 0$) but not marked under $\mathbf{m}$ ($\mathbf{y} \cdot \mathbf{m} = 0$)

- In every home state of a live system every trap is marked: If $\mathbf{m}$ is a home state then the following system must be infeasible:

$$\mathbf{y} \cdot \mathbf{C}_\Theta \geq \mathbf{0} \wedge \mathbf{y} \geq \mathbf{0} \wedge \mathbf{y} \cdot \mathbf{m} = 0 \tag{19}$$

Notice that not every trap (or siphon) property can be expressed as a single system of linear inequalities. For instance, we cannot express that *when a trap becomes marked* it remains marked.

4 Analysis of Properties Using the State Equation

4.1 Overview

As we indicated in Subsection 3.1 the linear description — more precisely, relaxation — of the state space (and fireable sequences) provided by the state equation can be readily used for the analysis: the state equation is incorporated in the set of (linear) constraints of a linear (or integer) programming problem, where the property to analyse is either part of the linear constraints or appears in the cost function. We gave in (4) a straightforward application of this method to analyse reachability of a given marking. Owing to the possible presence of spurious solutions, in general this kind of analysis allows to *semidecide* only; in the case of reachability, for instance, the structural condition is only necessary. For other properties, only sufficient conditions are obtained.

Using this method we can analyse properties stated as existence or non existence of markings and/or firing sequences that satisfy some restrictions expressed in terms of linear inequalities. For existence we obtain necessary conditions, and sufficient for non-existence. Also we can compute (bounds for) the maximum of a linear function, or semidecide whether it exists. Some examples are submarking reachability, boundedness, repetitiveness, implicitness of a place, mutual exclusion, or deadlock-freeness:

- Is $\mathbf{m} \geq \mathbf{m}_x$ reachable? Existence of $\mathbf{m}$ such that $\mathbf{m} \geq \mathbf{m}_x$.
- Is p k-bounded? Non-existence of $\mathbf{m}$ such that $\mathbf{m}[p] > k$.
- Bound of p? Maximise $\mathbf{m}[p]$.
- Is t repetitive? Unbounded maximisation of $\sigma[t]$.
- Is p implicit? Non-existence of $\mathbf{m}$ and t where $\mathbf{m}[p]$ is the only marking variable that prevents the firing of t, i.e., non-existence of $\mathbf{m}$ and t such that $\mathbf{m}[P - \{p\}] \geq \mathbf{Pre}[P - \{p\}, t]$ and $\mathbf{m}[p] < \mathbf{Pre}[p, t]$.
- Are the places in $\Pi \subseteq P$ in pairwise mutual exclusion? Non-existence of $\mathbf{m}$ that marks a pair of them.
- Deadlock-freeness? Non-existence of $\mathbf{m}$ where no transition is enabled.

In the following subsections we discuss the analysis of some of these properties in more detail, and we show how several classical results in structure theory of net systems are a rather direct consequence of this analysis, typically by making use of basic results on linear programming and duality theory.

It is quite apparent from the few examples of properties listed above that their expression can be more or less complicated. Formally, they are logic propositions where atoms are linear inequalities on marking and firing count variables. When these propositions are atomic or linked *conjunctively*, as it is the case of boundedness, repetitiveness, implicitness, or mutual exclusion between two places (Subsections 4.2 to 4.5) their inclusion in the set of linear constraints poses no particular problem.

When they are linked *disjunctively*, the verification requires checking a number of systems of linear inequalities for existence of solutions. The case of mutual exclusion is illustrative: For the mutual exclusion between two places, p and p', we check non-existence of $\mathbf{m}$ such that $\mathbf{m}[p] > 0$ and $\mathbf{m}[p'] > 0$; for the pairwise mutual exclusion between three places, p, p', and p'', we must check, in principle, non-existence of $\mathbf{m}$ such that $\mathbf{m}[p] > 0$ and $\mathbf{m}[p'] > 0$, non-existence of $\mathbf{m}$ such that $\mathbf{m}[p] > 0$ and $\mathbf{m}[p''] > 0$, and non-existence of $\mathbf{m}$ such that $\mathbf{m}[p'] > 0$ and $\mathbf{m}[p''] > 0$ (in general, for n places, non-existence of solution to $n(n-1)/2$ systems of linear inequalities). If the places were known to be safe, for instance, it would suffice to check just one system, because we could express the property as non-existence of $\mathbf{m}$ such that $\mathbf{m}[p] + \mathbf{m}[p'] + \mathbf{m}[p''] > 1$. Subsection 4.6 studies the case of deadlock-freeness, and illustrates how this problem with disjunctions can be palliated, even overcome in most practical cases.

Properties discussed so far are *safety properties* expressed as first order logic predicates where the domain is defined by only one type of quantifier (either existential or universal). One may conceive the analysis of other properties where the domain is defined by a combination of quantifiers of both types. However, since we can only obtain necessary conditions for existentially quantified predicates, and sufficient conditions for universally quantified ones, due to the possible presence of spurious solutions, as a result we would not even semidecide on such properties — leaving apart the case where one of the quantifications can be unrolled because the elements of the definition domain are finite and known a priori. Consider reversibility: $\mathbf{m}_0$ reachable from every reachable marking, i.e., *for all* $\mathbf{m}$ reachable from $\mathbf{m}_0$ *there exists* a sequence σ leading from $\mathbf{m}$ to $\mathbf{m}_0$.

If we used the state equation relaxation, we would analyse:

$$(\forall \mathbf{m})(\exists \boldsymbol{\sigma})(\mathbf{m} = \mathbf{m_0} + \mathbf{C} \cdot \boldsymbol{\sigma}' \geq \mathbf{0} \wedge \mathbf{m_0} = \mathbf{m} + \mathbf{C} \cdot \boldsymbol{\sigma} \geq \mathbf{0} \wedge \boldsymbol{\sigma}, \boldsymbol{\sigma}' \geq \mathbf{0}) \quad (20)$$

Validity of the above predicate is neither sufficient for reversibility, because the $\boldsymbol{\sigma}$'s may not be fireable, nor necessary, because an $\mathbf{m}$ invalidating it may be spurious. Liveness (of a transition) suffers from the very same problem. Of course, in net subclasses where these "difficult" properties are equivalent to others that can be analysed using the state equation method, the problem is solved. For instance, liveness of bounded strongly connected equal conflict systems is equivalent to deadlock-freeness [92], and live equal conflict systems do not have spurious deadlocks, so absence of deadlock markings which are solution to the state equation proves liveness (see Remark 35).

As a last comment, many of the properties that we consider here are particular — but specially relevant — cases of general *synchronic properties*. For instance, the *synchronic lead* of a subset of transitions with respect to another accounts for the maximum difference between the (possibly weighted) number of firings of the former and the latter. Boundedness of a place is a matter of boundedness of the synchronic lead of its input transitions with respect to its output transitions, and mutual exclusion between two safe places is equivalent to the synchronic lead of their input transitions with respect to their output transitions taking value one. The structural analysis of synchronic properties, which is presented in [83], is a more compact or abstract view of the analysis of many safety properties that we describe here.

4.2 Place Marking Bounds and Structural Boundedness

The marking bound of a place p in a net system $\mathcal{S}$ is defined as:

$$\mathbf{b}[p] = \max\{\mathbf{m}[p] \mid \mathbf{m} \in \mathrm{RS}(\mathcal{S})\} \quad (21)$$

When this bound is finite, the place is said to be bounded. Using the state equation, and writing $\mathbf{m}[p]$ as $\mathbf{1}_p \cdot \mathbf{m}$, we define the *structural bound* of a place p as:

$$\mathbf{sb}[p] = \max\{\mathbf{1}_p \cdot \mathbf{m} \mid \mathbf{m} - \mathbf{C} \cdot \boldsymbol{\sigma} = \mathbf{m_0} \wedge \mathbf{m}, \boldsymbol{\sigma} \geq \mathbf{0}\} \quad (22)$$

In principle, Equation (22) is an integer programming problem. If integrality constraints (on $\mathbf{m}$ and $\boldsymbol{\sigma}$) are disregarded then (22) is a linear programming problem, that can be solved in polynomial time.

According to Theorem 11, the marking for which the structural bound is reached could be spurious, so clearly, in general, we have $\mathbf{sb}[p] \geq \mathbf{b}[p]$. Therefore, if we were investigating the k-boundedness of p (i.e., is $\mathbf{m}[p] \leq k$?), an efficient *sufficient* condition is $\mathbf{sb}[p] \leq k$. We insist that the condition is only sufficient, and even it is possible that the structural bound for a bounded place does not exist (i.e., the programming problem is unbounded; we note at this point that, although the structural bound using integer programming can be more accurate than using linear programming, when this particular linear programming problem is unbounded so it is the integer version — see Appendix). For instance, place p in the live and safe system of Figure 8 is bounded, but the linear programming problem (22) is unbounded.

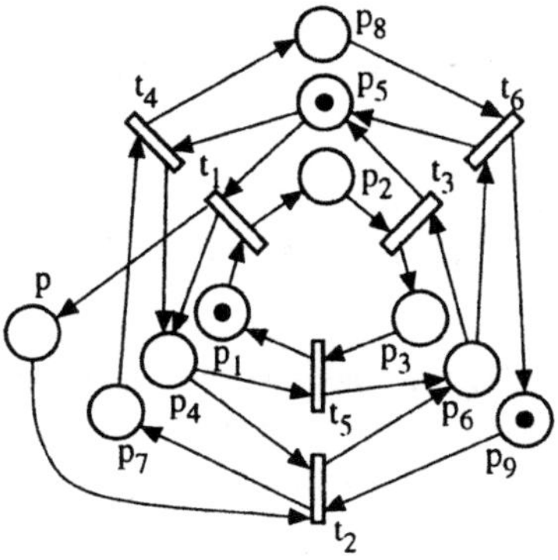

Fig. 8. Place p is bounded but not structurally bounded.

In the sequel we apply results from duality theory of linear programming. Although duality results are available for integer programming, we concentrate on linear programming, partly owing to the objective of re-encountering some classical results in net theory, and also to the pragmatic reason that the derived algorithms are more efficient, actually polynomial time. The dual linear programming problem of (22) is:

$$\mathbf{sb}[p]' = \min\{\mathbf{y} \cdot \mathbf{m_0} \mid \mathbf{y} \cdot \mathbf{C} \leq \mathbf{0} \wedge \mathbf{y} \geq \mathbf{1}_p\} \tag{23}$$

Since (22) has always a feasible solution ($\mathbf{m} = \mathbf{m_0}$, $\boldsymbol{\sigma} = \mathbf{0}$), both problems (22) and (23) are bounded iff a feasible solution for (23) exists, according to the duality and unboundedness theorems. In other words, if $\mathbf{y} \geq \mathbf{1}_p$ exists such that $\mathbf{y} \cdot \mathbf{C} \leq \mathbf{0}$, then p is structurally bounded, i.e., bounded for every initial marking. (Moreover, in such case $\mathbf{sb}[p] = \mathbf{sb}[p]'$.)

Remark 14. The linear programming problem (23) performs in polynomial time a *search* for the vector $\mathbf{y}$ that allows to obtain the best structural bound for p among all the P-semiflows and other vectors $\mathbf{y} \geq \mathbf{0}$ such that $\mathbf{y} \cdot \mathbf{C} \leq \mathbf{0}$. Clearly, it gives a more accurate bound than considering the P-semiflows only (besides being more efficient!) □

By the alternatives theorem, existence of $\mathbf{y} \geq \mathbf{1}_p$ such that $\mathbf{y} \cdot \mathbf{C} \leq \mathbf{0}$ is equivalent to non existence of $\mathbf{x} \geq \mathbf{0}$ such that $\mathbf{C} \cdot \mathbf{x} \geq \mathbf{1}_p$. Observe that, if such $\mathbf{x}$ exists, then we can find a sufficiently large initial marking $\mathbf{m_0}$ allowing to fire once and again a sequence with firing count vector $\mathbf{x}$, what shows that p is unbounded. Therefore, non existence of $\mathbf{x} \geq \mathbf{0}$ such that $\mathbf{C} \cdot \mathbf{x} \geq \mathbf{1}_p$ is necessary for structural boundedness of p. In the example of Figure 8, $\mathbf{x} = t_1 + t_3 + t_5$ disproves structural boundedness of p. Notice that no sequence with such firing count vector is fireable under the marking shown in the figure; nevertheless, $t_1\, t_3\, t_5$ becomes fireable with $\mathbf{m_0} = p_1 + p_5 + p_6 + p_9$, and p becomes unbounded. (Interpreting boundedness as ***stability*** of the dynamic system, it can be said that non structurally bounded systems which are bounded are only ***conditionally stable***, in the sense that an increment of the initial marking may lead to unstable behaviour.)

In summary:

Theorem 15. *Let $\mathcal{N}$ be a P/T net, and p one of its places. The following three statements are equivalent:*

1. *p is structurally bounded, i.e., bounded for every $\mathbf{m_0}$.*
2. *There exists $\mathbf{y} \geq \mathbf{1}_p$ such that $\mathbf{y} \cdot \mathbf{C} \leq \mathbf{0}$.*
3. *There does not exist $\mathbf{x} \geq \mathbf{0}$ such that $\mathbf{C} \cdot \mathbf{x} \geq \mathbf{1}_p$.*

The above statement, the same as several others in the sequel, contains two dual perspectives of a property, one ***place-based*** (item 2) and the other ***transition-based*** (item 3). We remark that the ***duality*** in this kind of net properties is rooted on ***duality*** theory in linear programming.

Applying the above characterisation of structural boundedness of p to every place of $\mathcal{N}$:

Corollary 16. *Let $\mathcal{N}$ be a P/T net. The following three statements are equivalent:*

1. *$\mathcal{N}$ is structurally bounded, i.e., every place is bounded for every $\mathbf{m_0}$.*
2. *There exists $\mathbf{y} > \mathbf{0}$ such that $\mathbf{y} \cdot \mathbf{C} \leq \mathbf{0}$.*
3. *There does not exist $\mathbf{x} \geq \mathbf{0}$ such that $\mathbf{C} \cdot \mathbf{x} \gneq \mathbf{0}$.*

Remark 17. It is well known that nets with *inhibitor arcs* — which is a widely used extension of P/T systems — can be simulated with plain P/T systems using the *complementary place* construction in the case that the inhibiting places are bounded. This is particularly true for structurally bounded places, with the additional advantage that the transformation can be done for whichever initial marking. In the sequel, it goes without saying that all the results can be extended to nets with inhibitor arcs in the case of structural boundedness. □

4.3 Transition Fireability Bounds and Structural Repetitiveness

The fireability or repetitiveness bound of a transition t in a net system $\mathcal{S}$ is defined as:

$$\mathbf{r}[t] = \max\{\boldsymbol{\sigma}[t] \mid \sigma \in \mathrm{L}(\mathcal{S})\} \tag{24}$$

If this bound is zero, then the transition is *dead* (or *0-live*). If the bound takes a finite positive value, the transition can only be fired a finite number of times (it is *1-live*). When the bound does not exist, transition t is said to be ***repetitive*** (or ***2-live***), and this is necessary for liveness of t.

Using the state equation, we define the ***structural repetitiveness bound*** of a transition t as:

$$\mathbf{sr}[t] = \max\{\mathbf{1}_t \cdot \boldsymbol{\sigma} \mid \mathbf{m} - \mathbf{C} \cdot \boldsymbol{\sigma} = \mathbf{m}_0 \wedge \mathbf{m}, \boldsymbol{\sigma} \geq 0\} \tag{25}$$

Clearly, if the transition is repetitive, then (25) is unbounded, since $\mathbf{sr}[t] \geq \mathbf{r}[t]$. (The converse is not true because the $\boldsymbol{\sigma}$'s that make the problem unbounded may not correspond to actually fireable sequences: all the transitions in the nets of Figure 2 are structurally repetitive and none of them is repetitive.)

The dual linear programming problem of (25) is:

$$\mathbf{sr}[t]' = \min\{\mathbf{y} \cdot \mathbf{m}_0 \mid \mathbf{y} \cdot \mathbf{C} \leq -\mathbf{1}_t \wedge \mathbf{y} \geq 0\} \tag{26}$$

Since (25) has always a feasible solution ($\mathbf{m} = \mathbf{m}_0$, $\boldsymbol{\sigma} = \mathbf{0}$), it is unbounded iff (26) is infeasible, according to the duality and unboundedness theorems. In other words, if some $\mathbf{y} \geq \mathbf{0}$ exists such that $\mathbf{y} \cdot \mathbf{C} \leq -\mathbf{1}_t$, then t is not structurally repetitive, i.e., not repetitive for any initial marking.

By the alternatives theorem, non existence of $\mathbf{y} \geq \mathbf{0}$ such that $\mathbf{y} \cdot \mathbf{C} \leq -\mathbf{1}_t$ is equivalent to existence of $\mathbf{x} \geq \mathbf{1}_t$ such that $\mathbf{C} \cdot \mathbf{x} \geq \mathbf{0}$. Observe that, if such $\mathbf{x}$ exists, then we can find a sufficiently large initial marking $\mathbf{m}_0$ allowing to fire once and again a sequence with firing count vector $\mathbf{x}$, what shows that t is repetitive. Therefore, existence of $\mathbf{x} \geq \mathbf{1}_t$ such that $\mathbf{C} \cdot \mathbf{x} \geq \mathbf{0}$ is sufficient for structural repetitiveness of t. In summary:

Theorem 18. *Let $\mathcal{N}$ be a P/T net, and t one of its transitions. The following three statements are equivalent:*

1. *t is structurally repetitive, i.e., repetitive for some $\mathbf{m}_0$.*
2. *There does not exist $\mathbf{y} \geq \mathbf{0}$ such that $\mathbf{y} \cdot \mathbf{C} \leq -\mathbf{1}_t$.*
3. *There exists $\mathbf{x} \geq \mathbf{1}_t$ such that $\mathbf{C} \cdot \mathbf{x} \geq \mathbf{0}$.*

Applying the above characterisation of structural repetitiveness of t to every transition of $\mathcal{N}$:

Corollary 19. *Let $\mathcal{N}$ be a P/T net. The following three statements are equivalent:*

1. *$\mathcal{N}$ is structurally repetitive, i.e., every transition is repetitive for some $\mathbf{m}_0$.*
2. *There does not exist $\mathbf{y} \geq \mathbf{0}$ such that $\mathbf{y} \cdot \mathbf{C} \lneq \mathbf{0}$.*
3. *There exists $\mathbf{x} > \mathbf{0}$ such that $\mathbf{C} \cdot \mathbf{x} \geq \mathbf{0}$.*

Comparing Corollaries 16 and 19 (or Theorems 15 and 18) it is quite apparent that structural boundedness and structural repetitiveness are dual notions: a net with incidence matrix $\mathbf{C}$ is structurally repetitive iff the reverse dual, with incidence matrix $-\mathbf{C}^{\perp}$, is structurally bounded. Both properties together are equivalent to conservativeness and consistency: Let $\mathbf{x} > \mathbf{0}$ and $\mathbf{y} > \mathbf{0}$ be the vectors such that $\mathbf{C}\cdot\mathbf{x} \geq \mathbf{0}$ and $\mathbf{y}\cdot\mathbf{C} \leq \mathbf{0}$, respectively. By $\mathbf{C}\cdot\mathbf{x} \geq \mathbf{0}$, $\mathbf{y}\cdot(\mathbf{C}\cdot\mathbf{x}) \geq \mathbf{0}$, while by $\mathbf{y}\cdot\mathbf{C} \leq \mathbf{0}$, $(\mathbf{y}\cdot\mathbf{C})\cdot\mathbf{x} \leq \mathbf{0}$. Therefore $\mathbf{y}\cdot\mathbf{C}\cdot\mathbf{x} = \mathbf{0}$, hence $\mathbf{C}\cdot\mathbf{x} = \mathbf{0}$ and $\mathbf{y}\cdot\mathbf{C} = \mathbf{0}$. Since (structural) repetitiveness is necessary for (structural) liveness, i.e., liveness *for some* initial marking, see Proposition 10, we can state:

Theorem 20. *Let $\mathcal{N}$ be a P/T net. If $\mathcal{N}$ is structurally live and structurally bounded, then (equivalently):*

1. *$\mathcal{N}$ is structurally repetitive and structurally bounded.*
2. *$\mathcal{N}$ is conservative and consistent.*

Taking into account Proposition 12:

Corollary 21. *Let $\mathcal{N}$ be a P/T net. If $\mathcal{N}$ is structurally live and structurally bounded, then for every $\mathbf{m_0}$:*

$$\mathrm{LRS}^{\mathrm{SEIR}}(\mathcal{N}, \mathbf{m_0}) = \mathrm{LRS}^{\mathrm{Pf}}(\mathcal{N}, \mathbf{m_0}) = \mathrm{LRS}^{\mathrm{Psf}}(\mathcal{N}, \mathbf{m_0})$$

4.4 Implicit Places and Structurally Implicit Places

In general, places impose constraints on the firing of their output transitions. When they never do, they could be removed without affecting the behaviour of the rest of the system. (However, even being redundant, they might still be useful — one such application is developed in Section 5.)

A place whose removal does not affect the behaviour of the system is called an *implicit place*. Here, by behaviour we understand the *interleaving semantics*, i.e., the sequential observations or language, although the notion of implicit place can be directly extended to cope with a *step semantics* (see Remark 27).

Definition 22. Let $\mathcal{S} = \langle P \cup \{p\}, T, \mathbf{Pre}, \mathbf{Post}, \mathbf{m_0}\rangle$ be a P/T system. The place p is *implicit* iff $\mathrm{L}(\mathcal{S}) = \mathrm{L}(\langle P, T, \mathbf{Pre}[P,T], \mathbf{Post}[P,T], \mathbf{m_0}[P]\rangle)$.

In other words, p is never the unique place that prevents the firing of a transition, i.e., $\mathbf{m} \geq \mathbf{Pre}[P,t] \Rightarrow \mathbf{m}[p] \geq \mathbf{Pre}[p,t]$ for all $t \in p^\bullet$, so it produces fictitious synchronisations in its output transitions. For instance, in Figure 9 (a), t_4 seems to represent a synchronisation, but whenever p_3 is marked so it is p, so p is implicit. It is worth noticing at this point that, in general, $\mathbf{m}[p]$ may not be computable from the marking of other places. For instance, the marking of p in Figure 9 (a) cannot be deduced from the marking of the other places, it depends also on the number of occurrences of t_1. When the marking of an implicit place can be computed from the marking of other places, we say that it is *marking implicit.* In Figure 9 (b), both p_1 and p_4 are marking implicit: $\mathbf{m}[p_1] = \mathbf{m}[p] + \mathbf{m}[p_3]$ and $\mathbf{m}[p_4] = \mathbf{m}[p] + \mathbf{m}[p_2]$. Therefore, they can be removed without affecting the behaviour of the system. The net in Figure 9 (b) shows also that, naturally, being implicit or not is sometimes a matter of the initial marking. With $p_1 + p_4 + 2p$ as initial marking, p is marking implicit instead of p_1 and p_4, and $\mathbf{m}[p] = \mathbf{m}[p_1] + \mathbf{m}[p_4]$: the behaviour of the system is that of

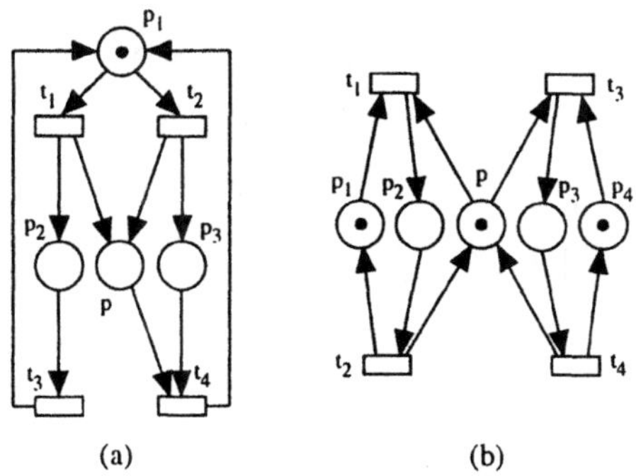

Fig. 9. Implicit and marking implicit places. Place p is implicit but not marking implicit in (a). Places p_1 and p_4 are marking implicit in (b). If we add a token to p in (b), then it becomes marking implicit instead.

two cyclic processes, $(t_1\, t_2)^*$ and $(t_3\, t_4)^*$ that evolve in parallel independently, in spite of the apparent synchronisation introduced by p.

Notice that once the initial marking of the rest of the system is fixed, implicitness of a place is *monotonic* with respect to its initial marking (i.e., if the place is implicit for some initial marking, it is also implicit with a greater one).

To decide whether a given p is implicit, in principle, we should check that no reachable marking exists such that for some t we have $\mathbf{m}[P] \geq \mathbf{Pre}[P,t]$ and $\mathbf{m}[p] < \mathbf{Pre}[p,t]$. Of course, it is necessary that p is not the only input place of its output transitions. This syntactical check rapidly tells that, for instance, p_2 and p_3 cannot be implicit in the net of Figure 9 (b).

The condition that no reachable marking exists such that for some t (actually, $t \in p^\bullet$) we have $\mathbf{m}[P] \geq \mathbf{Pre}[P,t]$ and $\mathbf{m}[p] < \mathbf{Pre}[p,t]$ can be expressed conveniently using a "transition selector" $\mathbf{s}$, i.e., the characteristic vector of one transition ($\mathbf{s} \geq \mathbf{0}$ and $\mathbf{1} \cdot \mathbf{s} = 1$): Place p is implicit in $\mathcal{S} = \langle P \cup \{p\}, T, \mathbf{Pre}, \mathbf{Post}, \mathbf{m_0}\rangle$ iff there are no $\mathbf{m}$ and $\mathbf{s}$ which are solution to:

$$\begin{aligned} \mathbf{m} &\in \mathrm{RS}(\mathcal{S}) \\ \mathbf{m}[P] - \mathbf{Pre}[P,p^\bullet] \cdot \mathbf{s} &\geq \mathbf{0} \\ \mathbf{m}[p] - \mathbf{Pre}[p,p^\bullet] \cdot \mathbf{s} &< 0 \\ \mathbf{1} \cdot \mathbf{s} &= 1 \\ \mathbf{s} &\geq \mathbf{0} \end{aligned} \tag{27}$$

where $\mathbf{m}[P] \geq \mathbf{Pre}[P,p^\bullet] \cdot \mathbf{s}$ states that $\|\mathbf{s}\|$ is enabled by P, and $\mathbf{m}[p] < \mathbf{Pre}[p,p^\bullet] \cdot \mathbf{s}$ states that it is not enabled by p.

Replacing for the first condition the state equation, as usual, we obtain a structural *sufficient* condition for p implicit, in terms of non existence of solution $\mathbf{m}$, $\boldsymbol{\sigma}$, and $\mathbf{s}$ to:

$$
\begin{aligned}
\mathbf{m} - \mathbf{C} \cdot \boldsymbol{\sigma} &= \mathbf{m_0} \\
\mathbf{m}[P] - \mathbf{Pre}[P, p^\bullet] \cdot \mathbf{s} &\geq \mathbf{0} \\
\mathbf{m}[p] - \mathbf{Pre}[p, p^\bullet] \cdot \mathbf{s} &< 0 \\
\mathbf{1} \cdot \mathbf{s} &= 1 \\
\mathbf{m}, \boldsymbol{\sigma}, \mathbf{s} &\geq \mathbf{0}
\end{aligned}
\tag{28}
$$

Since $\mathbf{m}[p] = \mathbf{m_0}[p] + \mathbf{C}[p, T] \cdot \boldsymbol{\sigma}$, the above is equivalent to $\mathbf{m_0}[p]$ being greater than or equal to the optimal value of the following linear programming problem:

$$
\begin{aligned}
\max\{\mathbf{Pre}[p, p^\bullet] \cdot \mathbf{s} - \mathbf{C}[p, T] \cdot \boldsymbol{\sigma} \mid \;& \mathbf{m}[P] - \mathbf{C}[P, T] \cdot \boldsymbol{\sigma} = \mathbf{m_0}[P] \\
& \mathbf{m}[P] - \mathbf{Pre}[P, p^\bullet] \cdot \mathbf{s} \geq \mathbf{0} \\
& \mathbf{1} \cdot \mathbf{s} = 1 \\
& \mathbf{m}, \boldsymbol{\sigma}, \mathbf{s} \geq \mathbf{0}\}
\end{aligned}
\tag{29}
$$

The initial marking does not appear in the constraints of the dual problem of (29), which is:

$$
\begin{aligned}
\min\{\mathbf{y} \cdot \mathbf{m_0} + \mu \mid \;& \mathbf{y} \cdot \mathbf{C}[P, T] \leq \mathbf{C}[p, T] \\
& \mathbf{j} \cdot \mathbf{Pre}[P, p^\bullet] + \mu \mathbf{1} \geq \mathbf{Pre}[p, p^\bullet] \\
& \mathbf{y} \geq \mathbf{j} \geq \mathbf{0}\}
\end{aligned}
\tag{30}
$$

Considering $\mathbf{j} = \mathbf{y}$ does not affect the solution, leading to the following sufficient condition for p implicit:

Theorem 23. *Let $\mathcal{S} = \langle P \cup \{p\}, T, \mathbf{Pre}, \mathbf{Post}, \mathbf{m_0} \rangle$ be a P/T system. If $\mathbf{m_0}[p]$ is greater than or equal to the optimal value of the following linear programming problem:*

$$
\begin{aligned}
\min\{\mathbf{y} \cdot \mathbf{m_0}[P] + \mu \mid \;& \mathbf{y} \cdot \mathbf{C}[P, T] \leq \mathbf{C}[p, T] \\
& \mathbf{y} \cdot \mathbf{Pre}[P, p^\bullet] + \mu \mathbf{1} \geq \mathbf{Pre}[p, p^\bullet] \\
& \mathbf{y} \geq \mathbf{0}\}
\end{aligned}
\tag{31}
$$

then p is implicit.

Let us consider the places which can be made implicit *for every* initial marking of the rest of the system (i.e., we are abstracting from the initial marking, as for other structural versions of properties). We call such places *structurtally implicit*:

Definition 24. Let $\mathcal{N} = \langle P \cup \{p\}, T, \mathbf{Pre}, \mathbf{Post} \rangle$ be a P/T net. The place p is *structurally implicit* iff for every $\mathbf{m_0}[P]$, a $\mathbf{m_0}[p]$ exists such that p is implicit in $\mathcal{S} = \langle P \cup \{p\}, T, \mathbf{Pre}, \mathbf{Post}, \mathbf{m_0} \rangle$.

The linear programming problem in Theorem 23 is feasible iff a $\mathbf{y} \geq \mathbf{0}$ exists such that $\mathbf{C}[p, T] \geq \mathbf{y} \cdot \mathbf{C}[P, T]$, because the constraint containing μ can always be satisfied taking a sufficiently large μ. When a place fulfills such condition (or its alternative formulation), it becomes implicit with a sufficiently large initial marking, at least with that given by the linear programming problem, so it is structurally implicit. (Notice that it may become implicit also with a smaller marking in some cases, see later Remark 39.)

On the other hand, consider a net $\mathcal{N}$ and a place p for which no $\mathbf{y} \geq \mathbf{0}$ exists such that $\mathbf{C}[p, T] \geq \mathbf{y} \cdot \mathbf{C}[P, T]$. By the alternatives theorem, a $\mathbf{x} \geq \mathbf{0}$ exists such that $\mathbf{C}[P, T] \cdot \mathbf{x} \geq \mathbf{0}$ and $\mathbf{C}[p, T] \cdot \mathbf{x} < 0$. With a sufficiently large initial marking $\mathbf{m_0}[P]$, a sequence with firing count vector $\mathbf{x}$ can be fired repeatedly in $\langle \mathcal{N}, \mathbf{m_0}[P] \rangle$, and this would empty p for whichever $\mathbf{m_0}[p]$, so p is not structurally implicit.

In summary, a characterisation of structurally implicit places is given by the following result:

Theorem 25. *Let $\mathcal{N} = \langle P \cup \{p\}, T, \mathbf{Pre}, \mathbf{Post} \rangle$. The place p is structurally implicit iff (equivalently):*

1. *A $\mathbf{y} \geq \mathbf{0}$ exists such that $\mathbf{C}[p, T] \geq \mathbf{y} \cdot \mathbf{C}[P, T]$*
2. *No $\mathbf{x} \geq \mathbf{0}$ exists such that $\mathbf{C}[P, T] \cdot \mathbf{x} \geq \mathbf{0}$ and $\mathbf{C}[p, T] \cdot \mathbf{x} < 0$*

Remark 26. Similarly as structural boundedness and structural repetitiveness are dual notions, the dual notion of a structural implicit place can be defined. The resulting objects are called *structural bypass transitions*, which have also a behavioural interpretation (see [73]). □

When $\mathbf{C}[p, T] = \mathbf{y} \cdot \mathbf{C}[P, T]$ in Theorem 25.1, the place p is *marking structurally implicit* and its marking can be computed from the marking of other places:

$$\begin{aligned} \mathbf{m}[p] &= \mathbf{m_0}[p] + \mathbf{C}[p, T] \cdot \boldsymbol{\sigma} \\ &= \mathbf{m_0}[p] + \mathbf{y} \cdot \mathbf{C}[P, T] \cdot \boldsymbol{\sigma} \\ &= \mathbf{m_0}[p] + \mathbf{y} \cdot (\mathbf{m}[P] - \mathbf{m_0}[P]) \end{aligned} \quad (32)$$

As an example of (marking) structurally implicit place, consider p with $\mathbf{C}[p, T] = t_3 - t_5$ added to the net system in Figure 1. The optimal solution to the linear programming problem in Theorem 23 is zero, and it is obtained for $\mathbf{y} = p_1 + p_2 + p_5$ and $\mu = -1$. Therefore, p is (marking) structurally implicit ($\mathbf{C}[p, T] = \mathbf{y} \cdot \mathbf{C}[P, T]$), and with $\mathbf{m_0}[p] = 0$ it is (marking) implicit — as can be easily checked. It is interesting to observe in this case that, once p is added, the places p_1, p_3, p_4, and p are the support of a *new* P-semiflow, which induces the new marking invariant $\mathbf{m}[p_1] + \mathbf{m}[p_3] + \mathbf{m}[p_4] + \mathbf{m}[p] = 1$. This invariant reveals, in particular, that p_3 and p_4 are safe and in mutual exclusion, what could not be proven using the state equation of the original net, where $2p_3$, $2p_4$, and $p_3 + p_4$ were spurious marking. The addition of an implicit place, which *does not affect the behaviour*, may be useful to allow proving properties by structural techniques!

An example of implicit place which is not structurally implicit, effectively showing that the structural condition is only sufficient, is p in Figure 8. (In practice, though, specially under liveness and boundedness, implicit places are most often structurally implicit.)

Remark 27. A place is ***concurrent implicit*** when it preserves not only the transition firing sequences but also the ***step*** firing sequences. Any concurrent implicit place is of course (sequential) implicit, and (sequential) implicit places without self-loops are also concurrent implicit — notice that they differ only when a transition is enabled ***several times*** by the marking of the other places while it is merely enabled by the implicit place.

The sufficient condition for a place p to be implicit given by Theorem 23 can be generalised to cope with concurrent implicit places by substituting the cost function in (31) by $\mathbf{y} \cdot \mathbf{m_0}[P] + \alpha \cdot \mu$, where α is a positive integer, not necessarily one. A sensible choice for α must be greater than or equal to the maximum ***enabling bound*** of the transitions in $p^\bullet$, where the enabling bound of a transition is the maximum number of times that it can occur in a step. A structural enabling bound for a transition t, possibly greater than the actual enabling bound, can be computed as:

$$\mathbf{seb}[t] = \max\{k \mid \mathbf{m} \geq k\mathbf{Pre}[P,t] \wedge \mathbf{m} - \mathbf{C} \cdot \boldsymbol{\sigma} = \mathbf{m_0} \wedge \mathbf{m}, \boldsymbol{\sigma} \geq 0\} \tag{33}$$

□

4.5 Mutual Exclusion and Concurrency Relations

We consider now the verification of mutual exclusion between two places, which is the basic property to analyse in order to investigate pairwise mutual exclusions, as was discussed in Subsection 4.1.

The problem of pairwise concurrency of transitions is closely related. Even it can be transformed into non mutual exclusion between places by splitting the analysed transitions into paths transition → place → transition. More directly, for transitions that are in pairwise concurrency relation we can find, for each pair t and t', a marking $\mathbf{m}$ such that $\mathbf{m} \geq \mathbf{Pre}[P,t] + \mathbf{Pre}[P,t']$.

We concentrate on the mutual exclusion of two places, p and p':

Theorem 28. *Let S be a P/T system.*
If there is no solution to

$$\mathbf{m} - \mathbf{C} \cdot \boldsymbol{\sigma} = \mathbf{m_0} \wedge \mathbf{m} \geq \mathbf{1}_{\{p,p'\}} \wedge \boldsymbol{\sigma} \geq 0 \tag{34}$$

or (equivalently) there exists solution to

$$\mathbf{y} \cdot \mathbf{C} \leq \mathbf{0} \wedge \mathbf{y} \cdot \mathbf{1}_{\{p,p'\}} - \mathbf{y} \cdot \mathbf{m_0} > 0 \wedge \mathbf{y} \geq \mathbf{0} \tag{35}$$

then p and p' are in mutual exclusion.

Remark 29. As usual, the alternative formulation given by (35) is only equivalent when considering (34) over the reals. For instance, consider the net on the right and top in Figure 2 initially marked with $2p_2$ instead of $p_1 + p_2$ (which now becomes a spurious solution). While (35) does not allow to prove mutual exclusion of p_1 and p_2, (34) does if it is interpreted over the integers. □

The interpretation of (35) is the following: From $\mathbf{y} \cdot \mathbf{m}_0 < \mathbf{y} \cdot \mathbf{1}_{\{p,p'\}}$ and $\mathbf{y} \cdot \mathbf{C} \leq \mathbf{0}$ it follows that in every reachable marking $\mathbf{y} \cdot \mathbf{m} < \mathbf{y} \cdot \mathbf{1}_{\{p,p'\}}$, so $\mathbf{m} \geq \mathbf{1}_{\{p,p'\}}$ is impossible, what proves mutual exclusion. (In structurally repetitive nets, the condition $\mathbf{y} \cdot \mathbf{C} \leq \mathbf{0}$ can be replaced by $\mathbf{y} \cdot \mathbf{C} = \mathbf{0}$, because, according to Proposition 19, it is never the case that $\mathbf{y} \cdot \mathbf{C} \lneq \mathbf{0}$.)

The dual formulation of the mutual exclusion problem given by (35) proceeds in the same way as the dual formulation of the boundedness problem, efficiently searching for a suitable vector $\mathbf{y}$ to prove the property (see Remark 14). An additional advantage of this formulation compared to (34) is that the obtained $\mathbf{y}$ provides an *explanation* of the property and may be useful to derive other mutual exclusion relations by the way. As an example, consider again the net system in Figure 1 after adding the place p with $\mathbf{C}[p,T] = t_3 - t_5$, and suppose that we are trying to prove the mutual exclusion between p_1 and p_4 applying (35). A solution $\mathbf{y} = p_1 + p_2 + p_4 + p$ is obtained. The corresponding marking invariant, $\mathbf{m}[p_1] + \mathbf{m}[p_3] + \mathbf{m}[p_4] + \mathbf{m}[p] = 1$, proves not only mutual exclusion of p_1 and p_4, but also mutual exclusion of any other pair, i.e., it proves at once the pairwise mutual exclusion of p_1, p_3, p_4, and p.

4.6 Deadlock-freeness and Termination Properties

Proving that a concurrent system cannot reach a deadlock condition (i.e., no activity because every process is waiting for some other to continue) may be difficult when the size of the system becomes large and its structure intricate due to complex and perhaps subtle interactions. For instance, (flexible) manufacturing systems are indeed a characteristic domain where the study of deadlocks is specially relevant [3,34,94,95]. The tasking behaviour of Ada programs is another example [64].

We apply here the state equation method to analyse the absence of deadlocks in P/T models. The approach is conceptually simple but, since the condition for a marking to be a deadlock is relatively complex, it still requires a considerable computational effort. We present here several techniques to improve the performance, often reducing the problem to checking a single system of linear inequalities for existence of solutions.

In net terms, a deadlock corresponds to a marking from which no transition is fireable. That t is disabled at $\mathbf{m}$ can be expressed:

$$\bigvee_{p \in {}^\bullet t} \mathbf{m}[p] < \mathbf{Pre}[p,t] \tag{36}$$

Clearly, every reachable deadlock is a solution to the state equation where every transition is disabled, what leads to the following basic general sufficient condition for deadlock-freeness:

Theorem 30. *Let S be a P/T system.*

If there is no (integer) solution to

$$\begin{array}{c} \mathbf{m} - \mathbf{C} \cdot \boldsymbol{\sigma} = \mathbf{m_0} \\ \mathbf{m}, \boldsymbol{\sigma} \geq \mathbf{0} \\ \bigvee_{p \in {}^\bullet t} \mathbf{m}[p] < \mathbf{Pre}[p, t] \quad \forall t \in T \end{array} \tag{37}$$

then $\mathcal{S}$ is deadlock-free.

In general, the disabledness conditions are not linear, because they consist of linear inequalities linked disjunctively. Anyway we can rewrite (37) as a set of systems of linear inequalities/equations, i.e., as a logic sum of products, applying the distributive property to the actual product of sums: If for every mapping $\alpha : T \to P$ that assigns to each transition one of its input places there is no solution to:

$$\begin{array}{c} \mathbf{m} - \mathbf{C} \cdot \boldsymbol{\sigma} = \mathbf{m_0} \\ \mathbf{m}, \boldsymbol{\sigma} \geq \mathbf{0} \\ \mathbf{m}[\alpha(t)] < \mathbf{Pre}[\alpha(t), t] \quad \forall t \in T \end{array} \tag{38}$$

then $\mathcal{S}$ is deadlock-free.

The problem now is that we have to check (38) for *every* mapping α, so we have to check $\prod_{t \in T} |{}^\bullet t|$ linear systems. In general this number might be large if there are many synchronisations (join transitions). We aim at reducing this number as much as we can, preserving the decision power, that is, the set of *integer* solutions to (37).

Firstly, we can obviously remove from (37) a disabledness condition if there is another disabledness condition that is "weaker". In other words, if $\mathbf{Pre}[P, t] \leq \mathbf{Pre}[P, t']$ for some t and t', the disabledness condition for t' can be removed without affecting the set of solutions to (37), even over the reals.

It is also quite obvious that we can remove from (37) the disabledness conditions of transitions that are known to be *dead* (or *0-live*, they do not appear in any sequence from $\mathbf{m_0}$) because whenever all the other transitions are disabled, we are certainly in a deadlock situation. This does not affect either the set of real solutions to (37).

Clearly, a sufficient condition for t dead is non-existence of solution to:

$$\mathbf{m} - \mathbf{C} \cdot \boldsymbol{\sigma} = \mathbf{m_0} \wedge \boldsymbol{\sigma} \geq \mathbf{0} \wedge \mathbf{m} \geq \mathbf{Pre}[P, t] \tag{39}$$

Considering (39) over the reals, and applying the alternatives theorem, we obtain an alternative sufficient condition for t dead, in terms of *existence* of solution to:

$$\mathbf{y} \cdot \mathbf{C} \leq \mathbf{0} \wedge \mathbf{y} \geq \mathbf{0} \wedge \mathbf{y} \cdot (\mathbf{Pre}[P, t] - \mathbf{m_0}) > 0 \tag{40}$$

If there were a solution $\mathbf{y}$, it would induce the marking invariant $\mathbf{y} \cdot \mathbf{m} \leq \mathbf{y} \cdot \mathbf{m_0}$. If some $\mathbf{m}$ enabled t, then $\mathbf{m} \geq \mathbf{Pre}[P, t]$, so $\mathbf{y} \cdot \mathbf{m} \geq \mathbf{y} \cdot \mathbf{Pre}[P, t] > \mathbf{y} \cdot \mathbf{m_0}$, contradiction.

A simpler (and weaker) sufficient condition for t dead is that there exists $p \in {}^\bullet t$ such that $\mathbf{sb}[p] < \mathbf{Pre}[p,t]$. For the application of the following results, the knowledge of $\mathbf{sb}$ is assumed, so this weaker condition can be applied with no extra computational effort.

Remark 31. Termination properties can often be rephrased in terms of fireability (or ***quasi-liveness***) of an artificial transition that removes the tokens from the desired final state and restores the initial marking (e.g., see [10]). Proving that such transition is dead, e.g., using (39) or (40), is sufficient — of course not necessary — to disprove termination. □

Although disregarding some transitions applying the above arguments may be helpful, typically the more drastic reduction in the number of systems to check is produced by the results that we present in the sequel. They provide rules to rewrite the disabledness condition of a transition in a less complex way while preserving the set of *integer* solutions to (37). (Notice that if the systems were finally checked disregarding the integrality of variables, these rules might diminish the decision power.)

Proposition 32. *Let t be a transition such that for every $p \in \pi \subseteq {}^\bullet t$ the following holds: $\mathbf{sb}[p] \leq \mathbf{Pre}[p,t]$. Replacing in (37) for the disabledness condition corresponding to transition t the following (less complex) condition:*

$$\left(\sum_{p\in\pi} \mathbf{m}[p] < \sum_{p\in\pi} \mathbf{Pre}[p,t]\right) \vee \left(\bigvee_{p\in{}^\bullet t\setminus\pi} \mathbf{m}[p] < \mathbf{Pre}[p,t]\right)$$

the set of integer *solutions is preserved.*

Proof. From $\mathbf{m}[p] \leq \mathbf{sb}[p]$ for every $p \in P$ and $\mathbf{m} \in \mathrm{LRS}^{\mathrm{SE}}(\mathcal{S})$, and $\mathbf{sb}[p] \leq \mathbf{Pre}[p,t]$ for every $p \in \pi$, it follows that $\sum_{p\in\pi}(\mathbf{m}[p] - \mathbf{Pre}[p,t]) \leq 0$ for every $\mathbf{m} \in \mathrm{LRS}^{\mathrm{SE}}(\mathcal{S})$. Since all the addends are non-positive, the "$<$" holds for the sum iff it holds for one of them. □

By the application of this result to a transition t, the number of linear systems to be solved is divided by $\frac{|{}^\bullet t|}{|{}^\bullet t|-|\pi|+1}$, what is deduced from the ratio between the number of input places to the transition and the actual number of them that need being considered separately.

In the particular case where $\pi = {}^\bullet t$, the disabledness condition is reduced to a linear inequality. Therefore a *single* system of linear inequalities is needed for structurally safe systems (i.e., those having all places with structural bound equal to one), for instance.

Also when *all but one* of the input places of a transition are such that their structural bound equals to the weight of the arc, the disabledness condition of the transition can be reduced to a linear inequality, applying the following result, which generalises Proposition 32:

Proposition 33. *Let t be a transition such that ${}^\bullet t = \pi \cup \{p'\}$, where* $\mathbf{sb}[p'] > 0$ *and* $\mathbf{sb}[p] \leq \mathbf{Pre}[p,t]$ *for every* $p \in \pi$. *Replacing in (37) for the disabledness condition corresponding to transition t the following (less complex) condition:*

$$\mathbf{sb}[p'] \cdot \sum_{p\in\pi} \mathbf{m}[p] + \mathbf{m}[p'] < \mathbf{sb}[p'] \cdot \sum_{p\in\pi} \mathbf{Pre}[p,t] + \mathbf{Pre}[p',t] \tag{41}$$

the set of integer *solutions is preserved.*

Proof. Let us rewrite first the condition in (37) for t disabled:

$$\left(\bigvee_{p\in\pi} \mathbf{m}[p] < \mathbf{Pre}[p,t]\right) \vee (\mathbf{m}[p'] < \mathbf{Pre}[p',t]) \tag{42}$$

It must be shown that (42) $\Leftrightarrow$ (41).

For "$\Rightarrow$", we distinguish two cases for t disabled:

1. Some $p'' \in \pi$ is such that $\mathbf{m}[p''] < \mathbf{Pre}[p'',t]$, so $\mathbf{m}[p''] \leq \mathbf{Pre}[p'',t] - 1$. Using also that $\mathbf{m}[p] \leq \mathbf{sb}[p]$ for every $p \in P$, and that $\mathbf{sb}[p] \leq \mathbf{Pre}[p,t]$ for every $p \in \pi$, we get:

$$\begin{aligned}\mathbf{sb}[p'] \cdot \textstyle\sum_{p\in\pi} \mathbf{m}[p] + \mathbf{m}[p'] =\\ \mathbf{sb}[p'] \cdot \textstyle\sum_{p\in\pi\setminus\{p''\}} \mathbf{m}[p] + \mathbf{sb}[p'] \cdot \mathbf{m}[p''] + \mathbf{m}[p'] \leq\\ \mathbf{sb}[p'] \cdot \textstyle\sum_{p\in\pi\setminus\{p''\}} \mathbf{Pre}[p,t] + \mathbf{sb}[p'] \cdot (\mathbf{Pre}[p'',t] - 1) + \mathbf{m}[p'] =\\ \mathbf{sb}[p'] \cdot \textstyle\sum_{p\in\pi} \mathbf{Pre}[p,t] - (\mathbf{sb}[p'] - \mathbf{m}[p']) \leq\\ \mathbf{sb}[p'] \cdot \textstyle\sum_{p\in\pi} \mathbf{Pre}[p,t] <\\ \mathbf{sb}[p'] \cdot \textstyle\sum_{p\in\pi} \mathbf{Pre}[p,t] + \mathbf{Pre}[p',t]\end{aligned}$$

2. Now p' is such that $\mathbf{m}[p'] < \mathbf{Pre}[p',t]$. Using that $\mathbf{m}[p] \leq \mathbf{sb}[p]$ for every $p \in P$, and that $\mathbf{sb}[p] \leq \mathbf{Pre}[p,t]$ for every $p \in \pi$, the result follows.

For "$\Leftarrow$", let us rewrite first (41):

$$\sum_{p\in\pi} (\mathbf{m}[p] - \mathbf{Pre}[p,t]) < \frac{\mathbf{Pre}[p',t] - \mathbf{m}[p']}{\mathbf{sb}[p']} \tag{43}$$

Assume contrary: (43) holds and t is enabled. In particular, $\mathbf{m}[p'] \geq \mathbf{Pre}[p',t]$, hence $\sum_{p\in\pi}(\mathbf{m}[p] - \mathbf{Pre}[p,t]) < 0$. Since $\mathbf{m}[p] - \mathbf{Pre}[p,t] \leq 0$ for every $p \in \pi$, it follows that there exists $p \in \pi$ such that $\mathbf{m}[p] - \mathbf{Pre}[p,t] < 0$. Thus, t is not enabled, against the hypothesis. □

By the application of this result to a transition t, the number of linear systems to be solved is obviously divided by $|{}^\bullet t|$.

Let us illustrate this latter rule. The idea is to describe the set of solutions to the state equation for which t is disabled by means of a linear inequality. In the case of Figure 10, the markings for which t is disabled are those for which $\mathbf{m}[p] < 3$ and those for which $\mathbf{m}[p'] < 1$. Observe that, taking into account that

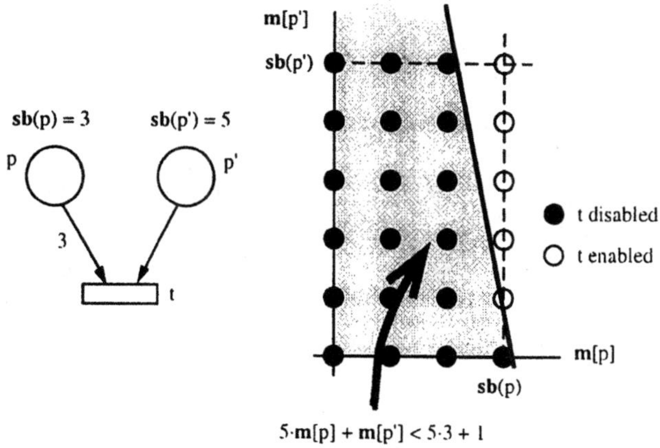

Fig. 10. Illustration of Proposition 33.

the solutions to the state equation respect, in particular, the structural bounds, all the solutions for which t is disabled are among the integer points in the region described by $5 \cdot \mathbf{m}[p] + \mathbf{m}[p'] < 5 \cdot 3 + 1$. Notice that the slope of the boundary of such region could be greater than that given by $\mathbf{sb}[p']$, provided it is not vertical. (Therefore, in case we finally check non-existence of solutions using integer programming, it makes no difference to use any greater value.)

In summary, up to now, apart from disregarding transitions with a stronger or equal precondition than others and dead transitions, we are able to write as a linear inequality the disabledness condition for all transitions $t \in T$ that have *at most* one input place p such that $\mathbf{sb}[p] > \mathbf{Pre}[p, t]$.

Proposition 32 can also be applied to reduce the number of terms when $|\pi - {}^\bullet t| > 1$, and $|\pi| > 1$, although in this case the disabledness condition is not reduced to a single term.

We can still further reduce the number of systems to solve by preapplying a transformation to the system that preserves deadlock-freeness (actually, it preserves the projected language). The transformation, illustrated in Figure 11, can be applied as needed to every place p with homogeneous weighting. After the transformation, we have one more transition ($t^{(p)}$ in the figure), the disabledness condition of which can be written as a linear inequality because the structural bound of $p^{(c)}$ is one. On the other hand, the structural bound of $p^{(b)}$ is also one, thus we have in each transition in $p^\bullet$ one input place less with structural bound greater than the weight (perhaps only one, or even none, remains, and then the disabledness condition for such transition can be written as a linear inequality too).

After the presented results, clearly the state equation based verification of deadlock-freeness reduces to checking non-existence of (integer) solution to a *single* linear system of inequalities in the case of structurally bounded P/T systems with homogeneous weighting — in particular ordinary and equal conflict

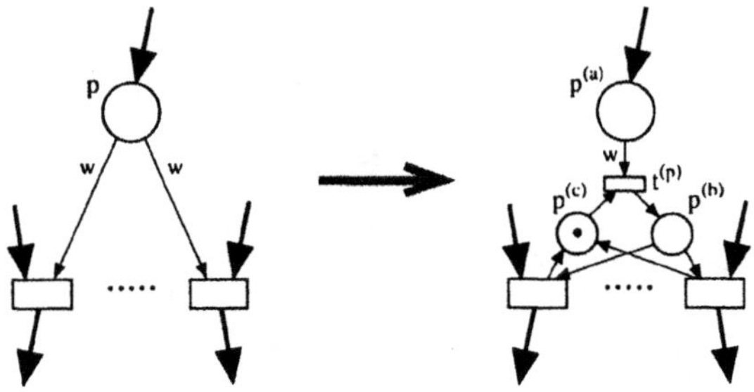

Fig. 11. A transformation preserving the projected language (in particular, preserving deadlock-freeness).

systems — because the transformation in Figure 11 can be applied as necessary to enable Propositions 32 or 33. Moreover, since every P/T system can be simulated by another with homogeneous weighting preserving the projected language (see Figure 12 for an illustrative example of the kind of transforma-

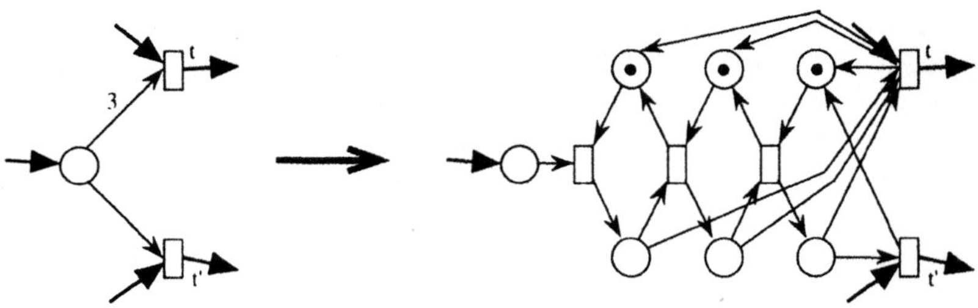

Fig. 12. Simulating weights with ordinary nets preserving the projected language.

tion used; more compact transformations exist for particular cases), it follows that *every* structurally bounded P/T system (or merely known to be k-bounded, because these can always be made structurally bounded using the complementary place construction) can be transformed to require a single linear system of inequalities:

Theorem 34. *Let S be a structurally bounded P/T system. Then (37) in the sufficient condition for deadlock-freeness given by Theorem 30 can be rewritten as a single system of linear inequalities.*

Remark 35. Since live equal conflict systems do not have "killing" spurious solutions (i.e., spurious deadlocks), existence of a solution guarantees non liveness. If no deadlock solution is found the system is proven deadlock-free, what in bounded strongly connected equal conflict systems implies liveness. Observe that applying the transformation illustrated in Figure 11 to an equal conflict system leads to another equal conflict system. This allows in this class to characterise liveness, in the presence of boundedness, through non-existence of integer solution to a single system of linear inequalities [92]. □

Let us illustrate the application of some of the above techniques to a simple example. Consider the system in Figure 13 (a), where all the structural marking

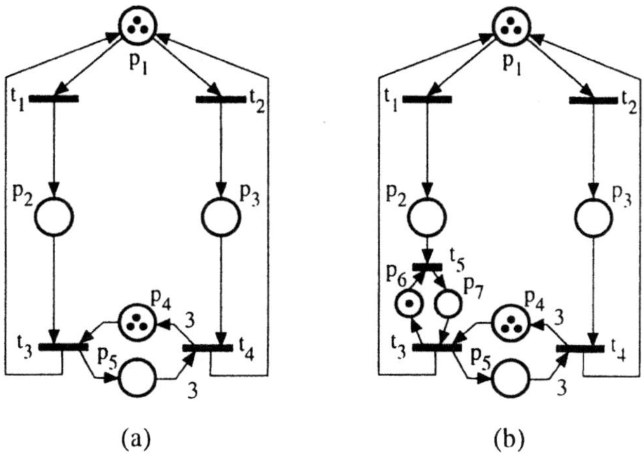

Fig. 13. Example of deadlock-freeness analysis.

bounds are three. The disabledness condition for t_2 can be disregarded. Transition t_1 has only one input place, so its disabledness condition is linear. Transition t_4 has only one input place (p_3) with structural bound greater than the weight, so its disabledness condition can be written linearly applying Proposition 33. The structural bounds of both input places of t_3 are greater than the corresponding weights, but we can apply the transformation in Figure 11 to one of them, e.g., to p_2, see Figure 13 (b). Now the disabledness conditions of t_3 and t_5 can both be written linearly.

Therefore, finally, the analysis can be done using a single linear system. That is, if $\mathbf{m}$ is a deadlock then the following linear system has a solution:

$$
\begin{aligned}
\mathbf{m} - \mathbf{C} \cdot \sigma = \mathbf{m_0} \wedge \mathbf{m}, \sigma \geq \mathbf{0} & \quad \text{(state eq.)} \\
\mathbf{m}[p_1] \leq 0 & \quad (t_1) \\
3 \cdot \mathbf{m}[p_5] + \mathbf{m}[p_3] \leq 9 & \quad (t_4) \\
3 \cdot \mathbf{m}[p_7] + \mathbf{m}[p_4] \leq 3 & \quad (t_3) \\
3 \cdot \mathbf{m}[p_6] + \mathbf{m}[p_2] \leq 3 & \quad (t_5)
\end{aligned}
$$

Solving (with the simplex algorithm) we obtain, for instance, the following solution: $\mathbf{m} = 3p_3 + 3p_4 + p_6$, $\sigma = 3t_2$, which in this case is actually reachable.

Remark 36. Notice that Proposition 33 and transformations like that illustrated in Figure 11 are intended to express the disabledness of a transition linearly.

Therefore, they are not only useful for deadlock-freeness analysis but for every analysis that requires checking the disabledness of some transitions. In particular, they can be applied to analyse the non fireability of *facts* [37]. □

5 Improving the State Equation

A major limitation of the analysis methods based on the state equation that we have described in the previous sections stems from the existence of ***spurious solutions***: only necessary or sufficient conditions for the analysed properties are obtained. As an example, for the net system in Figure 1, due to the spurious solutions, the state equation based analysis does not allow to prove safeness, deadlock-freeness, or pairwise mutual exclusion between p_2, p_3, and p_4. This limitation motivates the interest of the techniques that we present in this section to remove some, ideally all, of the spurious solutions (with respect to $\mathrm{LRS}^{\mathrm{SE}\mathbb{R}}$).

In Subsection 3.5 we observed that traps and siphons induce new invariant laws that add information to the state equation. We concentrate here on traps. Assume Θ is an initially marked trap of $\mathcal{S}$. To fix ideas, consider $\Theta = \{p_1, p_2, p_5\}$ of the net in Figure 1. An initially marked trap induces a trap invariant as an inequality, which in the example is:

$$\mathbf{m}[p_1] + \mathbf{m}[p_2] + \mathbf{m}[p_5] \geq 1 \tag{44}$$

Now assume that we have a P-semiflow $\mathbf{y}$ whose support includes Θ, which induces another marking invariant concerning the places of Θ. The P-semiflow $\mathbf{y} = 2p_1 + p_2 + p_3 + p_4 + p_5$ in the example, induces the marking invariant:

$$2\mathbf{m}[p_1] + \mathbf{m}[p_2] + \mathbf{m}[p_3] + \mathbf{m}[p_4] + \mathbf{m}[p_5] = 2 \tag{45}$$

By subtracting (44) from (45) we obtain another inequality invariant:

$$\mathbf{m}[p_1] + \mathbf{m}[p_3] + \mathbf{m}[p_4] \leq 1 \tag{46}$$

By introducing a *slack* variable, we can transform the latter inequality into an equality. The non-negative slack variable can be interpreted as the marking of a new place, p_Θ:

$$\mathbf{m}[p_1] + \mathbf{m}[p_3] + \mathbf{m}[p_4] + \mathbf{m}[p_\Theta] = 1 \tag{47}$$

Observe that the new place is marking (structurally) implicit. In the example, and taking into account (45):

$$\begin{aligned} \mathbf{m}[p_\Theta] &= 1 - (\mathbf{m}[p_1] + \mathbf{m}[p_3] + \mathbf{m}[p_4]) \\ &= 1 - (2 - (\mathbf{m}[p_1] + \mathbf{m}[p_2] + \mathbf{m}[p_5])) \\ &= \mathbf{m}[p_1] + \mathbf{m}[p_2] + \mathbf{m}[p_5] - 1 \end{aligned} \tag{48}$$

The place p_Θ coincides with the place p that we considered in Subsection 4.4 after Theorem 25. Remember that, when marked initially with the optimum value of (31), which is zero in this case, the addition of this place, which is (marking) implicit, removes some spurious solutions. Actually, from (48) it is not difficult to see that the removed solutions are precisely those that violate the trap invariant (44): both perspectives, in terms of trap invariants and marking (structurally) implicit places, are "dual" ways of interpreting the same removal of spurious solutions from the state equation.

The above reasoning shows that basic trap invariants, i.e., the information that a set of places must remain marked, can be *coded* in a marking structurally implicit place that is added to the original net system. The converse is also true: when a marking structurally implicit place removes, or *cuts*, spurious solutions, the cut can be interpreted in terms of trap invariants, as we show in Subsection 5.1. Although not complete, this leads to a procedure to add cutting implicit places to improve the accuracy of a given state equation, that we describe in Subsection 5.2. Making use of the linear algebraic formulation of the property that initially marked traps remain marked given in (18), it is also possible to improve the state equation by directly adding a generator of trap invariants in order to require this property, what is presented in Subsection 5.3.

5.1 Cutting Implicit Places

After realising that structural implicit places can improve the state equation by cutting spurious solutions, several questions naturally arise:

- Which structural implicit places do cut?
- Which spurious solutions are cut?
- Is it possible to eliminate in this way *any* spurious solution?

We devote this subsection to answering them.

Let p be a structurally implicit place with $\mathbf{m_0}[p]$ equal to the optimal value of (31), and let $\mathbf{y}$ and μ be a corresponding optimal solution. If $\mu \geq 0$, then

$$\mathbf{m}[p] = \mathbf{m_0}[p] + \mathbf{C}[p,T] \cdot \boldsymbol{\sigma} \geq \mathbf{y} \cdot (\mathbf{m_0}[P] + \mathbf{C}[P,T] \cdot \boldsymbol{\sigma}) = \mathbf{y} \cdot \mathbf{m}[P] \geq 0 \quad (49)$$

so the inequality $\mathbf{m}[p] \geq 0$ becomes *redundant* in the state equation ($\mathbf{m}[p]$ is a *non-extremal variable* in the terminology of convex geometry), hence its addition or removal does not affect $\mathrm{LRS}^{\mathrm{SE}\mathbb{R}}$. (In the particular case of live marked graphs, where $\mathrm{RS} = \mathrm{LRS}^{\mathrm{SE}\mathbb{R}}$, every implicit place is of this kind.) On the contrary, when $\mu < 0$ in every optimal solution, the constraint on non-negativity of $\mathbf{m}[p]$ is not redundant, so the state equation with p has less solutions that the state equation without it; since p is implicit, the difference are spurious solutions, precisely those where $\mathbf{m}[p]$ would have been negative. Taking into account that

$$\begin{aligned} \mathbf{m}[p] &= \mathbf{y} \cdot \mathbf{m_0}[P] + \mu + \mathbf{C}[p,T] \cdot \boldsymbol{\sigma} \\ &= \mathbf{y} \cdot (\mathbf{m}[P] - \mathbf{C}[P,T] \cdot \boldsymbol{\sigma}) + \mu + \mathbf{C}[p,T] \cdot \boldsymbol{\sigma} \end{aligned} \quad (50)$$

we can state that:

Theorem 37. *Let S be a P/T system. Let p be a structurally implicit place, and let $\mathbf{m_0}[p]$ be the optimal value of (31) corresponding to the optimal solution $\mathbf{y} \geq \mathbf{0}$ and $\mu < 0$. The addition of p to S cuts at least one spurious solution (with respect to* LRS^{SEIR}*), i.e., p is a* cutting implicit place, *iff there is no solution to*

$$\begin{array}{c} \mathbf{y}' \cdot \mathbf{C}[P,T] \leq \mathbf{C}[p,T] \\ \mathbf{y}' \cdot \mathbf{Pre}[P,p^\bullet] + \mu'\mathbf{1} \geq \mathbf{Pre}[p,p^\bullet] \\ \mathbf{y}' \cdot \mathbf{m_0}[P] + \mu' = \mathbf{m_0}[p] \\ \mathbf{y}', \mu' \geq \mathbf{0} \end{array} \tag{51}$$

The spurious solutions that are cut are those that fulfill:

$$\mathbf{y} \cdot \mathbf{m}[P] + (\mathbf{C}[p,T] - \mathbf{y} \cdot \mathbf{C}[P,T]) \cdot \boldsymbol{\sigma} < -\mu \tag{52}$$

In other words, the addition of the cutting implicit place p expresses the additional restriction that in every reachable marking the following holds:

$$\mathbf{y} \cdot \mathbf{m}[P] \geq -\mu - (\mathbf{C}[p,T] - \mathbf{y} \cdot \mathbf{C}[P,T]) \cdot \boldsymbol{\sigma} \tag{53}$$

Observe that $(\mathbf{C}[p,T] - \mathbf{y} \cdot \mathbf{C}[P,T]) \cdot \boldsymbol{\sigma} \geq 0$ and that it depends on $\boldsymbol{\sigma}$. For the case of a marking structurally implicit place, $\mathbf{C}[p,T] = \mathbf{y} \cdot \mathbf{C}[P,T]$, and $\mathbf{y} \cdot \mathbf{m}[P] \geq -\mu$. This reminds us a trap invariant, where $\|\mathbf{y}\|$ is the trap and $-\mu$ is the minimal (weighted) token content. To demonstrate this fact, rewriting $\mathbf{C}$ as $\mathbf{Post} - \mathbf{Pre}$ we get:

$$\mathbf{y} \cdot \mathbf{Post}[P,T] - \mathbf{y} \cdot \mathbf{Pre}[P,T] = \mathbf{Post}[p,T] - \mathbf{Pre}[p,T] \tag{54}$$

Assume without loss of generality that p is pure, i.e., $p^\bullet \cap {}^\bullet p = \emptyset$. (If p was not pure, the same place cancelling the self-loops would be also marking structurally implicit, possibly with a lesser value of μ, i.e., with a "more negative" μ.) Considering separately $p^\bullet$, ${}^\bullet p$, and $T - (p^\bullet \cup {}^\bullet p)$, we have:

$$\mathbf{y} \cdot \mathbf{Post}[P,p^\bullet] - \mathbf{y} \cdot \mathbf{Pre}[P,p^\bullet] = -\mathbf{Pre}[p,p^\bullet] < \mathbf{0} \tag{55}$$

$$\mathbf{y} \cdot \mathbf{Post}[P,{}^\bullet p] - \mathbf{y} \cdot \mathbf{Pre}[P,{}^\bullet p] = \mathbf{Post}[p,{}^\bullet p] > \mathbf{0} \tag{56}$$

$$\mathbf{y} \cdot \mathbf{Post}[P,T - (p^\bullet \cup {}^\bullet p)] - \mathbf{y} \cdot \mathbf{Pre}[P,T - (p^\bullet \cup {}^\bullet p)] = \mathbf{0} \tag{57}$$

Since $\mathbf{y} \cdot \mathbf{Pre}[P,p^\bullet] + \mu\mathbf{1} \geq \mathbf{Pre}[p,p^\bullet]$, because $\mathbf{y}$, μ is a solution to (31), Equation (55) becomes:

$$\mathbf{y} \cdot \mathbf{Post}[P,p^\bullet] \geq -\mu\mathbf{1} > \mathbf{0} \tag{58}$$

From (56–58), for every transition t we have $\mathbf{y} \cdot (\mathbf{Post}[P,t] - \mathbf{Pre}[P,t]) \geq 0$. It follows that if $\mathbf{y} \cdot \mathbf{Pre}[P,t] \neq 0$, then $\mathbf{y} \cdot \mathbf{Post}[P,t] \neq 0$, i.e., if $t \in \|\mathbf{y}\|^\bullet$ then $t \in {}^\bullet\|\mathbf{y}\|$, so $\|\mathbf{y}\|$ is effectively a trap. In summary:

Theorem 38. *Let S be a P/T system. Let p be a marking structurally implicit place, and let $\mathbf{m_0}[p]$ be the optimal value of (31) corresponding to the optimal solution $\mathbf{y} \geq \mathbf{0}$ and $\mu < 0$. If p is a cutting implicit place (according to Theorem 37) then $\|\mathbf{y}\|$ is a trap of $\mathcal{N}$ such that its weighted token content, defined by $\mathbf{y} \cdot \mathbf{m}[P]$, is never less than $-\mu$.*

In order to further illustrate the "dual" interpretation of the cut in terms of implicit places and traps, and also to introduce the limitations of this approach, consider the example shown in Figure 14, which is a most spare PN representation of a prototype distributed mutual exclusion algorithm. Let $q = r = 1$, to

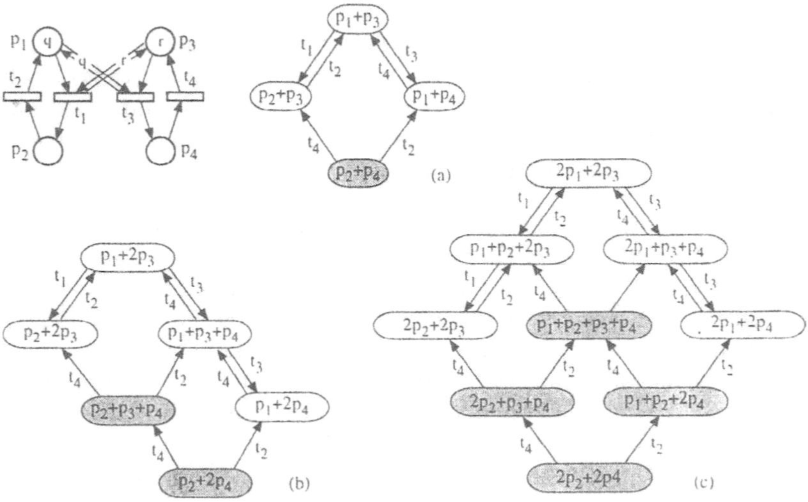

Fig. 14. A family of P/T systems where p_2 and p_4 are in mutual exclusion, and the LRG^{SE} for (a) $q = r = 1$, (b) $q = 1, r = 2$, and (c) $q = r = 2$.

start with. The marking $p_2 + p_4$ is a (spurious) solution to the state equation, which prevents proving mutual exclusion of p_2 and p_4 using the state equation (notice that the same happens in the net where the self-looped transitions are splitted, which is merely more cumbersome). The place p: $\mathbf{C}[p, T] = t_2 + t_4 - t_1 - t_3$, initially marked with one token, is marking implicit; an optimal solution to (31) is $\mathbf{y} = p_1 + p_3$, $\mu = -1$. Its addition cuts the markings where $\mathbf{y} \cdot \mathbf{m} = \mathbf{m}[p_1] + \mathbf{m}[p_3] < 1 = -\mu$, that is, the markings where the initially marked trap $\{p_1, p_3\}$ is unmarked. In this case the only spurious marking is cut, so mutual exclusion can be proven using the state equation method in the net with the implicit place.

Let $q = 1$ and $r = 2$. Now the markings $p_2 + p_3 + p_4$ and $p_2 + 2p_4$ are spurious (and prevent us from proving mutual exclusion the same as before). The place p used above, marked with two tokens, is marking implicit, but does not allow to prove mutual exclusion because it does not cut the spurious marking $p_2 + p_3 + p_4$. Fortunately, a different choice of the weights of the arcs helps. The place p': $\mathbf{C}[p', T] = 2t_2 + t_4 - 2t_1 - t_3$, initially marked with two tokens, is marking implicit; an optimal solution to (31) is $\mathbf{y} = 2p_1 + p_3$, $\mu = -2$. Its addition cuts the markings where $2\mathbf{m}[p_1] + \mathbf{m}[p_3] < 2$, that is, the markings where the weighted token content of the initially marked trap $\{p_1, p_3\}$ is *less than two*. (Again in this case all the spurious markings are cut.) What is remarkable in this case is the fact that we are removing even spurious markings, namely $p_2 + p_3 + p_4$, in which the trap *is marked* (with a weighted token content below a minimum, though). We shall come back to this case in Subsection 5.2.

Finally, let $q = r = 2$. Although the addition of the place p: $\mathbf{C}[p,T] = t_2 + t_4 - t_1 - t_3$, initially marked with two tokens, cuts the markings where $\mathbf{m}[p_1] + \mathbf{m}[p_3] < 2$, the spurious marking $p_1 + p_2 + p_3 + p_4$ is not cut. Notice that this spurious marking *cannot* be cut by any other place because it is a positive linear combination of two reachable markings: $2p_2 + 2p_3$ and $2p_1 + 2p_4$ (the reachability set is *not convex* in this case). This example manifests the incompleteness of the addition of cutting implicit places in order to remove spurious solutions: they *improve* the linear description, but *not always* completely.

Remark 39. The incompleteness of the addition of cutting implicit places refers to applying this method *alone*. It is conceivable that the net to be analysed can be transformed in such a way that the properties under study are preserved and in the transformed net spurious solutions can be removed by cutting implicit places, e.g., with the transformation rule shown in Figure 11. For instance, mutual exclusion of p_2 and p_4 of the net system in Figure 14 with $q = r = 2$ is equivalent to mutual exclusion of p'_2 and p_4 in the net system in Figure 15 (a). Place p shown in (b) with $\mathbf{m_0}[p] = 2$ is marking (structurally) implicit and removes the spurious solutions where p'_2 and p_4 are simultaneously marked, what allows to conclude through state equation analysis that p_2 and p_4 were in mutual exclusion in the original system.

It is remarkable that in this case the initial marking of p computed from (31) is four instead of two, so it *is not* the minimal marking that makes p implicit, showing that Theorem 23 is only sufficient for p implicit. □

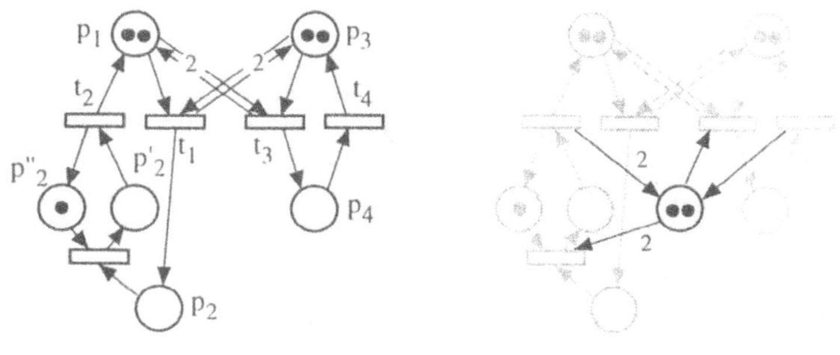

(a) (b)

Fig. 15. Mutual exclusion of p_2 and p_4 of the net system in Figure 14 with $q = r = 2$ is equivalent to mutual exclusion of p'_2 and p_4 in the net system (a). Adding the implicit place p shown in (b) allows to prove mutual exclusion.

5.2 Improving the State Equation with Implicit Places

When faced to a net system the state equation of which is to be improved by adding cutting implicit places, a question arises: how can we select the candidate places? We outline here some indications to answer this question.

From the association of cutting implicit places and initially marked traps that are insufficiently marked in some spurious solutions, the following procedure naturally comes to mind:

1. Compute (minimal) initially marked traps.
2. For each trap Θ a marking structurally implicit place is obtained as $\mathbf{C}[p_\Theta, T] = \mathbf{1}_\Theta \cdot \mathbf{C}[P, T]$, that is, taking $\mathbf{y} = \mathbf{1}_\Theta$ in Theorem 25.
3. Let the initial marking for p_Θ be the optimal value of (31) fixing $\mathbf{y} = \mathbf{1}_\Theta$. If $\mu < 0$ in the optimal solution, then apply Theorem 37 to verify that it cuts and to obtain the expression of the achieved cut, and add the place if it cuts.

The above procedure removes all the spurious solutions that can be proven non reachable observing that the *non weighted* token content of an initially marked trap is less than a minimum value, particularly those where the trap is unmarked. For instance, to improve the state equation of the net system in Figure 1, we would consider the initially marked traps $\Theta_1 = \{p_1, p_2, p_5\}$, $\Theta_2 = \{p_1, p_3, p_5\}$, and $\Theta_3 = \{p_1, p_2, p_3, p_4\}$, that lead to the corresponding marking structurally implicit places shown in Figure 16 (a). Their initial markings,

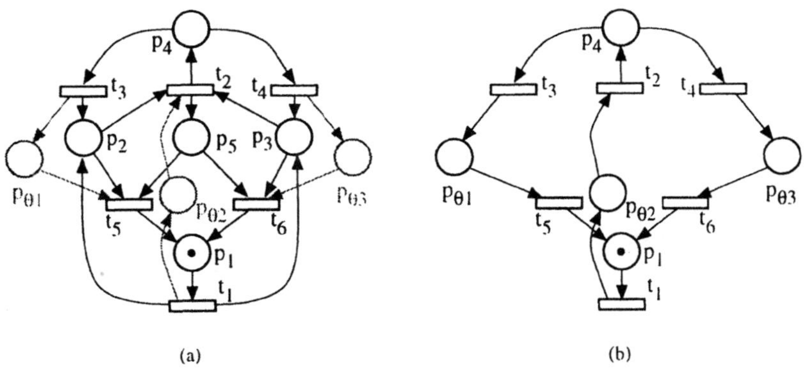

Fig. 16. Adding implicit places to the net system in Figure 1 cuts every spurious solution. Some original places become implicit when the cutting implicit places are added.

obtained from (31), are all zero, and all of them cut spurious solutions (we have already discussed in detail the case of p_{Θ_1}). Actually, in this case, the reader can easily check that *every* spurious solution is cut. It can also be checked that places p_2, p_3, and p_5 are implicit in the net system of Figure 16 (a). Their removal leads to the system in Figure 16 (b), which is a live and safe state machine isomorphic to the reachability graph.

Remark 40. If a different **y** is taken in the above procedure to compute cutting implicit places from given traps, different solutions appear. For the example in Figure 1 and 16, an alternative set of cutting implicit places is shown in [23]. □

Let us come back to the example of Figure 14 with $q = 1$ and $r = 2$. If we apply the above procedure, from the initially marked trap $\Theta_1 = \{p_1, p_3\}$ we obtain the marking structurally implicit place p_{Θ_1}: $\mathbf{C}[p_{\Theta_1}] = t_2 + t_4 - t_1 - t_3$. The initial marking, obtained from (31), is two, corresponding to $\mu = -1$, so the addition of p_{Θ_1} introduces the new P-invariant $\mathbf{m}[p_2]+\mathbf{m}[p_4]+\mathbf{m}[p_{\Theta_1}] = 2$, which cuts the spurious solution $p_2 + 2p_4$ but not $p_2 + p_3 + p_4$. In principle, it seems that the latter spurious solution cannot be cut because, as we discussed, the matter is the *weighted* token content of the trap, not merely the token content. It may be surprising at first sight that it is cut if the above procedure *is applied iteratively*. A little thought reveals that this is quite natural, since the addition of new places originates new traps, so new chances for improving. (Of course, only traps including newly added places are useful now.) Effectively, besides other traps, after the addition of p_{Θ_1}, $\Theta_2 = \{p_1, p_{\Theta_1}\}$ is an initially marked trap, that leads to p_{Θ_2}: $\mathbf{C}[p_{\Theta_2}] = 2t_2 + t_4 - 2t_1 - t_3$. As we saw, when initially marked with two tokens, this place removes all the spurious solutions. (In fact, once p_{Θ_2} is added, p_{Θ_1} becomes redundant and can be removed.)

We can proceed until no spurious solution is removed, i.e., no cutting implicit place is added, in an iteration. (Naturally, the procedure stops after a finite number of iterations in structurally bounded nets, because the number of possible spurious solutions is finite.) We insist that this *does not* guarantee that every spurious solution has been removed.

Remark 41. An alternative way of finding a cutting implicit place, in the case that we are trying to disprove reachability of some (sub)markings that we suspect are spurious, is postulating a *monitor place* that forbids reaching them, and then analysing whether it is implicit or not. Consider again the example of Figure 14 with $q = 1$ and $r = 2$. As we want to prove that p_2 and p_4 are in mutual exclusion, but the state equation does not allow us to conclude, we introduce a place that forces the mutual exclusion of p_2 and p_4, that is a place p such that $2\mathbf{m}[p_2] + \mathbf{m}[p_4] + \mathbf{m}[p] = 2$ becomes a (new) P-invariant. As we know, the place is p: $\mathbf{C}[p, T] = 2t_2 + t_4 - 2t_1 - t_3$, initially marked with two tokens, which is effectively implicit, so we are done.

This approach is specially suitable for structurally safe systems, where the place that forbids a given (sub)marking $\mathbf{m} = \mathbf{1}_{\|\mathbf{m}\|}$ is simply p: $\mathbf{C}[p, T] = -\mathbf{m} \cdot \mathbf{C}[P, T]$, initially marked with $\|\mathbf{m}\| - 1 - \mathbf{m} \cdot \mathbf{m_0}[P]$ tokens (this is the kind of *constraints* introduced by Patil, see [67]). Actually, since doing so we can remove any one precise marking, if the spurious solutions were known, all of them could be removed. This implies that for any given structurally repetitive and structurally safe system a "place completed" version with the same behaviour exists such that $\mathrm{RS} = \mathrm{LRS}^{\mathrm{Psf}}$. □

5.3 Improving the State Equation with a Generator of Trap Invariants

Equation (18) expresses linearly that initially marked traps remain marked. In order to add this condition to the state equation, it must be stated in terms of existence of solution, what can be done using the alternatives theorem. Doing so, we obtain that initially marked traps are marked under $\mathbf{m}$ iff there is a solution to:

$$\mathbf{C}_\Theta \cdot \mathbf{x}' + x'\mathbf{m}_0 - \alpha\mathbf{m} \leq \mathbf{0} \wedge \mathbf{x}' \geq \mathbf{0} \wedge x' > 0 \wedge \alpha \geq 0 \tag{59}$$

If a solution with $\alpha = 0$ exists, then so does a solution with $\alpha > 0$. Dividing (59) by α we get:

$$\mathbf{C}_\Theta \cdot \mathbf{x} + x\mathbf{m}_0 - \mathbf{m} \leq \mathbf{0} \wedge \mathbf{x} \geq \mathbf{0} \wedge x > 0 \tag{60}$$

This condition can be interpreted as a generator of trap invariants. It can be added to the state equation leading to the following *improved state equation* (integrality of $\boldsymbol{\sigma}$ can be required for better accuracy, while $\mathbf{x}$ and x are rational because they originate in the transformation of (18) by the alternatives theorem):

Theorem 42. *Let $\mathcal{S}$ be a P/T system. If $\mathbf{m} \in \mathrm{RS}(\mathcal{S})$ then it is a solution to:*

$$\begin{aligned} \mathbf{m} - \mathbf{C} \cdot \boldsymbol{\sigma} &= \mathbf{m}_0 \\ \mathbf{m} - x\mathbf{m}_0 - \mathbf{C}_\Theta \cdot \mathbf{x} &\geq \mathbf{0} \\ \mathbf{m}, \boldsymbol{\sigma}, \mathbf{x} &\geq \mathbf{0} \\ x &> 0 \end{aligned} \tag{61}$$

where $\mathbf{C}_\Theta$ is as defined in Theorem 13.

Remark 43. Since in live, bounded, and reversible free choice systems the reachable markings are the vectors in $\mathrm{LRS}^{\mathrm{Psf}}$ that mark every trap [30], Theorem 42 characterises reachability. □

This method has the advantage that no previous computation of traps is required. (It can be said that the method based on implicit places is "compiled", in the sense that trap invariants obtained in some previous or off-line computation are incorporated or coded as part of the — transformed — net. Following the same analogy, the method based on a generator of trap invariants is "interpreted", in the sense that markings that are solution to the state equation but violate the trap condition are eliminated on-line.)

Nevertheless, the gain in efficiency of this method is paid by the loss in accuracy compared to the method based on implicit places: only markings where an initially marked trap is unmarked are cut. For instance, the marking $p_2 + p_3 + p_4$ in the net system of Figure 14 with $q = 1$ and $r = 2$ is still spurious with respect to the improved state equation because no trap is unmarked.

The advantages of both methods can be combined by (iteratively) adding cutting implicit places to a given net up to a certain point, and then using the improved state equation (i.e., incorporating the generator of trap invariants) instead of the plain state equation for the analysis of properties.

6 Structural Liveness and the Rank Theorems

We have encountered that the state equation method is best suited to the analysis of properties formulated in terms of existence or non-existence of markings (and firing vectors), such as marking bounds, mutual exclusion, or even deadlock-freeness, for which it provides at least necessary or sufficient conditions. For other properties the statement of which combines existential and universal quantifiers, e.g., liveness: *for every* reachable marking *there exists* a reachable successor that enables t, the direct application of the state equation method does not allow to reach any conclusion.

Of course, the sufficient condition for deadlock-freeness is useful for liveness analysis (deadlock-freeness is necessary for liveness; even in some cases, e.g., bounded strongly connected equal conflict systems, it is sufficient). But this is not the only way in which the incidence matrix of a net can be exploited to facilitate the analysis of liveness. In this section we describe in some detail efficient tests — polynomial time — that give necessary conditions for the existence of an initial marking that makes a given net live (and bounded). Such necessary conditions are also sufficient in some net subclasses.

6.1 The Rank Theorem: A General Necessary Condition for Liveness and Boundedness

A well-known polynomial time necessary condition for liveness and boundedness of a net system, based solely on purely structural properties, is strong connectedness [77] and consistency (Proposition 10) of the net. (As stated by Theorem 20, for structural boundedness and structural liveness, conservativeness and consistency are necessary.)

These conditions are very useful to discard models that are not correct before undertaking a more costly analysis. Unfortunately they are only necessary: there are strongly connected and consistent nets that cannot be lively and boundedly marked (see Figure 17).

We present here an improved — and still polynomial time — necessary condition, that incorporates an upper bound for the rank of the incidence matrix, namely that it must be less than the number of equal conflict sets of the net. We make use of *circuit arbiters*, which are a particular class of the *regulation nets* of [47] that we use to regulate non-trivial equal conflicts. For an equal conflict set $e \in$ SEQS, its circuit arbiter is defined as follows:

Definition 44. Let $\mathcal{N}$ be a P/T net, and let $e \in$ SEQS such that $|e| > 1$. A net $\mathcal{A}_e = \langle P_e, e, \mathbf{Pre}_e, \mathbf{Post}_e \rangle$ is an *(ordinary) circuit arbiter* for the equal conflict set e iff $\mathcal{A}_e$ is an ordinary net such that $P_e \cap P = \emptyset$ and its underlying graph is an elementary circuit.

Some straightforward properties of these arbiters are: being circuits, they have the same number of places and transitions, i.e., $|e|$; the set of places of a

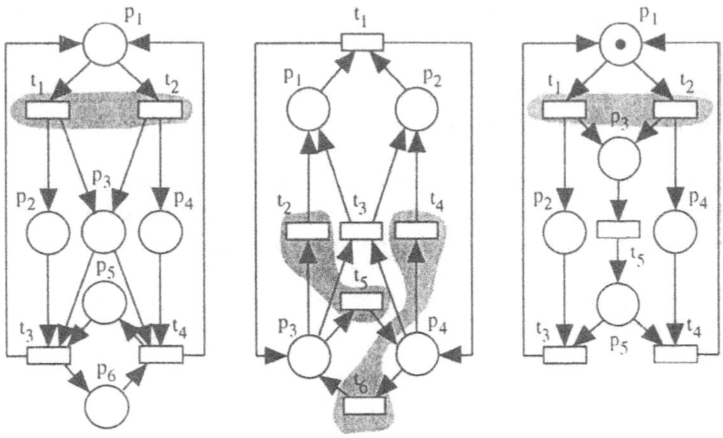

Fig. 17. Three conservative and consistent nets with rank(**C**) = 3. Their (non-trivial) equal conflict sets are shaded. The nets (a) and (b) cannot be lively and boundedly marked, while (c) is live and bounded with the marking shown.

circuit arbiter in a net is the support of a minimal P-semiflow; with every non-empty initial marking, a circuit arbiter is live, bounded, and reversible. Figure 18 represents a circuit arbiter (shaded places) merged on an equal conflict set.

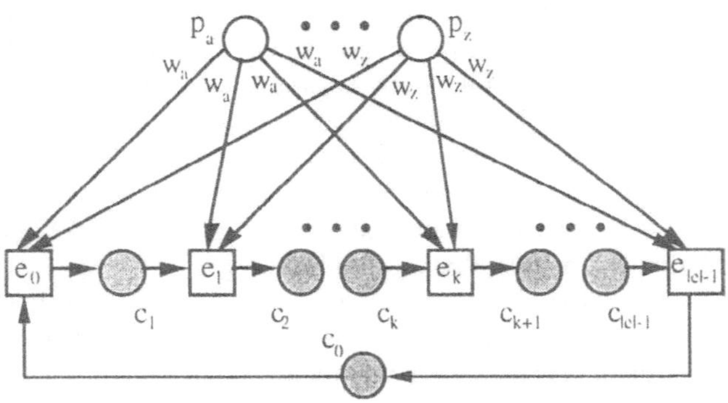

Fig. 18. A circuit arbiter (shaded places) merged on an equal conflict set.

Theorem 45 (The rank theorem). *If* S *is a live and bounded P/T system, then* $\mathcal{N}$ *is strongly connected, consistent, and* $\text{rank}(\mathbf{C}) < |\text{SEQS}|$.

A weaker but more "symmetric" statement, that clearly shows the extension of Theorem 20, is:

Corollary 46. *If $\mathcal{N}$ is a structurally live and structurally bounded P/T net, then $\mathcal{N}$ is conservative, consistent, and* $\text{rank}(\mathbf{C}) < |\text{SEQS}|$.

Lemma 47. *Let $\mathcal{S}$ be a P/T net, and let $e \in$ SEQS such that $|e| > 1$. Let $\mathcal{A}_e$ be a circuit arbiter for e, and let $\mathcal{N}'$ be the net $\mathcal{N}$ merged with the circuit arbiter $\mathcal{A}_e$ sharing the transitions in e. If $\mathcal{S}$ is live and bounded then*

1. *A marking $\mathbf{m_0}'$ with $\mathbf{m_0}'[P] = \mathbf{m_0}$ such that $\mathcal{S}' = \langle \mathcal{N}', \mathbf{m_0}' \rangle$ is live and bounded exists.*
2. $\text{rank}(\mathbf{C}') = \text{rank}(\mathbf{C}) + |e| - 1$

Proof (of Lemma 47). For Part 1, boundedness of $\mathcal{S}$ and conservativeness of $\mathcal{A}_e$ guarantee boundedness of $\mathcal{S}'$ for every $\mathbf{m_0}'$ with $\mathbf{m_0}'[P] = \mathbf{m_0}$. Since $\mathcal{S}$ is live and bounded, then the number $r_e = \max\{\min\{\#(e, \sigma) \mid \sigma t \in L(\mathcal{N}, \mathbf{m})\} \mid t \in T \wedge \mathbf{m} \in RS(\mathcal{S})\}$ is well-defined. This is a bound for the number of firings of transitions in e that are *required* to enable an arbitrary transition from an arbitrary reachable marking. We put r_e tokens in each place in P_e, what completes the definition of $\mathbf{m_0}'$ and now we prove that $\mathcal{S}'$ is live. Let $\mathbf{m}' \in RS(\mathcal{S}')$ and $t \in T$. We shall prove that t can ultimately be enabled from $\mathbf{m}'$. We claim that there exists a marking $\mathbf{m}'' \in RS(\mathcal{N}', \mathbf{m}')$ such that $\mathbf{m}''[P_e] = \mathbf{m_0}'[P_e]$. In that case, since (1) $\mathcal{S}$ is live, (2) $\mathbf{m}''[P] \in RS(\mathcal{S})$, and (3) $\mathbf{m_0}'[P_e]$ has been defined in a way that it does not interfere when firing a sequence to enable an arbitrary t from an arbitrary reachable marking, then we can fire in $\langle \mathcal{N}', \mathbf{m}'' \rangle$ the same sequence that we could fire in $\langle \mathcal{N}, \mathbf{m}''[P] \rangle$ in order to enable t. To prove the claim, let $\sigma_e = e_{i_1} e_{i_2} \cdots e_{i_k} \in L(\mathcal{A}_e, \mathbf{m}'[P_e])$ be such that $\mathbf{m}'[P_e] \xrightarrow{\sigma_e} \mathbf{m_0}'[P_e]$, i.e., a sequence in the circuit arbiter returning to the initial marking. It is easy to see that a sequence such that its projection on e is σ_e can be fired in $\langle \mathcal{N}', \mathbf{m}' \rangle$. The idea is firing transitions not in e, which does not affect the marking of places in P_e, until e are P-enabled (their input places in P have enough tokens, no matter how many tokens are there in other places), which will eventually happen thanks to liveness of $\langle \mathcal{N}, \mathbf{m}'[P] \rangle$, then firing e_{i_1} which is also P_e-enabled according to our definition of σ_e, then firing more transitions not in e until e are P-enabled again, then firing e_{i_2} which is also P_e-enabled, etc.

To prove Part 2, for $\text{rank}(\mathbf{C}') = \text{rank}(\mathbf{C}) + |e| - 1$, we shall prove that $|e| - 1$ out of the $|e|$ rows corresponding to the places of the circuit arbiter are linearly independent. Let us fix a notation for the equal conflict set and the circuit arbiter (see Figure 18):

- $e = \{e_0, e_1, \ldots, e_k, \ldots, e_{|e|-1}\}$
- $P_e = \{c_0, c_1, \ldots, c_k, \ldots, c_{|e|-1}\}$
- $\mathbf{Pre}_e[c_i, e_i] = 1$ and $\mathbf{Post}_e[c_{i \oplus 1}, e_i] = 1$ (otherwise zeros), where $\oplus$ represents the sum modulo $|e|$.

It is clear that there is one row being a linear combination of the rest, for instance $\mathbf{C}[c_0, T] = -\sum_{p \in P_e \setminus \{c_0\}} \mathbf{C}[p, T]$, so we remove it and then we prove that the rows corresponding to places in $P_e \setminus \{c_0\}$ are all linearly independent. Assume, on the contrary, that c_k, where $1 \leq k \leq |e|-1$, is a linear combination of the other places (let *the other places* be denoted by $OP = P \cup P_e \setminus \{c_0, c_k\}$):

$$\mathbf{C}[c_k, T] = \sum_{p \in OP} \boldsymbol{\lambda}[p] \cdot \mathbf{C}[p, T] = \boldsymbol{\lambda} \cdot \mathbf{C}[OP, T] \tag{62}$$

Thus, the marking increment produced by a sequence σ should also be a linear combination of the marking increment of the other places:

$$\Delta\mathbf{m}[c_k] = \mathbf{C}[c_k, T] \cdot \boldsymbol{\sigma} \stackrel{by\ (62)}{=} \boldsymbol{\lambda} \cdot \mathbf{C}[OP, T] \cdot \boldsymbol{\sigma} = \boldsymbol{\lambda} \cdot \Delta\mathbf{m}[OP] \tag{63}$$

Clearly, it is possible to fire in $\mathcal{S}$ a sequence σ such that $\#(e_i, \sigma) =$ **if** $i < k$ **then** ω **else** 0, where ω is arbitrarily large. In that case $\Delta\mathbf{m}[c_k] = \mathbf{C}[c_k, T] \cdot \boldsymbol{\sigma} = \omega$ is arbitrarily large, while all the entries in $\Delta\mathbf{m}[OP]$ are finite, what contradicts (63). □

Proof (of Theorem 45). Only the rank condition needs to be proven. Let $\mathcal{N}'$ be the net $\mathcal{N}$ together with circuit arbiters merged to *every* non-trivial equal conflict set. Applying Lemma 47.2 repeatedly after each circuit arbiter is merged, what can be done thanks to Lemma 47.1, it follows that:

$$|T| - 1 \geq \text{rank}(\mathbf{C}') = \text{rank}(\mathbf{C}) + \sum_{e \in \text{SEQS}} (|e| - 1)$$

Rearranging the above inequality we obtain a bound for the rank:

$$\text{rank}(\mathbf{C}) \leq |T| - \sum_{e \in \text{SEQS}} (|e| - 1) - 1$$

Since $\sum_{e \in \text{SEQS}} |e| = |T|$, this bound is $|\text{SEQS}| - 1$, so the result follows. □

In the example of Figure 17 (a), Theorem 45 allows to decide that the net cannot be lively and boundedly marked. Both nets in Figure 17 (b) and (c) "pass" the test of Theorem 45, although only the latter can be lively and boundedly marked, e.g., with the marking shown. In summary, the rank condition in Theorem 45 allows indeed to discriminate some cases but, unfortunately, not all of them, that is, it is not sufficient.

In the next subsection we seek for cases where the sufficiency holds, but before let us briefly discuss on the kind of situations leading to non structural liveness that the rank theorem detects. Intuitively, from the proof of Theorem 45, it becomes apparent that the rank condition fails when some individual "choices between alternatives" are not independent from the rest of the system (synchronisations, other choices, etc.). Apart from "flow problems" (i.e., absence of consistency), absence of such independence indicates that a wrong decision taken in an individual choice may affect the rest of the system (to the point of "killing" it).

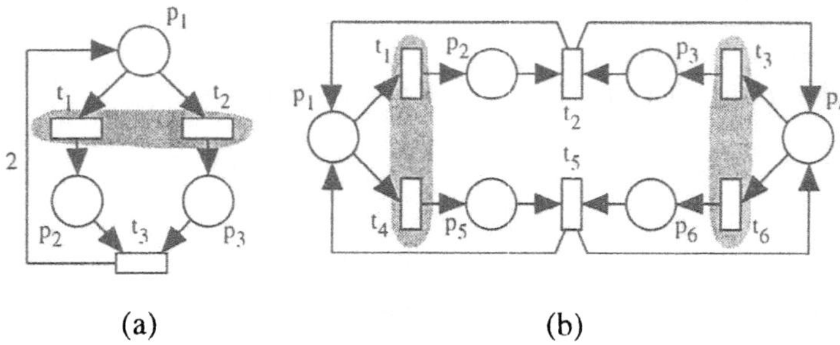

Fig. 19. Two conservative and consistent nets where the rank theorem detects non structural liveness. The (non-trivial) equal conflicts are shaded.

Let us illustrate these ideas with a couple of simple examples where the rank theorem detects non structural liveness. In the net of Figure 19 (a), the only minimal T-semiflow is **1**, and it is also a basis of T-flows, so rank(**C**) $= 2 =$ |SEQS|. Notice that the fact that t_1 and t_2 are together in every T-semiflow — what is due to the synchronisation or join transition t_3 — means that in every infinite sequence they should be fired in a fixed proportion (one to one in this case). Nevertheless, since the choice between t_1 and t_2 is free, the net does not prevent that this proportion is violated. This mismatch between conflicts and synchronisations is what the rank theorem detects. Observe that if we merge a circuit arbiter on t_1 and t_2, say c_0 from t_2 to t_1 and c_1 from t_1 to t_2, the rank is not increased: one place is clearly a linear combination of the other, say $\mathbf{C}[c_0, T] = -\mathbf{C}[c_1, T]$; but also c_1 is a linear combination of other places, namely $\mathbf{C}[c_1, T] = \mathbf{C}[p_1, T] + 2\mathbf{C}[p_2, T]$, what reveals the problem. In terms of implicit places, both places of the circuit arbiter are structurally implicit, so they can be made implicit with a large enough initial marking. Implicitness of the arbiter reveals that the choice is not free. (Remarkably, for $\mathbf{m_0} = 2p_1$, when the arbiter is marked with only one token the system with the arbiter is live. Notice that in such case the arbiter places are *not* implicit, actually they affect the behaviour avoiding the deadlocks. Increasing the marking of the arbiter places eventually makes them implicit, while it destroys liveness — liveness is not monotonic.)

In the net of Figure 19 (b), the minimal T-semiflows are $t_1 + t_2 + t_3$ and $t_4 + t_5 + t_6$, and they are also a basis of T-flows, so rank(**C**) $= 4 =$ |SEQS|. Now the synchronisations (t_2 and t_5) do not impose a given resolution of each conflict to allow infinite activity (the outcomes of each conflict are in different minimal T-semiflows), but they impose that each conflict is solved according to the other, what is again not guaranteed by the net structure where the choices are free. The rank theorem detects also this mismatch. If we merge a circuit arbiter, say on t_1 and t_4, it increases the rank. Now the net with the arbiter has a unique minimal T-semiflow, **1**, and the second circuit arbiter does not increase the rank, in the same way as in the previous example (after merging an arbiter on one conflict, a proportion between the outcomes of the other conflict has been fixed).

6.2 The Rank Theorem for Some Subclasses

For certain subclasses, the general necessary condition for structural liveness (Theorem 45) has been proven to be sufficient too. Loosely speaking, these subclasses have in common that their syntactical constraints leave only conflicts that are essentially "choices between alternatives" (equal conflicts), so representing competition or resource sharing is very limited when not forbidden. This is particularly the case of equal conflict systems (it is also trivially the case of state machines or marked graphs):

Theorem 48. *Let $\mathcal{N}$ be an equal conflict net.*

A marking $\mathbf{m_0}$ *such that $\mathcal{S}$ is live and bounded exists iff $\mathcal{N}$ is strongly connected, consistent, and* $\text{rank}(\mathbf{C}) = |\text{SEQS}| - 1$.

Proof (Sketch). The necessity part is the general rank theorem, where $\text{rank}(\mathbf{C}) < |\text{SEQS}|$ reduces to $\text{rank}(\mathbf{C}) = |\text{SEQS}| - 1$ because a live and bounded equal conflict net where circuit arbiters have been merged to every equal conflict set has a unique minimal T-semiflow. For existence of this T-semiflow, notice that the "arbitered" net must be consistent. For unicity, it suffices to show that the support of every T-semiflow is the whole T. Let $\mathbf{x}$ be a T-semiflow of the arbitered net, and let $t \in \|\mathbf{x}\|$. All the transitions in CCS(t) are also in $\|\mathbf{x}\|$ because the places in the circuit arbiters have only one output transition. Every output place of the transitions in CCS(t) must have at least one output transition in $\|\mathbf{x}\|$, so we can apply repeatedly the same argument and, by strong connectedness, all the transitions are shown to be in $\|\mathbf{x}\|$.

The sufficiency part requires a closer investigation of the structural properties of the net that is out of the scope of this work, but we outline here the proof, referring to [92]. Consistency and $\text{rank}(\mathbf{C}) = |\text{SEQS}| - 1$ imply *P-allocatability* (see [92, Theorem 20]). Strong connectedness and P-allocatability imply conservativeness (see [92, Theorem 24]), hence boundedness for every initial marking. Moreover, strong connectedness and P-allocatability imply that liveness of the whole net is guaranteed by liveness of every *P-component* (see [92, Theorem 27.2]). A marking $\mathbf{m_0}$ such that $\mathbf{m_0}[p] = \mathbf{Pre}[p, t]$ for every p, where $t \in p^\bullet$, is sufficient to make every P-component live. □

From the proof of the sufficiency part, it follows in particular that live and bounded equal conflict systems are structurally bounded, thus conservative taking consistency into account:

Corollary 49. *If $\mathcal{S} = \langle \mathcal{N}, \mathbf{m_0} \rangle$ is a live equal conflict system, then $\mathcal{S}$ is bounded iff $\mathcal{N}$ is conservative (hence structurally bounded).*

In the case of free choice nets [39], the P-components are strongly connected state machine P-subnets. Strongly connected state machines are live iff they are marked, so in the ordinary case the above statement can be made stronger, particularly showing the polynomial complexity of the liveness and boundedness problem for free choice systems:

Corollary 50. *A free choice system $\mathcal{S}$ is live and bounded iff $\mathcal{N}$ is strongly connected, consistent, and* $\text{rank}(\mathbf{C}) = |\text{SEQS}| - 1$, *and no P-semiflow* $\mathbf{y}$ *such that* $\mathbf{y} \cdot \mathbf{m_0} = 0$ *exists.*

Remark 51. Some results from the classical free choice theory can be deduced easily from the above rank theorem, particularly the *duality theorem* [39], which states that a free choice net is structurally live and structurally bounded iff its reverse-dual net (which is also free choice) is. □

The rank based characterisation of structural liveness and boundedness has been extended to larger subclasses of nets. In particular, for the class of *deterministically synchronised sequential processes (DSSP)* [71,74], that is intended for the modular modelling of sequential agents that cooperate through buffers, the corresponding result is proven in [71].

A DSSP is a net system formed by a collection of *sequential agents* interconnected in a restricted way through *buffers*. The sequential agents are live and safe state machines. The buffers are places whose outputs are in one sequential agent (i.e., buffers are *destination private*) and that do not condition the resolution of the conflicts of their destination (i.e., all the outcomes of a conflict in a sequential agent have the same precondition). Under interleaving semantics, DSSP are a *strict generalisation* of equal conflict systems. In other words, provided that only sequential observations are relevant, equal conflict systems can be *simulated* by DSSP. The construction is simple (see Figure 20): add self-loop

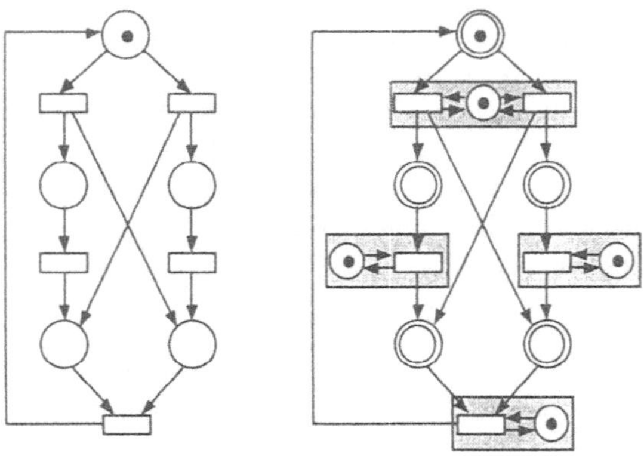

Fig. 20. Simulation of equal conflict systems by DSSP.

places marked with one token around each equal conflict set of a given equal conflict system. These self-loop places (with their adjacent transitions) are the sequential agents, and the original places of the equal conflict system play the role of buffers.

Other results for equal conflict systems have been extended to DSSP (see [74]), including the equivalence of liveness and deadlock-freeness (under boundedness and strong connectedness), the existence of home states (under liveness and boundedness), or the absence of spurious deadlocks (under liveness and consistency). Therefore, in particular, it is possible to verify liveness using integer programming, the same as in equal conflict systems (the deadlock-freeness condition can be written as a single system of linear inequalities preserving the class membership also in this case using a particular transformation rule).

Extending the DSSP definition recursively, the class of $\{\mathrm{SC}\}^*\mathrm{EQS}$ is defined, for which also a rank based characterisation of structural liveness and boundedness holds [72].

6.3 Application of the Rank Theorems for Subclasses to General Nets

Given a P/T net for which the available rank theorems do not allow to decide on its structural liveness, e.g., the nets in Figure 17 (b,c), it is sometimes possible, using certain net *transformation rules* (e.g., removal of bypass transitions or implicit places, other classical reduction rules, equalisation, release, etc.) to obtain another net where the corresponding property preservation of the rules together with the available results allow to decide. This topic deserves a closer investigation that is out of the scope of this work (see [73]), but we illustrate it with some examples.

In the net of Figure 17 (b), transition t_3 is a linear combination of t_2 and t_4: $\mathbf{C}[P,t_3] = \mathbf{C}[P,t_2] + \mathbf{C}[P,t_4]$, so the effect of firing t_3 is the same as the effect of firing t_2 and t_4. Moreover, ${}^\bullet\{t_2,t_4\} \cap \{t_2,t_4\}^\bullet = \emptyset$, so t_3 can only be fired when both t_2 and t_4 can be fired in one step (t_3 is a particular case of ***bypass*** transition, its occurrence "bypasses" the occurrence of the step $t_2 + t_4$ or any of its interleavings). Clearly, by removing t_3 from Figure 17 (b) we could not destroy liveness, i.e., liveness with t_3 ensures liveness without it. But in the net that we obtain after removing t_3, $\mathrm{rank}(\mathbf{C}) = 3$ and $\mathrm{SEQS} = \{\{t_1\},\{t_2,t_5\},\{t_4,t_6\}\}$, so Theorem 45 shows that it cannot be lively and boundedly marked, hence the net of Figure 17 (b) is proven not to be structurally live.

In the net of Figure 17 (c), the path $p_3 \to t_5 \to p_5$ can be substituted by a (macro)place p_{35}. This place is implicit, hence it can be removed. Since the resulting net can be proven (structurally) live, so it is the original one — actually, in this case we need not apply the rank theorem, since the net is simply a state machine.

Similarly, in the conservative and consistent net of Figure 3, where $\mathrm{rank}(\mathbf{C}) = 9 = |\mathrm{SEQS}| - 1$ (i.e., the necessary condition for structural liveness holds), place R is structurally implicit. Since the removal of R leads to a net where structural liveness can be proven, we are done — again in this case we need not apply the rank theorem, since the net without R is simply a marked graph.

A particular transformation, called *equalisation*, allows to obtain a general sufficient condition for structural liveness and boundedness. *Total equalisation* of a net consists in adding arcs or increasing weights as needed to make every coupled conflict set equal without changing $\mathbf{C}$ (if we add an arc from p to t we must add another from t to p). Figure 21 shows a net and the net obtained

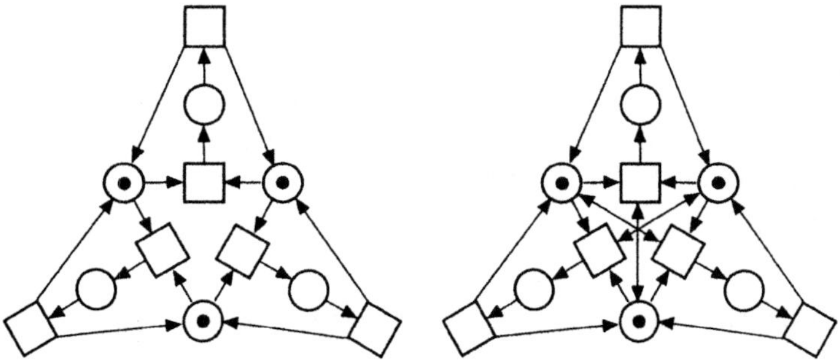

Fig. 21. Total equalisation of a net.

by total equalisation. The resulting net is equal conflict; if it can be proven structurally live and bounded (using Theorem 48), then the original net is proven structurally live and bounded too:

Theorem 52. *Let $\mathcal{N}$ be a P/T net.*

If $\mathcal{N}$ is strongly connected, consistent, and $\text{rank}(\mathbf{C}) = |\text{SCCS}| - 1$, *then* $\mathbf{m_0}$ *such that $\mathcal{S}$ is live and bounded exists.*

Proof. After total equalisation we obtain an equal conflict net $\mathcal{N}'$ such that $\mathbf{C}' = \mathbf{C}$ and $\text{SEQS}' = \text{SCCS}' = \text{SCCS}$. It follows that $\mathcal{N}'$ is strongly connected, consistent, and $\text{rank}(\mathbf{C}') = |\text{SEQS}'| - 1$, so, by Theorem 48, a marking $\mathbf{m_0}$ exists such that $\langle \mathcal{N}', \mathbf{m_0} \rangle$ is live and bounded. Since live and bounded equal conflict systems are conservative (see Corollary 49), $\mathcal{N}'$ is conservative too, hence so it is $\mathcal{N}$, and then $\mathcal{S}$ is bounded. Assume $\mathcal{S}$ non live. Then $t \in T$ and $\mathbf{m}_t \in \text{RS}(\mathcal{S})$ exist such that t cannot be fired from any $\mathbf{m} \in \text{RS}(\mathcal{N}, \mathbf{m}_t)$. Clearly, $\mathbf{m}_t \in \text{LRS}^{\text{SE}}(\mathcal{N}', \mathbf{m_0})$, thus a marking $\mathbf{m}_1$ exists such that $\mathbf{m}_1 \in \text{RS}(\mathcal{N}', \mathbf{m}_t) \cap \text{RS}(\mathcal{N}', \mathbf{m_0})$ (see Subsection 3.4). Since $\langle \mathcal{N}', \mathbf{m_0} \rangle$ is live, $\mathbf{m}_2 \in \text{RS}(\mathcal{N}', \mathbf{m}_1)$ exists such that t is enabled. Contradiction, since $\mathbf{m}_2 \in \text{RS}(\mathcal{N}, \mathbf{m}_t)$. □

In the case of ordinary nets, basically the same argument allows to make use of Corollary 50, leading to:

Corollary 53. *Let $\mathcal{S}$ be an ordinary P/T system.*

If $\mathcal{N}$ is strongly connected, consistent, and $\text{rank}(\mathbf{C}) = |\text{SCCS}| - 1$, *and no P-semiflow* $\mathbf{y}$ *such that* $\mathbf{y} \cdot \mathbf{m_0} = 0$ *exists, then $\mathcal{S}$ is live and bounded.*

The above result(s) allow, for instance, to prove (structural) liveness and boundedness of the system (net) in Figure 21.

In summary, given an arbitrary P/T net $\mathcal{N}$ that is strongly connected and consistent (otherwise it cannot be lively and boundedly marked), only when $|\text{SEQS}| - 1 \geq \text{rank}(\mathbf{C}) \geq |\text{SCCS}| - 1$ a marking $\mathbf{m_0}$ such that $\mathcal{S}$ is live and bounded exists, what is guaranteed in case $\text{rank}(\mathbf{C}) = |\text{SCCS}| - 1$. (The inequality $\text{rank}(\mathbf{C}) \geq |\text{SCCS}| - 1$ can be deduced as follows: given $\mathcal{N}$ strongly connected and consistent, by total equalisation we obtain an equal conflict net $\mathcal{N}'$ that is strongly connected and consistent too. Merging arbiters in every equal conflict set as in the proof of Theorem 48, the resulting net $\mathcal{N}''$ may have either one minimal T-semiflow or none. Being consistent, a basis of T-flows can be made up of T-semiflows only — dual of Proposition 12.1 — so the dimension of the space of right annullers of $\mathbf{C}''$ is at most one.)

7 Bibliographical Remarks

Linear algebra has been used in net theory at different net levels (e.g., P/T, or high level) and with different purposes (e.g., logical analysis, performance evaluation, controller design, or net synthesis). In logical/correctness analysis — which is the topic of this paper — other properties not considered here have also been studied (e.g., home states [44], or fairness [84]). For *performance evaluation and optimization*, linear techniques have been applied for the computation of performance bounds [78,69,17], for approximation techniques [14], or for initial marking optimization [15]. For the design of logic controllers for a plant modelled with P/T, linear techniques have been applied within the so called *supervisory control* theory [68] (e.g., [38,53,54,40]). The synthesis of P/T systems from an automata using the *theory of regions* applies also linear algebraic techniques [2]. In the case of high level PN, linear algebraic techniques have been developed mainly for the computation of P- (and T-) invariants (most relevant works are collected in [43]).

The use of integer linear algebra for the correctness analysis of PN dates back to the seventies [52], where the invariant method is introduced. Other pioneering works, using linear algebra in the real domain, introduce the notions of consistency and conservativeness [70,55]. The alternatives theorem (or Farkas lemma) is applied in [59,79] to provide dual perspectives of structural boundedness and repetitiveness, laying a first bridge between net theory and convex geometry. Taking into account that in live marked graphs reachability is linearly characterised, and that the incidence matrix of a marked graph is unimodular, linear programming can be used for the analysis, as it was firstly done in [36]. After realising that the evolution equation of a net system is a state equation in control theory sense, [63] tries to lay a bridge between nets and classical linear control theories. Despite the great conceptual interest of this bridge, integrality and non-negativity constraints, and the existence of spurious solutions, limit its strength.

The topic was surveyed in both previous Advanced Courses on PN: In [60] emphasis is given to boundedness, repetitiveness, and duality, [51] is an introductory tutorial on linear algebraic techniques for P/T nets, and [61] mainly overviews the invariant method for high level nets.

The basic idea in early works was the intensive exploitation of the P- (and T-) invariants — the so called invariant method — focusing on minimality, decomposition, and applicability to prove properties. In essence, the idea is to find the appropriate invariants to (dis)prove the interesting properties, for what the computation of the fundamental set is important. A need for this kind of computations has been encountered — and solutions re-discovered — quite often in several disciplines (dating back to Fourier! see [22] for more details). Within the PN field, [57] gives a first algorithm to compute the fundamental set of semiflows, taking advantage from a rank based property to remove non minimal semiflows before their computation is completed. In order to reduce the computational complexity of the algorithm, some heuristics have been proposed (e.g., [57,1]). In [22] the interpretation of semiflows as directions of a cone is explicited, and the existing algorithms are reviewed, improved, and their performance is analysed. Taking into account integrality constraints reveals that $\mathbf{y} \cdot \mathbf{C} = \mathbf{0}$ is an homogeneous linear Diophantine system, and different solutions are investigated in [48]. Actually, in principle, the state equation comprises integrality constraints. If only integer solutions are to be considered to prove non-reachability of a given marking, then we should solve it using integer linear algebraic methods, e.g., by means of the Smith normal form [45,76]. Integrality constraints in linear equation systems can be treated from a modulo-arithmetic perspective. By applying this approach to the state equation, modulo-invariants are obtained in [32], generalising the notion of P-flows.

Traps and siphons, which lead to different marking invariants, have been extensively used in the structure theory of (mainly ordinary) P/T, particularly in the case of some net subclasses (e.g., [39,41,10,4,31,5]). Regarding the computation of traps and siphons using linear algebraic techniques, the initial attempts try to translate the logic conditions defining the corresponding objects into a set of linear inequalities (e.g. [1,81]). A new line of thinking was opened in [50], where the computation of traps and siphons is carried out through the computation of semiflows in a transformed net. This approach was used with slight improvements in [33]. A similar approach, where only the weighting of the problem net is possibly transformed, was introduced in [35]. (A particular instance of this method has been presented here.)

In summary, the classical method consists in computing some structural objects (semiflows, traps, etc.) and then using the corresponding invariants in order to prove properties. The point of view adopted here considers directly the state equation, possibly improved, e.g., by taking into account the information provided by traps. This method was introduced in [83], where the analysis of several synchronic properties of general P/T systems is carried out through linear programming problems based on the state equation, and it was developed in detail in [19]. The idea of using implicit places to cut spurious solutions and their relation with trap invariants was presented in [23]. In [87] the method is implemented, and it is observed that the iterative application can remove more spurious solutions because the added implicit places lead to new traps. The improvement of the linear description is not only helpful for correctness analysis, but also for performance evaluation [16]. Monitor places have been used in order to forbid reaching some markings within supervisory control theory [38,93]. The observation that they could be used also to remove given spurious solutions, and that this method allows to remove all of them in structurally safe systems appears in [21]. The idea of incorporating a generator of trap invariants into the state equation appears in [58]. Also in [23] a totally different improvement method is introduced, consisting in removing spurious solutions without predecessors, for what partial enumeration is required (by the way, this method removes the spurious solutions of the example system in Figure 14 with $q = r = 2$).

Implicit places were introduced in [6]. Actually, in this seminal work, only implicit places the marking equation of which is redundant were considered, i.e., $\mu \geq 0$, and they were called redundant places. It was observed in [81] that redundancy (in a convex geometry sense) is not necessary for the place to be implicit (in the sense that it does not affect the behaviour), i.e., allowing $\mu < 0$. Implicit places were revisited in [23], where the structural ones were derived, using duality theory, from the linear inequalities expressing the condition that the behaviour is not affected. Moreover, a sufficient condition for a place to be implicit in terms of a linear programming problem was introduced. Besides their interest for reduction techniques and improvement of the state equation, implicit places play an important role in implementation techniques. On one hand, the addition of implicit places increases the Hamming distance of the code defined by marking vectors, what is interesting for fault-tolerant (error detecting and correcting) implementations [80,85,86]. On the other hand, since new semiflows appear after the addition of implicit places, the possibilities for decomposition are increased, what is useful for distributed implementations [81,24]. The new possibilities for decomposition are also interesting for approximate performance evaluation techniques [14,66] and exact performance evaluation [18].

Non-existence of solution to the $|T|$ linear systems of equations for t dead of the form (40) is essentially the necessary condition for liveness presented in [60,49]. In this sense, non-existence of solution to (37) is a greatly more accurate necessary condition for liveness. Some techniques to improve the performance of verifying non-existence of solution to (37) were presented in [89], and they have been recalled and improved here.

It was early realised that the state equation (or the invariants that can be deduced from it) is in general insufficient to analyse liveness or similar properties. Quoting from [51]: "Token counting in P-semiflows is by far not subtle enough to solve general liveness problems." One way of approaching the problem from structure theory is to investigate the conditions under which the net structure allows a live marking, i.e., structural liveness. Presently, the best linear conditions are given by the so called rank theorems. The rank theorem for free choice systems was conceived from the problem of computability of visit ratios in stochastic free choice nets in [13]. A proof, based on the Commoner's theorem [39], so limited to the free choice case, was published in [28]. In order to extend the applicability of the result, the necessity part was developed for general P/T nets [20], and the sufficiency part was developed for equal conflict systems [91], independently of the classical free choice theory. (Both results, that have been recalled here with minor modifications, can be found in [92].) This allowed to obtain rank based characterisations of structural liveness and boundedness in larger subclasses, namely DSSP [71] and $\{SC\}^*EQS$ [72]. The idea of using the rank theorem for free choice — applying equalisation — to obtain a general sufficient condition for liveness and boundedness in ordinary P/T systems (actually, to define a subclass, the *regular marked nets*, that are always live and bounded) appears in [29]. The extension to general P/T systems is contained in [71]. Generalising this approach by means of other transformation rules increases the decision power of rank theorems [73].

Acknowledgements

The authors wish to thank the careful reading and valuable suggestions of E. Badouel, J. Ezpeleta, C. Girault, K. Lautenbach, T. Murata, L. Ojala, L. Pomello, and L. Recalde.

References

1. H. Alaiwan and J. M. Toudic. Recherche des semi-flots, des verrous et des trappes dans les réseaux de Petri. *Technique et Science Informatiques*, 4(1):103–112, 1985.
2. E. Badouel and P. Darondeau. A survey on net synthesis. In Borne et al. [9], pages 309–316.
3. Z. A. Banaszak and B. H. Krogh. Deadlock avoidance in flexible manufacturing systems with concurrently competing process flows. *IEEE Trans. on Robotics and Automation*, 6(6):724–734, 1990.
4. K. Barkaoui and M. Minoux. A polynomial-time graph algorithm to decide liveness of some basic classes of bounded Petri nets. In Jensen [42], pages 62–75.
5. K. Barkaoui and J. F. Pradat-Peyre. On liveness and controlled siphons in Petri nets. In Billington and Reisig [8], pages 57–72.
6. G. Berthelot and G. Roucairol. Reduction of Petri netss. In *Procs. of the Symposium on MFCS '76*, volume 45 of *Lecture Notes in Computer Science*, pages 202–209. Springer, 1976.
7. E. Best, J. Desel, and J. Esparza. Traps characterize home states in free choice systems. *Theoretical Computer Science*, 101:161–176, 1993.
8. J. Billington and W. Reisig, editors. *Application and Theory of Petri Nets 1996*, volume 1091 of *Lecture Notes in Computer Science*. Springer, 1996.
9. P. Borne, J. C. Gentina, E. Craye, and S. El Khattabi, editors. *Symposium on Discrete Events and Manufacturing Systems. CESA '96 IMACS Multiconference*, Lille, France, July 1996.
10. G. W. BRAMS. *Réseaux de Petri: Théorie et Pratique*. Masson, 1983.
11. W. Brauer, editor. *Net Theory and Applications*, volume 84 of *Lecture Notes in Computer Science*. Springer, 1980.

12. W. Brauer, W. Reisig, and G. Rozenberg, editors. *Petri Nets: Central Models and their Properties. Advances in Petri Nets 1986, Part I*, volume 254 of *Lecture Notes in Computer Science*. Springer, 1987.
13. J. Campos, G. Chiola, and M. Silva. Properties and performance bounds for closed free choice synchronized monoclass queueing networks. *IEEE Trans. on Automatic Control*, 36(12):1368–1382, 1991.
14. J. Campos, J. M. Colom, H. Jungnitz, and M. Silva. Approximate throughput computation of stochastic marked graphs. *IEEE Trans. on Software Engineering*, 20(7):526–535, 1994.
15. J. Campos, J. M. Colom, and M. Silva. Performance evaluation of repetitive automated manufacturing systems. In *Procs. 2nd Int. Conf. on Computer Integrated Manufacturing and Automation Technology (CIMAT '90)*, pages 74–81. IEEE-Computer Society Press, 1990.
16. J. Campos, J. M. Colom, and M. Silva. Improving throughput upper bounds for net based models. In S. G. Tzafestas and J. C. Gentina, editors, *Robotics and Flexible Manufacturing Systems*, pages 281–294. Elsevier, 1992.
17. J. Campos and M. Silva. Structural techniques and performance bounds of stochastic Petri net models. In G. Rozenberg, editor, *Advances in Petri Nets 1992*, volume 609 of *Lecture Notes in Computer Science*, pages 352–391. Springer, 1992.
18. J. Campos, M. Silva, and S. Donatelli. Structured solution of stochastic DSSP systems. In *Procs. of the 7th Int. Workshop on Petri Nets and Performance Models (PNPM97)*. IEEE Computer Society Press, 1997.
19. J. M. Colom. *Análisis Estructural de Redes de Petri. Programación Lineal y Geometría Convexa.* PhD thesis, DIEI. Univ. Zaragoza, June 1989.
20. J. M. Colom, J. Campos, and M. Silva. On liveness analysis through linear algebraic techniques. In *Procs. of the AGM of Esprit BRA 3148 (DEMON)*, 1990.
21. J. M. Colom et al. Linear algebraic characterisation of structurally safe P/T systems. Technical report, DIIS. Univ. Zaragoza, 1997. In preparation.
22. J. M. Colom and M. Silva. Convex geometry and semiflows in P/T nets. A comparative study of algorithms for computation of minimal P-semiflows. In Rozenberg [75], pages 79–112.
23. J. M. Colom and M. Silva. Improving the linearly based characterization of P/T nets. In Rozenberg [75], pages 113–145.
24. J. M. Colom, M. Silva, and J. L. Villarroel. On software implementation of Petri nets and colored Petri nets using high-level concurrent languages. In *Proc. 7th European Workshop on Application and Theory of Petri Nets*, pages 207–241, Oxford, England, July 1986.
25. F. Commoner, A. W. Holt, S. Even, and A. Pnueli. Marked directed graphs. *Journal on Computer Systems Science*, 5:72–79, 1971.
26. R. G. Coyle. *Management System Dynamics.* Wiley, 1977.
27. R. David and H. Alla. *Petri Nets and Grafcet.* Prentice-Hall, 1992.
28. J. Desel. A proof of the rank theorem for extended free choice nets. In Jensen [42], pages 134–153.
29. J. Desel. Regular marked Petri nets. In J. Leeuwen, editor, *WG' 93: 19th Int. Workshop on Graph-Theoretic Concepts in Computer Science*, volume 790 of *Lecture Notes in Computer Science*, pages 264–275. Springer, 1993.
30. J. Desel and J. Esparza. Reachability in cyclic extended free choice systems. *Theoretical Computer Science*, 114:93–118, 1993.
31. J. Desel and J. Esparza. *Free Choice Petri Nets*, volume 40 of *Cambridge Tracts in Theoretical Computer Science.* Cambridge University Press, 1995.

32. J. Desel, K. P. Neuendorf, and M. D. Radola. Proving nonreachability by modulo-invariants. *Theoretical Computer Science*, 153(1-2):49–64, 1996.
33. J. Esparza and M. Silva. A polynomial time algorithm to decide liveness of bounded free choice nets. *Theoretical Computer Science*, 102:185–205, 1992.
34. J. Ezpeleta, J. M. Colom, and J. Martínez. A Petri net based deadlock prevention policy for flexible manufacturing systems. *IEEE Trans. on Robotics and Automation*, 11(2):173–184, 1995.
35. J. Ezpeleta, J. M. Couvreur, and M. Silva. A new technique for finding a generating family of siphons, traps and ST-components. application to coloured Petri nets. In G. Rozenberg, editor, *Advances in Petri Nets 1993*, volume 674 of *Lecture Notes in Computer Science*, pages 126–147. Springer, 1993.
36. H. J. Genrich and K. Lautenbach. Synchronisationgraphen. *Acta Informatica*, 2:143–161, 1973.
37. H. J. Genrich, K. Lautenbach, and P. S. Thiagarajan. Elements of general net theory. In Brauer [11], pages 21–163.
38. A. Giua, F. DiCesare, and M. Silva. Generalized mutual exclusion constraints on nets with uncontrollable transitions. In *IEEE Int. Conf. on Systems, Man, and Cybernetics*, Chicago, IL, USA, October 1992.
39. M. H. T. Hack. Analysis of production schemata by Petri nets. Master's thesis, M.I.T., Cambridge, MA, USA, 1972. (Corrections in *Computation Structures Note* 17, 1974).
40. L. E. Holloway, B. H. Krogh, and A. Giua. Petri nets for the control of discrete event systems: A tutorial survey. In *Supervisory Control of Discrete Event Systems*. Laboratoire d'Automatique de Grenoble, INPG, September 1995.
41. M. Jantzen and R. Valk. Formal properties of Place/Transition nets. In Brauer [11], pages 165–212.
42. K. Jensen, editor. *Application and Theory of Petri Nets 1992*, volume 616 of *Lecture Notes in Computer Science*. Springer, 1992.
43. K. Jensen and G. Rozenberg, editors. *High-level Petri Nets*. Springer, 1991.
44. C. Johnen. Algorithmic verification of home spaces in P/T systems. In *Procs. IMACS 1988*, 12th *World Congress on Scientific Computation*, pages 491–493, 1988.
45. R. Kannan and A. Bachem. Polynomial algorithms for computing the Smith and Hermite normal forms of an integer matrix. *SIAM Journal on Computing*, 8:499–507, 1979.
46. P. Kemper. Linear time algorithm to find a minimal deadlock in a strongly connected free-choice net. In M. Ajmone Marsan, editor, *Application and Theory of Petri Nets 1993*, volume 691 of *Lecture Notes in Computer Science*, pages 319–338. Springer, 1993.
47. W. E. Kluge and K. Lautenbach. The orderly resolution of memory access conflicts among competing channel processes. *IEEE Trans. on Computers*. 31(3):194–207, 1982.
48. F. Krückeberg and M. Jaxy. Mathematical methods for calculating invariants in Petri nets. In G. Rozenberg, editor, *Advances in Petri Nets 1987*, volume 266 of *Lecture Notes in Computer Science*, pages 104–131. Springer, 1987.
49. J. B. Lasserre and P. Mahey. Using linear programming in Petri net analysis. *Operations Research*, 26(1):43–50, 1989.
50. K. Lautenbach. Linear algebraic calculation of deadlocks and traps. In K. Voss et al., editors, *Concurrency and Nets*, pages 315–336. Springer, 1987.
51. K. Lautenbach. Linear algebraic techniques for Place/Transition nets. In Brauer et al. [12], pages 142–167.

52. K. Lautenbach and H. A. Schmid. Use of Petri nets for proving correctness of concurrent process systems. In *Procs. IFIP Congress 74*, pages 187–191. North-Holland, 1974.
53. Y. Li and W. M. Wonham. Control of vector discrete event systems I - the base model. *IEEE Trans. on Automatic Control*, 38(8):1214–1227, 1993.
54. Y. Li and W. M. Wonham. Control of vector discrete event systems II - controller synthesis. *IEEE Trans. on Automatic Control*, 39(3):512–531, 1994.
55. Y. E. Lien. Termination properties of generalized Petri nets. *SIAM Journal on Computing*, 5(2):251–265, 1976.
56. D. G. Luenberger. *Introduction to Linear and Non Linear Programming*. Addison Wesley, 1972.
57. J. Martínez and M. Silva. A simple and fast algorithm to obtain all invariants of a generalized Petri net. In C. Girault and W. Reisig, editors, *Application and Theory of Petri Nets*, pages 301–310. Springer, 1982.
58. S. Melzer and J. Esparza. Checking system properties via integer programming. In H. R. Nielsen, editor, *Proceedings of ESOP '96*, volume 1058 of *Lecture Notes in Computer Science*, pages 250–265. Springer, 1996.
59. G. Memmi. Applications of the semiflow notion to the boundedness and liveness problems in Petri net theory. In *Conf. on Information Sciences and Systems*. John Hopkins University, 1978.
60. G. Memmi and G. Roucairol. Linear algebra in net theory. In Brauer [11], pages 213–223.
61. G. Memmi and J. Vautherin. Analysing nets by the invariant method. In Brauer et al. [12], pages 300–336. Collected in [43].
62. T. Murata. Circuit theoretic analysis and synthesis of marked graphs. *IEEE Trans. on Circuits and Systems*, 24(7):400–405, 1977.
63. T. Murata. State equation, controllability, and maximal matchings of Petri nets. *IEEE Trans. on Automatic Control*, 22(3):412–416, 1977.
64. T. Murata, B. Shenker, and S. M. Shatz. Detection of Ada static deadlocks using Petri net invariants. *IEEE Trans. on Software Engineering*, 15(3):314–326, 1989.
65. K. G. Murty. *Linear Programming*. Wiley and Sons, 1983.
66. C. J. Pérez, J. Campos, and M. Silva. On approximate performace evaluation of manufacturing systems modelled with weighted T-systems. In Borne et al. [9], pages 201–207.
67. J. L. Peterson. *Petri Net Theory and the Modeling of Systems*. Prentice-Hall, 1981.
68. P. J. G. Ramadge and W. M. Wonham. The control of discrete event systems. *Proceedings of the IEEE*, 77(1):81–98, 1989.
69. C. V. Ramamoorthy and G. S. Ho. Performance evaluation of asynchronous concurrent systems using Petri nets. *IEEE Trans. on Software Engineering*, 6(5):440–449, 1980.
70. C. Ramchandani. Analysis of asynchronous concurrent systems by Petri nets. Technical Report Project MAC, TR-120, M.I.T., Cambridge, MA, USA, 1974.
71. L. Recalde, E. Teruel, and M. Silva. On well-formedness analysis: The case of deterministic systems of sequential processes. In J. Desel, editor, *Proc. of the Int. Workshop on Structures in Concurrency Theory (STRICT)*, Workshops in Computing, pages 279–293. Springer, 1995.
72. L. Recalde, E. Teruel, and M. Silva. {SC}*ECS: A class of modular and hierarchical cooperating systems. In Billington and Reisig [8], pages 440–459.
73. L. Recalde, E. Teruel, and M. Silva. Improving the decision power of rank theorems. In *IEEE Int. Conf. on Systems, Man, and Cybernetics*, Orlando, Florida, USA, October 1997. To appear.

74. L. Recalde, E. Teruel, and M. Silva. Modeling and analysis of sequential processes that cooperate through buffers. Technical report, DIIS. Univ. Zaragoza, 1997. Submitted paper.
75. G. Rozenberg, editor. *Advances in Petri Nets 1990*, volume 483 of *Lecture Notes in Computer Science*. Springer, 1991.
76. A. Schrijver. *Theory of Linear and Integer Programming*. Wiley, 1986.
77. M. W. Shields. *An Introduction to Automata Theory*. Blackwell Scientific Publications, 1987.
78. J. Sifakis. Uses of Petri nets for performance evaluation. In *Measuring, Modelling, and Evaluating Computer Systems*, pages 75–93. North-Holland, 1977.
79. J. Sifakis. Structural properties of Petri nets. In J. Winkowski, editor, *Mathematical Foundations of Computer Science 1978*, pages 474–483. Springer, 1978.
80. J. Sifakis. Realization of fault-tolerant systems by coding Petri nets. *Design Automation and Fault-Tolerant Computing*, 3(2):93–107, 1979.
81. M. Silva. *Las Redes de Petri: en la Automática y la Informática*. AC, 1985.
82. M. Silva. Introducing Petri nets. In *Practice of Petri Nets in Manufacturing*, pages 1–62. Chapman & Hall, 1993.
83. M. Silva and J. M. Colom. On the computation of structural synchronic invariants in P/T nets. In G. Rozenberg, editor, *Advances in Petri Nets 1988*, volume 340 of *Lecture Notes in Computer Science*, pages 387–417. Springer, 1988.
84. M. Silva and T. Murata. B-fairness and structural b-fairness in Petri net models of concurrent systems. *Journal of Computer and System Sciences*, 44(3):447–477, 1992.
85. M. Silva and S. Velilla. Detección y corrección de errores mediante la codificación de redes de Petri. In *II Simposium Nacional IFAC: Automática en la Industria*, pages 491–500, Zaragoza, Spain, November 1984.
86. M. Silva and S. Velilla. Error detection and correction on Petri net models of discrete event control systems. In *Proc. ISCAS 85*, pages 921–924, 1985.
87. E. Teruel. Programa para la mejora de la descripción lineal de redes de Petri mediante adición de lugares implícitos secantes. Master's thesis, DIEI. Univ. Zaragoza, 1990.
88. E. Teruel, P. Chrzastowski, J. M. Colom, and M. Silva. On weighted T-systems. In Jensen [42], pages 348–367.
89. E. Teruel, J. M. Colom, and M. Silva. Linear analysis of deadlock-freeness of Petri net models. In *Procs. of the* 2nd *European Control Conference*, volume 2, pages 513–518. North-Holland, 1993.
90. E. Teruel, J. M. Colom, and M. Silva. Choice-free Petri nets: A model for deterministic concurrent systems with bulk services and arrivals. *IEEE Trans. on Systems, Man, and Cybernetics*, 27(1):73–83, 1997.
91. E. Teruel and M. Silva. Well-formedness of equal conflict systems. In R. Valette, editor, *Application and Theory of Petri Nets 1994*, volume 815 of *Lecture Notes in Computer Science*, pages 491–510. Springer, 1994.
92. E. Teruel and M. Silva. Structure theory of equal conflict systems. *Theoretical Computer Science*, 153(1-2):271–300, 1996.
93. F. Tricas and J. Martínez. An extension of the liveness theory for concurrent sequential processes competing for shared resources. In *IEEE Int. Conf. on Systems, Man, and Cybernetics*, pages 4119–4124, Vancouver, Canada, October 1995.
94. N. Viswanadham, Y. Narahari, and T. L. Johnson. Deadlock prevention and deadlock avoidance in flexible manufacturing systems using Petri net models. *IEEE Trans. on Robotics and Automation*, 6(6):713–723, 1990.

95. M. C. Zhou, F. DiCesare, and A. A. Desrochers. A hybrid methodology for synthesis of Petri nets for manufacturing systems. *IEEE Trans. on Robotics and Automation*, 8(3):350–361, 1992.

Elements of Linear Programming and Duality Theory

Many text books cover linear programming (see, for instance, [56,65]). Here we recall a few definitions and results that are used in the paper.

Any *linear programming problem* can be written in standard form (possibly requiring the incorporation of *slack variables* to transform inequalities into equations) as:

$$z = \max\{\mathbf{c} \cdot \mathbf{x} \mid \ \mathbf{A} \cdot \mathbf{x} = \mathbf{b} \wedge \mathbf{x} \geq \mathbf{0}\} \tag{64}$$

where $\mathbf{x}$ are the (real valued) variables, $\mathbf{c} \cdot \mathbf{x}$ is the *cost function* to optimise, and $\mathbf{A} \cdot \mathbf{x} = \mathbf{b} \wedge \mathbf{x} \geq \mathbf{0}$ are the *linear constraints*. The computational complexity of linear programming problems is polynomial time. They are usually solved using the *simplex* algorithm, which, among other advantages compared to polynomial algorithms, performs most often in linear time in spite of its worst case exponential complexity.

Depending on the existence of solutions to the linear constraints and the value of the objective function, a linear programming problem can be:

- *Non feasible*: No solution to the linear constraints.
- *Unbounded*: The value of the cost function can be increased arbitrarily.
- *Bounded*: There are *optimal solutions* $\mathbf{x}$ that maximise the value of the cost function.

The *dual* of the *primal* problem (64) is:

$$z' = \min\{\mathbf{b} \cdot \mathbf{y} \mid \ \mathbf{y} \cdot \mathbf{A} \geq \mathbf{c}\} \tag{65}$$

Note that the dual of the primal problem (65) is (64).

The *weak duality* theorem states that, if $\mathbf{x}$ and $\mathbf{y}$ are feasible solutions to (64) and (65), respectively, then $\mathbf{c} \cdot \mathbf{x} \leq \mathbf{b} \cdot \mathbf{y}$.

The *duality* theorem states that, if both (64) and (65) are feasible, then both are bounded and $z = z'$.

The *unboundedness* theorem states that, if only one of (64) or (65) is feasible, then it is unbounded.

These theorems allow to prove the *alternatives* theorem (for homogeneous or non-homogeneous systems). Two out of the many formulations of this theorem are the following:

- One and only one of the following systems is feasible:

$$\mathbf{A} \cdot \mathbf{x} \geq \mathbf{b} \tag{66}$$

$$\mathbf{y} \geq \mathbf{0} \wedge \mathbf{y} \cdot \mathbf{b} > 0 \wedge \mathbf{y} \cdot \mathbf{A} = \mathbf{0} \tag{67}$$

– One and only one of the following systems is feasible:

$$\mathbf{A} \cdot \mathbf{x} \geq \mathbf{0} \wedge \mathbf{x} > \mathbf{0} \tag{68}$$

$$\mathbf{y} \cdot \mathbf{A} \lneq \mathbf{0} \wedge \mathbf{y} \geq \mathbf{0} \tag{69}$$

If $\mathbf{x}$ is restricted to be integer in (64), then it is an *integer programming problem*, which is NP-complete. (A typical algorithm is *branch and bound* based on linear programming.) In the particular case that $\mathbf{A} = [\mathbf{I}\ \mathbf{A}']$, a property that we use is that boundedness of the integer programming problem is equivalent to boundedness of the corresponding linear programming problem where integrality is disregarded, although the optimal value may not coincide.

Decidability and Complexity of Petri Net Problems – An Introduction*

Javier Esparza

Institut für Informatik, Technische Universität München,
Arcisstr. 21, D-80290 München, Germany,
e-mail: esparza@informatik.tu-muenchen.de

Abstract. A collection of 10 "rules of thumb" is presented that helps to determine the decidability and complexity of a large number of Petri net problems.

1 Introduction

The topic of this paper is the decidability and complexity of verification problems for Petri nets. I provide answers to questions like "is there an algorithm to decide if two Petri nets are bisimilar?", or "how much time is it needed (in the worst case) to decide if a 1-safe Petri net is deadlock-free?"

My intended audience are people who work on the development of algorithms and tools for the analysis of Petri net models and have some basic understanding of complexity theory. More precisely, I assume that the reader is familiar with the notion of undecidable problem, with the definitions of deterministic and nondeterministic complexity classes like NP or PSPACE, with the notion of hard and complete problems for a complexity class, and with the use of reductions to prove hardness and completeness results. Theoreticians acquainted with the topic of this paper are warned: They won't find much in it that they didn't know before.[2] On the other hand, they might be interested in the paper's unified view of complexity questions for 1-safe and general Petri nets, and in a few simplifications in the presentation of some proofs.

When I was invited to write this paper, I hesitated for a while. I remembered the statement of the Greek scepticist Gorgias:

> Nothing exists;
> if anything does exist, it is unknowable;
> if anything can be known, knowledge of it is incommunicable.

and imagined a Greek chorus advising me not to write the paper because, in their opinion:

---

* Work partially supported by the Sonderforschungsbereich 342 "Werkzeuge und Methoden für die Nutzung paralleler Rechnerarchitekturen".

[2] Only one result has not been published before, namely a PSPACE algorithm for the model-checking problem of CTL and 1-safe Petri nets, presented in Section 4.

All results about decidability and complexity of Petri nets were already obtained in the early eighties;
if there are new results, you have included them for sure in the paper "Decidability issues for Petri nets – a survey" you wrote with Mogens Nielsen in 1994 [10];
if you haven't included them in the survey, they are only of interest for specialists; moreover, these results just show that all interesting problems are intractable – finer classifications, like NP-, PSPACE- or EXPSPACE-hardness have no practical relevance.

Since, as you can see, I still decided to write the paper, I would like to anticipate my answer to these three possible criticisms.

• *There have been important recent developments about decidability and complexity questions, of interest for the whole Petri net community.*

During the late seventies and early eighties there was an outburst of theoretical work on decidability and complexity problems for (Place/Transition) Petri nets. Well-known computer scientists, like Rabin, Rackoff, Lipton, Mayr, Meyer, and Kosaraju, just to mention a few, obtained a very impressive collection of results. The decidability of most problems, like boundedness, liveness, reachability, language equivalence, etc. was settled, and in many cases tight complexity bounds were obtained.

However, while these results were being obtained, two developments in computer science opened new problems:

• In the late seventies, temporal logic was proposed as a query language for the specification of reactive and distributed systems; a few years later, model-checking was introduced as a technique for the verification of arbitrary temporal properties. Howell, Rosier, and Yen were the first to study the decidability and complexity of model-checking problems for Petri nets in the second half of the eighties [17, 19, 20]. Today most questions in this research field have been answered [9, 14].

• In the early eighties, process algebras were introduced for the formal description of concurrent and reactive systems. It was seen that language equivalence was not an adequate equivalence notion for this class of systems, since for instance it may consider deadlock-free systems as equivalent to systems with deadlocks. New equivalence relations were introduced, like bisimulation and failures equivalence. In the early nineties, the decidability of these equivalences for systems with infinite state spaces started to receive a lot of attention, and led to renewed interest in Petri nets. Jančar proved only a few years ago a fundamental result showing the undecidability for Petri nets of all equivalence notions described in the literature [22, 21].

These two developments still had another effect. During the eighties, many researchers started to study the relationship of process algebras to Petri nets. Net models in which a place can carry at most one token, like condition/event systems or elementary net systems, turned out to be particularly useful for these studies. These nets, which have by definition a finite number of states, became even more interesting after the introduction of automatic model-checkers, when it was realised that they could be used to model a large number of interesting systems which were within the reach of automatic verification. The questions that had been asked and mostly solved for Place/Transition nets were now asked again for these models. In the last years the complexity of classical properties (reachability, liveness ...), model-checking problems for different temporal logics, and equivalence problems for different equivalence notions, has been completely determined [2, 23, 31].

• *This paper has a different approach than the '94 survey paper, and has been written to complement it.*

Research on the decidability and complexity of verification problems for Petri nets has produced well over 100 papers, maybe even 150. Many of them have been published in well-known journals, and are thus available in any good library. My survey paper with Mogens Nielsen [10] summarises many results, and provides a rather comprehensive list of references.

Petri net researchers often need information about the complexity of a particular problem (the Petri net mailing list receives now and then postings with this kind of requests). In most cases, a similar problem has already been studied in the literature, and pointers to relevant papers can be found in [10]. If one is familiar with a number of basic techniques, it is easy to apply these existing results to the new problem. However, acquiring this familiarity is at the moment a rather hard task, specially for Ph. D. students: one has to go through many papers and distill an understanding which is not explicitly contained in the papers themselves. The purpose of these pages is to make this task a bit easier. Instead of listing results and references, I concentrate on a few general results of broad applicability. I also provide "rules of thumb", which I think can be more useful than formal theorems.

• *All researchers interested in the development and implementation of analysis algorithms for Petri nets can greatly profit from some basic knowledge on the computational complexity of analysis problems.*

All researchers are regularly confronted with the problem of having to prove or disprove a conjecture. Should one first try to find a proof or a counterexample? The wrong choice can make one lose precious time. Complexity theory can often help by showing that the truth or falsity of the conjecture implies an unlikely fact, like P=NP or NP=PSPACE. I present here some examples in the form of three stories taken from my personal experience:

Story I. After graduating in Physics, I became a Ph. D. student of computer science. At that time I knew very little about theoretical computer science, and there were no theoreticians in my environment. I started to work on the analysis of free-choice Petri nets, a net class for which there was hope of finding efficient verification algorithms, and more precisely I began to investigate the liveness problem. My hope was to efficiently transform the problem into a set of linear inequations that could be solved using linear programming. 'Efficiently' meant for me that the number and size of the equations should grow quadratically, say, in the size of the net.

During the next four months I could not find any encoding, but I read some textbooks on theoretical computer science. I came across Garey and Johnson's book on the theory of NP-completeness [12], and I found the problem I was working on (more precisely, its complement) in the list of NP-complete problems at the end of the book. Since there exist polynomial algorithms for Linear Programming but the complement of the liveness problem for free-choice nets was NP-complete, the existence of an efficient encoding would imply P=NP, and so it was highly unlikely.

The NP-completeness of the non-liveness problem for free-choice Petri nets is proved in Section 10.

Story II. Some years ago I refereed a paper submitted to the Petri net conference. The paper contained a conjecture on the reachability problem for Petri nets that can be stated as follows. Let $\mathcal{N}$ be a net, and let M_0 and M be markings of $\mathcal{N}$ such that M is reachable from M_0. Conjecture: M can be reached from M_0 through a sequence of transition firings which only visits intermediate markings of size $O(n + m_0 + m)$, where n, m_0, m are the sizes of $\mathcal{N}$, M_0 and M, respectively. The author of the paper had constructed a random generator of nets and markings and had tested the conjecture in one thousand cases, always with a positive answer.

It is certainly possible to disprove the conjecture by exhibiting a counterexample, but it is faster to use a complexity argument. I show this argument in Section 7.

Story III. I have recently come across a paper containing a characterisation of the set of reachable markings of 1-safe Petri nets. A simple complexity analysis shows that the characterization is most probably wrong, although I haven't found a counterexample yet. In order to formulate the characterisation we need some definitions and notations. A *siphon* of a net is a subset of places R satisfying ${}^\bullet R \subseteq R^\bullet$. A *trap* is a subset of places R satisfying $R^\bullet \subseteq {}^\bullet R$. Given a net $\mathcal{N} = (S, T, F)$ and a set $U \subseteq T$, we define the net $\mathcal{N}_U$ as the result of first removing all transitions of $\mathcal{N}$ not belonging to U, and then removing all places that are not connected to any transition anymore.

Now, let $\mathcal{N} = (S, T, F)$ be a net, and let M_0 and M be markings of $\mathcal{N}$ such that the Petri net $(\mathcal{N}, M_0)$ is 1-safe. The characterization states M is reachable from M_0 if and only if there exists a mapping $X: T \to \mathbb{N}$ satisfying the following three properties:

(1) for every place s, $M(s) = M_0(s) + \sum_{t \in T}(F(t,s) - F(s,t)) \cdot X(t)$,
(2) every nonempty siphon of $\mathcal{N}_{TX}$ is marked at M_0, and
(3) every nonempty trap of $\mathcal{N}_{TX}$ is marked at M.

where TX is the set of transitions t such that $X(t) > 0$.

I strongly believe that the proof of this result contains a mistake, and that a counterexample exists. I show why in Section 3.[3]

• *The classification of a problem as NP-, PSPACE- or EXPSPACE-hard does have practical relevance*

The complexity of Petri nets was first studied in the seventies, when NP-complete problems were really intractable: computer scientists were unable to deal even with very small instances due to the lack of computing power and of good theoretical results. At that time it probably didn't make so much difference for a practitioner whether a problem was PSPACE-hard or only NP-complete. In my opinion, today's picture is very different:

- NP-complete problems are no longer "intractable". It is certainly true that all known algorithms that solve them have exponential worst-case complexity. However, today there exist commercial systems for standard NP-complete problems, like satisfiability of propositional logic formulas or integer linear programming problems, that routinely solve instances of large size.
- The last years have witnessed a proliferation of model-checking tools, like COSPAN, PEP, PROD, SMV, SPIN, and others (see [11] and [30] for comprehensive information). Although the problems they solve are PSPACE-complete, they have been successfully applied to the verification of many interesting finite state systems. Commercial versions are starting to appear.
- Experimental tools for the analysis of timed-systems are starting to emerge. Examples are Hy-Tech, KRONOS, UPPAAL [11]. Many of the problems solved by these tools are EXPSPACE-complete. The size of the instances they can handle is certainly much smaller than in the case of model-checkers, but the results are very promising.
- Theorem provers like HOL, Isabelle, PVS, and others are being applied with good success to the verification of systems with infinite state spaces. They use heuristics to try to solve particular instances of undecidable analysis problems.

My conclusion is that the old "tractable – intractable" classification has become too rough. A finer analysis provides very valuable information about the size of instances that can be handled by automatic tools, and about the possibility of applying existing tools to a particular problem.

Organisation of the paper

The paper is divided into two parts. The first is devoted to 1-safe Petri nets, which are Place/Transition Petri nets having the property that no reachable marking puts more than one token in any place. Nearly all results hold for n-safe Petri nets (at most n tokens on a place) too, assuming that the algorithms

[3] After I wrote this paper, but before its publication, Stephan Melzer found a counterexample with 5 places and 3 transitions.

receive n as part of the input, which implies in particular that n must be known in advance. The second part is devoted to general Place/Transition nets. Both parts are divided into the same four sections. Each section contains one or more "rules of thumb". These are general informal statements which try to summarise a number of formal results in a concise, necessarily informal, but informative way. They could also be called "useful lies": statements which do not tell all the truth and nothing but the truth, but are more useful than a complicated formal theorem with many ifs and buts. There is a total of 10 rules of thumb in the paper; with their help I can solve most of the complexity questions I come across in my own research.

Rules of thumb are displayed in the text like this:

> Rule of thumb 0:
> To find the rules of thumb, look for pieces of text within a box.

This is only a rule of thumb, because other pieces of text are also surrounded by a box, in fact by a *double* box. They are fundamental formal results used to derive the rules of thumb.

> Fundamental results are displayed within a double box.

The first section contains a universal lower bound for "interesting" Petri net problems. The second section deals with upper bounds: for 1-safe Petri nets it is possible to give an almost universal upper bound, whereas the case of general Petri nets is more delicate. The third section deals with equivalence problems: are two given nets equivalent with respect to a given equivalence notion? Upper and lower bounds are considered simultaneously. Finally, the fourth section gives information about how far one can go with polynomial time algorithms.

Only some of the results mentioned in the paper are proved; for others the reader is referred to the literature. The results with a proof are those fulfilling two conditions: they are very general, applicable to a variety of problems, and admit relatively simple, non-technical proofs. I have devoted special effort to presenting proofs in the simplest possible way. My goal was to produce a paper that could be read straight through from beginning to end. I don't know if the goal has been achieved, but I tried my best.

Table of Contents

2 Preliminaries

We assume that the reader is acquainted with the basic notions of net theory, like firing rule, reachable marking, liveness, boundedness, etc., and also with other basic computation models like Turing machines. This section just fixes some notations.

Petri nets. A *net* is a triple $\mathcal{N} = (S, T, F)$, where S and T are finite sets of *places* and *transitions*, and $F \subseteq (S \times T) \cup (T \times S)$ is the *flow relation.* We identify F with its characteristic function $(S \times T) \cup (T \times S) \to \{0, 1\}$. The *preset* and *postset* of a place or transition x are denoted by ${}^\bullet x$ and $x^\bullet$, respectively. Given a set $X \subseteq S \cup T$, we denote ${}^\bullet X = \bigcup_{x \in X} {}^\bullet x$ and $X^\bullet = \bigcup_{x \in X} x^\bullet$. A *marking* is a mapping $M: S \to I\!N$. A *(Place/Transition) Petri net* is a pair $N = (\mathcal{N}, M_0)$, where $\mathcal{N}$ is a net and M_0 is the initial marking. A transition t is *enabled* at a marking M if $M(s) > 0$ for every $s \in {}^\bullet t$. If t is enabled at M, then it can *fire* or *occur*, and its firing leads to the successor marking M' which is defined for every place s by

$$M'(s) = M(s) + F(t, s) - F(s, t)$$

The expression $M \xrightarrow{t} M'$ denotes that M enables transition t, and that the marking reached by the occurrence of t is M'. A finite or infinite sequence

$M_0 \xrightarrow{t_1} M_1 \xrightarrow{t_2} M_2 \cdots$ is called a *firing sequence*. The maximal firing sequences of a Petri net (i.e., the infinite firing sequences plus the finite firing sequences which end with a marking that does not enable any transition) are called *runs*. Given a sequence $\sigma = t_1 t_2 \ldots t_n$, $M \xrightarrow{\sigma} M'$ denotes that there exist markings $M_1, M_2, \ldots, M_{n-1}$ such that $M \xrightarrow{t_1} M_1 \ldots M_{n-1} \xrightarrow{t_n} M'$.

A Petri net is *1-safe* if $M(s) \leq 1$ for every place s and every reachable marking M.

We encode a net (S, T, F) as two $|S| \times |T|$ binary matrices *Pre* and *Post*. The entry $Pre(s,t)$ is 1 if there is an arc from s to t, and 0 otherwise. The entry $Post(s,t)$ is 1 if there is an arc from t to s, and 0 otherwise. The *size of a net* is the number of bits needed to write down these two matrices, and is therefore $O(|S| \cdot |T|)$. The *size of a Petri net* is the size of the net plus the size of its initial marking. Markings are encoded as vectors of natural numbers. The *size of a marking* is defined as the number of bits needed to write it down as a vector, where each component is written in binary. Observe that the size of a 1-safe Petri net is $O(|S| \cdot |T|)$, since the initial marking has size $O(|S|)$.

A *labelled net* is a fourtuple (S, T, F, λ), where (S, T, F) is a net and λ is a mapping that associates to each transition t a label $\lambda(t)$ taken from some given set of actions Act. Given $a \in Act$, we denote by $M \xrightarrow{a} M'$ that there is some transition t such that $M \xrightarrow{t} M'$ and $\lambda(t) = a$. A *labelled Petri net* is a pair $(\mathcal{N}, M_0)$, where $\mathcal{N}$ is a labelled net and M_0 is the initial marking.

Turing machines. In the paper we use single tape Turing machines with one-way infinite tapes, i.e., the tape has a first but not a last cell. For our purposes it suffices to consider Turing machines starting on empty tape, i.e., on tape containing only blank symbols. So we define a *(nondeterministic) Turing machine* as a tuple $M = (Q, \Gamma, \delta, q_0, F)$, where Q is the set of states, Γ the set of tape symbols (containing a special *blank* symbol), $\delta: (Q \times \Gamma) \rightarrow \mathcal{P}(Q \times \Gamma \times \{R, L\})$ the transition function, q_0 the initial state, and F the set of final states. The *size of a Turing machine* is the number of bits needed to encode its transition relation.

Linearly and exponentially bounded automata. We work several times with Turing machines that can only use a finite tape fragment, or equivalently, with Turing machines whose tape has both a first and a last cell. We call them *bounded automata.* If a bounded automaton tries to move to the right from the last tape cell it just stays in the last cell.

A function $f: \mathbb{N} \rightarrow \mathbb{N}$ induces the class of *$f(n)$-bounded automata*, which contains for all $k \geq 0$ the bounded automata of size k that can use $f(k)$ tape cells. Notice that we deviate from the standard definition, which says that an automaton is $f(n)$-bounded if it can use at most $f(k)$ tape cells for an *input word* of length k. Since we only consider bounded automata working on empty tape, the standard definition is not appropriate for us. When $f(n) = n$ and $f(n) = 2^n$ we get the classes of *linearly bounded* and *exponentially bounded automata*, respectively.

Complexity classes and reductions. In the paper we use some of the most basic complexity classes, like P, NP, and PSPACE. We also use the class EXPSPACE, defined by[4]

$$\mathrm{EXPSPACE} = \bigcup_{k \geq 0} \mathrm{DSPACE}(2^{n^k})$$

We always work with polynomial reductions, i.e., given an instance x of a problem A we construct in polynomial time an instance y of a problem B. Many of the results also hold for logspace reductions, or even log-lin reductions, but we do not address this point.

PartI

1-safe Petri nets

We study the complexity of analysis problems for 1-safe Petri nets. Given a 1-safe Petri net $(\mathcal{N}, M_0)$, where $\mathcal{N} = (S, T, F)$, we say that the *possible markings of* $\mathcal{N}$ or just the *markings of* $\mathcal{N}$ are the set of markings that put at most one token in a place. Clearly, there are $2^{|S|}$ possible markings. Each of the markings can be identified with the set of places marked at it. Observe that the size of a marking is linear in the size of the net.

3 A universal lower bound

In this section we obtain a universal lower bound for the complexity of deciding whether a 1-safe Petri net satisfies an interesting behavioural property:

Rule of thumb 1: All interesting questions about the behaviour of 1-safe Petri nets are PSPACE-hard.

Notice that a rule of thumb is not a theorem. There are behavioural properties of 1-safe Petri nets that can be solved in polynomial time. For instance, the question "Is the initial marking a deadlock?" can be answered very efficiently; however, it is so trivial that hardly anybody would consider it really interesting. So a more careful formulation of the rule of thumb would be that all questions described in the literature as interesting are at least PSPACE-hard. Here are 14 examples:

- Is the Petri net live?
- Is some reachable marking a deadlock?
- Is a given marking reachable from the initial marking?

[4] Notice that some books (for instance [1]) define $\mathrm{EXPSPACE} = \bigcup_{k \geq 0} \mathrm{DSPACE}(k \cdot 2^n)$.

- Is there a reachable marking that puts a token in a given place?
- Is there a reachable marking that does not put a token in a given place?
- Is there a reachable marking that enables a given transition?
- Is there a reachable marking that enables more than one transition?
- Is the initial marking reachable from every reachable marking?
- Is there an infinite run?
- Is there exactly one run?
- Is there a run containing a given transition?
- Is there a run that does not contain a given transition?
- Is there a run containing a given transition infinitely often?
- Is there a run which enables a transition infinitely often but contains it only finitely often?

The PSPACE-hardness of all these problems is a consequence of one single fundamental fact, first observed by Jones, Landweber and Lien in 1977 [24]:

> A linearly bounded automaton of size n can be simulated by a 1-safe Petri net of size $O(n^2)$. Moreover, there is a polynomial time procedure which constructs this net.

The notion of simulation used here is very strong: a 1-safe Petri net simulates a Turing machine if there is bijection f between configurations of the machine and markings of the net such that the machine can move from a configuration c_1 to a configuration c_2 in one step if and only if the Petri net can move from the marking $f(c_1)$ to the marking $f(c_2)$ through the firing of exactly one transition.

Let $A = (Q, \Gamma, \Sigma, \delta, q_0, F)$ be a linearly bounded automaton of size n. The computations of M visit at most the cells $c_1, \ldots, c_n$. Let C be this set of cells. The simulating Petri net $N(A)$ contains a place $s(q)$ for each state $q \in Q$, a place $s(c)$ for each cell $c \in C$, and a place $s(a, c)$ for each symbol $a \in \Gamma$ and for each cell $c \in C$. A token on $s(q)$ signals that the machine is in state q. A token on $s(c)$ signals that the machine reads the cell c. A token on $s(a, c)$ signals that the cell c contains the symbol a. The total number of places is $|Q| + n \cdot (1 + |\Sigma|)$.

The transitions of $N(A)$ are determined by the state transition relation of A. If $(q', a', R) \in \delta(q, a)$, then we have for each cell c a transition $t(q, a, c)$ whose input places are $s(q)$, $s(c)$, and $s(a, c)$ and whose output places are $s(q')$, $s(a', c)$ and $s(c')$, where c' is the cell to the right of c (this signals that the tape head has moved to the right) unless c is the last cell, in which case $c' = c$. The last cell is an exception, because by assumption the machine cannot move to the right from there. If $(q', a', L) \in \delta(q, a)$ then we add a similar set of transitions; this time the first cell is the exception. The total number of transitions is at most $2 \cdot |Q|^2 \cdot |\Gamma|^2 \cdot n$, and so $O(n^2)$, because the size of A is $O(|Q|^2 \cdot |\Gamma|^2)$.

The initial marking of $N(A)$ puts one token on $s(q_0)$, on $s(c_1)$, and on the place $s(B, c_i)$ for $1 \leq i \leq n$, where B denotes the blank symbol. The total size of the Petri net is $O(n^2)$.

It follows immediately from this definition that each move of A corresponds to the firing of one transition. The configurations reached by A along a computation correspond to the markings reached along its corresponding run. These markings put one token in exactly one of the places $\{s(q) \mid q \in Q\}$, in exactly one of the places $\{s(c) \mid c \in C\}$, and in exactly one of the places $\{s(a,c) \mid a \in \Sigma\}$ for each cell $c \in C$. So $N(A)$ is 1-safe.

In order to answer a question about a linearly bounded automaton A we can construct the net $N(A)$, which is only polynomially larger than A, and solve the corresponding question about the runs of A. For instance, the question "does any of the computations of A terminate?" corresponds to "has the Petri net $N(A)$ a deadlock?"

It turns out that most questions about the computations of linearly bounded automata are PSPACE-hard. To begin with, the *(empty tape) acceptance problem* is PSPACE-complete:

Given: a linearly bounded automaton A.
To decide: if A accepts the empty input.

Moreover, the PSPACE-hardness of this problem is very robust: it remains PSPACE-complete if we restrict it to

- deterministic bounded automata,
- bounded automata having one single accepting state,
- bounded automata having one single accepting configuration.

Many other problems can be easily reduced to the acceptance problem in polynomial time, and so are PSPACE-hard too. Examples are:

- does A halt?,
- does A visit a given state?,
- does A visit a given configuration?
- does A visit a given configuration infinitely often?

We obtain in this way a large variety of PSPACE-hard problems. Since $N(A)$ is only polynomially larger than A, all the corresponding Petri net problems are PSPACE-hard as well. For instance, a reduction from the problem "does A ever visit a given configuration?" proves PSPACE-hardness of the reachability problem for 1-safe Petri nets. Furthermore, once we have some PSPACE-hard problems for 1-safe Petri nets we can use them to obtain new ones by reduction. For instance, the following problems can be easily reduced to the problem of deciding if there is a reachable marking that puts a token on a given place:

- is there a reachable marking that concurrently enables two given transitions t_1 and t_2?
- can a given transition t ever occur?
- is there a run containing a given transition t infinitely often?

13 out of the 14 problems at the beginning of the section (and many others) can be easily proved PSPACE-hard using these techniques. The liveness problem, the first in our list, is a bit more complicated. The interested reader can find the reduction in [2].

The solution to Story III

Recall the conjecture of Story III: Let $\mathcal{N} = (S, T, F)$ be a net, and let M_0 and M be markings of $\mathcal{N}$ such that the Petri net$(\mathcal{N}, M_0)$ is 1-safe. M is reachable from M_0 in $\mathcal{N}$ if and only if there exists a mapping $X: T \rightarrow \mathbb{N}$ satisfying the following three properties:

(1) for every place s, $M(s) = M_0(s) + \sum_{t \in T}(F(t,s) - F(s,t)) \cdot X(t)$,
(2) every nonempty siphon of $\mathcal{N}_{TX}$ is marked at M_0, and
(3) every nonempty trap of $\mathcal{N}_{TX}$ is marked at M.

where TX is the set of transitions t such that $X(t) > 0$.

We show that if the conjecture is true, then the reachability problem for 1-safe Petri nets belongs to NP. Since we know that this problem is PSPACE-hard, the truth of the conjecture implies NP=PSPACE, which is highly unlikely. So, very probably, the conjecture is false; one should look for a counterexample instead of trying to prove it.

We need a well-known result (see for instance [16]):

> There is a polynomial time nondeterministic algorithm Feasible(S) for the problem of deciding if a system of linear equations S with integer coefficients has a solution in the natural numbers.

It is easy to decide if every siphon of a net $\mathcal{N}$ is marked at a given marking M. The following (deterministic) algorithm, due to Starke [33, 5], does it for you. It first computes the largest siphon R contained in the set of places not marked at M. Clearly, all nonempty siphons are marked at M if and only if R is empty.

Algorithm All_Siphons_Marked($\mathcal{N}$, M):

variable: R of type set of places;

begin
    $R :=$ set of places of N unmarked under M;
    while there is $s \in R$ and $t \in {}^\bullet s$ such that $t \notin R^\bullet$ **do**
        $R := R \setminus \{s\}$
    od;
    if $R = \emptyset$ **then return true**
    else return false
end

The algorithm All_Traps_Marked is very similar: just change the loop condition to: there is $s \in R$ and $t \in s^\bullet$ such that $t \notin {}^\bullet R$. Clearly, these two algorithms run in polynomial time.

The following nondeterministic algorithm checks conditions (1), (2) and (3). It first guesses the set TX of transitions, and checks that (2) and (3) hold. Then, it checks if condition (1) holds for a vector X such that $TX = \{t \in T \mid X(t) > 0\}$. For that, it checks if the system of equations $\mathcal{S}$ containing the equations of condition (1) plus the equation $X(t) \geq 1$ for every $t \in TX$, and the equation $X(t) = 0$ for every $t \in T \setminus TX$ has a solution.

Algorithm Check_Conditions($\mathcal{N}$, M_0, M):

begin
  guess a subset of transitions TX of $\mathcal{N}$;
  if All_Siphons_Marked($\mathcal{N}_{TX}$, M_0)
    and All_Traps_Marked($\mathcal{N}_{TX}$, M)
    and Feasible($\mathcal{S}$)
  then return true fi
end

Since the system of equations $\mathcal{S}$ has linear size in the net N, Feasible($\mathcal{S}$) runs in polynomial time in the size of the net. So Check_Conditions runs in polynomial time, and the problem of checking if conditions (1), (2), and (3) hold belongs to NP.

Remark Even if we didn't know about the All_Siphons_Marked algorithm, we could still conclude that the conjecture is probably false. Only from the existence of the procedure Feasible($\mathcal{S}$) we can already conclude that the reachability problem for 1-safe nets belongs to Σ_2^P, the second level of the polynomial-time hierarchy (see for instance [1]). The general opinion of complexity theorists is that Σ_2^P = PSPACE is almost as unlikely as NP=PSPACE.

4 A nearly universal upper bound

In this section we obtain a nearly universal upper bound matching the PSPACE-hard lower bound of the last section:

> Rule of thumb 2:
> Nearly all interesting questions about the behaviour of 1-safe Petri nets can be decided in polynomial space.

Observe that the rule of thumb says "nearly all" and no longer "all". The reason is that the literature contains at least one interesting question requiring more than polynomial space. This exception to the rule is described at the end of the section.

We substantiate the rule of thumb with the help of temporal logics. Since their first application to computer science in the late seventies by Pnueli and others, temporal logics have become the standard query languages used to express properties of reactive and distributed systems. A good introduction to the application of temporal logics to computer science can be found in [6].

Temporal logics can be *linear-time* and *branching-time*: linear-time logics are interpreted on the single computations of a system, while branching-time logics are interpreted on the tree of all its possible computations. The most popular linear and branching-time temporal logics are LTL (*linear-time propositional temporal logic*) and CTL (*computation tree logic*). Most of the safety and liveness properties of interest for practitioners, like deadlock-freedom, reachability, liveness (in the Petri net sense), starvation-freedom, strong and weak fairness, etc. can be expressed in LTL or in CTL (often in both).

We show that all the properties expressible in LTL and CTL can be decided in polynomial space. Actually, we even show that they can be *uniformly* decided in polynomial space, i.e., we prove that the degree of the polynomial does not depend on the property we consider. More precisely, let $|N|$ denote the size of a Petri net N, and let $|\phi|$ denote the length of a formula ϕ (its number of symbols). For each of LTL and CTL we give an algorithm that accepts as input a Petri net N and a formula ϕ, and answers "yes" or "no" according to whether the net satisfies the formula or not; the algorithm uses $O(p(|N| + |\phi|))$ space, where p is a polynomial independent of N and ϕ.

4.1 Linear-time propositional temporal logic

The formulas of LTL are built from a set *Prop* of atomic propositions, and have the following syntax:

$$\begin{array}{lll} \phi ::= & p \in \mathit{Prop} & \\ & \neg\phi & \\ & \phi_1 \wedge \phi_2 & \\ & X\phi & (\phi \text{ holds at the next state}) \\ & \phi_1 U \phi_2 & (\phi_1 \text{ holds until } \phi_2 \text{ holds}) \end{array}$$

Usual abbreviations are $\mathit{true} = p \vee \neg p$, $F\phi = \mathit{true}U\phi$ (eventually ϕ), and $G\phi = \neg F \neg \phi$ (always ϕ).

LTL formulas are interpreted on *computations*. A computation is a finite or infinite sequence $\pi = P(0)P(1)P(2)\ldots$ of sets of atomic propositions. Intuitively, $P(i)$ is the set of propositions that hold in the computation after i steps. For a computation π and a point i in the computation, we have that:

$\pi, i \models p$	iff	$p \in P(i)$
$\pi, i \models \neg\phi$	iff	$\text{not}(\pi, i \models \phi)$
$\pi, i \models \phi_1 \wedge \phi_2$	iff	$\pi, i \models \phi_1$ and $\pi, i \models \phi_2$
$\pi, i \models X\phi$	iff	there exists a point $i+1$ in the computation, and $\pi, i+1 \models \phi$
$\pi, i \models \phi_1 U \phi_2$	iff	for some $j \geq i$, we have $\pi, j \models \phi_2$ and for all k, $i \leq k < j$, we have $\pi, k \models \phi_1$

We say that a computation π satisfies a formula ϕ, denoted $\pi \models \phi$, if $\pi, 0 \models \phi$.

The atomic propositions are intended to be propositions on the states of a system. They can only be chosen after the class of systems on which the logic is to be applied has been fixed. In the case of 1-safe Petri nets the states of the system are the markings, and so the atomic propositions are predicates on the possible markings of the net. It is then natural to have one atomic proposition per place. The markings satisfying the atomic proposition s are those that put a token in s. Observe that a computation is now a sequence of sets of places, and so a sequence of markings. In particular, the sequences of markings obtained from the runs of N by removing the intermediate transitions are computations. Abusing language, we also call these particular computations runs. We now define that a Petri net N satisfies ϕ if *all* its runs satisfy ϕ. Here are some LTL formulas that can be interpreted on the Petri net of Figure 1, which models a variation of Lamport's 1-bit mutual exclusion algorithm for two processes [26]:

(1) All runs are infinite (true for the net of Figure 1): $GX\mathit{true}$.
(2) All runs mark place cs_1 infinitely often (false): $GFcs_1$.
(3) In all runs, if place req_1 becomes marked then place cs_1 will eventually become marked (true): $G(req_1 \Rightarrow Fcs_1)$.

Formula (1) expresses deadlock-freedom; formula (3) expresses that the requests of the first process to the critical section are eventually granted.

The model-checking problem for LTL and 1-safe Petri nets consists of, given a 1-safe Petri net N and a formula ϕ, deciding whether N satisfies ϕ or not.

The solution to the model-checking problem we give here makes use of automata theory. We have to introduce automata on infinite words. Let $A = (\Sigma, Q, q_0, \delta, F)$ be a nondeterministic automaton, where Σ is a finite alphabet, Q is a finite set of states, q_0 is the initial state, $\delta \subseteq Q \times \Sigma \times Q$ is the transition relation, and F is a set of finite states. The language of A, denoted by $L(A)$, is defined as the set of finite words accepted by A. We define now the language of *infinite words* accepted by A, which we denote by $L_\omega(A)$. A word $w = a_0 a_1 a_2 \ldots$ belongs to $L_\omega(A)$ if there is an infinite sequence of states $q_0 q_1 q_2 \ldots$ such that $(q_i a_i q_{i+1}) \in \delta$ for every $i \geq 0$.

When we are interested in the language of infinite words of an automaton, then we call it *Büchi automaton*.

We have the following important result:

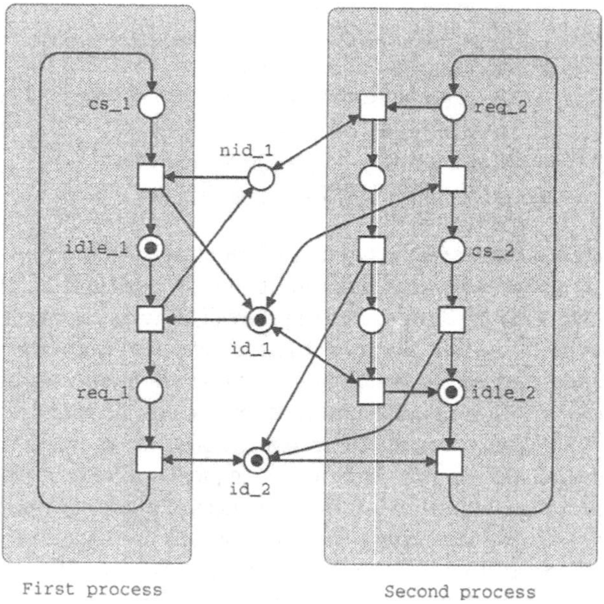

Fig. 1. A Petri net model of Lamport's 1-bit mutex algorithm

> Given an LTL formula ϕ, one can build a finite automaton A_ϕ and a Büchi automaton B_ϕ such that $L(A_\phi) \cup L_\omega(B_\phi)$ is exactly the set of computations satisfying the formula ϕ.

Since computations are sequences of sets of atomic propositions, the alphabet of the automata A_ϕ and B_ϕ is the set 2^{Prop}. In our case *Prop* is the set of places of the net, and so the alphabet of the automata is the set of all markings.

The construction of A_ϕ and B_ϕ exceeds the scope of this paper (see for instance [37]). For our purposes, it suffices to know the following facts:

- The states of A_ϕ are sets of subformulas of ϕ; the states of B_ϕ are pairs of sets of subformulas of ϕ. Since there are exponentially many sets of subformulas, A_ϕ and B_ϕ may have exponentially many states in $|\phi|$.
- Given two states q_1, q_2 of A_ϕ or B_ϕ and a marking M, there is an algorithm which decides using polynomial space whether $(q_1, M, q_2) \in \delta_\phi$.

We also need two automata $A_N = (2^S, Q_N, q_{0N}, \delta_N, F_N^A)$ and $B_N = (2^S, Q_N, q_{0N}, \delta_N, F_N^B)$ obtained from the Petri net N, as follows:

- Q_N is the set of reachable markings of N;
- q_{0N} is the initial marking M_0;
- δ_N contains the triples of markings (M_1, M_1, M_2) such that $M_1 \xrightarrow{t} M_2$ for some transition t;

- F_N^A is the set of deadlocked reachable markings of N;
- $F_N^B = Q$, i.e., F_N^B is the set of reachable markings of N.

Loosely speaking, both automata correspond to the reachability graph of N, with the peculiarity that edges are labelled with the marking they come from. A_N and B_N differ only in their final states. Clearly, $L(A_N)$ is the set of all finite runs of N, and $L_\omega(B_N)$ the set of all infinite runs.

In order to solve the model-checking problem for input N, ϕ, let A be the product of the automata $A_{\neg\phi}$ and A_N, and let B be the product of the automata $B_{\neg\phi}$ and B_N, where the product $(\Sigma, Q, q_0, \delta, F)$ of two automata $(\Sigma, Q_1, q_{01}, \delta_1, F_1)$ and $(\Sigma, Q_2, q_{02}, \delta_2, F_2)$ is defined in the usual way:

$$\begin{aligned} Q &= Q_1 \times Q_2 \\ q_0 &= (q_{01}, q_{02}) \\ \delta &= \{((q_1, q_2), a, (q_1', q_2')) \mid (q_1, a, q_1') \in \delta_1 \text{ and } (q_2, a, q_2') \in \delta_2\} \\ F &= F_1 \times F_2 \end{aligned}$$

Clearly, we have $L(A) = L(A_{\neg\phi}) \cap L(A_N)$ and $L_\omega(B) = L_\omega(B_{\neg\phi}) \cap L_\omega(B_N)$.[5] So the union of $L(A)$ and $L_\omega(B)$ is the set of runs of N that do *not* satisfy ϕ; in other words, N satisfies ϕ if and only if $L(A) = \emptyset$ and $L_\omega(B) = \emptyset$.

We have reduced the model checking problem to the following one: Given N and ϕ, decide if $L(A)$ and $L_\omega(B)$ are empty. We have to solve this problem using only polynomial storage space in the size of N and ϕ. The first natural idea is to construct A and B, and then use the standard algorithms for emptiness of automata for finite and infinite words. Unfortunately, both A and B may have exponentially many states in $|N|$ and $|\phi|$.

At this point, complexity theory helps us by means of Savitch's construction. Recall that a nondeterministic decision procedure for a problem is an algorithm which can return "yes" or fail, and satisfies the following property: the answer to the problem is "yes" if and only if *some* (not necessarily all) execution of the algorithm returns "yes". A deterministic decision procedure always answers "yes" or "no".

Savitch's construction: Given a nondeterministic decision procedure for a given problem using $f(n)$ space, Savitch's construction yields a deterministic procedure for the same problem using $f^2(n)$ space.

This construction makes our life easier: it suffices to give a *nondeterministic* algorithm for the emptiness problem of A and B running in polynomial space. Actually, it also suffices to give a nondeterministic algorithm for the *nonemptiness* problem: by Savitch's construction there exists a deterministic algorithm

[5] The product of two Büchi automata doesn't always accept the intersection of the languages, but this is so in our case.

for the nonemptiness problem, and by reversing the answer of this algorithm we obtain another one for the emptiness problem.

The nondeterministic algorithm for the nonemptiness problem constructs A and B "on the fly". The algorithm keeps track of a current state of A or B, which is initially set to the initial state. The algorithm repeatedly guesses a next state, checks that there is a transition leading from the current state to the next state, and updates the current state. In the case of A, the algorithm returns "true" when (and if) it reaches a final state:

Algorithm Nonempty_A(N, ϕ)

variables: q of type state of $A_{\neg\phi}$;
M of type state of A_N (i.e., of type marking);

begin
$(q, M) := (q_{0\neg\phi}, M_0)$;
while (q, M) is not a final state of A **do**
choose a state q' of $A_{\neg\phi}$ such that $(q, M, q') \in \delta_{\neg\phi}$
and a marking M' such that $M \xrightarrow{t} M'$ for some transition t;
$(q, M) := (q', M')$;
od;
return true
end

In order to estimate the space used by Nonempty_A, observe that all the operations and tests can be performed in polynomial space. For that, recall that given two states $q_1, q_2 \in Q_{\neg\phi}$ and $M \in 2^S$, there is an algorithm which decides using polynomial space whether $(q_1, M, q_2) \in \delta_{\neg\phi}$. The algorithm needs to store one state q of $A_{\neg\phi}$ and a marking M of N. Since the states of $A_{\neg\phi}$ are sets of subformulas of ϕ, q has quadratic size in $|\phi|$. Since M has linear size in $|N|$, polynomial space suffices.

The case of B is a bit more complicated. Since B has finitely many states, $L_\omega(B)$ is nonempty if and only if there exists a reachable final state q such that there is a loop from q to itself. So the algorithm proceeds as in the case of A, but, at some point, it guesses that the current final state will be revisited; it then stores the current state to be able to check if the guess is true. The rest of the algorithm checks the guess nondeterministically.

Algorithm Nonempty_B(N, ϕ):

variables: M, M_r of type state of B_N (i.e., of type marking);
q, q_r of type state of $B_{\neg\phi}$;
flag of type boolean;

begin
$(q, M) := (q_{0\neg\phi}, M_0)$; $flag := false$;
while $flag = false$ **do**
choose a state q' of $A_{\neg\phi}$ such that $(q, M, q') \in \delta_{\neg\phi}$
and a marking M' such that $M \xrightarrow{t} M'$ for some t;
$(q, M) := (q', M')$;
if (q, M) is a final state **then**
choose between $flag := false$ and $flag := true$
fi
od;
$(q_r, M_r) := (q, M)$;
repeat
choose a state q' of $A_{\neg\phi}$ such that $(q, M, q') \in \delta_{\neg\phi}$
and a marking M' such that $M \xrightarrow{t} M'$ for some t;
$(q, M) := (q', M')$
until $(q, M) = (q_r, M_r)$;
return true
end

Again, Nonempty_B(N, ϕ) uses only polynomial space. Since the deterministic algorithm obtained after the application of Savitch's construction to Nonempty_A and Nonempty_B also needs polynomial space, the model-checking problem for LTL belongs to PSPACE.

Observe that the only properties of 1-safe nets we have used in order to obtain this result are:

- a state has polynomial size (actually, even linear) in $|N|$, and
- given two markings M, M', it can be decided in polynomial space if $M \xrightarrow{t} M'$ for some transition t.

These conditions are very weak, and so the PSPACE result can be extended to a number of other models. As observed in [35], conditions (1) and (2) hold for other Petri net classes, like *condition/event systems*, *elementary net systems*, but also for process algebras with certain limitations to recursion, and for several other models based on a finite number of state machines communicating by finite means. The conditions also hold for bounded Petri nets, assuming that the bound is also given to Nonempty_A and Nonempty_B as part of the input. This assumption is necessary, because the bound of a bounded Petri net (the maximal number of tokens a place can contain under a reachable marking) can be much bigger than the size of the net, and so we may need more than polynomial space in order to just write down a reachable marking.

The PSPACE result can also be extended to more general logics, like the linear-time mu-calculus, for which the translation into automata still works (see for instance [4]).

4.2 Computation Tree Logic

Some interesting properties of Petri nets cannot be expressed in LTL. An example is liveness (in the Petri net sense). Recall that a transition is live if it can always occur again. One possibility to express this to allow existential or universal quantification on the set of computations starting at a marking. CTL introduces this quantification on top of LTL's syntax The syntax of CTL is

$$
\begin{array}{rll}
\phi ::= & p \in \mathit{Prop} & \\
 & \neg\phi & \\
 & \phi_1 \wedge \phi_2 & \\
 & EX\phi & \text{existential next operator} \\
 & AX\phi & \text{universal next operator} \\
 & E[\phi_1 U \phi_2] & \text{existential until operator} \\
 & A[\phi_1 U \phi_2] & \text{universal until operator}
\end{array}
$$

Disjunction and implication are defined as usual. Other abbreviations are $\mathit{true} = p \vee \neg p$, $EF\phi = E[\mathit{true}U\phi]$ (possibly ϕ), $AG\phi = \neg EF \neg\phi$ (always ϕ), $AF\phi = A[\mathit{true}U\phi]$ (eventually ϕ) and $EG\phi = \neg AF\neg\phi$ (ϕ holds at every state of some computation).

CTL formulas are interpreted on *computation trees*, which are possibly infinite trees where each node n is labelled with a set of atomic propositions $P(n)$. A path of a computation tree that cannot be extended to a larger path is called a computation; notice that it is a computation in the LTL sense. The intuition is that the nodes of the tree correspond to the states of a system; a state may have an arbitrary number of successors, corresponding to different computations. $P(n)$ is the set of atomic propositions that hold at node (state) n. For a tree τ and a node n we have that:

$$
\begin{array}{lll}
\tau, n \models p & \text{iff} & p \in P(n) \\
\tau, n \models \neg\phi & \text{iff} & \text{not}(\tau, n \models \phi) \\
\tau, n \models \phi_1 \wedge \phi_2 & \text{iff} & \tau, n \models \phi_1 \text{ and } \tau, n \models \phi_2 \\
\tau, n \models AX\phi & \text{iff} & \text{for every child } n' \text{ of } n,\ \tau, n' \models \phi \\
\tau, n \models EX\phi & \text{iff} & \text{for some child } n' \text{ of } n,\ \tau, n' \models \phi \\
 & & (n \text{ must have at least one child}) \\
\tau, n \models A[\phi_1 U \phi_2] & \text{iff} & \text{for all computations } n = n_0 n_1 n_2 \ldots \\
 & & \text{there exists } i \geq 0 \text{ such that } n_i \models \phi_2 \\
 & & \text{and for every } j,\ 0 \leq j < i,\ n_j \models \phi_1 \\
\tau, n \models E[\phi_1 U \phi_2] & \text{iff} & \text{for some computation } n = n_0 n_1 n_2 \ldots \\
 & & \text{there exists } i \geq 0 \text{ such that } n_i \models \phi_2 \\
 & & \text{and for every } j,\ 0 \leq j < i,\ n_j \models \phi_1
\end{array}
$$

If the tree τ is clear from the context we shorten $\tau, n \models \phi$ to $n \models \phi$. We say that a tree τ satisfies a formula ϕ if $\mathit{root}(\tau) \models \phi$.

Observe that $AX\phi$ is equivalent to $\neg EX \neg\phi$, i.e., EX and AX are dual operators. So actually we could remove AX from the syntax without losing expressive power. It might seem that the existential and universal until operators are also dual of each other, but this is not true. The dual operator of the universal until is the existential *weak* until, with syntax $E[\phi_1\, WU \phi_2]$, and the following semantics:

$$\tau, n \models E[\phi_1\, WU \phi_2] \;\text{ iff }\; \tau, n \models E[\phi_1 U \phi_2] \vee EG(\phi_1)$$

It holds that

$$A[\phi_1 U \phi_2] = \neg E[\neg\phi_2\, WU \neg\phi_1]$$

In order to use CTL to specify properties of a 1-safe Petri net N, we choose again the places of N as atomic propositions. With this choice a computation tree is a tree of sets of places, and so a set of markings. We can associate to N a computation tree τ_N as follows: the root is labelled with the initial marking M_0; the children of a node labelled by M are labelled with the markings M' such that $M \xrightarrow{t} M'$ for some transition t. We say that N satisfies ϕ if the tree τ_N satisfies ϕ.

The computation tree corresponding to the the net of Figure 1 is shown in Figure 2. Essentially, the tree is just the unfolding into a tree of the reachability graph of the net. Different nodes in the tree can be labelled with the same

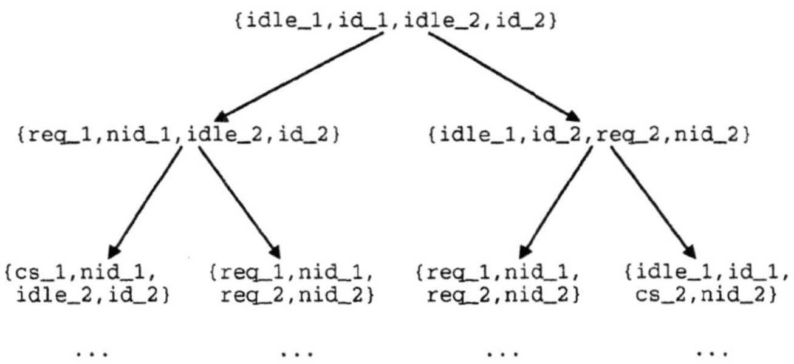

Fig. 2. Computation tree of the Petri net of Figure 1

marking, but all subtrees whose roots are labelled with the same marking are isomorphic. Given a formula ϕ and a marking M, either all or none of the nodes labelled by M satisfy ϕ. So it makes sense to say that M satisfies ϕ, meaning that all nodes labelled by M satisfy ϕ.

Here are some CTL queries on the Petri net of Figure 1:

- No reachable marking puts tokens in cs_1 and cs_2 (true): $AG(\neg cs_1 \vee \neg cs_2)$.
- The output transition of the place req_1 is live (true): $AGEF(req_1 \wedge id_2)$.
- The initial marking is reachable from every reachable marking (true): $AGEF(idle_1 \wedge id_1 \wedge id_2 \wedge idle_2))$
- Eventually place cs_1 becomes marked (false): $AFcs_1$
- There is a run that never marks cs_2 (true): $EG\neg cs_2$
- If req_2 becomes marked, then eventually cs_2 becomes marked (false): $AG(req_2 \Rightarrow AFcs_2)$

We show that the model checking problem for CTL is in PSPACE. It follows from the discussion above that it suffices to give a polynomial space algorithm for the syntax

$$\phi ::= s \mid \neg\phi_1 \mid \phi_1 \wedge \phi_2 \mid EX\phi \mid E[\phi_1 U \phi_2] \mid E[\phi_1\, WU \phi_2]$$

We give a (deterministic) algorithm Check(M, ϕ) with a marking M and a formula ϕ as parameters which answers "true" if M satisfies ϕ, and "false" otherwise. The model-checking problem is then solved by Check(M_0, ϕ).

Check(M, ϕ) is a recursive procedure on the structure of ϕ, i.e., Check(M, $Op(\phi_1, \ldots, \phi_n)$), where Op is some operator of the logic, calls Check(M, ϕ_1), ..., Check(M, ϕ_n).

Algorithm Check(M, ϕ):

begin
  if $\phi = s$ **then**
    if $M(s) = 1$ **then return true else return false fi**
  elseif $\phi = \neg\phi_1$ **then return not** Check(M, ϕ_1)
  elseif $\phi = \phi_1 \wedge \phi_2$ **then return** Check(M, ϕ_1) **and** Check(M, ϕ_2))
  elseif $\phi = EX\phi_1$ **then**
    for every M' such that $M \xrightarrow{t} M'$ for some transition t **do**
      if Check(M', ϕ_1) **then return true fi**
    od
  elseif $\phi = E[\phi_1 U \phi_2]$ **then return** EU(M, ϕ_1, ϕ_2)
  elseif $\phi = E[\phi_1\, WU \phi_2]$ **then return** EWU(M, ϕ_1, ϕ_2)
  fi
end

It remains to define the procedures EU(M, ϕ_1, ϕ_2) and EWU(M, ϕ_1, ϕ_2). We start with EU(M, ϕ_1, ϕ_2).

It is not possible to deterministically explore the infinitely many computations starting at M, and check directly if one of them satisfies $\phi_1 U \phi_2$. The reader might feel tempted to give a nondeterministic algorithm which explores one of the computations, and then apply Savitch's technique. This seems to be a good idea, but in fact doesn't work! There is a rather subtle problem. Consider the formulas

$$\phi_n = E[E \ldots E[s_0 U s_1] \ldots]U s_{n-1}]U s_n]$$

where $s_1, \ldots, s_n$ are places. We obtain a checking algorithm ϕ_n through n applications of Savitch's technique. It is easy to give a $\Omega(|N|)$-space nondeterministic algorithm for $E[s_0 U s_1]$. Unfortunately, the deterministic algorithm obtained by Savitch's technique requires $\Omega(|N|^2)$ space, the algorithm for $E[E[s_0 U s_1] U s_2]$ $\Omega(|N|^4)$ space, and the algorithm for ϕ_n no less than $\Omega(|N|^{2n})$ space. So the degree of the polynomial in $|N|$ depends on the formula we are considering.

We proceed in a different way. In a fist step we reduce the problem to the exploration of a finite number of finite paths. We extend the syntax of CTL with new operators $E[\phi_1 U_b \phi_2]$, one for each natural number b. Loosely speaking, a node satisfies $E[\phi_1 U_b \phi_2]$ if in at least one of the computations starting at it we find a node satisfying ϕ_2 ***after at most b steps***, and all nodes before it satisfy ϕ_1. Formally:

$$\begin{array}{ll} \tau, n \models E[\phi_1 U_b \phi_2] \text{ iff } & \text{for some computation } n = n_0 n_1 n_2 \ldots \\ & \text{there exists } i,\ 0 \leq i \leq b-1 \text{ such that} \\ & n_i \models \phi_2 \text{ and } n_j \models \phi_1 \text{ for every } j,\ 0 \leq j < i \end{array}$$

It follows immediately from this definition that if τ, n satisfies $E[\phi_1 U_b \phi_2]$ for some number b then it also satisfies $E[\phi_1 U \phi_2]$.

Now, let n be an arbitrary node of τ_N, and let k be the number of places of N. We prove

$$n \models E[\phi_1 U \phi_2] \iff E[\phi_1 U_{2^k} \phi_2]$$

It suffices to prove that $n \models E[\phi_1 U \phi_2]$ implies $n \models E[\phi_1 U_{2^k} \phi_2]$. Assume that n satisfies $E[\phi_1 U \phi_2]$. Then, τ_N contains a computation $n = n_0 n_1 n_2 \ldots$ satisfying $\phi_1 U \phi_2$: $n_i \models \phi_1$ for some $i \geq 0$ and $n_j \models \phi_1$ for every j, $0 \leq j < i$. If $i \leq 2^k - 1$, then this computation satisfies $\phi_1 U_{2^k} \phi_2$, and so $n \models \phi_1 U_{2^k} \phi_2$. Let us now consider the case $i \geq 2^k$. Let $M_0 M_1 M_2 \ldots$ be the sequence of markings corresponding to $n_0 n_1 n_2 \ldots$. Since N is 1-safe and has k places, it has at most 2^k reachable markings. So there are indices j_1 and j_2, $0 \leq j_1 < j_2 \leq i$, such that $M_{j_1} = M_{j_2}$. Since the markings labelling the successors of a node are completely determined but the marking labelling the node itself, τ_N contains another computation starting at n_0 and labelled by

$$M_0 \ldots M_{j_1} M_{j_2+1} M_{j_2+2} \ldots$$

Loosely speaking, the sequence of markings of the new computation is obtained from the old sequence by "cutting out" the piece $M_{j_1+1} \ldots M_{j_2}$ and "glueing" the two ends M_{j_1} and M_{j_2+1}. In this new sequence the marking M_i appears at the position $i - (j_2 - j_1)$, and so closer to M_0 than in the original computation. We now iterate the "cutting and glueing" procedure until M_i appears before the 2^k -th position. The computation so obtained satisfies $\phi_1 U_{2^k} \phi_2$, and so $n \models \phi_1 U_{2^k} \phi_2$.

So we have solved our first problem: instead of a potentially infinite number of computations, it suffices to explore finitely many paths containing at most 2^k nodes, and check that at least one of them satisfies $\phi_1 U_{2^k} \phi_2$ (more precisely, that at least one of them can be extended to a computation satisfying $\phi_1 U_{2^k} \phi_2$).

We construct EU(M, ϕ_1, ϕ_2) with the help of another algorithm Path(M, M', ϕ, ψ, l), still to be designed, with the following specification:

Path(M, M', ϕ, ψ, l) returns "true" if and only if τ_N has a path $n_0 \ldots n_l$ such that

- n_0 is labelled by M and n_l is labelled by M',
- $n_i \models \phi$ for every i, $0 \leq i < l$, and
- $n_l \models \psi$.

We can take:

```
Algorithm EU(M, φ1, φ2)

constant: k = number of places of N;

begin
   for every marking M' of N and every 0 ≤ l < 2^k do
      if Path(M, M', φ1, φ2, l) then return true
   od;
   return false
end
```

Since each iteration of the **for** loop can reuse the same space, the space used by EU(M, ϕ_1, ϕ_2) is the space used by Path(M, M', ϕ_1, l) plus the space needed to store M' and l. So Path(M, M', ϕ_1, l) should use at most polynomial space for every $l < 2^k$. A backtracking algorithm, which would be the obvious choice, does not meet this requirement, because it stores all the nodes of the computation being currently explored having still unexplored branches, and there can be exponentially many of those.

A trick frequently applied in complexity theory[6] helps us out of the problem. Loosely speaking, for each reachable marking M'', we explore all paths leading from M to M'' and containing $\lceil \frac{l}{2} \rceil + 1$ nodes, and then, *reusing the same space*, all paths leading from M'' to M' and containing $\lfloor \frac{l}{2} \rfloor + 1$ nodes. This trick of splitting the paths into two parts is applied recursively until paths having at most 2 nodes are reached.

```
Algorithm Path(M, M', φ, ψ, l)

constant: k = number of places of N;

begin
   if l = 0 then
      if M = M' and Check(M, ψ)
      then return true fi
   fi;
   if l = 1 then
      if M --t--> M' for some transition t
         and Check(M, φ) and Check(M', ψ)
      then return true fi
   fi;
   for every marking M'' of N do
      if Path(M, M'', φ, true, ⌈l/2⌉) and Path(M'', M, φ, ψ, ⌊l/2⌋)
      then return true fi
   od;
   return false
end
```

In order to estimate the space complexity of Path(M, M', ϕ, l), let $c(\phi)$ be the maximum over all markings M of the space needed by Check(M, ϕ), and let

[6] In fact, this trick lies at the heart of Savitch's technique.

$p(\phi, \psi, l)$ be the maximum over all pairs of markings M, M' of the space needed by Path(M, M', ϕ, ψ, l). Then we have

$$
\begin{aligned}
p(\phi, \psi, 0) &= O(c(\psi)) \\
p(\phi, \psi, 1) &= O(\max\{c(\phi), c(\psi)\}|N|) \\
p(\phi, \psi, l) &= O(\max\{p(\phi, \psi, \lceil \tfrac{l}{2} \rceil), p(\phi, \psi, \lfloor \tfrac{l}{2} \rfloor)\}|N|)
\end{aligned}
$$

and so, in particular

$$p(\phi, \psi, 2^k) = O(\max\{c(\phi), c(\psi)\} + k \cdot |N|) = O(\max\{c(\phi), c(\psi)\} + |N|^2)$$

It remains to construct EWU(M, ϕ_1, ϕ_2). The interested reader can easily prove that for every node n of τ_N

$$n \models E[\phi_1 \, WU \phi_2] \iff E[\phi_1 \, WU_{2^k} \phi_2]$$

where the semantics of $E[\phi_1 \, WU_b \phi_2]$ is given by

$$
\begin{aligned}
\tau, n \models E[\phi_1 \, WU_b \phi_2] \text{ iff } & n \models E[\phi_1 U_b \phi_2] \text{ or} \\
& \text{there exists a path } n = n_0 n_1 n_2 \ldots n_b \\
& \text{such that } n_i \models \phi_1 \text{ for every } 0 \le i \le b
\end{aligned}
$$

So we can take

```
Algorithm EWU(M, φ1, φ2)

constant: k = number of places of N;

begin
  if EU(M, φ1, φ2) then return true
  else
    for every marking M' of N do
      if Path(M, M', φ1, true, 2^k) then return true
    od;
  return false
end
```

This completes the definition of Check(M, ϕ). It is easy to see that it runs in polynomial space in $|N|$ and $|\phi|$, but let us determine the space complexity a bit more precisely. We have:

$$
\begin{aligned}
c(s) &= O(|N|) \\
c(\phi_1 \wedge \phi_2) &= O(\max\{c(\phi_1), c(\phi_2)\} + |N|) \\
c(\neg\phi) &= O(c(\phi)) \\
c(E[\phi_1 U \phi_2]) &= O(p(\phi_1, \phi_2, 2^k) + |N|) \\
&= O(\max\{c(\phi_1), c(\phi_2)\} + |N|^2) \\
c(E[\phi_1 U_w \phi_2]) &= O(\max\{c(E[\phi_1 U \phi_2]), p(\phi_1, true, 2^k)\} + |N|) \\
&= O(\max\{c(\phi_1), c(\phi_2)\} + |N|^2)
\end{aligned}
$$

and so we finally get $c(\phi) = O(|\phi| \cdot |N|^2)$.

4.3 An exception

The most interesting exception to Rule of Thumb 2 is the *controllability* property. Let T_0 be a subset of transitions of a 1-safe Petri net $N = (S, T, F, M_0)$, and let $t \in T \setminus T_0$. We say that T_0 *controls t by a sequence* $\sigma \in T_0^*$ if for every occurrence sequence $M_0 \xrightarrow{\tau} M$ such that the projection of τ onto T_0 is σ, the transition t cannot occur at M. The intuition is that T_0 can control t in the sense that once the sequence σ has occurred, possibly interleaved with transitions of $T \setminus T_0$, t cannot occur until transitions of T_0 occur again. We say that T_0 *can control* t if T_0 can control t by at least one sequence σ.

The controllability problem is defined as follows:

Given: a 1-safe Petri net with a set T of transitions, $T_0 \subseteq T$, $t \in T \setminus T_0$
To decide: if T_0 can control t.

Jones, Landweber and Lien show in [24] that controllability is EXPSPACE-complete.

4.4 A remark on action-based temporal logics

We have defined LTL and CTL as *state-based* logics, because in order to know if a run satisfies a property one only needs information about the states – the markings – visited during its execution, and not about which transitions lead from a marking to the next. It is possible to define *action-based* versions of these logics, in which the identities of the markings visited during the execution of a run is irrelevant, while the information is carried by the sequence of transitions that occur. These action-based versions are particularly useful for labelled Petri nets.

The action-based version of LTL – tailored for labelled Petri nets – looks as follows: the set of basic propositions contains only one element, namely the proposition *true*. The operators X and U are replaced by a set of *relativised* operators X_K, U_K, where k is a subset of a certain finite set of actions *Act*. A *computation* is now a finite or infinite sequence $\pi = a_0 a_1 a_2 \ldots$ of actions. Let $\pi^{(i)} = a_i a_{i+1} \ldots$. We have:

$$
\begin{array}{lll}
\pi \models \mathit{true} & & \text{always} \\
\pi \models X_K \phi & \text{iff} & \pi \neq \epsilon,\ a_0 \in k, \text{ and } \pi^{(1)} \models \phi \\
\pi \models \phi_1 U_K \phi_2 & \text{iff} & \text{for some } j \geq 0 \text{ we have } \pi^{(j)} \models \phi_2 \text{ and} \\
& & \text{for all } k,\ 0 \leq k \leq j, \text{ we have } a_i \in k \text{ and } \pi^{(k)} \models \phi_1
\end{array}
$$

In order to interpret the logic on a 1-safe labelled Petri net N, we choose *Act* as the set of labels carried by the transitions of N. We say that N satisfies a formula ϕ if *all* the sequences of transition labels obtained from the runs of N by removing the markings satisfy ϕ.

Similarly, in the action-based version of CTL the operators of the logic EX, AX, $E[\ldots U \ldots]$, and $A[\ldots U \ldots]$ are replaced by sets of relativised operators

EX_K, AX_K, $E[\ldots U_K \ldots]$, and $A[\ldots U_K \ldots]$. Computation trees are now trees whose edges are labelled with actions. The semantics is exactly what one expects.

It is easy to prove that the model-checking problem for these two new logics can be reduced to the model-checking problem for their state-based versions. More precisely: given a labelled 1-safe Petri net N and a formula ϕ of action-based LTL (CTL), one can construct in polynomial time an unlabelled 1-safe Petri net N' and a formula ϕ' of state-based LTL (CTL) such that N satisfies ϕ if and only if N' satisfies ϕ'. It follows that the model-checking problem for the action-based LTL and CTL is also in PSPACE.

In Section 8 we study the model checking problems for temporal logics and arbitrary Petri nets. There, the distinction between state-based and action-based logics plays a much more important rôle.

5 Deciding equivalences

In this section we investigate the complexity of deciding if two labelled 1-safe Petri nets are equivalent with respect to a given equivalence notion.

Since the early eighties many different equivalence notions have been presented in the literature. Van Glabbeek has classified them in several papers, e.g. [36]. Most of these equivalences fit between the so-called *trace* equivalence, which is a process theory counterpart of the classical language equivalence used in formal language theory, and *bisimulation* equivalence. An equivalence notion X fits between trace and bisimulation equivalence if bisimilar systems are X-equivalent, and X-equivalent systems are trace equivalent.

Trace and bisimulation equivalences are defined as follows. Let N be a labelled Petri net, where transitions are labelled with the elements of a set of actions *Act*. The set of *traces* of N, denoted by $\mathcal{T}(N)$ is the set of words $a_1 \ldots a_n \in Act^*$ such that there exist markings $M_1, \ldots M_n$ satisfying $M_0 \xrightarrow{a_1} M_1 \xrightarrow{a_2} \ldots \xrightarrow{a_n} M_n$[7]. Two Petri nets N_1 and N_2 are *trace equivalent* if $\mathcal{T}(N_1) = \mathcal{T}(N_2)$.

A relation $\mathcal{R}$ between the sets of markings of two nets is a *(strong) bisimulation* if for every pair $(M_1, M_2) \in \mathcal{R}$ and for every action $a \in Act$,

- if $M_1 \xrightarrow{a} M_1'$, then $M_2 \xrightarrow{a} M_2'$ for some marking M_2' such that $(M_1', M_2') \in \mathcal{R}$, and
- if $M_2 \xrightarrow{a} M_2'$, then $M_1 \xrightarrow{a} M_1'$ for some marking M_1' such that $(M_1', M_2') \in \mathcal{R}$.

Two Petri nets N_1 and N_2 are *(strongly) bisimilar* if there exists a (strong) bisimulation $\mathcal{R}$ containing the pair (M_{01}, M_{02}) of initial markings of N_1 and N_2.

We have the following

[7] Recall: $M \xrightarrow{a} M'$ denotes that there is a transition t labelled by a such that $M \xrightarrow{t} M'$.

Rule of thumb 3:
Equivalence problems for 1-safe Petri nets are harder to solve than model-checking problems, but they need at most exponential space.

We provide a first piece of evidence for this rule of thumb by showing that the equivalence problem for 1-safe Petri nets and any equivalence notion fitting between trace and bisimulation equivalence is PSPACE-hard. It turns out that all the concrete equivalences mentioned in the literature have at least DEXPTIME-hard equivalence problems, and so this general PSPACE-hardness lower bound can possibly be improved.

We proceed by reduction from the following PSPACE-hard problem

Given: a 1-safe Petri net N, a place s of N
To decide: if some reachable marking of N puts a token on s.

We start by labelling each transition of N with the same label, say a. N is now a labelled net. We put N side by side with the labelled net N' consisting of a loop containing one single place marked with one token and one single transition labelled by a. We denote the resulting Petri net by $N \parallel N'$.

Now, we consider two labelled nets. The first one is $N \parallel N'$; the second is a small modification of it obtained by adding a new output transition to the place s of N. The new transition has s as unique input place, no output places, and carries a label different from a, say b.

The following holds:

- If some reachable marking puts a token on s, then the two nets are not trace equivalent: the second one can do a b, while the first one can't.
- If no reachable marking puts a token on s, then the two nets are bisimilar: the relation containing all pairs (M_1, M_2), where M_1 is a reachable marking of the first net and M_2 a reachable marking of the second net, is clearly a bisimulation.

Therefore, given any equivalence notion X fitting between trace and bisimulation equivalence, we can solve the PSPACE-hard problem above by constructing the two nets and deciding if they are X-equivalent. So the equivalence problem for any such notion is PSPACE-hard.

Apart from this little result, the real evidence supporting the rule of thumb above is the work of Rabinovich [31] and Jategaonkar and Meyer [23]. This last paper contains a table with the complexity of 18 equivalence notions. Bisimilarity and many variants of it are DEXPTIME-complete, while trace equivalence, failures equivalence, and several variants of them are EXPSPACE-complete. They also consider so-called *partial order equivalences*, for which the concurrent execution of two actions is not equivalent to their interleaved execution (i.e., a system that executes a and b in parallel is not considered to be equivalent to a system which chooses between executing a and then b, or b and then a). The complexity results (up to some open problems) are similar.

6 Can anything be done in polynomial time?

We have seen that all interesting problems for arbitrary 1-safe Petri nets are at least PSPACE-hard, and so that there is very little hope of finding polynomial algorithms for them. The natural question to ask is if there are important subclasses of 1-safe Petri nets for which one could solve at least some problems in polynomial time. In this section we get some general answers in the form of rules of thumb.

A first rule, which tends to be surprising for many people is

> Rule of thumb 4:
> Most interesting questions about the behaviour of *acyclic* 1-safe Petri nets are NP-hard.

Here, as in Section 3, a word of warning is required about the meaning of "interesting". Liveness is certainly an interesting question for arbitrary 1-safe nets, but not for the acyclic ones: 1-safe acyclic Petri nets are always non-live, because no transition can fire more than once. Interesting questions for 1-safe acyclic Petri nets, all of them NP-hard, are

- Is a given marking reachable from the initial marking?
- Is there a reachable marking which marks a given place?
- Is there a reachable marking which does not mark a given place?
- Is there a reachable marking which enables a given transition?
- Is the initial marking reachable from every reachable marking?
- Is there a run containing a given transition?
- Is there a run that does not contain a given transition?

Let us prove NP-hardness of the second problem: Is there a reachable marking which marks a given place? We present a polynomial time construction which associates to a boolean formula in conjunctive normal form an acyclic 1-safe Petri net. The net nondeterministically selects a truth assignment for the variables of the formula, and then checks if the formula is true under the assignment. The construction is illustrated in Figure 3 by means of an example.

It seems[8] that in order to obtain classes with polynomial decision algorithms one has to impose *local* constraints on the net's structure. Here "local constraint" means a constraint which can be shown not to hold by looking at only a small part of the net. For instance, "every transition has exactly one input place" is a local constraint; if the constraint does not hold, then one can always point at a particular transition in the net, together with its input places, and show that the constraint is not satisfied because of this transition. A constraint like "the net is acyclic" is not local, because the smallest circuit of the net may be the net itself.

The two following local constraints have been very intensely studied in the literature:

[8] Although I don't know of any formal proof.

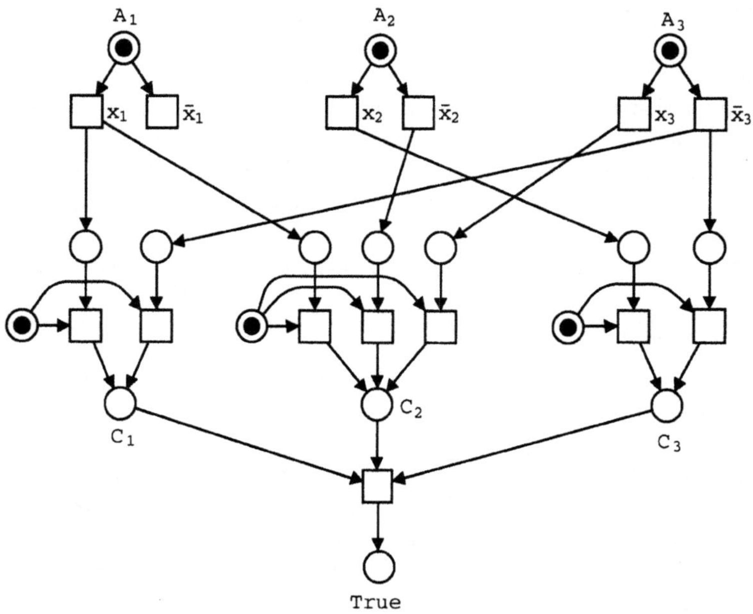

Fig. 3. Acyclic net corresponding to the formula $(x_1 \vee \overline{x}_3) \wedge (x_1 \vee \overline{x}_2 \vee x_3) \wedge (x_2 \vee \overline{x}_3)$

- the *conflict-freeness* constraint: $s^\bullet \subseteq {}^\bullet s$ for every place s with more than one output transition; in the case of 1-safe Petri nets this constraint is equivalent to "every place has at most one output transition" for nearly all purposes;
- the *free-choice* constraint: if (s,t) is an arc from a place to a transition, then so is (s',t') for every place $s' \in {}^\bullet t$ and for every transition $t' \in s^\bullet$.

Unfortunately, it is not possible to summarise the results of the research on conflict-free and free-choice Petri nets in a concise and general rule of thumb. But we can still say:

Rule of thumb 5:
Many interesting questions about 1-safe conflict-free Petri nets are solvable in polynomial time.
Some interesting questions about *live* 1-safe free-choice Petri nets are solvable in polynomial time (and liveness of 1-safe free-choice Petri nets is decidable in polynomial time too).
Almost no interesting questions for 1-safe net classes substantially larger than free-choice Petri nets are solvable in polynomial time.

Among the "many" interesting polynomial questions for conflict-free nets are all those that can be expressed in the fragment of CTL with syntax

$$\phi ::= s \mid \neg\phi \mid \phi_1 \wedge \phi_2 \mid EX\phi \mid EF\phi$$

(see [7]). Among the "some" interesting polynomial questions for live free-choice nets are the following [5]:

- Is there a reachable marking which marks a given place?
- Is there a reachable marking which does not mark a given place?
- Is there a reachable marking which enables a given transition?
- Is the initial marking reachable from every reachable marking?
- Is there a run that does not contain a given transition?

Interestingly, the reachability problem for 1-safe live free-choice nets is NP-complete [8], and so it is unlikely that it will ever be added to this list.

PartII

General Petri nets

In this second part of the paper we consider arbitrary (finite) Place/Transition Petri nets. The *possible markings of a net $\mathcal{N}$* or just the *markings of $\mathcal{N}$* are now the set of all mappings $S \to \mathbb{N}$, where S is the set of places of $\mathcal{N}$. Observe that, contrary to the 1-safe case, there is no a priori relation between the size of a net and the size of its markings. Notice also that the set of reachable markings may be infinite.

7 A universal lower bound

This section is the counterpart of Section 3 for Place/Transition Petri nets. The rule of thumb is now:

> Rule of thumb 6:
> All interesting questions about the behaviour of (Place/Transition) Petri nets are EXPSPACE-hard. More precisely, they require at least $2^{O(\sqrt{n})}$-space.

In particular, all the questions we asked about 1-safe Petri nets can be reformulated for Petri nets, and turn out to have at least this space complexity. As in the case of 1-safe Petri nets, this is a consequence of one single fundamental fact:

> A deterministic, exponentially bounded automaton of size n can be simulated by a Petri net of size $O(n^2)$. Moreover, there is a polynomial time procedure which constructs this net.

In order to answer a question about the computation of an exponentially space bounded automaton A, we can construct the net that simulates A, which has size $O(n^2)$, and solve the corresponding question. If the original question requires 2^n space, as is the case for many properties, then the corresponding question about nets requires at least $2^{O(\sqrt{n})}$-space.

The fundamental fact above was first proved by Lipton [27]. Mayr and Meyer proved in [29] that it is possible to make the simulating net *reversible* (a net is reversible if for each transition t there is a reverse transition $\overline{t}$ which "undoes" the effect of t). Since reversible nets are equivalent to commutative semigroups, the construction by Mayr and Meyer has important applications in mathematics.

Since Mayr and Meyer's construction is more involved than Lipton's, and since reversibility is not a main concern for this paper, we consider Lipton's construction in detail. It would have been easier to refer to Lipton's paper, but unfortunately it only exists as an old Yale report, quite difficult to find.

Bounded automata and general Place/Transition Petri nets do not "fit" well. It is not appropriate to model a cell of a bounded automaton as a place, as we did in the 1-safe case, because the cell contains one out of a *finite* number of possible symbols, while the place can contain infinitely many tokens, and so the same information as a nonnegative integer variable. So we use an intermediate model, namely *counter programs*. It is well-known that so-called bounded counter programs can simulate bounded automata (see below), and we show that Petri nets can simulate bounded counter programs.

A counter program is a sequence of *labelled commands* separated by semicolons. Basic commands have the following form, where l, l_1, l_2 are *labels* or *addresses* taken from some arbitrary set, for instance the natural numbers, and x is a variable over the natural numbers, also called a *counter*:

l: $x := x + 1$
l: $x := x - 1$
l: **goto** l_1 — unconditional jump
l: **if** $x = 0$ **then goto** l_1 **else goto** l_2 — conditional jump
l: **halt**

A program is syntactically correct if the labels of commands are pairwise different, and if the destinations of jumps correspond to existing labels. For convenience we can also require the last command to be a **halt** command.

A program can only be executed once its variables have received initial values. In this paper we assume that the initial values are always 0. The semantics of programs is that suggested by the syntax. The only point to be remarked is that the command $l : x := x - 1$ *fails* if $x = 0$, and causes abortion of the program. Abortion must be distinguished from proper termination, which corresponds to the execution of a **halt** command. Observe in particular that counter programs are deterministic.

A counter program C is k-bounded if after any step in its unique execution the contents of all counters are smaller than or equal to k. We make use of a well known construction of computability theory:

There is a polynomial time procedure which accepts a deterministic bounded automaton A of size n and returns a counter program C with $O(n)$ commands simulating the computation of A on empty tape; in particular, A halts if and only if C halts. Moreover, if A is exponentially bounded, then C is 2^{2^n}-bounded.

Now, it suffices to show that a 2^{2^n}-bounded counter program of size $O(n)$ can be simulated by a Petri net of size $O(n^2)$. This is the goal of the rest of this section.

Since a direct description of the sets of places and transitions of the simulating net would be very confusing, we introduce a net programming notation with a very simple net semantics. It is very easy to obtain the net corresponding to a program, and execution of a command corresponds exactly to the firing of a transition. So we can and will look at the programming notation as a compact description language for Petri nets.

A *net program* is rather similar to a counter program, but does not have the possibility to branch on zero; it can only branch nondeterministically. However, it has the possibility of transferring control to a subroutine. The basic commands are as follows:

l: $x := x + 1$	
l: $x := x - 1$	
l: **goto** l_1	unconditional jump
l: **goto** l_1 **or goto** l_2	nondeterministic jump
l: **gosub** l_1	subroutine call
l: **return**	end of subroutine
l: **halt**	

Syntactical correctness is defined as for counter programs. We also assume that programs are well-structured. Loosely speaking, a program is *well-structured* if it can be decomposed into a main program that only calls first-level subroutines, which in turn only call second-level subroutines, etc., and the jump commands in a subroutine can only have commands of the same subroutine as destinations.[9] We do not formally define well-structured programs, it suffices to know that all the programs of this section are well-structured.

We sketch a (Place/Transition) Petri net semantics of well-structured net programs. The Petri net corresponding to a program has a place for each label, a place for each variable, a distinguished *halt* place, and some additional places used to store the calling address of a subroutine call. There is a transition for each assignment and for each unconditional jump, and two transitions for each nondeterministic jump, as shown in Figure 4. We illustrate the semantics of the subroutine command by means of the program

[9] Here we consider the main program as a zero-level subroutine, i.e., jump commands in the main program can only have commands of the main program as destinations.

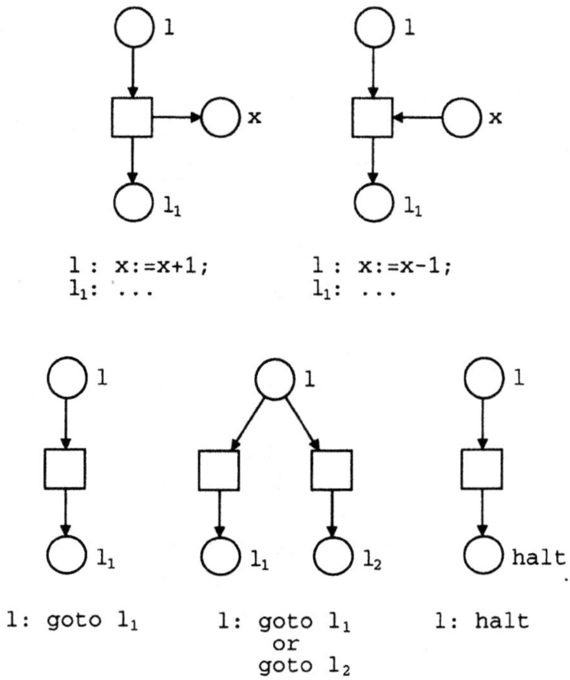

Fig. 4. Net semantics of assignments and jumps

1: **gosub** 4;
2: **gosub** 4;
3: **halt**;
4: **goto** 5 **or goto** 6;
5: **return**;
6: **return**

The corresponding Petri net is shown in Figure 5. Observe that the places *1_calls_4* and *2_calls_4* are used to remember the address from which the subroutine was called.

Clearly, the Petri net corresponding to a net program with k commands has $O(k)$ places and $O(k)$ transitions, and its initial marking has size $O(k)$. So it is of size $O(k^2)$.

Let C be a 2^{2^n}-bounded counter program with $O(n)$ commands. We show that C can be simulated by a net program $N(C)$ with $O(n)$ commands, which corresponds to a Petri net of size $O(n^2)$. Unfortunately, the construction of $N(C)$ requires quite a bit of low-level programming. But the reward is worth the hacking effort.

The notion of simulation is not as strong as in the case of 1-safe Petri nets. In particular, net programs are nondeterministic, while counter programs are deterministic. A net program N simulates a counter program C if the follow-

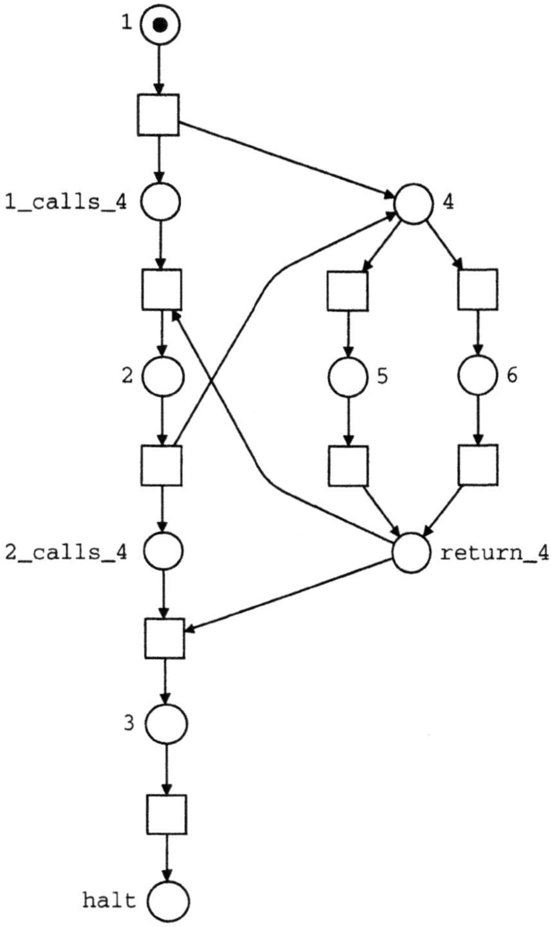

Fig. 5. Net semantics of subroutines

ing property holds: C halts (executes the command **halt**) if and only if *some* computation of N halts (other computations may fail).

Each variable x of N (be it a variable from C or an auxiliary variable) has an auxiliary *complement variable* $\overline{x}$. N takes care of setting $\overline{x} = 2^{2^n}$ at the beginning of the program. We call the code that takes care of this $N_{init}(C)$.[10] The rest of $N(C)$, called $N_{sim}(C)$, simulates C and takes care of keeping the invariant $\overline{x} = 2^{2^n} - x$.

We design $N_{sim}(C)$ first. This program is obtained through replacement of each command of C by an adequate net program. Commands of the form $x := x + 1$ ($x := x - 1$) are replaced by the net program $x := x + 1; \overline{x} := \overline{x} - 1$

[10] Recall that by definition all variables of N have initial value 0. Therefore, if we need $\overline{x} = 2^{2^n}$ initially, then we have to design preprocessing code for it.

$(x := x - 1; \overline{x} := \overline{x} + 1)$. Unconditional jumps are replaced by themselves. Let us now design a program

$$\text{Test}_n(x, \texttt{ZERO}, \texttt{NONZERO})$$

to replace a conditional jump of the form

l: **if** $x = 0$ **then goto** ZERO
else goto NONZERO

The specification of Test_n is as follows:

If $x = 0$ ($1 \leq x \leq 2^{2^n}$), then some execution of the program leads to ZERO (NONZERO), and no computation leads to NONZERO (ZERO); moreover the program has no side-effects: after any execution leading to ZERO or NONZERO no variable has changed its value.

Actually, it is easier to design a program $\text{Test}'_n(x, \texttt{ZERO}, \texttt{NONZERO})$ with the same specification but a *side-effect*: after an execution leading to ZERO, the values of x and $\overline{x}$ are swapped.[11] Once Test'_n has been designed, we can take:

```
Program Test_n(x, ZERO, NONZERO):
          Test'_n(x, continue, NONZERO);
continue: Test'_n(x̄, ZERO, NONZERO)
```

because the values of x and $\overline{x}$ are swapped 0 times if $x > 0$ or twice if $x = 0$, and so Test_n has no side effects.

The key to the design of Test'_n lies in the following observation: Since x never exceeds 2^{2^n}, testing $x = 0$ can be replaced by nondeterministically choosing

- to decrease x by 1, and if we succeed then we know that $x > 0$, or
- to decrease $\overline{x}$ by 2^{2^n}, and if we succeed then we know that $\overline{x} = 2^{2^n}$, and so $x = 0$.

If we choose wrongly, that is, if for instance $x = 0$ holds and we try to decrease x by 1, then the program fails; this is not a problem, because we only have to guarantee that the program *may* (not must!) terminate, and that if it terminates then it provides the right answer.

Decreasing x by 1 is easy. Decreasing $\overline{x}$ by 2^{2^n} is the difficult part. We leave it for a routine Dec_n to be designed, which must satisfy the following specification:

If the initial value of s is smaller than 2^{2^n}, then every execution of Dec_n fails. If the value of s is greater than or equal to 2^{2^n}, then all executions terminating with a **return** command have the same effect as $s := s - 2^{2^n}; \overline{s} := \overline{s} + 2^{2^n}$; in particular, there are no side-effects. All other executions fail.

[11] Executions leading to NONZERO must still be free of side-effects.

Test'_n proceeds by transferring the value of x to a special variable s, and then calling the routine Dec_n, which decreases s by 2^{2^n}. In this way we need one single routine Dec_n, instead of one for each different variable to be decreased, which leads to a smaller net program.

Program $\text{Test}'_n(x,$ ZERO, NONZERO):

** initially $s = 0$ and $\overline{s} = 2^{2^n}$ **
    goto nonzero **or goto** loop;
nonzero: $x := x - 1$; $x := x + 1$; goto NONZERO;
loop: $\overline{x} := \overline{x} - 1$; $x := x + 1$; $s := s + 1$; $\overline{s} := \overline{s} - 1$;
    goto exit **or goto** loop
exit: **gosub dec$_n$**; **goto** ZERO
** the routine called at dec$_n$ is $\text{Dec}_n(s)$ **

It is easy to see that Test'_n meets its specification: if $x > 0$, then we may choose the nonzero branch and reach NONZERO. If $x = 0$, then $\overline{x} = 2^{2^n}$. After looping 2^{2^n} times on loop the values of x, $\overline{x}$ and s, $\overline{s}$ have been swapped. The values of s and $\overline{s}$ are swapped again by the subroutine Dec_n, and then the program moves to ZERO. Moreover, if $x = 0$ then no execution reaches the NONZERO branch, because the program fails at $x := x - 1$. If $x > 0$, then no execution reaches the ZERO branch, because s cannot reach the value 2^{2^n}, and so Dec_n fails.

The next step is to design Dec_n. We proceed by induction on n, starting with Dec_0. This is easy, because it suffices to decrease s by $2^{2^0} = 2$. So we can take

Subroutine $\text{Dec}_0(s)$:

$s := s - 1$; $\overline{s} := \overline{s} + 1$;
$s := s - 1$; $\overline{s} := \overline{s} + 1$;
return

Now we design Dec_{i+1} under the assumption that Dec_i is already known. The definition of Dec_{i+1} contains two copies of a program Test'_i, called with different parameters. We define this program by substituting i for n everywhere in Test'_n. Test'_i calls the routine Dec_i at the address *dec$_i$*. Notice that this is correct, because we are assuming that the routine Dec_i has already been defined.

The key to the design of Dec_{i+1} is that decreasing by $2^{2^{i+1}}$ amounts to decreasing 2^{2^i} times by 2^{2^i}, because

$$2^{2^{i+1}} = (2^{2^i})^2 = 2^{2^i} \cdot 2^{2^i}$$

So decreasing by $2^{2^{i+1}}$ can be implemented by two nested loops, each of which is executed 2^{2^i} times, such that the body of the inner loop decreases s by 1. The loop variables have initial values 2^{2^i}, and termination of the loops is detected by testing the loop variables for 0. This is done by the Test'_i programs.

Subroutine $\text{Dec}_{i+1}(s)$:

```
** Initially y_i = 2^(2^i) = z_i, ȳ_i = 0 = z̄_i **
** The initialisation is carried out by N_init **
outer_loop: y_i := y_i - 1; ȳ_i := ȳ_i + 1;
inner_loop: z_i := z_i - 1; z̄_i := z̄_i + 1;
            s := s - 1; s̄ := s̄ + 1;
            Test'_i(z_i, inner_exit , inner_loop);
inner_exit: Test'_i(y_i, outer_exit, outer_loop);
outer_exit: return
```

Observe also that both instances of Test'_i call the same routine at the same label.

It could seem that Dec_{i+1} swaps the values of $y_i, \overline{y}_i$ and $z_i, \overline{z}_i$, which would be a side-effect contrary to the specification. But this is not the case. These swaps are compensated by the side-effects of the ZERO branches of the Test'_i programs! Notice that these branches are now the inner_exit and outer_exit branches. When the program leaves the inner loop, Test'_i swaps the values of z_i and $\overline{z}_i$. When the program leaves the outer loop, Test'_i swaps the values of y_i and $\overline{y}_i$.

This concludes the description of the program Test_n, and so the description of the program $N_{sim}(C)$. It remains to design $N_{init}(C)$. Let us first make a list of the initialisations that have to be carried out. $N_{sim}(C)$ contains

- the variables $x_1, \ldots, x_l$ of C with initial value 0; their complementary variables $\overline{x}_1, \ldots, \overline{x}_l$ with initial value 2^{2^n};
- a variable s with initial value 0; its complementary variable $\overline{s}$ with initial value 2^{2^n};
- two variables y_i, z_i for each i, $0 \le i \le n-1$, with initial value 2^{2^i}; their complementary variables $\overline{y}_i, \overline{z}_i$ for each i, $0 \le i \le n-1$, with initial value 0.

Now, the specification of $N_{init}(C)$ is simple

> $N_{init}(C)$ uses only the variables in the list above; every successful execution leads to a state in which the variables have the correct initial values.

$N_{init}(C)$ calls programs $\text{Inc}_i(v_1, \ldots, v_m)$ with the following specification:

> All successful executions have the same effect as
>
> $$v_1 := v_1 + 2^{2^i}; \quad \ldots; \quad v_m := v_m + 2^{2^i}$$
>
> In particular, there are no side-effects.

These programs are defined by induction on i, and are very similar to the family of Dec_i programs. We start with Inc_0:

Program $\mathrm{Inc}_0(v_1, \ldots, v_m)$:

$$v_1 := v_1 + 1;\ v_1 := v_1 + 1;$$
$$\ldots$$
$$v_m := v_m + 1;\ v_m := v_m + 1$$

and now give the inductive definition of Inc_{i+1}:

Program $\mathrm{Inc}_{i+1}(v_1, \ldots, v_m)$:

```
** Initially y_i = 2^(2^i) = z_i, ȳ_i = 0 = z̄_i **
outer_loop: y_i := y_i - 1; ȳ_i := ȳ_i + 1;
inner_loop: z_i := z_i - 1; z̄_i := z̄_i + 1;
            v_1 := v_1 + 1;
            ...
            v_m := v_m + 1;
            Test'_i(z_i, inner_exit , inner_loop);
inner_exit: Test'_i(y_i, outer_exit, outer_loop);
outer_exit: ...
```

It is easy to see that these programs satisfy their specifications. Now, let us consider $\mathrm{N}_{init}(C)$. Apparently, we face a problem: in order to initialise the variables $v_1, \ldots, v_m$ to $2^{2^{i+1}}$ the variables y_i and z_i must have already been initialised to 2^{2^i}! Fortunately, we find a solution by just carrying out the initialisations in the right order:

Program $\mathrm{N}_{init}(C)$:

$$\mathrm{Inc}_0(y_0, z_0);$$
$$\mathrm{Inc}_1(y_1, z_1);$$
$$\ldots$$
$$\mathrm{Inc}_{n-1}(y_{n-1}, z_{n-1});$$
$$\mathrm{Inc}_n(\overline{s}, \overline{x}_1, \ldots, \overline{x}_l)$$

This concludes the description of $N(C)$, and it is now time to analyse its size. Consider $N_{sim}(C)$ first. It contains two assignments for each assignment of C, an unconditional jump for each unconditional jump in C, and a different instance of Test_k for each conditional jump. Moreover, it contains (one single instance of) the routines Dec_n, Dec_{n-1}, ..., Dec_0 (notice that Test_n calls Dec_n, which calls Dec_{n-1}, etc.). Both Test_n and the routines have constant length. So the number of commands of $N_{sim}(C)$ is $O(n)$.

$N_{init}(C)$ contains (one single instance of) the programs Inc_i $1 \leq i \leq n$. The programs Inc_1, ..., Inc_{n-1} have constant size, since they initialise a constant number of variables. The number of commands of Inc_n is $O(n)$, since it initialises $O(n)$ variables.

So we have proved that $N(C)$ contains $O(n)$ commands. It follows that its corresponding Petri net has size $O(n^2)$, which concludes our presentation of Lipton's result.

The solution to Story II

Recall the conjecture of Story II: given a net $\mathcal{N}$ and two markings M_1 and M_2, if M_2 is reachable from M_1 then it is reachable from M_1 through a sequence $M_0 \xrightarrow{t_0} M_1 \xrightarrow{t_1} \cdots \xrightarrow{t_n} M_n = M$ such that all the markings $M_1, \ldots, M_n$ have size $O(n + m_0 + m)$, where n, m_0, m are the sizes of $\mathcal{N}$, M_0 and M respectively.

Let c be the constant such that $M_0, \ldots, M_n$ have size at most $c \cdot (n + m_0 + m)$. If the conjecture is true, then the following nondeterministic algorithm solves the reachability problem, since it may always answer "true" when M is reachable:

Algorithm Reachable($\mathcal{N}$, M_0, M):

variable: M' of type marking;

begin
  $M' := M_0$;
  while $M' \neq M$ **do**
    choose a marking M'' of size at most $c \cdot (n + m_0 + m)$
    such that $M' \xrightarrow{t} M''$ for some transition t;
    if there is no such marking then stop;
    $M' := M''$;
  od;
  return true
end

Since the algorithm only visits markings of size $c \cdot (n + m_0 + m)$, it runs in linear space. By Savitch's construction there is a deterministic algorithm which uses quadratic space. Since the reachability problem requires exponential space, the conjecture is false.

8 Upper bounds

The general exponential space lower bound of the last section is almost the best we can hope for, because Rackoff gave in [32] an almost matching exponential space upper bound for the covering and boundedness problems for Petri nets. More precisely, the upper bound is $2^{O(n \log n)}$ space, very close to the $2^{O(\sqrt{n})}$ lower bound. The covering problem consists of deciding if there exists a reachable marking M such that $M \geq M'$ for a given marking M', i.e., if there exists a reachable marking M *covering* M'; the boundedness problem consists of deciding if the number of reachable markings is finite.

Yen showed some years later in [38] that the same upper bound holds for the problem of deciding if there exists a firing sequence

$$M_0 \xrightarrow{\sigma_1} M_1 \xrightarrow{\sigma_2} \cdots \xrightarrow{\sigma_k} M_k$$

satisfying a given predicate $F(M_1, \ldots, M_k, \sigma_1, \ldots, \sigma_k)$ constructed using the following syntax:[12]

$$\begin{aligned} F ::= \; & M_i(s) \geq c \mid M_i(s) > c \\ & M_i(s) \leq M_j(s') \mid M_i(s) \geq M_j(s') \\ & \#_{\sigma_i}(t) \leq c \mid \#_{\sigma_i}(t) \geq c \\ & \#_{\sigma_i}(t) \leq \#_{\sigma_j}(t') \mid \#_{\sigma_i}(t) \geq \#_{\sigma_j}(t') \\ & F_1 \wedge F_2 \mid F_1 \vee F_2 \end{aligned}$$

where s and s' are places, t and t' are transitions, c is a constant, and $\#_\sigma(t)$ denotes the number of times that t occurs in σ. Both the covering and the boundedness problem can be reduced to Yen's problem. The covering problem for a marking $M = (m_1, \ldots, m_n)$ corresponds to deciding if there exists a firing sequence $M_0 \xrightarrow{\sigma_1} M_1$ such that $M_1(s_1) \geq m_1 \wedge \ldots \wedge M_1(s_n) \geq m_n$. The boundedness problem can be easily shown to be equivalent to the problem of deciding if there exists a sequence $M_0 \xrightarrow{\sigma_1} M_1 \xrightarrow{\sigma_2} M_2$ such that $M_1(s_1) \geq M_2(s_1) \wedge \ldots \wedge M_1(s_n) \geq M_2(s_n)$. Observe however that the reachability problem *cannot* be reduced to Yen's problem, because the predicate $M(s) = c$ does not belong to the syntax. The reachability problem was shown to be decidable by Mayr [28] and shortly after with a simpler proof by Kosaraju [25], but all known algorithms are non-primitive recursive. Closing the gap between the exponential space lower bound and the non-primitive recursive upper bound is one of the most relevant open problems of net theory.

Is it possible to give more general results about the properties that are decidable, and the properties that are decidable in exponential space? In particular, we would like to show that all the properties of a certain temporal logic are decidable, or decidable in exponential space. As we are going to see, there is a very significant difference between state-based logics and action-based logics, and so we consider them separately.

8.1 The state-based case

We have the following very general rule of thumb:

Rule of thumb 7: The model-checking problems of all interesting state-based logics are undecidable.

As in the 1-safe case, we first have to choose a set of atomic propositions. We take again $Prop = S$, i.e., the atomic propositions are the places of N. We say that a marking M satisfies the proposition s if M is marked at s. Observe that a computation is ***no longer*** a sequence of markings; a computation is a sequence of

[12] The syntax is actually a bit more general, see [38] for the details.

sets of places, as in the 1-safe case, but the markings of general Place/transition nets are not sets of places anymore.

With this choice of atomic propositions we can only express that a place is marked or not; we can say nothing about the number of tokens it contains. Unfortunately, even with this restricted expressive power the model checking problems for LTL and CTL turn out to be undecidable.

The proof is in both cases by reduction from the following problem, which is known to be undecidable:

Given: a counter program C with counters initialised to 0.
To decide: if C halts.

We simulate once again counter programs by net programs. Given a counter program C, we obtain a net program $N'(C)$ through replacement of each counter command

$$\text{l: } \textbf{if} \;\; x = 0 \;\; \textbf{then goto} \;\; \text{l}_1 \; \textbf{else goto} \;\; \text{l}_2$$

by the net program

$$\begin{aligned} \text{l: } & \textbf{goto } \texttt{test\_l}_1 \;\; \textbf{or goto} \;\; \texttt{test\_l}_2; \\ \texttt{test\_l}_1\text{: } & \textbf{goto} \;\; \text{l}_1; \\ \texttt{test\_l}_2\text{: } & \textbf{goto} \;\; \text{l}_2 \end{aligned}$$

while other commands are replaced by themselves.

The net program $N'(C)$ simulates C in a much weaker sense than that of Section 7. $N'(C)$ has a *honest* run that exactly mimics the (unique) execution of C: whenever C executes the command l, $N'(C)$ chooses the same branch as C. However, it also has many other runs that "cheat", i.e., runs that at some point choose the wrong branch. The labels $\texttt{test\_l}_1$ and $\texttt{test\_l}_2$ correspond to two places of $N'(C)$ which can be used to test if the program has cheated or not when executing the conditional jump.

Suppose that there exists a temporal logic formula *Halt* with the following property:

$N'(C)$ satisfies *Halt* if and only if the honest execution of $N'(C)$ halts.[13]

Since the honest run exactly mimics the execution of the counter program C, $N'(C)$ satisfies *Halt* if and only if C halts. Therefore, the problem of deciding if *Halt* is satisfied by a given Petri net N is undecidable. It follows that the model-checking problem of those logics in which *Halt* was expressed is undecidable as well.

We construct in CTL and LTL very simple formulas *LTL-Halt* and *CTL-Halt.* We first define a formula *Cheat* without temporal operators. *Cheat* is the conjunction over all conditional jumps l: **if** $x = 0$ **then goto** l_1 **else goto** l_2 of the formulas:

[13] Since $N'(C)$ is just a shorthand description of a Petri net, it makes sense to ask if $N'(C)$ satisfies a property formalised as a temporal formula.

$$(\texttt{test\_l}_1 \wedge x) \vee (\texttt{test\_l}_2 \wedge \neg x)$$

If a run visits a marking satisfying *Cheat*, then we know that it is dishonest: if the marking satisfies $(\texttt{test\_l}_1 \wedge x)$, then at some conditional jump the run has taken the l_1 branch even though $x > 0$; if $(\texttt{test\_l}_2 \wedge \neg x)$, then the run has taken the l_2 branch even though $x = 0$. Now, we define

$$LTL - Halt = F(Cheat \vee halt)$$

where *halt* is the place in the net semantics corresponding to all the **halt** commands. A run satisfies *LTL-Halt* if at some point it cheats or it halts. $N'(C)$ satisfies *LTL-Halt* iff every run satisfies *LTL-Halt*. Since the honest run is the only one that doesn't cheat, $N'(C)$ satisfies *LTL-Halt* iff the honest run halts.

The formula *CTL-Halt* is :

$$CTL - Halt = AF(Cheat \vee halt)$$

It follows immediately from the semantics of formulae that $N'(C)$ satisfies *CTL-Halt* if and only if it satisfies *LTL-Halt*.

Since the formula *CTL-Halt* only contains the operator AF, the fragment of CTL that extends propositional logic with the operators EF and its dual AG could still be decidable. Unfortunately, a different proof [9] shows that this is not the case.

8.2 The action-based case

As mentioned above, the action-based case is very different from the state-based case:

> Rule of thumb 8:
> The model-checking problems of all interesting branching-time, action-based logics are undecidable. The model-checking problems of all interesting linear-time, action-based logics are decidable.

The undecidability of branching-time logics in the action-based case is an immediate consequence of the following fact: given an unlabelled Petri net N and a formula ϕ of state-based CTL there is a labelled net N' and a formula ϕ' of action-based CTL such that N satisfies ϕ if and only if N' satisfies ϕ'.

The net N' is obtained by labelling the transitions of N with some label, say a, and then adding for each place s a new transition t_s having s as only input place, no output place at all, and labelled by s. The formula ϕ' is obtained through replacement of each atomic proposition s by $EX_s true$, and of each temporal operator EX, AX, $E[\ldots U \ldots]$, $A[\ldots U \ldots]$ by $EX_{\{a\}}$, $AX_{\{a\}}$, $E[\ldots U_{\{a\}\ldots]}$, and $A[\ldots U_{\{a\}} \ldots]$, respectively. Observe that s holds iff the transition t_s can occur, i.e., iff $EX_s true$ holds.

We cannot use the same technique to prove the undecidability of the model-checking problem for LTL, because the problem is decidable! As in the 1-safe case, the model-checking algorithm is based on automata theory. Given an LTL formula ϕ, one can build a finite automaton A_ϕ and a Büchi automaton B_ϕ such that $L(A_\phi) \cup L_\omega(B_\phi)$ is exactly the set of computations satisfying the formula ϕ. In the action-based case both A_ϕ and B_ϕ are automata over the alphabet *Act*.

In the 1-safe case, given a net N and a formula ϕ, we first constructed two automata $A_{\neg\phi}$ and $B_{\neg\phi}$ such that $L(A_{\neg\phi}) \cup L_\omega(B_{\neg\phi})$ is exactly the set of computations *violating* the formula ϕ. In the general case we proceed exactly in the same way. The second step was to construct two finite automata A_N and B_N from the Petri net N, which were both essentially equal to the reachability graph of the net. Here we have a problem: the automata A_N and B_N can be defined just as in the 1-safe case, but since N may now have infinitely many reachable markings, they are not guaranteed to be finite.

The solution to this problem is easy: instead of constructing two automata A_N and B_N out of the Petri net N, we construct two labelled Petri nets $NA_{\neg\phi}$ and $NB_{\neg\phi}$ out of the automata $A_{\neg\phi}$ and $B_{\neg\phi}$ in the following obvious way:

- the places of NA_ϕ are the states of A_ϕ;
- for each transition $q \xrightarrow{a} q'$ in A_ϕ add a transition to NA_ϕ, labelled by a, with q and q' as input and output place.

NB_ϕ is constructed analogously. Now we construct the products $N \times NA_\phi$ and $N \times NB_\phi$, where the product $N_1 \times N_2$ of two Petri nets N_1 and N_2 is another Petri net defined in the following way:

- the set of places of N is the union of the sets of places of N_1 and N_2;
- for each pair of transitions t_1 of N_1 and t_2 of N_2 labelled by a same action a, the product N contains a transition (t_1, t_2) also labelled by a; the input (output) places of (t_1, t_2) are the union of the input (output) places of t_1 and t_2.

The two following results are easy to prove:

- $L_\omega(B_N) \cap L(B_\phi) \neq \emptyset$ holds if and only if the Petri net $N \times NB_\phi$ has a run which marks some place corresponding to a final state of B_ϕ infinitely often.
- $L(A_N) \cap L(A_\phi) \neq \emptyset$ holds if and only if the Petri net $N \times NA_\phi$ has a reachable dead marking which marks some place corresponding to a final state of A_ϕ.

Finding a run of $N \times NB_\phi$ that marks some place from a given set FS of final places infinitely often is equivalent to deciding if there exists a firing sequence $M_0 \xrightarrow{\sigma_1} M_1 \xrightarrow{\sigma_2} M_2 \xrightarrow{\sigma_3} M_3$ in the net $N \times NB_\phi$ such that

$$(\bigwedge_{s \in S} M_3(s) \geq M_1(s)) \wedge (\bigvee_{s \in FS} M_2(s) \geq 1)$$

where S denotes the set of all places. By Yen's result, introduced at the beginning of this section, the problem can be solved in exponential space in the size of

$N \times NB_\phi$. In a more detailed analysis [14], Habermehl shows that this problem is EXPSPACE-complete in the size of N and PSPACE-complete in the length of ϕ.

Finding a dead reachable marking of $N \times NA_\phi$ that marks some place from a given set FS of final places can be reduced to and is at least as hard as the reachability problem. Therefore, there exist so far no primitive recursive algorithms for it.

As in the 1-safe case, these results can be generalised to any logic for which the translation into automata theory holds [9].

9 All equivalence problems are undecidable

This section's rule of thumb has a rather negative flavour:

> Rule of thumb 9:
> All equivalence problems for Petri nets are undecidable.

This rule is supported by a recent and very nice result due to Jančar, showing that every equivalence notion between trace and bisimulation equivalence is undecidable for Petri nets.[14] Jančar himself has presented his result very clearly in [22]; here we do it in a slightly different way. We proceed by reduction from the problem

Given: a counter program C,
To decide: if C halts (recall that all counters are initialised to 0).

which is known to be undecidable.

Although the result can be presented directly by constructing two Petri nets out of C (and this is the way the proof in [22] goes), we prefer to use again a net programming language with a very simple net semantics, this time a language of *guarded commands.* A program is a sequence of instructions, and instructions are expressions of the form

$$\begin{aligned} \mathrm{l} : [\ & \square\ \mathit{guard}_1 \xrightarrow{\mathit{action}_1} \mathit{command}_1 \\ & \square\ \mathit{guard}_2 \xrightarrow{\mathit{action}_2} \mathit{command}_2 \\ & \dots \\ & \square\ \mathit{guard}_n \xrightarrow{\mathit{action}_n} \mathit{command}_n\ ] \end{aligned}$$

where l is a label, $\mathit{action}_1, \dots, \mathit{action}_n$ are actions, a *guard* is either the special string **true** or a conjunction of expressions of the form $x > 0$ (no guards of the form $x = 0$ are allowed), and the possible *commands* are

[14] Actually, the result is a bit stronger, since bisimulation can be replaced by an even finer equivalence.

$$\textbf{skip}\ ,\ x := x+1\ ,\ x := x-1\ ,\ \textbf{goto}\ \mathrm{l}\ ,\ \textbf{halt}$$

Operationally, an instruction is executed as follows: one of the guards that evaluate to true at the current state is nondeterministically selected (if no guard evaluates to true, the program aborts). Then, two things happen: the action of the selected guard is sent to the environment, and its command is executed (if the command is $x := x - 1$ and $x = 0$ holds, then the program aborts). If the command is a jump **goto** l, then execution continues at the instruction with label l. If the command is **skip** or an assignment, then execution continues with the next instruction. An observer can only see the actions executed by the program, but not the values of its variables, or the label of the instruction being currently executed.

Guarded command programs can be easily translated into labelled Petri nets. Figure 6 shows the labelled net corresponding to the instruction

$$\begin{array}{ll} 1 : [& \square\ x > 0 \xrightarrow{a} x := x - 1 \\ & \square\ \mathit{true} \xrightarrow{b} x := x + 1 \\ & \square\ x > 0 \wedge y > 0 \xrightarrow{a} \textbf{goto}\ 3 \\ & \square\ \mathit{true} \xrightarrow{c} \textbf{halt}\ ] \end{array}$$

(where we assume that the instruction following 1 in the program is labelled by 2). There is a place for each variable and each label, plus a special place *halt*. There is a transition for each alternative, labelled by the alternative's action. The semantics of a program is obtained by merging places of the nets corresponding to its instructions carrying the same label. We identify a program with its corresponding labelled Petri net. In particular, two programs are trace or bisimulation equivalent if their corresponding labelled nets are.

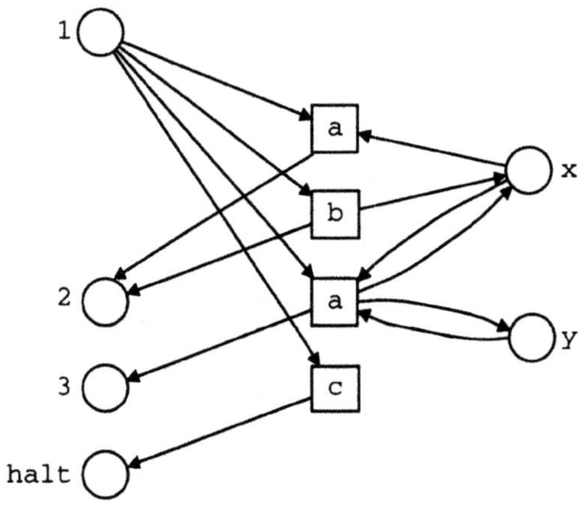

Fig. 6. Net corresponding to an instruction

Given a counter program C, we construct two net programs $N_1(C)$ and $N_2(C)$ satisfying the following two properties:

(1) if C halts, then $N_1(C)$ and $N_2(C)$ are not trace equivalent, and
(2) if C does not halt, then $N_1(C)$ and $N_2(C)$ are bisimilar.

For the proof of these properties it is very useful to characterise trace and bisimulation equivalences in terms of *two-person games.* We describe first the features common to both trace and the bisimulation games. The board of the games are the two programs $N_1(C)$ and $N_2(C)$ in their initial states. The games are played by two players, Alice and Bob, who alternate moves. Alice makes the first move. A move is the execution of (one of the alternatives of) an instruction in either $N_1(C)$ or $N_2(C)$, and is named after the action corresponding to the executed alternative. That is, an a-move is the execution of an alternative of the form *guard* $\xrightarrow{a}$ *command.* If Alice makes an a-move in one of the programs, then Bob can only answer with an a-move in the other program. It may help your intuition to imagine that Alice wishes the programs to be non-equivalent, while Bob wishes them to be equivalent. The winner of a game is decided as follows:

- if Alice has no move available, then Bob wins;
- if Bob cannot answer to Alice's move, then Alice wins;
- if the game does not terminate, then Bob wins.

If you find the idea of a non-terminating game awkward, think of chess without the 50-move rule. If a position with only the two kings on the board is reached, then the game goes on forever. In the trace and bisimulation games a situation like this is not a draw, but a win for Bob. Bob only wins after infinite time, which can make the game rather tedious, but that's his problem: the winning condition is well defined, and every game has a winner.

We describe now the differences between the trace and bisimulation games, which are surprisingly small. In a trace game, Alice chooses one of the programs *at the beginning of the game*, and makes *all her moves* in this program; Bob must make all his moves in the other program. In a bisimulation game, Alice chooses one of the programs *before each move*, and makes *her next move* in this program. For instance, in the bisimulation game Alice can make her first move in the first program (Bob must answer in the second), and her second move in the second program (Bob must answer in the first).

A *strategy* for a player is a function which gets the list of moves played so far and yields the player's next move. A strategy is *winning* if a player that sticks to it wins *all* games. We have the following nice result (see for instance [34]), which at least in the case of the trace game is intuitively very plausible:

In the trace and bisimulation games for $N_1(C)$ and $N_2(C)$: if Alice has a winning strategy, then the two programs are equivalent; if Bob has a winning strategy, then the two programs are not equivalent.

So the properties (1) and (2) that $N_1(C)$ and $N_2(C)$ – both to be constructed – have to satisfy can be reformulated as follows:

(1) if C halts, then Alice has a winning strategy in the trace game, and
(2) if C does not halt, then Bob has a winning strategy in the bisimulation game.

It is time to start with the definition of $N_1(C)$ and $N_2(C)$. To make things a bit simpler, assume without loss of generality that the counter program C contains one single **halt** instruction, and that this instruction is the last one.[15] The programs $N_1(C)$ and $N_2(C)$ look as follows:

Program $N_1(C)$:

$$\begin{array}{l} \texttt{start:}[\ \textbf{true} \xrightarrow{start} y := y+1\ ];\\ \quad N'(C);\\ \texttt{halt:}\ [\ y>0 \xrightarrow{halt} \textbf{halt}\ ] \end{array}$$

Program $N_2(C)$:

$$\begin{array}{l} \texttt{start}\{\ \textbf{true} \xrightarrow{start} \textbf{skip}\ ];\\ \quad N'(C);\\ \texttt{halt:}\ [\ y>0 \xrightarrow{halt} \textbf{halt}\ ] \end{array}$$

where the program $N'(C)$ still has to be defined. Observe that the two programs differ only in the first instruction, and that after this instruction is executed, the variable y has the value 1 in $N_1(C)$ and the value 0 in $N_2(C)$.

The program $N'(C)$ is obtained by replacing each command of C but the unique **halt** command through an instruction of the new language. The instructions corresponding to assignments and jumps are:

$$\begin{array}{lll} \text{l: } x := x+1 & \text{is replaced by} & \text{l: } [\textbf{true} \xrightarrow{inc} x := x+1]\\ \text{l: } x := x-1 & \text{is replaced by} & \text{l: } [\textbf{true} \xrightarrow{dec} x := x-1]\\ \text{l: } \textbf{goto}\ \text{l}_1 & \text{is replaced by} & \text{l: } [\textbf{true} \xrightarrow{jump} \textbf{goto}\ \text{l}_1] \end{array}$$

Conditional jumps are the delicate part. A command of the form

$$\begin{array}{l} \text{l: } \textbf{if}\ \ x=0\ \ \textbf{then goto}\ \ \texttt{ZERO}\\ \quad\ \textbf{else goto}\ \ \texttt{NONZERO} \end{array}$$

is replaced by the following sequence of two instructions:

$$\begin{array}{l} \text{l}: [\ \Box\ x>0 \xrightarrow{nonzero} \textbf{goto}\ \texttt{NONZERO}\\ \quad\ \ \Box\ \textbf{true} \xrightarrow{zero} \textbf{skip}\\ \quad\ \ \Box\ x>0 \wedge y>0 \xrightarrow{zero} y := y-1];\\ \text{l}': [\ \textbf{true} \xrightarrow{zero} \textbf{goto}\ \texttt{ZERO}\ ] \end{array}$$

This completes the description of $N_1(C)$ and $N_2(C)$. Before going on, we observe that the program $N'(C)$ has an *honest* run that mimics the execution of C, and looks as follows: whenever C executes a command, $N'(C)$ executes its

[15] If there are several **halt** instructions, we can replace them by jumps to a new label at the end of the program, and place there a unique **halt** command.

corresponding instruction. If the command is a conditional jump and C takes the NONZERO-branch, then $N'(C)$ chooses the *nonzero* alternative of the corresponding instruction; if C takes the *ZERO* branch, then $N'(C)$ chooses the *first* of the two *zero* alternatives, namely **true** $\xrightarrow{zero}$ **skip**, and then it executes the **goto** ZERO instruction.

There is an important difference between $N_1(C)$ and $N_2(C)$. Assume that in both $N_1(C)$ and $N_2(C)$ we execute the *start* action, followed by the honest execution of $N'(C)$. If and when the honest execution terminates, we can execute the *halt* action in $N_1(C)$, because y has been set to 1 by the *start* action, but we *cannot* execute it in $N_2(C)$, because y still has the value 0 there.

We are now ready to describe the winning strategies for Alice and Bob in the different games.

Assume that C halts. Here is the strategy for Alice in the trace game. Alice chooses to play on $N_1(C)$, and so Bob is forced to play on $N_2(C)$. Alice sticks to the following sequence of moves, completely disregarding Bob's answers: she plays the *start*-move, continues with the moves of the honest execution of $N'(C)$, and – if the honest run terminates – finishes with a *halt*-move.

We show in the first place that, if Alice follows this strategy, then from the second move on Bob is forced to play *exactly the same moves* as Alice (i.e., exactly the same alternatives in the same commands). When Alice plays a *nonzero* move, Bob can only answer with a unique *nonzero* move, so this case is easy. When Alice plays a *zero* move, it seem as if Bob can choose between two *zero*-answers, namely

$$\mathbf{true} \xrightarrow{zero} \mathbf{skip} \quad \text{and} \quad x > 0 \land y > 0 \xrightarrow{zero} y := y - 1$$

But remember: Alice is playing the honest run, and so she only plays a *zero*-move when $x = 0$. So, whenever Alice plays a *zero* move, Bob observes that the guard $x > 0 \land y > 0$ evaluates to false, and so that his only move is **true** $\xrightarrow{zero}$ **skip**.

Let us now see that Alice's strategy is winning. Since C halts, the honest run terminates, and so eventually Alice plays a *halt* move.[16] All along the game Bob has patiently repeated Alice's moves, waiting for a chance, but his efforts are in vain: he cannot reply to Alice's *halt* move, because in his program $N_2(C)$ the variable y has the value 0, and so the guard $y > 0$ of the *halt* move evaluates to false. So Bob loses.

Assume that C does not halt. Here is the strategy for Bob in the bisimulation game. Alice has to play the *start* move in one of the two programs, and Bob just replies with the *start* move in the other program. Then, as long as Alice plays the honest run of $N'(C)$ (possibly switching between the two programs), Bob patiently repeats her moves in the other program.[17] The first time (if at all) that Alice deviates from the honest run by playing

[16] Incidentally, observe that Alice can indeed play *halt*, because she set y to 1 with her *start* move, and she never touched y during the honest execution.

[17] He has no choice anyway!

$$x > 0 \wedge y > 0 \xrightarrow{zero} y := y - 1$$

in one of the programs, Bob replies with

$$\mathbf{true} \xrightarrow{zero} \mathbf{skip}$$

in the other program. After this move, Bob goes on playing exactly the same moves as Alice.

Let us see that Bob wins all games. If Alice sticks to the honest execution, then, since C does not halt, she never plays a *halt*-move. Since all other moves can be mimicked by Bob without problems, the game never terminates: a win for Bob. So Alice's only chance to win is to deviate from the honest run at some point by playing $x > 0 \wedge y > 0 \xrightarrow{zero} y := y - 1$ at a marking in which $x > 0$ – a cheat. But with this cheat she digs her own grave: she sets y to 0, and now *all* variables have exactly the same value in $N_1(C)$ and $N_2(C)$! Bob replays $\mathbf{true} \xrightarrow{zero} \mathbf{skip}$, and after his move both programs are in exactly the same state. So Bob wins by playing the same moves as Alice.

9.1 Partial-order equivalences are also undecidable

As we mentioned in Section 5, the literature contains many so-called ***partial-order*** equivalence notions which do not fit between trace and bisimulation equivalence. So Jančar's result might seem not to apply for them. But it does. Say that two transitions t_1 and t_2 are *concurrently enabled* at a marking M if $M(s) \geq F(s,t_1) + F(s,t_2)$ for every place s, and say that a Petri net is ***sequential*** if no reachable marking enables two transitions concurrently. It is easy to see that the Petri nets $N_1(C)$ and $N_2(C)$ we have constructed above are sequential. So, actually, we have just proved that any equivalence relation which fits between trace and bisimulation equivalence *for the class of sequential Petri nets* is undecidable. Partial-order equivalences turn out to fit between trace and bisimulation equivalence for sequential nets. Actually, this is what one would expect: partial-order equivalences should distinguish concurrency from interleaving, but if there is no concurrency at all then there is also nothing to distinguish.

10 Can anything be done in polynomial time?

The general EXSPACE-hardness bound of Section 7 raises the question if there are better results (PSPACE, NP, polynomial problems) for classes of Place/Transition Petri nets. Since a complete treatment of this question is out of the scope of this paper, we concentrate on how far can one go with polynomial algorithms. Obviously, we cannot expect to go further than for 1-safe Petri nets. So the first question is if at least some problems for conflict-free nets and free-choice nets that are not necessarily 1-safe can still be solved in polynomial time. The answer is a qualified "no". Even though [18, 39] contain some polynomial algorithms for conflict-free Petri nets, most of the important problems for these two classes become at least NP-hard. For instance, the reachability problem for conflict-free Petri nets is NP-complete [8], and the liveness problem for free-choice Petri nets is co-NP-complete (i.e., it is the complement of an NP-complete problem) [24, 5] (the proof is sketched below as the solution to Story I). Notice that the liveness and reachability problems for arbitrary Petri nets are much harder, and so these NP-completeness results can also be seen as positive results.

Is there any interesting constraint leading to polynomial algorithms for many problems? There seems to be essentially a single non-trivial one: every place has exactly one input transition and exactly one output transition ("exactly" can also be generalised to "at most") The Petri nets satisfying this constraint have been called *marked graphs*, *synchronisation graphs*, and *T-systems*. Two of the oldest papers in net theory show that many problems for these nets can be solved using simple graph algorithms or linear programming [3, 13]. So let us formulate our last rule of thumb:

> Rule of thumb 10:
> Many interesting problems about marked graphs are solvable in polynomial time. Almost no interesting problems about Petri net classes substantially larger than marked graphs are solvable in polynomial time.

The solution to Story I

The non-liveness problem for free-choice Petri nets can be formulated as follows:

Given: a free-choice Petri net N,
To decide: if N is non-live.

Membership in NP is non-trivial; it follows from Commoner's theorem [15, 5]. NP-hardness, on the contrary, is very easy to prove by a reduction, first presented in [24], from the satisfiability problem for boolean formulas in conjunctive normal form.[18]. Figure 7 shows the Petri net corresponding to the formula

$$(x_1 \vee \overline{x}_3) \wedge (x_1 \vee \overline{x}_2 \vee x_3) \wedge (x_2 \vee \overline{x}_3)$$

and we explain the construction on this example. Loosely speaking, the Petri net works as follows: first, the variables are nondeterministically assigned truth values by firing either the transition x_i or $\overline{x}_i$ for each variable x_i. Once all variables have been assigned a value, a transition C_j is enabled if and only if the assignment makes the clause C_j false. For instance, C_2 is enabled if and only if the transitions $\overline{x}_1, x_2, \overline{x}_3$ have fired; this corresponds to the assignment $x_1 :=$ *false*, $x_2 :=$ *true*, $x_3 :=$ *false*, which is the only assignment making C_2 false. So we have that the place *False* gets tokens if and only if the formula is false under the assignment. If the formula is satisfiable, then there is an assignment making the formula true, and for this assignment the place *False* never gets marked. So the Petri net is not live. On the contrary, if the formula is unsatisfiable, then the place *False* can always get marked again, and the net is live.

[18] It is interesting to compare this reduction with the one of Section 6.

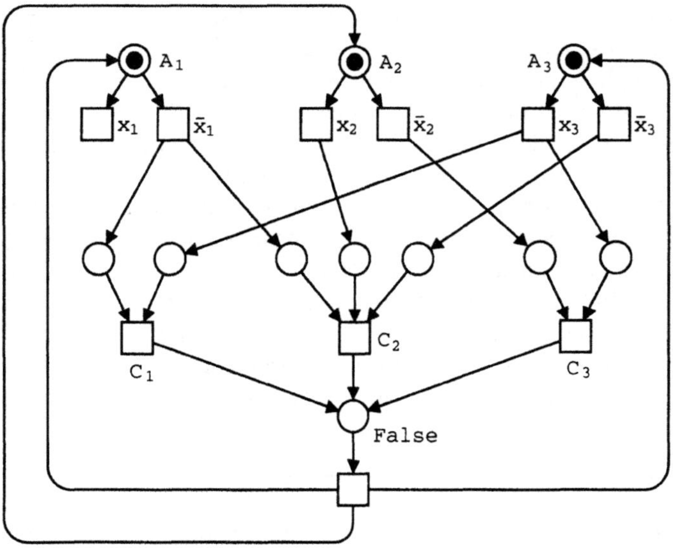

Fig. 7. Petri net corresponding to the formula $(x_1 \vee \overline{x}_3) \wedge (x_1 \vee \overline{x}_2 \vee x_3) \wedge (x_2 \vee \overline{x}_3)$

Since the formula is satisfiable, the Petri net of Figure 7 is non-live.

11 Conclusions

I'd like to conclude by listing the 10 rules of thumb of the paper. You can find them in Table 11. I've allowed myself to suppress the word "interesting" from all the rules, since it should no longer lead to confusion.

Acknowledgments

Many thanks to Eike Best, Peter Habermehl, Ernst Mayr, Richard Mayr, Peter Rosmanith, P.S. Thiagarajan, Antti Valmari and Frank Wallner for helpful suggestions, discussions, and informations. The PSPACE-algorithm for CTL of Section 4 is joint work with Peter Rossmanith.

References

1. J.L. Balcázar, J. Díaz, and J. Gabarró. *Structural Complexity I*, volume 11 of *Monographs in theoretical Computer Science*. Springer-Verlag, 1988.
2. A. Cheng, J. Esparza, and J. Palsberg. Complexity Results for 1-safe Nets. *Theoretical Computer Science*, 147:117–136, 1995.
3. F. Commoner, A. W. Holt, S. Even, and A. Pnueli. Marked Directed Graphs. *Journal of Computer and System Sciences*, 5:511–523, 1971.

The 10 Rules of Thumb

1. All questions about the behaviour of 1-safe Petri nets are PSPACE-hard.
2. Nearly all questions about the behaviour of 1-safe Petri nets can be solved in polynomial space.
3. Equivalence problems for 1-safe Petri nets are harder to solve than model-checking problems. They need at most exponential space.
4. Most questions about the behaviour of acyclic 1-safe Petri nets are NP-hard.
5. Many questions about 1-safe conflict-free Petri nets are solvable in polynomial time.
   Some questions about live 1-safe free-choice Petri nets are solvable in polynomial time (and liveness of 1-safe free-choice Petri nets is decidable in polynomial time too).
   Almost no questions for 1-safe net classes substantially larger than free-choice Petri nets are solvable in polynomial time.
6. All questions about the behaviour of Petri nets are EXPSPACE-hard.
7. The model-checking problems for Petri nets and all state-based logics are undecidable.
8. The model-checking problems for Petri nets and all branching-time, action-based logics are undecidable.
   The model-checking problems for Petri nets and all linear-time, action-based logics are decidable.
9. All equivalence problems for Petri nets are undecidable.
10. Many questions about marked graphs are solvable in polynomial time.
    Almost no questions about Petri net classes substantially larger than marked graphs are solvable in polynomial time.

Table 1.

4. M. Dam. Fixpoints of Büchi automata. In *Proceedings of the 12th International Conference of Foundations of Software Technology and Theoretical Computer Science*, volume 652 of *Lecture Notes in Computer Science*, pages 39–50, 1992, Also: LFCS Report, ECS-LFCS-92-224, University of Edinburgh.
5. J. Desel and J. Esparza. *Free-choice Petri Nets*, volume 40 of *Cambridge Tracts in Theoretical Computer Science*. Cambridge University Press, 1995.
6. E. A. Emerson. Temporal and Modal Logic. In *Handbook of Theoretical Computer Science Volume B*, pages 995–1027, 1990.
7. J. Esparza. Model Checking Using Net Unfoldings. *Science of Computer Programming*, 23:151–195, 1994.
8. J. Esparza. Reachability in Live and Safe Free-Choice Petri Nets is NP-Complete. Technical Report SFB-Bericht Nr. 342/12/96 A, Technische Universität München, 1996. To appear in Theoretical Computer Science.
9. J. Esparza. Decidability of Model-Checking for Infinite-State Concurrent Systems. *Acta Informatica*, 34:85–107, 1997.
10. J. Esparza and M. Nielsen. Decidability Issues for Petri Nets – a Survey. In *Bulletin of the EATCS*, volume 52, pages 245–262, 1994
    Also: Journal of Information Processing and Cybernetics 30(3):143–160, 1995.

11. Formal methods page of the WWW Virtual Library at http://www.comlab.ox.ac.uk/archive/formal-methods.html#notations.
12. M. R. Garey and D. S. Johnson. *Computers and Intractability: A Guide to the Theory of NP-completeness*. Freeman, 1979.
13. H. J. Genrich and K. Lautenbach. Synchronisationsgraphen. *Acta Informatica*, 2:143–161, 1973.
14. P. Habermehl. On the Complexity of the Linear-Time Mu-Calculus for Petri Nets. In P. Azéma and G. Balbo, editors, *Application and Theory of Petri Nets*, volume 1248 of *Lecture Notes in Computer Science*, pages 102–116. Springer-Verlag, 1997.
15. M. H. T. Hack. *Analysis of Production Schemata by Petri Nets*. M.s. thesis, Cambridge, Mass.: MIT, Dept. Electronical Engineering, 1972.
16. J. E. Hopcroft and J. Ullman. *Introduction to Automata Theory, Languages and Computation*. Addison-Wesley, 1979.
17. R. R. Howell and L. Rosier. On Questions of Fairness and Temporal Logic for Conflict-free Petri Nets. In G. Rozenberg, editor, *Advances in Petri Nets*, volume 340 of *Lecture Notes in Computer Science*, pages 200–220, 1988.
18. R. R. Howell and L. E. Rosier. An $O(n^{1.5})$ Algorithm to Decide Boundedness for Conflict-free Vector Replacement Systems. *Information Processing Letters*, 25(1):27–33, 1987.
19. R. R. Howell and L. E. Rosier. Problems Concerning Fairness and Temporal Logic for Conflict-free Petri Nets. *Theoretical Computer Science*, 64:305–329, 1989.
20. R. R. Howell, L. E. Rosier, and H. Yen. A Taxonomy of Fairness and Temporal Logic Problems for Petri Nets. *Theoretical Computer Science*, 82:341–372, 1991.
21. P. Jančar. All Action-based Behavioural Equivalences are Undecidable for Labelled Petri Nets. *Bulletin of EATCS*, 56:86–88, 1995.
22. P. Jančar. Undecidability of Bisimilarity for Petri Nets and Some Related Problems. *Theoretical Computer Science*, 148:281–301, 1995.
23. L. Jategaonkar and A. Meyer. Deciding True Concurrency Equivalences on Safe, Finite Nets. *Theoretical Computer Science*, 154(1):107–143, 1996.
24. N. D. Jones, L. H. Landweber, and Y. E. Lien. Complexity of Some Problems in Petri Nets. *Theoretical Computer Science*, 4:277–299, 1977.
25. S.R. Kosaraju. Decidability of Reachability in Vector Addition Systems. In *14th Annual ACM Symposium on Theory of Computing*, pages 267–281, San Francisco, 1982.
26. L. Lamport. The Mutual Exclusion Problem. Part II – Statement and Solutions. *Journal of the ACM*, 33(2), 1986.
27. R. Lipton. The Reachability Problem Requires Exponential Space. Technical Report 62, Yale University, 1976.
28. E. W. Mayr. An Algorithm for the General Petri Net Reachability Problem. *SIAM Journal on Computing*, 13:441–460, 1984.
29. E.W. Mayr and A.R. Meyer. The Complexity of the Word Problems for Commutative Semigroups and Polynomial Ideals. *Advances in Mathematics*, 46:305–329, 1982.
30. WWW page on Petri net tools at http://www.daimi.aau.dk/petrinets/tools/.
31. A. Rabinovich. Complexity of Equivalence Problems for Concurrent Systems of Finite Agents. *Information and Computation*, 127(2):164–185, 1997.
32. C. Rackoff. The Covering and Boundedness Problem for Vector Addition Systems. *Theoretical Computer Science*, 6:223–231, 1978.
33. P. Starke. *Analyse von Petri-Netz-Modellen*. Teubner, 1990.

34. C. Stirling. Bisimulation, Model Checking and Other Games. Notes for Mathfit instructional meeting on games and computation, Edinburgh, June 1977. Available at http://www.dcs.ed.ac.uk/home/cps/.
35. A. Valmari. *State Space Generation: Efficiency and Practicality.* Phd thesis, Tampere University of Technology, 1988.
36. R. J. van Glabbeek. The Linear Time – Branching Time Spectrum. In *Proceedings of CONCUR '90*, volume 458 of *Lecture Notes in Computer Science*, pages 278–297, 1990.
37. M. Vardi. An Automata-Theoretic Approach to Linear temporal Logic. In *Logics for Concurrency: Structure versus Automata*, volume 1043 of *Lecture Notes in Computer Science*, pages 238–265, 1996.
38. H. C. Yen. A Unified Approach for Deciding the Existence of Certain Petri Nets Paths. *Information and Computation*, 96(1):119–137, 1992.
39. H. C. Yen. A Polynomial Time Algorithm to Decide Pairwise Concurrency of Transitions for 1-Bounded Conflict Free Petri Nets. *Information Processing Letters*, 38:71–76, 1991.

The State Explosion Problem

Antti Valmari

Tampere University of Technology, Software Systems Laboratory,
PO Box 553, FIN-33101 Tampere, FINLAND,
email: ava@cs.tut.fi

Abstract. *State space methods* are one of the most important approaches to computer-aided analysis and verification of the behaviour of concurrent systems. In their basic form, they consist of enumerating and analysing the set of the states the system can ever reach. Unfortunately, the number of states of even a relatively small system is often far greater than can be handled in a realistic computer. The goal of this article is to analyse this *state explosion problem* from several perspectives. Many advanced state space methods alleviate the problem by using a subset or an abstraction of the set of states. Unfortunately, their use tends to restrict the set of analysis or verification questions that can be answered, making it impossible to discuss the methods without some taxonomy of the questions. Therefore, the article contains a lengthy discussion on alternative ways of stating analysis and verification questions, and algorithms for answering them. After that, many advanced state space methods are briefly described. The state explosion problem is investigated also from the computational complexity point of view.

1 Introduction

There are two main approaches to checking that a concurrent system is correct with respect to a formal specification: *theorem proving* and *state space* methods.

Theorem proving is based on formulating the correctness claim as a mathematical theorem. The theorem is then proven either manually or with the help of a theorem proving tool. The proof usually uses several *invariants* and *variant functions* or *bound functions*. An invariant states a property that all states that the system can reach must have. Invariants are used for showing that the system does only correct or "acceptable" things. A variant states an upper limit to the number of times that something can happen before something else happens. Variants are used to show that a system makes progress, and eventually does the things it should do. Because verification of concurrent systems by theorem proving is a major paradigm, the amount of literature on it is huge and cannot be surveyed here. As a starting point for reading one can use the textbooks and surveys [10, 19, 29, 56, 60], for instance.

Because manual proving of theorems takes time, requires highly skilled personnel, and is prone to human errors, several more or less automatic *theorem prover* tools have been developed. Although such a tool is sometimes able to prove the correctness of an invariant or variant fully automatically, it is more common that lots of human assistance is needed. Furthermore, the invariants and variants must almost always be provided by the user. Theorem-proving-based verification is thus very seldom fully automatic. In practice, theorem prover tools can be used only by specially trained personnel, and the construction of a proof may take lots of time, even weeks.

Another problem with theorem proving is that it is geared towards proving correctness. As a consequence, it is not particularly good in providing *debugging information*, that is, information on the nature and location of errors. Basically, an error manifests itself by causing all proof attempts to fail. The error can often be traced by locating the invariant or variant that could not be proven, and analysing where attempts to prove it fail. This is, however, an indirect and sometimes unpleasant way of debugging. A related problem is that theorem proving is clumsy in the *analysis* of system behaviour. That is, it is not good in answering questions of the type "how does the system behave" instead of "is it certain that the system behaves so-and-so". These features make theorem proving somewhat ill-suited to development work, such as experimenting with new system design ideas, and fixing an incorrect design.

The other side of the coin is that theorem proving is very generally applicable. For instance, unbounded data types (such as the ordinary integers) can often be handled without particular difficulties. To give another example, theorem proving techniques work well with systems whose process structure evolves, that is, where new processes are added and old processes aborted during an execution of the system.

State space methods aim at more automatic analysis and verification of the behaviour of systems. In their basic form, they are based on constructing a structure that consists of all states that a system can reach, and all transitions that the system can make between those states. This structure is often called the *state space*. The construction of state spaces can be fully automated. Furthermore, practical algorithms are known for answering various verification and analysis questions, given the state space of a system. No invariants or variants need be provided by the user. Therefore, at best, the "only" task that the user of a state space method has to be able to accomplish is to formulate an analysis or verification question and start a tool. As a consequence, state space methods can be used by less trained personnel than theorem proving. When state space methods apply, they are also usually fast to use.

Many state space methods are capable of producing good debugging information if the system proves incorrect. Furthermore, if a system cannot be fully investigated for some reason, state space methods are often able to give partial answers. Although an incomplete analysis cannot prove the correctness of a correct system, it can point out an error in an incorrect system. These properties make state space methods good for experimenting and as an advanced form of testing. Another advantage of state space methods is that they are flexible: a large set of analysis and verification questions of many different kinds can be answered from a single state space. This is different from, say, *place invariants* and *structure theory* [73, 17] of Petri nets, that can answer only certain kinds of questions (although sometimes very efficiently). State space methods also support the analysis of behaviour well.

State space methods are thus free from most of the problems of theorem proving, and sound like an almost ideal behavioural analysis and verification technique. Unfortunately, they suffer from one problem that is so big and fundamental that it has led many to believe that state space methods will never work well enough for large-scale practical use: *state explosion*. The number of states of almost any system of interest is huge. To get an appreciation of how serious the state explosion problem is, it suffices to investigate the number of states of some rather trivial systems:

- The system consisting of n non-interacting processes, each with k local states, has k^n states.
- The classic dining philosophers system with n 4-state philosophers (Figure 1 left) has $3^n - 1$ states.
- A simple token ring protocol described in [38] (Figure 1 right) has $9n2^{n-2}$ states, where n is the number of stations.

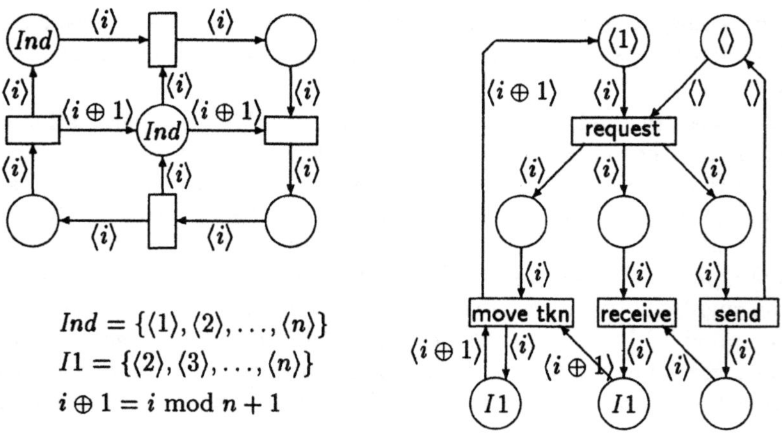

Fig. 1. The dining philosophers system (left) and the Graf-Steffen token ring (right) as high-level nets.

These examples reveal a common pattern: in a family of similar systems of different size, the number of states tends to grow exponentially in the parameter that describes the size ("n" in the above examples). More generally, intuitively speaking, the size of a state space of a system tends to grow exponentially in the number of its processes and variables, where the base of the exponentiation depends on the number of local states a process has and the number of values a variable may store, and on some kind of "tightness" of the connection between the components of systems (that is, the extent to which the local states of components are determined by the local states of other components). In the above examples, the bases are small because the systems are simple and rather tightly coupled. In many practical analysis or verification tasks the situation is much worse.

At first sight, the state explosion problem looks so formidable that it seems to make state space methods useless for analysis and verification of systems in practice. However, the great advantages of state space methods have motivated researchers to try to find ways of alleviating the problem. During the last decade, many methods have been suggested that reduce the number of states that need to be constructed for answering certain verification or analysis questions. Such enhanced state space methods increase substantially the size of systems that can be analysed or verified, while preserving most of the advantages of state space methods. The goal of this article is to discuss many issues arising in attempts to cure the state explosion problem, and to briefly introduce several enhanced state space methods.

Unfortunately, most state space reduction techniques sacrifice one or another of the advantages of state space methods. Most often, an advanced state space method can answer only certain kinds of analysis or verification questions without losing its ability to reduce the number of states. In some cases the algorithm for constructing the reduced state space takes advantage of details of the particular verification question, and may run simultaneously and interact with the algorithm that answers the question. As a consequence, advanced state space methods cannot be properly discussed without first making a classification of different kinds of analysis and verification questions, and investigating a bit the algorithms with which they can be solved.

The next section introduces many commonly used concepts that relate to state spaces, system behaviour, and the abstractions used for extracting the latter from the former. After that, the structure of typical analysis and verification problems is examined in Section 3. Section 4 is devoted to a discussion of various kinds of analysis and verification questions, and state space algorithms for answering them. Then the state explosion problem is analysed from the complexity theory point of view in Section 5. Some strategies to state space reduction are discussed in Section 6, and many individual reduction techniques in Section 7. The conclusions are in Section 8.

2 Basic Concepts

2.1 State Spaces and Executions

To capture those aspects of state spaces that are used by the state space methods discussed later, we define a *state space* as the tuple (S, T, Δ, S_I), where

- S is a set of *states*.
- T is a set of *structural transitions*.

– Δ is a set of *semantic transitions* or *edges*. It satisfies $\Delta \subseteq S \times T \times S$.
– S_I is a set of *initial states*. It has the properties $S_I \subseteq S$ and $S_I \neq \emptyset$.

"States" are simply (a subset of) the global states of the system under analysis. Often, but not always, only those states are included that the system can reach during an execution that starts in an initial state. In the case of Petri nets, states are called *markings*, and they are usually denoted with the letter M (perhaps with extensions, like M', M_{i+1}).

"Structural transitions" are entities whose execution causes the system under analysis to change its state. They may be, for instance, atomic statements in a concurrent programming language, or Petri net transitions. They are thus present in the program or system under analysis.

"Semantic transitions", on the other hand, model actual changes of state by the system. They relate to the *behaviour* of the system, not the system itself. They can be called *occurrences* of structural transitions. Formally, a semantic transition is a triple (s, t, s') consisting of a *start state* $s \in S$, structural transition $t \in T$, and *end state* $s' \in S$. As we will see below, sometimes the middle component t is omitted or replaced by a more abstract transition label. To help reading, specific notation is often used for saying that a triple is a semantic transition, such as $M[t\rangle M'$ in the case of Petri nets, and $s \xrightarrow{a} s'$ or $s -a\rightarrow s'$ in process algebras. In other words, $s -t\rightarrow s' \overset{\text{def}}{\iff} (s, t, s') \in \Delta$.

"Initial states" are the states the system may be in when its execution starts. It is customary to define only one initial state. Petri nets, for instance, are typically defined with one initial marking, which we will denote with M_I. However, the possibility of more than one initial state is sometimes important. For instance, one might want to show that a system is correct independently of the initial value of some variable. Then one can use a set of initial states, consisting of one initial state for each possible value of the variable.

A state space with many initial states can be transformed to a state space with one initial state by adding a new state s_{new}, declaring it as the initial state, and adding the semantic transition (s_{new}, t, s) for every transition (s_I, t, s) that starts in any initial state s_I. If there is an initial state that has no incoming transitions, it can be used instead of s_{new}. It is intuitively easy to accept that this transformation does not change the "behaviour" of the state space, and it is also mathematically true for almost all precise notions of "behaviour" or "behavioural property" in this article (home state properties are an exception). Although this description was at the level of the global state space, this construction produces the desired result also when applied to each individual component of the system. Therefore, it is often implementable in a verification model in practice.

A state space is *finite* if and only if S and T are finite — then also Δ and S_I are finite. For talking about the computational complexity of state space algorithms it is handy to define the *size* of a finite state space as $|S| + |T| + |\Delta|$. In all but exceptional cases $|S| + |T| + |\Delta| = \Theta(|\Delta|)$, and the presence of $|S|$ and $|T|$ in the measure of size simplifies the treatment of the exceptional cases. The state space is *finitely branching* if and only if each state has a finite number of output edges; that is, for every $s \in S$, the set $(\{s\} \times T \times S) \cap \Delta$ is finite.

The notion of the "universe" U of "syntactically possible states" is often useful when discussing the construction of state spaces from a description of the system in some modelling formalism, such as Petri nets or a parallel programming language. In such a description, the (global) state of the system is a combination of the local states of its components, values in its variables, etc. The universe is simply the set of all such combinations. It is common that it contains many states that the system cannot reach. The set S of the states of a state space is always a subset of U.

Sequences of semantic transitions are often abbreviated, and the structural transitions that label them can be omitted:

$s_0 -t_1\!\!\rightarrow s_1 -t_2\!\!\rightarrow \cdots -t_n\!\!\rightarrow s_n$ means that $s_0 -t_1\!\!\rightarrow s_1$, $s_1 -t_2\!\!\rightarrow s_2$, ..., and $s_{n-1} -t_n\!\!\rightarrow s_n$.

$s_0 -t_1 t_2 \cdots t_n\!\!\rightarrow s_n$ claims that there are states $s_1, s_2, \ldots, s_{n-1}$ such that $s_0 -t_1\!\!\rightarrow s_1 -t_2\!\!\rightarrow \cdots -t_n\!\!\rightarrow s_n$. By choosing $n = 0$ we see that $s -\varepsilon\!\!\rightarrow s$ holds for every $s \in S$, where ε denotes the empty sequence.

$s \rightarrow s'$ says that there is some $t \in T$ such that $s -t\!\!\rightarrow s'$.

$s \rightarrow^* s'$ holds if and only if there is a (possibly empty) sequence $t_1 t_2 \cdots t_n$ of structural transitions such that $s -t_1 t_2 \cdots t_n\!\!\rightarrow s'$.

Structural transitions are not needed by all state space methods. If they are absent, then a state space is defined as the triple (S, Δ, S_I), where $\Delta \subseteq S \times S$ and $\emptyset \neq S_I \subseteq S$. A semantic transition $(s, s') \in \Delta$ may then be written as $s \rightarrow s'$, and sequences of zero or more semantic transitions as $s \rightarrow^* s'$.

A structural transition t is *enabled* in a state s, if and only if there is a state s' such that $(s, t, s') \in \Delta$. This is often written in process algebras as $s -t\!\!\rightarrow$, and in Petri nets as $M[t\rangle$. If $s -t\!\!\rightarrow s'$, then it is said that t may *occur* in s yielding s'. We say that a state is a *deadlock* if no structural transition is enabled in it.

A structural transition t is *deterministic* if and only if the state resulting from its occurrence in any given state is unique, that is, $\forall s, s_1, s_2 \in S : (s -t\!\!\rightarrow s_1 \wedge s -t\!\!\rightarrow s_2 \Rightarrow s_1 = s_2)$. If determinism is not guaranteed, then structural transitions are said to be *nondeterministic*. For example, place/transition net (that is, "ordinary" Petri net) transitions are deterministic, but transition labels in process algebras are nondeterministic.

When modelling a system at a low level of abstraction, it is usually possible to get rid of any nondeterministic structural transition t by replacing it with a set of deterministic structural transitions, which together represent exactly the possible outcomes of the occurrences of t. At higher levels of abstraction this is not always possible. For instance, assume that a communication protocol may reply ok or error to a transmission request. At a detailed level of abstraction, the choice between ok and error is determined by a loss or no loss of a message in a channel. If we look at the *service* provided by the protocol (that is, how its users see it), the transitions corresponding to the operation of the channel are not shown, so the choice between ok and error looks nondeterministic. Because many advanced state space methods rely on these kinds of abstractions, we cannot avoid nondeterministic transitions in the sequel.

An *execution* of a system is a finite or infinite sequence $\langle s_0, t_1, s_1, t_2, \ldots, t_n, s_n\rangle$ or $\langle s_0, t_1, s_1, t_2, \ldots\rangle$ such that $s_0 \in S_I$ and $s_0 -t_1\!\rightarrow s_1 -t_2\!\rightarrow \cdots -t_n\!\rightarrow s_n$ (or $s_0 -t_1\!\rightarrow s_1 -t_2\!\rightarrow \cdots$). A finite execution may be *incomplete* in the sense that it may end in a state with enabled structural transitions. Thus every prefix of an execution that ends with a state (instead of a transition) is an execution. In particular, a single state s is an execution if and only if $s \in S_I$. An execution is *deadlocking*, if and only if it is finite and its last state is a deadlock state. An execution is *complete*, if and only if it is infinite or deadlocking. The set of executions of a system is the same as the set of all prefixes of the complete executions of the system. We denote the set of complete executions of the system in question by CEx.

We can talk about executions that *start at some given state* s. The definitions are otherwise the same as above, but the requirement $s_0 \in S_I$ is replaced by $s_0 = s$.

A state s' is *reachable from the state* s, if and only if it is the last state of some finite execution starting at s, that is, $s \rightarrow^* s'$. The set of Petri net markings reachable from the marking M is denoted by $[M\rangle$. A semantic transition is reachable from s, if and only if its start state is — then also its end state is. A state or semantic transition is *reachable*, if and only if it is reachable from some initial state. By the *reachable part* of a state space (S, T, Δ, S_I) we mean the tuple (S', T, Δ', S_I), where S' is the set of reachable states and Δ' is the set of reachable semantic transitions. Most state space tools construct only the reachable part of the state space,[1] and it is intuitively clear that the "behaviour" of a system depends only on it. Indeed, all formal definitions of behavioural properties in this article take only the reachable part into account.

A state space represents an *interleaving* semantics of a system. That is, it does not model the possibility that two or more structural transitions occur simultaneously. Semantic structures where this possibility is modelled are often called *true concurrency* models. An obvious way of extending state spaces to a true concurrency model would be to add semantic transitions labelled by nonempty sets of simultaneously occurring structural transitions. The corresponding semantics is often called *step semantics*.

Opinions differ as to whether interleaving or true concurrency semantics should be used in verification. If the property in question is inherently truly concurrent, then, of course, interleaving models are not appropriate. On the other hand, deadlocks, livelocks, formulae in typical temporal logics — perhaps most of the properties that people want to analyse or verify — are insensitive to the difference between interleaving and truly concurrent models.[2]

[1] Not all, though. Symbolic model checking with BDDs is one exception.

[2] This is one of the statements that loses its validity in weird enough situations. Consider a parallel composition of infinitely many non-interacting processes that make one step each and then terminate. In a truly concurrent model the system as a whole can terminate, but not in the interleaving model, because in it only finitely many transitions can be executed in a finite time.

Truly concurrent models are sometimes suggested as a way of avoiding state explosion. Step semantics does not seem promising in this respect, because it does not affect the number of states and increases the number of semantic transitions. However, it is not the only possible truly concurrent semantics of systems. Indeed, we will see in Section 7.2 that states can be reduced with another truly concurrent semantic model, namely the ***unfoldings*** of Petri nets. Interestingly, concurrency can also be taken advantage of when constructing a reduced interleaving state space, as will be explained in Section 7.4.

2.2 Abstractions of States and Transitions

It is often reasonable to define the correctness of a system at a higher level of abstraction than the system itself. (As a matter of fact, this is what the famous software engineering principle of keeping "what" and "how" apart from each other recommends.) Instead of formulating correctness claims and analysis questions in terms of the details of individual states and structural transitions, more abstract notions may be defined that capture those properties of states and structural transitions that are relevant for the user of the system. For instance, it is more pleasant to require "processes 1 and 2 should not be in their critical sections simultaneously" than "processes 1 and 2 should not be in line 7 simultaneously".

At first sight this kind of abstraction might seem a minor user-friendliness issue, but it is actually crucial for many advanced state space methods. This is because they are based on throwing away information on those aspects of system behaviour that are not relevant for the specification or analysis questions. Without abstraction, an analysis question can potentially refer to just anything, leaving the state space method little or no room for obtaining reduction. In the presence of the abstraction, one can stipulate that the question may refer to the system *only* with the abstract concepts, and the state space method can take advantage of this fact. (This is an example of "information hiding", another famous principle in software engineering.)

For the purpose of discussing typical abstraction mechanisms, we employ two sets of "observables", Π and Σ, together with a special symbol "τ". The intention is that the properties of the system may be referred to only with these observables.

Π is a set of (atomic) *propositions*. Propositions refer to the properties of individual states. More formally, a proposition φ is a function from S to the set $\{\mathsf{False}, \mathsf{True}\}$ of truth values. Examples: Critical2 $\overset{\text{def}}{\Longleftrightarrow}$ "Process 2 is in its critical section", AB_empty $\overset{\text{def}}{\Longleftrightarrow}$ "There are no messages in the channel AB", p1_is_good $\overset{\text{def}}{\Longleftrightarrow}$ $M(p_1) \geq M(p_2)$.

Σ is a set of *observable transition labels*, also called *observable* or *visible actions*. The set Σ is often called the *alphabet*. Transition labels are, in essence, user-oriented names of individual (structural or semantic) transitions or groups of them. Examples: "press_start_button", "send_message⟨12345⟩".

τ is a special *unobservable* or *invisible action*. It is used to label those transitions that the specification should not talk about, because they are intended to model implementation details and be internal to the system. The formula $\tau \notin \Sigma$ is assumed to hold.

The values of the observables on a given state space (S, T, Δ, S_I) are determined by the following two evaluation functions.

$\mathcal{E}_\Pi : S \mapsto 2^\Pi$ assigns to each $s \in S$ the set $\mathcal{E}_\Pi(s) \subseteq \Pi$ of propositions that hold in s. That is, $\varphi(s) = \mathsf{True}$ (i.e. φ holds in s, often written as $s \models \varphi$) if and only if $\varphi \in \mathcal{E}_\Pi(s)$.

$\mathcal{E}_\Sigma : T \mapsto \Sigma \cup \{\tau\}$ gives new names to structural transitions. Several different structural transitions may have the same name. The structural transitions whose occurrences are intended to be unobservable are given the name τ, that is, $\mathcal{E}_\Sigma(t) = \tau$.

One might ask for the reason of having both structural transitions T, and labels of transitions $\Sigma \cup \{\tau\}$. The two play a different role. Structural transitions come from the formalism used for describing the system — they tell something about how the system has been put together. Transition labels, on the other hand, reflect the meaning of a semantic transition at some higher level of abstraction, or from the point of view of the user of the system. It is common that an operation is modelled with more than one structural transition, especially if the system description formalism is not rich.

For instance, the assignment of zero tokens to a place/transition net place with capacity n in one transition occurrence requires $n+1$ alternative transitions, one for each possible number of tokens before the assignment. This is because a place/transition net transition can change the number of tokens only by a constant value, and the operation requires $n+1$ different changes to be possible. However, all the alternative transitions model the same operation, so it is good if they can be given a common name such as "reset_x".

Another reason for having the two levels is that some advanced verification methods can take advantage of the low-level information provided by structural transitions, while answering verification questions stated in terms of transition labels. The CSP-preserving stubborn set method in Section 7.4 is an example. It uses Σ to determine whether the swapping of successive transition occurrences affects the property that is being verified, and T to find out whether both orderings will lead to the same future states.

If transition labels are needed but structural transitions are not, then it is customary to replace T in the definition of state spaces by Σ or $\Sigma \cup \{\tau\}$, yielding (S, Σ, Δ, S_I). Then $\Delta \subseteq S \times (\Sigma \cup \{\tau\}) \times S$, and $\mathcal{E}_\Sigma$ becomes the identity function and is discarded. This is the usual case with process algebras, for instance.

The functions $\mathcal{E}_\Pi$ and $\mathcal{E}_\Sigma$ can be extended to executions in a natural way:

- $\mathcal{E}_\Pi(\langle s_0, t_1, s_1, t_2, \ldots, t_n, s_n \rangle) = \langle \mathcal{E}_\Pi(s_0), \mathcal{E}_\Pi(s_1), \ldots, \mathcal{E}_\Pi(s_n) \rangle$,
- $\mathcal{E}_\Sigma(\langle s_0, t_1, s_1, t_2, \ldots, t_n, s_n \rangle) = \langle \mathcal{E}_\Sigma(t_1), \mathcal{E}_\Sigma(t_2), \ldots, \mathcal{E}_\Sigma(t_n) \rangle$, and

- $\mathcal{E}_{\Pi+\Sigma}(\langle s_0, t_1, s_1, t_2, \ldots, t_n, s_n\rangle) = \langle \mathcal{E}_\Pi(s_0), \mathcal{E}_\Sigma(t_1), \mathcal{E}_\Pi(s_1), \mathcal{E}_\Sigma(t_2), \ldots, \mathcal{E}_\Sigma(t_n), \mathcal{E}_\Pi(s_n)\rangle$.

Stuttering. An important issue arising with $\mathcal{E}$-abstracted executions is that of *stuttering*. Stuttering means that a semantic transition in an execution has no observable effect. Let $\xi = \langle s_0, t_1, s_1, t_2, \ldots, t_n, s_n\rangle$ be a finite or infinite execution. When Π is used stuttering means that $\mathcal{E}_\Pi(s_{i+1}) = \mathcal{E}_\Pi(s_i)$ for some i, and with Σ that $\mathcal{E}_\Sigma(t_i) = \tau$ for some i.

The exact number of transitions a concurrent system takes to accomplish some task is usually considered irrelevant in verification. For instance, we do not usually care whether it took 5 or 10 internal transitions for a protocol to deliver a message. Therefore, executions that differ only in the amount of stuttering are usually considered equivalent in verification, with the exception that infinite stuttering (that is, $\mathcal{E}_\Pi(s_j) = \mathcal{E}_\Pi(s_i)$ or $\mathcal{E}_\Sigma(t_j) = \tau$ for every $j \geq i$) is commonly distinguished from finite stuttering.

A property is *stuttering-insensitive* if and only if its truth value never changes when finite stuttering is added to or removed from a system. Most properties we want to verify are stuttering-insensitive. When verifying such properties, if $\mathcal{E}_\Pi(\xi) = \langle P_0, P_1, \ldots, P_n\rangle$ is finite, we could remove all P_i that satisfy $P_i = P_{i-1}$ without modifying verification results. From an infinite $\langle P_0, P_1, \ldots\rangle$ we can remove all P_i such that $P_i = P_{i-1}$ and there is $j > i$ such that $P_i \neq P_j$. Similarly, we could remove all τ-symbols from $\mathcal{E}_\Sigma(\xi)$, as long as we do not remove any infinite suffix consisting only of τs.

State-based and action-based formalisms. In many formalisms, one can refer to the properties of a system only with the elements of Π. Because Σ is not used, structural transitions cannot be referred to, and semantic transitions can be referred to only indirectly, as changes of state. Such formalisms can be called *state-based.* Most (but not all) temporal logics are in this category. In *action-based* formalisms Π is not used but Σ is. Most, if not all, process algebras are action-based. So are also some temporal logics, most notably those that have been intended to be used in connection with process algebras.

It would be possible to use both Π and Σ in the same abstraction formalism, but such formalisms seem to be rare. One reason for this is that in many cases, information about states is redundant if information about actions (i.e. transition labels) is available, and vice versa. Let $s \xrightarrow{t} s'$. State information can be encoded into actions by replacing $\mathcal{E}_\Sigma(t)$ with the pair $(\mathcal{E}_\Sigma(t), \mathcal{E}_\Pi(s'))$, and storing $\mathcal{E}_\Pi(s_I)$ separately for each initial state s_I. This approach has, unfortunately, a problem: it is not clear which pairs could take the role of the invisible action, because the set $\mathcal{E}_\Pi(s')$ may be important in the pair $(\tau, \mathcal{E}_\Pi(s'))$. Furthermore, this idea is difficult to implement at the level of the modelling formalism (such as Petri nets), because the global state is needed for computing $\mathcal{E}_\Pi(s')$.

These problems can be solved by using the triples $(\mathcal{E}_\Sigma(t), P_{\mathsf{on}}, P_{\mathsf{off}})$ where $P_{\mathsf{on}} = \mathcal{E}_\Pi(s') - \mathcal{E}_\Pi(s)$ and $P_{\mathsf{off}} = \mathcal{E}_\Pi(s) - \mathcal{E}_\Pi(s')$ as the new transition labels. The idea is that P_{on} and P_{off} describe the *change* in the $\mathcal{E}_\Pi$-abstracted state caused by the occurrence of t, so the triple $(\tau, \emptyset, \emptyset)$ can be used as the invisible action. Furthermore, the information that P_{on} and P_{off} contain is local in the sense that it can be computed based on knowledge of the neighbourhood of t. It is worth noticing that if originally all transitions are labelled with τ, then this mapping transforms state stuttering (i.e. repetition of the same $\mathcal{E}_\Pi$-abstracted state) into action stuttering (i.e. occurrence of the invisible action), and vice versa.

Similarly, action information can be encoded into states by adding the label of the most recently executed transition into each state. This may divide the state to several copies, one for each possible input transition label, but this is usually pretty harmless. If insensitivity to stuttering is desired, then one may store the most recent *observable* transition label together with a bit that alternates its value each time a transition with an observable label occurs. The purpose of the bit is to make it possible to distinguish repeated occurrences of the same visible action. Repetition of the same visible action is not stuttering and should not be ignored in verification. For instance, if a protocol may execute, in a row, two receive$\langle msg \rangle$-actions with the same message msg, then we would probably want to know it, because it may be a sign of erroneous duplication of a message. Also this mapping makes state-based and action-based stuttering match each other.

The above encodings of state information to transitions and transition information to states are important, because they allow the use of state-based methods for action-based verification tasks and vice versa. For instance, the stubborn set methods (Section 7.4) are inherently action-based because they rely on analysing relations between structural transitions, but thanks to the above mapping they can be applied to state-based linear temporal logic.

2.3 Linear and Branching Time

Let us assume that we have decided to use an abstraction mechanism consisting of Π and $\mathcal{E}_\Pi$, Σ and $\mathcal{E}_\Sigma$, or both. The mechanism specifies what we can say about the properties of *individual* states and transitions. Another important question is: What do we want to be able to say about their *relations over time*?

Linear time. One popular answer is that it suffices to look separately at each complete execution of the system or, more precisely, what can be seen of it through the abstraction mechanism. In other words, "$\xi \models \varphi$" (what it means for an execution ξ to satisfy a property φ) is defined as $\mathcal{E}_\Pi(\xi) \models \varphi$ or $\mathcal{E}_\Sigma(\xi) \models \varphi$ or $\mathcal{E}_{\Pi+\Sigma}(\xi) \models \varphi$, and the system has the property φ if and only if $\xi \models \varphi$ for every $\xi \in CEx$. Any property whose validity is defined in this way is a *linear-time* property.

For instance, as we will soon demonstrate, reachability of a deadlock, and 4-boundedness of a Petri net place p (that is, $\forall M \in [M_I\rangle : M(p) \leq 4$) are linear-time properties. So is the property that at any instant of time, the sequence of messages that have been output from an alleged fifo queue is a prefix of the input sequence; and the property that the length of the output sequence will eventually be at least the same as the present length of the input sequence.

Reachability of a deadlock can be checked (in the mathematical, not necessarily computational, sense) by including the proposition is_deadlock $\overset{\text{def}}{\Longleftrightarrow} \forall t \in T : \neg M[t\rangle$ in Π, going through all abstracted states in the sequences in $\mathcal{E}_\Pi(CEx) = \{\, \mathcal{E}_\Pi(\xi) \mid \xi \in CEx \,\}$, and checking whether is_deadlock holds in any of them. The same principle applies to 4-boundedness. To analyse the properties of the fifo, one may include in Σ the actions input$\langle x \rangle$ and output$\langle x \rangle$ for every possible message type x. Then the "prefix" property can be checked by searching $\mathcal{E}_\Sigma(CEx) = \{\, \mathcal{E}_\Sigma(\xi) \mid \xi \in CEx \,\}$ for a prefix of an abstracted execution that violates the property. Finally, the "length" property is violated if and only if $\mathcal{E}_\Sigma(CEx)$ contains an abstracted complete execution with a finite number of output messages that is smaller than the number of input messages (the latter may be infinite).

As an example of a property that, making reasonable assumptions about the abstraction (we will return to this at the end of this section), is *not* linear-time we may take "liveness" of structural transitions in the Petri net sense of the word. A structural transition $t \in T$ is *Petri-net-live*, if and only if $\forall M \in [M_I\rangle :$ $\exists M' \in [M\rangle : M'[t\rangle$ (or, in the CTL logic that will be introduced in Section 4.3, AG EF "t is enabled").

Figure 2 shows a Petri net and its state space abstracted such that $\Pi = \{t\_\text{enab}\}$, where t_enab holds in exactly those states where t is enabled. The transition t is not Petri-net-live, but if the dashed transition is removed from the net, then t becomes Petri-net-live. The set $\mathcal{E}_\Pi(CEx)$ of complete abstracted executions of the net consists of the sequences $\langle P_0, P_1, P_2, \ldots \rangle$, where $P_i = \emptyset$ whenever $i = 0$ or i is odd, and each of the remaining P_i is either $\{t\_\text{enab}\}$ or $\emptyset$. The same set is obtained if the dashed transition is removed. So we see that the removal of the dashed transition does not affect the linear-time properties of the net with respect to the abstraction Π, although it affects the Petri-net-liveness of t. Therefore, Petri-net-liveness of t cannot be reasoned from $\mathcal{E}_\Pi(CEx)$, and Petri-net-liveness is not a linear-time property.

It is worth noticing that the fact that Π refers to the marking of only one place (namely the input place of t) was not important in the above example. The example remains valid for every Π that contains t_enab and does not refer to $M(p_1)$.

Branching time. On the other hand, Petri-net-liveness can be determined if t_enab $\in \Pi$ and all *execution trees* of the system are known. Just like the definition of an execution, the definition of an execution tree requires that one initial state $s_I \in S_I$ is chosen as the starting point. Consider the part of the state space that consists of the states and semantic transitions that are reachable from s_I. An execution tree is an unfolding of that part into a tree. It represents all executions that start at s_I, and it also records all the positions where any two executions separate from each other.

An execution tree of the state space (S, T, Δ, S_I) with the start state $s_I \in S_I$ may be defined formally as the rooted edge-labelled graph (V, E, s_I) such that

– V is the set of finite executions $\langle s_0, t_1, \ldots, t_n, s_n \rangle$ of the system where $s_0 = s_I$

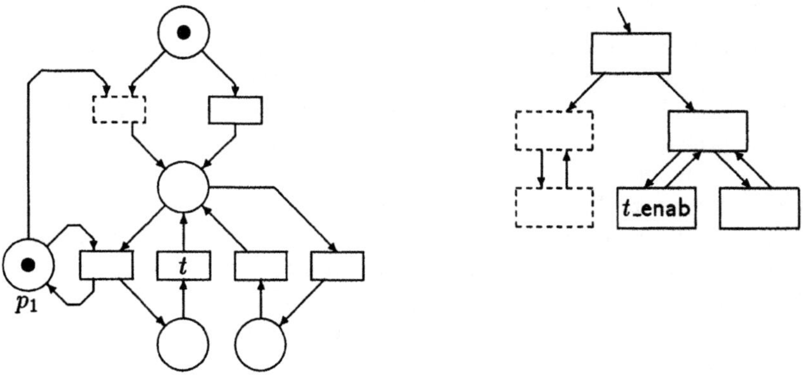

Fig. 2. A Petri-net-liveness example.

- $E = \{ (\langle s_I, \ldots, s \rangle, t, \langle s_I, \ldots, s, t, s' \rangle) \mid \langle s_I, \ldots, s \rangle \in V \wedge (s, t, s') \in \Delta \}$

Although the nodes of the execution tree are defined above as executions and thus each node contains complete information of its history, one should consider nodes as anonymous and void of any information other than that provided by $\mathcal{E}_\Pi$. Unifying nodes with their histories is just a mathematical trick that ensures that the result is a tree. The edges of an execution tree may be defined also as

$E = \{ (\langle s_I, \ldots, s \rangle, \mathcal{E}_\Sigma(t), \langle s_I, \ldots, s, t, s' \rangle) \mid \langle s_I, \ldots, s \rangle \in V \wedge (s, t, s') \in \Delta \}$
if structural transitions are abstracted with Σ, or
$E = \{ (\langle s_I, \ldots, s \rangle, \langle s_I, \ldots, s, t, s' \rangle) \mid \langle s_I, \ldots, s \rangle \in V \wedge \exists t \in T : (s, t, s') \in \Delta \}$
if structural transitions are abstracted away totally.

Any property whose validity is defined on $\mathcal{E}$-abstracted execution trees is a *branching-time* property. We may call a branching-time property *proper* if it is not also a linear-time property. Thus Petri-net-liveness is a proper branching-time property.

Even the set of branching-time properties does not cover all properties of interest. Consider the net in Figure 3, and assume that $\Pi = \{\mathsf{p1}, \mathsf{p2}, \mathsf{p3}, \mathsf{p4}\}$, where $\mathsf{p}i \overset{\text{def}}{\Longleftrightarrow} M(p_i) = 1$. A marking M_H is a *home marking* if and only if it is reachable from all reachable markings, that is, $\forall M \in [M_I\rangle : M_H \in [M\rangle$, or AG EF "the marking is M_H". The example net does not have home markings. However, if p_5 and its adjacent arc are removed, then $M(p_4) = 1 \wedge M(p_1) = M(p_2) = M(p_3) = 0$ becomes a home marking.

On the other hand, because we assumed that $\mathsf{p5} \notin \Pi$, the $\mathcal{E}_{\Pi+\Sigma}$-abstracted execution trees of the net with and without p_5 are the same. As a consequence, under this abstraction the property "M_H is a home marking" is not a branching-time property. That it was possible to describe it with the above CTL formula is because the formula implicitly refers to p_5 in "the marking is M_H".

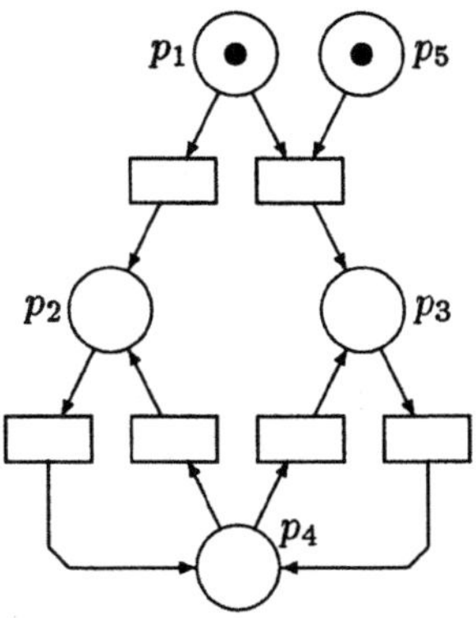

Fig. 3. A home marking example.

In the above examples we assumed that an abstraction mechanism was used that prevented us from reading the marking of one place of the net. This was necessary, because if no abstraction mechanism is used, then the distinction between linear-time, branching-time and even more general state space properties disappears in a certain theoretical sense. Namely, then we would have access to the properties of the states and transitions in full detail, and could reconstruct the reachable part of the state space from the set of complete executions by merging all occurrences of the same state. As a consequence, any state space property that does not depend on unreachable states could be checked from the set of complete executions. In the presence of an abstraction mechanism this trick does not work, because then we cannot be sure that s_1 and s_2 are the same state even if we know that $\mathcal{E}_\Pi(s_1) = \mathcal{E}_\Pi(s_2)$.

2.4 Safety, Liveness, and Fairness

In the verification of sequential programs the distinction between *partial correctness* and *termination* has proven useful. Partial correctness means that the program does not ever do anything illegal such as deliver incorrect results or crash. Partial correctness requirements depend on the task of the program, and are often formulated in some programming logic. The requirement of termination is the same for all programs, and it is simple to state: the program should not run forever. Partial correctness is proven with *weakest preconditions*, *loop invariants*, etc., while proofs of termination are usually based on *bound* or *variant functions* whose values decrease as the program continues execution but cannot decrease without limit. Thus the techniques used for stating and proving termination are very different from the techniques used for stating and proving partial correctness.

A similar distinction has been made in the world of concurrent systems between *safety* and *liveness* (or *progress*) properties [55]. Safety plays the role of partial correctness and liveness that of termination. Since concurrent programs are usually not intended to terminate, liveness consists of a set of requirements of the kind "these things should eventually happen". Unlike termination, liveness properties are system-specific and often difficult to formulate.

Linear-time safety properties can be defined as those properties of $\mathcal{E}$-abstracted executions that have *finite* counterexamples. If a prefix of an execution matches any one of the counterexamples, then the execution violates the property, no matter how it continues. A system has a safety property, if and only if all its executions have it. The "prefix" property of a fifo queue and 4-boundedness of a Petri net place (Section 2.3) are safety properties. Namely, to demonstrate that either one of them does not hold, it suffices to give an example of a finite, perhaps incomplete execution in which a message is output that has not been input, or the place contains at some point more than four tokens. Also the property "the program will not terminate" is a safety property, if we assume that $\mathcal{E}_\Pi$ contains the proposition **stopped** $\overset{\text{def}}{\Longleftrightarrow} \forall t \in T : \neg(s -t\rightarrow)$.

The precise meanings of the terms "liveness" and "progress" vary. In this article we roughly follow [2, 59] and define linear-time liveness properties as the properties such that $\mathcal{E}$-abstractions of only *complete* executions qualify as counterexamples. Furthermore, if the complete execution is finite and its $\mathcal{E}$-abstraction is $\langle P_0, a_1, P_1, a_2, \ldots, a_n, P_n \rangle$, then also its infinite completion should be a counterexample, where the completion is $\langle P_0, a_1, P_1, a_2, \ldots, a_n, P_n, \tau, P_n, \tau, P_n, \ldots \rangle$ (that is, the last abstracted state is repeated forever with invisible transitions in between). This extra condition ensures that no property is simultaneously a safety and liveness property.

For instance, the property "the program will eventually terminate" is a liveness property, because to demonstrate that it does not hold, an infinite, and thus complete, execution is needed. If we have observed the execution of a system only a finite time and the system has not terminated, we cannot claim that the execution violates any liveness property, because it is possible that the execution continues in a way that makes the property valid. The "length" property of fifo queues is a liveness property. This is because if n messages have been input to the queue and less than n have been output, it is possible that the remaining messages will come out if we wait long enough.

The definition of safety can be naturally applied to branching time by replacing "execution trees" for "executions". Because an execution tree may be simultaneously infinite and incomplete, it is not immediately clear how branching-time liveness should be defined. For instance, a counterexample for the Petri-net-liveness of a transition t has to contain a complete branch where t is never enabled, but other branches of the tree may be discarded.

If a linear-time safety property does not hold, then there is a finite execution that works as a counterexample. A state space tool can print that execution, and the user can use it to trace the reason of the error. If a linear-time liveness property does not hold, then counterexamples are usually infinite. An infinite execution cannot be printed in full, but almost always a counterexample can be made to consist of a finite prefix followed by an infinitely repeating finite cycle, and the prefix and the cycle can be printed. Counterexamples for proper branching-time properties are more complicated. Let us again use Petri-net-liveness as an example. It is easy to print an execution that leads to a state after which t cannot any more become enabled, but it is not clear what the tool should print to show that t indeed stays disabled from that state on.

Safety properties are easier to handle in verification than liveness properties in at least two ways. First, there are verification approaches and algorithms that work for safety properties, but not for liveness properties. We will see examples of this in Section 7. Second, validity of a liveness property tends to depend on subtle assumptions about the scheduling policy of the system and liveness properties of its components, while safety properties are totally insensitive to them. For instance, a mutual exclusion algorithm cannot guarantee that each customer that has requested for the shared resource will eventually get it, unless it is assumed that each customer who gets the resource will eventually release it. (The usefulness of Petri-net-liveness in verification is increased by the fact that although it fulfills to some extent similar verification needs as liveness properties, it does not depend on fairness assumptions.)

Fairness is a particular type of assumption that is very often needed for ensuring liveness. Numerous different notions of fairness have been defined (see the book [28], for instance). Perhaps the following two from [59] are the most well known. *Weak fairness* (also known as *justice*) towards a structural transition t promises that if t is enabled in every state from some point on, then it will eventually occur. The definition rules out executions where, after some state, t never occurs although it is always ready to occur. Thus weak fairness is suitable for modelling assumptions such that each process of a system gets processor time.

Strong fairness (or *compassion*) requires that if t is enabled infinitely many times, then it should also occur infinitely many times. With strong fairness one can specify, for instance, that a server does not systematically disfavour any one of its clients.

3 Structure of the Analysis or Verification Problem

State space methods are good both in the *verification*, *analysis*, *validation* and *error detection* of the behaviour of systems. In this article, these terms have the following meanings.

Verification is the act of proving or checking that a formal system has a formally stated property. In the strictest interpretation of the word, the goal is just to find rigorous evidence to the claim that the system is correct in the sense of having the property. If the system is not correct, then it suffices that a verification technique fails either by not terminating or by giving a failure report that leaves the correctness question unanswered. This kind of a verification technique may be called *one-sided* — it can answer "yes, the system is correct" and perhaps also "cannot say", but it cannot answer "no, it is incorrect".

In a more permissive interpretation, a verification technique may also give the answer "incorrect" and, better still, give some debugging information, such as an example of an execution that violates the specification. A verification technique that, given enough time and memory, is guaranteed to eventually terminate with the answer "yes" or "no" is a *verification algorithm*. A *one-sided verification algorithm* will eventually terminate and answer "yes" or "cannot say".

Analysis means finding answers to formal questions about the behaviour of a system. Analysis differs from verification in that

- the question is not necessarily a "yes"/"no" question, and
- even if it is, the answer "yes" is not given priority over "no".

An *analysis algorithm* is guaranteed to eventually terminate with a correct answer (other than "cannot say"), unless it runs out of computer resources. Examples of analysis questions: "What is the maximum number of messages simultaneously in this queue?", "Give me a list of those Petri net places that have more than one token in some reachable marking."

Validation is the process of obtaining confidence that a system behaves as intended. The behaviour of the system is compared to the expectations of the human being who is validating the system. Validation is thus always inherently informal.

Verification does not guarantee that a system behaves as we want, because we might have made an error when formulating the correctness criteria (not to mention the problem that the object of verification is never the real system, but only a formal model of it, that may or may not represent the system accurately enough). Formal analysis or verification may, however, be used as a means of validation. Knowing that a system does not deadlock and that its input and output are related in a certain way may increase our reliance on it quite a lot. Some other, widely used forms of validation are ordinary testing and reviews of program code.

Error detection is, in a sense, opposite to verification and validation: its purpose is to find errors. In most cases the goal of making a flawless system is beyond reach. Therefore, the less ambitious goal of finding as many of the remaining errors as possible within reasonable time and costs is taken. Ordinary testing is widely used for this purpose. Analysis and verification algorithms are also useful, because they are powerful in detecting errors, and they tend to find different errors than testing.

With this attitude, the failure of one individual analysis or verification run because of state explosion is a pity, but not a catastrophe — the search for errors may be continued with another, perhaps simpler analysis question. The verification system is successful if it reveals errors, even if it falls far short of full coverage of the state space. Furthermore, the capability of finding errors is more important than the ability of proving correctness when trying to convince engineers of the benefits of verification algorithms and tools. An engineer does not necessarily know what verification means and is perhaps not much impressed when told "I verified your design and it is correct", but pointing out an error that the engineer was not aware of gives the engineer concrete evidence of the power and usefulness of state space techniques.

To some extent, the difference between verification and analysis is a matter of taste, or point of view. For instance, a verification algorithm (in the above sense) is also an analysis algorithm, and the production of debugging information can be thought of as analysis. One practical difference in the use of these terms is that in analysis, one does most of the thinking *after* running the tool, while in verification most of the thinking is done *before* running the tool. Formulating a reasonable correctness claim for verification takes a lot of effort, whereas the answer can be interpreted easily, especially if it happens to be "yes". In analysis, on the other hand, a question may be thrown in just to see what happens, and interpreting the answer may take quite some time.

A typical framework for computer-aided analysis or verification has the following four components:

1. **A formalism for modelling the system.** It may be, for instance, a parallel programming language, some class of Petri nets, a process-algebraic language, or a formalism consisting of finite-state automata communicating via first-in-first-out queues. In the sequel we will call it *modelling formalism*.
2. **A formalism for stating analysis questions or properties for verification.** We will call the former *query formalism* and the latter *specification formalism*. Examples are state space query languages and temporal logics. A specification or query formalism may also contain aspects that are related to the modelling formalism, such as, in the case of Petri nets, fact transitions[3] and lists of places that are expected to be bounded. A query formalism may be rudimentary, and may reside in part or in full in the names and options of the commands used for invoking analysis tools. A document in a specification formalism is a *specification*.
3. **A formal meaning for the relation "the system has the properties".** The system is given in a modelling formalism, and "the properties" mean either the specification, or the results of an analysis. This relation is known as *satisfies* in the remainder of this article. In the case of temporal logics the satisfies-relation is usually the "is-a-model-of"-relation denoted by "$\models$". With process algebras it is commonly some process equivalence or preorder. A Petri net satisfies a fact transition if and only if the fact transition is disabled in all reachable markings.
4. **An algorithm for checking whether a given system satisfies a given specification.** Even if we restrict ourselves to state-space-based algorithms, the algorithm and its computational complexity depend heavily on the specification formalism. Sections 6 and 7 are devoted to a discussion of various ways of organising a verification algorithm, and of techniques that can be used to avoid or alleviate the effects of state explosion.

[3] In Petri net terminology, a *fact transition* is a structural transition that is never expected to be enabled. Thus the negation of its enabling condition is expected to always hold, to be a persistent fact.

Some major specification formalisms and the corresponding satisfies-relations and algorithms will be discussed in Section 4.

4 Specification and Query Formalisms

4.1 Statistics at the Level of Modelling Formalism

Many state space tools can produce various statistics on the state space that are formulated in the terminology of the modelling formalism. A Petri net tool, for instance, may generate lists of deadlock markings, maximum numbers of tokens in each place, transitions that never occur, transitions that are not Petri-net-live, etc. To get the lists, the user needs not do more than tell the tool to print them. On the other hand, after the lists have been printed, the user has to think a lot to check whether the results are acceptable and the system is working correctly. So this approach is more oriented towards analysis than verification.

Because a typical state space contains vast amounts of information, lists of the above kind may be rather long. Therefore, and also to facilitate more detailed investigation of the state space, some state space tools allow the user to make various queries on the state space. For instance, the user may ask for a list of markings where a certain transition is enabled. Advanced state space query languages allow the user to talk about the properties of a marking in great detail, investigate the immediate predecessor and successor markings of a given marking, pose questions about the reachability relation, etc.

State space statistics can usually be produced and queries of this kind answered efficiently with rather simple algorithms. The properties that can be checked with a typical state space query language concentrate on linear-time safety properties. Home markings, questions about the reachability relation, etc. allow the analysis of some other types of properties, but these facilities tend to be somewhat ad-hoc and clumsy. This observation has an explanation: we will see in Section 4.3 that it is very unlikely that simple, efficient verification algorithms exist for a certain powerful specification formalism that is capable of expressing liveness and branching-time properties (namely CTL*). A formalism for specifying a wide range of properties has thus to be chosen carefully, in order to make it expressive enough without making it too expensive to use.

Another problem with statistics and queries that are at the level of the modelling formalism is that they are sometimes not "semantic" enough. They do not use a terminology that is relevant for the system, but the terminology of the modelling formalism. For instance, a tool can produce a list of halted states, but cannot divide them into illegal deadlock and legal termination states unless the notion of "legal" is somehow explained to the tool. Furthermore, sometimes the statistics answer a slightly wrong question. As an example we will discuss home states.

As was defined in Section 2.3, a *home state* is a state that is reachable from all reachable states of the system. If a system has a home state, then it has some mode of operation that it can always enter but never exit. The existence of a home state is usually a good sign, because it reveals that the system cannot do anything irrevocable after an initialisation stage; whatever is the service it provides in and around the home state, it can and will provide it forever. It is, however, possible that the home state belongs to a livelock or even deadlock, in which case the intended service is provided in the "initialisation stage", if at all. This is not possible if all initial states are home states, which is, therefore, considered as a particularly good sign. Home states are a special case of the analysis of strongly connected components of the state space.

Consider any Petri net that has a home state. Let us add to it the isolated net fragment that is shown on the left in Figure 4. This addition has absolutely no effect on the behaviour of the rest of the net. The only thing it does is to introduce exactly one extra semantic transition that is chosen from two possibilities, may take place at any instant of time, and occurs only once. Therefore, making the usual assumption that stuttering is insignificant, the addition does not affect the correctness of the net. However, after the addition the net has no home states. So we see that home states are "fragile" in the sense that a modification that is totally irrelevant for the correct functioning of the net can make a significant change to its home state properties.

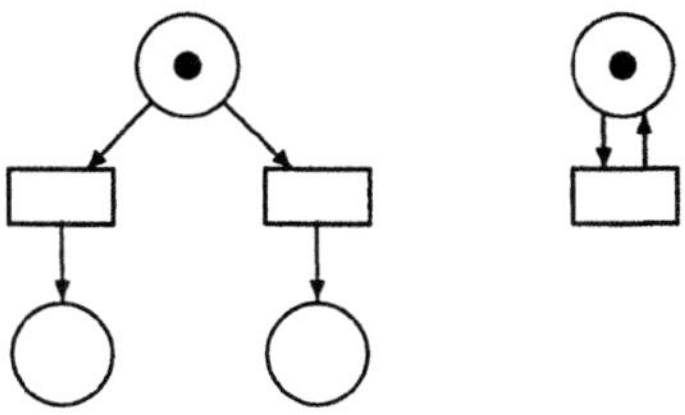

Fig. 4. Two isolated net fragments.

From the point of view of this article, comprehensive statistics and versatile query languages have yet another drawback. Namely, they do not go together well with techniques for alleviating state explosion. Many state space reduction methods are based on throwing some information away. To obtain correct results, a method should preserve those pieces of information that may be needed for producing the answers. When comprehensive statistics are produced or versatile query languages are used, the method should preserve lots of information from all over the net and all over its state space. As a consequence, good reduction results cannot be obtained.

On the other hand, the user seldom needs all that information. If we restrict ourselves to certain parts of the net by, say, letting the analysis questions only refer to certain places and transitions, then the reduction methods can throw away information on the uninteresting places and transitions and yield better results. This can be done with the abstraction mechanisms of Section 2.2.

4.2 Instrumenting the Model

Software modules are often tested in a test bed that sends them input, and receives and checks their output. The same idea can be applied to verification. For instance, a high-level Petri net verification model of a communication protocol may be augmented with the net fragment in Figure 5. As long as the protocol works correctly, the test bed keeps on sending it new messages for transmission. If the protocol ever delivers a wrong message, then t_3 occurs, and if it duplicates a message, then t_4 may occur. If it fails to deliver a message then the system either deadlocks or livelocks depending on whether the protocol part stops.

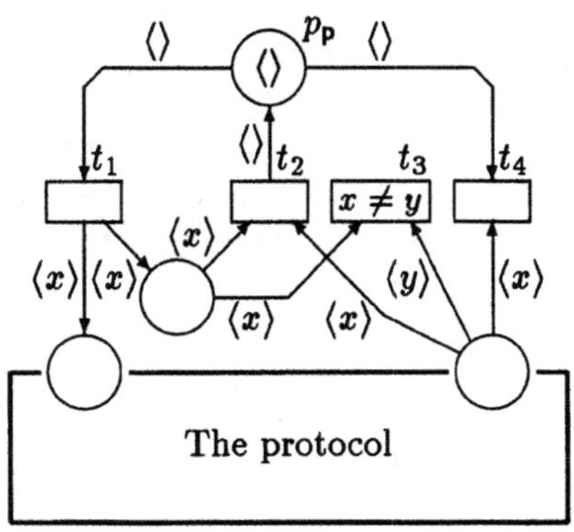

Fig. 5. A protocol test bed.

Deadlock and the occurrence of t_3 or t_4 can be easily checked from typical state space statistics of the kind in Section 4.1. If the initial states of the system as a whole are home states, then it is certain that the protocol cannot enter a livelock that it could not exit. On the other hand, the absence of home states does not necessarily mean that the protocol is incorrect in this respect.

The absence of livelocks that the protocol can exit cannot be verified with home states, occurrence checks and Petri-net-liveness checks. To see this, it suffices to take a net without such livelocks and add to it the net fragment on the right in Figure 4. (If we consider that this addition should not cause a livelock because the added transition cannot prevent the rest of the net from making progress, then this example demonstrates the need for fairness assumptions.)

On-the-fly verification of linear-time safety properties. In the above example, t_3 and t_4 were used as *fact transitions*, that is, Petri net transitions which are never expected to be enabled. Because the enabling condition of a fact transition may refer to several places all over the net, and because one can add extra places to keep track of useful information, fact transitions offer a rather versatile way of specifying verification questions.

It is actually easy to see that fact transitions can be used to check any linear-time property whose counterexamples can be expressed as a regular language over T or $\Sigma \cup \{\tau\}$. For every such property there is a finite automaton that accepts exactly the finite sequences of transition occurrences that violate the property. This automaton can be easily built from Petri net places and transitions and connected to the system with transition fusion.[4] The arrival of a token to a place that corresponds to an acceptance state of the automaton can be detected with a fact transition.

If the properties of interest are stuttering-insensitive, then τ-transitions of the system need not be fused with any transitions of the automaton. This is important for some advanced state space methods, such as the compositional analysis of Section 7.3.

There is an alternative way of connecting the test automaton to the system that is handy for state-based specifications, and used especially with the Büchi automata that are described later in this section (Section 4.2). In it, the test automaton makes its first transition according to what elements of Π are valid in the initial state of the system. Then, for each semantic transition (s, t, s') of the system, the test automaton makes a transition according to what elements of Π are valid in s'. (A more formal definition will be given when Büchi automata are discussed in more detail.) This way of connection requires support from the state space construction tool, as it cannot be implemented in typical modelling formalisms.

Can fact transitions and finite test automata express *all* linear-time safety properties? There is a classic simple argument from set theory that answers "no, no formalism can". Let $\Sigma = \{a, b\}$. Consider any $X \subseteq \{0, 1, 2, \ldots\}$, and let φ_X be the property "the number of a-transitions before the first b-transition is in X". The number of such properties is clearly uncountable. On the other hand, any object or property we ever specify in any formalism must have a finite description in that formalism (although the object itself may be infinite), otherwise we cannot write the specification in full. A finite description is a finite string of characters, and there are only countably many of them. Therefore, there are much more properties than specifications.

Because of the above, it is more meaningful to ask whether all those linear-time safety properties can be specified with fact transitions and finite test automata that can with some other given formalism. The answer is obviously "yes" for all process-algebraic semantics of Section 4.4 provided that the specification LTS is finite, because the automata-theoretic complement of the specification LTS can be used as the test automaton. The answer is also "yes" for all the linear-time safety properties that can be expressed in the temporal logics of Section 4.3, assuming either that the modelling formalism allows the encoding of

[4] *Transition fusion* means that all transitions t_S in the system and t_A in the automaton such that $\mathcal{E}_\Sigma(t_S) = \mathcal{E}_\Sigma(t_A)$ are replaced by transitions t_{SA}, one for each possible (t_S, t_A)-pair. The input and output arcs of t_S and t_A are copied to t_{SA} with their possible arc inscriptions, and the guard of t_{SA} is the conjunction of the guards of t_S and t_A.

atomic propositions on states into Σ in the sense of Section 2.2, or that the automaton is connected to the system with the alternative state-based mechanism discussed above. So we see that fact transitions and finite test automata are actually quite expressive (although they have also limitations, as will be discussed below).

A larger set of properties is obtained if the test automaton is allowed to be unbounded — any Turing machine, for instance. However, although many modelling formalisms allow the implementation of infinite-state automata, such automata tend to cause serious problems to verification algorithms and tools.

The fact that fact transitions allow the verification of any property whose counterexamples can be expressed as a regular language over T or Σ does not imply that this is always practical, or compatible with an advanced verification method one plans to use, or sufficient for verifying even simple properties stated in terms of Π. Representing the condition "no structural transition is enabled" as the disjunction of the enabling conditions of a set of fact transitions is possible in the case of Petri nets where each place may store at most one token at any instant of time, but it may be awkward and require an exponential number of fact transitions. With ordinary Petri nets it is not even possible, because it would make it possible to implement a transition that is enabled exactly when a certain place is empty. This, in turn, would make it possible to simulate two-counter machines, but that is impossible, because two-counter machines can simulate Turing machines, and ordinary Petri nets cannot. If an advanced verification method relies on the assumption that the model has a certain special property — that it is a *free-choice* Petri net, for instance — then it may be that the addition of the fact transition destroys this property, making it impossible to use the method.

When a fact transition becomes enabled it is known that the system is not correct. A state space tool may then immediately stop and print a message. This may save lots of analysis time. Such use of fact transitions is a rudimentary form of *on-the-fly verification*, a powerful idea that we will investigate in more detail in Section 6.

On-the-fly livelock detection. Fact transitions can be used only for linear-time safety properties. We mentioned earlier that livelocks that the protocol can exit cannot be checked with them. In general, a livelock corresponds to an infinite execution that produces no useful result.

In the protocol example, it is reasonable to expect that for each message given for transmission, the protocol computes only a finite amount of time. As a consequence, every infinite execution of the system should have an infinite number of transmission requests, that is, occurrences of t_1. This is an example of a general technique for specifying absence of livelocks: a set of *progress transitions* is specified, and a livelock is reported if and only if the system has an infinite execution that contains only a finite number of occurrences of progress transitions. If the state space is finite, then a livelock of this kind corresponds to a cycle where no edge is labelled by a progress transition.

Instead of progress transitions, it is possible to define *progress states* as the states that satisfy some condition, and interpret livelocks as cycles which do not contain progress states. Such a cycle is called *non-progress cycle* in [43]. In the protocol test bed example, one can define that progress states are those where p_{p} is marked. At first it might seem that this technique would lose those livelocks where the protocol is not transmitting a message, because then p_{p} is continuously marked. This is not the case, however, because in such a situation t_1 is enabled, and the state space tool will investigate also the sequence consisting of the livelock preceded by an occurrence of t_1. In this execution p_{p} is empty, so the livelock is detected.

We see from this example that a test bed need not necessarily be deterministic: it suffices that at least one nondeterministic alternative leads to the detection of the error, and no alternative gives a false alarm. A deterministic test bed is, however, often better than a nondeterministic one in that with it, the state space tool will report an error immediately when enough of the behaviour of the system has been investigated. When a nondeterministic test bed is being used, it may be that an alternative that does not give an alarm is investigated first. Then it is possible that erroneous behaviour and after it many more states are investigated before an error-detecting branch of the test bed is finally chosen and the error is detected.

In the action-based case it is natural to declare all transitions with a visible label as progress transitions. Then livelocks correspond to $\mathcal{E}_\Sigma$-abstracted infinite executions that end up with an infinite sequence of τ-transitions. It is not necessarily the case that all such executions are errors. If the error executions are taken and all τ-transitions are removed from them, a language consisting of finite strings and representing some stuttering-insensitive liveness property is obtained. If that language is regular, it can be represented as a finite automaton.

In [87] that automaton was connected to the system with transition fusion just like above with safety properties. Non-progress cycles were reported if and only if the automaton was in an acceptance state. Thus progress transitions (visible transitions) were used simultaneously with progress states (global states where the automaton is not in an acceptance state) for efficient on-the-fly detection of a large set of livelock errors. We may call finite automata that are used in this way *livelock detection automata.* They are particularly useful for certain process-algebraic verification tasks (Section 4.4).

Cycles of a finite directed graph can be detected easily and efficiently with the ordinary depth-first search. If the first search does not cover the graph in full, then a new search is initiated in some ignored vertex and so on, until all vertices have been investigated. Following the colour encoding in [15], let us call a vertex *gray* if the search has entered it but not yet backtracked from it. The graph has a cycle if and only if the search ever encounters an edge from the current vertex to a grey vertex. Non-progress cycles can be found from a finite state space by first removing all edges that correspond to occurrences of progress transitions or are adjacent to progress states, and then seeking for cycles.

It is also possible to integrate the detection of non-progress cycles with the construction of the state space by processing states and their output (semantic) transitions in a certain order, as was explained in [87]. The basic idea is to investigate progress transitions or semantic transitions that start in a progress state only when there are no other uninvestigated semantic transitions, and investigate all other semantic transitions in depth-first order. This yields an on-the-fly algorithm for detecting livelocks.[5]

Progress transitions or states can also be used in the presence of certain fairness assumptions. For instance, we may want to check that the protocol always eventually delivers the correct message, assuming that the channel loses only a finite number of messages in a row. This can be done by declaring both t_2 and t_{loss} as progress transitions, where t_{loss} is the transition that corresponds to loss of messages in the channel. Then a non-progress cycle corresponds to an execution where the last message is not delivered, although only a finite number of messages is lost. The remaining executions either contain an infinite number of deliveries, or violate the fairness assumption.

On-the-fly error detection with Büchi automata and tester processes. With progress transitions or states, an error is declared if an infinite execution contains only a finite number of them. There is an alternative, even more versatile way of detecting errors: states that are visited or transitions that occur an infinite number of times in an erroneous execution. A *Büchi automaton* $\mathcal{B} = (Q, \Sigma, \Delta, q_I, F)$ is otherwise like a finite automaton (thus $\Delta \subseteq Q \times \Sigma \times Q$, $q_I \in Q$ and $F \subseteq Q$), but the notion of "acceptance" has been defined in a different way. Namely, $\mathcal{B}$ accepts an infinite string $a_1a_2a_3\cdots$ of elements of Σ if and only if there is an infinite sequence $q_0q_1q_2\cdots$ of states such that $q_0 = q_I$, $(q_{i-1}, a_i, q_i) \in \Delta$ for $i \geq 1$, and $q_i \in F$ for infinitely many values of i.

A Büchi automaton $\mathcal{B}$ is usually connected to the system such that the alphabet of $\mathcal{B}$ is 2^Π, that is, the set of all subsets of the set of atomic propositions. Initially and immediately after each semantic transition of the system, $\mathcal{B}$ makes a transition according to the propositions that hold in the system state. More formally, the joint state space of a state space (S, T, Δ, S_I) and the Büchi automaton $(Q, 2^\Pi, \Delta_{\mathcal{B}}, q_I, F)$ has

- $\{ (s_I, q) \mid s_I \in S_I \wedge (q_I, \mathcal{E}_\Pi(s_I), q) \in \Delta_{\mathcal{B}} \}$ as its initial states, and
- the rule $(s, q) -t\rightarrow (s', q') \overset{\text{def}}{\iff} (s, t, s') \in \Delta \wedge (q, \mathcal{E}_\Pi(s'), q') \in \Delta_{\mathcal{B}}$ determines the set of transitions.[6]

[5] Also [43] describes an algorithm for the task, but it is based on investigating each state twice. The algorithm in [87] does it only once, and is thus usually more efficient. "Usually", because the two algorithms differ in the order in which the state space is investigated, making it possible for the [43] algorithm to detect an error earlier in some cases.

[6] The connection may be defined also such that $\mathcal{B}$ determines its move according to s instead of s'. Then the initial states would be $\{ (s_I, q_I) \mid s_I \in S_I \}$. This formulation does not, however, work appropriately in the presence of deadlock states.

Because 2^{Π} might be big and not all propositions $P \in \Pi$ are always relevant, the transitions of $\mathcal{B}$ may be labelled also by two sets $P_{\mathsf{on}} \subseteq \Pi$ and $P_{\mathsf{off}} \subseteq \Pi$. The transition $(q, P_{\mathsf{on}}, P_{\mathsf{off}}, q')$ is then an abbreviation of the set of the transitions (q, P, q') such that $P_{\mathsf{on}} \subseteq P$ and $P_{\mathsf{off}} \cap P = \emptyset$. That is, $\mathcal{B}$ may make the transition $(q, P_{\mathsf{on}}, P_{\mathsf{off}}, q')$, if and only if at least the propositions in P_{on} evaluate to True and at least the propositions in P_{off} evaluate to False in the system state according to which $\mathcal{B}$ chooses its transition.

Because a Büchi automaton expects an infinite input string, deadlock states have to be handled as a special case. The usual assumption is to add a τ-transition from the deadlock state to itself, or simulate such a transition in the state space construction tool or the algorithm that checks whether the automaton accepts. We saw a similar trick already in Section 2.4 when defining liveness properties.

With this approach, the detection of an error corresponds to acceptance by the Büchi automaton. This acceptance can be checked by seeking for a strongly connected component in the state space where in at least one of the states the local state of the Büchi automaton is an acceptance state. The cost of doing this is proportional to the size of the joint state space of the system and the Büchi automaton which, in turn, is in the worst case proportional to the product of the size of the system state space and the size of the Büchi automaton. The checking of a state space against a Büchi automaton is thus inexpensive.

There is also a nice and efficient algorithm for detecting Büchi acceptance on the fly [16]. It is based on investigating the state space in depth-first order, and starting a second search on already investigated states each time the primary search is about to backtrack from a Büchi acceptance state.

A Büchi automaton may be connected to the system also with other methods, such as transition fusion. The automaton may be connected to all or just visible transitions. In the latter case only stuttering-insensitive properties can be checked. On the other hand, fusion with all transitions is bad for some important methods of alleviating state explosion.

Fusion with only visible transitions makes it possible that an execution is infinite, but the Büchi automaton does not participate it from some point on. This raises a question: if the Büchi automaton stops in an acceptance state, should the corresponding execution be declared as erroneous? It may be reasonable to do so. On the other hand, this is exactly the situation for which we introduced livelock detection automata above. Therefore, it may also be reasonable to let the Büchi automaton ignore all executions where it stops, and handle them with a separate livelock detection automaton.

Of course, the same automaton may have acceptance states of both kinds, and even more. In [87] the use of four kinds of acceptance states was suggested. An error is declared if

- the automaton ever reaches a *reject state*,
- the system stops while the automaton is in a *deadlock monitor state*,
- the system livelocks while the automaton is in a *livelock monitor state*, or

– the automaton passes infinitely many times through an *infinite trace monitor state*.

So these automata combine the acceptance states of Büchi automata, livelock detection automata, and the finite automata that were used above for on-the-fly verification of safety properties, and treat deadlocks as a special case. They were called *tester processes* in [87] because of their intended application in process-algebraic verification (Section 4.4).

It is easy to add error detection by reject and deadlock monitor states to any state space construction algorithm, including the above-mentioned on-the-fly algorithms for detecting non-progress cycles and Büchi acceptance. Thus the first, the second, and one of the latter two kinds of errors can be detected during the same construction of the state space. Unfortunately, the on-the-fly algorithms for non-progress cycles and Büchi acceptance require the construction of the state space in different order, and are thus difficult to combine. On the other hand, we will see in Section 4.4 that the combination of the first three kinds of acceptance states suits well for process-algebraic verification.

It is known that everything that can be specified in the linear temporal logic of Section 4.3 can be checked with Büchi automata. (Actually, Büchi automata are strictly more expressive than the logic.) One may therefore ask: what need is there for reject and livelock monitor states? In the case of reject states the answer lies in the fact that with them, errors are found much earlier than with the above on-the-fly algorithm for Büchi acceptance. The situation with livelock monitor states is less straightforward, but there is a heuristic argument that suggests that also then Büchi acceptance tends to be significantly slower. It is as follows.

The above on-the-fly algorithm for Büchi acceptance is based on a depth-first search, and finds errors when backtracking. Thus an error cannot be found before some branch of the depth-first search tree has been constructed in full. On the other hand, the above on-the-fly algorithm for non-progress cycle detection detects an error immediately when it has constructed all states and semantic transitions of a non-progress cycle. Furthermore, because it disfavours progress states and transitions, it will complete any non-progress cycle rather soon after finding a state in it. One has to notice, however, that this reasoning does not take into account all factors that affect the relative speeds of the algorithms. Deeper theoretical analysis or experimental results would be needed for checking whether non-progress cycle detection really is faster in practice, but for the time being none is available.

The theoretical properties of Büchi and many other types of automata on infinite objects are surveyed in [80].

4.3 Temporal Logics

Temporal logics are ordinary logics augmented with operators for specifying temporal relationships. The temporal logics used in the verification of systems are usually state-based, but also action-based logics have been suggested, especially in connection with process algebras. Temporal logics are often very expressive, but specifications written in them tend to be somewhat cryptic.

The syntax of a typical temporal logic consists of the following components (we restrict ourselves to *propositional* temporal logics, that is, logics without variable symbols and the quantifiers "$\forall$", "$\exists$"):

- atomic propositions,
- ordinary propositional operators, and
- temporal operators.

The set of atomic propositions of a state-based logic is what we called Π in Section 2.2. It is the link that connects the logic to the application domain. From the point of view of the logic Π is just some abstract set whose content is not important. From the point of view of a user of a temporal logic verification tool the contents of Π are important, because they specify the elementary terms in which the user can talk about the properties of the system. This makes it reasonable to divide a temporal logic verification tool into two parts:

- A modelling-formalism-dependent part that gives the user versatile facilities for writing expressions that describe the properties of individual states, such as $\neg M[t_1\rangle \wedge M(p_2) > M(p_3)$.
- A temporal part that treats the expressions of the above part as atomic propositions, and analyses the temporal relations of states.

The ordinary propositional operators are and "$\wedge$", or "$\vee$", not "$\neg$", implies "$\Rightarrow$", etc. They have their usual meanings.

The first temporal operators that we will discuss, namely the *path operators*, talk about the properties of infinite sequences $\sigma = \langle P_0, P_1, P_2, \ldots\rangle$ of subsets of Π. In a typical application, the sequence σ is the $\mathcal{E}_\Pi$-abstraction of a complete execution of the system in question. If the execution is finite, it is completed to infinite by letting its last state repeat forever. The most important path operators and their meanings are:

If $\varphi \in \Pi$, then $\langle P_0, P_1, P_2, \ldots\rangle \models \varphi$ if and only if $\varphi \in P_0$ (i.e. φ holds in P_0).

$\langle P_0, P_1, P_2, \ldots\rangle \models \Box\varphi$ if and only if $\langle P_i, P_{i+1}, P_{i+2}, \ldots\rangle \models \varphi$ for every $i \geq 0$. "$\Box\varphi$" thus means that φ holds continuously from now on, and is pronounced "always φ" or "henceforth φ".

$\langle P_0, P_1, P_2, \ldots\rangle \models \Diamond\varphi$ if and only if $\langle P_i, P_{i+1}, P_{i+2}, \ldots\rangle \models \varphi$ for at least one $i \geq 0$. "$\Diamond\varphi$" thus means that φ holds now or will hold at least once in the future, and is pronounced "eventually φ" or "sometimes φ". The formula $\Diamond\varphi \Leftrightarrow \neg\Box\neg\varphi$ holds.

$\langle P_0, P_1, P_2, \ldots\rangle \models \varphi\,\mathcal{U}\,\psi$ if and only if there is $i \geq 0$ such that $\langle P_i, P_{i+1}, P_{i+2}, \ldots\rangle \models \psi$, and $\langle P_j, P_{j+1}, P_{j+2}, \ldots\rangle \models \varphi$ for every j such that $0 \leq j < i$. "$\varphi\,\mathcal{U}\,\psi$" is pronounced "$\varphi$ until ψ" and means that ψ will eventually hold, and φ holds until then. The formula $\Diamond\varphi \Leftrightarrow \mathsf{True}\,\mathcal{U}\,\varphi$ holds.

$\langle P_0, P_1, P_2, \ldots\rangle \models \bigcirc\varphi$ if and only if $\langle P_1, P_2, P_3, \ldots\rangle \models \varphi$. "$\bigcirc\varphi$" thus means that φ will hold in the next state. The use of this operator in specification is often discouraged, because, unlike with "$\Box$", "$\Diamond$" and "$\mathcal{U}$", with it one can specify properties that are sensitive to stuttering. However, "$\bigcirc$" is important for the development of temporal logic verification algorithms, because formulas such as $\varphi \,\mathcal{U}\, \psi \Leftrightarrow \psi \vee \varphi \wedge \bigcirc(\varphi \,\mathcal{U}\, \psi)$ use it and have proven useful in that work.

Many other path operators have been defined in the literature, such as "weak until" that is equivalent to $(\varphi \,\mathcal{U}\, \psi) \vee \Box\varphi$, "$\leadsto$" or "leads to" $\Box(\varphi \Rightarrow \Diamond\psi)$, and operators that refer to the past instead of the future.

The infinite repetition of the last state of a finite complete execution guarantees that $\bigcirc\varphi$ has a well-defined meaning also in the last state of the execution. The meanings of "$\Box$", "$\Diamond$" and "$\mathcal{U}$" are not affected by it. Terminating executions were extended to infinite also in the definition of liveness properties in Section 2.4, and of acceptance of Büchi automata in Section 4.2.

The extension has the consequence that unless Π contains enough information for characterising deadlock states, the temporal logics in this section do not suffice for distinguishing between a deadlock and livelock. At first this might seem a deficiency, but it is actually in harmony with the "philosophy" of abstraction: all that matters is when and how the $\mathcal{E}_\Pi$-abstracted state may change, and if it does not change from some point on, it is not important whether the system has terminated or is running an endless invisible loop.

LTL. So far we have defined what it means for a *complete execution* to satisfy a temporal logic formula. Now we need to define it for a *system*. The simplest way to do that is to say that a system satisfies a formula built of the above operators, if and only if all of its complete executions satisfy it. With this definition, the logic consisting of the above operators can be only used to specify properties that are linear-time in the sense of Section 2.3. Correspondingly, it is called *(propositional) linear temporal logic*, and often abbreviated as "LTL".

LTL has been studied extensively in numerous articles, and the textbooks [59, 60]. The sublogic of LTL where the use of "$\bigcirc$" is forbidden is denoted by $\mathrm{LTL}_{-\mathrm{X}}$ in this article. With it, only stuttering-insensitive properties can be specified.

If φ is an LTL formula and σ is an infinite sequence of subsets of Π, then σ satisfies either φ or $\neg\varphi$. However, the same is not true for systems: it is possible that neither φ nor $\neg\varphi$ holds in a given system. This is because the system may have one execution that satisfies φ thus violating $\neg\varphi$, and another one that violates φ. This induces a natural preorder relation between systems for any fixed Π: Sys_1 *implements* or *is more deterministic than* Sys_2, if and only if $\varphi \models Sys_2$ implies $\varphi \models Sys_1$ for every LTL-formula φ whose atomic propositions are from Π.

A *model checking* algorithm inputs a state space and a temporal logic formula φ and checks whether φ is a true statement about the state space. LTL has a model checking algorithm whose worst-case time consumption is linear in the size of the state space, and exponential in the length of the formula [57]. It is thus feasible for short LTL formulae, but may run very long with long formulae. Fortunately, the time consumption of the algorithm depends on what operators and how are used in the formula, so not every long formula causes the algorithm to slow down. Furthermore, the LTL formulae needed in verification are often rather short. It is very unlikely that a worst-case polynomial time LTL model checking algorithm could ever be found, because the LTL model checking problem is known to be PSPACE-complete [78].

The above-mentioned model checking algorithm is not particularly popular in automatic verification as such, but its basic techniques have been used in an alternative, more popular approach. Namely, every LTL formula can be compiled to a Büchi automaton that accepts exactly the executions described by the formula. One advanced practical algorithm for this was given in [32]. As was told in Section 4.2, Büchi automata can be efficiently and even on the fly used for verifying that no execution has the property described by the automaton. Therefore, the validity of any LTL formula φ can be checked by constructing a Büchi automaton for $\neg\varphi$ and checking the system with it. This technique is called the *automata-theoretic approach to model checking* [97], and it suits well both ordinary and on-the-fly verification.

Unfortunately, the PSPACE-hardness of LTL model checking must manifest itself somewhere also in the automata-theoretic approach: the size of a Büchi automaton may be exponential compared to the size of the LTL formula from which it was constructed.

Assume that t_enab denotes that the structural transition t is enabled, and t_occ that it has just occurred. Weak fairness towards t can be expressed with the LTL formula $\Diamond\Box t\_\mathsf{enab} \Rightarrow \Box\Diamond t\_\mathsf{occ}$, and strong fairness with $\Box\Diamond t\_\mathsf{enab} \Rightarrow \Box\Diamond t\_\mathsf{occ}$. As a consequence, a fairness assumption could be taken into account in the verification of the LTL formula φ by expressing the assumption with an LTL formula γ, and checking the validity of $\gamma \Rightarrow \varphi$. However, because fairness assumptions are very common, special techniques for handling them have been developed and integrated to LTL model checking algorithms.

CTL and CTL*. Another possibility of extending the logic to apply to systems relies on the following two new operators. They say whether a formula should be valid in one or all possible executions that start at a given state s:

$s \models \mathsf{A}\varphi$, if and only if φ holds in all paths of the state space that start at s.
$s \models \mathsf{E}\varphi$, if and only if φ holds in at least one path of the state space that starts at s.

The system satisfies a formula, if and only if all of its initial states satisfy it.

Because the validity of A- and E-formulae has been defined for every individual state of the state space, it is meaningful to apply path operators on formulae built with A and E. In this way one gets formulae like $\mathsf{A}\,\Box\,\mathsf{E}\,\Diamond\,t\_\mathsf{enab}$, which says that in all possible futures, there is always a future where eventually

t_enab holds. If t_enab means that t is enabled, then this formula expresses the Petri-net-liveness of t. In this context the operators "$\Box$", "$\Diamond$", "$\mathcal{U}$" and "$\bigcirc$" are usually written as "G", "F", "U" and "X", respectively. Thus the above formula is more commonly written as A G E F t_enab.

This logic was developed by Emerson and Halpern [20] on the basis of LTL and the earlier CTL logic of Clarke and Emerson [11], and it is known as *CTL** ("CTL" stands for *computation tree logic*). The validity of CTL*-formulae could also be defined on execution trees instead of state spaces, but that would not change the meaning of the logic. Properties expressible with CTL* are thus branching-time in the sense of Section 2.3. Any LTL formula φ can be represented as the CTL* formula Aφ, so CTL* is an extension of LTL. As a consequence, also CTL* model checking is PSPACE-hard. It is not harder, though, as was proven in [21].

The older logic CTL is a restriction of CTL*. It has a fast model checking algorithm [11]. In CTL, every path operator must be immediately preceded by A or E. Thus A G E F t_enab is a valid CTL formula (and more commonly written as AG EF t_enab), but E(φ U Gψ) is not. CTL has become very popular in automatic verification. Although not all LTL formulae can be expressed in CTL, CTL seems to have enough expressive power for many verification tasks, and its efficient model checking algorithm is definitely an advantage. Like with LTL, the use of the next state operator is sometimes forbidden, yielding the logics $\mathrm{CTL}_{-\mathsf{X}}$ and $\mathrm{CTL}^*_{-\mathsf{X}}$.

The basic idea of the CTL model checking algorithm can be illustrated with the case of the AU operator; the remaining operators can be handled with similar methods. Let us say that a state is *φ-marked*, if and only if the algorithm has found out that φ holds in it. Consider a formula of the form $\theta = \mathsf{A}(\varphi \mathbin{\mathsf{U}} \psi)$. The algorithm is first run recursively to φ-mark every state where φ holds, and similarly with ψ. Then every ψ-marked state is marked with θ. Finally the following is repeated as long as possible: if there is a state s that is φ-marked but not θ-marked, and each of its immediate successor states is θ-marked, then s is marked with θ. This last stage can be implemented efficiently with a suitable backwards search.

In order to compare CTL* and $\mathrm{CTL}^*_{-\mathsf{X}}$ to the branching-time process-algebraic semantic models discussed in Section 4.4, we define two equivalence notions between the state spaces $Sys_1 = (S_1, \Delta_1, S_{I1})$ and $Sys_2 = (S_2, \Delta_2, S_{I2})$ [7]:

1. Sys_1 and Sys_2 are *Π-bisimilar* if and only if there is a relation "$\sim$" $\subseteq S_1 \times S_2$ such that the following hold for every $s_1, s_1' \in S_1$ and $s_2, s_2' \in S_2$ (the prefix "Π" was added to the name of the relation to avoid confusion with the "strong bisimilarity" relation in Section 4.4):
   - If $s_1 \sim s_2$ and $\pi \in \Pi$, then $s_1 \models \pi$ if and only if $s_2 \models \pi$.
   - If $s_1 \in S_{I1}$, then there is $s \in S_{I2}$ such that $s_1 \sim s$.
   - If $s_2 \in S_{I2}$, then there is $s \in S_{I1}$ such that $s \sim s_2$.
   - If $s_1 \sim s_2$ and $(s_1, s_1') \in \Delta_1$, then there is s such that $s_1' \sim s$ and $(s_2, s) \in \Delta_2$.

- If $s_1 \sim s_2$ and $(s_2, s_2') \in \Delta_2$, then there is s such that $s \sim s_2'$ and $(s_1, s) \in \Delta_1$.

2. The complete executions $\langle s_{1,0}, s_{1,1}, s_{1,2}, \ldots \rangle$ and $\langle s_{2,0}, s_{2,1}, s_{2,2}, \ldots \rangle$ *stutter-simulate* each other according to the relation "$\simeq$" $\subseteq S_1 \times S_2$, if and only if $\langle s_{1,0}, s_{1,1}, s_{1,2}, \ldots \rangle$ has a partition $B_{1,0}, B_{1,1}, B_{1,2}, \ldots$ and $\langle s_{2,0}, s_{2,1}, s_{2,2}, \ldots \rangle$ has a partition $B_{2,0}, B_{2,1}, B_{2,2}, \ldots$ such that for every $i \geq 0$ the following hold: $0 < |B_{1,i}| < \infty$, $0 < |B_{2,i}| < \infty$, and $\forall s_1 \in B_{1,i} : \forall s_2 \in B_{2,i} : s_1 \simeq s_2$.
   The state spaces Sys_1 and Sys_2 are *stuttering equivalent* if and only if there is a relation "$\simeq$" $\subseteq S_1 \times S_2$ such that the following hold for every $s_1 \in S_1$ and $s_2 \in S_2$:
   - If $s_1 \simeq s_2$ and $\pi \in \Pi$, then $s_1 \models \pi$ if and only if $s_2 \models \pi$.
   - If $s_1 \in S_{I1}$, then there is $s \in S_{I2}$ such that $s_1 \simeq s$.
   - If $s_2 \in S_{I2}$, then there is $s \in S_{I1}$ such that $s \simeq s_2$.
   - If $s_1 \simeq s_2$, then for every complete execution of Sys_1 that starts at s_1 there is a complete execution of Sys_2 starting at s_2 such that the two executions stutter-simulate each other according to "$\simeq$".
   - If $s_1 \simeq s_2$, then for every complete execution of Sys_2 that starts at s_2 there is a complete execution of Sys_1 starting at s_1 such that the two executions stutter-simulate each other according to "$\simeq$".

It is relatively easy to see that if Sys_1 and Sys_2 are Π-bisimilar and Sys_1 satisfies a CTL* formula φ, then also $Sys_2 \models \varphi$. This was proven in [7] together with a reverse result where CTL (without the "*") suffices: if Sys_1 and Sys_2 are finite and satisfy the same CTL formulae, then they are Π-bisimilar. A similar pair of results holds for stuttering equivalence, $\text{CTL}^*_{-\text{X}}$, and $\text{CTL}_{-\text{X}}$.

The article [19] is an excellent survey on various temporal logics and their theoretical properties, including model checking.

4.4 Process-Algebraic Semantics

A process algebra, such as the *Calculus of Communicating Systems* (*CCS*) [65] and *Communicating Sequential Processes* (*CSP*) [6, 42, 75] consists of a language for specifying systems, and a theory of the behaviour of the systems specified in that language. Most (or perhaps all) process algebras are action-based. Instead of structural transitions, emphasis is put on *actions* that are $\mathcal{E}_\Sigma$-abstractions (Section 2.2) of structural transitions. Correspondingly, in the context of process algebras, state spaces are usually defined as (S, Σ, Δ, S_I), where Σ does not contain the invisible action symbol τ, and $\Delta \subseteq S \times (\Sigma \cup \{\tau\}) \times S$. Such structures are called *labelled transition systems*, abbreviated *LTS*. Occurrences of actions (i.e. semantic transitions) are sometimes called *events*.

Process composition operators. The language of a process algebra is a "modelling formalism" in the sense of Section 3. It practically always contains some operators for *parallel composition* of processes and for *hiding* of actions. (The CCS language does not have a separate hiding operator, but its parallel composition is capable of hiding.) Communication between parallel processes is synchronous, like Ada rendez-vous. The meanings of operators are usually defined in terms of expressions of the language, and their details vary, but the basic ideas can be illustrated with state-space-level definitions.

The following definition corresponds to a popular version of parallel composition. Let $L_1 = (S_1, \Sigma_1, \Delta_1, S_{I1})$ and $L_2 = (S_2, \Sigma_2, \Delta_2, S_{I2})$ be LTSs. Their parallel composition $L_1 || L_2$ is the LTS (S, Σ, Δ, S_I) such that

- $S = S_1 \times S_2$
- $\Sigma = \Sigma_1 \cup \Sigma_2$
- $((s_1, s_2), a, (s_1', s_2')) \in \Delta$ if and only if either
  - $(s_1, a, s_1') \in \Delta_1$, $a \notin \Sigma_2$, and $s_2' = s_2$,
  - $a \notin \Sigma_1$, $s_1' = s_1$, and $(s_2, a, s_2') \in \Delta_2$, or
  - $(s_1, a, s_1') \in \Delta_1$, $(s_2, a, s_2') \in \Delta_2$, and $a \neq \tau$.
- $S_I = S_{I1} \times S_{I2}$

Since τ is never a member of Σ, this definition implies that a component of a parallel composition can do its τ-transitions independently of and without affecting the state of the other component, and this is the only way in which τ-transitions can be executed. The same holds for transitions that are labelled with a visible action that is not in the alphabet of the other component, because $(s, a, s') \in \Delta$ implies $a \in \Sigma$ or $a = \tau$. If a visible action is in the alphabet of both components, then a corresponding transition must be executed by both components simultaneously. This parallel composition is essentially the same thing as the transition fusion of Section 4.2.

This definition extends easily to more than two components. Alternative definitions of parallel composition may specify the synchronisation of visible actions in a different way, but it is probably universal that τ-transitions do not synchronise. A parallel composition computed according to the above definition contains usually many unreachable states. Since their presence is most of the time either insignificant or harmful, the definition of parallel composition is often formulated in such a way that unreachable states and their adjacent semantic transitions are not generated or are removed from the result. This is not an important issue, however, because, with reasonable notions of "behaviour", the removal of unreachable states does not affect the behaviour of the system.

The hiding operator converts visible actions into τ. A popular version of it can be defined as **hide** A **in** $(S, \Sigma, \Delta, S_I) = (S, \Sigma', \Delta', S_I)$, where A is some set of actions, and

- $\Sigma' = \Sigma - A$, and
- $\Delta' = \{ (s, a, s') \mid (s, a, s') \in \Delta \wedge a \notin A \}$
  $\quad \cup \{ (s, \tau, s') \mid \exists a \in A : (s, a, s') \in \Delta \}$.

Strong bisimilarity. As was mentioned above, a theory of the behaviour of systems is an essential component of a process algebra. Because the theory of CCS was originally presented in the context of the CCS language and similarly for CSP, it is common to associate each theory to a particular language. This is not necessary, however, and we can and will discuss the theories of behaviour at the level of state spaces in the sequel.

To jump from a language to state spaces it suffices that an operational semantics for process-algebraic expressions is defined which, given any well-defined expression in the language, produces a labelled transition system that represents the behaviour of the object described by the expression. For a reason that will be described later in this section, there is significant freedom in the choice of the technical details of this semantics. For instance, even if it is intuitively clear that the system has only a finite number of states, there is no compelling reason to require that the corresponding LTS must be finite.

Behavioural equivalence is a central notion in process-algebraic theories of behaviour. Behavioural equivalences can be divided to two categories according to how they treat τ-transitions. In the first category τ is handled in the same way as all visible actions. This category contains only one important equivalence: *strong bisimilarity* [66]. The LTSs $L_1 = (S_1, \Sigma, \Delta_1, S_{I1})$ and $L_2 = (S_2, \Sigma, \Delta_2, S_{I2})$ that have the same alphabet are *(strongly) bisimilar*, denoted in this article by $L_1 \simeq_{\mathsf{sb}} L_2$, if and only if there is a relation "$\sim$" $\subseteq S_1 \times S_2$ such that the following hold for every $s_1, s_1' \in S_1$, $s_2, s_2' \in S_2$, and $a \in \Sigma \cup \{\tau\}$:

- If $s_1 \in S_{I1}$, then there is $s \in S_{I2}$ such that $s_1 \sim s$.
- If $s_2 \in S_{I2}$, then there is $s \in S_{I1}$ such that $s \sim s_2$.
- If $s_1 \sim s_2$ and $(s_1, a, s_1') \in \Delta_1$,
  then there is s such that $s_1' \sim s$ and $(s_2, a, s) \in \Delta_2$.
- If $s_1 \sim s_2$ and $(s_2, a, s_2') \in \Delta_2$,
  then there is s such that $s \sim s_2'$ and $(s_1, a, s) \in \Delta_1$.

The relation "$\sim$" is called a *strong bisimulation*. The basic idea of this definition is that if the LTSs are in strongly bisimilar states and one of them makes a transition, the other can simulate it with a transition of its own such that the labels of the transitions are the same, and after both transitions the systems are again in strongly bisimilar states. Since it is assumed that every initial state of each LTS is simulated by some initial state of the other LTSs, a simple induction argument reveals that each system can simulate everything the other does. It is difficult to think of any property of LTSs that could reasonably be called "behavioural" and that would hold in one and not in the other of two strongly bisimilar LTSs. (Two strongly bisimilar LTSs can have different numbers of states, for instance, but the number of states can hardly be called "behavioural".)

Two strongly bisimilar systems thus have "the same behaviour" in a very strong sense. Indeed, strong bisimilarity is the strongest equivalence that is in common use in process algebras. It is useful for technical purposes, such as abstracting away from irrelevant details of the mapping that associates to each process-algebraic expression the LTS that represents its behaviour. It is needed because it is common that two process-algebraic expressions that, according to our intuition, have obviously the same behaviour are nevertheless formally different because of syntactic details, and cannot thus be unified outright. For instance, according to the above definition, $L_1||L_2$ is not the same as $L_2||L_1$, because their states list their components in opposite order.

The *Lotos* specification language [44, 4] provides another, more fundamental example. Its operational semantics is defined in such a way that a cyclic process "remembers" the number of times the cycle has been executed. As a consequence, the state after the cycle is formally different from the state before the cycle although intuitively it is "the same state", and the resulting LTS is not a cycle but an infinite chain. However, the chain can be folded into a cycle on the basis that the start states of each round are strongly bisimilar. Strong bisimilarity gives a sound mathematical notion for unifying states.[7]

Algorithms for strong bisimilarity. We may say that two states s_1 and s_2 of an LTS $L = (S, \Sigma, \Delta, S_I)$ are strongly bisimilar, if and only if there is a strong bisimulation relation between $(S, \Sigma, \Delta, \{s_1\})$ and $(S, \Sigma, \Delta, \{s_2\})$. Let $[[s]]$ denote the set of states that are strongly bisimilar with s, and let $L_{\mathsf{min}} = (S', \Sigma, \Delta', S'_I)$, where $S' = \{ [[s]] \mid s \in S \}$, $S'_I = \{ [[s]] \mid s \in S_I \}$, and $([[s]], a, [[s']]) \in \Delta'$ if and only if there is $s'' \in [[s']]$ such that $(s, a, s'') \in \Delta$ (this definition of Δ' is independent of the choices of the states s that span the sets $[[s]]$). Then $L \simeq_{\mathsf{sb}} L_{\mathsf{min}}$.

Furthermore, it is possible to show that if L is finite, then L_{min} is minimal (that is, has the smallest possible number of states and transitions) among the LTSs that are strongly bisimilar to L, and all LTSs that are strongly bisimilar to L and have as few states as L_{min} are the same as L_{min} except for the names of states. A similar result holds also when L is infinite, although then one has to be careful with what is meant by "minimal".

There is a very efficient algorithm that constructs L_{min} from any given finite LTS [26]. It resembles the classic "block splitting" algorithm for the minimisation of deterministic finite automata. It can also be used for checking strong bisimilarity of two finite LTSs simply by minimising their disjoint union. The LTSs are strongly bisimilar, if and only if each block that contains an initial state of one of them also contains an initial state of the other.

Strong bisimilarity is almost the same as the Π-bisimilarity of Section 4.3. The only difference is that now the labels of matching transitions are compared instead of the sets of propositions that hold in matching states. This difference can be easily taken into account in the minimisation of an LTS with block splitting. As a consequence, algorithms for strong bisimilarity can be taken advantage of in the verification of CTL* formulae.

Abstract process equivalences. The second category of behavioural equivalences consists of those where most information about τ-transitions is abstracted away in one way or another. We will call them *abstract process equivalences*. It

[7] Mathematicians often use isomorphism for these kinds of tasks. Isomorphism is a strictly stronger notion than strong bisimilarity. It is too strong for the above Lotos example, because it does not unify a cycle with the infinite chain that is its unfolding.

is often desirable to preserve some information on the indirect consequences of τ-transitions. However, opinions differ regarding how much and what kind of information should be preserved, which has led to the development of many different equivalences. Furthermore, the desire of making equivalences to have certain mathematical properties that are useful either for theory development (such as unique minimal solutions to recursive equations) or for applications has further increased the number of equivalences that have been presented in the literature. The survey [95] lists 155. In this article we can mention only some. Fortunately, many of the others are small variations of what we will present.

An important requirement, especially from the point of view of advanced verification methods, is that the equivalence should be a *congruence* with respect to parallel composition, hiding, and whatever other process operators are used. This means that if "$\simeq$" denotes the equivalence, then $L_1 \simeq L_2$ should guarantee that **hide** A **in** $L_1 \simeq$ **hide** A **in** L_2, $L_1||L \simeq L_2||L$, $L||L_1 \simeq L||L_2$, and similarly for the other operators, where L_1, L_2 and L are arbitrary LTSs. The congruence property guarantees that any LTS that is a component of a bigger system can freely be replaced by an equivalent LTS, and the behaviour of the bigger system does not change. Although the congruence property seems a natural one, it is difficult to obtain, and many otherwise nice equivalences lack it.

Trace semantics and verification with a process equivalence. The simplest widely used abstract process equivalence is *trace equivalence*. In process algebras, a *trace* is the sequence of visible actions obtained from a finite execution by removing all states and all τ-symbols. In other words, it is the $\mathcal{E}_\Sigma$-abstraction of a finite execution with all stuttering removed. (Process-algebraic traces have thus nothing to do with Mazurkiewicz traces [61].)

The set of traces of an LTS (which we will denote with $Tr(L)$) is called its *trace semantics*, and two LTSs that have a common alphabet are *trace equivalent* (written as $L_1 \simeq_{\text{tr}} L_2$ in the sequel) if and only if they have the same trace semantics. *Trace preorder* "$\sqsupseteq_{\text{tr}}$" is defined by $L_1 \sqsupseteq_{\text{tr}} L_2$ if and only if $Tr(L_1) \subseteq Tr(L_2)$.[8] Trace equivalence is a congruence with respect to most, or perhaps all, process operators that have been suggested in the literature.

Trace semantics can be used in the verification of a system *Sys* in at least three different ways. For the sake of examples, let us assume that *Sys* is a model of a communication protocol implementation that consists of a protocol sender process and a protocol receiver process connected to each other via a bidirectional channel, such that all actions are hidden except those with which the protocol sender inputs transmission requests from and reports success or failure to the sending client, and those with which the protocol receiver delivers arrived messages to the receiving client. Assume further that *Spec* is a single LTS that models the service that the protocol is supposed to provide to the clients. We presume that *Spec* has the same alphabet as *Sys*; if necessary, the hiding operator is first used to convert the extra actions in either alphabet to τ.

[8] This direction of "$\sqsupseteq$" seems to be more common in the literature, although also the opposite direction has supporters.

1. One can compare whether $Sys \simeq_{\mathrm{tr}} Spec$. If not, then Sys either has an illegal sequence of communication with the clients of the protocol (such as the delivery of a wrong message), or it lacks some desired sequence (for instance, fails to deliver a message). If yes, then Sys provides exactly the required service as far as one can say on the basis of the trace semantics.
2. One can compare whether $Sys \sqsupseteq_{\mathrm{tr}} Spec$. It is common that a specification does not define a system in full, but only states minimal requirements that the system must satisfy. In the case of a protocol, *Spec* may allow the protocol sender to give up and report failure to the sending client, if it does not receive an acknowledgement from the protocol receiver within a specified time. However, if the channel and protocol receiver are reliable and fast enough, then the acknowledgement is never lost or delayed, and *Sys* never reports failure. In that case *Sys* lacks some traces that *Spec* has, but for a very acceptable reason. In this kind of a situation the requirement that $Sys \simeq_{\mathrm{tr}} Spec$ is too stringent, but $Sys \sqsupseteq_{\mathrm{tr}} Spec$ works well (except that for many applications, it is not stringent enough, but other process equivalences that are described later in this section will solve this problem).
3. One can construct a small LTS that is trace-equivalent to *Sys* and investigate its properties, for example, with model checking tools. Of course, one should investigate only those properties φ that trace equivalence *preserves* in the sense that if $L_1 \models \varphi$ and $L_1 \simeq_{\mathrm{tr}} L_2$, then $L_2 \models \varphi$.

In the terminology of Section 3, *Spec* is the "specification". As a consequence, the "specification formalism" is the formalism in which *Spec* was written, that is, a process-algebraic language or LTSs. It may very well be the same formalism in which *Sys* was written! The "satisfies" relation that says what it means for the system to satisfy its specification is $\simeq_{\mathrm{tr}}$ in the first case above, and $\sqsupseteq_{\mathrm{tr}}$ in the second case.

Let φ be any stuttering-insensitive linear-time safety property over Σ. Let L_φ be an LTS whose finite executions are exactly those which do not violate φ. The LTS L_φ is not necessarily finite, but it exists, at least in a theoretical sense. Now, an LTS L has the property φ, which we may write as $L \models \varphi$, if and only if $L \sqsupseteq_{\mathrm{tr}} L_\varphi$. We see that preorder checking is conceptually close to model checking. This fact has useful consequences in automatic verification: thanks to it, similar techniques can be used for model checking and preorder checking in many cases.

Because trace semantics is defined on the basis of finite executions, it preserves only stuttering-insensitive linear-time safety properties. On the other hand, as the following simple argument shows, it preserves all of them. If L violates a stuttering-insensitive linear-time safety property φ, then it has a finite execution ξ that acts as a counterexample. Let σ be the trace corresponding to that execution. If $L' \simeq_{\mathrm{tr}} L$, then also L' has a finite execution that has σ as its trace; call it ξ'. The executions ξ and ξ' differ only in the amount of finite stuttering. Thus also ξ' is a valid counterexample to φ, because φ is stuttering-insensitive. So L' violates φ. This argument is valid also for state-based properties, if the transformation in Section 2.2 is used for encoding state information into actions.

In Section 4.2 we pointed out that no formalism can specify all safety properties, because there are uncountably many of them. This might seem to be in contradiction with the above claim that trace semantics preserves all safety properties, because trace semantics was used above in specification by describing an LTS *Spec* and requiring that $Sys \simeq_{\mathsf{tr}} Spec$ or $Sys \sqsupseteq_{\mathsf{tr}} Spec$. There is no contradiction, however, because trace semantics was not used as the specification formalism but as an abstract mathematical concept that serves as the "satisfies" relation. The specification formalism is the formalism used for describing *Spec*, and it allows only countably many specifications to be written.

Algorithms for trace semantics. Every finite LTS L can be interpreted as a finite automaton in the classic automata-theoretic sense — we only need to specify which states are accepting. If we declare that all states are accepting, then $Tr(L)$ is the same as the language accepted by the automaton. As a consequence, classic algorithms can be used for determinising LTSs, and minimising deterministic LTSs and comparing the languages they accept. Unfortunately, these algorithms consume exponential time (and space) in the worst case, because the minimal deterministic finite automaton that accepts the same language as a given finite automaton may be exponentially larger than the latter.

That $Sys \sqsupseteq_{\mathsf{tr}} Spec$ holds can be checked on the fly as follows. First *Spec* is determinised yielding *DSpec*. Then the LTS $Sys||DSpec$ is computed. If the result has at least one state where *Sys* is ready to execute a visible action that *DSpec* is not ready to execute, then $Sys \sqsupseteq_{\mathsf{tr}} Spec$ does not hold, otherwise it does.

Instead of comparing the next possible actions of *Sys* and *DSpec* in each joint state, the errors can also be detected with the machinery of tester processes in Section 4.2. It suffices to add one extra reject state s_R, and extra transitions (s, a, s_R) to *DSpec* for those $s \in S_{DSpec}$ and $a \in \Sigma$ such that *DSpec* has no a-transition out of s. Then violations against $Sys \sqsupseteq_{\mathsf{tr}} Spec$ are detected as $Sys||DSpec$ reaching any state of the form (s, s_R). This algorithm consumes exponential time and memory in the size of *Spec*, but only linear time in the size of *Sys*. If *Spec* is deterministic, then this algorithm is linear in the sizes of both of its arguments.

It is known well that the problem of checking whether a finite automaton $\mathcal{A} = (Q, \Sigma, \Delta, q_I, F)$ accepts all strings in Σ^* is PSPACE-complete (see e.g., [1, 30]). Consider the LTS $L = (S, \Sigma', \Delta', S_I)$ that is obtained from $\mathcal{A}$ by choosing a new state $s_{\mathsf{new}} \notin Q$ and a new action $\delta \notin \Delta$ and letting $S = Q \cup \{s_{\mathsf{new}}\}$, $\Sigma' = \Sigma \cup \{\delta\}$, $\Delta' = \Delta \cup \{ (s, \delta, s_{\mathsf{new}}) \mid s \in F \} \cup \{ (s_{\mathsf{new}}, a, s_{\mathsf{new}}) \mid a \in \Sigma' \}$, and $S_I = \{q_I\}$. The automaton $\mathcal{A}$ accepts all strings in Σ^* if and only if $Tr(L) = \Sigma'^*$. Thus the question "are all finite strings over the alphabet of a given LTS traces of that LTS" is PSPACE-hard.

The problem "does $Sys \sqsupseteq_{\mathsf{tr}} Spec$ hold" is in PSPACE, because an execution σ of *Sys* can be guessed that produces a trace that is not in $Tr(Spec)$, and "$\sigma \notin Tr(Spec)$" can be verified by maintaining the set of all the states where *Spec* could be when simulating the trace. Polynomial space does not necessarily suffice for storing the trace, but it need not be stored, if simulation is interleaved with the guessing of the execution.

From the above results we can reason that the problems "does $Sys \simeq_{\text{tr}} Spec$ hold" and "does $Sys \sqsupseteq_{\text{tr}} Spec$ hold" are PSPACE-complete, even if Sys is replaced by the one-state LTS All_{Σ} that has a transition from that state to itself for every member of Σ. Namely, the question "does $All_{\Sigma} \sqsupseteq_{\text{tr}} Spec$ hold" is the same as "is $Tr(Spec) = \Sigma^*$". Therefore, checking "$Sys \sqsupseteq_{\text{tr}} Spec$" is PSPACE-complete in the size of $Spec$. On the other hand, the above algorithm checks it in time that is linear in the size of Sys. Indeed, the question "does $Sys \sqsupseteq_{\text{tr}} All_{\Sigma}$ hold" is computationally very easy: its answer is always "yes".

In conclusion, we can check $Sys \sqsupseteq_{\text{tr}} Spec$ inexpensively in terms of the size of Sys, but expensively in the size of $Spec$. It is common that Sys, representing the implementation, consists of several parallel processes and has a huge state space, while $Spec$ consists of only one process and has a small state space. This is fortunate, because it means that expensive operations are done to small and inexpensive operations to big objects. This observation generalises to the verification of linear-time properties in many formalisms: we already saw a similar situation with LTL (Section 4.3), and we will soon see it again with certain other linear-time abstract process semantics. If we wanted to verify $Sys \simeq_{\text{tr}} Spec$ or $Spec \sqsupseteq_{\text{tr}} Sys$, then we would have to do expensive operations also to Sys, leading soon to the exhaustion of computational resources. This suggests that we can attack the state explosion problem much better if we prefer "$\sqsupseteq$"-type notions of "system satisfies specification" over "$\simeq$"-type.

Another operation used frequently in process-algebraic verification is *LTS reduction*. It means the construction of an LTS that is equivalent to the input LTS, but as small as the algorithm can produce. If the result is the smallest possible equivalent LTS, then this operation is known as *minimisation* of the LTS. Unfortunately, as the counterexample in Figure 6 shows, trace semantics does not guarantee the existence of a unique minimal equivalent LTS. Even the problem of finding some equivalent LTS with the smallest possible number of states is PSPACE-hard, because its solution can obviously be easily used to solve "does $Sys \simeq_{\text{tr}} All_{\Sigma}$ hold". (It is also in PSPACE because it can be solved by constructing LTSs of increasing size until an equivalent one is found.)

Fig. 6. Two non-isomorphic trace-equivalent minimal LTSs.

Fortunately, various heuristics can be used for trace-equivalence-preserving

LTS reduction. Examples include minimisation with respect to weak bisimilarity, and the construction of the minimal equivalent deterministic LTS with rejection of the result if it is not smaller than the input LTS. These kinds of techniques have proven to work reasonably well in practice.

Refusals, failures and divergence. Because trace semantics does not preserve any liveness properties, other semantics have been developed. Unlike in temporal logics, in process algebras it is common (but not ubiquitous) to distinguish between deadlock and livelock. This is partly due to a reason illustrated by the following example. If L is ready to do an a-transition and only it, a is not in the alphabet of either L_1 or L_2, and L_1 is in deadlock and L_2 in livelock, then $L||L_1$ is guaranteed to eventually do an a-transition, but $L||L_2$ is not. In terms of a single-processor multi-process computer system, the system running $L||L_1$ will eventually execute L because L_1 is not executable, but the system running $L||L_2$ might spend all processor time on L_2 unless we make some fairness assumption about the scheduling policy. So we see that although we cannot detect directly whether an LTS has deadlocked or livelocked, we can find it out by putting the LTS into a suitable context.

A livelock corresponds to an infinite execution that has only a finite number of visible transitions. The removal of all τ-symbols from the $\mathcal{E}_\Sigma$-abstraction of the execution yields a finite sequence. That sequence is called a *divergence trace.* We will denote the set of the divergence traces of an LTS L with $\mathit{Divtr}(L)$.

To obtain a congruence that handles deadlocks properly, a more general notion of *refusal* has been defined in process algebras. The need for more generality can be illustrated with the following example. Consider two LTSs L_1 and L_2, both of which have $\{a, b\}$ as their alphabet. It may be that in some situation, L_1 is ready to execute an a-transition but not any b-transition, while L_2 is ready for b but not a. Then $L_1||L_2$ is in deadlock, although neither process would be in deadlock if it were alone.

Let $L = (S, \Sigma, \Delta, S_I)$ be an LTS, $A \subseteq \Sigma$, and $s \in S$. We say that s is *stable*, or L *is stable in* s, if and only if $\neg(s -\tau\rightarrow)$, that is, s has no outgoing τ-transitions. Furthermore, s *refuses* A, or L *refuses* A *in* s, if and only if s is stable and has no outgoing transitions labelled with an element of A.[9] This definition has the consequence that $L_1||L_2$ deadlocks in (s_1, s_2) if and only if s_1 and s_2 are stable, and there are sets A_1 and A_2 such that $A_1 \cup A_2 = \Sigma_1 \cup \Sigma_2$ and L_i refuses A_i in s_i for $i \in \{1, 2\}$. It is worth noticing that if a belongs to both Σ_1 and Σ_2, then it suffices that a is in one of A_1 and A_2; that is, one of the LTSs is not ready for a.

A *stable failure* of an LTS $L = (S, \Sigma, \Delta, s_I)$ is any pair (σ, A) such that L has a stable state s that refuses A, and a finite execution that ends at s and has σ as the corresponding trace. The set of the stable failures of L is denoted by $\mathit{Sfail}(L)$. The set of traces that lead to a deadlock is $\{\, \sigma \mid (\sigma, \Sigma) \in \mathit{Sfail}(L) \,\}$, and we have the identity $\mathit{Tr}(L) = \mathit{Divtr}(L) \cup \{\, \sigma \mid (\sigma, \emptyset) \in \mathit{Sfail}(L) \,\}$.

[9] Not all authors require the stability of s in this definition.

Finally, consider an infinite execution that contains an infinite number of occurrences of visible transitions. The result of the removal of all τ-symbols from its $\mathcal{E}_\Sigma$-abstraction is an infinite sequence of elements of Σ, and it is called an *infinite trace*. The notation $\mathit{Inftr}(L)$ denotes the set of the infinite traces of L. Let $\mathit{Tr}\omega(L) = \{ \omega \in \Sigma^\omega \mid \forall \sigma < \omega : \sigma \in \mathit{Tr}(L) \}$, that is, $\mathit{Tr}\omega(L)$ is the set of those infinite sequences of elements of Σ whose every proper prefix is a trace of L. It is clear that $\mathit{Inftr}(L) \subseteq \mathit{Tr}\omega(L)$. If L is finite, then also $\mathit{Tr}\omega(L) \subseteq \mathit{Inftr}(L)$.

Because $\mathit{Tr}(L)$, $\mathit{Divtr}(L)$ and $\mathit{Inftr}(L)$ are defined on the basis of executions, they are linear-time notions. The set $\mathit{Sfail}(L)$ can be thought of as having some branching-time aspect, because its elements talk about all possible next visible actions after a particular execution. However, whether or not a state refuses a set of actions can be thought of as a property of the state, and thus a member of Π. In this interpretation, the property "L has the stable failure (σ, A)" is clearly a linear-time property. Therefore, we will classify as linear-time all abstract process semantics that are only built from $\mathit{Tr}(L)$, $\mathit{Sfail}(L)$, $\mathit{Divtr}(L)$ and $\mathit{Inftr}(L)$.

Failure-based semantic models. The most well-established linear-time abstract process semantics that preserves some liveness properties is certainly the standard *failures–divergences* model of CSP [6, 42, 75]. Because of reasons that at least partly depend on the mathematical approach used in the development of the theory of CSP, this semantics uses different notions of failures and divergences from what we defined above.

Let $\mathit{CSPdivtr}(L) = \{ \sigma \in \Sigma^* \mid \exists \rho : \rho \leq \sigma \wedge \rho \in \mathit{Divtr}(L) \}$. That is, a *CSP-divergence trace* of L is any divergence trace of L continued with any finite sequence of actions. *CSP-failures* of L are all stable failures of L, and all pairs (σ, A) where σ is a CSP-divergence trace of L and $A \subseteq \Sigma$. That is, $\mathit{CSPfail}(L) = \mathit{Sfail}(L) \cup (\mathit{CSPdivtr}(L) \times 2^\Sigma)$. The *CSP-semantics* of L is the pair $(\mathit{CSPfail}(L), \mathit{CSPdivtr}(L))$, and CSP-preorder and CSP-equivalence of two LTSs that have the same alphabet are defined as

$$L_1 \sqsupseteq_{\mathsf{CSP}} L_2 \overset{\mathsf{def}}{\Longleftrightarrow}$$
$$\mathit{CSPfail}(L_1) \subseteq \mathit{CSPfail}(L_2) \wedge \mathit{CSPdivtr}(L_1) \subseteq \mathit{CSPdivtr}(L_2), \text{ and}$$
$$L_1 \simeq_{\mathsf{CSP}} L_2 \overset{\mathsf{def}}{\Longleftrightarrow} L_1 \sqsupseteq_{\mathsf{CSP}} L_2 \wedge L_2 \sqsupseteq_{\mathsf{CSP}} L_1.$$

The CSP-semantics of L has the property that if σ is a divergence trace, then $\sigma\rho$ is a CSP-divergence trace and $(\sigma\rho, A)$ is a CSP-failure, for every $\rho \in \Sigma^*$ and $A \subseteq \Sigma$. This implies that CSP-semantics does not preserve any information about the behaviour of the system after it has executed σ. This property of CSP-semantics is called *catastrophic divergence*. Any system C that has ε as a divergence trace has $\mathit{CSPdivtr}(C) = \Sigma^*$ and $\mathit{CSPfail}(C) = \Sigma^* \times 2^\Sigma$, and is called *chaos*. The catastrophic divergence property makes CSP-semantics useless for analysing the behaviour of a system after a divergence trace. Despite this, CSP-semantics has been very successful both in terms of theory development and practical use. The textbook [75] is a thorough treatment of the CSP language, semantic theory of CSP, and automatic verification in the context of CSP.

The catastrophic divergence problem has motivated researchers to develop alternative semantic models based on some kinds of failures and/or divergence traces. Among them, the *Chaos-Free Failures Divergences* (*CFFD*) semantics [94] is interesting because of its special relation to $\mathrm{LTL}_{-\mathsf{X}}$ (Section 4.3). It is the triple $(\mathit{Sfail}(L), \mathit{Divtr}(L), \mathit{Inftr}(L))$. (Originally it had also a fourth *initial stability* component. It was needed for ensuring the congruence property in the presence of the so-called *choice* process composition operator.) CFFD-preorder and -equivalence are defined just like the corresponding CSP notions. If the LTSs are finite, then the sets of infinite traces need not be compared, because then $\mathit{Inftr}(L) = \mathit{Tr}\omega(L)$ and $\mathit{Tr}\omega(L)$ is uniquely determined by $\mathit{Sfail}(L)$ and $\mathit{Divtr}(L)$.

CFFD-equivalence implies CSP-equivalence, trace equivalence and the NDFD-equivalence mentioned below in the sense that if $L_1 \simeq_{\mathsf{CFFD}} L_2$, then $L_1 \simeq_{\mathsf{CSP}} L_2$, $L_1 \simeq_{\mathsf{tr}} L_2$ and $L_1 \simeq_{\mathsf{NDFD}} L_2$. Furthermore, NDFD-equivalence implies CSP- and trace equivalences.

Assume that Π is encoded into Σ as was discussed in Section 2.2. Then CFFD-equivalence is the weakest possible (that is, makes least distinctions between systems) congruence that (1) preserves the validity of formulae written in $\mathrm{LTL}_{-\mathsf{X}}$, and (2) distinguishes between deadlock and livelock. Furthermore, a slight modification of CFFD-equivalence called *nondivergent failures divergences* (*NDFD*, it takes into account only those stable failures (σ, A) where σ is not a divergence trace) equivalence is the weakest congruence that has the property (1). These facts were proven in [47], and elaborated a bit further in [89]. Due to them, CFFD- and NDFD-semantics provide a means for applying process-algebraic verification methods to the verification of $\mathrm{LTL}_{-\mathsf{X}}$ formulae.

CFFD- and NDFD-equivalences do, however, suffer from a problem that to some extent restricts their applicability to the verification of liveness properties in general and $\mathrm{LTL}_{-\mathsf{X}}$ formulae in particular: they do not handle fairness assumptions in a satisfactory way. As was mentioned in Section 4.3, fairness assumptions can be encoded in the formula whose validity is being verified. However, doing that leads to a big Σ, and that is bad for process-algebraic methods of alleviating state explosion. This problem is not a deficiency of only the CFFD- and NDFD-equivalences, but is present in most or all abstract process semantics that are both congruences and strong enough for handling liveness properties. Some ways of working around it have been found [71, 92], but more work needs to be done in this field.

Algorithms for failure-based semantics. Most complexity results concerning trace semantics generalise relatively easily to CSP-, CFFD- and NDFD-semantics. For instance, the LTSs in Figure 6 are CSP-, CFFD- and NDFD-equivalent, thus also these equivalences fail to have unique minimal LTSs.

The paper [14] shows how algorithms for strong bisimilarity can be used for failure-based semantic models by first transforming the LTSs into *acceptance graphs* that are based on the *acceptance trees* of [41]. Acceptance graphs are, in essence, deterministic LTSs augmented with a special representation for refusal and divergence information. Acceptance graphs were applied in a slightly different way in [93], where algorithms for LTS equivalence comparison and reduction according to CFFD semantics were developed. The results of [93] can be adapted to CSP- and NDFD-semantics.

The *tester processes* presented towards the end of Section 4.2 can be used for checking CFFD-preorder, and even without infinite trace monitor states [87]. The basic idea is that illegal traces are caught with reject states, illegal stable failures with deadlock monitor states, and illegal divergence traces with livelock monitor states. Illegal infinite traces need not be worried of, because the presence of such a trace in a finite LTS implies the presence of an illegal stable failure or illegal divergence trace. Reject states are not absolutely necessary because every illegal trace can be detected as an illegal deadlock or illegal divergence trace. Reject states are, however, very easy to implement efficiently in a state space tool, and they detect an error immediately when the error trace has been executed, while deadlock and livelock monitor states require continuation to a deadlock or livelock. Reject states thus improve efficiency.

There is an algorithm that, given an LTS *Spec*, produces a tester process that checks $Sys \sqsupseteq_{\mathsf{CFFD}} Spec$. It is based on the construction of the *DSpec* with an extra reject state s_R that was defined in "Algorithms for trace semantics", the taking of "mirror images" of refusal sets as has been described in [5], and the marking of those states as deadlock and livelock monitor states which correspond to traces after which *Spec* cannot deadlock or livelock, respectively. The algorithm consumes exponential time and space in the worst case, both because the number of states may grow exponentially, and because the handling of refusal sets may be expensive with pathological inputs. In practice, however, the algorithm often runs reasonably fast. This approach can be adapted to NDFD- and CSP-preorders with small changes. Like with trace semantics, the use of tester processes is inexpensive in the size of *Sys*.

Another on-the-fly method for checking CSP-preorder was described in [74, 75]. It also relies on constructing *DSpec*, but from then on it works differently from the previous one. It augments *DSpec* with an acceptance-tree-type representation of refusal and divergence information, making it possible to compare on the fly each state of *Sys* directly with the corresponding state in *DSpec*. (The "corresponding state" is the unique state of *DSpec* that is reached with the same trace as with which the state of *Sys* was reached.) Again, small changes suffice to apply this algorithm to CFFD- and NDFD-preorders.

Branching-time abstract semantics. *Weak bisimilarity* or *observation equivalence* [65] is certainly the most well known abstract branching-time semantic model in process algebras. Its definition resembles that of strong bisimilarity, but has the difference that when simulating a transition, the simulating LTS may do any number of transitions (including zero) as long as the resulting sequence of the visible actions is the same in both sides.

Let $s -\tau^*\rightarrow_i s'$ denote that s' can be reached from s with zero or more τ-transitions in the LTS in question. The LTSs $L_1 = (S_1, \Sigma, \Delta_1, S_{I1})$ and $L_2 = (S_2, \Sigma, \Delta_2, S_{I2})$ are *weakly bisimilar*, denoted here $L_1 \simeq_{\mathsf{wb}} L_2$, if and only if there is a relation "$\sim$" $\subseteq S_1 \times S_2$ such that for every $s_1, s_1' \in S_1$, $s_2, s_2' \in S_2$, and $a \in \Sigma$:

- If $s_1 \in S_{I1}$, then there is $s \in S_{I2}$ such that $s_1 \sim s$.
- If $s_2 \in S_{I2}$, then there is $s \in S_{I1}$ such that $s \sim s_2$.
- If $s_1 \sim s_2$ and $(s_1, \tau, s_1') \in \Delta_1$, then there is s such that $s_2 -\tau^*\rightarrow_2 s$ and $s_1' \sim s$.
- If $s_1 \sim s_2$ and $(s_2, \tau, s_2') \in \Delta_2$, then there is s such that $s_1 -\tau^*\rightarrow_1 s$ and $s \sim s_2'$.
- If $s_1 \sim s_2$ and $(s_1, a, s_1') \in \Delta_1$, then there are s, s' and s'' such that $s_2 -\tau^*\rightarrow_2 s'$, $(s', a, s'') \in \Delta_2$, $s'' -\tau^*\rightarrow_2 s$, and $s_1' \sim s$.
- If $s_1 \sim s_2$ and $(s_2, a, s_2') \in \Delta_2$, then there are s, s' and s'' such that $s_1 -\tau^*\rightarrow_1 s'$, $(s', a, s'') \in \Delta_1$, $s'' -\tau^*\rightarrow_1 s$, and $s \sim s_2'$.

The definition allows the simulation of a local τ-loop (that is, a transition of the form (s, τ, s)) by no transition at all. As a consequence, weak bisimilarity does not distinguish deadlock from livelock. If this is considered a deficiency, an extra requirement may be added saying that if $s_1 \sim s_2$, then $s_1 -\tau^\omega\rightarrow_1 \Leftrightarrow s_2 -\tau^\omega\rightarrow_2$, where $s -\tau^\omega\rightarrow$ means that there is an infinite execution that starts in s and contains only τ-transitions. This modification of weak bisimilarity implies CFFD-equivalence (excluding the "initial stability" component). Its theory and algorithms were investigated in [18].

Assume that an LTS $L = (S, \Sigma, \Delta, S_I)$ is converted to $Sat(L) = (S, \Sigma, \Delta', S_I)$ such that $\Delta' = \Delta \cup Sc_1 \cup Sc_2$, where *Sat* and *Sc* stand for "saturate" and "shortcut", $Sc_1 = \{ (s, a, s') \mid (s \neq s' \vee a \neq \tau) \wedge \exists s_1, s_2 : s -\tau^*\rightarrow s_1 \wedge (s_1, a, s_2) \in \Delta \wedge s_2 -\tau^*\rightarrow s \}$, and $Sc_2 = \{ (s, \tau, s) \mid s \in S \}$. Clearly $L' \simeq_{\mathsf{wb}} L$. What is more, $L_1 \simeq_{\mathsf{wb}} L_2 \Leftrightarrow Sat(L_1) \simeq_{\mathsf{sb}} Sat(L_2)$. The computation of $Sat(L)$ is essentially the problem of computing the transitive closure of a relation. It can be solved in cubic time (even faster, but the significance of the known faster algorithms is mostly theoretical).

In this way minimisation and comparison of LTSs according to weak bisimilarity can be converted to the related problems with strong bisimilarity, which, as we have seen, have very efficient algorithms. This approach is from [48], and yields cubic equivalence checking and minimisation algorithms for weak bisimilarity. Unfortunately, the number of semantic transitions in a saturated LTS tends to be high, so this approach consumes a lot of memory. This can be avoided by simulating the "$-\tau^*\rightarrow$"-relation as needed instead of storing it in an explicit form, but then the algorithm becomes slower.

It is clear that weak bisimilarity does not preserve the validity of all CTL$_{-\mathsf{X}}$-formulae, because it discards livelock information. However, as the following example demonstrates, weak bisimilarity does not even preserve the validity of all those CTL$_{-\mathsf{X}}$ properties that do not depend on livelocks. The LTSs in Figure 7 are weakly bisimilar, but the formula $\mathsf{EG}(\neg\alpha \wedge (\mathsf{EF}\alpha \vee \neg\mathsf{EF}\beta))$ holds only in the rightmost one.

In Section 4.3 a *stuttering equivalence* was defined that preserves the validity of CTL$^*_{-\mathsf{X}}$-formulae. It requires that each state of an execution at one side is

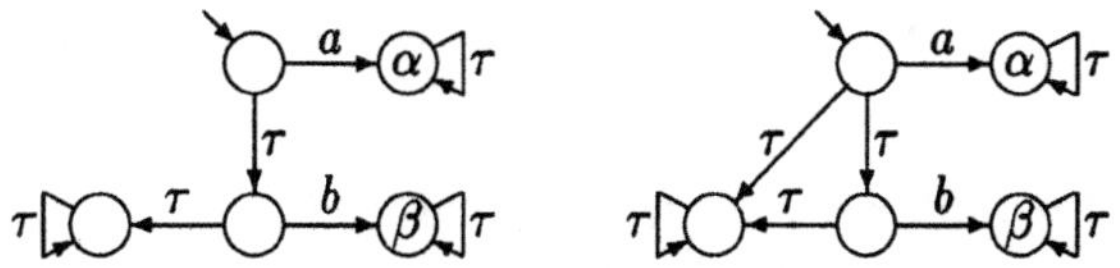

Fig. 7. Weak bisimilarity does not preserve CTL.

simulated by one or more states at the opposite side. Therefore, according to it, the execution where the rightmost LTS in Figure 7 jumps to the bottom left state without visiting the bottom middle state cannot be stutter-simulated by the leftmost LTS, because this execution has no state that could be the pair of the bottom middle state.

Branching bisimilarity [96] is a process-algebraic equivalence that is between weak bisimilarity and stuttering equivalence. Branching bisimilarity implies weak bisimilarity, and differentiates the LTSs in Figure 7 from each other. It, too, discards livelock information, and therefore fails to preserve the validity of $\mathrm{CTL}_{-\mathrm{X}}$ formulae. It can be modified to preserve $\mathrm{CTL}_{-\mathrm{X}}$, but then it becomes essentially the same as stuttering equivalence. One attractive feature of branching bisimilarity is that its algorithms are similar to the algorithms for weak bisimilarity, but need less memory, because the saturation of LTSs is not needed.

The article [90] is a tutorial on the basic ideas of process-algebraic semantic models and automatic verification with them.

5 The Complexity of Verification

5.1 Complexity in Terms of the State Space

In the previous section we mentioned the computational complexity of some verification tasks as a function of the size of the state space $|\mathcal{S}|$ and the size of the property or specification $|\varphi|$:

- The checking of various individual properties, such as the absence of deadlocks, 4-boundedness of a Petri net place and Petri-net-liveness of a transition are linear in $|\mathcal{S}|$.
- LTL and CTL* model checking are linear in $|\mathcal{S}|$ and PSPACE-complete in $|\varphi|$.
- CTL model checking is linear in $|\mathcal{S}|$ and $|\varphi|$.
- Minimisation with respect to trace, CSP-, CFFD- and NDFD-semantics is not well-defined, because there is no unique minimum. Finding some equivalent LTS with a minimal number of states is PSPACE-complete.
- Trace, CSP-, CFFD- and NDFD-equivalence checking are PSPACE-complete.

- Trace, CSP-, CFFD- and NDFD-preorder checking are linear in the alleged smaller LTS, and PSPACE-complete in the other argument.
- An LTS can be minimised with respect to weak or branching bisimilarity in $O(|\mathcal{S}|^3)$ time.
- Equivalence according to weak or branching bisimilarity can be checked in $O(|\mathcal{S}_1|^3 + |\mathcal{S}_2|^3)$ time.

Excluding CTL* model checking, the above results suggest a general tendency: verification tasks related to linear-time properties tend to be PSPACE-complete, whereas similar tasks for branching time can be performed in low-order polynomial time. This may feel counterintuitive, because branching time seems more complicated than linear time and, for instance, weak bisimilarity implies trace equivalence. The case of LTL vs. CTL has an easy explanation: CTL restricts seriously what one can say about the properties of individual executions, and when those restrictions are removed (yielding CTL*), model checking becomes PSPACE-hard again. In the case of weak bisimilarity vs. trace semantics the explanation lies in the fact that checking equivalence under two different semantics is two distinct problems that do not necessarily have much in common, even if one of the equivalences implies the other. We will now illustrate this with an example.

Consider the following seven different equivalences, each of which implies the next: identity ($(S_1, \Sigma_1, \Delta_1, S_{I1}) = (S_2, \Sigma_2, \Delta_2, S_{I2})$ if and only if $S_1 = S_2$, $\Sigma_1 = \Sigma_2$, $\Delta_1 = \Delta_2$, and $S_{I1} = S_{I2}$), isomorphism, strong bisimilarity, branching bisimilarity, weak bisimilarity, trace equivalence, and the "universal" equivalence that says that all LTSs are equivalent. Identity is easy: it can be checked in linear time on the average by inputting the LTSs into a hash table, and in worst-case linear time by replacing the hash table with the (impractical) data structure in exercise 12.1-4 of [15]. Isomorphism is clearly in NP, but has not been proven NP-complete — as a matter of fact, it is a strong candidate for a problem that is not solvable in polynomial time but is not NP-hard either [30]. It is thus rather complex. Strong bisimilarity, despite of its resemblance to isomorphism, can be checked in low-order polynomial time. Known algorithms for branching bisimilarity take some more time and weak bisimilarity still some more, but they are still within cubic time. Trace equivalence is PSPACE-complete and thus (apparently) the hardest in this list. On the other hand, the universal equivalence is the easiest of all: the algorithm **print** "yes" solves it in constant time.

We see that the strength of an equivalence and the complexity of its equivalence checking have low correlation. The fact that branching-time semantic models have fast equivalence checking algorithms seems thus coincidental. Intuitively, their low complexity compared to equivalence checking in linear-time models is due to the fact that branching-time tasks rely on *local* information, while linear-time verification requires simultaneous information about all possible executions that have the same $\mathcal{E}$-abstraction.

We will see in Section 5.3 that there is, however, a sense in which the weakness of an equivalence is beneficial regarding the cost of verification.

5.2 Complexity in Terms of the Modelling Formalism

In the previous section we discussed the complexity of verification as a function of the size of the state space of the system. In reality, the system is usually not originally given as a state space, but as a program in some concurrent programming language, Petri net, parallel composition of LTSs, etc. This representation is typically much, much smaller than the state space — after all, that is what the state explosion problem is about. As a consequence, the complexity of a problem as a function of the size of the state space does not give complete picture of the cost of verification.

Numerous results exist about the complexity of the verification of some property from a system represented in some formalism. One can derive more with the following technique. (More detail can be found in [89], for instance.)

A *linear bounded automaton* is a nondeterministic single-tape Turing machine whose use of the tape is restricted as follows. The input string is surrounded by two endmarkers that the machine cannot write over or bypass. The tape alphabet may be much larger than that used in the input string, and the machine is allowed to write over the input. Therefore, the machine can use the part of the tape that originally contains the input also as working storage. As a consequence, the working storage of the machine is linear in the size of the input string (while with the ordinary nondeterministic Turing machine it is unbounded).

The question whether a given linear bounded automaton accepts a given string is PSPACE-complete ([30], p. 265). A linear bounded automaton is basically a finite automaton that has access to a fixed amount of memory that it can address only by jumping from a memory cell to the next or previous cell. Given a typical formalism that is intended for modelling concurrent systems, it is usually relatively easy to design a system that simulates an arbitrary linear bounded automaton with an arbitrary input, and whose size is polynomial in the size of the description of the automaton and input. The simulator can be made such that when the automaton reaches an acceptance state, the simulator does something that it otherwise cannot do, such as stops or executes a certain transition. Checking the ability of doing "that something" is thus PSPACE-hard for the formalism in question. In this way it is possible to show that the detection of deadlocks, checking whether a given structural transition may ever become enabled, and so on are PSPACE-hard for Petri nets, parallel labelled transition systems, and so on.

To prove that something is PSPACE-complete instead of just PSPACE-hard, it suffices to additionally show that an execution leading to a deadlock, ending up with the occurrence of a given structural transition, etc. can be simulated using at most a polynomial amount of space as a function of the size of the description of the system. Because of the famous result by Savitch that PSPACE = NPSPACE ([76], [30] p. 176, [1] p. 395), the simulator needs not "know" what step it should take next if there are many choices; we may assume that it can always "guess" it. However, when making proofs of this kind, one has to remember that polynomial space does not usually suffice for storing the guesses or the simulated sequence, so the sequence can be used only once in the proof.

The detection of deadlocks, checking whether a given structural transition may ever become enabled, and so on are indeed PSPACE-complete for parallel labelled transition systems and many other formalisms. The result is not valid for ordinary Petri nets because polynomial space does not suffice for storing the marking, as the number of tokens in a place may grow very high. It becomes valid if a fixed upper bound to the number of tokens is enforced for each place.

We can make the conclusion from the above that an interesting verification problem for an interesting modelling formalism can be easier than PSPACE-hard only in exceptional cases. Excluding those exceptional cases (and ignoring the very unlikely possibility that PSPACE = P), *every* verification algorithm for an interesting task must have bad worst-case performance. Thus each verification method must contain at least one potentially expensive step. With ordinary state spaces, this step is obviously the construction of the state space. The same holds for most advanced methods discussed in Section 7. We will see in Section 7.2 that there are methods that can sometimes pack full state space information into a very small representation, but with them the problem of extracting eventual answers from the packed state space tends to be complicated.

Fortunately, the above complexity results apply only to the worst case. They do not prevent a verification method from being reasonably efficient in a large set of practical cases. They do, however, imply that for each method there are systems with which the method becomes very slow (unless PSPACE = P).

5.3 How Much Information to Preserve?

An advanced state space method can reduce the size of the state space either by throwing some information away, or by representing the information in a denser form. Throwing information away implies that some verification questions cannot any more be answered. In theory, packing the information more densely does not rule out the ability to answer any verification question. In practice, both the above complexity results and experience with advanced state space methods suggest that if the packed representation is significantly denser than the ordinary state space, then extracting the answers to certain verification questions becomes so difficult that it becomes a bottleneck. It is thus reasonable to accept the idea that state explosion cannot be significantly alleviated without losing some analysis capability.

Consider two sets Γ_1 and Γ_2 of analysis questions such that $\Gamma_1 \subseteq \Gamma_2$ and $\Gamma_1 \neq \Gamma_2$. Any method that is capable of answering all questions in Γ_2 can be used also for Γ_1. Furthermore, there may be advanced methods that are legal for Γ_1, but fail to answer correctly some verification question in $\Gamma_2 - \Gamma_1$. Therefore, at least in principle, Γ_1 allows the use of at least the same and may allow more tools for attacking the state explosion problem than Γ_2. Therefore, *the more information we are willing to give away, the more we can do to avoid state explosion.*

As an example of the above principle, we will now explain how the fact that some equivalence (say, trace equivalence) preserves strictly less information than some other (weak bisimilarity, for instance) can be used to reduce the cost of verification, although equivalence checking and minimisation are expensive for trace equivalence and cheap for weak bisimilarity. To compare the uses of two different equivalences in the same verification task it is necessary to assume that both equivalences apply to that task, that is, they preserve the property φ in question. For instance, if φ is any linear-time safety property that is stated in terms of Σ, then both trace equivalence and weak bisimilarity preserve it. In such a situation the equivalences are just tools in the verification of φ, not goals in themselves.

As will be discussed in more detail in Section 7.3, equivalences can be used to avoid the construction of the full state space of the system, and produce a smaller but equivalent state space instead. Because any two weakly bisimilar state spaces are also trace-equivalent but the opposite does not hold, trace semantics allows the production of smaller state spaces than weak bisimilarity. Computing the smallest possible trace-equivalent state space is expensive, but it is not necessary in this kind of an application. It is perfectly legal to first run the inexpensive algorithm that produces the unique minimal weakly bisimilar state space, and then apply some inexpensive trace-equivalence-preserving heuristics to reduce the result further still, and perhaps minimise again with respect to weak bisimilarity. In this way a state space is constructed that is never bigger and may be much smaller than the smallest weakly bisimilar state space, but is still valid for the verification of any linear-time safety property φ.

5.4 How Much Memory is Really Needed?

Consider the problem of detecting deadlocks from an ordinary Petri net that has the property that no place ever contains more than one token. This problem can be easily solved by constructing the state space, if enough time and memory is available. The number of the reachable states of the net may be exponential, so this method uses exponential space and consequently also exponential time in the worst case. Even so, unless the net belongs to some special subclass, state spaces are the most practical deadlock detection method known today. We saw above, however, that deadlock detection is PSPACE-complete for this type of Petri nets. Deadlocks can thus be detected in polynomial space, at least in theory. Why are polynomial space deadlock detection algorithms not used in practice?

The result that deadlocks can be detected in polynomial space relies on a theorem by Savitch ([76], [30] p. 176, [1] p. 370). From the proof that Savitch gave to his theorem, it is possible to derive the following polynomial space algorithm "is_reachable(M_1, M_2, k)" for checking whether M_2 is reachable from M_1 in at most k steps.

If $k = 0$ the algorithm returns True if $M_1 = M_2$, and False otherwise. If $k = 1$ the algorithm returns True if there is a structural transition t such that $M_1 [t\rangle M_2$, and False otherwise. For bigger values of k, the algorithm constructs and tests one at a time every possible marking M that assigns to each place either 0 or 1 tokens. A marking is tested by calling "is_reachable(M_1, M, $\lfloor \frac{k}{2} \rfloor$)" and "is_reachable($M$, M_2, $\lceil \frac{k}{2} \rceil$)". If both calls return True, then it is known that there is a path from M_1 to M_2 that goes through M, so the algorithm returns

True. If either call returns False, the test of M fails, and the algorithm proceeds to the next value of M. If the test fails for all values of M, then the algorithm returns False. This is correct, because if there were a path from M_1 to M_2 whose length is at most k, then the middlemost marking of the path would make both "is_reachable(M_1, M, $\lfloor \frac{k}{2} \rfloor$)" and "is_reachable($M$, M_2, $\lceil \frac{k}{2} \rceil$)" to return True.

Let the number of the places in the net be n. The net has at most 2^n reachable markings, because we assumed that $M(p) \leq 1$ for every place p and reachable marking M. Therefore, if M_2 is reachable from M_1, it is reachable via a path whose length is less than 2^n. As a consequence, reachability of a deadlock can be checked by letting M_1 scan through all initial markings and M_2 through all markings, and by calling "is_reachable(M_1, M_2, 2^n)" for each M_2 that is a deadlock. Each such call creates a tree of recursive calls of "is_reachable" whose height is at most $n+1$. Each invocation of "is_reachable(M_1, M_2, k)" consumes only $\Theta(n)$ bits of memory, because each of M, M_1, M_2 and k fits $n+1$ bits. Thus the total memory consumption is $O(n^2)$, which is low-order polynomial in n.

The memory consumption of the above algorithm is actually quite reasonable. Even a straightforward non-optimised stack-based (instead of recursive) implementation survives with $4n + 2$ bits of memory per recursion level. If, for instance, $n = 1000$, then roughly 4 000 000 bits or 500 kilobytes of memory suffices for checking the reachability of a deadlock.

On the other hand, the time consumption of the above algorithm is woeful. If no marking is reachable from another, then the bottom level of recursion is called 2^{n^2} times. That is $2^{1\,000\,000}$ or roughly $10^{300\,000}$ times for $n = 1000$. The estimated age of the universe is only about 10^{27} nanoseconds.

The reason why the above algorithm consumes so little space and so much time is that it throws away almost all subresults it has computed, and recomputes them again and again. Significant savings in time may be obtained by storing the subresults and fetching them from the store as needed instead of recomputing them. But that would require exponential memory. This observation is valid for PSPACE-complete problems in general: although they can be solved in polynomial space in theory, in practice exponential space is used because it is much faster.

In conclusion, it is in theory possible to solve interesting verification tasks in relatively small memory. However, the known algorithms that do that consume unimaginable amounts of time. The possibilities of making a big enough improvement to the speed of the algorithms while keeping the memory consumption polynomial seem minuscule. On the other hand, a much, much faster algorithm is obtained, one that is often fast enough in practice, if one is willing to use much more memory and store intermediate results in it. This is exactly what the state space methods do.

6 Reduction Strategies

An advanced verification algorithm that tries to alleviate state explosion may be organised in many different ways. In this section we discuss the most important of them.

Transparent construction-time reduction. Perhaps the easiest way of connecting a state explosion alleviation method to verification with state spaces is *transparent construction-time reduction.* In it, instead of the full state space, a reduced state space is constructed that gives the same answers to a predefined class of verification questions as the full state space would give. Specifications may be checked and analysis questions may be answered with the same algorithms and tools as with full state spaces.

In order to avoid incorrect answers, the user has to know what properties the reduced state space has preserved, or the analysis tool must refuse to answer illegal questions. The latter may be implemented by adding to the state space some information on what properties it preserves.

The basic stubborn set method that preserves deadlocks and non-termination (Section 7.4) is an example of transparent construction-time reduction. So is also the compositional LTS construction method that is commonly used in the verification of process-algebraic equivalences between systems (Section 7.3).

Guided construction-time reduction. *Guided construction-time reduction* is an elaboration of transparent construction-time reduction, where information about the analysis or verification question or special knowledge on the system is used for guiding the construction of the reduced state space. The abstraction mechanisms of Section 2.2 are one simple way of doing this: the set Π or Σ is specified before the construction of the reduced state space, the state space reduction algorithm uses knowledge of that set to decide where it may reduce, and only the elements in the set can be used as basic components in the verification questions. When the user wants to ask a question that refers to details that are not present in Π or Σ, a new state space has to be constructed.

For instance, the LTL_{-X} and CTL^*_{-X}-preserving stubborn set methods (Section 7.4) work in this way. Also compositional LTS construction falls into this category when it is used as an aid in the verification of individual properties (such as mutual exclusion) instead of checking process equivalence.

If the user specifies a small Π or Σ, then the state space can answer only a small set of verification questions. On the other hand, if Π or Σ is made larger, then also the state space becomes larger and may soon exceed the capacity of the verification tools. The user is thus in a trade-off situation.

The user may have special knowledge of the system that is useful in alleviating state explosion. For instance, the user may know that a correct token-ring system contains at most one token at any instant of time. Because an incorrect system does not necessarily have this property, one has to be careful when taking advantage of this kind of information in verification in order to avoid circular reasoning. In Sections 7.1 and 7.3 we will see examples of how this can be done (the elimination of remnant variable values with **uninitialise**-statements, and the interface processes of Graf and Steffen).

Preprocessing the model. State explosion can be alleviated also by modifying the system description before starting the construction of the state space, or by taking the needs of state space methods into account already when modelling the system. For instance, it is customary to use in a verification model as few variables as possible and to restrict their types to as small as possible. It is also customary to make the degree of atomicity of transitions as coarse as possible.

The correctness of abstractions of this kind is in the responsibility of the author of the verification model. It is also possible to apply sound theories and automatic tools for preprocessing the model such that it is guaranteed that the answers to certain verification questions are not changed. This is the topic of Section 7.1.

Packed state spaces. A *packed state space* is a nonstandard, dense way of storing the state space or a part of it. The packed state space may contain full or incomplete information on the interleaved (or even true concurrency) behaviour of the system.

The analysis algorithms and tools have to be modified to work on packed state spaces instead of ordinary ones. Because the information on states and semantic transitions is to some extent implicit in a packed state space, extracting an answer to a verification question is sometimes hard. This leads to slow and complicated algorithms for certain verification questions. As a consequence, methods based on packed state spaces tend to work well only for certain types of analysis questions, even if no information is lost in the packing. Fortunately, the most important types of packed state spaces support well quite large types of analysis questions.

The symmetry method, the unfolding method, and BDDs are examples of packed state space methods that preserve full information. Petri net coverability graphs preserve incomplete information, and allow the verification of only a restricted class of properties. Also Holzmann's supertrace algorithm throws information away. It allows the handling of a rather large set of properties, but gives only approximate answers. It is thus not a verification method but a validation and error detection method. All these methods are described in Section 7.2.

The construction of an ordinary state space is based on the execution of individual structural transitions in individual states, that is, the construction of semantic transitions one at a time. With packed state spaces this scheme has to be modified to at least some extent, because the processing of semantic transitions one by one would lead to a total amount of work of the order of the size of the full state space, which would largely destroy the benefits of packing. With the symmetry, coverability graph and supertrace methods, the construction of semantic transitions does not differ much from the construction of ordinary state spaces (except that semantic transitions are not stored when supertrace is used). On the other hand, when the set of reachable states is represented with a BDD, semantic transitions are handled with an entirely different technique that will be discussed in Section 7.2.

A packed state space that preserves full information on the behaviour of a system must rely on some regularity in the behaviour, otherwise the packed representation could not be essentially smaller than an explicit representation. The symmetry method, for instance, relies on the assumption that the system consists of several components that are identical except the name or index of the component. Therefore, intuitively speaking, the storing of many behaviours can be replaced by a remark that says that those behaviours are the same as certain stored behaviours except the naming of entities.

The regularity on which a packed state space relies needs not be well-specified. For example, the type of regularity that BDDs rely on is very difficult to characterise.

On-the-fly verification. In *on-the-fly verification* the algorithm that checks the validity of a property is integrated to the algorithm that constructs the state space. The construction of the state space is stopped immediately when an error against the property is found. The state space is usually thrown away when the algorithm has finished. We have seen several on-the-fly algorithms and techniques already in Sections 4.2, 4.3 and 4.4.

To appreciate the importance of on-the-fly verification, let us consider the introduction of an error to an originally correct system. In some cases the error makes the state space of the system smaller — this happens, for example, if it causes the system to deadlock right at the start. It is more common, however, that the error greatly increases the size of the state space. This is because a correct system almost always obeys some state invariant that significantly restricts the number of states that it can reach, and is violated by the broken system.

For instance, the alternating bit values of messages in a fifo channel of the well-known alternating bit protocol are not arbitrary, but they can change in at most one point within the fifo. If the capacity of the channel is k and the actual contents of the message are not modelled, arbitrary values of the alternating bits allow $2^0 + 2^1 + \cdots + 2^k = 2^{k+1} - 1$ different contents of the channel, whereas only $1 + 2 \cdot 1 + 2 \cdot 2 + \cdots + 2 \cdot k = k^2 + k + 1$ value combinations satisfy the invariant. Once the broken system has entered a state that violates the invariant, usually nothing stops it from slowly corrupting the state more and more, and thus eventually reaching a great number of states that are disallowed by the invariant.

It is obvious that stopping when an error is found does not speed up the verification of a correct system. On the other hand, we argued above that an incorrect system tends to have a much larger state space than the corresponding correct system. If the validity of the above invariant is monitored on the fly with a method that detects violations without delay, then analysis is stopped at the latest when one more state has been constructed than what the invariant permits. On-the-fly verification thus tends to reduce the number of states of the erroneous system back to roughly the same level with the correct system. This is of great help when testing design ideas during system development.

According to the above reasoning, the ability of an on-the-fly method to keep the state space small depends on how soon the method detects the error after all states of an erroneous execution have been constructed. Because of this, we recommended in Section 4.2 the use of reject states and livelock monitor states for checking linear-time safety properties and livelocks, even if they could be checked also with Büchi automata (or infinite trace monitor states).

On-the-fly verification may reduce the number of states in also another way, by avoiding the construction of those parts of the state space that are not relevant for the property. This may be thought of as an instance of guided construction-time reduction, and it reduces states also when analysing a correct system. For the sake of an example, consider a tester process (Section 4.2) that is connected to the system with transition fusion. Assume that the alphabet of the system is $\Sigma = \{a, b, c\}$, and the property to be checked is "if the first visible action is a, then the next one is b". The 3-state tester 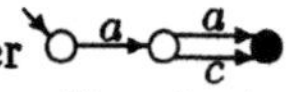 (the black state is a reject state) suffices for checking the property. Due to transition fusion, the tester prevents the system from executing b or c as the first visible action, thus pruning a potentially big part of the state space.

On-the-fly verification can often be combined to advanced state space methods of other kinds, and some advanced methods (most notably Holzmann's supertrace, Section 7.2) actually require the use of on-the-fly verification. However, the addition of extra components to the system that is required by most on-the-fly methods sometimes more or less destroys the ability of an advanced state space method to reduce states.

For instance, the addition of a Büchi automaton by synchronising with every structural transition of the system makes stubborn-set-type methods (Section 7.4) and process-algebraic compositionality (Section 7.3) more or less useless, because that requires the making of every transition visible, and (significant) reduction is obtained only from invisible transitions. This problem can be avoided by using an automaton that represents a stuttering-insensitive property, and either connecting it to only visible transitions (then one has to be careful with infinite stuttering, because the automaton does not even see stuttering), or synchronising it to each state and using a method that preserves the property without getting guidance from the automaton.

The lesson to be learnt here is that although it is often technically possible to use more than one advanced verification paradigm or method simultaneously, it may lead to bad results if done without good enough understanding of the situation.

7 Advanced State Space Methods

Many different methods for coping with state explosion have been suggested in the literature, and the field is still evolving rapidly. In this section the basic ideas of a dozen or so methods are described. The discussion concentrates on methods and ideas whose importance has already been proven by practical use, influence on ongoing research, or some other way. Unfortunately, it would have required too much effort to carefully assess the merits of all the numerous approaches suggested in the literature. It is thus possible that some important ideas may be missing from this section. Stubborn-set-type methods are presented in more detail than other methods because there was a wish that this article would also serve as a stubborn set tutorial.

7.1 Preprocessing Methods

Every experienced user of state space tools knows that the way in which the system is modelled for the tool may have a great effect on the size of the state space. Some of the tricks that (assuming sufficient tool support) the modeller of the system can use to alleviate state explosion without affecting verification results are discussed in this section.

Elimination of remnant variable values. The state of a system is determined by the history of the system, and usually most of the state information affects the future behaviour of the system. With some systems it is, however, possible that some variable may contain several different values due to different histories, but all values lead to the same possible future behaviours. We might then say that the variable contains a *remnant value*.

A good example of remnant values can be found from within a communication protocol with a bounded number of retransmissions. The protocol sender process contains a counter for keeping track of transmission attempts. After the sender has received an acknowledgement, the counter will not be touched until the sender receives a new transmission request from the sending client. Then it resets the counter to zero. Thus the value of the retransmission counter is unnecessary between the reception of the acknowledgement and the start of the transmission of a new message.

Although remnant values do not affect future behaviour, they contribute to the global state of the system, and thus increase the number of reachable states. This effect can be prevented by adding to the verification model of the system statements that reset the variable to a specified value immediately after the value of the variable becomes unnecessary.

If the reset statements are added manually, there is the risk that a variable is reset although its value is not unnecessary, which may lead to incorrect verification results. Fortunately, this risk can be avoided if the input language of the verification tool contains the following feature. In addition to the "normal" values, the type of each variable should contain a special value undefined, and there should be a statement **uninitialise**(v) that assigns undefined to the variable v. Each attempt to use the value of v when it is undefined is immediately reported by the verification tool as a run-time error (like division by zero). Thus the modeller can add **uninitialise**-statements wherever the value of the variable is believed to be remnant, and the tool reveals if it was not remnant after all.

Coarsening of atomicity. The sizes of atomic actions have a great effect on the size of the state space of a system. As a consequence, system modellers often try to make the actions as large as possible. Unfortunately, careless coarsening of atomicity often removes important behavioural errors that we would like to find with verification algorithms. A widely used way of avoiding this problem is to adhere to the following "critical reference" rule.

Let us say that a reference to a variable (or any other block of memory) that is shared by two or more parallel processes is *critical*, if and only if some other process can write to that variable (that is, change its value), or the reference is a writing reference and some other process has access to the variable. If a sequence of statements contains at most one critical reference, may be blocked only at the beginning, and does not contain a potentially non-terminating loop, then the sequence may be collapsed to one atomic action without affecting the answer to almost any verification question. There are special cases where this rule can be made more liberal. For instance, if only one process can write to an unbounded fifo and no process can test the emptiness of the fifo, then the writing reference needs not be considered critical.

The coarsening of atomicity often requires support from the modelling language. For instance, the *Promela* modelling language of the *Spin* tool contains an "**atomic**" statement with which an arbitrarily long sequence of successive transitions can be glued into one atomic transition [43]. If such a sequence is deterministic, then coarsening may be implemented in a state space construction tool by executing the whole sequence as one semantic transition. If the sequence contains a nondeterministic choice, then it may simplify the implementation to store the state where the choice is made. In this "intermediate" state no other transitions than those that belong to the process in question are investigated.

A simple technique that facilitates both the implementation of the above intermediate states and some other modelling tricks is to assign priorities to structural transitions. A low-priority transition may occur only if no higher-priority transition is enabled. An example of other uses of priorities is the modelling of the assumption that a timer does not occur prematurely, that is, while some other processing is still going on. This is achieved be giving the expiration of the timer lower priority than anything else.

The stubborn-set-type methods of Section 7.4 have a similar but usually much stronger effect than the coarsening of atomicity. Coarsening is thus more or less unnecessary when those methods are used. Coarsening is, however, much easier to implement, and, unlike stubborn-set-type methods, can be used with almost any verification question. Only questions that refer to states in the middle of a coarsened action or are sensitive to the disappearance of such states cannot be answered when actions are coarsened.

System transformations. The effect of the elimination of remnant variable values and coarsening of atomicity may to some extent be obtained by applying suitable transformations to the system model before the construction of the state space. Examples of such transformations of Petri nets are given in [40]. Unfortunately, system transformations tend to either preserve only a restricted set of behavioural properties, or contain complicated conditions on their use. An example of good use of net reductions is contained in [77].

System transformations work at the structural level. They cannot thus take advantage of situations where, for instance, a variable contains a remnant value (Section 7.1) but this fact cannot be seen by investigating the structure of the system. Process-algebraic equivalence-preserving reductions of state spaces (Section 7.3) can almost always do the same as system transformations and much more. Their use, however, requires that the state space of a suitably chosen part of the system is first constructed. Sometimes such a part cannot be identified, or has too big a state space. In such a situation process-algebraic reductions do not make system transformations unnecessary.

Data-independence. Many systems are *data-independent* in the sense that they just move data around without looking at it or modifying it. Most obvious examples of this are communication protocols and cache memories. In such a case it is clearly unnecessary for verification to model the data in full detail. However, if the model contains only one data value, then errors such as the swapping of messages in a protocol cannot be detected. This raises the question: what is the minimum (or at least a sufficient) number of data values for detecting all behavioural errors of a given data-independent system?

This question is usually answered with relatively simple manual reasoning. For instance, two different values suffice for detecting the swapping of two successive messages in a data-independent protocol. To see this, assume that the protocol may swap the nth and $(n+1)$th message in a sequence of m messages, where m may be infinite or any finite integer greater than n. (We have to give n and m this much freedom, unless we know that the possibility of swapping two messages does not depend on the total length of the sequence and the location of the swapping within it.)

Because of data-independence, the swapping may occur also in the sequence that consists of n messages of type m_1 and then $m - n$ (or infinitely many, if $m = \infty$) messages of type m_2. In that case the output of the protocol contains at least one instance of m_2 being immediately followed by m_1. On the other hand, such an output is impossible for the given input sequence if the protocol never swaps messages.

Swapping may thus be detected by adding to the system a data source process that gives the protocol an arbitrary sequence of the above kind, and a tester process that checks the output of the protocol. Only two values of messages are needed. State explosion is further reduced by the fact that the two values are not sent in arbitrary combinations. The error is detected on-the-fly which, as was discussed in Section 6, also helps to alleviate state explosion.

The theory of data-independence was carefully investigated in [99]. Recently, several theorems giving sufficient numbers of data values in CSP-type preorder verification were developed by Ranko Lazić; these results are summarised in [75]. A theory that allows the postponing of the instantiation of data values in process-algebraic compositional verification (Section 7.3) until doing it causes less state explosion was developed in [51].

Further remarks on modelling a system for verification. The size of a state space of a system is at worst proportional to the product of the numbers of different local states and values that its component processes and variables may have. For simplicity, we will discuss only the values of variables in the sequel. This does not imply loss of generality, because the set of local states of a process can be thought of as a variable whose value specifies the current local state.

The more a variable v contains "arbitrary" information that does not depend on the values of other variables, the bigger is the contribution of v to the size of the state space. In particular, if the value of v can be uniquely determined from the values of the other variables, then v is harmless from the point of view of state explosion. (Due to this, tester processes used in on-the-fly verification like in Section 4.2 do not usually increase the size of the state space very much.) Therefore, it is not essential to keep the number of variables small; what is essential is to ensure that the variables do not store arbitrary information.

It is common that a system contains more than one variable of some data type, and that the values of those variables may be partially unrelated. For instance, one of the variables may contain the value that is currently being processed, while another contains the previously processed value as a remnant value. In such a situation, making the data type grow affects the size of the state space more than linearly. Because of this, it is often important for state space methods to keep data types very small. This also explains why elimination of remnant values sometimes helps a lot in fighting state explosion.

The power of coarsening of atomic actions has a similar explanation. Coarsening reduces the extent to which component processes may proceed independently of each other and without affecting non-local variable values. It thus reduces significantly the amount of arbitrary information in component processes.

Many systems contain one or more parameters which may assume different values. Examples of such parameters are the number of philosophers in the dining philosophers system, and the window size, maximum number of retransmissions and queue capacities in a sliding window protocol. Strictly speaking, such "systems" are actually infinite *families* of similar systems of different size. Most state space methods require that all such parameters are given a constant value, and usually the values must be small to avoid state explosion. Fortunately, it seems that the majority of design errors manifest themselves already with small values of parameters. Therefore, verification with small parameter values can be used as an error detection and validation method for systems with large parameter values.

7.2 Methods Based on Packed State Spaces

Holzmann's supertrace. Most modelling formalisms allow the representation of the state of the system as a bit vector of fixed length, at least if integers, queues, etc. are replaced with bounded data types such as 8-bit integers and queues of fixed capacity. A bit vector of length k has 2^k different value combinations. These combinations comprise a "universe" U of all "syntactically possible" states, of which the set S of reachable states is a subset. It is common that only a small fragment of syntactically possible states are reachable. Consider the n dining philosophers system in Figure 1. The state of a philosopher can be encoded to two bits, and the state of a fork to one bit. This makes a total of $3n$ bits and thus $2^{3n} = 8^n$ value combinations. However, only $3^n - 1$ of them are reachable. We can say that S is usually *sparse*.

The most memory-efficient way of storing an entirely arbitrary subset X of a set $U = \{u_1, u_2, \ldots, u_n\}$ of n elements is a bit vector of length n, where the value of the ith bit tells whether $u_i \in X$. To avoid confusion with bit vectors that represent states, we will use the term *bitset* of a bit vector that stores a set. With sparse sets, the bitset approach leads to the storing of numerous "no" bits, making the explicit enumeration of all elements of X as the keys of the records of some data structure (such as a linked list, hash table, or binary tree) much more memory-efficient. For instance, the state of the 10 philosophers system fits 30 bits or 4 bytes. If we assume that the data structure requires two additional 4-byte pointers per record, the total amount of memory becomes 59 048 states $\times$ 12 bytes/state $\approx$ 700 kilobytes. The bitset representation requires $2^{30}/8 =$ 128 megabytes.

Because of this, the majority of traditional state space tools store states in explicit form. The use of bitsets is possible only if the state can be encoded to a number of bits that is at most the base 2 logarithm of the amount of available memory in bits.

The classic state space construction algorithm needs to store two sets of states: the states *Found* that have been found so far, and its subset *Incomplete* consisting of those states that have been found but not yet fully processed. When the algorithm is run, *Found* keeps on growing until it contains all reachable states, while *Incomplete* alternates between growing and shrinking. It is rather common that the size of *Incomplete* stays small compared to the set of reachable states. Thus the bottleneck is the storing of *Found*.

The basic idea of Holzmann's *supertrace* algorithm [43] is to replace *Found* with an approximation that is obtained by artificially mapping the set U of all syntactically possible states to a small enough set H, and store the approximation as a bitset $H[\,]$. The mapping is implemented as a hash function $h : U \mapsto H$ that takes the original representation of a state s as a long bit vector and produces a short enough bit vector $h(s)$. The test $s \in \mathit{Found}$ is replaced by the test $H[h(s)] = 1$.

Because $|H| < |U|$ (and $|H| << |U|$), the risk of two different reachable states mapping to the same element of H (hash collisions, in other words) cannot be eliminated. As a consequence, every now and then the supertrace algorithm treats a newly found state as an old state and fails to add it to *Incomplete*. This causes the algorithm to ignore the output transitions of some or many states, and leave a part of the state space uninvestigated. Supertrace is thus *not* a verification algorithm, but a validation and error detection algorithm.

The bitset approximation of *Found* is not particularly useful after supertrace has terminated, because it is neither a lower (an unreachable state may hash to a marked bit) nor an upper (supertrace did not investigate all reachable states) approximation of the set of reachable states. It is difficult to think of any way how supertrace could store reasonable additional information on the sets of states and semantic transitions that it has investigated without increasing the memory requirements at least by an order of magnitude. Therefore, with supertrace, errors must be detected on the fly. Fortunately, as we saw in Section 4.2, violations against linear-time safety properties, linear-time livelock properties and even arbitrary LTL properties can be detected on-the-fly. Supertrace has thus wide applicability.

Perhaps the most important advantage of supertrace is that it can be used almost independently of the size of the state space of the system in question and the amount of available memory. Supertrace never investigates more states than fits the bitset. This also sets an upper limit to the time consumption of supertrace. If the bitset is small, hash collisions start to occur early in the analysis, and only a small fraction of the state space is investigated. If more memory is given, the bitset may be made larger, leading to the investigation of a bigger part of the state space.

In this way supertrace almost always gives an answer within the resources given to it, and the quality of the answer improves if the resources are increased. Furthermore, set membership can be tested and a new element added very efficiently with bitsets, so supertrace uses the resources given to it in an exceptionally efficient way. It may, however, run out of memory, because the set *Incomplete* is stored in the ordinary way, and its maximum size may be difficult to predict in advance.

Supertrace is thus not a verification technique, but it is a very pragmatic validation and error detection technique. In addition, it is easy to implement.

Petri net coverability graphs. A marking M' of a place/transition net *covers* a marking M, denoted by $M' \geq M$, if and only if $M'(p) \geq M(p)$ for every place p. The notation $M' > M$ means that $M' \geq M$ and $M' \neq M$. If M' covers M and $M\,[t\rangle\,M_1$, then there is a marking M_1' such that $M'\,[t\rangle\,M_1'$ and M_1' covers M_1. As a matter of fact, $M_1' - M_1 = M' - M$, where $M_x - M_y$ denotes the function from places to integers such that $(M_x - M_y)(p) = M_x(p) - M_y(p)$ for every place p. This fact implies that if $M_0\,[t_1t_2\ldots t_n\rangle\,M_1$ and $M_1 \geq M_0$, then $M_1\,[t_1t_2\ldots t_n\rangle\,M_2\,[t_1t_2\ldots t_n\rangle\,M_3\,[t_1t_2\ldots t_n\rangle\cdots$, where $M_k = M_0 + k(M_1 - M_0)$. If, furthermore, $M_0 \in [M_I\rangle$ and p is a place such that $M_1(p) > M_0(p)$, then the number of tokens in p may grow without limit, and $[M_I\rangle$ is infinite.

On the other hand, if a place/transition net has a finite number of places and transitions but an infinite number of reachable markings, then it can be proven to have an infinite execution with infinitely many different markings. Furthermore, any such execution has a marking M_0 and a later marking M_1 such that $M_1 > M_0$.

Let us use the term *ω-marking* of any function from the set of places to $\{0, 1, 2, \ldots\} \cup \{\omega\}$, where ω is a special symbol that denotes unboundedness. The notion of covering can be extended to ω-markings by defining that $i < \omega$ (and thus $i \leq \omega$) for every integer i. An ω-marking M_ω denotes a set $[[M_\omega]]$ of ordinary markings such that

- for every place p and every $M \in [[M_\omega]]$, if $M_\omega(p) \neq \omega$, then $M(p) = M_\omega(p)$; and
- for every ordinary marking M such that $M \leq M_\omega$, $[[M_\omega]]$ contains an ordinary marking M' such that $M \leq M'$.

The set denoted by an ω-marking M_ω that contains no ω-symbols is thus $\{M_\omega\}$. If M_ω contains ω-symbols, $[[M_\omega]]$ is not uniquely defined, but it is certainly infinite. If M and M' are ordinary markings, then $M \leq M_\omega \Leftrightarrow \exists M' \in [[M_\omega]] : M \leq M'$.

What it means for a transition to be enabled in an ω-marking, and what is the result of its occurrence are defined like with place/transition nets, except that a place marked with ω always contains enough tokens, and is marked with ω also after the transition occurrence. This implies that if M_ω contains at least one ω-symbol, then $M_\omega\, [t\rangle\, M'_\omega$ represents an infinite number of occurrences of t from a marking in $[[M_\omega]]$ to a marking in $[[M'_\omega]]$.

A *coverability graph* of a place/transition net is constructed like the ordinary state space with the following exception. Each time it is found out that a newly constructed ω-marking (remember that also ordinary markings are ω-markings) M' covers and is reachable from an older ω-marking M, M' is replaced by the unique ω-marking M_ω such that $M \in [[M_\omega]]$, $M' \in [[M_\omega]]$, and M_ω contains as few ω-symbols as possible. The details of when and to which ω-markings M' is compared may vary, and it is also possible to replace M_ω for M instead of or in addition to M'. To guarantee termination it suffices that each newly constructed M' is compared to each ω-marking in the path along which M' was found. Figure 8 shows an example of a Petri net and its coverability graph.

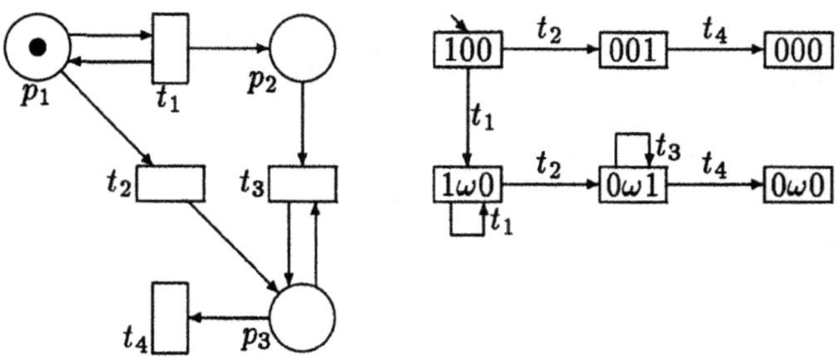

Fig. 8. A coverability graph example.

Coverability graphs can be easily used to detect unbounded places of a place/transition net. Because the set represented by an ω-marking is not uniquely defined, coverability graphs cannot be used for checking the reachability of a marking. For instance, if the weight of the arc from t_1 to p_2 were changed to 2 in Figure 8, then the coverability graph would not change, but the marking 110 would become unreachable. This implies that there are simple state-based properties that cannot be verified from coverability graphs. On the other hand, coverability graphs can be used for checking reachability of a state where a transition is enabled. As was discussed in Section 4.2, this suffices for checking a large set of action-based linear-time safety properties.

Regarding liveness properties, the example net cannot execute t_3 an infinite number of times. However, if a new arc were drawn from t_3 to p_2, then t_3 would become infinitely executable without any change in the coverability graph. Thus coverability graphs throw away essential information regarding the verification of liveness properties.

Coverability graphs are a technique for handling infinite state spaces. They are of no help if the state space is known to be finite. A coverability graph of a Petri net with a finite number of reachable markings is the same as its state space.

With coverability graphs it is possible to answer many questions that are undecidable for Turing machines, such as the possibility of non-termination. This implies that place/transition nets are computationally strictly weaker than Turing machines, and that the coverability graph construction cannot be fully transformed to Turing-strong formalisms such as automata that communicate through unbounded fifo-queues. On the other hand, the ideas of coverability graphs could perhaps be of use also with fifo systems. They provide an incomplete test for detecting that the state space is infinite, and allow some reasoning from an infinite state space. Because of undecidability, the coverability graph of a fifo system will every now and then be infinite and thus impossible to construct, but state spaces that are too big to be constructed have always been a part of everyday life with state space methods.

More information on coverability graphs can be found in many books on the theory or analysis of Petri nets. The usual definition of coverability graphs is somewhat liberal, making it possible to obtain different coverability graphs from the same Petri net. The article [27] defines and analyses *minimal* coverability graphs and shows how they can be constructed. The minimal coverability graph of a Petri net is unique.

The symmetry method. Many systems exhibit symmetry in one form or another. For instance, a system may contain several identical components that are coupled to each other and the rest of the system in a regular way. Consider a system with the syntactically possible states U and the state space (S, T, Δ, S_I). The notion of symmetry may be formalised with the aid of a set of bijections $f : (U \cup T) \mapsto (U \cup T)$ that have for every states s and s' and structural transition t the properties that $f(s) \in U$, $f(t) \in T$ and $s -t\rightarrow s' \Leftrightarrow f(s) -f(t)\rightarrow f(s)'$. Furthermore, if $s \in S_I$, then also $f(s) \in S_I$ and vice versa. We may call such bijections *symmetry bijections*.

It is clear that the identity function $Id : (U \cup T) \mapsto (U \cup T) : Id(x) = x$

is a symmetry bijection. Furthermore, if f and g are symmetry bijections, then so are also the inverse f^{-1} of f and the function composition $g \circ f$ defined by $(g \circ f)(x) = g(f(x))$. Let B be a nonempty set of symmetry bijections, and assume that B^* is the set of all symmetry bijections that can be constructed from the elements of B by repeated application of the inverse and/or function composition operations. For instance, in the case of a token-ring protocol, if B contains only one element and that element is the rotation of the ring one step to the right, then B^* is the set of all rotations of the protocol. If B consists of all possible ways of swapping two clients in a star-shaped client-server system with one server and many clients, then B^* is the set of all permutations of the client processes.

The pair $(B^*, \circ)$ is a mathematical *group*, and the relation $s \simeq s' \overset{\text{def}}{\Longleftrightarrow} \exists f \in B^* : s' = f(s)$ is an equivalence. Let us denote its equivalence classes with $[[s]]$. That is, $[[s]] = \{ f(s) \mid f \in B^* \}$. It follows from the above definitions that if any state in $[[s]]$ is reachable, then every state in $[[s]]$ is reachable. Then we may say that $[[s]]$ is reachable.

The basic idea of the reduction of state spaces with symmetries is to store only one state from each reachable $[[s]]$. We may call that state the *representative* for $[[s]]$. If s has been stored and $s -t\rightarrow s'$, then an edge is stored that starts from s, is labelled with t, and ends in the state s_r that represents $[[s']]$. (Some versions of the symmetry method store as an additional label of the edge the symmetry bijection h or h^{-1} such that $s_r = h(s')$.) It may or may not be the case that $s_r = s'$. This means that for every reachable semantic transition (s, t, s'), the symmetry method stores the semantic transition $\big(f(s), f(t), g(s')\big)$, where f and g are some symmetry bijections. Furthermore, if (s, t, s') is an edge in the symmetry state space, then there is a symmetry bijection g such that $\big(f(s), f(t), f(g(s'))\big)$ is a reachable semantic transition for every symmetry bijection f. Therefore, assuming that the set of symmetry bijections is known, the symmetry method preserves full information on the reachable part of the state space.

We pointed out in Section 5.3 that even if a packed state space preserves full information on the behaviour, it is not guaranteed that answers to verification questions can be extracted efficiently enough. Fortunately, if each $[[s]]$ is finite (which is certainly the case if the ordinary state space of the system is finite), then the symmetry method can be used to verify all LTL properties with the on-the-fly techniques in Section 4.2. In order to not break the symmetry and thus invalidate the symmetry method, one has to add to the system the necessary number of symmetric copies of the fact transition, Büchi automaton, or tester process with which the property is represented. Then errors can be detected with reject states, deadlock monitor states, livelock monitor states and Büchi acceptance states (or infinite trace monitor states) exactly like with ordinary state spaces. We will now show that this is the case with Büchi acceptance; the other forms of error detection can be handled similarly (and do not need the finiteness assumption in the case of reject and deadlock monitor states).

Consider an infinite execution $s_0 -t_1\rightarrow s_1 -t_2\rightarrow \cdots$ that goes infinitely many times through some Büchi acceptance state s_B. The symmetry state space contains the path $f_0(s_0)$ $-f_0(t_1)\rightarrow$ $f_1(s_1)$ $-f_1(t_2)\rightarrow$ $\cdots$, where f_0, f_1, ... are some symmetry bijections. This path contains infinitely many occurrences of elements of $[[s_B]]$. Because $[[s_B]]$ is finite, at least one element in it occurs infinitely many times, and the path $f_0(s_0)$ $-f_0(t_1)\rightarrow$ $f_1(s_1)$ $-f_1(t_2)\rightarrow$ $\cdots$ is Büchi-accepted. If, on the other hand, the symmetry state space contains a path s_0 $-t_1\rightarrow$ s_1 $-t_2\rightarrow$ $\cdots$ where some Büchi acceptance state s_B occurs infinitely many times, then $f_0(s_0)$ $-f_0(t_1)\rightarrow$ $f_1(s_1)$ $-f_1(t_2)\rightarrow$ $\cdots$ is an execution for at least some symmetry bijections f_0, f_1, ... Due to the same reason as above, some $s \in [[s_B]]$ occurs infinitely many times in that execution. Let f be the symmetry bijection that maps s to its corresponding acceptance state in the Büchi automaton that represents the original property. Then $f(f_0(s_0))$ $-f(f_0(t_1))\rightarrow$ $f(f_1(s_1))$ $-f(f_1(t_2))\rightarrow$ $\cdots$ is an execution that violates the property.

Efficient verification of proper branching-time properties with the symmetry method is possible, but less straightforward. The problem can be illustrated with the net 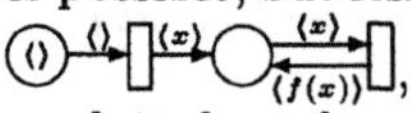
, where x may get the values 0, 1, 2 and 3. If $f(x) = (x + 1) \bmod 4$, then the occurrence mode $x = 1$ of the rightmost transition is Petri-net-live, but it is not if $f(x) = (x + 2) \bmod 4$. However, in both cases the symmetry state space has the structure 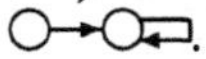. To decide Petri-net-liveness it is thus necessary to look at the actual symmetry bijections in addition to the structure of the symmetry state space, making verification more difficult.

The state space reduction power of the symmetry method depends crucially on the set of symmetry bijections that is used. The application of rotation symmetry to a ring-like system of n components can divide the number of states by at most n, which is usually not sufficient for curing state explosion. On the other hand, the symmetry consisting of all permutations can give much better reduction results.

The addition of (all symmetric copies of) a Büchi automaton or tester process to a system increases the amount of information that is stored of each symmetric component. States that would be symmetric without the extra information may become asymmetric in the presence of the extra information. Because of this effect, on-the-fly verification with symmetries does not always work well. This effect can be fought against when designing the automaton by paying careful attention to the amount and nature of the information that it stores.

The construction of symmetry state spaces and their use in the verification of Petri nets has been discussed in [46]. The symmetry method was applied to CTL* model checking in [12, 22], of which [12] used also BDDs. The use of symmetries when model checking with Büchi automata in the presence of fairness assumptions was investigated in [23, 39]. The combination of the symmetry method with the stubborn set method (Section 7.4) was investigated in [84, 53] with a small number of case studies, and was found to be better than either method alone.

The unfolding method. Consider an ordinary Petri net $\mathcal{N}$ that has the property that $M(p) \leq 1$ for every place p in every reachable marking M. Let M_I be an initial marking of $\mathcal{N}$. The *unfolding* of $\mathcal{N}$ from M_I is a (usually infinite) unmarked acyclic Petri net $\mathcal{U}$ that represents the truly concurrent behaviour of $\mathcal{N}$ in a certain way. Each place b of $\mathcal{U}$ corresponds to some place $fb(b)$ (b "folded back") of $\mathcal{N}$, and similarly for transitions. The unfolding of $\mathcal{N}$ is obtained as follows.

To start with, a place b_i^I is added to $\mathcal{U}$ for each place p_i of $\mathcal{N}$ that is marked in M_I, and, naturally, $fb(b_i^I)$ is chosen to be p_i. We will call these places the *initial places* of $\mathcal{U}$. Then new places and transitions are added according to the following rule as long as possible. Let t be a transition of $\mathcal{N}$, and let the numbers of input and output places of t be denoted by k and h. If $b_1, \ldots, b_k$ are places of $\mathcal{U}$ such that they are *concurrent* in the sense explained soon, and $fb(b_1), \ldots, fb(b_k)$ are exactly the input places of t, then a new transition e and new places $b'_1, \ldots, b'_h$ are added, and arcs are drawn from each b_i to e and from e to each b'_j, provided that such a transition and places have not already been added. The function fb is extended to the newly added elements such that $fb(e) = t$ and $\{fb(b'_1), \ldots, fb(b'_h)\}$ is the set of the output places of t.

In the unfolding, a place can never have more than one input transition. Furthermore, every place is reachable from at least one initial place. Let b_1 and b_2 be two different places of $\mathcal{U}$. They are *causally related*, if and only if there is a path from one of them to the other. They are in *conflict*, if and only if $\mathcal{U}$ has a place b such that $b \neq b_1$, $b \neq b_2$, and there is a path from b to b_1 and another path from b to b_2 such that the paths have nothing else in common than b. Finally, the places $b_1, \ldots, b_k$ of $\mathcal{U}$ are *concurrent*, if and only if no two of them are causally related or in conflict. It follows from the construction of $\mathcal{U}$ that no two of its places can simultaneously be causally related and in conflict.

Let $\mathcal{M}_I$ be the marking of $\mathcal{U}$, where each initial place contains exactly one token, and the remaining places contain no tokens. Consider the set of markings that $\mathcal{U}$ can reach starting from $\mathcal{M}_I$. If $\mathcal{M} \in [\mathcal{M}_I\rangle$ is the marking where exactly the places $b_1, \ldots, b_n$ are marked, then we let $fb(\mathcal{M})$ denote the marking of $\mathcal{N}$ where exactly $fb(b_1), \ldots, fb(b_n)$ are marked. For instance, $fb(\mathcal{M}_I) = M_I$ (that is, the initial marking of $\mathcal{N}$). The construction of $\mathcal{U}$ guarantees that if $\mathcal{M} \in [\mathcal{M}_I\rangle$, then $fb(\mathcal{M})$ is a reachable marking of $\mathcal{N}$. If, furthermore, $\mathcal{M}\,[e\rangle\,\mathcal{M}'$, then $fb(\mathcal{M})\,[fb(e)\rangle\,fb(\mathcal{M}')$.

In the reverse direction, if M is a reachable marking of $\mathcal{N}$, then $\mathcal{U}$ has a reachable marking $\mathcal{M}$ such that $fb(\mathcal{M}) = M$. If, furthermore, $M\,[t\rangle\,M'$, then $\mathcal{U}$ has a transition e and a marking $\mathcal{M}'$ such that $\mathcal{M}\,[e\rangle\,\mathcal{M}'$, $fb(e) = t$ and $fb(\mathcal{M}') = M'$. (To see that this holds it is helpful to first notice that the property "all simultaneously marked places are concurrent" is an invariant property of $\mathcal{U}$ with the initial marking $\mathcal{M}_I$.) Thus $\mathcal{U}$ contains complete information on the reachable part of the state space of $\mathcal{N}$.

The unfolding $\mathcal{U}$ is usually infinite and cannot be constructed as such. However, if $\mathcal{N}$ is finite, then $\mathcal{U}$ has at least one finite prefix that represents all reachable states of $\mathcal{N}$. An algorithm for constructing one such prefix was presented in [62]. Its result — let us call it the *finite unfolding* and denote it with $\mathcal{U}^{\text{fin}}$ — may be much smaller than the state space of $\mathcal{N}$. Some properties of $\mathcal{N}$ can be checked easily from $\mathcal{U}^{\text{fin}}$, such as the reachability of a state where a certain transition is enabled. Unfortunately, the checking of many seemingly simple properties from $\mathcal{U}^{\text{fin}}$ is quite hard, as the following example from [62] demonstrates.

Let $\varphi = \varphi_1 \wedge \cdots \wedge \varphi_n$ be a Boolean formula in 3-conjunctive normal form (that is, each φ_j is a disjunction of three different *literals*, where a literal is either a variable symbol or a negated variable symbol). Let the variables that occur in φ be $v_1, \ldots, v_m$. We will now construct a Petri net $\mathcal{N}_\varphi$ from φ. For each variable v_i the net contains two transitions t_i^{F} and t_i^{T} and three places p_i^I, p_i^{F} and p_i^{T}. There is an arc from p_i^I to t_i^{F}, from t_i^{F} to p_i^{F}, from p_i^I to t_i^{T}, and from t_i^{T} to p_i^{T}. For each conjunct $\varphi_j = l_{j,1} \vee l_{j,2} \vee l_{j,3}$ there are two transitions t_j^φ and t_j^∞ and a place p_j^φ. The place p_j^φ has input arcs from both t_j^φ and t_j^∞, and an output arc to t_j^∞. Finally, for each $k \in \{1,2,3\}$, t_j^φ has an input arc from p_i^{F} if and only if $l_{j,k} = v_i$, and p_i^{T} if and only if $l_{j,k} = \neg v_i$ (here "$\neg$" indeed goes together with "T"). The places p_i^I are initially marked, and the other places are not.

The unfolding of $\mathcal{N}_\varphi$ is the same as $\mathcal{N}_\varphi$ without the initial marking, except that each p_j^φ-t_j^∞-pair is replaced by an infinite linear chain of places and transitions. By cutting these chains after the second place, a finite unfolding $\mathcal{U}_\varphi^{\text{fin}}$ is obtained that represents all reachable markings of $\mathcal{N}_\varphi$. It can thus be constructed from φ in polynomial time.

Consider any marking where for each i, either p_i^{F} or p_i^{T} is marked, and the remaining places are empty. This marking is clearly reachable from the initial marking of $\mathcal{N}_\varphi$. If that marking corresponds to an assignment of truth values to $v_1, \ldots, v_m$ that makes φ false, then at least one φ_j is false, and t_j^φ is enabled. Otherwise all φ_j are true, and $\mathcal{N}_\varphi$ is in a deadlock. It is clear that in all other reachable markings some t_i^{T} or some t_j^∞ is enabled. Therefore, $\mathcal{N}_\varphi$ may deadlock if and only if φ is satisfiable. As a consequence, detecting deadlock from $\mathcal{N}_\varphi$, and thus also from $\mathcal{U}_\varphi^{\text{fin}}$, is NP-hard. So the existence of a worst-case polynomial time algorithm for deadlock detection from the finite unfolding is very unlikely.

On the other hand, it is trivial to check reachability of a deadlock from the state space of a system in linear time — linear *in the size of the state space*. In the case of $\mathcal{N}_\varphi$ (and many other systems, for that matter), linear in the size of the state space is exponential in the size of the system, because the state space clearly has an exponential number of states. So we see that the detection of deadlocks (most likely) contains an exponential step both with unfoldings and with state spaces, it is just that with unfoldings this step is after and with state spaces during the construction of the object that represents the behaviour of the system. As a matter of fact, making the very likely assumption that NP $\neq$ PSPACE, *every* deadlock detection method must contain a step that cannot be performed in (even nondeterministic) polynomial time.

The difficulty of finding a deadlock from the finite unfolding is thus a consequence of the facts that finding a deadlock is hard with *any* method, and, for a significant class of systems, the unfolding is small and can be constructed in polynomial time. We can interpret this by saying that the finite unfolding is an "intermediate" representation of behaviour that is between a Petri net and its state space. Because it is sometimes much smaller than the state space, and not everything is hard to check from it, it provides answers to some questions on some systems with much less time and memory than state spaces do. There are, however, many systems for which the finite unfolding is of exponential size. This is the case at least when an exponential number of transition occurrences is needed before a certain transition becomes enabled. A binary counter is a simple example.

For the practical use of unfoldings, it is important to distinguish properties that can be checked easily and efficiently enough. Although deadlock detection is hard in the worst case, [62] presents a heuristic algorithm that works often well. An alternative algorithm that uses integer linear-algebraic techniques was given in [64].

An algorithm for checking the validity of formulae in a certain simple temporal logic from the finite unfolding was developed in [24]. The logic consists of atomic propositions of the form "$M(p) = 1$", the Boolean connectives "$\wedge$" and "$\neg$", and the CTL operator "EF" (from which "$\vee$" and "AG" can be constructed). The publication [24] also distinguishes a non-trivial class of Petri nets for which the finite unfolding is always small and constructible in polynomial time, yielding polynomial time verification algorithms for formulae of the above logic that are in a certain form. (Of course, the binary counter cannot be modelled with such Petri nets, and the conversion of an arbitrary formula to the required form may increase its length exponentially.)

The original algorithm [62] for constructing a finite unfolding may produce an unnecessarily large result if the Petri net contains lots of conflicts, leading sometimes to finite unfoldings that are significantly bigger than the corresponding state spaces. An improved algorithm that avoids this problem was presented in [25].

Binary decision diagrams. An *(ordered) binary decision diagram* or *(O)BDD* [8] is a data structure for representing a set of bit vectors of equal length or, equivalently, a Boolean formula. It is a directed acyclic graph whose each vertex has either zero or exactly two successor vertices. Vertices with no output edges are labelled by "F" or "T". Each of the remaining vertices is labelled by a variable, and its output edges are labelled by "0" and "1". Exactly one vertex, the *root*, has no incoming edge. Figure 9 shows a BDD that represents the set $\{0011, 0111, 1011, 1100, 1101, 1110, 1111\}$, or the formula $(v_1 \wedge v_2) \vee (v_3 \wedge v_4)$. A vector $v_1 v_2 v_3 v_4$ is in the set if and only if the corresponding path from the root down through the BDD ends with "T", where the "corresponding" path is the one where the output edge from each node v_i is selected according to the value of v_i.

BDDs are usually made as small as possible by merging any two nodes that are roots of isomorphic sub-BDDs, and removing all nodes whose both output edges lead to isomorphic sub-BDDs. This can be done very efficiently by working bottom-up. The ordering in which the variables occur in a BDD is fixed.

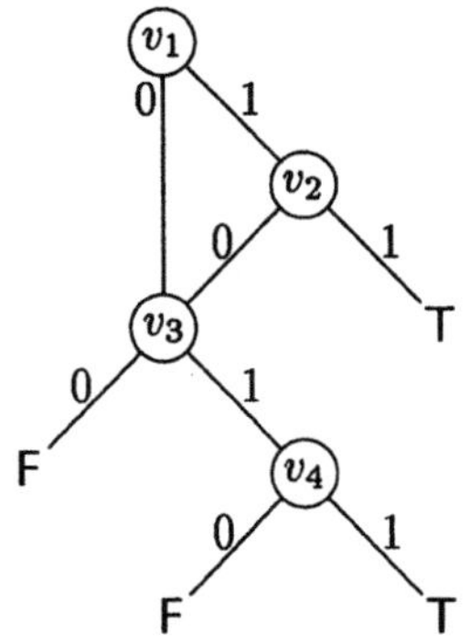

Fig. 9. A BDD example: $(v_1 \wedge v_2) \vee (v_3 \wedge v_4)$.

This has the consequence that each set has a unique fully reduced BDD representation and, furthermore, basic set operations including union, intersection, complementation, and equivalence test can be done efficiently with fully reduced BDDs. On the other hand, the size of a BDD (that is, the number of nodes in it) may depend crucially on the ordering of the variables, and it is difficult to know in advance whether a particular ordering would be good.

Several BDDs over the same set of variables and with the same ordering of those variables can be represented efficiently in one data structure by sharing identical sub-BDDs. This is useful in algorithms that manipulate BDDs.

With BDDs, it is natural to think of each structural transition t as defining a relation in $S \times S$ or, if the structural transitions are deterministic, a partial function $S \mapsto S$. This relation can be represented as a Boolean formula $R_t(s, s')$ where s and s' are the state before and after the occurrence of t. In these terms, the enabling condition of t is $\exists s' : R_t(s, s')$. The set of all structural transitions corresponds to the formula $R(s, s') \overset{\text{def}}{\Longleftrightarrow} \bigvee_{t \in T} R_t(s, s')$.

Assuming that the formula $S(s)$ describes some set of states, the formula $\exists s' : S(s') \wedge R(s', s)$ describes the set of those states that are reachable from the present set with one transition occurrence. A formula representing the set of all reachable states may be constructed by starting with a formula $S(s)$ that represents the set of initial states, and repeating the operation $S(s) := S(s) \vee \exists s' : S(s') \wedge R(s', s)$ until $S(s)$ does not change any more. (Alternatively, one can restrict the application of $R(s', s)$ to those reachable s' to which it has not yet been applied, or those that the previous application of $R(s', s)$ produced.) All these formulae can be represented with BDDs, and the operations needed in this approach can be performed efficiently in the sizes of the BDDs.

When the set $S(s)$ of reachable states is obtained, one can check whether it contains a forbidden state by describing forbidden states with a formula $F(s)$, computing $S(s) \wedge F(s)$, and checking whether the result is False. Other properties may be checked with more complicated algorithms, such as the algorithm for Petri net liveness in [67]. Perhaps the most important such algorithm is the *symbolic model checking* algorithm for CTL, LTL, weak bisimilarity and some other properties described in [9]. In it, the construction of the BDD proceeds backwards guided by the formula that is being checked. A BDD representing the set $G(s)$ of syntactically possible states that satisfy the formula is obtained as a result. One can then compute $I(s) \wedge \neg G(s)$ where $I(s)$ models the set of initial states to verify that the system satisfies the formula.

As was mentioned in Section 6, to obtain a small BDD it is necessary that the set of the reachable states of a system is regular in some informal and not well understood sense. (In the case of symbolic model checking it is the set of syntactically possible states that satisfy the formula that must be regular.) On the other hand, the total number of states is seldom an important factor of the BDD size. As a consequence, intermediate BDDs may be much larger than the final BDD in the above algorithm for constructing a BDD for the set of reachable states. For instance, [67] contains case studies where the biggest BDD has more than ten times as many nodes as the final BDD. Furthermore, systems exist which have no small BDDs. A combinatorial multiplier circuit is an often cited example. This problem has been attacked by developing variants of BDDs.

It is unclear whether it is useful to combine BDDs to other state space reduction methods. Most methods discussed in this article aim at reducing the number of states, but the number of states is not important with BDDs. On the other hand, the simultaneous use of another method may well pay off, if it makes the set of states more regular in the above informal sense. For instance, [3] contains a successful example of the combination of BDDs to a stubborn-set-type method.

BDDs have found lots of use especially in circuit design, as can be seen from the survey [63].

7.3 Methods Based on Process-Algebraic Compositionality

Compositional LTS construction. The goal of *compositional LTS construction* is to construct a reduced state space that is equivalent to the full state space in the sense of some process equivalence. The method is usually applied to process-algebraic verification of the kind in Section 4.4, but it can also be used for state-based verification.

The method can be illustrated with the example in Figure 10. The ordinary LTS of the example system can be computed according to the process-algebraic expression

$$P = \mathbf{hide}\, u, v, w, x, y\, \mathbf{in}(P_1 || P_2 || P_3 || P_4).$$

Due to the properties of hiding and parallel composition, the expression

$$\mathbf{hide}\, w, x\, \mathbf{in}\, (\mathbf{hide}\, u, v\, \mathbf{in}(P_1 || P_3) \;||\; \mathbf{hide}\, y\, \mathbf{in}(P_2 || P_4))$$

produces the same result (excluding the names of states). Let $\simeq$ be a process equivalence that is a congruence with respect to "**hide**" and "||". If *Red* is an algorithm that reduces an LTS but preserves $\simeq$, then $P \simeq P_{\mathsf{red}}$, where

$$P_{\mathsf{red}} = \textbf{hide}\, w, x\, \textbf{in}\big(Red(\textbf{hide}\, u, v\, \textbf{in}(P_1 || P_3)) \,||\, Red(\textbf{hide}\, y\, \textbf{in}(P_2 || P_4))\big).$$

Thus P_{red} can be used for verifying all the properties that are preserved by $\simeq$. On the other hand, the LTS P_{red} may be much smaller than the LTS P. (Suitable equivalences $\simeq$and reduction algorithms *Red* were discussed in Section 4.4).

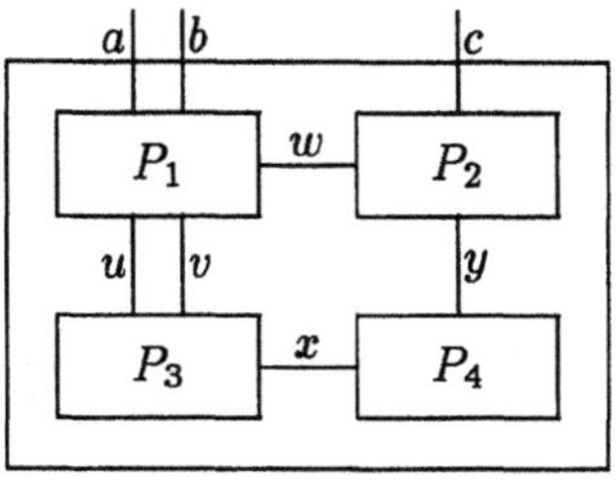

Fig. 10. A system consisting of four processes.

Compositional LTS construction can be applied hierarchically. For instance, if P_1 consists of several processes, then a compositionally constructed reduced LTS of it can be used in its place. Even if P_1 is a single process, it may be worthwhile to replace it by its reduced LTS.

Compositionality is inherent in process-algebraic theories, so it is difficult to give anyone credit of inventing compositional LTS construction. An early article where its use was explicitly suggested in a modern form is [58]. The article [90] is a tutorial on compositional LTS construction.

Compositional LTS construction is most often used for the verification of action-based properties of systems with synchronous interprocess communication, but it can be used also for state-based properties and with shared-variable communication. Shared variables can be handled by thinking of them as processes in their own right. Statements that read, test or modify the values of a shared variable are interpreted as actions that synchronise with the variable. When all processes that touch a shared variable have been added to the system, a special *initialisation* operator is used to give the variable its initial value and hide it from the interface of the (sub)system. This trick was applied to Petri net places in [88].

State-based properties can be handled by encoding state properties into action names as was described in Section 2.2. It is also possible to proceed as follows [88, 89]. Consider a state-based temporal logic formula φ such that its atomic propositions refer to the local states of processes $P_1, \ldots, P_n$, but not to the local states of processes $Q_1, \ldots, Q_m$. Then the validity of φ on $P_1 || \ldots || P_n ||$

$Q_1||\dots||Q_m$ can be checked by compositionally constructing a reduced LTS Q_{red} of $Q_1||\dots||Q_m$, and checking the validity of φ on $P_1||\dots||P_n \; || \; Q_{red}$. Of course, the equivalence that is preserved by the reduction must preserve the temporal logic in which φ was written. NDFD-equivalence preserves LTL_{-X} and branching bisimilarity has a variant that preserves CTL_{-X} and even CTL^*_{-X}.

The interface processes of Graf and Steffen. The ability of compositional LTS construction to save effort relies on the assumption that the intermediate LTSs are smaller than the full LTS of the whole system. This is not always the case. A typical example is a token-ring protocol where one token circulates in a ring consisting of n processes. To keep the example simple to discuss, we assume that a process without a token can be in one and the process with the token in two different states. Then the system as a whole has only $2n$ different states. The number of states is that small, because the system maintains the property that there is only one token in the ring. Any process that is not in the possession of the token is willing to input the token, but in all but one case the previous process in the ring does not have the token and is thus unable to give it.

Consider then an open-ended segment of the token ring consisting of $k < n$ processes. The input actions of the first process of the segment are now connected to nowhere. Therefore, if the first process does not have a token, then it can input one even if some other process in the segment has a token. As a consequence, the segment may contain up to k tokens, and may be in 3^k different states. Even for relatively small values of k, 3^k is much bigger than $2n$.

The problem in this example is that a subsystem that is isolated from its proper context may exhibit lots of "spurious" behaviour that it does not have when it is a part of the system as a whole. This problem was pointed out in [38], and essentially the following solution was suggested. The fact that the full system allows the segment to contain only one token at a time is modelled by augmenting the segment with an extra process I that monitors the input and output of the segment, and lets the segment contain at most one token at a time. That process might be called an *interface process*, because it monitors and restricts the interface of the subsystem. The process I allows the segment to input a token even if it already contains one, but then the input action leads to a specially marked "cut" state where no actions are possible. This restricts the number of states of the segment to $2k + 2$ — one state with no tokens, $2k$ states with one token and the special cut state with two tokens. Whenever the segment contains at most one token, I does not prevent any actions of the segment.

An LTS L of the system as a whole is constructed compositionally in the usual way, except that wherever needed, subsystems are augmented with suitable interface processes, and the special cut states are treated during LTS reduction as equivalent to each other and different from all other states. In L, inputting a second token should be impossible and the cut states should thus have disappeared. If this did not happen, then the user knows that some I was incorrect, and L is not valid for verifying the system. If the cut states did disappear, then the interface processes have no effect on the behaviour of the complete system, and L is equivalent to what would have been obtained without them.

Instead of cut states, [38] developed and used a theory of "undefinedness" predicates. The theory allows the merging of any states whose S_i-components are the same during the computation of the parallel composition of a subsystem S_i and its interface process I_i.

The use of compositional LTS construction in this example is clearly an overkill, because the system as a whole has so few states. The purpose of the example was only to illustrate the spurious behaviour problem, and the use of interface processes for avoiding it. In [38] the token ring in Figure 1 was used as an example. It has $9n2^{n-2}$ states, but if all actions except move tkn are hidden and the full LTS is reduced preserving weak bisimilarity, then the result contains only n states. Any attempt to use the ordinary compositional LTS method leads to LTSs that are bigger than the full LTS, but the biggest LTS encountered with interface processes has (according to the numerical experiments in [38]) only $4n + 4$ states.

The interface processes resemble on-the-fly verification in that they allow the user to give the verification algorithm information about the *expected* (not necessarily *real*) behaviour of the system that helps to keep the number of states small. The user cannot fool the algorithm to give wrong answers by giving incorrect information, because verification results are obtained only if cut states disappear and the information has thus proven correct.

Behavioural fixed points and network invariants. Sometimes process-algebraic compositionality makes it possible to verify an infinite family of similar systems of increasing size. The dining philosophers system provides a simple example. It was observed in [93] that when a CFFD-semantics-preserving reduced LTS was constructed for it with the compositional approach, the open-ended segment PF_4 with four philosophers and forks where only the interface actions at the ends are visible is CFFD-equivalent to the segment PF_3 consisting of three philosophers and forks. By adding k philosophers and forks to PF_4 and similarly to PF_3 we get the result that $PF_{k+4} \simeq_{\mathsf{CFFD}} PF_{k+3}$, which implies that $PF_i \simeq_{\mathsf{CFFD}} PF_3$ for any $i \geq 3$.

The system of n dining philosophers can be composed of one philosopher and fork that are connected to both ends of PF_{n-1}. As a consequence, the analysis results obtained from the four-philosopher system are valid also for all larger philosopher systems, provided that the analysis questions refer only to the state of one philosopher and to the fork on her left side (the fork that he takes first). However, that suffices for detecting deadlock and starvation in the philosopher system.

The situation where the behaviour of a system does not change when more components are added to it may be called a *behavioural fixed point*. The stronger a behavioural equivalence is, the less likely it is to yield a behavioural fixed point in a given system. One would thus expect behavioural fixed points to be found more often with trace, CFFD- and CSP-equivalences than with weak bisimilarity.

The behavioural fixed point method can also be used for the verification of systems with integer parameters that represent some upper bounds, as was demonstrated in [92]. The maximum number of transmission attempts that a sender process of a communication protocol will make before giving up is an example of such an upper bound. The basic idea is to model a counter that counts towards the upper bound as a stack of processes that may be added to the system one by one, thus increasing the upper bound. If a behavioural fixed point is found when the bound is k, then the system is independent of the value of the bound provided that it is at least k. This method was also used in [92] to model a certain fairness assumption by letting the counter count the number of times a certain unfavourable action (namely the loss of a message in a channel) has been chosen in a row instead of a favourable action.

Other methods that make it possible to verify an infinite family of systems of similar structure include [13, 54, 100].

7.4 Methods Based on Commutativity

Introduction to stubborn-set-type methods. The total effect of a set of concurrent actions is independent of the order in which the actions occur. As is obvious from the trivial "n non-interacting k-state processes" example of Section 1, the fact that the ordinary state space contains all possible orderings of concurrent actions is a major source of state explosion.

This observation has led many researchers to develop advanced state space methods where only some of the orderings are investigated, ideally only one ordering for each set of concurrent actions. In these methods, at each state that has been added to the reduced state space, only a subset of the semantic transitions out of that state are investigated. In other words, only a subset of structural transitions is used when constructing output edges for the state. The subset is chosen such that the occurrences of the remaining structural transitions can be postponed, or perhaps ignored altogether, without modifying the answers to the verification questions at hand.

Although the above idea is natural, a number of problems arise when trying to turn it into a working verification method. We will illustrate them with the aid of Figure 11.

- An enabled structural transition t may be in conflict with another structural transition t' that is disabled at the moment ("conflict" means, roughly speaking, that they are not concurrent even when both of them are enabled). This is the situation, for instance, with t_2 and t_3 of the example net in the marking $M_{2,4,7}$ where exactly p_2, p_4 and p_7 are marked. Executions where t' becomes enabled and occurs (perhaps disabling t) before the occurrence of t should not be ignored. Therefore, t' must somehow be taken into account in that state although it is disabled.
- Sometimes only one of two concurrent actions is taken in an execution that is relevant for verification. This happens, for instance, if we want to check the reachability of the above-mentioned marking $M_{2,4,7}$. Namely, initially t_1 and t_4 are concurrent, but only t_1 should occur to reach $M_{2,4,7}$. The method

must thus be able to favour one transition over another if that is important for the verification question.

- Some verification tasks require that the ordering of concurrent actions is partially preserved. For instance, to preserve the set of the process-algebraic traces of a system, the ordering of all transitions t such that $\mathcal{E}_\Sigma(t) \neq \tau$ must be preserved.
- The transitions that will be investigated in a state must be chosen reasonably efficiently, and the choice cannot rely on information on states that have not yet been investigated. For instance, it is obvious for a human who knows the net well that in $M_{2,4,7}$ it suffices to investigate t_4 alone but it is not sufficient to investigate t_2 alone, although both are enabled. An algorithm is needed for this kind of an analysis.
- The *ignorance* problem: If the investigation of some structural transition is postponed in each state of a cycle, then there is the risk that the transition is postponed forever. As an extreme example, if only t_6 is investigated in the initial marking of the example net, then no other markings will be constructed, and most of the behaviour of the net will be ignored.

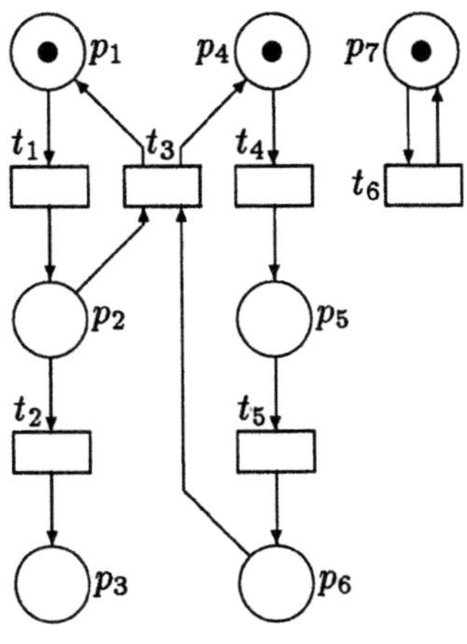

Fig. 11. A stubborn set example.

Different verification tasks require different answers to these problems, and different researchers have fine-tuned the details in different ways. This has led to the development of several methods under the names *stubborn sets* (since 1988 [81, 82]), *persistent sets* (since 1990 [33], although the term "persistent" is slightly more recent) and *ample sets* (since 1993 [68], but the idea was used in manual verification already in [49]). The persistent set approach is explained in detail in [34], and ample sets in [70]. These methods are so similar that it would perhaps be a good idea to think of them as variants of the same method. In this article we call them *stubborn-set-type methods*. We use mostly the basic technical definitions in [72, 98], because they both cover or generalise most of the definitions in the literature, and are intuitively reasonably clear.

Stubborn-set-type methods are often classified as *partial-order methods*, but in the opinion of the present author that is somewhat misleading. "Partial order" refers to the truly concurrent semantics known as *Mazurkiewicz traces* [61], in which the ordering of actions is, indeed, partial. It would be tempting to think of stubborn-set-type methods as attempting to construct one linearisation of each Mazurkiewicz trace of the system, taking advantage of the symmetric *independency* relation in the theory of Mazurkiewicz traces.

In reality, however, the methods construct *representatives* for executions of the system, where a representative is not necessarily a member of the same Mazurkiewicz trace as an execution that it represents. The construction of representatives is based on an asymmetric relation that is more liberal than the independency relation. Furthermore, a Mazurkiewicz trace that is not relevant for the verification question at hand may be left without representatives. We will return to this issue after presenting the basic stubborn set method.

Basic stubborn set method. A set $T_s \subseteq T$ of structural transitions is *dynamically stubborn* at state s_0, if and only if the following hold:

D1 If $t \in T_s$, $t_1, \ldots, t_n \notin T_s$, $s_0 -t_1 t_2 \cdots t_n\rightarrow s_n$, and $s_n -t\rightarrow s_n'$, then there is s_0' such that $s_0 -t\rightarrow s_0'$ and $s_0' -t_1 t_2 \cdots t_n \rightarrow s_n'$.

D2 There is at least one transition $t_k \in T_s$ such that if $t_1, \ldots, t_n \notin T_s$ and $s_0 -t_1 t_2 \cdots t_n \rightarrow s_n$, then $s_n -t_k\rightarrow$. The transition t_k is called a *key transition* of T_s at s.

By letting $n = 0$ in D2 we see that a key transition is enabled. A set of structural transitions is *strongly* dynamically stubborn at state s, if and only if it is dynamically stubborn at s, and all of its enabled transitions are key transitions, that is, they qualify as the t_k in D2. Strongly dynamically stubborn sets are an important subclass of dynamically stubborn sets, because most (but not all) known algorithms for stubborn set construction produce them, and some analysis algorithms require them.

Because a dynamically stubborn set must contain a key transition, deadlock states have no dynamically stubborn sets. (A deadlock state is a state with no enabled structural transitions.) If s is not a deadlock state, then the set T of all structural transitions is dynamically stubborn and even strongly dynamically stubborn at s.

Let $T_s : S \mapsto 2^T$ be a function that assigns to each non-deadlock state s a dynamically stubborn set $T_s(s)$. The *basic stubborn set method* starts with the initial states, and constructs for each state s that has been found so far only those output edges and immediate successor states that are obtained by firing the enabled structural transitions in $T_s(s)$.

Let s_d be a deadlock state and s any state in the reduced state space constructed with the basic stubborn set method. If $s -t_1 t_2 \cdots t_n \rightarrow s_d$, then D2 implies that at least one of $t_1, t_2, \ldots, t_n$ must belong to $T_s(s)$, because otherwise all key transitions of $T_s(s)$ would be enabled at s_d. Let i be as small as possible such that $t_i \in T_s(s)$. D1 implies that there is s' such that $s -t_i\rightarrow s'$, and s_d is reachable from s' with $n-1$ steps, namely by firing $t_1, \ldots, t_{i-1}, t_{i+1}, \ldots, t_n$. If s is an initial state, then by repeating this argument in s' and so on a total of n times we see that the reduced state space contains s_d. We have shown the hardest part of the following theorem:

A reduced state space that has been constructed with the basic stubborn set method (that is, for each state s in it, $T_s(s)$ satisfies D1 and D2) contains all deadlock states of the system that are reachable from the initial states. Furthermore, all deadlock states of the reduced state space are deadlock states of the system.

If the transitions are deterministic, then the following can be proven from D1 and D2:

D3 If T_s is dynamically stubborn at s, $t_1, t_2, \ldots \notin T_s$, t_k is a key transition of T_s at s, and $s -t_1 t_2 \cdots\rightarrow$, then there is s' such that $s -t_k\rightarrow s'$ and $s' -t_1 t_2 \cdots\rightarrow$.

From D1, D2 and D3 it is not difficult to prove that if the system has an infinite execution, then also its reduced state space obtained with the basic stubborn set method contains an infinite execution. Therefore, the basic stubborn set method can be used for checking whether a system with deterministic transitions may fail to terminate. The result can be applied also when transitions are not deterministic by first showing that the stubborn sets that are used satisfy D3.

A reduced state space that has been constructed with the basic stubborn set method contains an infinite execution if and only if the full state space contains an infinite execution, provided that either all structural transitions of the system are deterministic, or $T_s(s)$ satisfies D3 in each state s of the reduced state space.

On alternative definitions of stubborn sets. Several different definitions of stubborn (or ample or persistent) sets of different strength have appeared in the literature. We say that a definition of stubborn sets is *weaker* than another definition, if every set that is stubborn according to the latter is stubborn also according to the former, but not necessarily vice versa.

The weaker a definition of stubborn sets is, the more sets it classifies as stubborn, and the better are the chances that a stubborn set with very few enabled transitions may be found. As a consequence, weak definitions of stubborn sets have more potential for good reduction results than strictly stronger definitions. This has motivated researchers to try to find as weak definitions as possible without making the theory too complicated [36, 50, 72, 81, 83]. The conditions

D1 to D3 (essentially from [72]) were chosen for this article, because they are simple and among the weakest that have appeared in the literature.

As an example of the above, consider the situation where a process is ready to read from a non-empty fifo queue that other processes can access only by writing to it. Assume that the process has no alternative actions to the reading. Let t_{read} be the structural transition that corresponds to the reading. The conditions D1 to D3 classify the set $\{t_{\mathsf{read}}\}$ as dynamically stubborn (and even strongly). On the other hand, writing to and reading from a fifo are not independent in the "classic" sense used, for instance, in the theory of Mazurkiewicz traces, because sometimes the writing transition enables the reading transition. Therefore, $\{t_{\mathsf{read}}\}$ could not be stubborn according to the classic notion of independency. This problem does not affect D1 to D3, because we applied them in a state where t_{read} is known to be enabled.

As a consequence, a theory of stubborn sets based on the classic notion of independency would yield worse reduction results than the theory presented in this article. So the theory of Mazurkiewicz traces is not an optimal starting point for the development of stubborn-set-type methods, although it provides good background intuition.

Unlike an ample set [70] or persistent set [34], a stubborn set may contain disabled transitions. Disabled transitions might seem unnecessary, because they do not contribute to the set of output edges of a state. However, their presence simplifies the formulation of both the conditions V and L2 that will be presented later, and the notion of static stubborn sets that is useful for developing stubborn set construction algorithms. It was shown in [98] that if transitions are deterministic, then the non-empty persistent sets of [34] correspond precisely to the sets of enabled transitions of strongly dynamically stubborn sets.

Static definitions of stubborn sets. To implement the basic stubborn set method, it is necessary to design an algorithm that, given a non-deadlock state, produces the enabled transitions in a dynamically stubborn set of transitions. The set of all structural transitions is stubborn at every non-deadlock state, but it should be returned only as a last resort, because it will not yield any reduction of the number of output edges of the state. The definition of dynamical stubbornness does not directly lead to an algorithm, because it refers to states that are in the future of s and are thus not yet available.

To solve this problem, a static notion of a *stubborn set* of structural transitions has been defined. (Historically, stubborn sets were defined before dynamically stubborn sets.) This notion depends on the formalism used for modelling the system, and can be given several different definitions even for the same formalism, depending on how much effort one is willing to put into the analysis of the dependencies between transitions. What is important is that one must be able to prove for the chosen notion of stubborn sets that each stubborn set is also a dynamically stubborn set. A definition of "stubborn sets" is thus nothing but a static sufficient condition for a set being dynamically stubborn. Furthermore, if transitions are not deterministic and one wants to use stubborn sets for the detection of failure of termination, then one should show also that the sets satisfy D3.

In the case of ordinary Petri nets, the following is a possible simple definition of stubborn sets:[10]

- If $t \in T_s$ and $\neg M[t\rangle$, then there is $p \in \bullet t$ such that $M(p) < W(p,t)$ and $\bullet p \subseteq T_s$.
- If $t \in T_s$ and $M[t\rangle$, then $(\bullet t)\bullet \subseteq T_s$ or $\bullet(\bullet t) \subseteq T_s$.
- T_s contains a transition t_k such that $M[t_k\rangle$ and $(\bullet t_k)\bullet \subseteq T_s$.

This definition can be converted to a more restricted definition that allows only strongly dynamically stubborn sets simply by removing "or $\bullet(\bullet t) \subseteq T_s$".

Also the following, more complicated definition from [83] (with a small improvement from [98]) implies D1 and D2. It represents a more careful analysis of the dependencies between transitions. It is implied by the above simple definition but is not equivalent. Therefore, it accepts more sets as stubborn. So it can yield smaller stubborn sets than the above simple definition and thus produce better reduction results.

- If $t \in T_s$ and $\neg M[t\rangle$, then there is $p \in \bullet t$ such that $M(p) < W(p,t)$ and $\{ t' \mid W(p,t') < W(t',p) \wedge W(p,t') \leq M(p) \} \subseteq T_s$.
- If $t \in T_s$ and $M[t\rangle$, then for every $p \in \bullet t$, either $\{ t' \mid \min(W(t,p), W(t',p)) < \min(W(p,t), W(p,t')) \} \subseteq T_s$, or $\{ t' \mid \min(W(t,p), W(p,t')) < \min(W(p,t), W(t',p)) \} \subseteq T_s$.
- T_s contains a transition t_k such that $M[t_k\rangle$ and for every $p \in \bullet t_k$ $\{ t' \mid W(t',p) < \min(W(p,t_k), W(p,t')) \} \subseteq T_s$.

A strong version is obtained by replacing everything after and including the "or" with "T_s contains a transition t_k such that $M[t_k\rangle$."

The following is an example of a definition of stubborn sets with nondeterministic transitions, stated in the language of labelled transition systems (Section 4.4). It applies to the state $(s_1, \ldots, s_n)$ of $L = L_1 || \cdots || L_n$, where $L_i = (S_i, \Sigma_i, \Delta_i, S_{Ii})$ for $1 \leq i \leq n$, and "$s_i -a\rightarrow_i s_i'$" denotes that $(s_i, a, s_i') \in \Delta_i$. It produces only strongly dynamically stubborn sets, and guarantees also D3. To simplify the presentation of the definition, it is assumed that each τ-action is subscribed by the index of the component process that executes it. The stubborn set T_s is a subset of $\Sigma \cup \{\tau_1, \ldots, \tau_n\}$.

- If $a \in T_s$ and $\neg(s -a\rightarrow\ )$, then there is $1 \leq i \leq n$ such that $a \in \Sigma_i \cup \{\tau_i\}$, $\neg(s_i -a\rightarrow_i)$, and $\{ b \mid s_i -b\rightarrow_i \} \subseteq T_s$.
- If $a \in T_s$ and $s -a\rightarrow$, then for every $1 \leq i \leq n$, either $a \notin \Sigma_i \cup \{\tau_i\}$, or $\{ b \mid s_i -b\rightarrow_i \} \subseteq T_s$.

[10] "$\bullet x$" denotes the set of the input places or transitions of the transition or place x, "$x\bullet$" is the similar notion for output, and "$\bullet x$" and "$x\bullet$" are extended to sets by taking the union of the results for each member of the set. $W(p,t)$ and $W(t,p)$ are the numbers of tokens that the transition t consumes from and produces for the place p when it occurs.

– There is $a_k \in T_s$ such that $s -a_k\rightarrow$.

As was emphasised above, each of the above definitions is *static* in the sense that, unlike the earlier definition of dynamically stubborn sets, it refers to only one state. Given a state and a set of transitions, it is possible (and easy) to check whether the set satisfies the definition in the state.

When constructing a reduced state space with stubborn-set-type methods, it is necessary at each state to construct a "good" stubborn set, given only the state. It is not known what stubborn set would be the best regarding reduction results. For instance, [82] contains an example showing that always choosing the stubborn set with the smallest number of enabled transitions does not necessarily produce the smallest reduced state space. Even so, it is easy to believe that trying to keep the number of enabled transitions small is a reasonable heuristics. In particular, if the set of enabled transitions in T_{s1} is a proper subset of the set of enabled transitions in T_{s2}, then T_{s1} is preferable, because with it the basic stubborn set method cannot produce a bigger but may produce a smaller reduced state space than with T_{s2}.

Algorithms for constructing stubborn sets. Each of the above static definitions of stubborn sets can be thought of as consisting of a requirement of the form "T_s must contain a transition t_k with a certain property k", and of conditions of the form "if $t \in T_s$, then all transitions in either $f_1(s,t)$ or $f_2(s,t)$ or ...or $f_k(s,t)$ must be in T_s", where $f_1, \ldots, f_k$ are functions of the state and t. If we are ready to give up some of the generality of the above definitions, we can introduce a more or less arbitrary rule for choosing one of the f_i, given s and t. Then the above conditions span a binary relation between transitions that we will denote with "$\rightsquigarrow$": $t_1 \rightsquigarrow t_2 \overset{\text{def}}{\Longleftrightarrow} t_2 \in f_i(s,t_1)$.

For instance, in the case of the first of the above definitions, we may choose $t_1 \rightsquigarrow t_2 \overset{\text{def}}{\Longleftrightarrow} \bullet t_1 \cap \bullet t_2 \neq \emptyset$ when $M[t_1\rangle$, and $t_1 \rightsquigarrow t_2 \overset{\text{def}}{\Longleftrightarrow} t_2 \in \bullet p(M,t_1)$ when $\neg M[t_1\rangle$, where $p(M,t)$ is the smallest-numbered input place of t according to some fixed numbering of places such that $M\big(p(M,t)\big) < W\big(p(M,t),t\big)$.

The closure algorithm. The above notions lead to a simple sufficient graph-theoretic condition for stubborn sets: if a set is closed under "$\rightsquigarrow$" and it contains a transition that has the property k, then it is stubborn. Therefore, a stubborn set may be constructed by picking a transition t_k that satisfies k, and then performing a graph search (such as the depth-first search) in the graph $(T, \rightsquigarrow)$ starting at t_k.

This algorithm is easy to implement and reasonably fast: it consumes $O(|T| + |\rightsquigarrow|)$ time (but see the comment on speed below). The size of "$\rightsquigarrow$" depends on the modelling formalism. In the case of Petri nets it is at most $c^2|T|$, where c is the maximum number of input or output places or transitions of a transition or place. It can be reduced to $2c|T|$ by letting also the places act as vertices of the graph, yielding $O(c|T|)$ worst-case time.

Unfortunately, the quality of the stubborn sets this algorithm produces depends a lot on the choices of the start transitions.

The strong component algorithm. A better stubborn set is obtained by searching from $(T, \rightsquigarrow)$ for a maximal strongly connected component such that it contains a transition that satisfies k, and no other maximal strongly connected component reachable from it contains such a transition [81]. The stubborn set consists of the transitions in that component and all components reachable from it.

If the f_i are chosen such that each enabled transition automatically satisfies k, then this reduces to a search for a maximal strongly connected component that contains an enabled transition but its descendants do not (and then the resulting sets are strongly dynamically stubborn). For instance, with the above simple definition this means that if t is enabled, $(\bullet t)\bullet$ is used and not $\bullet(\bullet t)$.

In this application, it is better to search for the component with Tarjan's algorithm [79, 1] than with the more modern one described in [15]. This is because the former can be easily adapted to stop when a suitable component has been found even if most of $(T, \rightsquigarrow)$ has not been touched, whereas the latter requires that $(T, \rightsquigarrow)$ is searched completely through at least once. The former algorithm has thus much better best-case performance than the latter. Also this stubborn set construction algorithm consumes $O(|T| + |\rightsquigarrow|)$ or $O(c|T|)$ time in the worst case.

Instead of stopping immediately when the first suitable component has been found, the algorithm can be continued until it has found all suitable components. Then the one resulting in the smallest number of enabled transitions in the stubborn set may be chosen.

The deletion algorithm. The need to choose artificially one of the above-mentioned functions f_i can be avoided by spending more time in the construction of stubborn sets as follows [82]. One can take the different f_i into account by replacing $(T, \rightsquigarrow)$ by a more complicated graph containing "and"- and "or"-vertices. For instance, in the case of the above simple definition of stubborn sets for Petri nets, a disabled transition t corresponds to an or-vertex such that $t \rightsquigarrow p$ for every p such that $M(p) < W(p,t)$, and each such p corresponds to an and-vertex such that $p \rightsquigarrow t'$ for each $t' \in \bullet p$.

The algorithm starts with the full graph, and performs "removal searches". Each removal search starts by removing an enabled transition. When any vertex of the graph is removed, also those of its immediate "$\rightsquigarrow$"-predecessor vertices are removed that either are and-vertices, or are or-vertices and have lost all of their immediate "$\rightsquigarrow$"-successors. If the graph does not any more correspond to a stubborn set after a removal search, then the effects of the search are cancelled. Removal searches are continued as long as possible. The worst-case time consumption of this algorithm is $O(|T||\rightsquigarrow|)$. With the above Petri net definitions this makes $O(c|T|^2)$.

The time spent in constructing a stubborn set is an important factor in the time taken by the stubborn set method, because a stubborn set is constructed in every state that is investigated. Although all of the above algorithms have good asymptotic performance, they require a lot of analysis of "$\rightsquigarrow$".

In particular, the above definitions of stubborn sets of Petri nets may cause the algorithms to investigate long chains consisting of a disabled transition, its empty input place, a disabled input transition of that place, and so on. The above process-algebraic definition is better in this respect, because it causes the analysis to jump directly to the location where the control of the process is at the moment, instead of stepping one disabled transition at a time. It can be further improved by precomputing information on the reachability between the states of the same component LTS, and not making the jump if the start point of the jump is not reachable from the end point of the jump.

It may sometimes be more efficient to use simple heuristics for constructing a stubborn set rapidly, and use the set of all transitions when the heuristics fail. This may lead to worse stubborn sets and a bigger reduced state space, but may pay back in time consumption if the heuristics are fast enough. For instance, in [34] an algorithm is investigated that, in essence, computes the set of transitions that are reachable from an enabled transition with "$\leadsto$", until either the set is ready, or a disabled transition is encountered. In the latter case, the algorithm gives up and returns the set of all transitions.

When the stubborn set method is combined to packed state space methods such as the symmetry method, the problem of constructing good stubborn sets fast enough becomes more challenging. For instance, stubborn sets for a coloured Petri net may be constructed by first unfolding the net to an ordinary Petri net and then using the above methods, but one would not like to do this, because the ordinary Petri net may be much bigger than the original coloured Petri net. Unfortunately, in [53] it is shown that unfolding or something equally expensive is sometimes unavoidable in the construction of "good" stubborn sets for coloured Petri nets. It is, however, possible to find "good" stubborn sets in time that is proportional to the size of the coloured Petri net, if the net is given additional structure [53].

Another result of this kind is the definition of stubborn sets in [51]. It works for parallel labelled transition systems that are extended with symbolic data.

Stubborn set methods for safety properties. Because of the above-mentioned "ignorance" problem, the basic stubborn set method is not good for checking other properties than those that directly relate to the termination of the system. The addition of the following assumption makes the basic stubborn set method capable of analysing various safety properties.

S For every state s in the reduced state space and every $t \in T$, if $s -t\rightarrow$, then there is an execution $s_0 -t_1\rightarrow s_1 -t_2\rightarrow \cdots -t_n\rightarrow s_n$ such that $s_0 = s$, $t \in T_s(s_n)$, and t_i is a key transition of $T_s(s_{i-1})$ for $1 \leq i \leq n$.

We say that a structural transition t is *ignored* at s in the reduced state space, if a violation of S occurs with s and t. Ignoring a structural transition resembles being weakly unfair to it, but is not the same thing. Weak fairness states a requirement for all futures, whereas t is not ignored if there is at least one future of a particular kind.

If the conditions D1, D2 and S hold in a reduced state space, then the following claims hold [83]. They make it possible to check a number of safety properties from a reduced state space. The last three of the claims are corollaries of the first one. By a *terminal component* we mean a maximal strongly connected component such that each edge that starts in the component also ends in it.

- If $s -t_1 t_2 \cdots t_n \rightarrow s'$ is a finite execution that starts at a state s that is in the reduced state space, then there are structural transitions $t_{n+1}, \ldots, t_{n+k}$ ($k \geq 0$) and a state s'' such that $s' -t_{n+1} \cdots t_{n+k} \rightarrow s''$, and the reduced state space contains the execution $s -t_{\pi(1)} t_{\pi(2)} \cdots t_{\pi(n+k)} \rightarrow s''$ for some permutation $t_{\pi(1)}, t_{\pi(2)}, \ldots, t_{\pi(n+k)}$ of $t_1, t_2, \ldots, t_{n+k}$.
- A structural transition t labels an edge in the reduced state space if and only if t labels an edge in the full state space.
- A structural transition t is Petri-net-live (Section 2.3) if and only if t is Petri-net-live in the reduced state space.
- If the full state space is finite and contains a terminal component C, then the reduced state space contains a terminal component C_{red} such that $C_{\mathsf{red}} \subseteq C$, and a structural transition t occurs in C_{red} if and only if it occurs in C. In the reverse direction, each terminal component C_{red} of the reduced state space is a subset of some terminal component C of the full state space such that t occurs in C_{red} if and only if it occurs in C.

The second of the above results makes it possible to verify various linear-time safety properties with the fact transition technique and tester processes described in Section 4.2. In this approach it is important that the tester process synchronises only with a subset of the transitions of the system. This is because the addition of the tester introduces new dependencies between the transitions it synchronises with. If the tester is synchronised with every transition, then it is almost always the case that the stubborn set must contain all enabled transitions, leading to no reduction in the size of the state space.

A strong stubborn set algorithm for ensuring the condition S. In the case of strongly stubborn sets, the condition S reduces to "if s is in the reduced state space and $s -t\rightarrow$, then t is investigated in at least one of the states that are reachable from s in the reduced state space". This observation leads to an efficient algorithm for ensuring that S holds [83]. In it, the reduced state space is constructed in depth-first order. Tarjan's algorithm is used during the construction to recognise terminal components of the state space.

Assume that a terminal component has just been completed, and the construction of the reduced state space is just about to backtrack from a state s in the component to a state outside the component (or is about to terminate, if s has no predecessor states in the depth-first search tree). We will say that the structural transition t is *ignored in the component,* if it is enabled in s but occurs nowhere in the component. The algorithm checks that no structural transition is ignored in the component. If this does not hold, the stubborn set $T_s(s)$ that was used in s is extended such that at least one structural transition that was ignored in the component is included into the extended set, and the extended set is stubborn.

From the point of view of the depth-first search discipline and Tarjan's algorithm, the extension of $T_s(s)$ adds new output edges to s, and does that exactly when the original output edges have been scanned through. Because the depth-first search has not yet made any actions after investigating the original output edges of s, from its point of view the newly added output edges could have been there all the time. Therefore, the extension of $T_s(s)$ does not confuse the depth-first search discipline or Tarjan's algorithm.

One possibility of computing the extension of $T_s(s)$ is to compute a new stubborn set that contains a transition that is ignored in the component. This is correct, because it is easy to check from D1 to D3 that the union of two stubborn sets is stubborn. If both the original $T_s(s)$ and its extension are constructed with the closure or strong component algorithm using the same relation "$\leadsto$", then the construction of the extension needs not enter the original stubborn set.

More details of this algorithm are given in [83].

Preserving process-algebraic traces. The above algorithm makes it possible to check on the fly whether $L_1 \sqsupseteq_{\mathsf{tr}} L_2$ with the tester processes described in Section 4.2. It is, however, sometimes beneficial to have an algorithm that preserves process-algebraic traces and is *transparent* in the sense of Section 6. This is the case, for instance, if the LTS is intended to be used as a component in the compositional LTS construction method of Section 7.3. This goal is obtained by stating an additional requirement for the stubborn sets $T_s(s)$ that are used in the states s of the reduced state space:

V If $T_s(s)$ contains a structural transition t such that $s -t\rightarrow$ and $\mathcal{E}_\Sigma(t) \neq \tau$, then $T_s(s)$ contains all those structural transitions t' such that $\mathcal{E}_\Sigma(t') \neq \tau$ (even if t' is disabled).

The condition is labelled "V" because it talks about the handling of *visible* transitions, that is, those structural transitions whose $\mathcal{E}_\Sigma$-abstraction is not τ. It ensures that when the stubborn set method produces a representative for an execution, it does not change the relative ordering of visible transitions. Assuming that D1, D2, S and V are satisfied, then the reduced state space is trace-equivalent with the full state space.

There are at least two practical ways of implementing V. The more complicated one consists of first trying to find a stubborn set which contains no enabled visible transitions with, for instance, the strong component algorithm. If that fails, then the closure algorithm is applied to every visible transition and the union of the results is taken. The simpler way consists of just adding to "$\leadsto$" an edge from t_1 to t_2 for every enabled visible t_1 and visible t_2 before attempting to construct a stubborn set.

If all transitions are visible, then V forces the reduced state space to be the same as the full state space. Therefore, no reduction is obtained unless there are invisible transitions.

The above-mentioned on-the-fly method of checking "$\sqsupseteq_{\rm tr}$" with tester processes does not need the condition V. Therefore, it might seem to allow for better stubborn sets. Sometimes it indeed does that, but perhaps not as often as one might think, because the addition of the tester process to the system introduces new dependencies between transitions. In particular, if the tester process is synchronised only with the visible actions and it never refuses any visible actions, then each state $s_{\rm t}$ of the tester process has for every $a \in \Sigma$ an output edge that is labelled with a. In most (but not all) cases this implies that if one output action of $s_{\rm t}$ is taken into the stubborn set, then all of them must be taken, which is equivalent to the extension of "$\leadsto$" described above. If the tester is synchronised with all actions instead of just visible ones, then the situation is even worse, because then almost all actions become dependent of each other.

It is essential for V that a stubborn set may contain disabled transitions. Instead of V, [70] uses the strictly stronger condition V′ that either $T_{\mathbf{s}}(s)$ contains no enabled visible transitions, or $T_{\mathbf{s}}(s) = T$. This condition leads to worse reduction results than V, but is easier to implement.

Other ways of handling safety properties. An alternative approach to the verification of safety properties that is based on a transformation of the system description was presented in [37]. Safety properties can be analysed also with the CFFD-, $\mathrm{LTL}_{-\mathrm{X}}$-, and branching-time-preserving stubborn set methods that are described below.

Stubborn set methods for liveness properties. The above stubborn set methods for safety properties suffice for the verification of various properties whose validity depend on finite executions. However, they do not suffice for properties that depend on infinite executions, as the example in Figure 12 demonstrates. In the example, a and b are visible actions, and τ_1, τ_2 and τ_3 are invisible. If the dashed edges are removed, the resulting reduced state space satisfies D1, D2, D3, S and V. However, according to it it is guaranteed that if a is executed, then also b will be executed, although this is not true in the full state space.

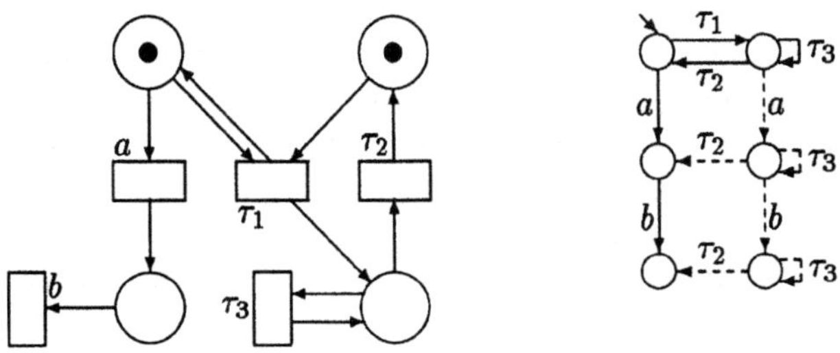

Fig. 12. Safety-preserving stubborn set methods do not suffice for liveness.

The following two conditions make the stubborn set method capable of handling various linear-time liveness properties. They replace the condition S.

L1 If s is a state in the reduced state space such that there is $t \in T$ such that $\mathcal{E}_\Sigma(t) = \tau$ and $s -t\rightarrow$ in the full state space, then $T_s(s)$ has a key transition t_k such that $\mathcal{E}_\Sigma(t_k) = \tau$.

L2 If $s_0 -t_1\rightarrow s_1 -t_2\rightarrow \cdots$ is an infinite execution in the reduced state space starting at any state s_0, then for each t_v such that $\mathcal{E}_\Sigma(t_v) \neq \tau$, there is $i \geq 0$ such that $t_v \in T_s(s_i)$. (This implies that there are actually infinitely many such i.)

Regarding the execution $M_I -\tau_1 a \tau_3^\omega\rightarrow$ of the system in Figure 12, L2 ensures that a is investigated after τ_1 (the infinite execution in L2 is $M_I -\tau_1\tau_3^\omega\rightarrow$). L1 ensures that τ_3 is investigated after a.

The following theorem is slightly strengthened from a similar one in [86]:

- If s is a state in a reduced state space that satisfies the conditions D1, D2, D3, V, L1 and L2, and if $s -t_1t_2\cdots\rightarrow$ is an infinite execution that starts at s, then the reduced state space contains an infinite execution $s -t_1't_2'\cdots\rightarrow$ such that $\mathcal{E}_\Sigma(t_1't_2'\cdots) = \mathcal{E}_\Sigma(t_1t_2\cdots)$.
- If s is a state in a reduced state space that satisfies the conditions D1, D2 and V, and if $s -t_1t_2\cdots t_n\rightarrow s_d$ is a finite execution that starts at s and ends in a deadlock state, then the reduced state space contains a finite execution $s -t_1't_2'\cdots t_n'\rightarrow s_d$ such that $\mathcal{E}_\Sigma(t_1't_2'\cdots t_n') = \mathcal{E}_\Sigma(t_1t_2\cdots t_n)$, and $t_1't_2'\cdots t_n'$ is a permutation of $t_1t_2\cdots t_n$.

The condition L1 can be implemented by first constructing a stubborn set without worrying about L1, and then extending L1 to contain an invisible key transition if necessary. The closure and strong component algorithms are handy for computing the extension. If the condition V is replaced by the stronger condition V' mentioned above, then L1 holds automatically and needs not be worried of.

With finite state spaces, L2 is equivalent to the condition that for each visible transition t_v, each cycle of the state space must contain at least one state s such that $T_s(s)$ contains t_v. Perhaps surprisingly, t_v needs not be enabled in s; it suffices that it affects the construction of the stubborn set. This condition is difficult to implement in its full generality. It is, however, easily implementable, if we do not mind using the same s for each t_v. Then cycles that violate it can be recognised with the non-progress cycle detection algorithm that was explained in Section 4.2. The stubborn set of the current state is extended with the "$\leadsto$"-closure of visible transitions if necessary. More detail can be found in [85, 86].

Another possibility is to construct the reduced state space in depth-first order, and extend the stubborn set each time the current state has an output edge to any state in the depth-first stack [68, 70]. (Only the edges in the reduced state space are taken into account in this test.) This algorithm is much simpler than the previous one, but extends stubborn sets more often.

Like with V and V′, it is possible to simplify the implementation of L2 at the cost of less reduction by requiring that $T_s(s) = T$ in the states s whose stubborn sets are extended [68, 70]. We will call this condition L2′.

Stubborn set methods for LTL_{-X}. The above theorem yields a transparent construction-time reduction method that preserves the validity of LTL_{-X}-formulae (or stuttering-insensitive LTL formulae). Let Π be the set of atomic propositions that appear in the formulae. We say that a structural transition t *affects* a proposition $P \in \Pi$ if and only if the reachable part of the state space contains states s and s' such that $s -t\rightarrow s'$, and either $s \models P$ and $s' \models \neg P$, or $s \models \neg P$ and $s' \models P$. That is, t affects P if and only if some occurrence of t changes the truth value of P.

Let T_Π be a set of structural transitions such that it contains *at least* those transitions that affect any $P \in \Pi$. The set T_Π is allowed to be larger than the precise set of transitions that affect members of Π, because the precise set may be difficult to find, and the use of a larger set does not sacrifice correctness (although it may lead to worse reduction results). The theorem that underlies the LTL_{-X}-preserving stubborn set method is as follows:

> If all transitions in T_Π are treated as visible and a reduced state space is constructed such that the conditions D1, D2, D3, V, L1 and L2 are satisfied, then the truth value of any LTL_{-X}-formula whose atomic propositions belong to Π is the same in the reduced and full state spaces.

Due to the condition V, the larger Π is, the larger will the reduced state space usually be. If φ is of the form $\varphi_1 \wedge \cdots \wedge \varphi_k$, then this effect may be fought against by verifying each φ_i separately. Then more than one reduced state space is constructed. However, with some luck each φ_i uses a smaller set of visible transitions and leads to a much smaller reduced state space than φ, so significant savings in total effort are possible (but not guaranteed). Unfortunately, this technique does not yield correct results for formulae of the form $\varphi_1 \vee \varphi_2$.

A more general technique for distributing the set of visible transitions was suggested in [68]. That paper assumed that the underlying model of computation is fair in a certain particular sense; we will return to this assumption a bit later. The LTL_{-X}-formula φ in question is converted into a form $B(\varphi_1, \ldots, \varphi_k)$, where $B(x_1, \ldots, x_k)$ represents some Boolean combination of $x_1, \ldots, x_k$, and each φ_i is as small as possible. For instance, $\Box(\varphi \wedge \Diamond\psi)$ can be converted to $(\Box\varphi) \wedge (\Box\Diamond\psi)$.

In this technique, only one reduced state space is constructed. Even so, instead of one set of visible transitions, each subformula has its own set, and the condition V is applied to each set separately. If t_1 only affects atomic propositions in φ_1 and t_2 only those in φ_2 and both t_1 and t_2 are enabled, then the original condition V insists that if t_1 is in the stubborn set, then also t_2 must be. On the other hand, this requirement disappears when V is applied to each subformula separately. This leads to smaller stubborn sets and better reduction results.

If V is applied to each subformula separately, then the proper treatment of the condition L1 becomes complicated. This problem did not arise in [68], because the fairness assumption in it made L1 unnecessary. On the other hand, the assumption rejects the execution $M_I -\tau_1 a \tau_3^{\omega}\rightarrow$ of the system in Figure 12 as unfair, so we are not always willing to make it. The thesis [98] solves this problem by showing how both V and L1 can be distributed to subformulae, so that the fairness assumption of [68] is not any more needed.

Handling fairness assumptions. The correctness of a system with respect to an LTL$_{-X}$-formula that expresses some liveness property often depends on some fairness assumptions (Section 2.4) about the system. As the following examples illustrate, customary fairness assumptions are not easy to handle with the LTL$_{-X}$-preserving stubborn set method, because the method may choose an unfair representative to a fair execution that violates a property.

Consider the alleged property $\Diamond(M(p_5) = 1)$ of the net on the left in Figure 13, when strong fairness towards t_5 is assumed. In the situation where exactly p_2 and p_3 are marked, a typical implementation of the LTL$_{-X}$-preserving stubborn set method would fire t_3 and only it, because $\{t_3\}$ is then stubborn, but no set not containing t_3 is stubborn. The situation is symmetric when exactly p_1 and p_4 are marked. As a consequence, in each infinite execution in the reduced state space, t_5 is enabled infinitely often. They are thus unfair and do not qualify as counterexamples to $\Diamond(M(p_5) = 1)$. So the reduced state space would lead to the conclusion that $\Diamond(M(p_5) = 1)$ holds. On the other hand, the full state space contains the fair counterexample $M_I -(t_1 t_2 t_3 t_4)^{\omega}\rightarrow$.

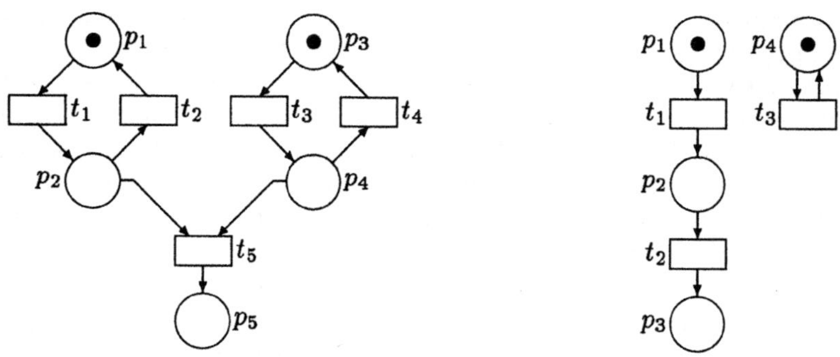

Fig. 13. Two fairness examples.

The net on the right in the figure exemplifies a similar problem with weak fairness. Let the formula be $\Diamond(M(p_3) = 1)$, and assume weak fairness towards t_2. The LTL$_{-X}$-preserving stubborn set method may choose $\{t_1\}$ as the stubborn set used in the initial marking. Unfortunately, if it does that, then it looses the only fair counterexample to the formula.

This problem can be avoided by representing the fairness assumptions as a part of the formula to be verified in the form "$(\mathsf{fair}_1 \wedge \cdots \wedge \mathsf{fair}_n) \Rightarrow \mathsf{property}$".

Then the ordinary $\mathrm{LTL}_{-\mathrm{X}}$-preserving stubborn set method is guaranteed to treat the fairness assumptions correctly, because they need no special treatment. This technique has, however, the problem that with it, very many transitions must be visible, which leads to bad reduction results. Fortunately, "$(\mathsf{fair}_1 \wedge \cdots \wedge \mathsf{fair}_n) \Rightarrow \mathsf{property}$" is a Boolean combination of "fair_1", ..., "fair_n", and "$\mathsf{property}$", so the above-mentioned techniques of [68, 98] can be used to alleviate the problem.

Another possibility is to use fairness assumptions that are insensitive to the difference between a sequence and its representative chosen by the stubborn-set-type method. With this goal in mind, in [68, 70] the following atypical but natural fairness assumption was suggested. Furthermore, the $\mathrm{LTL}_{-\mathrm{X}}$-preserving stubborn set method was proven correct when it and L2′ are assumed and the classic notion of "dependency" is used, even if the conditions L1 and D3 are dropped:

[68, 70]-fairness If a structural transition t is enabled continuously from some state on in an execution, then some transition that is dependent on t — perhaps t itself — occurs in the execution somewhere after that state. This is required from every $t \in T$.

Stubborn set methods for failure-based process semantics. A reduced state space that is constructed according to the $\mathrm{LTL}_{-\mathrm{X}}$-preserving stubborn set method — that is, obeying D1, D2, D3, V, L1 and L2 — is CFFD-equivalent (and thus also CSP- and NDFD-equivalent) with the full state space, assuming that all transitions t such that $\mathcal{E}_\Sigma(t) \neq \tau$ are treated as visible by the conditions V, L1 and L2. Therefore, the $\mathrm{LTL}_{-\mathrm{X}}$-preserving stubborn set method can also be used as a method of computing reduced parallel compositions when any of these three semantics is used.

When using the method, it is useful to notice that if the system is of the form

$$\mathbf{hide}\, a_1, \ldots, a_k \,\mathbf{in} (L_1 || \cdots || L_n),$$

then the transitions of $L_1 || \cdots || L_n$ labelled with a_1 or ... or a_k need not be considered visible by V, L1 and L2, because they are invisible at the system level, although they are visible from the point of view of an individual L_i.

The use of stubborn sets is beneficial even if the resulting state space will be further reduced with some process-algebraic LTS reduction algorithm. This is because the stubborn set method reduces the risk of the parallel composition being so big that it cannot be constructed or processed by the LTS reduction algorithm.

Worth mentioning is the fact that if a reduced state space obeys D1, D2, V and L1, then it contains all the stable states and all the stable failures of the original state space. Furthermore, if it obeys D1, D2, D3, V and L1, then it is CSP-equivalent with the full state space. In other words, L2 is not needed to preserve the CSP-semantics of a state space. This is an interesting consequence of the fact that, as was explained in Section 4.4, divergence is "catastrophic" in the CSP-semantics.

Stubborn set methods for process algebras were surveyed in [91].

On-the-fly methods. The use of automata-theoretic on-the-fly verification methods simultaneously with stubborn-set-type methods was discussed already in connection with safety-property-preserving stubborn set methods. Those Büchi automata that are connected to the system with the state synchronisation technique and do not restrict the transitions of the system (Section 4.2) are invisible to the stubborn set method. Therefore, Büchi automata can be used simultaneously with the LTL_{-X}-preserving stubborn set method. This idea has been developed further in [69].

An alternative approach is to connect the Büchi automaton (or tester process, Section 4.2) with transition fusion to only the visible transitions of the system. The automaton is now allowed to restrict the behaviour of the system. This idea was investigated in [87].

When verifying an LTL_{-X}-formula, it is often the case that an atomic proposition becomes irrelevant for the formula at some point of an execution. This is the case, for instance, with the formula $\Box(P \Rightarrow \Box\neg Q)$: when a state s such that $P \in \mathcal{E}_\Pi(s)$ has been seen, P does not matter any more, but the only interesting thing is whether Q may become true. An advanced on-the-fly method that exploits this fact was developed in [52].

A stubborn set method for branching time properties. A stubborn-set-type method that preserves the validity of CTL^*_{-X}-formulae was presented in [31], and adapted to nondeterministic structural transitions in [91]. To preserve the validity of CTL^*_{-X}-formulae, it is important that the points in time when decisions between different futures are made are located correctly with respect to each other and relative to $\mathcal{E}_\Pi$-abstracted states. This is exemplified by Figure 14 and the CTL-formula $\mathsf{EF}(\alpha \wedge (\mathsf{EF}\beta) \wedge (\mathsf{EG}\neg\beta))$. The formula holds in the net in the figure. The transitions t_2 and t_3 are invisible with respect to it, so the LTL_{-X}-preserving stubborn set method would allow to choose $\{t_2, t_3\}$ as the stubborn set that is used in the initial marking. That would, however, lead to a reduced state space where the formula does not hold, because in it the choice between t_2 and t_3 occurs too early.

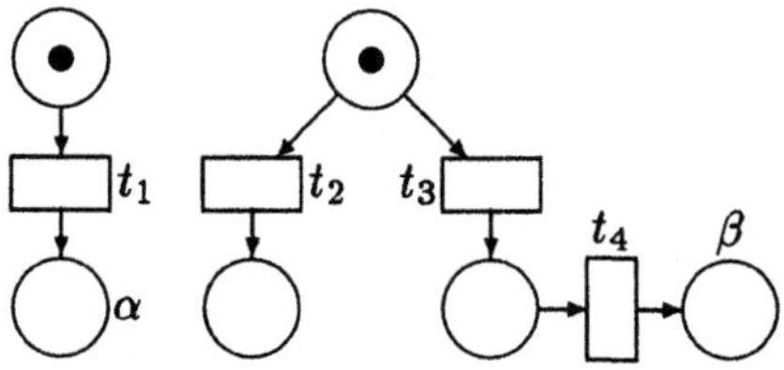

Fig. 14. A branching time example.

Because of the above problem, the CTL^*_{-X}-preserving stubborn set method is based on the following, rather strong condition:

B If s is a state in the reduced state space, then the stubborn set $T_s(s)$ either contains only one enabled structural transition, and that transition is invisible; or it contains all enabled structural transitions.

The following theorem expresses the correctness of the $\mathrm{CTL}^*_{-\mathrm{X}}$-preserving stubborn set method:

> If all structural transitions are deterministic, all structural transitions in T_Π are treated as visible, and a reduced state space is constructed such that the conditions D1, D2, S, and B are satisfied, then the truth value of any $\mathrm{CTL}^*_{-\mathrm{X}}$-formula whose atomic propositions belong to Π is the same in the reduced and full state spaces.

The construction of stubborn sets that satisfy D1, D2 and B is simple, because only the set of all structural transitions and the sets containing exactly one enabled structural transition need be taken into account. One may first scan through all enabled invisible structural transitions with the hope of finding one that, together with a suitable set of disabled structural transitions, satisfies D1 and D2. If that fails, then all structural transitions are taken.

Also the implementation of S is easy in this context, because states that do not have all enabled structural transitions in their stubborn sets have exactly one successor state. Such states may be constructed in a linear sequence. S is violated whenever the sequence enters any of its states anew. Then an arbitrary state of the sequence can be picked, and its output edges re-examined using this time the set of all transitions.

Allowing nondeterministic transitions. The condition B does not suffice with nondeterministic structural transitions, because it does not rule out the possibility that the same transition may occur in two different ways leading to two different futures. In [91] a stronger condition was suggested that is based on the notion of ***super-determinism***. We use the labelled transition system notation (Section 4.4) for explaining it, because nondeterministic transitions (or actions) are frequently used with them.

An action $a \in \Sigma$ is super-deterministic in the state s_0, if and only if the existence of an execution $s_0 -a_1\rightarrow \cdots -a_n\rightarrow s_n$ where $a_1, \ldots, a_n \neq a$ implies the existence of states $s'_0, \ldots, s'_n$ such that $s'_0 -a_1\rightarrow \cdots -a_n\rightarrow s'_n$, and for every $0 \leq i \leq n$: $s_i -a\rightarrow s'_i$ and if $s_i -a\rightarrow s''$, then $s'' = s'_i$. Less formally, a super-deterministic action is enabled, can yield only one result when it occurs even if other actions are executed before it, the execution of other actions cannot disable it, and its execution commutes with the execution of the other actions.

NB If s is a state in the reduced state space, then the stubborn set $T_s(s)$ either contains only one enabled structural transition, and that transition is invisible and super-deterministic; or it contains all enabled structural transitions.

The condition NB implies D1 and D2. Therefore, the $\text{CTL}^*_{-\text{X}}$-preserving stubborn set method can be formulated as follows in the presence of nondeterministic transitions:

> If all structural transitions in T_Π are treated as visible, and a reduced state space is constructed such that the conditions S and NB are satisfied, then the truth value of any $\text{CTL}^*_{-\text{X}}$-formula whose atomic propositions belong to Π is the same in the reduced and full state spaces.

If an LTS is constructed with the $\text{CTL}^*_{-\text{X}}$-preserving stubborn set method, then it is branching bisimilar and thus also weakly bisimilar with the ordinary LTS of the system, assuming that all transitions t such that $\mathcal{E}_\Sigma(t) \neq \tau$ are treated as visible.

The following sufficient condition for super-determinism [91] can be used in implementing the $\text{CTL}^*_{-\text{X}}$-preserving stubborn set method in the presence of nondeterministic transitions. Let $L_1 || \cdots || L_n$ be a parallel composition of LTSs, and $(s_1, \ldots, s_n)$ its state. Assume that $(s_1, \ldots, s_n) -a\rightarrow$. If for each $1 \leq i \leq n$, either $a \notin \Sigma_i \cup \{\tau_i\}$, or the local state s_i has only one output transition in the LTS L_i, then a is super-deterministic in $(s_1, \ldots, s_n)$. In other words, a is super-deterministic in $(s_1, \ldots, s_n)$, if each L_i that is interested in a-actions is ready to do only one transition, and that transition is labelled with a.

On the strength of the $\text{CTL}^*_{-\text{X}}$-preserving method. Stubborn set methods demonstrate nicely the principle that the more information one is willing to give away, the more powerful state space reduction methods are available. In this section we started with a method that is valid only for the verification of termination-related properties, and ended up with a method for full $\text{CTL}^*_{-\text{X}}$ and branching bisimilarity. Each time when we moved from a method to the next, we added one or more new restrictions on the construction of stubborn sets.

Although the relative strengths of the conditions underlying the methods are not always formally comparable, it is intuitively clear that B and NB prevent the use of a small stubborn set much more often than D3, V, L1 and L2, so the ability of the $\text{CTL}^*_{-\text{X}}$-preserving method of reducing state spaces is smaller than that of the other methods. On the other hand, it is easy to produce a fast implementation for the $\text{CTL}^*_{-\text{X}}$-preserving method without losing much of its generality, whereas the attempt to exploit the full power of the safety-property-preserving and $\text{LTL}_{-\text{X}}$-preserving methods leads to complicated and significantly slower algorithms. It is not known how significant the difference between the reduction power of the $\text{CTL}^*_{-\text{X}}$- and $\text{LTL}_{-\text{X}}$-preserving methods is in practice.

There are two important stuttering-insensitive branching-time specification formalisms that are strictly less powerful than $\text{CTL}^*_{-\text{X}}$ and branching bisimilarity, namely $\text{CTL}_{-\text{X}}$ and weak bisimilarity. It would be interesting to find a stubborn-set-type method that would preserve one of these, and would improve reduction results by not preserving $\text{CTL}^*_{-\text{X}}$ and branching bisimilarity. Unfortunately, the example in Figure 14 is valid also for $\text{CTL}_{-\text{X}}$ and weak bisimilarity, so the chances of finding such a method seem small.

Sleep sets. Sleep sets were first presented in [33], and perhaps the best source on them is [34]. Their theory has not been developed as far as that of stubborn sets. While stubborn-set-type methods save effort by postponing the investigation of structural transitions to future states, sleep sets avoid the investigation of transitions that have been investigated in the past states.

Assume that t_1 and t_2 are enabled and independent of each other in some state s, and neither of them has been put to *sleep* in the sense described below. Unlike stubborn-set-type methods, the sleep set method investigates both t_1 and t_2 at s. Assume that t_1 is investigated before t_2. Then, when investigating t_2, t_1 is put to sleep. Sleeping transitions will not be taken into account in future states until they are woken up. A sleeping transitions is woken up for certain reasons, such as the occurrence of a transition that is dependent on it.

As such, the sleep set method does not reduce the number of reachable states, it reduces only the number of edges. However, it can be used simultaneously with stubborn-set-type methods, and the combination gives better reduction results (in terms of reachable states) than the stubborn-set-type methods alone.

As was pointed out in [35], sleep sets can also be used to improve the *state space caching* method of [45]. The basic idea of state space caching is simply to start throwing reachable states away from the memory of the state space construction tool when the memory fills up. If a state has more than one input edge, as is very common in the presence of concurrency, and if it is thrown away before investigating all of its input edges, then it and at least a subset of its successor states will be constructed more than once. This is bearable if the full state space is not much bigger than what fits the memory, but leads to dramatic decrease of the performance of the algorithm if the full state space is big enough. Sleep sets reduce the number of times a state is entered, and therefore reduce the number of times a state is re-constructed. An impressive example of the power of sleep sets in this application is given in [34].

8 Conclusions

Many different techniques based on widely diverse principles have been suggested for alleviating the state explosion problem. In this article we have discussed a number of them. None of today's techniques solves the state explosion problem once and for all, and, because of the results regarding the computational complexity of typical verification tasks, it is unlikely that a perfect solution will be found in the near future. Even so, today's techniques facilitate the analysis, validation and verification of much bigger systems than ordinary state space construction. Although they are not strong and easy enough to be routinely used in all production of concurrent software and hardware, they can help a lot in finding serious design flaws in systems, and the proving of the correctness of various critical details.

An advanced state space method practically always restricts the set of properties that can be analysed or verified. This implies that the user of the method has to have at least some idea of the analysis questions before starting the tool.

If, during the analysis, new questions become relevant, it may be that a new reduced state space has to be constructed to answer them. This is a drawback compared to the use of full state spaces, where the same state space can be used for answering all analysis questions, even if the user invents new questions after obtaining the answers to the original ones. On the other hand, when the full state space of the system is too big to be constructed, it is better to get answers one at a time than to get no answers at all.

Because many advanced state space methods work well only with certain kinds of analysis questions, the user of the methods has to decide what kinds of questions will be asked. The smaller set of question types the user is happy with, the more methods are available for alleviating state explosion. Some arguments in favour of and against certain question types are given in the following.

Sensitivity to stuttering. Stuttering is usually not a problem for methods based on packed state spaces (perhaps excluding the unfolding method). In contrast, methods that are based on commutativity (Section 7.4) or process-algebraic compositionality (Section 7.3) are valid only for stuttering-insensitive properties.

Because the level of atomicity is often artificial to some extent, and because knowledge of how many actions of a parallel, independent process were scheduled between two actions of an interesting process in a one-processor machine is unimportant, many researchers consider stuttering as irrelevant in the context of concurrent systems.

Linear vs. branching time. Excluding CTL* and CTL^*_{-X}, the most common branching-time specification and query formalisms have fast algorithms for the last step of the verification of a property. With linear-time formalisms the algorithms for the last step tend to take exponential time in the length of the description of the property in the worst case (with the exception of preorder checking against a deterministic process-algebraic specification).

One may argue, however, that the bad worst-case performance of linear-time model and preorder checking is not an important problem, because the description of the property is often small, at least compared to the size of the system. Furthermore, linear-time formalisms allow the use of more powerful state space reduction methods than branching-time formalisms. For instance, we pointed out in Section 5.3 that labelled transition systems can be reduced more with failure-based semantic models than with weak bisimilarity. To give another example, the CTL^*_{-X}-preserving stubborn set method allows the use of a non-trivial stubborn set (that is, one that does not contain all enabled transitions) much less often than the LTL_{-X}-preserving method. Finally, it is difficult to think what a branching-time version of Holzmann's supertrace (Section 7.2) might look like.

As a consequence, although linear-time formalisms lose in the checking of the property from the reduced state space, they may win a lot during the construction of the reduced state space. It is thus difficult to say which approach is faster in practice. This question was analysed in more detail in [89].

Linear-time formalisms have also the advantage that if a property does not hold, then a counterexample that the user can easily simulate and analyse can be provided.

The verification of linear-time liveness properties needs often that fairness assumptions are made and taken into account. Fairness assumptions are problematic to certain advanced state space methods, including the process-algebraic compositional LTS construction in Section 7.3 and the $\mathrm{LTL}_{-\mathsf{X}}$-preserving stubborn set method in Section 7.4. Petri-net-liveness, or more generally the CTL "AG EF", gives to some extent similar information as linear-time liveness, although they are not precisely the same: linear-time liveness guarantees that, say, a structural transition t will occur in the future, whereas Petri-net-liveness only promises that the future occurrence of t is possible and cannot be made impossible.

Petri-net-liveness may, therefore, fall a bit short from what the user wants. It does not, however, need fairness assumptions; we saw in Section 7.4 that it is preserved by the safety-property-preserving stubborn set method; and it is within the scope of the unfolding method in Section 7.2. It is thus a particularly interesting branching-time operator from the point of view of alleviating state explosion.

State- vs. action-based. In many cases it is easy to switch from a state-based specification to an action-based one or vice versa with the mappings of Section 2.2. In those cases the choice between a state- or action-based formalism does not have much significance from the point of view of advanced verification methods.

There are, however, situations where advanced verification methods clearly favour one approach over the other. For instance, we saw in Section 7.2 that checking whether a certain transition can ever occur is easy for the unfolding method, but checking the reachability of a state with a certain property is difficult, unless the property can be handily encoded as the enabling condition of one or few transitions. It seems that the majority of advanced state space methods either handle arbitrary atomic propositions over states without difficulties (for instance, binary decision diagrams and Holzmann's supertrace), or require that the properties are ultimately stated in terms of actions (for instance, visible transitions in the stubborn-set-type methods). In the former case actions can be talked about by describing their enabling condition and the change of state they cause. It thus seems that action-based formalisms allow the use of more state space methods.

Action-based approaches often provide automatic insensitivity to stuttering. For instance, if a Büchi automaton is connected to the system with fusion with visible transitions instead of synchronisation to every state, then the automaton does not see stuttering. The acceptance condition of the automaton must then be modified so that executions where the automaton makes only a finite number of transitions are handled properly. This is not a big problem, because something similar has to be done in any case with executions that lead to deadlock states. We saw in Section 4.2 that the action-based approach allows faster on-the-fly detection of errors by facilitating the use of four different kinds of acceptance states for different types of erroneous executions.

We started this article by comparing state space methods to theorem proving. The goal of that was to highlight the advantages and disadvantages of state space methods. Despite the development of advanced state space methods, theorem proving is still much stronger in certain tasks, such as the proving of parameterised results. Sometimes best results are obtained by combining state space methods and theorem proving, for instance, by using a model checker as a sub-tool of a theorem prover, or debugging a system with state space methods and fixed parameter values, but conducting the final correctness proof with a theorem prover and symbolic parameters.

Acknowledgement Numerous researchers have contributed to this article by developing the original ideas, and communicating them to me through their books, papers, talks and discussions. I am afraid that here and there I was unable to trace the true source of an idea, and was unable to give the original inventor the recognition that (s)he deserves — please accept my apologies. Jaco Geldenhuys and Ilkka Kokkarinen helped me to reduce the number of errors in this article.

References

1. Aho, A. V., Hopcroft, J. E. & Ullman, J. D.: *The Design and Analysis of Computer Algorithms.* Addison-Wesley 1974, 470 p.
2. Alpern, B. & Schneider, F. B.: "Defining Liveness". *Information Processing Letters* 21(4), 1985, pp. 181–185.
3. Alur, R., Brayton, R. K., Henzinger, T. A., Qadeer, S. & Rajamani, S. K.: "Partial-Order Reduction in Symbolic State Space Exploration". *Proc. Computer Aided Verification (CAV) '97,* Lecture Notes in Computer Science 1254, Springer-Verlag 1997, pp. 340–351.
4. Bolognesi, T. & Brinksma, E.: "Introduction to the ISO Specification Language LOTOS". *Computer Networks and ISDN Systems* 14 (1987), pp. 25–59.
5. Brinksma, E.: "A Theory for the Derivation of Tests". *Protocol Specification, Testing and Verification VIII* (Proc. International IFIP WG 6.1 Symposium, 1988), North-Holland 1988, pp. 63–74.
6. Brookes, S. D., Hoare, C. A. R. & Roscoe, A. W.: "A Theory of Communicating Sequential Processes". *Journal of the ACM,* 31 (3) 1984, pp. 560–599.
7. Browne, M. C., Clarke, E. M. & Grumberg, O.: "Characterizing Finite Kripke Structures in Propositional Temporal Logic". *Theoretical Computer Science* 59, 1988, pp. 115–131.
8. Bryant, R. E.: "Graph-Based Algorithms for Boolean Function Manipulation". IEEE Transactions on Computers C-35 (8) 1986, pp. 677–691.
9. Burch, J. R., Clarke E. M., McMillan K. L., Dill D. L. & Hwang, L. J.: "Symbolic Model Checking: 10^{20} States and Beyond". Information and Computation 98 (2) 1992, pp. 142–170.
10. Chandy, K. M. & Misra, J.: *Parallel Program Design: A Foundation.* Addison-Wesley 1988, 516 p.

11. Clarke, E. M. & Emerson, E. A.: "Design and Synthesis of Synchronization Skeletons using Branching Time Temporal Logic". *Proc. Workshop on Logics of Programs*, Lecture Notes in Computer Science 131, Springer-Verlag 1981, pp. 52–71.
12. Clarke, E. M., Filkorn, T. & Jha, S.: "Exploiting Symmetry in Temporal Logic Model Checking". *Proc. Computer-Aided Verification (CAV) '93*, Lecture Notes in Computer Science 697, Springer-Verlag 1993, pp. 450–462.
13. Clarke, E. M., Grumberg, O. & Jha, S.: "Verifying Parameterized Networks using Abstraction and Regular Languages". *Proc. CONCUR '95, 6th International Conference on Concurrency Theory*, Lecture Notes in Computer Science 962, Springer-Verlag 1995, pp. 395–407.
14. Cleaveland, R. & Hennessy, M.: "Testing Equivalence as a Bisimulation Equivalence". *Formal Aspects of Computing*, 5 (1) 1993, pp. 1–20.
15. Cormen, T. H., Leiserson, C. E. & Rivest, R. L.: *Introduction to Algorithms*. The MIT Press, 1990, 1028 p.
16. Courcoubetis, C., Vardi, M., Wolper, P. & Yannakakis, M.: "Memory-Efficient Algorithms for the Verification of Temporal Properties", *Formal Methods in System Design* 1 (1992), pp. 275–288.
17. Desel, J. & Esparza, J.: *Free Choice Petri Nets*. Cambridge Tracts in Theoretical Computer Science 40, Cambridge University Press 1995, 244 p.
18. Eloranta, J., Tienari, M. & Valmari, A.: "Essential Transitions to Bisimulation Equivalences". *Theoretical Computer Science* 179 (1997) pp. 397–419.
19. Emerson, E. A.: "Temporal and Modal Logic". *Handbook of Theoretical Computer Science, Volume B: Formal Models and Semantics*, Elsevier Science Publishers 1990, pp. 995–1072.
20. Emerson, E. A., & Halpern, J. Y.: "'Sometimes' and 'Not Never' Revisited: on Branching Versus Linear Time Temporal Logic". *Journal of the ACM* 33 (1) 1986, pp. 151–178.
21. Emerson, E. A., & Lei, C.-L.: "Modalities for Model Checking: Branching Time Strikes Back". *Science of Computer Programming*, 8, 1987, pp. 275–306.
22. Emerson, E. A. & Sistla, A. P.: "Symmetry and Model Checking". *Proc. Computer-Aided Verification (CAV) '93*, Lecture Notes in Computer Science 697, Springer-Verlag 1993, pp. 463–477.
23. Emerson, E. A. & Sistla, A. P.: "Utilizing Symmetry when Model-Checking under Fairness Assumptions: An Automata-Theoretic Approach". *ACM Transactions on Programming Languages and Systems*, 19 (4) 1997, pp. 617–638.
24. Esparza, J.: "Model Checking Using Net Unfoldings". *Science of Computer Programming* (1994) 23: 151–195.
25. Esparza, J., Römer, S. & Vogler, W.: "An Improvement of McMillan's Unfolding Algorithm". *Proc. Tools and Algorithms for the Construction and Analysis of Systems '96*, Lecture Notes in Computer Science 1055, Springer-Verlag 1996, pp. 87–106.
26. Fernandez, J.-C.: "An Implementation of an Efficient Algorithm for Bisimulation Equivalence". *Science of Computer Programming* 13 (1989/90) pp. 219–236.
27. Finkel, A.: "The Minimal Coverability Graph for Petri Nets". *Advances in Petri Nets 1993*, Lecture Notes in Computer Science 674, pp. 210–243.
28. Francez, N.: *Fairness*. Springer-Verlag 1986, 295 p.
29. Francez, N.: *Program Verification*. Addison-Wesley 1992, 312 p.
30. Garey, M. R. & Johnson, D. S.: *Computers and Intractability: A Guide to the Theory of NP-Completeness*. W. H. Freeman and Company, 1979, 340 p.

31. Gerth, R., Kuiper, R., Peled, D. & Penczek, W.: "A Partial Order Approach to Branching Time Logic Model Checking". *Proc. Third Israel Symposium on the Theory of Computing and Systems*, IEEE 1995, pp. 130–139.
32. Gerth, R., Peled, D., Vardi, M. & Wolper, P.: "Simple On-the-fly Automatic Verification of Linear Temporal Logic". *Proc. Protocol Specification, Testing and Verification 1995*, Chapman & Hall 1995, pp. 3–18.
33. Godefroid, P.: "Using Partial Orders to Improve Automatic Verification Methods". *Proc. Computer-Aided Verification 90*, AMS-ACM DIMACS Series in Discrete Mathematics and Theoretical Computer Science, Vol. 3, 1991, pp. 321–340.
34. Godefroid, P.: *Partial-Order Methods for the Verification of Concurrent Systems, An Approach to the State-Explosion Problem.* Lecture Notes in Computer Science 1032, Springer-Verlag 1996, 143 p. (Earlier version: Ph.D. Thesis, University of Liège, 1994.)
35. Godefroid, P., Holzmann, G. J. & Pirottin, D.: "State Space Caching Revisited". *Proc. Computer-Aided Verification (CAV) '92*, Lecture Notes in Computer Science 663, Springer-Verlag 1993, pp. 178–191.
36. Godefroid, P. & Pirottin, D.: "Refining Dependencies Improves Partial-Order Verification Methods". *Proc. Computer-Aided Verification (CAV) '93*, Lecture Notes in Computer Science 697, Springer-Verlag 1993, pp. 438–449.
37. Godefroid, P., & Wolper, P.: "Using Partial Orders for the Efficient Verification of Deadlock Freedom and Safety Properties". *Proc. Computer Aided Verification (CAV) '91*, Lecture Notes in Computer Science 575, Springer-Verlag 1992, pp. 332–342.
38. Graf, S. & Steffen, B.: "Compositional Minimization of Finite State Processes". *Proc. Computer-Aided Verification '90*, AMS-ACM DIMACS Series in Discrete Mathematics and Theoretical Computer Science, Vol. 3, 1991, pp. 57–73.
39. Gyuris, V. & Sistla, P.: "On-the-Fly Model Checking Under Fairness That Exploits Symmetry". *Proc. Computer Aided Verification (CAV) '97*, Lecture Notes in Computer Science 1254, Springer-Verlag 1997, pp. 232–243.
40. Haddad, S.: "A Reduction Theory for Coloured Nets". *Advances in Petri Nets 1989*, Lecture Notes in Computer Science 424, Springer-Verlag 1990, pp. 209–235. Also in *High-level Petri Nets. Theory and Application*, Springer-Verlag 1991, pp. 399–425.
41. Hennessy, M.: "Acceptance Trees". *Journal of the ACM*, 32 (4) 1985, pp. 896–928.
42. Hoare, C. A. R.: *Communicating Sequential Processes.* Prentice-Hall 1985, 256 p.
43. Holzmann, G. J.: *Design and Validation of Computer Protocols.* Prentice-Hall 1991, 500 p.
44. *ISO 8807 International Standard: Information processing systems — Open Systems Interconnection — LOTOS — A formal description technique based on the temporal ordering of observational behaviour.* International Organization for Standardization 1989, 142 p.
45. Jard, C. & Jéron, T.: "Bounded-memory Algorithms for Verification On-the-fly". *Proc. Computer Aided Verification (CAV) '91*, Lecture Notes in Computer Science 575, Springer-Verlag 1992, pp. 192–202.
46. Jensen, K.: *Coloured Petri Nets. Volume 2, Analysis Methods.* Monographs in Theoretical Computer Science, Springer-Verlag 1995, 174 p.
47. Kaivola, R. & Valmari, A.: "The Weakest Compositional Semantic Equivalence Preserving Nexttime-less Linear Temporal Logic". *Proc. CONCUR '92, Third International Conference on Concurrency Theory*, Lecture Notes in Computer Science 630, Springer-Verlag 1992, pp. 207–221.

48. Kanellakis, P. C. & Smolka, S. A.: "CCS Expressions, Finite State Processes, and Three Problems of Equivalence". *Information and Computation* 86 (1990) pp. 43–68.
49. Katz, S. & Peled, D.: "An Efficient Verification Method for Parallel and Distributed Programs". *Proc. Linear Time, Branching Time and Partial Order in Logics and Models for Concurrency 1988*, Lecture Notes in Computer Science 354, Springer-Verlag 1989, pp. 489–507.
50. Katz, S. & Peled, D.: "Defining Conditional Independence Using Collapses". *Theoretical Computer Science* 101 (1992), pp. 337–359.
51. Kokkarinen, I.: *A Verification-Oriented Theory of Data in Labelled Transition Systems.* Ph.D. Thesis, Tampere University of Technology Publications 234, Tampere, Finland 1998, 105 p.
52. Kokkarinen, I., Peled, D. & Valmari, A.: "Relaxed Visibility Enhances Partial Order Reduction". *Proc. Computer Aided Verification (CAV) '97*, Lecture Notes in Computer Science 1254, Springer-Verlag 1997, pp. 328–339.
53. Kristensen, L. M. & Valmari, A.: "Finding Stubborn Sets of Coloured Petri Nets Without Unfolding". To appear in *Proc. International Conference on Application and Theory of Petri Nets*, 1998, 20 p.
54. Kurshan, R. P., Merritt, M., Orda, A. & Sachs, S. R.: "A Structural Linearization Principle for Processes". *Formal Methods in System Design* 5, 1994, pp. 227–244.
55. Lamport, L.: "Proving the Correctness of Multiprocess Programs". *IEEE Transactions on Software Engineering*, SE-3(2), 1977, pp. 125–143.
56. Lamport, L. & Lynch, N.: "Distributed Computing: Models and Methods". *Handbook of Theoretical Computer Science, Volume B: Formal Models and Semantics*, Elsevier Science Publishers 1990, pp. 1157–1199.
57. Lichtenstein, O. & Pnueli, A.: "Checking that Finite State Concurrent Programs Satisfy Their Linear Specifications". *Proc. 12th ACM Symposium on Principles of Programming Languages*, 1985, pp. 97–107.
58. Madelaine, E. & Vergamini, D.: "AUTO: A Verification Tool for Distributed Systems Using Reduction of Finite Automata Networks". *Proc. Formal Description Techniques II (FORTE '89)*, North-Holland 1990, pp. 61–66.
59. Manna, Z. & Pnueli, A.: *The Temporal Logic of Reactive and Concurrent Systems, Volume I: Specification.* Springer-Verlag 1992, 427 p.
60. Manna, Z. & Pnueli, A.: *Temporal Verification of Reactive Systems, Volume II: Safety.* Springer-Verlag 1995, 512 p.
61. Mazurkiewicz, A.: "Trace Theory". *Petri Nets: Applications and Relationships to Other Models of Concurrency*, Lecture Notes in Computer Science 255, Springer-Verlag 1987, pp. 279–324.
62. McMillan, K.: "Using Unfoldings to Avoid the State Explosion Problem in the Verification of Asynchronous Circuits". *Proc. Computer-Aided Verification (CAV) '92*, Lecture Notes in Computer Science 663, Springer-Verlag 1993, pp. 164–177.
63. Meinel, C. & Theobald, T.: "Ordered Binary Decision Diagrams and Their Significance in Computer-Aided Design of VLSI Circuits". *Bulletin of the European Association for Theoretical Computer Science* 64, 1998, pp. 171–187.
64. Melzer, S. & Römer, S.: "Deadlock Checking Using Net Unfoldings". *Proc. Computer Aided Verification (CAV) '97*, Lecture Notes in Computer Science 1254, Springer-Verlag 1997, pp. 352–363.
65. Milner, R.: *Communication and Concurrency.* Prentice-Hall 1989, 260 p.
66. Park, D.: "Concurrency and Automata on Infinite Sequences". *Theoretical Computer Science: 5th GI-Conference*, Lecture Notes in Computer Science 104, Springer-Verlag 1981, pp. 167–183.

67. Pastor, E., Roig, O., Cortadella, J. & Badia, R.: "Petri Net Analysis Using Boolean Manipulation". *Proc. Application and Theory of Petri Nets 1994*, Lecture Notes in Computer Science 815, Springer-Verlag 1994, pp. 416-435.
68. Peled, D.: "All from One, One for All: On Model Checking Using Representatives". *Proc. Computer-Aided Verification (CAV) '93*, Lecture Notes in Computer Science 697, Springer-Verlag 1993, pp. 409–423.
69. Peled, D.: "Combining Partial Order Reductions with On-the-fly Model-Checking". *Formal Methods in System Design* 8 (1) 1996: 39–64.
70. Peled, D.: "Partial Order Reduction: Linear and Branching Temporal Logics and Process Algebras". *Proc. POMIV'96, Workshop on Partial Order Methods in Verification*, DIMACS Series in Discrete Mathematics and Theoretical Computer Science Vol. 29, American Mathematical Society 1997, pp. 233–257.
71. Puhakka, A. & Valmari, A.: "Verification of Self-Synchronizing Alternating Bit Protocols with ARA". *Proc. Fifth Symposium on Programming Languages and Software Tools*, University of Helsinki, Department of Computer Science, Report C-1997-37, pp. 167–178.
72. Rauhamaa, M.: *A Comparative Study of Methods for Efficient Reachability Analysis.* Lic.Tech. Thesis, Helsinki University of Technology, Digital Systems Laboratory, Research Report A-14, Espoo, Finland 1990, 61 p.
73. Reisig, W.: *Petri Nets, An Introduction.* EATCS Monographs on Theoretical Computer Science, Vol. 4, Springer-Verlag 1985, 161 p.
74. Roscoe, A. W.: "Model-Checking CSP". *A Classical Mind: Essays in Honour of C. A. R. Hoare*, Prentice-Hall 1994, pp. 353–378.
75. Roscoe, A. W.: *The Theory and Practice of Concurrency.* Prentice-Hall 1998, 565 p.
76. Savitch, W. J.: "Relationships Between Nondeterministic and Deterministic Tape Complexities". *Journal of Computer and System Sciences* 4, 1970, pp. 177–192.
77. Shatz, S. M., Tu, S., Murata, T. & Duri, S.: "Application of Petri Net Reduction for Ada Tasking Deadlock Analysis". *IEEE Transactions on Parallel and Distributed Systems* 7 (12) 1996, pp. 1307–1322.
78. Sistla, A. P. & Clarke, E. M.: "The Complexity of Propositional Linear Temporal Logics". *Journal of the ACM* 32 (3) 1985, pp. 733–749.
79. Tarjan, R. E.: "Depth-first Search and Linear Graph Algorithms". *SIAM Journal on Computing*, 1 (2) 1972, pp. 146–160.
80. Thomas, W.: "Automata on Infinite Objects". *Handbook of Theoretical Computer Science, Volume B: Formal Models and Semantics*, Elsevier Science Publishers 1990, pp. 133–191.
81. Valmari, A.: "Error Detection by Reduced Reachability Graph Generation". *Proc. 9th European Workshop on Application and Theory of Petri Nets*, 1988, pp. 95–112.
82. Valmari, A.: *State Space Generation: Efficiency and Practicality.* Ph.D. Thesis, Tampere University of Technology Publications 55, Tampere, Finland 1988, 169 p.
83. Valmari, A.: "Stubborn Sets for Reduced State Space Generation". *Advances in Petri Nets 1990*, Lecture Notes in Computer Science 483, Springer-Verlag 1991, pp. 491–515.
84. Valmari, A.: "Stubborn Sets of Coloured Petri Nets". *Proc. 12th International Conference on Application and Theory of Petri Nets*, 1991, pp. 102–121.
85. Valmari, A.: *Alleviating State Explosion during Verification of Behavioural Equivalence.* Department of Computer Science, University of Helsinki, Report A-1992-4, Helsinki, Finland 1992, 57 p.

86. Valmari, A.: "A Stubborn Attack on State Explosion". *Formal Methods in System Design*, 1: 297–322 (1992).
87. Valmari, A.: "On-the-fly Verification with Stubborn Sets". *Proc. Computer-Aided Verification (CAV) '93*, Lecture Notes in Computer Science 697, Springer-Verlag 1993, pp. 397–408.
88. Valmari, A.: "Compositional Analysis with Place-Bordered Subnets". *Proc. Application and Theory of Petri Nets 1994*, Lecture Notes in Computer Science 815, Springer-Verlag 1994, pp. 531–547.
89. Valmari, A.: "Failure-based Equivalences Are Faster Than Many Believe". *Proc. Structures in Concurrency Theory 1995*, Springer-Verlag "Workshops in Computing" series, 1995, pp. 326–340.
90. Valmari, A.: "Compositionality in State Space Verification Methods". Invited talk, *Proc. Application and Theory of Petri Nets 1996*, Lecture Notes in Computer Science 1091, Springer-Verlag 1996, pp. 29–56.
91. Valmari, A.: "Stubborn Set Methods for Process Algebras". *Proc. POMIV'96, Workshop on Partial Order Methods in Verification*, DIMACS Series in Discrete Mathematics and Theoretical Computer Science Vol. 29, American Mathematical Society 1997, pp. 213–231.
92. Valmari, A. & Kokkarinen, I.: "Unbounded Verification Results by Finite-State Compositional Techniques: 10^{any} States and Beyond". *Proc. 1998 International Conference on Application of Concurrency to System Design*, IEEE Computer Society 1998, pp. 75–85.
93. Valmari, A. & Tienari, M.: "An Improved Failures Equivalence for Finite-State Systems with a Reduction Algorithm". *Proc. Protocol Specification, Testing and Verification XI*, North-Holland 1991, pp. 3–18.
94. Valmari, A. & Tienari, M.: "Compositional Failure-Based Semantic Models for Basic LOTOS". *Formal Aspects of Computing* (1995) 7: 440–468.
95. van Glabbeek, R.: "The Linear Time — Branching Time Spectrum II: The Semantics of Sequential Systems with Silent Moves". *Proc. CONCUR '93, Fourth International Conference on Concurrency Theory*, Lecture Notes in Computer Science 715, Springer-Verlag 1993, pp. 66–81.
96. van Glabbeek, R. & Weijland, W.: "Branching Time and Abstraction in Bisimulation Semantics (Extended Abstract)". *Proc. IFIP International Conference on Information Processing '89*, North-Holland 1989, pp. 613–618.
97. Vardi, M. Y. & Wolper, P.: "An Automata-Theoretic Approach to Automatic Program Verification". *Proc. IEEE Symposium on Logic in Computer Science*, 1986, pp. 332–344.
98. Varpaaniemi, K.: *On the Stubborn Set Method in Reduced State Space Generation.* Ph.D. Thesis, Helsinki University of Technology, Digital Systems Laboratory, Research Report A-51, Espoo, Finland 1998, 105 p.
99. Wolper, P.: "Expressing Interesting Properties of Programs in Propositional Temporal Logic". *Proc. 13th ACM Symposium on Principles of Programming Languages*, 1986, pp. 184–193.
100. Wolper, P. & Lovinfosse, V.: "Verifying Properties of Large Sets of Processes with Network Invariants". *Proc. Workshop on Automatic Verification Methods for Finite State Systems*, Lecture Notes in Computer Science 407, Springer-Verlag 1989, pp. 68–80.

Theory of Regions

Eric Badouel and Philippe Darondeau

Irisa, Campus de Beaulieu, F-35042 Rennes Cedex, France
E-mail : {Eric.Badouel,Philippe.Darondeau}@irisa.fr

Abstract. The synthesis problem for nets consists in deciding whether a given graph is isomorphic to the marking graph of some net and then constructing it. This problem has been solved in the literature for various types of nets ranging from elementary nets to Petri nets. The general principle for the synthesis is to inspect regions of graphs representing extensions of places of the likely generating nets. We follow in this survey the gradual development of the theory of regions from its foundation by Ehrenfeucht and Rozenberg, with a particular insistence on the abstract meaning of the theory, which is a general product decomposition of graphs into atomic components determined by a parameter called a type of nets, and on the derivation of efficient algorithms for net synthesis based on linear algebra.

Table of Contents:

1 Terminology of Graphs

Since the terminology on graph theory varies a lot from one author to the other, we found it necessary to begin by defining the terminology used in this document.

1.1 Graphs

A *graph* $G = (X, E)$ is a collection X of *vertices* or *nodes* together with a collection E of *edges*. The graph is said to be *finite* if it has finitely many vertices and edges. Each edge has either one endpoint: $\mathrm{end}(e) = \{x\}$ in which case e is termed a *loop* at vertex x, or two endpoints: $\mathrm{end}(e) = \{x, y\}$ in which case e is termed a *link* between vertices x and y. A graph is *simple* if it is *loop-free*: each edge is a link, and has no multiple edge: $\mathrm{end}(e_1) = \mathrm{end}(e_2) \Rightarrow e_1 = e_2$. Therefore an edge of a simple graph may be identified with the pair of its endpoints. The *incidence matrix* of a graph G is a matrix A with elements 0 and 1, where each row corresponds to a vertex, each column corresponds to an edge, and $A(x, e)$ is 1 if and only if x is an endpoint of e. A *chain* of length $n \geq 1$ with endpoints $\{x_1, x_{n+1}\}$ is a finite sequence $(x_1, e_1, x_2, \ldots, x_n, e_n, x_{n+1})$ of vertices and edges such that $\mathrm{end}(e_i) = \{x_i, x_{i+1}\}$ for all $1 \leq i \leq n$. We say that the chain *connects* its endpoints. For convenience, we consider that every vertex is connected to itself by an empty chain. The *connected component* of a vertex is the set of vertices connected to this vertex by some chain; the graph is *connected* if it has only one connected component. A non empty chain is said to be *simple* if all edges are distinct, a chain is said to be *elementary* if all the vertices but possibly the endpoints are pairwise distinct. A *cycle* is a simple chain whose endpoints coincide: $x_1 = x_{n+1}$. A *tree* is a graph with no cycle or alternatively a graph in which any two vertices are connected by a unique chain. $G' = (X', E')$ is a *subgraph* of $G = (X, E)$ if $X' \subseteq X$, $E' \subseteq E$, and the mappings that send an edge $e \in E'$ to its endpoints in G' and in G coincide. G' *spans* G if $X' = X$; a *spanning tree* of G is a subgraph which is a tree spanning G.

1.2 Directed Graphs

An *orientation* of an edge e is an ordered pair of vertices (x, y) such that $\mathrm{end}(e) = \{x, y\}$, thus a loop at x has only one possible orientation: (x, x), while a link between x and y has two possible orientations: (x, y) and (y, x). We let $e : (x, y)$ denote the assignement of the orientation (x, y) to the edge e; the vertices $x = \partial^0(e)$ and $y = \partial^1(e)$ are respectively called the *source* and *target* of edge e. An oriented edge is sometimes called an *arc*. A *directed* graph is a graph whose edges are given an orientation. A directed graph is *simple* if it is loop-free and has no multiple arc in the sense that two edges with the same endpoints are necessarily given opposite orientations: $(e_1 : (x, y) \;\wedge\; e_2 : (x, y)) \Rightarrow e_1 = e_2$. Therefore an edge of a simple directed graph may be identified with the ordered pair of its endpoints, and in that case we write $e = (x, y)$ when $\partial^0(e) = x$ and $\partial^1(e) = y$. Notice that the underlying graph of a simple oriented graph may not be simple as we can find two edges with the same endpoints but with opposite orientations. A subgraph of a directed graph G is a subgraph of the underlying graph with the orientations of edges inherited from G. The notions of chain, cycle, tree, spanning subgraph and spanning tree do not depend on the orientation of edges; therefore a chain (cycle, tree, ...) of a directed graph is a chain (cycle, tree, ...) of the underlying graph. The specific notions that take the orientation into account are the following. A *path* of length $n \geq 1$ from x_1 to x_{n+1} is a finite sequence $(x_1, e_1, x_2, \ldots, x_n, e_n, x_{n+1})$ of vertices and edges such that $\partial^0(e_i) = x_i$ and $\partial^1(e_i) = x_{i+1}$ for all $1 \leq i \leq n$. For convenience, we consider that there exists an empty path from any vertex to itself. A non empty path is said to be *simple* if all edges are distinct. A path is said to be *elementary* if all the vertices but possibly the endpoints are distinct. A *circuit* is a simple path whose endpoints coincide: $x_1 = x_{n+1}$. Thus paths and circuits are respectively chains and cycles of the underlying graph whose edges have compatible orientations. The *incidence matrix* of a directed graph is the matrix $A : X \times E \to \{-1;0;1\}$ given by $A(x,e) = \begin{cases} 1 & \text{if } \partial^0(e) = x \\ -1 & \text{if } \partial^1(e) = x \\ 0 & \text{otherwise} \end{cases}$.

2 Regional Representation of Partial 2-Structures

The theory of regions was founded by Ehrenfeucht and Rozenberg in [22] with the aim to obtain a set-theoretic representation of directed graphs (X, E), enriched with an equivalence $\equiv$ on edges. The resulting structures $(X, E, \equiv)$ are termed *partial 2-structures*. The representation problem for partial 2-structures consists in attaching properties p to nodes x so that the Kripke structure so obtained may be abstracted without loss of information to the data $\{x^* | \, x \in X\}$ and $\{e^* | \, e \in E\}$, where a node is encoded by the set $x^* = \{p \,|\, x \models p\}$ of properties it satisfies and an edge by the pair $e^* = (x^* \setminus y^*, y^* \setminus x^*)$ where x and y are the respective source and target of e. The main difficulty is to reconstruct the equivalence relation $\equiv$, and this cannot be done unless the considered properties are altered uniformly when passing along every edge in each equivalence class. These specific properties, seen as sets of nodes when identified with their extensions $\{x \,|\, x \models p\}$, are called *regions* in [22]. The presentation of regions in partial 2-structures given below is directly inspired from [22], where the proofs of the results may be found. The algorithmic aspects of elementary net synthesis will be examined in the next section.

2.1 Partial 2-Structures and their Regions

Definition 2.1 *A partial 2-structure is a triple $G = (X, E, \equiv)$ where X is a finite non empty set of nodes, $E \subseteq E_2(X) = \{(x_1, x_2) \in X \times X | \, x_1 \neq x_2\}$ is a set of 2-edges over X, and $\equiv$ is an equivalence relation on E. When $E = E_2(X)$ is the whole set of 2-edges over X, G is called a 2-structure.*

Partial 2-structures may be viewed as equivalence classes of labelled simple directed graphs, where two graphs are equivalent if their labelling functions have the same kernel. Of particular interest are the partial set 2-structures defined as follows.

Definition 2.2 *A partial set 2-structure of a finite set B is a partial 2-structure $G = (X, E, \equiv_\delta)$ where $X \subseteq \mathcal{P}(B)$ and $\equiv_\delta$ is the kernel of the function $\delta((M, M'))$ $= (M \setminus M', M' \setminus M)$ for $M, M' \in X$. Let $S2S(B)$ denote the (full) set 2-structure of B; i.e. when $X = \mathcal{P}(B)$ and $E = E_2(X)$.*

Thus in particular, any partial set 2-structure G of B is a substructure of $S2S(B)$. In notation, $G \leq S2S(B)$ where $(X_1, E_1, \equiv_1) \leq (X_2, E_2, \equiv_2)$ if $X_1 \subseteq X_2$, $E_1 \subseteq E_2$ and $\equiv_1$ is the restriction of $\equiv_2$ on $E_1 \times E_1$. The representation problem for partial 2-structures may be stated as follows.

> *Which partial 2-structures are isomorphic to substructures of $S2S(B)$ for some finite set B (of tokens)?*

The best way to grasp this problem is to examine the extents R_b of representation tokens $b \in B$ in the structure $S2S(B)$ itself, let $R_b = \{M \in \mathcal{P}(B) | \; b \in M\}$. So, $b \in M$ if and only if $M \in R_b$. The following may be observed.

1. For every pair of equivalent 2-edges (M_1, M_1') and (M_2, M_2'), and for every $b \in B$, $b \in M_1 \setminus M_1'$ entails $b \in M_2 \setminus M_2'$ and symmetrically $b \in M_1' \setminus M_1$ entails $b \in M_2' \setminus M_2$. This can also be expressed as follows:
   - $(M_1 \in R_b \wedge M_1' \notin R_b) \Rightarrow (M_2 \in R_b \wedge M_2' \notin R_b)$;
   - $(M_1 \notin R_b \wedge M_1' \in R_b) \Rightarrow (M_2 \notin R_b \wedge M_2' \in R_b)$.

   Thus, all the 2-edges in an equivalence class are incident to R_b outwards, or they are incident to R_b inwards, or they are not incident to R_b.
2. $\forall M_1, M_2 \in \mathcal{P}(B) \quad M_1 \neq M_2 \Rightarrow (\exists b \in B \; M_1 \in R_b \Leftrightarrow M_2 \notin R_b)$.
3. For every pair of inequivalent 2-edges (M_1, M_1') and (M_2, M_2'), there exists some token $b \in B$ such that one 2-edge is incident to R_b and the other is not, or one 2-edge is incident to R_b inwards and the other is incident to R_b outwards.

These properties are also valid for substructures $(X, E, \equiv_\delta)$ of $S2S(B)$, where R_b is the set $\{M \in X | \; b \in M\}$.

Definition 2.3 *A region in a partial 2-structure $G = (X, E, \equiv)$ is a subset of nodes $R \subseteq X$ such that for every pair of equivalent 2-edges (x_1, x_1') and (x_2, x_2') in E: $(x_1 \in R \wedge x_1' \notin R) \Rightarrow (x_2 \in R \wedge x_2' \notin R)$, and $(x_1 \notin R \wedge x_1' \in R) \Rightarrow (x_2 \notin R \wedge x_2' \in R)$. Let $\mathcal{R}_G$ denote the set of (non trivial) regions of G, and for $x \in X$, let $\mathcal{R}_G(x) = \{R \in \mathcal{R}_G | \; x \in R\}$.*

It is worth noting that the complement $X \setminus R$ of a region R is a region. In particular X and $\emptyset$ are regions (the trivial regions). Now the non trivial regions may serve as representation tokens for states, that is nodes, and at the same time for events, that is classes of equivalent 2-edges. One obtains in this way regional versions of partial 2-structures defined as follows.

Definition 2.4 *Given a partial 2-structure $G = (X, E, \equiv)$, the regional version of G is the partial set 2-structure* **regv**$(G) = (X', E', \equiv_\delta)$ *with components $X' = \{\mathcal{R}_G(x) | \; x \in X\}$ and $E' = \{(\mathcal{R}_G(x), \mathcal{R}_G(x')) | \; (x, x') \in E\}$.*

In this construction, illustrated in Fig. 1, a node x is mapped to the set $\mathcal{R}_G(x)$ of the regions which include x. It appears from Fig. 1, where equivalent edges bear a common label, that the map **regv** is not an equivalence of partial 2-structures . The following theorem states when **regv** maps a partial 2-structure isomorphically to a partial set 2-structure (the regional representation of the latter).

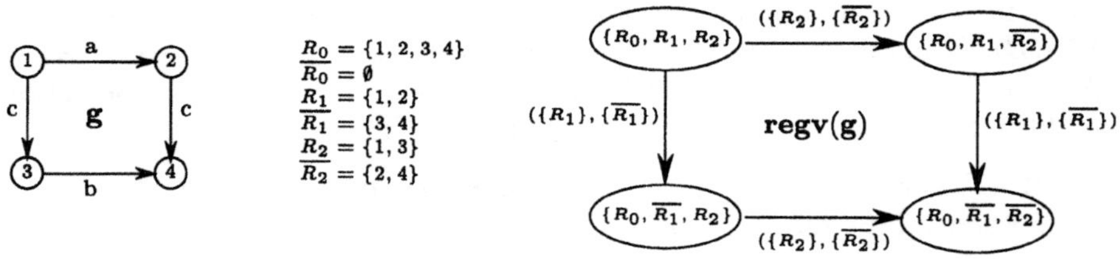

Fig. 1. a partial 2-structure and its regional version

Theorem 2.5 *A partial 2-structure $G = (X, E, \equiv)$ is isomorphic to a substructure of some set 2-structure if and only if $G \cong$ **regv**(G) (with $\mathcal{R}_G(\cdot)$ as the isomorphism) if and only if the following two axioms of separation are satisfied:*

1. STATES SEPARATION: $\forall x_1, x_2 \in X \;\; x_1 \neq x_2 \Rightarrow \exists R \in \mathcal{R}_G \; (x_1 \in R \Leftrightarrow x_2 \notin R)$.
2. EVENTS SEPARATION: *for all $(x_1, x_1'), (x_2, x_2') \in E$ with $(x_1, x_1') \not\equiv (x_2, x_2')$ there exists some region $R \in \mathcal{R}_G$ such that either (x_1, x_1') is incident to R outwards and (x_2, x_2') is not or (x_2, x_2') is incident to R outwards and (x_1, x_1') is not.*

There may exist nodes x_1, x_2, x_3 and x_4 such that $(x_1, x_2) \in E$, $\delta(x_1^*, x_2^*) = \delta(x_3^*, x_4^*)$, and $(x_3, x_4) \notin E$. Therefore **regv**(G) is not characterized by the sets $\{x^* | \; x \in X\}$ and $\{e^* | \; e \in E\}$. In order to reduce the mismatch, one should impose the additional axiom: $\forall (x_1, x_2) \in E \; \forall x_3, x_4 \in X \; \delta(\mathcal{R}_G(x_1), \mathcal{R}_G(x_2)) = \delta(\mathcal{R}_G(x_3), \mathcal{R}_G(x_4)) \Rightarrow (x_3, x_4) \in E$. Further on this way, one can even impose one or two stronger axioms:

FORWARD CLOSURE: $\forall (x_1, x_2) \in E \quad \forall x_3 \in X \;\; (\mathcal{R}_G(x_1) \setminus \mathcal{R}_G(x_2) \subseteq \mathcal{R}_G(x_3) \wedge \mathcal{R}_G(x_3) \cap \mathcal{R}_G(x_2) \setminus \mathcal{R}_G(x_1) = \emptyset) \Rightarrow \exists x_4 \in X \; (x_3, x_4) \in E \wedge \delta(\mathcal{R}_G(x_1), \mathcal{R}_G(x_2)) = \delta(\mathcal{R}_G(x_3), \mathcal{R}_G(x_4))$.

BACKWARD CLOSURE: $\forall (x_1, x_2) \in E \quad \forall x_4 \in X \;\; (\mathcal{R}_G(x_2) \setminus \mathcal{R}_G(x_1) \subseteq \mathcal{R}_G(x_4) \wedge \mathcal{R}_G(x_4) \cap \mathcal{R}_G(x_1) \setminus \mathcal{R}_G(x_2) = \emptyset) \Rightarrow \exists x_3 \in X \; (x_3, x_4) \in E \wedge \delta(\mathcal{R}_G(x_1), \mathcal{R}_G(x_2)) = \delta(\mathcal{R}_G(x_3), \mathcal{R}_G(x_4))$.

Partial 2-structures may be considered too general from a practical point of view, and one may prefer focusing on *reachable* partial 2-structures, such that all nodes can be reached by paths with a common origin. A familiar example of reachable partial set 2-structures is the class of sequential case graphs of elementary net systems.

Definition 2.6 *An elementary net is a directed bipartite graph $N = (P, E, F)$ such that* **dom**$(F) \cup$ **ran**$(F) = P \cup E$*. Elements of P, respectively E, are called conditions (or places), resp. events. Let $x \in {}^\bullet y$ and $y \in x^\bullet$ be alternative notations of $(x, y) \in F$. A case (or marking) of N is a subset of conditions $M \in \mathcal{P}(P)$. An event e has concession in case M (noted $M[e\rangle$) if and only if $({}^\bullet e, e^\bullet) = \delta(M, M')$ for some case M' (thus uniquely defined). The event e* *may then fire at M, resulting in the step $M[e{>}M'$. Thus, $M[e{>}$ if and only if $^\bullet e \subseteq M \wedge M \cap e^\bullet = \emptyset$, and then $M[e{>}M'$ where $M' = (M \setminus {}^\bullet e) \cup e^\bullet$.*

A net is *pure* if $\forall x \in P \cup E \quad x^\bullet \cap {}^\bullet x = \emptyset$; it is *simple* if $\forall x, y \in P \cup E \quad (x^\bullet = y^\bullet \;\wedge\; {}^\bullet x = {}^\bullet y) \Rightarrow x = y$. The elementary nets considered from now on are assumed to be pure and simple.

Definition 2.7 *An elementary net system is a structure $\mathcal{N} = (P, E, F, M_0)$ where $N = (P, E, F)$ is the underlying net and M_0 (in $\mathcal{P}(P)$) is the initial case. The sequential case graph of $\mathcal{N}$ is the partial set 2-structure* $\mathbf{scg}(\mathcal{N}) = (X', E', \equiv_\delta)$ *where $X' \subseteq \mathcal{P}(P)$ is the smallest set of cases reachable from M_0 by sequences of steps $M[e{>}M'$ and E' is the set of corresponding pairs (M, M').*

Lemma 2.8 *A partial set 2-structure $G = (X, E, \equiv_\delta)$ is the sequential case graph of an elementary net system if and only if it is reachable and the following property is satisfied: $\forall (x_1, x_2) \in E \quad \forall x_3 \in X \quad (x_1 \setminus x_2 \subseteq x_3 \;\wedge\; x_3 \cap x_2 \setminus x_1 = \emptyset) \Rightarrow \exists x_4 \in X \quad ((x_3, x_4) \in E \;\wedge\; \delta(x_1, x_2) = \delta(x_3, x_4))$.*

From Theo. 2.5 and Lem. 2.8, one obtains the following.

Corollary 2.9 *A partial 2-structure $G = (X, E, \equiv)$ is isomorphic to the sequential case graph of an elementary net system if and only if it is reachable and satisfies the axioms of states separation, events separation, and forward closure.*

The elementary net system in the above corollary is essentially the set of the ordered symmetric differences $\delta(\mathcal{R}_G(x), \mathcal{R}_G(y))$ for 2-edges $(x, y) \in E$. The representation problem for partial 2-structures set at the beginning of the section has in fact been given the solution $x^* = \mathcal{R}_G(x)$. The places of the net are the regions $r \in \mathcal{R}(G)$, the events are the equivalence classes of edges, and the flow relation is such that: $F([e]_\equiv, r) \Leftrightarrow r \in \mathcal{R}_G(y) \setminus \mathcal{R}_G(x)$ for some $(x, y) \in E$; and $F(r, [e]_\equiv) \Leftrightarrow r \in \mathcal{R}_G(x) \setminus \mathcal{R}_G(y)$ for some $(x, y) \in E$. The initial case of the net system is defined as $\mathcal{R}_G(x_0)$ for some $x_0 \in X$ such that every node of G is reachable from x_0.

2.2 Elementary Automata

The second part of the section paves the way for the algorithmic analysis of the region based correspondence between reachable graphs and elementary net systems. With this objective in mind, we recast the results obtained so far into the framework of transition systems, and illustrate the modified correspondence on a complete example.

Definition 2.10 *A (labelled) transition system is a triple $A = (S, E, T)$ with a set of states S, a set of events E, and a set of transitions $T \subseteq S \times E \times S$. Let $s \xrightarrow{e} s'$ be an equivalent notation for $(s, e, s') \in T$. An event e is enabled at state s (noted $s \xrightarrow{e}$) if $s \xrightarrow{e} s'$ for some s'. An event e is co-enabled at s' (noted $\xrightarrow{e} s'$) if $s \xrightarrow{e} s'$ for some s. An automaton is a structure $\mathcal{A} = (S, E, T, s_0)$ consisting of an underlying transition system $A = (S, E, T)$ and an initial state $s_0 \in S$.*

A partial 2-structure $G = (X, E, \equiv)$ may be identified with the transition system $(X, E/\equiv, T)$ where $x \stackrel{[e]_\equiv}{\rightarrow} x'$ if and only if $(x, x') \equiv e$. This transition system is *loopfree*: $s \stackrel{e}{\rightarrow} s' \Rightarrow s \neq s'$, *has no multiple arc*: $s \stackrel{e_1}{\rightarrow} s' \wedge s \stackrel{e_2}{\rightarrow} s' \Rightarrow e_1 = e_2$, and *it is reduced*: $\forall e \in E \quad \exists s, s' \in S \quad s \stackrel{e}{\rightarrow} s'$. The sequential case graphs of the reduced net systems defined hereafter fall in this subclass of transition systems.

Definition 2.11 *An elementary net system $\mathcal{N} = (P, E, F, M_0)$ is reduced if every event $e \in E$ has concession at some case M reachable from M_0, and for every two distinct conditions $p, p' \in P$ there exists some case M reachable from M_0 such that $p \in M \Leftrightarrow p' \notin M$. The dual of a reduced elementary net system $\mathcal{N}$ is the automaton $\mathcal{N}^* = (S, E, T, M_0)$ where S is the set of cases reachable from M_0 by sequences of steps $M[e{>}M'$ and T is the set of the corresponding transitions (M, e, M').*

Thus $\mathcal{N}^*$ is essentially the image of $\mathbf{scg}(\mathcal{N})$ through the map which sends the equivalence class of 2-edges $\{(M, M') |\ \delta(M, M') = ({}^\bullet e, e^\bullet)\}$ to the event e. Since $\mathcal{N}$ is simple and reduced, this map is one to one and onto. By construction, $\mathcal{N}^*$ is *reachable* from M_0, *deterministic*: $M \stackrel{e}{\rightarrow} M' \wedge M \stackrel{e}{\rightarrow} M'' \Rightarrow M' = M''$, and *co-deterministic*: $M' \stackrel{e}{\rightarrow} M \wedge M'' \stackrel{e}{\rightarrow} M \Rightarrow M' = M''$. The definition of regions may be carried to automata in the following form.

Definition 2.12 *A region in an automaton $\mathcal{A} = (S, E, T, s_0)$, or in the underlying transition system (S, E, T), is a subset of states $R \subseteq S$ such that*

$$\forall e \in E\ \forall s_1, s_2, s_3, s_4 \in S \quad s_1 \stackrel{e}{\rightarrow} s_2 \wedge s_3 \stackrel{e}{\rightarrow} s_4 \Rightarrow \begin{cases} s_1 \in R \wedge s_2 \notin R \Rightarrow s_3 \in R \wedge s_4 \notin R \\ s_1 \notin R \wedge s_2 \in R \Rightarrow s_3 \notin R \wedge s_4 \in R \end{cases}$$

Let $\mathcal{R}_\mathcal{A}$ denote the set of (non trivial) regions of $\mathcal{A}$, and for $s \in S$ let $\mathcal{R}_\mathcal{A}(s) = \{R \in \mathcal{R}_\mathcal{A} |\ s \in R\}$.

Thus, R is a region if and only if the label e of a transition suffices to determine whether the transition is incident to R inwards (R is then termed an output region for e, noted $e^\bullet R$), or it is incident to R outwards (R is then termed an input region for e, noted $R^\bullet e$), or it is not incident to R (it is internal to R or external to R). In particular, if $\mathcal{A}$ is reachable and reduced, the non trivial regions of $\mathcal{A}$ may be represented as maps $\eta_R : E \rightarrow \{-1, 0, 1\}$ such that $\eta_R(e) = 1$ if $e^\bullet R$, $\eta_R(e) = -1$ if $R^\bullet e$, and $\eta_R(e) = 0$ otherwise; the characteristic function of R, let $\chi_R : S \rightarrow \{0, 1\}$ where $\chi_R(s) = 1 \Leftrightarrow s \in R$, is then the unique map such that $s \stackrel{e}{\rightarrow} s' \Rightarrow \eta_R(e) = \chi_R(s') - \chi_R(s)$.

It is easily seen that for every condition p of a net system $\mathcal{N}$, the set of the reachable cases M that contain p is a region of $\mathcal{N}^*$. This region, denoted by p^* and called the extension of p, is such that $e^\bullet p^* \Leftrightarrow e \in {}^\bullet p$ and $p^{*\bullet} e \Leftrightarrow e \in p^\bullet$. Reversing the process which leads from net systems to sequential case graphs, let us recast the definition of regional versions in terms of nets and net systems.

Definition 2.13 *Given an automaton $\mathcal{A} = (S, E, T, s_0)$, the dual of $\mathcal{A}$ is the (reduced) elementary net system $\mathcal{A}^* = (\mathcal{R}_\mathcal{A}, (E/_\sim) \setminus \{\varepsilon\}, F, s_0^*)$ where: $\sim$ is the equivalence on E induced by regions, let*

$$e_1 \sim e_2 \Leftrightarrow (\forall R \in \mathcal{R}_\mathcal{A} \quad e_1{}^\bullet R \Leftrightarrow e_2{}^\bullet R \quad \wedge \quad R^\bullet e_1 \Leftrightarrow R^\bullet e_2);$$

ε is the equivalence class of the events which are inputless and outputless i.e. which are internal or external to all regions, if such events exist; F is the flow relation such that $F([e]_\sim, R) \Leftrightarrow e^\bullet R$ and $F(R, [e]_\sim) \Leftrightarrow R^\bullet e$; and $s_0^* = \{R \in \mathcal{R}_\mathcal{A} | \; s_0 \in R\}$.

The net system $\mathcal{A}^*$ is also called the saturated net version of $\mathcal{A}$ (for reasons explained in the sequel). The counterpart of Cor. 2.9 for automata is the following.

Theorem 2.14 *An automaton $\mathcal{A} = (S, E, T, s_0)$ is isomorphic to the dual $\mathcal{N}^*$ of an elementary net system if and only if $\mathcal{A} \cong \mathcal{A}^{**}$ if and only if $\mathcal{A}$ is simple (it has neither loop nor multiple arc), reduced, reachable and it satisfies the following properties of separation:*
SSP *(States Separation Property):*

$$\forall s, s' \in S \quad s \neq s' \Rightarrow \exists R \in \mathcal{R}_\mathcal{A} \quad (s \in R \Leftrightarrow s' \notin R)$$

ESP *(Events Separation Property):*

$$\forall e, e' \in E \quad e \neq e' \Rightarrow \exists R \in \mathcal{R}_\mathcal{A} \quad (R^\bullet e \wedge not(R^\bullet e')) \quad \vee \quad (e^\bullet R \wedge not(e'^\bullet R))$$

ESSP *(Events-States Separation Property):*

$$\forall e \in E \quad \forall s \in S \quad not(s \xrightarrow{e}) \quad \Rightarrow \exists R \in \mathcal{R}_\mathcal{A} \, (R^\bullet e \wedge s \notin R) \quad \vee \quad (e^\bullet R \wedge s \in R)$$

An automaton satisfying these conditions is termed an elementary automaton.

Observe that every event in an elementary automaton has input regions and output regions (from SSP), hence the map sending e to $[e]_\sim$ is a bijection between E and $(E/\sim) \setminus \{\varepsilon\}$ (from ESP). The isomorphism from $\mathcal{A}$ to $\mathcal{A}^{**}$ (the sequential case graph of the saturated net version of $\mathcal{A}$) maps e to $[e]_\sim$ and s to $s^* = \{R \in \mathcal{R}_\mathcal{A} | \; s \in R\}$. This isomorphism applies in particular to sequential case graphs, whence $\mathcal{N}^* \cong \mathcal{N}^{***}$ for every elementary net system. However, $\mathcal{N} = (P, E, F, M_0)$ is generally not isomorphic to its double dual $\mathcal{N}^{**}$. In fact, every condition p of $\mathcal{N}$ induces a corresponding region p^* of $\mathcal{N}^*$ which includes the reachable cases in which condition p holds, and $\mathcal{N}$ is isomorphic to the full subnet system of $\mathcal{N}^{**}$ with set of events $E/_\sim$ $(= E)$ and set of places $\{p^* | \; p \in P\}$. Thus, whenever $\mathcal{N}'^* \cong \mathcal{N}^*$, $\mathcal{N}'$ is isomorphic to a subnet system of $\mathcal{N}^{**}$ which is for that reason termed the saturated version of $\mathcal{N}$. Now, for an elementary automaton $\mathcal{A}$, $\mathcal{A} \cong \mathcal{A}^{**}$ entails that $\mathcal{A}^* \cong \mathcal{A}^{***}$, hence $\mathcal{A}^*$ is always a saturated net system. The aim of the next section is to optimize the synthesis process by looking at admissible subnets $\mathcal{N}$ of $\mathcal{A}^*$ such that $\mathcal{A} \cong \mathcal{N}^*$.

Before tackling the synthesis problem, we proceed to simplifying the presentation of elementary automata, and retrieve the usual presentation given in [11, 19, 34].

Proposition 2.15 *Let automaton $\mathcal{A}$ be simple, reduced and reachable, then $\mathcal{A}$ is elementary if and only if the separation properties* SSP *and* ESSP *are satisfied.*

Proof: Let $\mathcal{A} = (S, E, T, s_0)$, and assume for contradiction $e \neq e'$ and $\forall R \in \mathcal{R}_\mathcal{A}$ $(R^\bullet e \Leftrightarrow R^\bullet e') \;\wedge\; (e^\bullet R \Leftrightarrow e'^\bullet R)$. We show that $s \xrightarrow{e} s'$ entails $s \xrightarrow{e'} s'$ contradicting the assumption that $\mathcal{A}$ is simple. Assume $s \xrightarrow{e} s'$ and not $s \xrightarrow{e'}$, then by ESSP: $\exists R \in \mathcal{R}_\mathcal{A}$ $(R^\bullet e' \wedge s \notin R) \;\vee\; (e'^\bullet R \wedge s \in R)$ and the contradiction of $s \xrightarrow{e} s'$ follows from the definition of regions. Let $s'' \in S$ such that $s \xrightarrow{e'} s''$, then $\mathcal{R}_\mathcal{A}(s'') = \mathcal{R}_\mathcal{A}(s) \setminus {}^\bullet e' \cup e'^\bullet = \mathcal{R}_\mathcal{A}(s) \setminus {}^\bullet e \cup e^\bullet = \mathcal{R}_\mathcal{A}(s')$ and $s' = s''$ follows from ESP. ∎

For complete proofs of the results which have been stated in this subsection, the reader is referred to [19] where partial 2-structures are by-passed.

As an illustration, let us consider the elementary net system and the case graph given in Fig. 2. In Fig. 3 are displayed some of the non trivial regions of

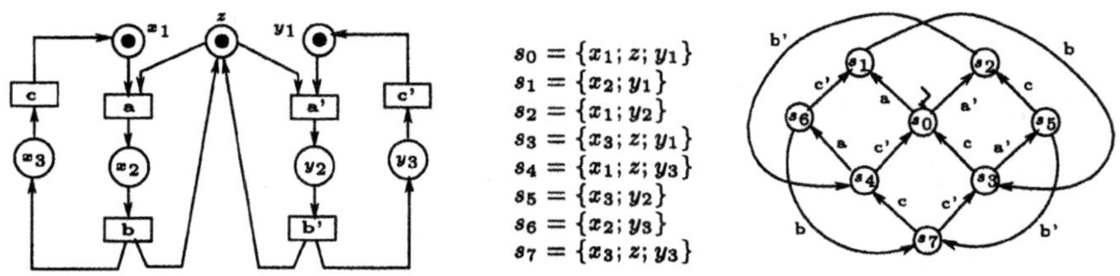

Fig. 2. an elementary net system and its case graph

this automaton. The missing items can be obtained by symmetry. Each drawing

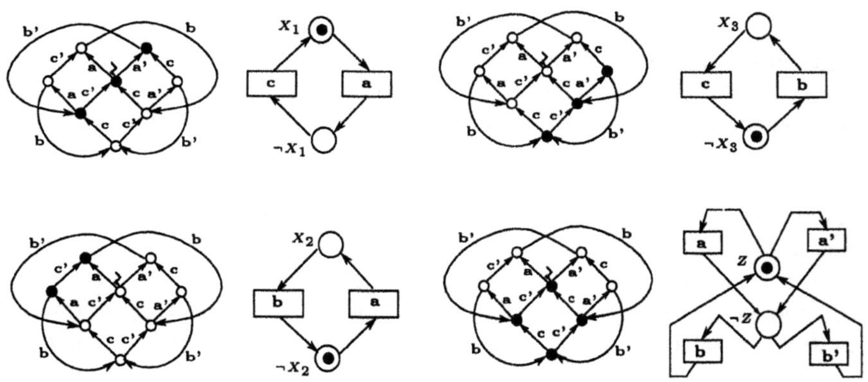

Fig. 3. some regions of the case graph of the elementary net system of Fig. 2 and their associated atomic net systems

represents a region R consisting of black states. The flow relations for the region R and for its complement $\neg R = S \setminus R$ are also represented pictorially; finally one token indicates which of these complementary regions contains the initial state.

We end up with the elementary net system of Fig. 4, which is the original net of Fig. 2 enriched with additional places (indicated by dashed lines) but with unchanged behaviour. The original net system is embedded into its saturated

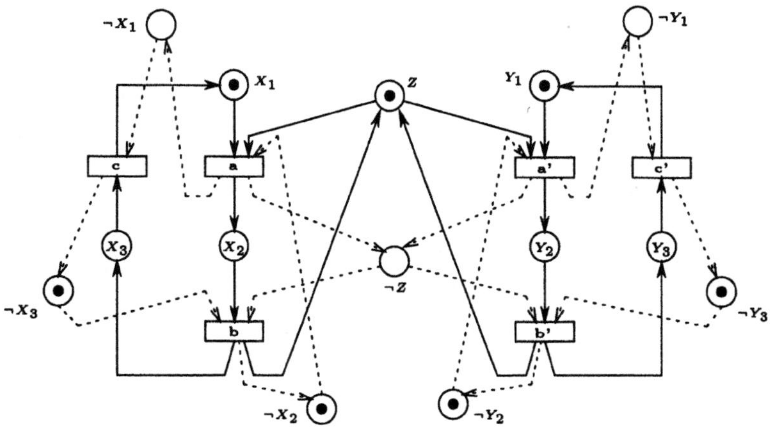

Fig. 4. the embedding of the elementary net system of Fig. 2 into its double dual

version by the map that sends a place x to its *extension* in the state graph i.e. the set of markings $\{M \in S|\ x \in M\}$.

3 The Synthesis of Elementary Net Systems

All automata considered in this section are assumed to be pre-elementary, i.e. simple, reachable and reduced. The synthesis problem of elementary net systems [19] is as follows:

> *Given a finite automaton $\mathcal{A} = (S, E, T, s_0)$, decide whether $\mathcal{A} \cong \mathcal{N}^*$ for some elementary net system $\mathcal{N}$ with the same set of events E, and if so, construct $\mathcal{N}$.*

Since the set $\mathcal{R}_\mathcal{A}$ of all the regions of $\mathcal{A}$ is finite, we already know from Prop. 2.15 that this problem can be decided in exponential time by simultaneously exploring $\mathcal{R}_\mathcal{A}$, for checking satisfaction of the separation properties ESP and ESSP, and constructing $\mathcal{N} = \mathcal{A}^*$. The aim of this section is to improve on this brute force solution. We review first Desel and Reisig's study of admissible sets of regions and their techniques for eliminating redundant regions. Next we account for Bernardinello's results on the synthesis of state machine decomposable net systems, based on the crucial remark that the minimal regions of an automaton form an admissible set, and for subsequent work by Cortadella et al. on the realization of automata by elementary nets up to some quotient of automata. We finally report the results obtained on the complexity of the synthesis problem in [25, 3].

3.1 Admissible sets of regions

In an elementary net system $\mathcal{N} = (P, E, F, M_0)$, each condition $p \in P$ determines an atomic subnet system of $\mathcal{N}$, let $\mathcal{N}_p = (\{p\}, E, F_p, M_{0,p})$ where F_p is the restriction of F and $M_{0,p}(p) = M_0(p)$. If we do not care about the isolated events in $\mathcal{N}_p$, these atomic subnet systems are elementary and $\mathcal{N}$ is just their sum $\sum_{p \in P} \mathcal{N}_p$, where nets are glued together on events $e \in E$. This decomposition may be used to isolate the contribution of each condition $p \in P$ to the global structure of the sequential case graph $\mathcal{N}^*$. This automaton may be seen as a deterministic recognizer of finite sequences, in which every state (i.e. case) is accepting. An automaton of this type is characterized up to isomorphism by the language $\mathcal{L}$ it accepts plus the equivalence $\equiv$ on $\mathcal{L}$ which identifies these sequences that lead to a common (accepting) state. Now in the case of $\mathcal{N}^*$, $\mathcal{L}$ and $\equiv$ are the intersections for p ranging over P of the respective languages and equivalences characteristic of $\mathcal{N}_p^*$: $\mathcal{L} = \bigcap_{p \in P} \mathcal{L}_p$ and $\equiv = \bigcap_{p \in P} \equiv_p$. Thus the role of each condition p is twofold: on the one hand, p cuts off sequences $u \cdot e$ such that $u \in \mathcal{L}$ but $u \cdot e \notin \mathcal{L}_p$, and on the other hand p separates pairs of words $u, v \in \mathcal{L}$ such that $u \not\equiv_p v$.

Returning to the synthesis problem, let us now clarify the relationship between automata and atomic net systems. Let $\mathcal{A} = (S, E, T, s_0)$ be a finite deterministic automaton, with language $\mathcal{L}$ and equivalence $\equiv$, and let $\mathcal{N}_p = (\{p\}, E, F_p, M_{0,p})$ be an atomic net system, inducing a dual automaton $\mathcal{N}_p^*$ with language $\mathcal{L}_p$ and equivalence $\equiv_p$. The automaton $\mathcal{N}_p^*$ has two states, $\emptyset$ and $\{p\}$, one of which is $M_{0,p}$, and it has transitions $\emptyset \stackrel{e}{\rightarrow} \{p\}$ if $F_p(e,p)$, $\{p\} \stackrel{e}{\rightarrow} \emptyset$ if $F_p(p,e)$, and otherwise $\emptyset \stackrel{e}{\rightarrow} \emptyset$ and $\{p\} \stackrel{e}{\rightarrow} \{p\}$. Suppose $\mathcal{L} \subseteq \mathcal{L}_p$ and $\equiv \subseteq \equiv_p$. Let R_p be the subset of states $s \in S$ such that $s_0 \stackrel{u}{\rightarrow} s$ in $\mathcal{A}$ and $M_{0,p} \stackrel{u}{\rightarrow} \{p\}$ in $\mathcal{N}_p^*$ for some sequence of events $u \in E^*$. Then R_p is a region of $\mathcal{A}$, $s_0 \in R_p \Leftrightarrow M_{0,p} = \{p\}$, and for every $e \in E$: $R_p{}^\bullet e \Leftrightarrow F_p(p,e)$ and $e^\bullet R_p \Leftrightarrow F_p(e,p)$. Conversely, for any region R_p of $\mathcal{A}$, the elementary net system $\mathcal{N}_p$ defined by the above relations induces a dual automaton $\mathcal{N}_p^*$ such that $\mathcal{L} \subseteq \mathcal{L}_p$ and $\equiv \subseteq \equiv_p$. Moreover, R_p separates two distinct states s' and s'' such that $s_0 \stackrel{u}{\rightarrow} s'$ and $s_0 \stackrel{v}{\rightarrow} s''$ in $\mathcal{A}$ if and only if $u \not\equiv_p v$, and R_p separates a state s such that $s_0 \stackrel{u}{\rightarrow} s$ from an event e such that not$(s \stackrel{e}{\rightarrow})$ if and only if $u \cdot e \notin \mathcal{L}_p$. Therefore, given a net system $\mathcal{N} = (P, E, F, M_0) = \sum_{p \in P} \mathcal{N}_p$, the dual automaton $\mathcal{N}^*$ is isomorphic to the automaton $\mathcal{A}$ if and only if $\mathcal{L} = \bigcap_{p \in P} \mathcal{L}_p$ and $\equiv = \bigcap_{p \in P} \equiv_p$, if and only if for all $p \in P$, $\mathcal{N}_p$ is an atomic net system defined from some corresponding region R_p in $\mathcal{A}$ and the following properties are satisfied:

SSP': $\forall u, v \in \mathcal{L} \quad u \not\equiv v \Rightarrow \exists p \in P \quad u \not\equiv_p v$,

ESSP': $\forall u \in \mathcal{L} \quad \forall e \in E \quad u \cdot e \notin \mathcal{L} \Rightarrow \exists p \in P \quad u \cdot e \notin \mathcal{L}_p$,

if and only if the family of regions $\{R_p \mid p \in P\}$ is admissible according to the following definition.

Definition 3.1 *Given an automaton $\mathcal{A} = (S, E, T, s_0)$, a subset of regions $\{R_p \mid p \in P\} \subseteq \mathcal{R}_\mathcal{A}$ is admissible if and only if it includes witnesses for the satisfaction of every instance of the following separation problems where $e \in E$ and $s, s', s'' \in S$ are such that $s' \neq s''$ and not$(s \stackrel{e}{\rightarrow})$:*

$\text{SSP}(s', s'')$: $\exists p \in P \quad s' \in R_p \Leftrightarrow s'' \notin R_p$,
$\text{ESSP}(s, e)$: $\exists p \in P \ (R_p{}^\bullet e \ \wedge \ s \notin R_p) \ \vee \ (e^\bullet R_p \ \wedge \ s \in R_p)$.

It is easily seen that problem $\text{SSP}(s', s'')$ cannot be solved positively in a non-deterministic automaton $\mathcal{A}$ where $s \xrightarrow{e} s'$ and $s \xrightarrow{e} s''$ for $s' \neq s''$. One rediscovers in this way a basic result established in [19].

Theorem 3.2 *An automaton $\mathcal{A} = (S, E, T, s_0)$ is isomorphic to $\mathcal{N}^*$ for $\mathcal{N} = (P, E, F, M_0)$ if and only if for every $p \in P$, the atomic subnet system $\mathcal{N}_p$ of $\mathcal{N}$ may be defined from some corresponding region R_p of $\mathcal{A}$, and the set of regions $\{R_p | \, p \in P\}$ is admissible.*

In view of Def. 3.1 and Theo. 3.2, the synthesis problem for $\mathcal{A} = (S, E, T, s_0)$ may be solved by considering at most $|S| \times (|S| + |E|)$ regions of $\mathcal{A}$. Nevertheless, this does not indicate how to select these regions from $\mathcal{R}_\mathcal{A}$. The purpose is to construct a subset of regions $\mathcal{R} \subseteq \mathcal{R}_\mathcal{A}$ as small as possible such that $\mathcal{R}$ is admissible if and only if the whole set of regions $\mathcal{R}_\mathcal{A}$ is admissible. Some structural rules are proposed in [19] for the stepwise elimination of redundant regions, starting from $\mathcal{R}_\mathcal{A}$.

Definition 3.3 *Let $\mathcal{R} \subset \mathcal{R}_\mathcal{A}$ be a set of regions. A region $R \in \mathcal{R}$ is redundant in $\mathcal{R}$ if the following assertions are equivalent:* (i) *$\mathcal{R}$ is admissible* (ii) *$\mathcal{R} \setminus \{R\}$ is admissible.*

Proposition 3.4 *Let $\mathcal{A} = (S, E, T, s_0)$ and $R \in \mathcal{R} \subset \mathcal{R}_\mathcal{A}$. In each of the following cases R is redundant in $\mathcal{R}$.*

1. $S \setminus R \in \mathcal{R}$,
2. $\exists R_1, R_2, R_3, R_4 \in \mathcal{R} \quad R = R_1 \cap R_2 \ \wedge \ S \setminus R = R_3 \cap R_4$,
3. $\exists R_1, R_2, R_3, R_4 \in \mathcal{R} \quad R = R_1 \cup R_2 \ \wedge \ S \setminus R = R_3 \cup R_4$,
4. $\exists R_1, R_2 \in \mathcal{R} \quad R = R_1 \cap R_2 \quad \wedge \quad \forall s \in R \quad \forall e \in E \quad \forall s' \in S \setminus R \quad s \xrightarrow{e} s' \Rightarrow s' \notin R_1 \cup R_2$.

Once a reduced set of regions $\mathcal{R}$ has been obtained from $\mathcal{R}_\mathcal{A}$, one can check directly from Def. 3.1 whether it is admissible, proving that $\mathcal{A}$ is elementary, and then extract from $\mathcal{R}$ a minimal subset $\{R_p | \, p \in P\}$ such that $\mathcal{A} \cong (\sum_{p \in P} \mathcal{N}_p)^*$. It is worth noting that there exists in general no least admissible set of regions. This fact is illustrated in Fig. 5 by the so-called "four seasons" example reproduced from [19]. The "four seasons" automaton may be realized by two minimal subnet systems of the dual saturated net system: one has four conditions and is contact-free while the other one has three conditions but is not contact-free.

Definition 3.5 *An elementary net system $\mathcal{N} = (P, E, F, M_0)$ is contact-free if ${}^\bullet e \subset M \Rightarrow M \cap e^\bullet = \emptyset$ for every event e and for every reachable case M.*

Thus, the subclass of elementary net systems which are contact-free and reduced coincides with the subclass of the reduced and one-safe Petri nets. Now, every saturated net system $\mathcal{N} = (P, E, F, M_0)$ is contact-free: every condition $p \in P$ induces two complementary regions R_p and $\overline{R_p}$ in $\mathcal{N}^*$, and since $\mathcal{N} \cong \mathcal{N}^{**}$ there

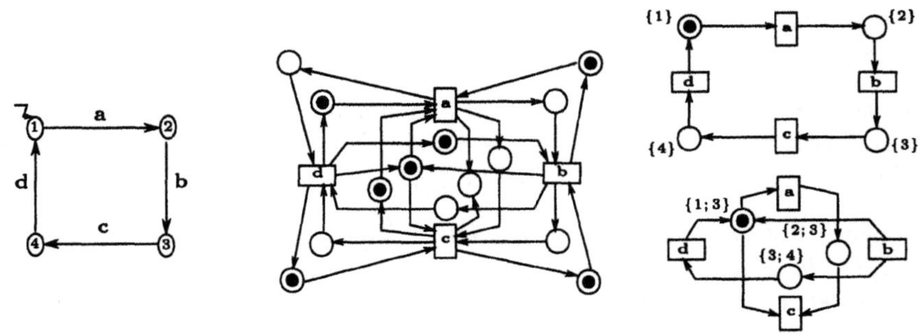

Fig. 5. the four seasons example: the automaton (on the left), the saturated net system (on the middle) and two elementary net systems corresponding to minimal sets of regions (on the right)

should exist some condition $\overline{p} \in P$ such that $R_{\overline{p}} = \overline{R_p}$. Therefore, every elementary automaton may be realized by a one-safe Petri net. The following adaptation of Theo. 3.2, based on the use of complementary regions, is established in [19]

Proposition 3.6 *An automaton* $\mathcal{A} = (S, E, T, s_0)$ *is isomorphic to* $\mathcal{N}^*$ *for a contact-free net system* $\mathcal{N} = (P, E, F, M_0) = \sum_{p \in P} \mathcal{N}_p$ *if and only if every atomic subnet system* $\mathcal{N}_p$ *of* $\mathcal{N}$ *may be defined from a corresponding region* $R_p \in \mathcal{R}_{\mathcal{A}}$ *and the following properties of separation are satisfied:*
SSP : $\forall s, s' \in S \quad s \neq s' \Rightarrow \exists p \in P \quad s \in R_p \Leftrightarrow s' \notin R_p$
ESSP$^\sharp$: $\forall e \in E \quad \forall s \in S \quad not(s \xrightarrow{e}) \quad \Rightarrow \exists p \in P \quad R_p{}^\bullet e \wedge s \notin R_p.$

3.2 Minimal Regions

Among the admissible sets of regions of an elementary automaton, the set of minimal regions plays a distinguished role because it leads naturally, as shown in [11], to a state machine decomposable (and hence contact-free) net system realizing the automaton.

Definition 3.7 *An elementary net system* $\mathcal{N} = (P, E, F, M_0)$ *is a state machine if its initial case is a singleton and every event has one precondition and one postcondition. A state machine component of* $\mathcal{N} = (P, E, F, M_0)$ *is a state machine* $\mathcal{N}' = (P', E', F', M_0')$ *such that* $P' \subseteq P$, $E' = \{e \in E | ({}^\bullet e \cup e^\bullet) \cap P' \neq \emptyset\}$, $F' = F \cap (E' \times P' \cup P' \times E')$, *and* $M_0' = M_0 \cap P'$. *A state machine decomposition of* $\mathcal{N} = (P, E, F, M_0)$ *is a family of state machines, let* $\mathcal{N}_i = (P_i, E_i, F_i, M_{0,i})$, *such that* $P = \cup_i P_i$, $E = \cup_i E_i$, $F = \cup_i F_i$, *and* $M_0 = \cup_i M_{0,i}$.

A state machine is nothing else than a reachable automaton, as can be seen from Fig. 6 where the elementary net system given in Fig. 2 is decomposed into three state machine components. The respective state machine components model sequential processes which are synchronized on their common events. In this example, the synchronization prevents the leftmost and rightmost processes from

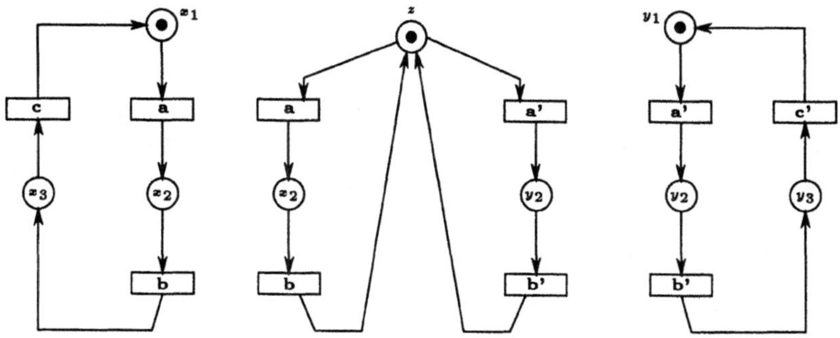

Fig. 6. three state machine components of the net system of Fig. 2

entering simultaneously the critical section figured by the mutually exclusive conditions x_2 and y_2. Each state machine component $\mathcal{N}_i$ of a net system $\mathcal{N} = \sum_i \mathcal{N}_i$ may be seen as a sequential observer of $\mathcal{N}^*$, projecting cases of $\mathcal{N}$ on observable conditions $p \in P_i$. By definition of state machine components, each case of $\mathcal{N}$ projects to one and exactly one condition $p \in P_i$, hence each case of $\mathcal{N}$ belongs to exactly one region R_p of $\mathcal{N}^*$ such that $p \in P_i$.

Proposition 3.8 *Every state machine component $\mathcal{N}_i = (P_i, E_i, F_i, M_{0,i})$ of an elementary net system $\mathcal{N} = \sum_i \mathcal{N}_i$ determines a regional partition $\{R_p |\ p \in P_i\}$ of the sequential case graph $\mathcal{N}^*$. Conversely, every regional partition $\{R_p |\ p \in P\}$ of $\mathcal{N}^*$ determines a state machine component of the saturated net system $\mathcal{N}^{**}$.*

Returning to the example, the regional partitions of $\mathcal{N}^*$ (Fig. 3) which determine the three state machine components shown in Fig. 6 are respectively $\{X_1, X_2, X_3\}$, $\{X_2, Z, Y_2\}$, and $\{Y_1, Y_2, Y_3\}$ where:

$$
\begin{aligned}
Z &= \{s_0; s_3; s_4; s_7\} \\
X_1 &= \{s_0; s_2; s_4\} \\
X_2 &= \{s_1; s_6\} \\
X_3 &= \{s_3; s_5; s_7\} \\
Y_1 &= \{s_0; s_1; s_3\} \\
Y_2 &= \{s_2; s_5\} \\
Y_3 &= \{s_4; s_6; s_7\}
\end{aligned}
$$

It may be observed that all these regions are minimal w.r.t. set inclusion in $\mathcal{R}_{\mathcal{N}^*}$. The particular interest of minimal regions for the net system realization of elementary automata is shown by the following proposition and corollaries.

Proposition 3.9 *Given an automaton $\mathcal{A} = (S, E, T, s_0)$, the following properties are satisfied by the set $\mathcal{R}_\mathcal{A}$ of regions of $\mathcal{A}$:*

1. *If R_1 and R_2 are disjoint regions then $R_1 \cup R_2$ is a region with*

$$
\begin{aligned}
{}^\bullet(R_1 \cup R_2) &= ({}^\bullet R_1 \cup {}^\bullet R_2) \setminus (({}^\bullet R_1 \cap R_2{}^\bullet) \cup ({}^\bullet R_2 \cap R_1{}^\bullet)) \\
(R_1 \cup R_2)^\bullet &= (R_1{}^\bullet \cup R_2{}^\bullet) \setminus (({}^\bullet R_1 \cap R_2{}^\bullet) \cup ({}^\bullet R_2 \cap R_1{}^\bullet)).
\end{aligned}
$$

2. *If R and R' are regions and $R' \subseteq R$ then $R \setminus R'$ is a region. If moreover R' is minimal then $e^\bullet(R \setminus R')$ for every event $e \in R'^\bullet$ which is not incident to R (i.e. such that $e \notin {}^\bullet R \cup R^\bullet$).*
3. *If R is a region and $s \in R$, then $s \in R'$ for some minimal region $R' \subseteq R$.*
4. *If R is a region and e an event such that $R^\bullet e$, then $R'^\bullet e$ for some minimal region $R' \subseteq R$; symmetrically if e is an event such that $e^\bullet R$, then $e^\bullet R'$ for some minimal region $R' \subseteq R$.*
5. *Every region is a disjoint union of minimal regions.*

Corollary 3.10 *A pre-elementary automaton is elementary if and only if its set of minimal regions is admissible.*

It may be further observed that the set of minimal regions of a pre-elementary automaton $\mathcal{A}$ is admissible w.r.t. the separation properties SSP and ESSP if and only if it is admissible w.r.t. the separation properties SSP and ESSP$^\sharp$. In fact, let $\{R_1, \ldots, R_n\}$ be any partition of the set of states of $\mathcal{A}$ into minimal regions, then each instance of the problem ESSP(s, e) solved by a region R_i such that $e^\bullet R_i$ and $s \in R_i$ can also be solved by a region R_j such that $R_j{}^\bullet e$ and $s \notin R_j$. Since the set of all partitions of the set of states of $\mathcal{A}$ into minimal regions induces a state machine decomposition of the net system $\sum_p \mathcal{N}_p$ defined from the set of all minimal regions R_p of $\mathcal{A}$, one deduces also the following.

Corollary 3.11 *Every elementary automaton may be realized by a state machine decomposable (and hence contact-free) elementary net system.*

An algorithm based on minimal regions has been proposed in [14] for a variant problem of realization of automata by net systems which may be stated as follows.

> *Given a pre-elementary automaton $\mathcal{A}$, decide whether exists and construct a (minimal) elementary net system $\mathcal{N}$ such that $\mathcal{N}^* \cong \mathcal{A}'$ for some quotient $\mathcal{A}'$ of $\mathcal{A}$.*

We recall that $\mathcal{A}' = (S', E, T', s_0')$ is a quotient of $\mathcal{A} = (S, E, T, s_0)$ if $s_1 \xrightarrow{e} s_2$ in $\mathcal{A}$ if and only if $\sigma(s_1) \xrightarrow{e} \sigma(s_2)$ in $\mathcal{A}'$ for some surjective map $\sigma : S \to S'$ such that $s_0' = \sigma(s_0)$. This problem is similar to the original synthesis problem, up to the fact that the states separation property SSP is ignored. Now the events-states separation property ESSP$^\sharp$ is valid in $\mathcal{A}$ if and only if for every event e the set of states $\{s \in S |\ s \xrightarrow{e}\}$ coincides with the intersection of the minimal regions R such that $R^\bullet e$. The algorithm starts from the sets $\{s \in S |\ s \xrightarrow{e}\}$ and increases them into minimal regions, which are generated until the validity of ESSP$^\sharp$ can be decided upon. The net $\mathcal{N}$ is then constructed from a minimal set of minimal regions admissible with respect to ESSP$^\sharp$. A variant form of this algorithm has been integrated to a software tool for the synthesis of asynchronous circuits [15].

It should be noted that the problem of realizing automata by nets up to a quotient differs significantly from the problem of realizing automata by nets up to behavioural equivalence (equality of the accepted languages). In order to make the difference visible, let us focus on finite and deterministic automata.

In this context, behavioural equivalence coincides with bisimilarity. Given a finite deterministic automaton $\mathcal{A}$, with language $\mathcal{L}$ and characteristic equivalence $\equiv$ on $\mathcal{L}$, the problem of realizing $\mathcal{A}$ up to behavioural equivalence consists in constructing an elementary net system $\mathcal{N}$ such that $\mathcal{N}^*$ recognizes $\mathcal{L}$. For the problem of realizing $\mathcal{A}$ up to a quotient, it is set as a further requirement that any two equivalent sequences in $\mathcal{L}$ lead to the same case when they are fired from the initial case of $\mathcal{N}$. In orther words, it is asked that $\equiv \subseteq \equiv_{\mathcal{N}^*}$. The reason why this constraint makes a notable difference is that the elementary automata are not closed under quotient. This counterfact is illustrated in Fig. 7: the automaton shown on the middle is isomorphic to the case graph of the net displayed on the left, but its minimized version shown on the right is not elementary (any region R such that $R^\bullet c$ must include state 3, hence the problem ESSP$(3, c)$ cannot be solved).

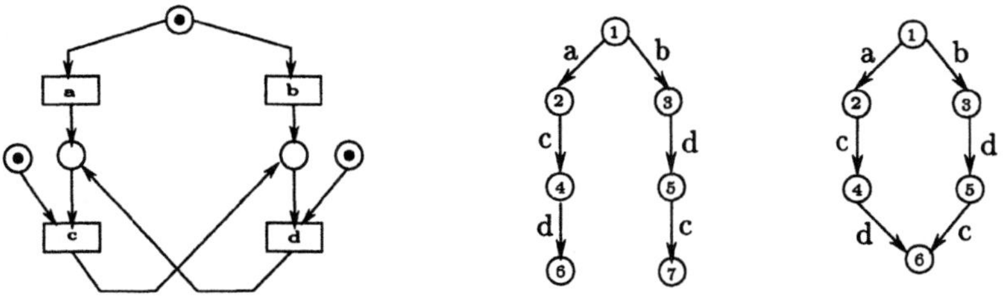

Fig. 7. elementary automata are not closed under quotient

3.3 Complexity Results

Hiraishi proved in [25] that the separation problems SSP(s, s') and ESSP$^\sharp(s, e)$ are NP-complete in the respective data $(\mathcal{A}, s, s')$ and $(\mathcal{A}, s, e)$. Since regions in $\mathcal{A}$ are closed under complementation, the problem ESSP(s, e) is also NP-complete. It does not follow therefrom that the synthesis problem for elementary net systems is NP-complete; however this is the case. The synthesis problem is obviously in NP since the total number of instances of separation problems in an automaton $\mathcal{A}$ is quadratic in the size of $\mathcal{A}$, and it can be checked in polynomial time whether a non-deterministically chosen subset of states is a region solving a fixed separation problem. Now a polynomial reduction of 3-SAT to the synthesis problem of elementary net systems was established in [3], showing NP-hardness since 3-SAT is NP-complete (see e.g. [23]). Recall that 3-SAT is the problem whether, given a finite set of boolean clauses over V, with three litterals per clause, there exists some truth assignment for V validating each clause. Each clausal system of this form is associated in [3] with an automaton such that the clausal system is satisfiable if and only if the automaton is elementary if and only if the separation property ESSP$^\sharp$ is valid. Therefore, the synthesis problem for elementary net systems is NP-complete, and so is the problem of realizing automata by nets up to a quotient. The problems of realizing automata by nets up to behavioural equivalence, or up to an unfolding (given $\mathcal{A}$ find $\mathcal{N}$ such that $\mathcal{A}$ is isomorphic to a quotient of $\mathcal{N}^*$) have unknown complexity.

4 Cutset Representation of Finite Graphs

We have seen that the region based synthesis of elementary net systems from initialized partial 2-structures $(X, E, \equiv, x_0)$ is a NP-complete problem. Nevertheless, this problem is trivial when the labelling equivalence is discrete: in that case, the partial 2-structure is essentially a state machine with set of places X; even better, this state machine is equivalent to a net system with $|X|-1$ places, whose case graph is a partial set 2-structure isomorphic to the given partial 2-structure. There exists a large variety of set-theoretic representations for an unlabelled graph (X, E), all of which using at most $|X|-1$ tokens. These representations, based on cuts and cutsets, may be computed by linear algebraic methods which are quite standard in applied graph theory. The purpose of this section is to review these methods, and thereby shed light on regions in two respects. First, we examine the close relationship between regions and cuts (this analogy was first pointed out to us by T. Murata). Second, we indicate the obstacles to using linear algebraic methods for the region based representation of labelled graphs. On account of this analysis, a variant definition of regions is proposed in the next section.

4.1 Cuts and Cutsets

Let $G = (X, E)$ be a finite, connected and simple directed graph with set of nodes $X = \{x_1, \ldots, x_n\}$ and set of 2-edges $E = \{e_1, \ldots, e_m\}$. So, G is free of loops multiple arcs, although a 2-edge $e = (x_l, x_k)$ may have an inverse $e^{-1} = (x_k, x_l)$ in E. A cutset of G is a *minimal* set of 2-edges whose removal increases the number of connected components by one. A cut of G is a cutset or an edge disjoint union of cutsets. Since G is connected, every cut or cutset $C \subseteq E$ determines two complementary subsets of nodes p and $X \setminus p$, both non empty, such that for every 2-edge $e = (x_k, x_l)$, $e \in C$ if and only if $x_k \in p \Leftrightarrow x_l \notin p$. Conversely, every non trivial subset $p \subseteq X$ determines a cut between p and $X \setminus p$, which is a cutset when both p and $X \setminus p$ are connected. An *orientation* of the cut C results from the choice of one of the two complementary subsets of nodes determined by the cut, let p. An oriented cut C may be coded by a vector $C \in \mathbb{R}^m$ such that for every 2-edge $e_i = (x_k, x_l)$, $C(i) = 1$ if $x_k \notin p$ and $x_l \in p$, $C(i) = -1$ if $x_k \in p$ and $x_l \notin p$, and $C(i) = 0$ if $x_k \in p \Leftrightarrow x_l \in p$.

Let $X = \{x_1, \ldots, x_n\}$ and $E = \{e_1, \ldots, e_m\}$. We will address the problem of constructing a variety of sets of properties $\{p_1, \ldots, p_{n-1}\}$ where $p_i \subseteq X$ such that the partial 2-structure $(\{x^* | \, x \in X\}, \{e^* | \, e \in E\}, \equiv_\delta)$ where $x^* = \{p_i | \, x \in p_i\}$, $(x_k, x_l)^* = (x_k^*, x_l^*)$, and $\delta(x_k^*, x_l^*) = (x_k^* \setminus x_l^*, x_l^* \setminus x_k^*)$ is isomorphic to G viewed as a partial 2-structure: $G = (X, E, id_E)$. Each family of tokens $\{p_1, \ldots, p_{n-1}\}$

will determine a corresponding set of (oriented) cuts $\{C_1, \ldots C_{n-1}\}$ which are linearly independent as vectors $C_i \in \mathbb{R}^m$.

The interesting fact here is that one can easily construct linear bases of cuts, given as sets of fundamental cutsets of G with respect to arbitrary spanning trees. Recall that a spanning tree is a set of edges $U \subseteq E$, free of cycles and connecting X. The fundamental cutsets w.r.t. U are the cuts which include exactly one branch of U. Each branch of U determines two connected components of U (and thus of G), with set of nodes p and $X \setminus p$, such that every other branch of U is internal either to p or to $X \setminus p$. The fundamental cutsets w.r.t. U may be computed by classical methods of linear algebra. These methods are recalled below, following the notations of [16].

4.2 Computing Cutsets

The graph $G = (X, E)$ is characterized up to isomorphism by its incidence matrix. We recall that this matrix $A = [a_{i,j}]$ is an $n \times m$ matrix with entries in $\{-1, 0, 1\}$, with $a_{i,j} = 0$ if edge e_j is not incident to node x_i, $a_{i,j} = 1$ if x_i is the source of e_j, and $a_{i,j} = -1$ if x_i is the target of e_j. Since every column contains exactly two non zero entries (1 and -1) every row can be computed from the other rows, and the matrix A has the same rank as the matrix A_1 obtained by erasing its last row. Let $A = \begin{bmatrix} A_1 \\ A_2 \end{bmatrix}$ where A_1 is an $(n-1) \times m$ matrix and A_2 is an $1 \times m$ matrix. Actually A_1 and A have rank $n-1$. Assume w.l.o.g. that the $(n-1)$ branches of the spanning tree U are the edges $e'_j = e_{j+(m-n+1)}$ for $j \in \{1, \ldots, n-1\}$. Then $A_1 = \begin{bmatrix} A_{11} \; A_{12} \end{bmatrix}$ where A_{12} is the $(n-1) \times (n-1)$ matrix corresponding to the edges of the tree (the *branches*) and A_{11} is the $(n-1) \times (m-n+1)$ matrix corresponding to the other edges (the *chords*).

The fundamental cutset C_i of G determined by the edge e'_i of the spanning tree is given by the ith row of the *fundamental cutset matrix* $Q_f = A_{12}^{-1} \cdot A_1$. This $(n-1) \times m$ matrix has the form $\begin{bmatrix} Q_{f11} \; I_{n-1} \end{bmatrix}$ where I_k is the identity matrix of rank k. The ith row of Q_f associated with the fundamental cutset C_i is an m vector with entries in $\{-1, 0, 1\}$. Let p_i and $X \setminus p_i$ be the two connected components of G separated by C_i, such that e'_i has its source in $X \setminus p_i$ and its target in p_i. Then for every $j \in \{1, \ldots, m\}$, $C_i(j) = 0$ if e_j is not in C_i, $C_i(j) = 1$ if e_j is oriented from $X \setminus p_i$ to p_i and $C_i(j) = -1$ if e_j is oriented from p_i to $X \setminus p_i$. A complete example is shown in Fig. 8.

It is worth noting that the matrix A_{12}^{-1} can be computed directly from G without inverting matrix A_{12}, for it coincides with the path matrix $P = [p_{i,j}]$ defined as follows. For each $j \in \{1, \ldots, n-1\}$, let Π_j be the unique chain (in the tree U) connecting x_j and the reference node x_n; then for $1 \leq i, j \leq n-1$, let $p_{i,j} = 0$ if e'_i does not belong to Π_j, $p_{i,j} = 1$ if e'_i belongs to Π_j and is oriented towards the reference node x_n, and $p_{i,j} = -1$ if e'_i belongs to Π_j and is oriented towards node x_j.

4.3 Cutset Representation of Graphs

The nodes of G may be coded injectively by $\{0,1\}$ vectors according to their membership to the properties p_j determined by the cuts C_j, resulting in an $n \times (n-1)$ matrix $S_a = [s_{i,j}]$, called the ***state matrix***, such that $s_{i,j} = 1$ if $x_i \in p_j$, and $s_{i,j} = 0$ if $x_i \notin p_j$. Let $S_a = [X_1 \cdots X_n]^t$, where the X_i are column vectors. The set $\{X_i^t | \, i \leq n\}$ of rows of S_a, representing nodes x_i, together with the set $\{C_i | \, i < n\}$ of rows of Q_f, representing fundamental cutsets, provide a representation of G. These data are also sufficient for retrieving the spanning tree. Actually, there is exactly one way to assemble the row vectors C_i into a matrix of the form $Q_f = \left[Q_{f11} \; I_{n-1}\right]$; and an ordered pair of vectors (X_k, X_l) represents an edge $e_j = (x_k, x_l)$ if and only if $X_l - X_k = Q_f(\cdot, j)$.

4.4 Variant Representations

A variant representation of G is given by the pair of matrices P and Q_{f11}. As a matter of fact, the reduced incidence matrix $A_1 = \left[A_{11} \; A_{12}\right]$ may be computed by $A_{12} = P^{-1}$ and $A_{11} = P^{-1} \cdot Q_{f11}$. The path matrix P can in turn be reconstructed from X_n and the reduced state matrix $S = [X_1, \ldots, X_{n-1}]^t$. Actually, for every $j < n$, $X_n = X_j + P_j$ where P_j is the jth column of P (coding the chain Π_j connecting x_j and x_n), hence the path matrix P and the reduced state matrix S are connected by the identity $S^t = \overbrace{[X_n, \ldots, X_n]}^{(n-1) \text{ times}}$ - P. In particular, $S = -P^t$ if all edges e'_j of U are oriented away from the reference node x_n.

4.5 Fundamental Cycles

It has some importance for the sequel to note that the information provided by the fundamental matrix Q_f is exactly the same as the information provided by the *fundamental cycle matrix* [1] B_f, defined as follows from the spanning tree U. Each chord (i.e. edge in $E \setminus U$) determines a cycle in G, consisting of this edge and the unique chain in U that connects its endpoints. This cycle may be represented by an m vector B_i with entries in $\{-1,0,1\}$ as follows: $B_i(j) = 0$ if e_j is not contained in the cycle, else $B_i(j) = 1$ or -1 depending on whether the orientation of e_j agrees with, or is opposite to the orientation of e_i within this cycle. The fundamental cycle matrix B_f is the $(m-n+1) \times m$ matrix defined by $B_f(i,j) = B_i(j)$. This matrix is of the form $B_f = \left[I_{m-n+1} \; B_{f12}\right]$, where $B_{f12} = -Q_{f11}^t$ (in particular, a branch belongs to the fundamental cycle defined by a chord if and only if the chord belongs to the fundamental cutset defined by the branch). Therefore, $B_f \cdot Q_f^t = 0$, and the vector spaces $\mathcal{V}_B$ and $\mathcal{V}_Q$ respectively generated over $\mathbb{R}$ by the fundamental cycles (rows of B_f) and by the fundamental cutsets (rows of Q_f) are orthogonal. These two vector spaces, which do not depend on the choice of the spanning tree, are indeed orthogonal complements of $\mathbb{R}^m$.

[1] called *fundamental circuit matrix* in [16]

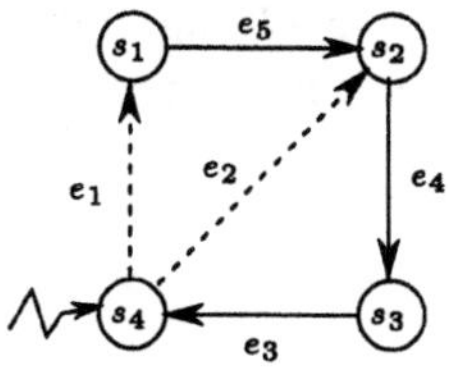

$n = 4$ vertices
$m = 5$ edges
$r = n - 1 = 3$ rank

Incidence Matrix:

$A =$

	e_1	e_2	e_3	e_4	e_5
s_1	−1	0	0	0	1
s_2	0	−1	0	1	−1
s_3	0	0	1	−1	0
s_4	1	1	−1	0	0

$$= \left[\begin{array}{c|c} A_{11} & A_{12} \\ \hline \multicolumn{2}{c}{A_2} \end{array}\right]$$

Path Matrix:

$P = A_{12}^{-1} =$

	s_1	s_2	s_3
e_3	1	0	0
e_4	1	1	0
e_5	1	1	1

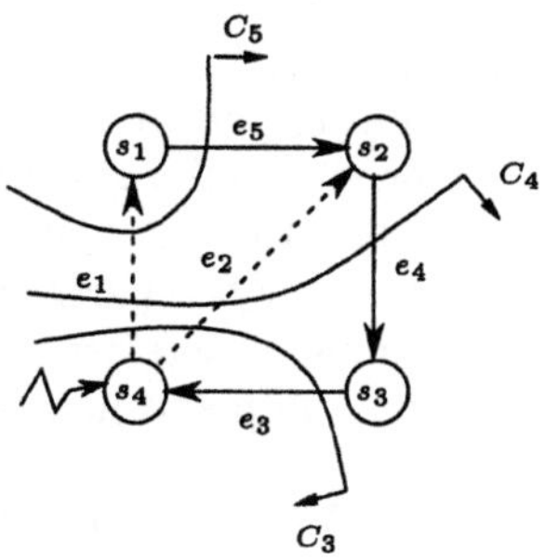

Fundamental Cutset Matrix:

$Q_f =$

	e_1	e_2	e_3	e_4	e_5
C_3	−1	−1	1	0	0
C_4	−1	−1	0	1	0
C_5	−1	0	0	0	1

$$= \left[\, Q_{f11} \;\; I_r \,\right]$$

State Matrix:

$S_a =$

	C_3	C_4	C_5
s_1	0	0	0
s_2	1	0	0
s_3	1	1	0
s_4	1	1	1

$$\boxed{Q_f = P \cdot A_1}$$

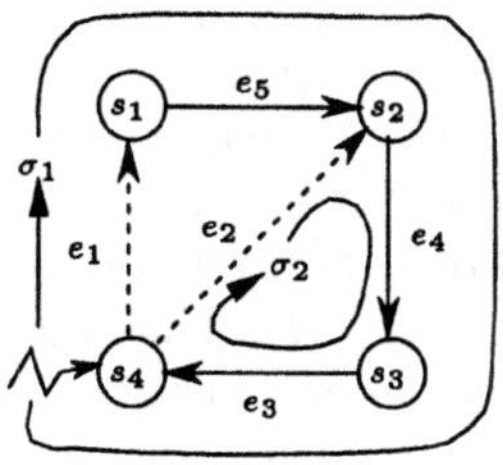

Fundamental Cycle Matrix:

$B_f =$

	e_1	e_2	e_3	e_4	e_5
σ_1	1	0	1	1	1
σ_2	0	1	1	1	0

$$= \left[\, I_{m-r} \;\; B_{f12} \,\right]$$

$$\boxed{B_{f12} = -Q_{f11}^t}$$

Fig. 8. fundamental cutsets and cycles

Every non null vector in $\mathcal{V}_B$ with entries -1, 0, and 1 is a sum of fundamental cycles and/or inverses of fundamental cycles, hence it is either a cycle or an edge disjoint union of cycles in vector form. Similarly, every non null vector C in $\mathcal{V}_Q$ with entries -1, 0, and 1 defines a cut $\{e_j | \ C(j) \neq 0\}$, but C may differ by the sign of its components from the vector which represents this cut (and also from the opposite of this vector). For a counterexample, let $C = (-1, 1)$ where e_1 and e_2 have the same target and distinct sources.

4.6 Back to set 2-Structures

We saw that G may be represented by a set of $\{0,1\}$ vectors expressing the set of properties of its nodes ($x_j \in p_i \Leftrightarrow X_j(i) = 1$), plus the set of the fundamental cutsets which define these properties (the cutset C_i defining p_i is given by the ith row of Q_f). A node x_j is then identified with the set of tokens $x_j^* = \{i | \ X_j(i) = 1\}$; similarly, an edge $e_j = (x_k, x_l)$ is identified with the ordered pair $e_j^* = (x_k^*, x_l^*)$. We show that the resulting partial 2-structure $G^* = (\{x^* | \ x \in X\}, \{e^* | \ e \in E\}, \equiv_\delta)$ is actually isomorphic to the given graph $G = (X, E, id_E)$. It is easily seen that the above representation is injective on nodes, since two different nodes of the spanning tree are always separated by a fundamental cutset. In order to prove that $G^* \cong G$, it suffices therefore to show that $\delta(e_j^*) = \delta(e_l^*)$ entails $e_j = e_l$. We establish a stronger property, namely:

Lemma 4.1 *Let $e_j = (x_k, x_l)$ be an edge of G, then for every pair of nodes x_p and x_q, $\delta(x_k^*, x_l^*) = \delta(x_p^*, x_q^*)$ entails that $x_p = x_k$ and $x_l = x_q$.*

Proof: Assuming the premises, let π be a chain connecting x_p and x_q in the spanning tree U, represented by a vector $\pi \in \{-1, 0, 1\}^m$ by "orienting" the chain from x_p to x_q. Suppose $\pi(j) = -1$, thus the edge e_j is oriented away from x_q and towards x_p in that chain. Let p_j be the property defined by the fundamental cutset which includes e_j, then necessarily $x_p, x_l \in p_j$ and $x_q, x_k \notin p_j$, hence $x_k^* \setminus x_l^* \neq x_p^* \setminus x_q^*$, contradicting our assumptions. Therefore, if we let 1_j denote the vector with a 1 at position j and 0 elsewhere, the vector $\pi - 1_j$ has all entries in $\{-1, 0, 1\}$. Since $Q_f \cdot \pi$ measures variations of properties along π, the assumption $\delta(x_k^*, x_l^*) = \delta(x_p^*, x_q^*)$ reads as $Q_f \cdot \pi = Q_f \cdot 1_j$. Thus the vector $\pi - 1_j$ lies in $\mathcal{V}_B$, and it is either a cycle or a disjoint union of cycles in vector form. Since there is no cycle in U, it follows that $\pi - 1_j$ is a cycle, hence $x_p = x_k$ and $x_q = x_l$ as was to show. ∎

Now, any set $\{p'_1, \ldots, p'_{n-1}\}$ of non trivial subsets of X determines a corresponding 2-structure $G^* = (\{x^* | \ x \in X\}, \{e^* | \ e \in E\}, \equiv_\delta)$, defined as above by setting $X_j^* = \{i | \ x_j \in p'_i\}$ and $(x_k, x_l)^* = (x_k^*, x_l^*)$. For $1 \leq i \leq n-1$, let C'_i denote the cut separating the complementary subsets $X \setminus p'_i$ and p'_i. We will show that $G^* \cong G$ whenever the corresponding vectors $C'_1, \ldots, C'_{n-1}$ are linearly independent. This is for instance the case when $p'_i = \{x_i\}$. Beware of the fact that G^* may be isomorphic to G even though $C'_1, \ldots, C'_{n-1}$ are not linearly independent. For an illustration, let $p'_1 = \{x_2, x_3\}$, $p'_2 = \{x_2, x_4\}$ and $p'_3 = \{x_1, x_3\}$ in

$G = (X, E)$ where $X = \{x_1, x_2, x_3, x_4\}$ and $E = \{e_1, e_2, e_3\}$ with $e_i = (x_1, x_{i+1})$, then $G^* \cong G$ but $C_2' + C_3' = 0$. Notice that in this representation of G the vectors e_1^*, e_2^*, and e_3^* are not linearly independent: $e_1^* + 2 \cdot e_2^* + e_3^* = 0$ even though there is no cycle in G.

Assuming that $C_1', \ldots, C_{n-1}'$ are linearly independent, let us prove that $G = (X, E, id_E)$ and $G^* = (\{x^* | \ x \in X\}, \{e^* | \ e \in E\}, \equiv_\delta)$ are isomorphic partial 2-structures. Let $x_k \neq x_l$ and assume for contradiction $x_k^* = x_l^*$. Let π be the chain in U connecting the vertices x_k and x_l. By construction of the cuts C_i', $\pi \cdot C_i' = 0$ for every $i \leq n-1$. Since $C_1', \ldots, C_{n-1}'$ are linearly independent, they span the vector space $\mathcal{V}_Q$ and π is a cycle, thus $x_k = x_l$. It remains to show that $\delta(e_j^*) = \delta(e_i^*)$ entails $e_j = e_i$.

Lemma 4.2 *Let $e_j = (x_k, x_l)$ be an edge of G, then for every pair of nodes x_p and x_q, $\delta(x_k^*, x_l^*) = \delta(x_p^*, x_q^*)$ entails that $x_p = x_k$ and $x_l = x_q$.*

Proof: Let π be a chain connecting x_p and x_q in the spanning tree, represented by a vector $\pi \in \{-1, 0, 1\}^m$ by "orienting" the chain from x_p to x_q. Suppose $\pi(j) = -1$, thus the edge e_j is in π and it is oriented away from x_q and towards x_p in that chain. From the assumption $\delta(x_k^*, x_l^*) = \delta(x_p^*, x_q^*)$ and by construction of the cuts C_i', it follows that $\pi \cdot C_i' = 1_j \cdot C_i'$ for all $i \leq n-1$, where 1_j denotes the vector with a 1 at position j and 0 elsewhere. Thus $(\pi - 1_j) \cdot C_i' = 0$ for all i, and since $C_1', \ldots, C_{n-1}'$ form a basis of the vector space $\mathcal{V}_Q$, it follows that $(\pi - 1_j) \cdot C_k = 0$ for all $k \leq n-1$ and in particular for $h = j - (m - n + 1)$. Now the first $m - n + 1$ entries of the vector $\pi - 1_j$ are zeros and the last $n - 1$ entries of C_h are zeros but $C_h(j)$ which is 1. Therefore, $\pi(j) = 1$ and we have reached a contradiction. Thus the vector $\pi - 1_j$ has all entries in $\{-1, 0, 1\}$. Since $(\pi - 1_j) \cdot C_i' = 0$ for all i, the vector $\pi - 1_j$ lies in $\mathcal{V}_B$, and it is either a cycle or a disjoint union of cycles in vector form. Since there is no cycle in U, it follows that $\pi - 1_j$ is a cycle, hence $x_p = x_k$ and $x_q = x_l$ as was to show. ∎

We now give an example (see Fig. 9) showing that the computation of cuts and cutsets cannot lead directly to a net representation of $G = (X, E)$. Let

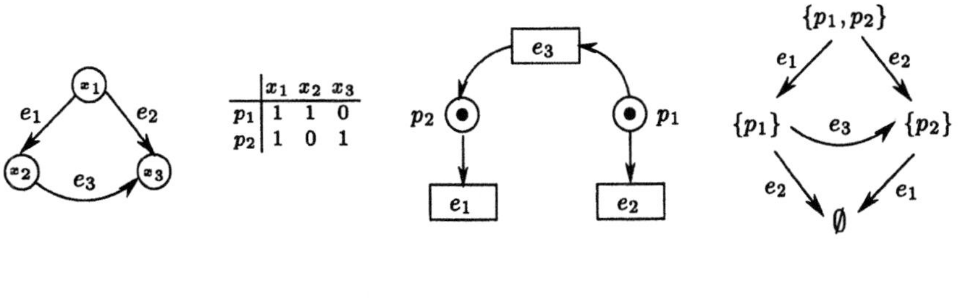

Fig. 9. elementary net system associated with a basis of cuts

$X = \{x_1, x_2, x_3\}$ and $E = \{e_1, e_2, e_3\}$ with $e_1 = (x_1, x_2)$, $e_2 = (x_1, x_3)$ and $e_3 = (x_2, x_3)$. A basis of cuts for G is given by the vectors $C_1 = (0, -1, -1)$

and $C_2 = (-1,0,1)$, inducing respective properties $p_1 = \{x_1, x_2\}$ and $p_2 = \{x_1, x_3\}$ such that $x_1^* = \{p_1, p_2\}$ and $x_2^* = \{p_1\}$ and $x_3^* = \{p_2\}$. Now let $\mathcal{N} = (\{p_1, p_2\}, E, F, x_1^*)$ be the elementary net system such that ${}^\bullet e_1 = x_1^* \setminus x_2^* = \{p_2\}$, $e_1{}^\bullet = x_2^* \setminus x_1^* = \emptyset$, ${}^\bullet e_2 = x_1^* \setminus x_3^* = \{p_1\}$, $e_2{}^\bullet = x_3^* \setminus x_1^* = \emptyset$, and ${}^\bullet e_3 = x_2^* \setminus x_3^* = \{p_1\}$, $e_3{}^\bullet = x_3^* \setminus x_2^* = \{p_2\}$. The case graph of $\mathcal{N}$ is not isomorphic to the initialized partial set 2-structure $(X^*, E^*, \equiv_\delta, x_1^*)$, due to the presence of two additional transitions $x_2^*[e_2 > \emptyset$ and $x_3^*[e_1 > \emptyset$.

4.7 Cuts and Regions

A non trivial region p of a partial 2-structure $(X, E, \equiv)$ always determines and is determined by a cut C of (X, E), which we may therefore call a ***regional cut.*** If we identify cuts C with the corresponding vectors $C : E \to \{-1,0,1\}$, then a cut is regional if and only if it is compatible with the equivalence $\equiv$ in the sense that $e \equiv e' \Rightarrow C(e) = C(e')$ for all $e, e' \in E$. In particular, all cuts are regional when $\equiv$ is the identity relation.

Let us adapt the above to transition systems. We saw that a non trivial region R of (S, E, T) is always determined from a corresponding map $\eta : E \to \{-1,0,1\}$ such that $\eta(e) = -1$ if $R^\bullet e$, $\eta(e) = 1$ if $e^\bullet R$, and $\eta(e) = 0$ otherwise ($\eta(e) = \chi_R(s') - \chi_R(s)$ when $s \xrightarrow{e} s'$). Let $\ell : T \to E$ be the labelling function such that $\ell(s \xrightarrow{e} s') = e$. Then a map $\eta : E \to \{-1,0,1\}$ determines a region in a pre-elementary transition system (S, E, T) if and only if the map $C : T \to \{-1,0,1\}$ defined by $C(t) = \eta(\ell(t))$ is a cut of the underlying graph (S, T).

On that basis, let us try to point out the obstacles to a polynomial synthesis of elementary net systems. On one hand, one can compute in polynomial time a linear basis for the real vector space $\mathcal{V}_Q$ which contains all cuts, but also elements which are not cuts even though all their entries are in $\{-1,0,1\}$. On the other hand, abstract regions are quotients of cuts, but it is not possible to derive a basis of abstract regions from a basis of cuts since abstract regions are not closed under summation. A well known recipe for getting rid of the first problem is to replace the real field $\mathbb{R}$ by the boolean field $\mathbf{2}$ in the definition of the vector space $\mathcal{V}_Q$. The second problem will then be overcome by a slight adaptation of the definition of regions, amounting to embed the elementary nets in a wider class of one-safe nets which have actually a polynomial time synthesis.

5 Flip Flop Nets and their Synthesis

We examine in this section extended regions in automata, defined as sets of states R such that all transitions with the same label are incident jointly to R, possibly inwards for some transitions and outwards for the others, or are not incident to R. A class of one-safe nets based on these regions, called ***flip flop nets*** and extending elementary nets, has been defined in [39]. We show that the synthesis problem for flip flop nets may be solved in polynomial time, following techniques of linear algebra based on cutsets. Pairs of complementary regions in an automaton may be identified with vectors $\eta : E \to \mathbf{2}$; these maps form a vector space over **2**, a basis of which is easily derived from any set of fundamental cutsets of the (undirected) graph underlying the automaton.

5.1 The Vector Space of Cuts

Let $\mathcal{A} = (S, E, T, s_n)$ be a loop-free, reachable and reduced finite automaton (not necessarily simple), with $S = \{s_1, \ldots, s_n\}$ and $T = \{t_1, \ldots, t_m\}$. Let $\partial^0(t) = s$, $\partial^1(t) = s'$ and $\ell(t) = e$ denote the respective source, target and label of a transition $t = s \xrightarrow{e} s'$. Let $A = [a_{i,j}]$ be the incidence matrix of the (undirected) graph (S, T). All definitions and results from section 4 carry to (undirected) graphs up to the replacement of $\mathbb{R}$ by **2**, see e.g. [32, 17]. Recall that a cut is a cutset or an edge-disjoint union of cutsets, where a cutset is a minimal set of edges whose removal increases the number of connected components by one. Cuts and cutsets are now represented as boolean vectors in $\mathbf{2} <T> \cong \mathbf{2}^m$, [2] and similarly for cycles and for edge-disjoint unions of cycles. The (pointwise) sum of two cuts is a cut, and similarly for two edge-disjoint unions of cycles. In other words, cuts and edge-disjoint union of cycles form vector spaces over **2**, let $\mathcal{V}_Q$ and $\mathcal{V}_B$. These vector spaces are respectively spanned by the rows of the fundamental cutset matrix Q_f and by the rows of the fundamental cycle matrix B_f, jointly computed from any spanning tree by the algorithms described in section 4, interpreted over the boolean field. Thus Q_f and B_f are $(n-1) \times m$ and $(m - n + 1) \times m$ matrices with boolean entries such that $B_f \cdot Q_f^t = 0$. Therefore $\mathcal{V}_Q$ and $\mathcal{V}_B$ form orthogonal complements in the boolean vector space $\mathbf{2} <T> \cong \mathbf{2}^m$.

5.2 The Vector Space of Abstract Regions

Our purpose is to transport the linear algebraic methods from the vector space $\mathbf{2} <T>$ to the vector space $\mathbf{2} <E>$ through the labelling function $\ell : T \to E$, which maps transitions to their labelling events.

Definition 5.1 *A cut $C = [c_j]$ is a regional cut if $\ell(t_j) = \ell(t_k) \Rightarrow c_j = c_k$ for all $1 \leq j, k \leq m$. An abstract region is a map $\eta : E \to \mathbf{2}$ such that $c_j = \eta(\ell(t_j))$ defines a regional cut $C = \eta \circ \ell$.*

By an abuse of notation, we make no distinction between vectors $C \in \mathbf{2} <T> \cong \mathbf{2}^m$ and the corresponding maps $C : T \to \mathbf{2}$. We make a similar confusion between maps $\eta : E \to \mathbf{2}$ and vectors $\eta \in \mathbf{2} <E> \cong \mathbf{2}^l$, where $E = \{e_1, \ldots, e_l\}$. Because $\mathcal{A}$ is reduced, regional cuts and abstract regions are in a bijective correspondence. Moreover, given any pair of regional cuts $C = \eta \circ \ell$ and $C' = \eta' \circ \ell$, their sum is a regional cut $C + C' = (\eta + \eta') \circ \ell$. Thus abstract regions form a subspace of the vector space $\mathbf{2} <E> \cong \mathbf{2}^l$. A method for computing a basis of abstract regions is indicated below.

[2] If K is a ring (or a field) and X a set (of generators) we let $K <X>$ denote the K-module (or vector space) freely generated by X.

Since the vector spaces $\mathcal{V}_Q$ and $\mathcal{V}_B$ are orthogonal complements, η is an abstract region if and only if $\eta \circ \ell \in \mathcal{V}_Q$ if and only if $C = \eta \circ \ell$ is orthogonal to all fundamental cycles B_i (rows of B_f). For any cycle $B = [b_j]$, let $\Pi(B) = [\pi_k] \in \mathbf{2}<E>$ be the *Parikh image* of B given by $\pi_k = \sum\{b_j |\ \ell(t_j) = e_k\}$. Otherwise stated $\pi_k = \sum_{j=1}^{m} \varphi_j(e_k)$ where $\varphi_j(e_k) = 1$ if $\ell(t_j) = e_k$ and $b_j = 1$, else 0. Then $C \cdot B = 0$ if and only if $\sum_{j=1}^{m} \eta(\ell(t_j)) \cdot b_j = 0$ if and only if $\sum_{k=1}^{l}(\eta(e_k) \cdot \sum_{j=1}^{m} \varphi_j(e_k)) = 0$ if and only if $\eta \cdot \Pi(B) = 0$. Since $\Pi(B + B') = \Pi(B) + \Pi(B')$, it follows that η is an abstract region if and only if η is orthogonal to the linear subspace of $\mathbf{2}<E>$ spanned by the Parikh images $\Pi(B_i)$ of the fundamental cycles B_i. Let $l - p$ be the dimension of the linear space $\Pi(\mathcal{V}_B)$. A basis of abstract regions $\{\eta_1, \ldots, \eta_p\}$ follows, e.g. by Gauss resolution.

5.3 Flip Flop Regions and Flip Flop Nets

Definition 5.2 *A flip flop region in* $\mathcal{A} = (S, E, T, s_n)$ *is a non trivial subset of states* $R \subset S$ *whose characteristic function* $\chi_R : S \to \mathbf{2}$ *satisfies* $\forall i \in \{1, \ldots, m\}$ $\chi_R(\partial^1(t_i)) = \chi_R(\partial^0(t_i)) + \eta(\ell(t_i))$ *for some abstract region* η.

Since $\mathcal{A}$ is reachable and reduced, abstract regions η are in bijective correspondence with pairs of complementary regions R and $S \setminus R$. Flip flop regions, represented as vectors $\chi_R \in \mathbf{2}<S> \cong \mathbf{2}^n$, form a linear subspace of $\mathbf{2}^n$, closed under the complementation operation $\chi_R + \mathbf{1} = \chi_{S\setminus R}$.

The definition of flip flop nets stems from the analysis of the possible crossing relations between a flip flop region R and the transitions bearing an identical label. All possible cases are covered by four relations:

$$\begin{array}{ll} R^\bullet e : \forall t \in T & \ell(t) = e \Rightarrow (\partial^0(t) \in R \ \wedge \ \partial^1(t) \notin R) \\ e^\bullet R : \forall t \in T & \ell(t) = e \Rightarrow (\partial^0(t) \notin R \ \wedge \ \partial^1(t) \in R) \\ e^\perp R : \forall t \in T & \ell(t) = e \Rightarrow (\partial^0(t) \in R \Leftrightarrow \partial^1(t) \in R) \\ e^\times R : \forall t \in T & \ell(t) = e \Rightarrow (\partial^0(t) \in R \Leftrightarrow \partial^1(t) \notin R) \end{array}$$

Conversely, any non trivial subset of states $R \subset S$ satisfying $R^\bullet e \vee e^\bullet R \vee e^\perp R \vee e^\times R$ for all events $e \in E$ is a flip flop region, associated with an abstract region η such that $\eta(e) = 0$ if and only if $e^\perp R$. It is now patent that flip flop regions are an extension of elementary regions, which must satisfy $R^\bullet e \vee e^\bullet R \vee e^\perp R$. Observe that $e^\bullet R \Rightarrow e^\times R$ and $R^\bullet e \Rightarrow e^\times R$. However these three relations play incomparable roles in flip flop nets, where they are called respectively *input* ($R^\bullet e$), *output* ($e^\bullet R$), and *swap* ($e^\times R$).

Definition 5.3 *A flip flop net is a triple* $N = (P, E, W)$ *where* P *is the set of places or conditions,* E *is the set of events, and* $W : P \times E \to \{\mathtt{input}, \mathtt{output}, \mathtt{nop}, \mathtt{swap}\}$ *is a matrix such that* $\forall e \in E \quad \exists p \in P \quad W(p, e) \neq \mathtt{nop}$. *A case of* N *is a map* $M : P \to \mathbf{2}$. *An event* e *has concession at* M *if and only if* $\forall p \in P$ $(W(p, e) = \mathtt{input} \Rightarrow M(p) = 1) \wedge (W(p, e) = \mathtt{output} \Rightarrow M(p) = 0)$. *The event* e *may then fire, resulting in a transition* $M[e\!>M'$ *where for every condition* p: $W(p, e) = \mathtt{nop} \Rightarrow M'(p) = M(p)$ *and* $W(p, e) \neq \mathtt{nop} \Rightarrow M'(p) = 1 + M(p)$.

Definition 5.4 *A flip flop net system is a structure $\mathcal{N} = (P, E, W, M_0)$, where M_0 is a case of the underlying flip flop net $N = (P, E, W)$. The sequential case graph of $\mathcal{N}$ is the automaton $\mathcal{N}^* = (S, E, T, M_0)$ where S is the set of cases reachable from M_0 by sequences of steps $M[e{>}M'$ and T is the subset of these steps in $S \times E \times S$.*

It follows that for every condition $p \in P$, the sets $\{M \in S|\ M(p) = 1\}$ and $\{M \in S|\ M(p) = 0\}$ are complementary flip flop regions of $\mathcal{N}^*$. So the sets of states $\{s_1, s_5, s_6\}$ and $\{s_2, s_3, s_4\}$ are flip flop regions in the example shown in Fig. 10.

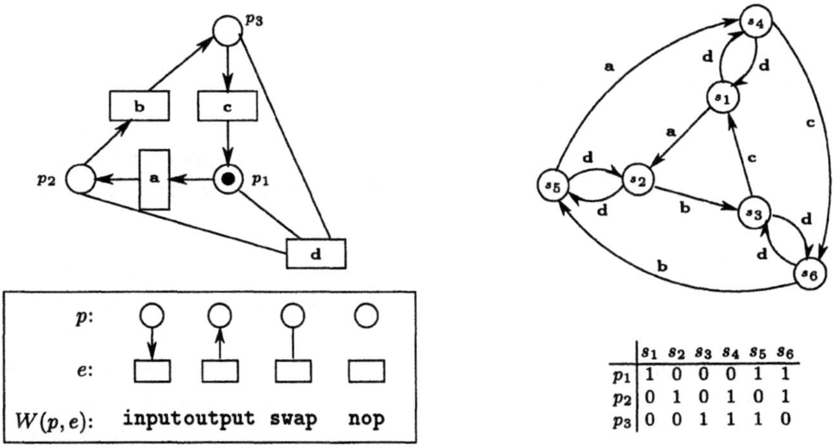

	s_1	s_2	s_3	s_4	s_5	s_6
p_1	1	0	0	0	1	1
p_2	0	1	0	1	0	1
p_3	0	0	1	1	1	0

Fig. 10. a flip flop net and its sequential case graph

5.4 Representation Result

The following result was established in [39].

Proposition 5.5 *A finite loop-free automaton $\mathcal{A} = (S, E, T, s_n)$, reachable from s_n and reduced, is isomorphic to the sequential case graph of a flip flop net system if and only if the following conditions are satisfied for R ranging over the set $\mathcal{R}_{\text{FFN}}(\mathcal{A})$ of flip flop regions of $\mathcal{A}$:*
SSP: $\forall s, s' \in S \quad s \neq s' \Rightarrow \exists R \quad (s \in R \Leftrightarrow s' \notin R)$.
ESSP: $\forall s \in S \quad \forall e \in E \quad \textit{not}\ s \xrightarrow{e} \quad \Rightarrow \exists R \quad (R^\bullet e \wedge s \notin R) \vee (e^\bullet R \wedge s \in R)$.

A synthesis algorithm follows easily. Let $\{\eta_1, \ldots, \eta_p\}$ be a basis of abstract regions of $\mathcal{A}$, computed from some spanning tree $U \subseteq T$. For each state $s_i \in S$, let p_i be the chain connecting s_i and s_n in the spanning tree. An instance $\text{SSP}(s_i, s_j)$ of the states separation problem can be solved if and only if $\eta_k \cdot (\Pi(p_i) + \Pi(p_j)) \neq 0$ for some $k \in \{1, \ldots, p\}$, where $\Pi(p) \in \mathbf{2}{<}E{>}$ is the

Parikh image of the chain $p \in \mathbf{2} <T>$. An instance of $\text{ESSP}(s_i, e)$ can be solved if and only if there exists a linear combination $\eta = \sum_{k=1}^{p} \alpha_k \cdot \eta_k$, where $\alpha_k \in \mathbb{Z}$, satisfying $\eta \cdot [\Pi(p_i) + \Pi(p_j)] = 1$ for every state s_j in which event e is enabled. When these conditions are satisfied, a net system $\mathcal{N} = (P, E, W, M_0)$ such that $\mathcal{A} \cong \mathcal{N}^*$ may be constructed by assembling the atomic net systems $\mathcal{N}_p$ defined from conditions p as follows:

1. for each instance $\text{SSP}(s_i, s_j)$ solved by η_k, let p be the condition such that $W(p, e) = \texttt{swap}$ if $\eta_k(e) = 1$ and $W(p, e) = \texttt{nop}$ if $\eta_k(e) = 0$, with $M_0(p)$ fixed arbitrarily to 0 or 1;
2. for each instance $\text{ESSP}(s_i, e)$ solved by an abstract region $\eta = \sum_{k=1}^{p} \alpha_k \cdot \eta_k$, let p be the condition such that $W(p, e) = \texttt{input}$, $M_0(p) = \eta \cdot \Pi(p_i)$, and for $e' \neq e$, $W(p, e') = \texttt{swap}$ if $\eta(e') = 1$, and $W(p, e') = \texttt{nop}$ if $\eta(e') = 0$.

A minimal system $\mathcal{N}'$ such that $\mathcal{A} \cong \mathcal{N}'^*$ may be obtained by eliminating from $\mathcal{N}$ redundant places. The following is proved in [39].

Proposition 5.6 *The synthesis problem for flip flop nets may be solved in time* $O(|S|^2 \times |E|^3)$*, where* S *and* E *are the respective sets of states and events of the automaton.*

The synthesis algorithm which has been suggested here is a simplified form of the synthesis algorithm for Petri nets proposed in [2] and presented in section 7 of this survey. The case of Petri nets is significantly more complex, to a limited extent because the integer module $\mathbb{Z} <E>$ is more complicated that the boolean vector space $\mathbf{2} <E>$, and to a large extent because combinatorial approximation techniques are needed for the synthesis of Petri nets, while they are useless for flip flop nets.

Before tackling the synthesis problem for Petri nets, we make a detour to show that the striking similarity of the representation results for elementary net systems and flip flop net systems is not incidental, and does not depend on the type of nets.

6 Regions for Arbitrary Types of Nets

The automata $\mathcal{A} = (S, E, T, s_0)$ considered in this section are always assumed to be reachable and deterministic, but they may not be simple, nor reduced, and they are not necessarily finite. The transition systems (S, E, T) are always assumed to be deterministic, but they are not necessarily connected. Recall that a morphism of transition systems $(\sigma, \eta) : (S, E, T) \to (S', E', T')$ is a pair of maps $\sigma : S \to S'$ and $\eta : E \to E'$ such that $s \stackrel{e}{\to} s'$ in T entails $\sigma(s) \stackrel{\eta(e)}{\to} \sigma(s')$ in T'; morphisms of automata are morphisms of the underlying transition systems which map the initial state to the initial state.

The extension of the concept of regions to arbitrary types of nets stems from the following observation. Let τ_{FFN} be the transition system given in Fig. 11. Solving the synthesis problem for $\mathcal{A} = (S, E, T, s_0)$ w.r.t. flip flop nets amounts

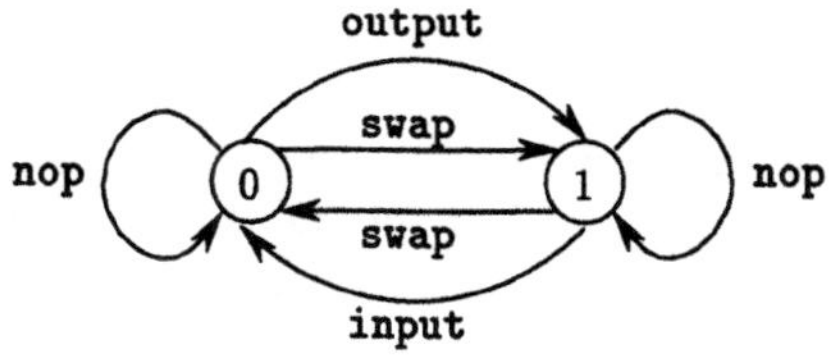

Fig. 11. the type τ_{FFN} of flip flop nets

to amalgamate on E a set of atomic net systems $\mathcal{N}_p = (\{p\}, E, W, M_0)$, defined from morphisms $p = (\sigma, \eta) : (S, E, T) \to \tau_{FFN}$ such that $W((\sigma, \eta), e) = \eta(e)$ and $M_0((\sigma, \eta)) = \sigma(s_0)$. The resulting net system $\mathcal{N} = \sum_{p \in P} \mathcal{N}_p$ has a case graph $\mathcal{N}^*$ isomorphic to $\mathcal{A}$ if and only if the family $\{\mathcal{N}_p | \, p = (\sigma, \eta) \in P\}$ is *admissible* in the sense that the following two separation conditions are satisfied:
SSP: $\forall s, s' \in S \quad s \neq s' \Rightarrow \exists (\sigma, \eta) \in P \quad \sigma(s) \neq \sigma(s')$.
ESSP: $\forall s \in S \quad \forall e \in E \quad \text{not } (s \xrightarrow{e}) \Rightarrow \exists (\sigma, \eta) \in P \quad \text{not } (\sigma(s) \xrightarrow{\eta(e)}) \quad \text{in} \quad \tau_{FFN}$.
Thus the concept of regions as sets of states may profitably be replaced by the richer concept of regions as morphisms, which is actually the central concept for the synthesis of net systems. The two concepts are not strictly equivalent for flip flop nets: several morphisms $(\sigma, \eta) : (S, E, T) \to \tau_{FFN}$ may actually determine the same set of states $\sigma^{-1}(\{1\})$, i.e. the same set theoretic region, because the component σ on states does not determine the component η (even though the transition system is connected and reduced).

Our aim is to show that the representation results which have been stated so far for elementary nets and for flip flop nets may be established at once for all possible types of nets, using the concept of regions as morphisms.

6.1 Types of Nets

For the sake of a uniform presentation, we depart here from the traditional definition of nets and adopt a parametric definition covering elementary nets, flip flop nets, and Petri nets as particular instances. The parameters of this general definition are called *types of nets* [6].

Definition 6.1 *A type of nets is a deterministic transition system $\tau = (LS, LE, \tau)$, where LS and LE are the respective sets of local states and local events, and $\tau \subseteq LS \times LE \times LS$ defines the partial action of local events on local states.*

Definition 6.2 *A net of type τ is a triple $N = (P, E, W)$ where P is a set of places, E is a set of events, and $W : P \times E \to LE$ is the weight matrix. A marking is a mapping $M : P \to LS$. A net system of type τ is a structure $\mathcal{N} = (P, E, W, M_0)$ where M_0, the initial marking, is a marking of the underlying net $N = (P, E, W)$.*

A net or net system is place simple if all rows of the weight matrix are different; it is event simple if all columns of the weight matrix are different. All nets and net systems considered in this section are assumed to be place simple but not necessarily event simple.

A net may be seen as an undirected complete bipartite graph whose edges are weighted by local events. As such, nets are of a static nature, but types (of nets) define their dynamics: the partial actions of events on markings may be inferred from the partial actions of local events on local states, using the weight matrix to control products of local events. The following definition extends in this way the usual *sequential firing rule.*

Definition 6.3 *Given a net $N = (P, E, W)$, of type $\tau = (LS, LE, \tau)$, the (sequential) marking graph of N is the transition system (LS^P, E, T) with set of transitions T defined by $(M \xrightarrow{e} M') \in T$ if and only if $\forall p \in P \ (M(p) \xrightarrow{W(p,e)} M'(p)) \in \tau$. Given a net system $\mathcal{N} = (P, E, W, M_0)$, the (sequential) marking graph of $\mathcal{N}$ is the (dual) automaton $\mathcal{N}^* = (S, E, T_S, M_0)$ where S is the inductive closure of $\{M_0\}$ w.r.t. forward transitions in T, and $T_S = T \cap (S \times E \times S)$.*

Thus an event has concession at marking M if and only if for every place p, the local event $W(p, e)$ is enabled at the local state $M(p)$ in the transition system τ (defining the type of the net). A net system is reduced if every event e has concession at some reachable marking and if for every pair of distinct places p and p', there exists some reachable marking M such that $M(p) \neq M(p')$. The net systems which we consider are generally not reduced. We now illustrate the above definitions on two classical examples, namely elementary nets and Petri nets.

Let τ_{EN} be the transition system shown in Fig. 12. The elementary nets

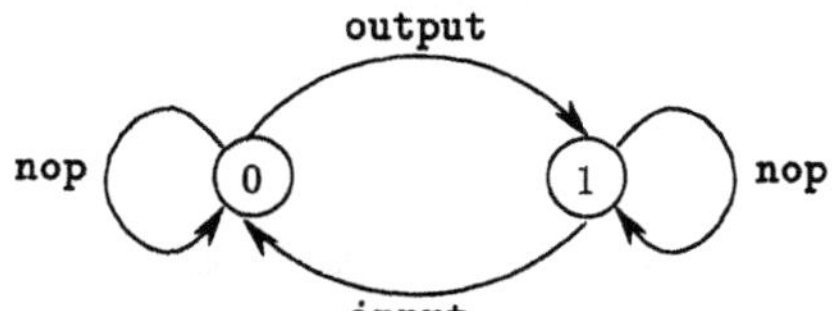

Fig. 12. the type τ_{EN} of elementary nets

(P, E, F) correspond bijectively with nets (P, E, W) of type τ_{EN}, with $W(p, e) =$ **input** $\Leftrightarrow F(p, e)$ and $W(p, e) =$ **output** $\Leftrightarrow F(e, p)$, and $W(p, e) =$ **nop** otherwise. One may easily verify that the corresponding nets have identical marking graphs.

Let us now recall the classical definition of Petri nets.

Definition 6.4 *A Petri net is a triple $N = (P, E, F)$ where P and E are disjoint sets of places and events, and F is a function, $F : (P \times E) \cup (E \times P) \to \mathbb{N}$. A marking of N is a map $M : P \to \mathbb{N}$. An event e has concession at M if and only if $\forall p \in P \quad F(p, e) \leq M(p)$. An event e which has concession at M may fire, resulting in a transition $M[e{>}M'$ where $\forall p \in P \quad M'(p) = M(p) - F(p, e) + F(e, p)$. A Petri net N is said to be pure if $\forall p \in P \ \forall e \in E \ \ F(p, e) \times F(e, p) = 0$.*

Let the type of pure Petri nets be the transition system $\tau_{PPN} = (\mathbb{N}, \mathbb{Z}, T)$ such that $n \xrightarrow{z} n'$ if and only if $n' = n + z$, i.e. τ_{PPN} is the full subgraph of the Cayley graph of $\mathbb{Z}$ induced by the restriction on the subset of nodes in $\mathbb{N}$. Pure Petri nets (P, E, F) are linked by a marking graph preserving bijective correspondence with nets (P, E, W) of type τ_{PPN} given by $W(p, e) = F(e, p) - F(p, e)$.

Let the type of Petri nets be the transition system $\tau_{PN} = (\mathbb{N}, \mathbb{N} \times \mathbb{N}, T)$ such that $n \xrightarrow{(p,q)} n'$ if and only if $n \geq p$ and $n' = (n - p) + q$. Petri nets are set in bijective correspondence with nets of type τ_{PN} by the relation $W(p, e) = (F(p, e), F(e, p))$. With this correspondence, the firing rule stated in Def. 6.3 reads actually as

$$M[e{>}M' \;\Leftrightarrow\; \forall p \in P \;\; M(p) \geq F(p,e) \wedge M'(p) = M(p) - F(p,e) + F(e,p)$$

The above Petri nets are a particular instance of the generalized Petri nets studied in [20]. In this paper, Droste and Shortt parametrize the classical definition of Petri nets (Def. 6.4), in which $\mathbb{N}$ is substituted for by the positive part G^+ of a partially ordered abelian group G. These authors further classify types of Petri nets over a fixed group G by the set of pairs $((F(p,e), F(e,p)) \in G^+ \times G^+$ occurring in associated subclasses of nets. For instance, condition-event nets are obtained by restricting nets over $\mathbb{Z}$ to the pairs $((F(p,e), F(e,p)) \in \{(0,0), (0,1), (1,0), (1,1)\}$. Note that all types of nets which we have defined so far can similarly be obtained from Cayley graphs (G, G, T) (i.e. $g' \xrightarrow{g} g''$ in T if and only if $g'' = g' + g$) by eliminating nodes and/or by restricting group actions to partially defined group actions. For instance, τ_{EN} is the Cayley graph of $\mathbb{Z}/_{3\mathbb{Z}}$ restricted on nodes 0 and 1, with $\texttt{nop} = 0$, $\texttt{output} = 1$, and $\texttt{input} = 2$. Similarly, τ_{FFN} is obtained from the Cayley graph of $\mathbb{Z}/_{2\mathbb{Z}}$ by identifying action 0 with `nop`, action 1 with `swap`, and the partial action 1 defined at node 0 (resp. at node 1) with `output` (resp. with `input`). Therefore, all nets considered so far are *reversible* in the sense that they have co-deterministic sequential marking graphs ($M_1[e{>}M$ and $M_2[e{>}M$ entail $M_1 = M_2$). Nevertheless, flip flop nets are not Petri nets over a group according to the definition of Droste and Shortt. The main reason why we do not stick here to types of nets based on groups is that we want to cover also non reversible nets, such as trace nets (see 6.5).

6.2 Regions as Morphisms

The firing rule for nets stated in Def. 6.3 tells us that for every place p in a net of type τ, the pair of maps (σ_p, η_p) defined by $\sigma_p(M) = M(p)$ and $\eta_p(e) = W(p, e)$ is a morphism of transition systems from the marking graph of the net to the type τ. Therefore, if we forget the internal structure of states in the marking graph, identified with any isomorphic transition system (S, E, T), and if we identify a place p with its extension (σ_p, η_p), we can rediscover the places of the net (and also discover implicit places) as morphisms $(\sigma, \eta) : (S, E, T) \to \tau$. This motivates the following definition of regions for arbitrary types of nets.

Definition 6.5 *Given a transition system $A = (S, E, T)$ and a type of nets $\tau = (LS, LE, \tau)$, the set $\mathcal{R}_\tau(A)$ of τ-type regions in A is the set of morphisms from A to τ.*

By an abuse of notation, we extend the above definition to automata by letting $\mathcal{R}_\tau(\mathcal{A}) = \mathcal{R}_\tau(A)$ where A is the transition system underlying the automaton $\mathcal{A}$. We now illustrate this definition on elementary nets and on Petri nets.

An elementary region in $A = (S, E, T)$ is a morphism $(\sigma, \eta) : A \to \tau_{EN}$. The map η classifies events $e \in E$ into three families according to their relationship with the property $R = \sigma^{-1}(\{1\})$: all events e such that $\eta(e) = \texttt{input}$ take R as an input condition and falsify R ($s \xrightarrow{e} s' \Rightarrow s \in R \wedge s' \notin R$), all events e such that $\eta(e) = \texttt{output}$ take the falsity of R as a precondition and establish R ($s \xrightarrow{e} s' \Rightarrow s \notin R \wedge s' \in R$), and the remaining events such that $\eta(e) = \texttt{nop}$ do not modify R ($s \xrightarrow{e} s' \Rightarrow (s \in R \Leftrightarrow s' \in R)$). One recognizes in $R = \sigma^{-1}(\{1\})$ a region according to the original definition of Ehrenfeucht and Rozenberg.

A pure Petri region in $A = (S, E, T)$ is a morphism $(\sigma, \eta) : A \to \tau_{PPN}$ (see Fig. 13 for an illustration). Here the map σ measures the availability of a

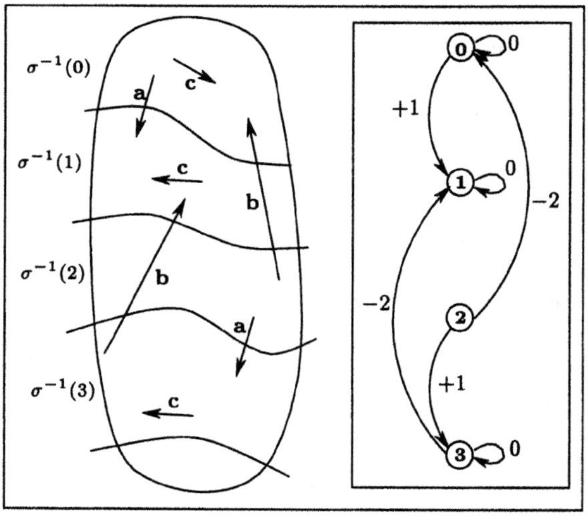

Fig. 13. a pure Petri region as a morphism: $A \xrightarrow{(\sigma,\eta)} \tau_{PPN}$

resource at each state $s \in S$, and the map η classifies events $e \in E$ according to the amount of resource which they produce (when $\eta(e) > 0$) or consume (when $\eta(e) < 0$) at each firing. When A is a finite transition system, the abstract regions η defined in this way are in bijective correspondence with weighted synchronic distances in A, measuring the relative degree of freedom of the respective subsets of events e such that $\eta(e) < 0$, resp. $\eta(e) > 0$ [12].

A Petri region in $A = (S, E, T)$ is a morphism $(\sigma, \eta) : A \to \tau_{PN}$. Here again the map σ measures the availability of a resource at each state. The map η

classifies events according to associated pairs $\eta(e) = ({}^\bullet\eta(e), \eta^\bullet(e))$ where ${}^\bullet\eta(e)$ measures the amount of resource consumed for triggering e while $\eta^\bullet(e)$ measures the amount of resource produced by e, amounting to a neat variation of resource $\eta^\bullet(e) - {}^\bullet\eta(e)$. These Petri regions coincide with the regions which have been defined by Mukund [33] (in the larger framework of step transition systems) and which have been adapted by Droste and Shortt [20] to Petri nets over partially ordered abelian groups.

6.3 A Galois Connection between Automata and Nets

We saw that regions may serve to reverse the production of marking graphs. The reversing process may also be applied to arbitrary transition systems, leading to the following definitions.

Definition 6.6 *Given a transition system $A = (S, E, T)$ and a type of nets τ, the dual of A is the net $A^* = (\mathcal{R}_\tau(A), E, W)$ with weights defined by $W((\sigma,\eta), e) = \eta(e)$. For any subset $\mathcal{R}$ of $\mathcal{R}_\tau(A)$, let $A^*_{\mathcal{R}}$ denote the subnet of A^* with restricted set of places $\mathcal{R}$.*

Definition 6.7 *Given an automaton $\mathcal{A}$ composed of a transition system A and an initial state s_0, and a type of nets τ, the dual of $\mathcal{A}$ is the net system $\mathcal{A}^*$ composed of the underlying net A^* and of the initial marking M_0 defined by $M_0(\sigma,\eta) = \sigma(s_0)$ for every $(\sigma,\eta) \in \mathcal{R}_\tau(\mathcal{A})$. For any subset $\mathcal{R}$ of $\mathcal{R}_\tau(A)$, let $\mathcal{A}^*_{\mathcal{R}}$ denote the subnet system of $\mathcal{A}^*$ with restricted set of places $\mathcal{R}$.*

We will show that the two $()^*$ operators mapping the automaton $\mathcal{A}$ to the net system $\mathcal{A}^*$ and the net system $\mathcal{N}$ to its marking graph $\mathcal{N}^*$ form a Galois connection: $\mathcal{A} \le \mathcal{N}^* \Leftrightarrow \mathcal{N} \le \mathcal{A}^*$. The main difficulty is to construct the appropriate order relations. One expects in particular $\mathcal{A} \le \mathcal{N}^*_p \Leftrightarrow \mathcal{N}_p \le \mathcal{A}^*$ for every region $p = (\sigma_p, \eta_p) \in \mathcal{R}_\tau(\mathcal{A})$ where $\mathcal{N}_p$ is the *atomic* subnet system of $\mathcal{A}^*$ with sole place p (i.e. $\mathcal{N}_p = \mathcal{A}^*_{\{p\}}$) and $\mathcal{N}^*_p$ is its marking graph. This particular case will help us to find out the order relation on automata. Since $\mathcal{N}_p$ is a subnet system of $\mathcal{A}^*$, both $\mathcal{N}_p \le \mathcal{A}^*$ and $\mathcal{A} \le \mathcal{N}^*_p$ are expected; by definition of regions, if E is the set of events of $\mathcal{A}$ then $(\sigma_p, 1_E)$ is an *event preserving* morphism from $\mathcal{A}$ to $\mathcal{N}^*_p$. Moreover, if there exists an event preserving morphism $(\sigma, 1_E) : \mathcal{A}_1 \to \mathcal{A}_2$ between two automata with set of events E, this morphism is necessarily unique owing to the strong properties of determinism and reachability we have assumed from all automata; therefore, if there exist morphisms $(\sigma_1, 1_E) : \mathcal{A}_1 \to \mathcal{A}_2$ and $(\sigma_2, 1_E) : \mathcal{A}_2 \to \mathcal{A}_1$, then $\mathcal{A}_1$ and $\mathcal{A}_2$ are identical up to the identity of states ($\mathcal{A}_1 =_E \mathcal{A}_2$). So let $\mathbf{Aut}(E)$ be the set of (deterministic and reachable) automata with fixed set of events E, quotiented by $=_E$, then

$$\mathcal{A}_1 \le \mathcal{A}_2 \;\; \textbf{if} \;\; \exists \sigma : \; (\sigma, 1_E) : \; \mathcal{A}_1 \to \mathcal{A}_2$$

is a partial order on $\mathbf{Aut}(E)$, such that $\mathcal{A} \le \mathcal{N}^*_p$ for every region $p \in \mathcal{R}_\tau(\mathcal{A})$. This partial order is a complete lattice, with greatest lower bounds computed as synchronized products. We remind the reader that the synchronized product $\bigwedge_{i\in I} \mathcal{A}_i$ of a family of automata $\mathcal{A}_i = (S_i, E, T_i, s_{0,i})$ indexed by $i \in I$ is the automaton (S, E, T, s_0) with components as follows: $s_0 = (s_{0,i})_{i\in I}$, S is the inductive closure of the set $\{s_0\}$ w.r.t. the *synchronized transition rule*

$$(s_i)_{i\in I} \xrightarrow{e} (s_i')_{i\in I} \quad \textbf{iff} \quad \forall i \in I \quad (s_i \xrightarrow{e} s_i') \in T_i$$

and T is the set of occurrences of this rule at states $(s_i)_{i\in I} \in S$. By definition of marking graphs, the automaton $\mathcal{N}^*$ dual to a net system $\mathcal{N} = (P, E, W, M_0)$ is actually the synchronized product $\bigwedge_{p\in P} \mathcal{N}_p^*$ of the marking graphs of its atomic subnet systems.

Concerning the order relation on net systems, the central assumption that $\mathcal{N}_p \leq \mathcal{A}^*$ for every region p of $\mathcal{A}$ leads to choose something akin to the substructure ordering: $\mathcal{N}_1 \leq_{sub} \mathcal{N}_2$ if $\mathcal{N}_1$ is $\mathcal{N}_2$ restricted on a subset of places. However *replicated* places may occur in a net system $\mathcal{N} = (P, E, W, M_0)$, i.e. places which the initial marking M_0 and the weight function W do not distinguish from one another, and we do not care about their degree of multiplicity nor about their identities. Let morphisms of net systems with fixed set of events be defined as follows: a morphism from $\mathcal{N}_1 = (P_1, E, W_1, M_{0,1})$ to $\mathcal{N}_2 = (P_2, E, W_2, M_{0,2})$ is a map $\beta : P_1 \to P_2$ such that $M_{0,1}(p) = M_{0,2}(p)$ and $W_1(p, e) = W_2(\beta(p), e)$ for all $p \in P_1$ and $e \in E$. Two net systems connected by morphisms in both directions are henceforth declared equivalent. Let **Nets**(E) denote the set of equivalence classes of net systems with set of events E (replication free nets are canonical representatives). One can equip **Nets**(E) with the partial order relation defined as:

$$\mathcal{N}_1 \leq \mathcal{N}_2 \quad \textbf{iff} \quad \exists \beta : \mathcal{N}_1 \to \mathcal{N}_2$$

This partial order is a complete lattice, with least upper bounds $\bigvee_{i\in I} \mathcal{N}_i$ of families of net systems computed by amalgamation of sets of places. Told in another way, if we identify a place p in a net system $\mathcal{N} = (P, E, W, M_0)$ with the pair $(M_0(p), \eta_p)$ such that $\eta_p(e) = W(p, e)$ for $e \in E$ then $\bigvee_{i\in I}(P_i, E, W_i, M_{0,i})$ $= (\bigcup_{i\in I} P_i, E, W, M_0)$ where $W(p, e) = W_i(p, e)$ and $M_0(p) = M_{0,i}(p)$ for $p \in P_i$. A net system $\mathcal{N}$ with set of places P is now the least upper bound $\bigvee_{p\in P} \mathcal{N}_p$ of its atomic subnet systems $\mathcal{N}_p$. In the particular case where $\mathcal{N} = \mathcal{A}^*$ is dual to the automaton $\mathcal{A}$, its set of places is the set of regions $\mathcal{R}_\tau(\mathcal{A})$, where τ is the type of $\mathcal{N}$, hence its atomic subnet systems $\mathcal{N}_p$ have the form $\mathcal{A}^*_{\{p\}}$ and we get the following.

Proposition 6.8 *Let* $\mathcal{R} \subseteq \mathcal{R}_\tau(\mathcal{A})$ *then* $\mathcal{A}^*_{\mathcal{R}} = (\bigvee_{p\in\mathcal{R}} \mathcal{A}^*_{\{p\}})$.

The key for the Galois connection between the ordered sets $(\mathbf{Aut}(E), \leq)$ and $(\mathbf{Nets}(E), \leq)$ is the following proposition, proved in [7]

Proposition 6.9 *Let* $\mathcal{N} = (\{p\}, E, W, M_0)$ *be an atomic net system of type* τ, *then* $\mathcal{A} \leq \mathcal{N}^*$ *if and only if* $M_0(p) = \sigma(s_0)$ *and* $\forall e \in E \quad W(p, e) = \eta(e)$ *for some region* $(\sigma, \eta) \in \mathcal{R}_\tau(\mathcal{A})$, *where* s_0 *is the initial state of the automaton* $\mathcal{A}$.

Thus any atomic net system $\mathcal{N}$ such that $\mathcal{A} \leq \mathcal{N}^*$ is isomorphic to $\mathcal{N}_p = \mathcal{A}^*_{\{p\}}$ for some region $p = (\sigma, \eta) \in \mathcal{R}_\tau(\mathcal{A})$. Since $\mathcal{A}^* = \bigvee\{\mathcal{A}^*_{\{p\}} | \, p \in \mathcal{R}_\tau(\mathcal{A})\}$, it follows that $\mathcal{N} \leq \mathcal{A}^*$. Conversely, by definition of the order relation on net systems, any atomic net system $\mathcal{N}$ such that $\mathcal{N} \leq \mathcal{A}^*$ is isomorphic to $\mathcal{N}_p = \mathcal{A}^*_{\{p\}}$ for some region $p = (\sigma, \eta) \in \mathcal{R}_\tau(\mathcal{A})$. Since $\mathcal{A} \leq \mathcal{N}^*_p$ by construction of the order relation on automata, it follows that $\mathcal{A} \leq \mathcal{N}^*$. Altogether, we obtain the following.

Proposition 6.10 *For any atomic net system $\mathcal{N}$, $\mathcal{A} \leq \mathcal{N}^* \Leftrightarrow \mathcal{N} \leq \mathcal{A}^*$.*

We are ready to establish the expected Galois connection between automata and net systems.

Proposition 6.11 *The two $()^*$ operators, mapping respectively the automaton $\mathcal{A}$ to the dual net system $\mathcal{A}^*$ and the net system $\mathcal{N}$ to its marking graph $\mathcal{N}^*$, constitute a Galois connection between the ordered sets* **Nets**(E) *and* **Aut**(E)*: $\mathcal{A} \leq \mathcal{N}^* \Leftrightarrow \mathcal{N} \leq \mathcal{A}^*$ for $\mathcal{A} \in$* **Aut**(E) *and $\mathcal{N} \in$* **Nets**(E).

Proof: By Prop. 6.10, $\mathcal{A} \leq \mathcal{N}^* \Leftrightarrow \mathcal{N} \leq \mathcal{A}^*$ if $\mathcal{N}$ is an *atomic* net system. Now for a net system $\mathcal{N} = \bigvee_{p \in P} \mathcal{N}_p$, where $\mathcal{N}_p$ is the atomic subnet system of $\mathcal{N}$ with the unique place p, $\mathcal{N}^* = \bigwedge_{p \in P} \mathcal{N}^*_p$ by definition of marking graphs. Thus $\mathcal{A} \leq \mathcal{N}^*$ *if and only if* $\mathcal{A} \leq \mathcal{N}^*_p$ for all $p \in P$ *if and only if* $\mathcal{N}_p \leq \mathcal{A}^*$ for all $p \in P$ (because $\mathcal{N}_p$ is atomic) *if and only if* $\mathcal{N} \leq \mathcal{A}^*$. ∎

The relations $\mathcal{A}_1 \leq \mathcal{A}_2 \Rightarrow \mathcal{A}^*_2 \leq \mathcal{A}^*_1$ (for $\mathcal{A}_1, \mathcal{A}_2 \in$ **Aut**(E)) and $\mathcal{N}_1 \leq \mathcal{N}_2 \Rightarrow \mathcal{N}^*_2 \leq \mathcal{N}^*_1$ (for $\mathcal{N}_1, \mathcal{N}_2 \in$ **Nets**(E)) follow immediately from the Galois connection. Another property of Galois connections is to produce *closure operators* by conjugated composition of the dual operators. Recall that an operator $\overline{()}$ on $(X, \leq)$, mapping x to $\overline{x}$, is a closure operator if it is increasing ($x_1 \leq x_2 \Rightarrow \overline{x_1} \leq \overline{x_2}$), extensive ($x \leq \overline{x}$), and idempotent ($\overline{\overline{x}} = x$). The double dual operators $()^{**}$ acting respectively on the ordered sets $($**Aut**$(E), \leq)$ and $($**Nets**$(E), \leq)$ are therefore closure operators. An automaton $\mathcal{A}$ equal to its closure $\mathcal{A}^{**}$ is said to be *separated* with respect to the fixed type of nets τ, while a net system $\mathcal{N}$ equal to its closure $\mathcal{N}^{**}$ is said to be *saturated.* Owing to the Galois connection, the lattices of separated automata and saturated net systems are dually order-isomorphic (i.e. isomorphic up to reversing the order).

6.4 Representation Results

By definition, an automaton separated with respect to type τ is isomorphic to the synchronized product of marking graphs $\mathcal{N}^*_p$ of atomic net systems $\mathcal{N}_p = \mathcal{A}^*_{\{p\}}$ derived from τ-regions p of $\mathcal{A}$ (in formulas: $\mathcal{A} \cong \bigwedge_{p \in \mathcal{R}_\tau(\mathcal{A})} \mathcal{N}^*_p$). Following [19], let us say that a subset of regions $\mathcal{R} \subseteq \mathcal{R}_\tau(\mathcal{A})$ is *admissible* if $\mathcal{A} \cong \bigwedge_{p \in R} \mathcal{N}^*_p$. So, $\mathcal{A}$ is separated if and only if $\mathcal{R}_\tau(\mathcal{A})$ is admissible, and of course every superset of an admissible set of regions is admissible. The marking graph $\mathcal{N}^*$ of a net system $\mathcal{N}$ is separated because $\mathcal{N}^* \cong \mathcal{N}^{***}$ follows from the Galois connection. In fact, the extensions (σ_p, η_p) of places p of $\mathcal{N}$ form an admissible set of regions of $\mathcal{N}^*$. The following criterion may be used to recognize admissible sets of regions, and consequently separated automata.

Theorem 6.12 *Given an automaton $\mathcal{A} = (S, E, T, s_0)$ and a type of nets τ, a set of regions $\mathcal{R} \subseteq \mathcal{R}_\tau(\mathcal{A})$ is admissible if and only if the following* separation properties *are satisfied for all states $s, s' \in S$ and for every event $e \in E$:*

(SSP) $s \neq s' \Rightarrow \exists(\sigma,\eta) \in \mathcal{R} : \sigma(s) \neq \sigma(s')$
(read: (σ,η) solves the states separation *problem at (s, s'))*

(ESSP) $s \overset{e}{\not\rightarrow} \Rightarrow \exists(\sigma,\eta) \in \mathcal{R} : \sigma(s) \overset{\eta(e)}{\not\longrightarrow}$ *w.r.t. τ*
(read: (σ,η) solves the event/state separation *problem at (s, e))*

When both properties are satisfied, $\mathcal{A} \cong (\mathcal{A}^*_\mathcal{R})^*$, where $\mathcal{A}^*_\mathcal{R}$ is the subnet system of $\mathcal{A}^*$ with restricted set of places $\mathcal{R}$ (also called the net synthesized *from $\mathcal{R}$).*

Proof: Let $\mathcal{N}_p = \mathcal{A}^*_{\{p\}}$ for $p \in \mathcal{R}$, and let $\mathcal{N}_\mathcal{R} = \mathcal{A}^*_\mathcal{R}$. Seeing that $\mathcal{A} \leq \mathcal{N}^*_p$ for every region p, $\mathcal{A} \leq \bigwedge_{p\in\mathcal{R}} \mathcal{N}^*_p = \mathcal{N}^*_\mathcal{R}$. Accordingly, there exists a morphism of automata $(\sigma, 1) : \mathcal{A} \to \mathcal{N}^*_\mathcal{R}$. Moreover this morphism is unique. On the other hand, every region $p = (\sigma_p, \eta_p)$ factors into $(\iota, \eta_p) \circ (\sigma_p, 1)$ where ι acts as the identity on the local states in its domain, and $(\sigma_p, 1)$ lifts to the unique event preserving morphism from $\mathcal{A}$ to $\mathcal{N}^*_p$. As $\mathcal{N}^*_\mathcal{R}$ is the synchronized product of $(\mathcal{N}^*_p)_{p\in\mathcal{R}}$, σ must be the map that sends each state s of $\mathcal{A}$ to the associated vector $\sigma(s) = (\sigma_p(s))_{p=(\sigma_p,\eta_p)\in\mathcal{R}}$ (the p-component is computed by evaluating region p at state s). Since $(\sigma, 1)$ is the unique morphism of this form from $\mathcal{A}$ to $\mathcal{N}^*_\mathcal{R}$, and seeing that all automata are accessible and deterministic, the assertion $\mathcal{A} \cong \mathcal{N}^*_\mathcal{R}$ is now equivalent to *(i)* σ is an injective map, and *(ii)* $s \overset{e}{\rightarrow}$ in $\mathcal{A}$ whenever $\sigma(s) \overset{e}{\rightarrow}$ in $\mathcal{N}^*_\mathcal{R}$. Now SSP is just another form of assertion *(i)*. By definition of the synchronized product, $\sigma(s) \overset{e}{\rightarrow}$ in $\mathcal{N}^*_\mathcal{R}$ entails $\sigma_p(s) \overset{e}{\rightarrow}$ in $\mathcal{N}^*_p$ for all $p \in \mathcal{R}$,hence $ESSP$ is just another form of assertion *(ii)*. ∎

Corollary 6.13 *Given an automaton $\mathcal{A} \in$ **Aut**(E) and a type of nets τ, $\mathcal{A} \cong \mathcal{N}^*$ for some net system $\mathcal{N} \in$ **Nets**(E) if and only if $\mathcal{A} \cong \mathcal{A}^{**}$ if and only if the conditions* SSP *and* ESSP *are valid in $\mathcal{A}$. Given an automaton $\mathcal{A} \in$ **Aut**(E), a type of nets τ, and a net system $\mathcal{N} \in$ **Nets**(E), $\mathcal{A} \cong \mathcal{N}^*$ if and only if $\mathcal{N}$ is isomorphic to subnet system of $\mathcal{A}^*$ determined from some admissible subset of regions $\mathcal{R} \subseteq \mathcal{R}_\tau(\mathcal{A})$.*

By setting $\tau = \tau_{EN}$, resp. $\tau = \tau_{FFN}$, in the above theorem and corollary, one retrieves the results of Ehrenfeucht and Rozenberg (Prop. 2.15) and Desel and Reisig (Prop. 3.6), resp. the result of Schmitt (Prop. 5.5). The application of the theorem to the types τ_{PPN} and τ_{PN} will be examined in section 7.

6.5 Some Applications to Safe Nets

We call *safe* nets all nets whose markings are defined as subsets of places $M \subseteq P$, or equivalently as maps $M : P \to \{0, 1\}$. Thus, a type of safe nets is a transition system $\tau = (LS, LE, \tau)$ whose set of local states is $LS = \{0, 1\}$. The largest type of safe nets, let τ_{safe}, is obtained by including in its transitions all the defined instances $s \overset{f}{\rightarrow} f(s)$ of partial functions $f : \{0, 1\} \to \{0, 1\}$.

Table 1. safe nets

	0	1	C/E nets	elementary nets	flip flop nets	trace nets
input	-	0	yes	yes	yes	yes
output	1	-	yes	yes	yes	yes
test=1	-	1	yes	no	no	yes
test=0	0	-	no	no	no	yes
set_0	0	0	no	no	no	yes
set_1	1	1	no	no	no	yes
nop	0	1	yes	yes	yes	yes
swap	1	0	no	no	yes	no

These functions, tabulated in Table 1, form a set LE_{safe}. Various types $\tau_X = (\{0,1\}, LE_X, \tau_X)$ follow as induced restrictions of τ_{safe} on particular subsets of local events $LE_X \subseteq LE_{safe}$ (see again Table 1 for some examples). Each type τ_X determines corresponding regions in transition systems $A = (S, E, T)$, defined as morphisms $(\sigma, \eta) : A \to \tau_X$. These morphisms define in turn set-theoretic regions $R = \sigma^{-1}(\{1\}) \in \mathcal{P}(S)$. We restrict our analysis of safe type to types τ_X larger than τ_{EN} and such that the complement of a region $\sigma^{-1}(\{1\})$ is a region. This amounts to set on LE_X the constraints $\{\mathtt{nop}, \mathtt{input}, \mathtt{output}\} \subseteq LE_X$ and $\mathtt{set\_0} \in LE_X \Leftrightarrow \mathtt{set\_1} \in LE_X$.

We declare equivalent, resp. weakly equivalent, two safe types τ_X which determine an identical family of separated automata, resp. identical families of set-theoretic regions in automata. With the above constraints, there are four classes of weakly equivalent types τ_X, each of which splitting into two equivalence classes of types.

The four possible concepts of set-theoretic regions are determined from five forbidding patterns displayed in Fig. 14. Each pattern represents a pair transi-

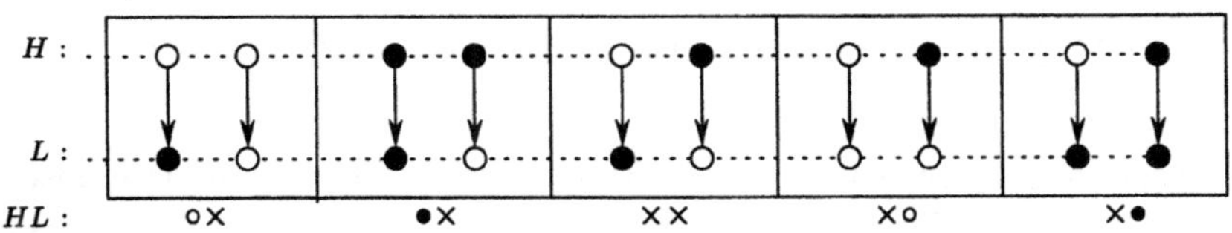

Fig. 14. five patterns for a pair of transitions with the same label $s_1 \xrightarrow{e} s_1'$ and $s_2 \xrightarrow{e} s_2'$ where $s \in R$ if and only if the corresponding node is coloured black

tions $s_1 \xrightarrow{e} s_1'$ and $s_2 \xrightarrow{e} s_2'$ with a common label e in a transition system $A = (S, E, T)$ whose states s are coloured black or white. Let $R \subseteq S$ be the subset of states coloured in black. The four possible concepts of **set-theoretic** regions are as follows.

1. R is an *elementary region* if and only if the patterns ○×, •×, ××, ×○, and ×• do not occur in A. A safe type τ_X induces the elementary regions if and

only if $LE_X \cap \{\texttt{swap}, \texttt{set\_0}, \texttt{set\_1}\} = \emptyset$. This case is met for τ_{EN} and for the type of C/E-nets, let τ_{CEN} where $LE_{CEN} = \{\texttt{nop}, \texttt{input}, \texttt{output}, \texttt{test=1}\}$.

2. R is a ***flip flop region*** if and only if the patterns $\circ\times$, $\bullet\times$, $\times\circ$, $\times\bullet$ do not occur in A. A safe type τ_X induces the flip flop regions if and only if $LE_X \cap \{\texttt{set\_0}, \texttt{set\_1}\} = \emptyset$. This case is met for τ_{FFN}.
3. R is a ***trace region*** if and only if the patterns $\circ\times$, $\bullet\times$, $\times\times$ do not occur in A. Trace regions have been introduced independently for ***trace nets*** in [4, 5] and for *chart nets* in [29] where they are called chart regions. A safe type τ_X induces the trace regions if and only if $\{\texttt{set\_0}, \texttt{set\_1}\} \subseteq LE_X$ and $\texttt{swap} \notin LE_X$. This case is met for the type of trace nets, let τ_{TRN} where $LE_{TRN} = \{\texttt{nop}, \texttt{input}, \texttt{output}, \texttt{test=1}, \texttt{test=0}, \texttt{set\_0}, \texttt{set\_1}\}$.
4. R is a ***safe region*** if and only if the patterns $\circ\times$ and $\bullet\times$ do not occur in A. A type τ_X induces the safe regions if and only if $\{\texttt{swap}, \texttt{set\_0}, \texttt{set\_1}\} \subseteq LE_X$.

Safe types may be classified further into pure types and impure types according to whether $LE_X \cap \{\texttt{test=0}, \texttt{test=1}\}$ is empty or not. One obtains in this way 8 classes of equivalent types. These classes may be ordered according to the inclusion of the associated sets of separated automata. All inclusions are shown in Fig. 15 together with representative automata showing they are strict.

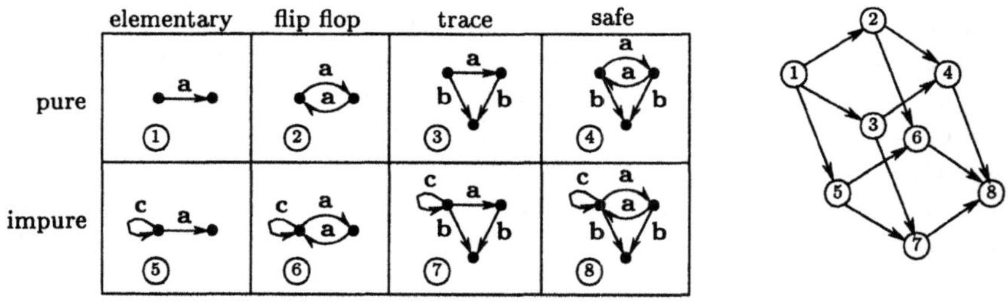

Fig. 15. classification of the equivalence classes of safe types

By specializing Theo. 6.12 to a particular type τ_X, one obtains an immediate characterization of the family of separated automata specific to its equivalence class. We have yet implicitly applied this technique to the type τ_{EN} of elementary nets and to the type τ_{FFN} of flip flop nets. Let us focus on the types τ_{CEN} (of C/E nets) and τ_{TRN} (of trace nets).

From Theo. 6.12 applied to τ_{CEN}, one retrieves Nielsen and Winskel's characterization of marking graphs of C/E nets [35]. Given an asynchronous automaton $\mathcal{A} = (S, E, \|, T, s_0)$ with an empty independence relation, the regions of $\mathcal{A}$ defined in [35] are actually in bijective correspondence with morphisms $(\sigma, \eta) : (S, E, T) \to \tau_{CEN}$ [1]. We recall that an asynchronous automaton according to the definition of Shields and Bednarczyk [9, 41] is a deterministic automaton $\mathcal{A} = (S, E, T, s_0)$, enriched with a symmetric and irreflexive relation of independence $\| \subseteq E \times E$ such that the following conditions are satisfied whenever $e_1 \| e_2$:

FORWARD DIAMOND PROPERTY: $s \xrightarrow{e_1} s_1 \wedge s \xrightarrow{e_2} s_2 \Rightarrow \exists s' \in S \;\; s_2 \xrightarrow{e_1} s' \wedge s_1 \xrightarrow{e_2} s'$.
COMMUTATION PROPERTY: $s \xrightarrow{e_1} s_1 \wedge s_1 \xrightarrow{e_2} s' \Rightarrow \exists s_2 \in S \;\; s \xrightarrow{e_2} s_2 \wedge s_2 \xrightarrow{e_1} s'$.

Following Nielsen and Winskel, let us define *asynchronous regions* in $\mathcal{A} = (S, E, \|, T, s_0)$ as the morphisms $(\sigma, \eta) : (S, E, T) \to \tau_{CEN}$ such that $\forall e_1, e_2 \in E \;\; e_1 \| e_2 \Rightarrow (\eta(e_1) = \mathbf{nop}) \vee (\eta(e_2) = \mathbf{nop})$. This is consistent with the usual definition of independence of events in C/E nets, according to which $e_1 \| e_2$ if and only if $({}^\bullet e_1 \cup e_1{}^\bullet) \cap ({}^\bullet e_2 \cup e_2{}^\bullet) = \emptyset$. Nielsen and Winskel show that the asynchronous automata which are generated from C/E net systems with this definition of independence are exactly those in which the separation properties SSP and ESSP are satisfied w.r.t. the asynchronous regions.

If we apply now Theo. 6.12 to the type τ_{TRN}, we retrieve the characterization of marking graphs of trace nets established in [5]. Given a trace automaton $\mathcal{A} = (S, E, \|, T, s_0)$ with an empty relation of independence, the trace regions of $\mathcal{A}$ defined in [5] are actually the morphisms $(\sigma, \eta) : (S, E, T) \to \tau_{TRN}$. We recall that a trace automaton according to the definition of Stark [42] is like an asynchronous automaton up to the removal of the commutation constraint. A typical trace automaton is shown in Fig. 16 together with a generating trace net. This trace automaton is not an asynchronous automaton since e.g. the

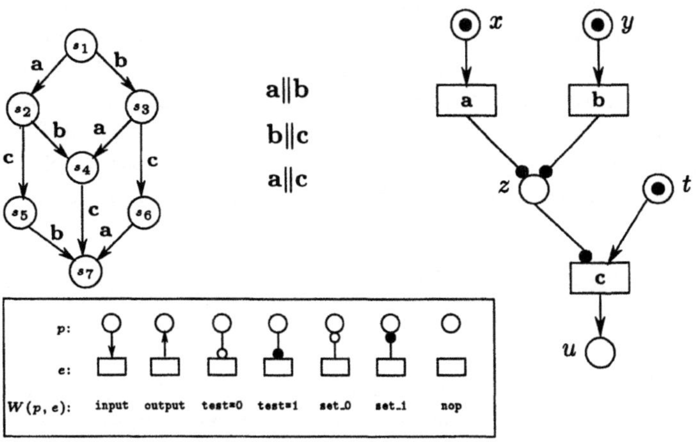

Fig. 16. a trace automaton and a generating trace net

sequence $a \cdot c$ can be fired from state s_1, but this is not the case with $c \cdot a$ although $a \| c$. The reader may verify that the separation problem ESSP(s_1, c) cannot be solved with elementary regions nor with asynchronous regions, but it is solved by the trace region corresponding to the place z of the trace net of Fig. 16. In the general case where the independence relation is not empty, a trace region of $\mathcal{A} = (S, E, \|, T, s_0)$ is defined as a trace region of the underlying automaton compatible with the independence of events in the sense that for any two independent events e_1 and e_2 one has *(i)* $\eta(e_1) \in \{\mathtt{input}, \mathtt{output}\} \Rightarrow \eta(e_2) = \mathtt{nop}$ and *(ii)* $\eta(e_1) \in \{\mathtt{test{=}1}, \mathtt{set\_1}\} \Rightarrow \eta(e_2) \neq \mathtt{set\_0}$. This is coherent with the independence of events in trace nets, defined similarly by $e_1 \| e_2$ if and only if for all places $p \in P$ *(i)* $W(p, e_1) \in \{\mathtt{input}, \mathtt{output}\} \Rightarrow W(p, e_2) = \mathtt{nop}$ and *(ii)* $W(p, e_1) \in \{\mathtt{test{=}1}, \mathtt{set\_1}\} \Rightarrow W(p, e_2) \neq \mathtt{set\_0}$. It is shown in [5] that the trace automata which are generated from trace net systems with this relation of independence are exactly those in which the separation properties SSP and ESSP are satisfied w.r.t. the trace regions. It is also shown that the finite trace automata which can be defined in the so-called *simple* format of Plotkin's Structural Operational Specification rules, with proofs of transitions as events and independence of proofs as independence of events, are exactly the finite and separated trace automata.

More will be said on the topic of independence in section 8.

6.6 Other Applications of Types

Types may serve alternatively to classify existing families of nets or to explore new families of nets. One may study hybrid types forged from existing ones by amalgamation, or by disjoint summation. One may study translations between classes of nets based on morphisms between their types. Theoretically speaking, this amounts to consider nets over a fixed set of events as a category indexed over the category of automata (their types). Compilation techniques for nets may also be defined on the following principle: let $N = (P, E, W)$ be a net of type τ, where τ is the marking graph of a net $N_\tau = (P_\tau, LE, W_\tau)$ of type τ', then N is equivalent to the net $N' = (P', E', W')$ of type τ' such that $P' = P \times P_\tau$, and $W'((p, p_\tau), e) = W_\tau(p_\tau, W(p, e))$. This amounts to consider nets as functors over automata, and composition of functors as compilation of nets.

7 Polynomial Time Algorithms for the Synthesis of Petri Nets

We present in this section the polynomial time algorithm proposed in [2] for the synthesis of pure Petri nets from finite automata. This algorithm has been implemented in the tool SYNET [13]. Next, we give a sketch of the variant algorithm for the synthesis of (general) Petri nets proposed in [7]. Finally, we indicate for both algorithms degenerated forms allowing to synthesize Petri nets from regular languages.

7.1 The Synthesis Problem for Pure Petri Nets

In the sequel, let $\mathcal{A} = (S, E, T, s_0)$ be a loop-free, reachable and reduced finite deterministic automaton, and let A denote the underlying transition system (S, E, T). The synthesis problem for pure Petri nets consists in *(i)* deciding whether an automaton $\mathcal{A}$, given as input, is isomorphic to the marking graph

$\mathcal{N}^*$ of some net system $\mathcal{N} = (P, E, W, M_0)$ of type τ_{PPN}, and if so *(ii)* producing as output a net system $\mathcal{N}$ such that $\mathcal{A} \cong \mathcal{N}^*$ and no proper subnet system of $\mathcal{N}$ satisfies this property. Recall that $\tau_{PPN} = (\mathbb{N}, \mathbb{Z}, \tau)$ with transitions $n \xrightarrow{z} n' \in \tau$ iff $n' = n + z$. On the grounds of Theo. 6.12, this amount to *(i)* deciding whether all instances of the separation problems in $\mathcal{A}$ can be solved by corresponding regions, and if so *(ii)* synthesizing the desired net system $\mathcal{N} = \mathcal{A}^*_{\mathcal{R}}$ from a minimal admissible subset of regions $\mathcal{R}$, where $\mathcal{A}^*_{\mathcal{R}} = (\mathcal{R}, E, W, M_0)$ with $W((\sigma, \eta), e) = \eta(e)$ and $M_0((\sigma, \eta)) = \sigma(s_0)$. Now, there is at most $|S|^2 - |S|$ possible inputs for the states separation problem:
$\text{SSP}_{\mathcal{A}}(s, s')$: "*construct from $\mathcal{A}$ and $s \neq s'$ a region (σ, η) s.t. $\sigma(s) \neq \sigma(s')$*"
and at most $|S| \times |E|$ instances of the event/state separation problem:
$\text{ESSP}_{\mathcal{A}}(s, e)$: "*construct from $\mathcal{A}$ and $(s \stackrel{e}{\not\rightarrow})$ a region (σ, η) s.t. $(\sigma(s) \stackrel{\eta(e)}{\not\longrightarrow})$*".
Part *(i)* of the problem will therefore be solved in time polynomial (in $|S|$ and $|E|$) as soon as $\text{SSP}_{\mathcal{A}}(s, s')$ and $\text{ESSP}_{\mathcal{A}}(s, e)$ are solved in polynomial time. Part *(ii)* consists in extracting from a set of regions with size polynomial in $|S|$ and $|E|$ a minimal admissible subset and this certainly can be done in polynomial time. So, a polynomial algorithm for the synthesis of pure Petri nets will follow if we succeed to construct procedures that solve in polynomial time $\text{SSP}_{\mathcal{A}}(s, s')$ and $\text{ESSP}_{\mathcal{A}}(s, e)$ with respect to the type τ_{PPN}. This is the main program of the section. The first stage of the program is to study the algebraic properties of the set of pure Petri regions of $\mathcal{A}$. The second stage of the program is to elaborate decision procedures based on these properties.

7.2 The Structure of Pure Petri Regions

Let $\mathcal{R}_{PPN}(\mathcal{A})$ denote the set of pure Petri regions of $\mathcal{A}$, i.e. the set of morphisms $(\sigma, \eta) : A \to \tau_{PPN}$. Before investigating the algebraic properties of $\mathcal{R}_{PPN}(\mathcal{A})$, let us recall some terminology borrowed from algebraic topology (see e.g. [31]). In the fixed transition system $A = (S, E, T)$, let $\partial^0, \partial^1 : T \to S$ and $\ell : T \to E$ denote the respective *source*, *target*, and *labelling* functions given by $\partial^0(t) = s$, $\partial^1(t) = s'$, and $\ell(t) = e$ for $t = s \xrightarrow{e} s' \in T$. A *0-chain* of A is a vector in the free $\mathbb{Z}$-module $C_0(A) = \mathbb{Z} <S>$ [3]. A *1-chain* of A is a vector in the free $\mathbb{Z}$-module $C_1(A) = \mathbb{Z} <T>$. The *boundaries* of the 1-chains are the 0-chains computed by the operator $\partial : C_1(A) \to C_0(A)$ such that $\partial(\Sigma z_j \cdot t_j) = \Sigma z_j \cdot (\partial^1(t_j) - \partial^0(t_j))$. The *co-boundaries* of the 0-chains are the 1-chains computed by the operator $\partial^* : C_0(A) \to C_1(A)$ such that $\partial^*(\sum z_i \cdot s_i) = \sum z_i \cdot \partial^*(s_i)$ where $\partial^*(s_i) = \sum\{t_j | \partial^1(t_j) = s_i\}$ - $\sum\{t_j | \partial^0(t_j) = s_i\}$. [4] A *cycle* of A is a 1-chain with a null boundary, and a *co-cycle* is a 0-chain with a null co-boundary. The cycles of A, resp. the

[3] we recall that the free $\mathbb{Z}$-module generated by a finite set $X = \{x_1, \ldots, x_n\}$ of generators is the set of maps α from X to $\mathbb{Z}$, viewed as vectors indexed by X with entries in $\mathbb{Z}$ and represented as formal sums $\alpha = \sum_i \alpha_i \cdot x_i$ where $\alpha(x_i) = \alpha_i$.

[4] the dual linear operators ∂ and ∂^* are associated respectively with $-A$ and its transpose $-A^t$, where A is the incidence matrix of the underlying graph. This change of sign is not technically significant and comes from different usages in the literature on graphs: the definition of the incidence matrix of a directed graph that we gave corresponds to the one used in [10, 16, 17], whereas Lefschetz [31] and Tutte [40]

co-boundaries of A, form submodules $\mathcal{V}_B$ resp. $\mathcal{V}_Q$ of $C_1(A)$ which are orthogonal complements. Linear bases for $\mathcal{V}_B$ and $\mathcal{V}_Q$ are supplied by the respective sets of fundamental cycles and fundamental cutsets of the underlying graph (S,T) w.r.t. some spanning tree $U \subseteq T$. Thus every cycle may be written as a linear combination $\sum z_i \cdot B_i$ of fundamental cycles $B_i : T \to \{-1,0,1\}$, and every co-boundary may be written as a linear combination $\sum z_i \cdot C_i$ of fundamental cutsets $C_i : T \to \{-1,0,1\}$, with integral coefficients $z_i \in \mathbb{Z}$. The Parikh images of the cycles form in turn a submodule of the free $\mathbb{Z}$-module $\mathbb{Z}<E>$, where the Parikh mapping $\pi : \mathbb{Z}<T> \to \mathbb{Z}<E>$ is the linear transformation given by $\pi(\sum z_i \cdot t_i) = \sum z_i \cdot \ell(t_i)$. In the sequel, the maps $\eta : E \to \mathbb{Z}$ are represented accordingly as formal sums $\eta = \sum z_i \cdot e_i$ where $z_i = \eta(e_i)$. For any two vectors $\alpha = \sum \alpha_i \cdot x_i$ and $\beta = \sum \beta_i \cdot x_i$ in a finite dimensional free $\mathbb{Z}$-module $\mathbb{Z}<X>$, we let $\alpha \cdot \beta$ denote the scalar product $\sum \alpha_i \cdot \beta_i \in \mathbb{Z}$.

Proposition 7.1 *$(\sigma,\eta) \in \mathcal{R}_{PPN}(A)$ if and only if $\sigma \cdot \partial(c) = \eta \cdot \pi(c)$ for all $c \in C_1(A)$.*

Proof: By linearity, the condition $\forall c \in C_1(A)\ \ \sigma \cdot \partial(c) = \eta \cdot \pi(c)$ is equivalent to the condition $\forall t \in T\ \ \sigma \cdot \partial(t) = \eta \cdot \pi(t)$ where t is identified with the chain $(1.t)$. Now the equation $\sigma \cdot \partial(t) = \eta \cdot \pi(t)$ is valid if and only if $\sigma(\partial^1(t))$ - $\sigma(\partial^0(t)) = \eta(\ell(t))$, if and only if $\sigma(\partial^0(t)) \stackrel{\eta(\ell(t))}{\longrightarrow} \sigma(\partial^1(t))$ w.r.t. the type τ_{PPN}, if and only if $(\sigma,\eta) \in \mathcal{R}_{PPN}(A)$ by definition of regions. ∎

Proposition 7.2 *A map $\eta : E \to \mathbb{Z}$ is the second projection of some region $(\sigma,\eta) \in \mathcal{R}_{PPN}(A)$ if and only if $\eta \cdot \pi(B) = 0$ for every cycle $B \in \mathcal{V}_B$; the regions $(\sigma,\eta) \in \mathcal{R}_{PPN}(A)$ which project on η are then characterized by the condition: $\sigma(s_0) + (\eta \cdot \pi(c)) \geq 0$ for every 1-chain $c \in C_1(A)$ such that $\partial(c) = s - s_0$ for some $s \in S$.*

Proof: From Prop. 7.1, the condition on η must hold and whenever it does, the scalar product $\eta \cdot \pi(c)$ takes an identical value for all 1-chains c with an identical boundary. From the definition of regions, the condition on $\sigma(s_0)$ must hold because the local states specified for the type of nets τ_{PPN} are the non negative integers. Now the two conditions taken together guarantee that one can always complete the data $(\sigma(s_0),\eta)$ to a pure region by selecting for each state $s \in S$ a corresponding 1-chain c_s such that $\partial(c_s) = s - s_0$ and then setting $\sigma(s) = \sigma(s_0) + \eta \cdot \pi(c_s)$, which is always possible since $\mathcal{A}$ is reachable. ∎

Let $\mathcal{R}_{abs}(A)$ denote the set of maps $\eta : E \to \mathbb{Z}$ characterized by Prop. 7.2, henceforth called *abstract regions*. It appears from this characterization that the abstract regions of A are in bijective correspondence with the co-boundaries of A which are compatible with the kernel of the labelling function $\ell : T \to E$. Actually, for every abstract region $\eta : E \to \mathbb{Z}$, the map $\kappa = \eta \circ \ell : T \to \mathbb{Z}$ is a

use the opposite matrix. In the same manner what we term co-boundary, following Lefschetz and Tutte and more generally those authors who identify graphs with 1-dimensional complexes, are called cocycles in many books on graph theory.

co-boundary of A, since for every cycle B, $\kappa \cdot B = \sum_t \eta(\ell(t)) \cdot B(t) = \sum_e \eta(e) \cdot \sum_{\ell(t)=e} B(t) = \eta \cdot \pi(B) = 0$. Conversely, every co-boundary $\kappa : T \to \mathbb{Z}$ such that $\ell(t) = \ell(t') \Rightarrow \kappa(t) = \kappa(t')$ determines a unique abstract region $\eta : E \to \mathbb{Z}$ such that $e = \ell(t) \Rightarrow \eta(e) = \kappa(t)$, since $\ell : T \to E$ is surjective and $\mathcal{A}$ is reduced.

An abstract region η determines a unique region (σ, η) such that $\sigma(s) = 0$ for some state s, called a *strict* region and given by $\sigma(s_0) = -\mathbf{min}\{\eta \cdot \pi(c) \mid \exists s \in S \;\; \partial(c) = s - s_0\}$, and an infinite family of non strict regions $(\sigma + h, \eta)$ for $h \in \mathbb{N} \setminus \{0\}$. Now any instance of the separation problems $\text{SSP}_{\mathcal{A}}(s, s')$ or $\text{ESSP}_{\mathcal{A}}(s, e)$ solved by $(\sigma + h, \eta)$ is also solved by (σ, η). For this reason, let us concentrate on strict regions, or equivalently on abstract regions.

The set $\mathcal{R}_{\text{abs}}(A)$ of abstract regions of A is obviously a $\mathbb{Z}$-module. From Prop. 7.2, a linear basis for this module may be computed as follows. Let $S = \{s_1, \ldots, s_n\}$, $T = \{t_1, \ldots, t_m\}$, and $E = \{e_1, \ldots, e_p\}$. Let $U \subseteq T$ be a spanning tree of the underlying graph $G = (S, T)$, and let $\{B_1, \ldots, B_{m-n+1}\}$ be the set of fundamental cycles of G w.r.t. U. Thus $\{B_1, \ldots, B_{m-n+1}\}$ is a basis for $\mathcal{V}_B$ and $\mathcal{R}_{\text{abs}}(A)$ is the kernel of the linear transformation $M_A : \mathbb{Z}^p \to \mathbb{Z}^{m-n+1}$ defined by the $(m - n + 1) \times p$ matrix M_A with integral coefficients

$$M_A(i, j) = \Sigma\{B_i(t_k) \mid \; 1 \leq k \leq m \;\; \wedge \;\; \ell(t_k) = e_j\}$$

Let k be the dimension of $Ker(M_A)$. The algorithm of von zur Gathen and Sieveking (see [38]), given M_A as input, produces in time polynomial in $m-n+1$ and p (or $|S| = n$ and $|E| = p$, because $m \leq n \times p$ follows from determinism of A) a basis $\{\eta_1, \ldots, \eta_k\}$ for $Ker(M_A) = \mathcal{R}_{abs}(A)$.

We have in hand all the elements needed for solving problems $SSP_A(s, s')$ and $ESSP_A(s, e)$ relatively to the type of pure Petri nets. The data needed are the spanning tree U, or more exactly the application $c_{(.)}$ that maps each state $s \in S$ to the unique chain c_s from s_0 to s in U, and the basis of abstract regions $\{\eta_1, \ldots, \eta_k\}$.

For the sake of illustration, let us exhibit these data for the automaton $\mathcal{A}$ shown in Fig. 17. Here $n = 8$, $m = 14$, and $p = 6$. The spanning tree U, indicated

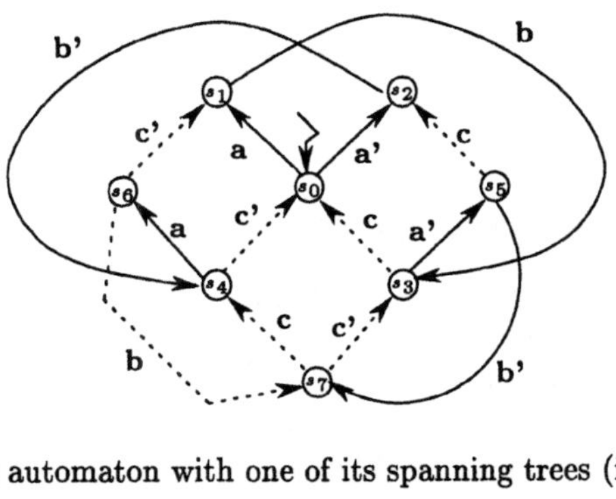

Fig. 17. an automaton with one of its spanning trees (in solid lines)

in solid lines, contains $n - 1 = 7$ transitions, The module $\mathcal{V}_B$ is generated from the $m-n+1=7$ fundamental cycles B_i defined by the respective chords t_i indicated in dashed lines, let $t_1 = s_5 \xrightarrow{c} s_2$, $t_2 = s_3 \xrightarrow{c} s_0$, $t_3 = s_7 \xrightarrow{c} s_4$, $t_4 = s_6 \xrightarrow{c'} s_1$, $t_5 = s_4 \xrightarrow{c'} s_0$, $t_6 = s_7 \xrightarrow{c'} s_3$, and $t_7 = s_6 \xrightarrow{b} s_7$. For instance, the chord t_1 defines the fundamental cycle

$$B_1 = (s_0 \xrightarrow{a} s_1) + (s_1 \xrightarrow{b} s_3) + (s_3 \xrightarrow{a'} s_5) + (s_5 \xrightarrow{c} s_2) - (s_0 \xrightarrow{a'} s_2)$$

whose Parikh image is $\pi(B_1) = a+b+c$. One can verify that $\pi(B_1)= \pi(B_2) = \pi(B_3) = a+b+c$, $\pi(B_4) = \pi(B_5) = \pi(B_6) = a'+b'+c'$, and $\pi(B_7)$ =0. The $\mathbb{Z}$-module of abstract regions consists of those vectors $\eta : E \to \mathbb{Z}$ such that:

$$\eta(a) + \eta(b) + \eta(c) = 0 \quad \text{and} \quad \eta(a') + \eta(b') + \eta(c') = 0$$

It is therefore a four dimensional $\mathbb{Z}$-module with basis as follows:

$$\eta_1 = a - c \ ; \ \eta_2 = b - c \ ; \ \eta_3 = a' - c' \ \text{and} \ \eta_4 = b' - c'$$

In this example, the spanning tree U is rooted at the initial state s_0 of the automaton. Let c_s denote the branch of U from s_0 to s and let $\pi_s = \pi(c_s)$ be its Parikh image. Thus, we have:

$$\begin{array}{llll} \pi_{s_0} = 0 & \pi_{s_1} = a & \pi_{s_2} = a' & \pi_{s_3} = a+b \\ \pi_{s_4} = a'+b' & \pi_{s_5} = a+b+a' & \pi_{s_6} = a'+b'+a & \pi_{s_7} = a+b+a'+b' \end{array}$$

The corresponding scalar products $\eta_i \cdot \pi_s$ are tabulated in Table 2

Table 2. states $s \in S$ represented by vectors $(\eta_i \cdot \pi_s)_i$ indexed by the set of basic abstract regions η_i

$\eta_i \cdot \pi_s$	π_{s_0}	π_{s_1}	π_{s_2}	π_{s_3}	π_{s_4}	π_{s_5}	π_{s_6}	π_{s_7}
η_1	0	1	0	1	0	1	1	1
η_2	0	0	0	1	0	1	0	1
η_3	0	0	1	0	1	1	1	1
η_4	0	0	0	0	1	0	1	1

7.3 Solving the separation problems

Let s and s' be two distinct states. From Prop. 7.1 and Prop. 7.2, $\text{SSP}_{\mathcal{A}}(s, s')$ has a solution in $\mathcal{R}_{PPN}(\mathcal{A})$ *iff* $\eta \cdot \pi(c_s - c_{s'}) \neq 0$ for some abstract region $\eta \in \mathcal{R}_{abs}(A)$ *iff* $\eta_i \cdot \pi(c_s - c_{s'}) \neq 0$ for some $i \in \{1, \cdots, k\}$, and the strict region (σ_i, η_i) determined from the basic abstract region η_i by setting $\sigma_i(s_0) = -\mathbf{min}\{\eta_i \cdot \pi(c_s) \mid s \in S\}$ is then a solution. Therefore, deciding whether $\text{SSP}_{\mathcal{A}}(s, s')$ has a solution and producing it takes time polynomial in $|S|$ and $|E|$.

In our running example all instances of the separation problem $\text{SSP}_{\mathcal{A}}(s, s')$ can be solved, because all the columns of table 2 are different.

Given $s' \in S$ and $e \in E$ such that $s' \overset{e}{\not\to}$, let us now consider the separation problem $\text{ESSP}_{\mathcal{A}}(s', e)$. From Prop. 7.2, this problem has a solution in $\mathcal{R}_{PPN}(\mathcal{A})$ *iff* there exists $\sigma(s_0) \in \mathbb{N}$ and $\eta \in \mathcal{R}_{abs}(A)$ such that

$$\forall s \in S \quad \sigma(s_0) + \eta \cdot \pi(c_s) \geq 0 \tag{1}$$

$$\sigma(s_0) + \eta \cdot \pi(c_{s'}) + \eta(e) < 0 \tag{2}$$

iff there exists $\eta \in \mathcal{R}_{abs}(A)$ satisfying the condition

$$\forall s \in S \quad \eta \cdot (\pi(c_{s'}) - \pi(c_s)) + \eta(e) < 0 \tag{3}$$

Whenever η satisfies condition 3, the strict region (σ, η) defined from η satisfies actually conditions 1 and 2 and therefore solves $\text{ESSP}_{\mathcal{A}}(s', e)$. Let $\eta = \sum_{i=1}^{k} x_i \cdot \eta_i$ where $\{\eta_1, \ldots, \eta_k\}$ is the basis of abstract regions, and $x_i \in \mathbb{Z}$. For every $s \in S$, let $\alpha_i^s = \eta_i \cdot (\pi(c_{s'}) - \pi(c_s)) + \eta_i(e)$. With these notations, condition 3 may be rewritten to the system of linear inequations $\{\sum_{i=1}^{k} \alpha_i^s \cdot x_i < 0 |\ s \in S\}$ in the variables $x_i \in \mathbb{Z}$. Now a system of linear inequations

$$MX \leq (-1)^n \tag{4}$$

where M is an integral matrix and $(-1)^n = < -1, \ldots, -1 >$ $(\in \mathbb{Z}^n)$ has an integral solution iff it has a rational solution. The method of Khachiyan (see [38] p.170) may be used to decide upon the feasability of (4) and to compute a rational solution, if it exists, in polynomial time. Thus, every instance of the problem $\text{ESSP}_{\mathcal{A}}(s', e)$ is solved up to a multiplicative factor, or shown unfeasible, in time polynomial in $|S|$ and $|E|$. In our running example, the system of linear inequations which express the separation problem $\text{ESSP}_{\mathcal{A}}(s_2, a)$ is the following:

$$\begin{array}{lr}
\eta \cdot (\pi_{s_2} - \pi_{s_0}) + \eta(a) < 0 : & x_1 + x_3 < 0 \\
\eta \cdot (\pi_{s_2} - \pi_{s_1}) + \eta(a) < 0 : & x_3 < 0 \\
\eta \cdot (\pi_{s_2} - \pi_{s_2}) + \eta(a) < 0 : & x_1 < 0 \\
\eta \cdot (\pi_{s_2} - \pi_{s_3}) + \eta(a) < 0 : & x_3 - x_2 < 0 \\
\eta \cdot (\pi_{s_2} - \pi_{s_4}) + \eta(a) < 0 : & x_1 - x_4 < 0 \\
\eta \cdot (\pi_{s_2} - \pi_{s_5}) + \eta(a) < 0 : & -x_2 < 0 \\
\eta \cdot (\pi_{s_2} - \pi_{s_6}) + \eta(a) < 0 : & -x_4 < 0 \\
\eta \cdot (\pi_{s_2} - \pi_{s_7}) + \eta(a) < 0 : & -x_2 - x_4 < 0
\end{array}$$

This system is solvable, and admits in particular the solution $x_1 = x_3 = -1$ and $x_2 = x_4 = 1$. Therefore, $\eta = -\eta_1 + \eta_2 - \eta_3 + \eta_4 = -a + b - a' + b'$ satisfies condition 3, and (σ, η) solves $\text{ESSP}_{\mathcal{A}}(s_2, a)$ with $\sigma(s_0) = 1$. The automaton of Fig. 17 is actually separated by the set of strict regions (σ_η, η) which are indicated in Table 3, computed from SYNET. The pure Petri net synthesized from this set of admissible regions is shown in Fig. 18. For full precision, it should be said that SYNET [13] does not relie on the method of Khachiyan but on the simplex method which has cubic complexity in the average (see [38]). A quite different solution to the synthesis problem of pure Petri nets from finite automata up to a quotient is described in [30]. This solution is based on the investigation of minimal regions.

Table 3. values taken on states by strict regions

$\sigma_\eta(s)$	s_0	s_1	s_2	s_3	s_4	s_5	s_6	s_7
$-\eta_1$	1	0	1	0	1	0	0	0
η_2	0	0	0	1	0	1	0	1
$-\eta_3$	1	1	0	1	0	0	0	0
η_4	0	0	0	0	1	0	1	1
$\eta_1 - \eta_2$	0	1	0	0	0	0	1	0
$\eta_3 - \eta_4$	0	0	1	0	0	1	0	0
$\eta_2 + \eta_4 - \eta_1 - \eta_3$	1	0	0	1	1	0	0	1

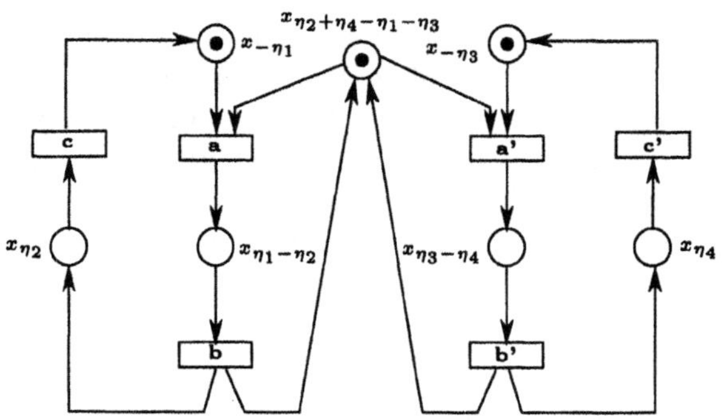

Fig. 18. the net synthesized from the admissible set of strict regions given in table 3

The key observation is the following: let $(\sigma_1, \eta_1), (\sigma_2, \eta_2) \in \mathcal{R}_{PPN}(\mathcal{A})$ such that $\forall s \in S \quad \sigma_1(s) \geq \sigma_2(s)$, then $(\sigma_1 - \sigma_2, \eta_1 - \eta_2)$ is a region of $\mathcal{A}$ and any instance of the separation problems which is solved by (σ_1, η_1) is solved either by (σ_2, η_2) or by $(\sigma_1 - \sigma_2, \eta_1 - \eta_2)$. Therefore, $\mathcal{A}$ is separated if and only if the set of its minimal regions is admissible.

7.4 The Case of General Petri Nets

A polynomial time algorithm for the synthesis of general Petri nets from finite automata was proposed in [7]. This algorithm is a modified form of the algorithm just described for pure Petri nets. We indicate below the main adaptations leading to the modified algorithm.

In the sequel, $\mathcal{A}$ is a reachable and reduced finite deterministic automaton, not necessarily simple. A region of $\mathcal{A}$ w.r.t. the type τ_{PN} of Petri nets is a morphism $(\sigma, ({}^\bullet\eta, \eta^\bullet)) : A \to \tau_{PN}$, called a Petri region, where ${}^\bullet\eta$ and $\eta^\bullet$ are maps from E to $\mathbb{N}$. As it was observed in [21], a Petri region is entirely determined from σ and ${}^\bullet\eta$ or alternatively from $\sigma(s_0)$, ${}^\bullet\eta$, and $\eta^\bullet$. Petri regions and pure Petri regions are connected by a pair of maps $J_{\mathcal{A}} : \mathcal{R}_{PN}(\mathcal{A}) \to \mathcal{R}_{PPN}(\mathcal{A})$ and $I_{\mathcal{A}} : \mathcal{R}_{PPN}(\mathcal{A}) \to \mathcal{R}_{PN}(\mathcal{A})$, such that $J_{\mathcal{A}}(\sigma, ({}^\bullet\eta, \eta^\bullet)) = (\sigma, \eta^\bullet - {}^\bullet\eta)$ and $I_{\mathcal{A}}(\sigma, \eta) = (\sigma, ({}^\bullet\eta, \eta^\bullet))$ where ${}^\bullet\eta(e) = \mathbf{max}\{0, -\eta(e)\}$ and $\eta^\bullet(e) = \mathbf{max}\{0, \eta(e)\}$.

Owing to this correspondence, an instance of the separation problem $\text{SSP}_{\mathcal{A}}(s, s')$ can be solved in $\mathcal{R}_{PN}(\mathcal{A})$ if and only if it can be solved in $\mathcal{R}_{PPN}(\mathcal{A})$. The respective solutions are actually connected by the maps $I_{\mathcal{A}}$ and $J_{\mathcal{A}}$.

The treatment of the event state separation problem is more delicate. An instance of $\text{ESSP}_{\mathcal{A}}(s', e)$ can be solved in $\mathcal{R}_{PN}(\mathcal{A})$ if and only if there exists $\sigma(s_0) \in \mathbb{N}$, ${}^\bullet\eta(e) \in \mathbb{N}$, and $\eta \in \mathcal{R}_{abs}(A)$ such that:

1. $\eta(e) + {}^\bullet\eta(e) \geq 0$
2. $\forall s \in S \quad \sigma(s_0) + \eta \cdot \pi(c_s) \geq 0$
3. $\forall s \in S \quad s \xrightarrow{e} \Rightarrow \sigma(s_0) + \eta \cdot \pi(c_s) \geq {}^\bullet\eta(e)$
4. $\sigma(s_0) + \eta \cdot \pi(c'_s) < {}^\bullet\eta(e)$

A solution is then given by the Petri region $(\sigma, ({}^\bullet\eta, \eta^\bullet))$ defined by ${}^\bullet\eta(e') = \mathbf{max}\{0, -\eta(e')\}$ for $e' \neq e$, and $\eta^\bullet(e') = \eta(e') + {}^\bullet\eta(e')$ for every $e' \in E$. Set $x = \sigma(s_0)$, $y = {}^\bullet\eta(e)$, and $\eta = \sum x_i \cdot \eta_i$ where $\{\eta_1, \ldots, \eta_k\}$ is the basis of $\mathcal{R}_{abs}(A)$ and $x_i \in \mathbb{Z}$. With these notations, the above conditions may be rewritten to a system of linear inequations in the $k+2$ variables x, y and x_i $(1 \leq i \leq k)$, where $w_{is} = \eta_i \cdot \pi(c_s)$:

1. $y + \sum z_i \eta_i(e) \geq 0$
2. $x + \sum z_i w_{is} \geq 0$ (one inequation for each $s \in S$)
3. $x - y + \sum z_i w_{is} \geq 0$ (one inequation for each $s \in S$ such that $s \xrightarrow{e}$)
4. $x - y + \sum z_i w_{is'} < 0$

This system, augmented with the constraints $x \geq 0$ and $y \geq 0$, is homogeneous and can therefore be solved or shown unfeasible in polynomial time following Khachiyan's method.

7.5 Synthesizing Bounded Net Systems up to Language Equivalence

A net system is termed bounded if its marking graph is finite. Given a reachable and reduced finite deterministic automaton $\mathcal{A}$, let $\mathcal{L}(\mathcal{A})$ denote the (prefix closed) language of words accepted by $\mathcal{A}$. We face now the problem of deciding whether $\mathcal{L}(\mathcal{A}) = \mathcal{L}(\mathcal{N}^*)$ for some bounded net system and if so constructing $\mathcal{N}$. This problem can be decided upon in polynomial time, for both types τ_{PPN} and τ_{PN}, when $\mathcal{A}$ is given in tree-like form.

Definition 7.3 *$\mathcal{A} = (S, E, T, s_0)$ is a tree-like automaton if there exists a spanning tree $U \subseteq T$ rooted at s_0, with all transitions in U directed away from s_0, such that for every chord $s \xrightarrow{e} s' \notin U$, s' is an ancestor of s in U.*

Suppose $\mathcal{L}(\mathcal{A}) = \mathcal{L}(\mathcal{N}^*)$, where $\mathcal{A}$ is tree-like and $\mathcal{N} = (P, E, W, M_0)$ is a bounded net system with type $\tau \in \{\tau_{PPN}, \tau_{PN}\}$. For each place $p \in P$, let $(\sigma_p, \eta_p) : \mathcal{N}^* \to \tau$ denote the associated region of $\mathcal{N}^*$. From the inclusion $\mathcal{L}(\mathcal{A}) \subseteq \mathcal{L}(\mathcal{N}^*)$ and the assumption of boundedness of $\mathcal{N}$, $\sigma'_p(M_0) = \sigma_p(M_0)$ and η_p define a region in $(\sigma'_p, \eta_p) : \mathcal{A} \to \tau$ (seeing that $s \xrightarrow{u} s$ [5] in $\mathcal{A}$ entails $\eta_p \cdot \pi(u) = 0$

[5] $s \xrightarrow{\varepsilon} s$ for every $s \in S$, where ε is the empty word, and $s \xrightarrow{u \cdot e} s' \Leftrightarrow \exists s'' \in S \quad s \xrightarrow{u} s'' \wedge s'' \xrightarrow{e} s'$.

for $\tau = \tau_{PPN}$). From the relation $\mathcal{L}(\mathcal{N}^*) = \mathcal{L}(\mathcal{A})$, every instance of the event state separation problem $\mathrm{ESSP}_{\mathcal{A}}(s', e)$ is solved by a region $(\sigma'_p, \eta_p) \in \mathcal{R}_\tau(\mathcal{A})$ defined as above from some corresponding place $p \in P$. Conversely, if the condition ESSP is valid in $\mathcal{A}$, then $\mathcal{L}(\mathcal{A}) = \mathcal{L}(\mathcal{N}^*)$ for any net system $\mathcal{N} = \sum_{p \in P} \mathcal{N}_p$ assembled from a subset of τ-regions of $\mathcal{A}$ admissible for ESSP. Therefore, in the restricted case of tree-like automata, the synthesis problem for pure or impure Petri nets up to language equivalence can be solved in polynomial time [2].

Now, every deterministic automaton $\mathcal{A} = (S, E, T, s_0)$ may be translated to an equivalent tree-like automaton $\mathcal{A}' = (S', E, T', (s_0, \varepsilon))$ with sets of states and transitions defined as follows.

- S' is the set of pairs $(s, u) \in S \times \mathcal{L}(\mathcal{A})$ such that $s_0 \xrightarrow{u} s$ in $\mathcal{A}$ and every two states of $\mathcal{A}$ visited in this path are different;
- $T' \subseteq S' \times E \times S'$ is the set of transitions $(s, u) \xrightarrow{e} (s', u')$ such that $s \xrightarrow{e} s'$ in $\mathcal{A}$ and $u' = u \cdot e$ or u' is a prefix of u.

It must be noted, however, that the size of the tree-like automaton $\mathcal{A}'$ constructed in this way is *exponential* in the size of $\mathcal{A}$ (so as the number of elementary circuits of $\mathcal{A}$, which shows that the case of general automata cannot be dealt with by polynomial algorithms).

8 Regions in Step Transition Systems

We leave now the classical frame of (sequential) transition systems for the more expressive frame of *step transition systems*, defined by Mukund so as to account fully for the independence of events in general Petri nets [33].

Definition 8.1 *A* step transition system (S, M, T) *over an abelian monoid M consists of a set of states S and a deterministic transition relation $T \subseteq S \times M \times S$, with distinguished* empty steps*: $s \xrightarrow{0} s'$ iff $s = s'$. A* step automaton *$\mathcal{A}$ is an initialized step transition system (S, M, T, s_0) with initial state $s_0 \in S$, such that every state $s \in S$ is reachable from s_0 in the underlying transition system $A = (S, M, T)$. The step automaton $\mathcal{A}$ is* finite *if the set of transitions T is finite. When $M = \langle E \rangle$ is the free abelian monoid freely generated from set E (the elements of M are then finite multisets over E), the step automaton $\mathcal{A}$ is said to be* reduced *if its* skeleton *$(S, E, T \cap (S \times E \times S), s_0)$ is a reduced automaton.*

This definition of step transition systems extends slightly Mukund's original definition, which was restricted to free abelian monoids. The extension allows to accomodate the idea of regions as morphisms to step transition systems which do not necessarily present the *intermediate state* property: $s \xrightarrow{\alpha+\beta} s' \Rightarrow \exists s'' \in S\ s \xrightarrow{\alpha} s'' \wedge s'' \xrightarrow{\beta} s'$. The definition of regions in step transition systems is parametric on enriched types of nets defined as follows.

Definition 8.2 *An enriched type of nets is a (deterministic) step transition system $\tau = (LS, LE, \tau)$, where LE is an abelian monoid $(LE, +, 0)$.*

For instance, the enriched type of Petri nets is just the type $\tau_{PN} = (\mathbb{N}, \mathbb{N} \times \mathbb{N}, \tau)$, where $n \stackrel{(i,j)}{\longrightarrow} n' \in \tau$ if and only if $n \geq i$ and $n' = n - i + j$, enriched with the operation of componentwise addition in $\mathbb{N} \times \mathbb{N}$. As a matter of fact, $(\mathbb{N} \times \mathbb{N}, +, (0,0))$ is the free abelian monoid generated from $(0,1)$ and $(1,0)$.

Each type of nets determines a specific concurrent firing rule and hence a specific construction of concurrent marking graphs.

Definition 8.3 *Given a net $N = (P, E, W)$ with (enriched) type $\tau = (LS, LE, \tau)$, the concurrent marking graph of N is the step transition system $(LS^P, <E>, T)$ with set of transitions T defined by :*

$$(M \stackrel{\alpha}{\rightarrow} M') \in T \Leftrightarrow \forall x \in P \ (M(x) \stackrel{W(x,\alpha)}{\longrightarrow} M'(x)) \in \tau \tag{5}$$

where $W(x, e_1 + \cdots + e_n) = W(x, e_1) + \cdots + W(x, e_n)$. Given a net system $\mathcal{N} = (P, E, W, M_0)$, the concurrent marking graph of $\mathcal{N}$ is the step automaton $\mathcal{N}^* = (S, <E>, T_S, M_0)$ where S is the inductive closure of the singleton set $\{M_0\}$ w.r.t. forward transitions in T, and $T_S = T \cap (S \times <E> \times S)$.

In order to illustrate this definition, let us inspect the relationship between the sequential and concurrent marking graphs of a Petri net. On the one hand, the sequential marking graph is the induced restriction of the concurrent marking graph on the subset of *atomic* steps, i.e. steps α such that $\Sigma_{e \in E}\, \alpha(e) = 1$. On the other hand, the concurrent marking graph cannot in general be reconstructed up to isomorphism from an arbitrary copy of the sequential marking graph, even though some additional informations are provided as in [20] by a binary relation of independence on events depending on markings, such that $e \,\|_M\, e'$ if and only if $M[\{e, e'\}>$. The example shown in Fig. 19, borrowed from [27], makes this fact clear. Regions may now be introduced, based on the following definition of morphisms of step transition systems.

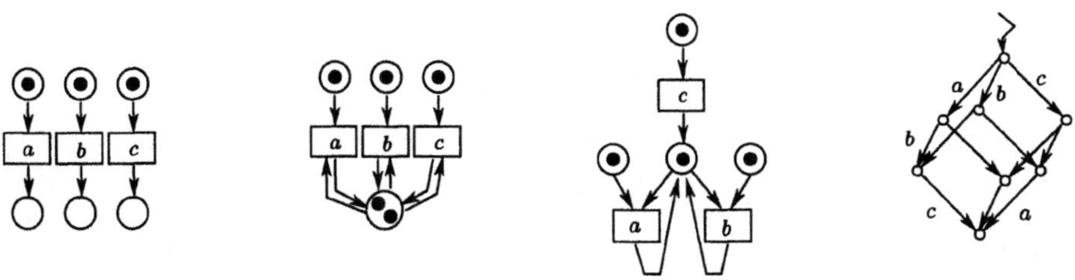

Fig. 19. three nets with an identical sequential marking graph but with different concurrent marking graphs: the three events a, b, and c are independent at the indicated marking in the first net whereas they are pairwise independent but not independent in the second net; the case of the third net is more involved: at the indicated marking the maximal sets of independent events are $\{a, c\}$ and $\{b, c\}$, but a and b become independent once c has been fired. Thus independence in Petri nets is marking dependent.

Definition 8.4 *A morphism of step transition systems from $A = (S, M, T)$ to $A' = (S', M', T')$ is a pair (σ, η), made of a map $\sigma : S \rightarrow S'$ and a monoid morphism $\eta : M \rightarrow M'$, such that $s \xrightarrow{\alpha} s' \Rightarrow \sigma(s) \xrightarrow{\eta(\alpha)} \sigma(s')$.* The morphisms of step automata *from $\mathcal{A}$ to $\mathcal{A}'$ are the morphisms from A to A' that preserve the initial state.*

Definition 8.5 *Given a step transition system $A = (S, M, T)$ and an enriched type of nets $\tau = (LS, LE, \tau)$, the set $\mathcal{R}_\tau(A)$ of τ-type (extended) regions of A is the set of morphisms of step transition systems from A to τ. The set of τ-type (extended) regions of a step automaton $\mathcal{A}$ is the set $\mathcal{R}_\tau(\mathcal{A}) = \mathcal{R}_\tau(A)$.*

By specializing this definition to the type τ_{PN}, one retrieves exactly the regions defined by Mukund in step transition systems over a free abelian monoid [33]. Special attention may be paid to the class of step transition systems $A = (S, M, T)$ derived from asynchronous transition systems $(S, E, \|, T')$ as follows: $M = <E>$ is the free abelian monoid generated by E (the elements of M are finite multisets of elements of E), $s \xrightarrow{\alpha} s'$ in T if and only if α is a sub*set* of pairwise independent events $\{e_1, \ldots, e_n\} \subseteq E$ (hence there is no auto-concurrency) and there exists in T' a sequence of transitions $s \xrightarrow{e_1} s_1 \xrightarrow{e_2} s_2 \ldots s_{n-1} \xrightarrow{e_n} s_n$ such that $s' = s_n$ (such sequences exist therefore for all permutations of $\{e_1, \ldots, e_n\}$). For this class of step transition systems, the regions $(\sigma, \eta) : A \rightarrow \tau_{PN}$ which are *safe* in the sense that $\sigma(s) \in \{0, 1\}$for all $s \in S$ are in bijective correspondence with the regions defined by Nielsen and Winskel for asynchronous transition systems [35].

The results about (ordinary) transition systems which have been presented in section 6 may be reproduced nearly intact in the richer setting of step transition systems over a free abelian monoid. In particular, Def. 6.6 and 6.7 and Prop. 6.11 may be extended to step transition systems, yielding a Galois connection $\mathcal{A} \leq \mathcal{N}^* \Leftrightarrow \mathcal{N} \leq \mathcal{A}^*$ between step automata $\mathcal{A} = (S, <E>, T, s_0)$ and net systems $\mathcal{N} = (P, E, W, M_0)$, for any enriched type of nets τ. The following counterpart to Theo. 6.12 for step transition systems appears in [7].

Theorem 8.6 *Given a step automaton over a free abelian monoid, let $\mathcal{A} = (S, <E>, T, s_0)$, and an enriched type of nets τ, a subset of extended regions $\mathcal{R} \subseteq \mathcal{R}_\tau(\mathcal{A})$ is admissible if and only if the following* separation properties *are satisfied for all states $s, s' \in S$ and for every multiset $\alpha \in <E>$:*

(SSP) $s \neq s' \;\Rightarrow\; \exists (\sigma, \eta) \in \mathcal{R} : \; \sigma(s) \neq \sigma(s')$

(ESSP) $s \stackrel{\alpha}{\not\rightarrow} \;\Rightarrow\; \exists (\sigma, \eta) \in \mathcal{R} : \; \sigma(s) \stackrel{\eta(\alpha)}{\not\longrightarrow}$ *in* τ

When both properties are satisfied, $\mathcal{A} \cong (\mathcal{A}^*_{\mathcal{R}})^*$, where $\mathcal{A}^*_{\mathcal{R}}$ is the subnet system of $\mathcal{A}^*$ with restricted set of places $\mathcal{R}$ (also called the net synthesized *from $\mathcal{R}$).*

Mukund's characterization of Petri net transition systems, established in [33], follows directly from Theo. 8.6 applied to the type τ_{PN}. Nielsen and Winskel's characterization of separated asynchronous automata, established in [35], follows therefrom as the subcase met when imposing on regions $(\sigma, \eta) \in \mathcal{R}$ the constraint that $\sigma(s) \in \{0, 1\}$ for every state s.

An algorithm for synthesizing Petri nets from finite step transition systems, based on Theo. 8.6, is proposed in [7]. This algorithm is an adaptation of the basic algorithm for pure Petri nets described in section 7. Let $\mathcal{A} = (S, <E>, T, s_0)$ be some finite and reduced step automaton. Seeing that the intermediate state property is always satisfied in the concurrent marking graph of a Petri net, we assume this property from $\mathcal{A}$. Thus, $(s \stackrel{\alpha+\beta}{\rightarrow} s') \in T \;\Rightarrow\; \exists s'' \in S \; (s \stackrel{\alpha}{\rightarrow} s'') \in T \wedge (s'' \stackrel{\beta}{\rightarrow} s') \in T$. The import is that we may assume a compact representation for $\mathcal{A}$, given by its skeleton and the set of maximal steps at each state $s \in S$. This makes sense since the set of steps of $\mathcal{A}$ is bounded, from the assumption that $\mathcal{A}$ is finite. As regards the event state separation problem, let us observe the following: if a region (σ, η) solves an instance $\text{ESSP}_{\mathcal{A}}(s, \alpha)$ of this problem, where α is a failure at s, then (σ, η) solves also every instance $\text{ESSP}_{\mathcal{A}}(s, \beta)$ such that $\alpha < \beta$. It is then sufficient to solve at each state s the instances $\text{ESSP}_{\mathcal{A}}(s, \alpha)$ such that α is a minimal failure in that state. From this remark and the assumed representation for $\mathcal{A}$, the following is proved in [7].

Theorem 8.7 *The synthesis problem for Petri net systems with the step firing rule, taking as inputs finite step transition systems, is polynomial in their numbers of states and events, in the size of the largest set of minimal failures in one state, and in the size of the largest set of maximal steps enabled in one state.*

Notice that the minimal failures are not determined at a given state by the maximal steps, as shown by the third net on Fig. 19 for which $Max\_steps(s_0) = \{a+c, b+c\}$ and $Min\_fails(s_0) = \{2a, a+b, 2b\}$ whence $Min\_fails(s) \not\subseteq \{\alpha + e \mid \alpha \in Max\_steps(s)\}$. Every step automaton may in fact be transformed to an ordinary automaton by splitting the alphabet of events: the states of the split automaton are the pairs $<s, \alpha>$ where α is a step with concession at s, and each transition $s \stackrel{e}{\rightarrow} s'$ gives rise to the pair of transitions $<s, \alpha> \stackrel{e^+}{\rightarrow} <s, \alpha + e>$ and $<s, \alpha + e> \stackrel{e^-}{\rightarrow} <s', \alpha>$ for every step $\beta = \alpha + e$ with concession at s. In [1] it is shown that the synthesis of Petri nets from step automata may be reduced to the synthesis of pure Petri nets from ordinary automata by splitting events, which yields a synthesis algorithm taking time polynomial in the number of higher-dimensional states. Now if $\beta \in Min\_fails(s)$ is a minimal failure, then $<s, \alpha> \stackrel{e^+}{\not\rightarrow}$ for any step α and event e such that $\beta = \alpha + e$, and the problem $\text{ESSP}_{\mathcal{A}}(s, \beta)$ is equivalent to the separation problem $\text{ESSP}_{\text{Split}(\mathcal{A})}(<s, \alpha>, e^+)$ for the event e^+ at the higher-dimensional state $<s, \alpha>$. There are (at most $|E|$ times) more instances of ESSP to be solved in the split automaton since β can be decomposed as $\beta = \alpha + e$ in several ways, but the total number of instances of the problem $\text{ESSP}_{\mathcal{A}}(s, \alpha)$ for minimal failure α in the step automaton is already exponential in the number of events.

In order to conclude with extended regions (σ, η) where η maps steps $\alpha \in <E>$ to (pairs of) integral weights, let us mention that such regions have also been used to solve different synthesis problems in the setting of generalized trace languages [26] and general event structures [27].

9 Adjunctions between Transition Systems and Nets

In section 6, a Galois connection $\mathcal{A} \leq \mathcal{N}^* \Leftrightarrow \mathcal{N} \leq \mathcal{A}^*$ between automata and net systems was established. However $\mathcal{A}^*$ and $\mathcal{N}^*$ are not constructed in a symmetrical way: $\mathcal{A}^*$ has been assembled from morphisms of transition systems from A to the type of nets τ, but $\mathcal{N}^*$ has not been constructed from net morphisms. We show in this section that $\mathcal{N}^*$ can equally well be assembled from net morphisms $\varphi : N \to \tau'$ where the places of τ' encode bijectively the transitions of τ.

Therefore types of nets are schizophrenic objects $<\tau, \tau'>$ living both in the category of transition systems and in the category of nets. Taking advantage of this fact, we adapt a work of Porst and Tholen [36] on concrete dualities induced by schizophrenic objects and construct dual adjunctions between transition systems and nets for any type of nets. We show in this way that the region based representation theorems for transition systems are a close analogue of the classical representation theorems for ordered algebras, which all arise from concrete dualities induced by schizophrenic objects based on the two element set $\mathbf{2} = \{0, 1\}$.

For the reader unfamiliar with schizophrenic objects, we review briefly some of the classical representation theorems. Birkhoff's duality between finite distributive lattices and finite partial orders relies on the schizophrenic object $\mathbf{2}$, viewed as a lattice and as an ordered set where $0 \leq 1$. The dual L^* of a distributive lattice L is the ordered set of its prime filters x whose characteristic functions are the lattice morphisms $\chi_x : L \to \mathbf{2}$. The dual X^* of an ordered set X is the lattice of its upwards closed subsets l whose characteristic functions are the morphisms of ordered sets $\chi_l : X \to \mathbf{2}$. Any ordered set is isomorphic to its double dual ($X \cong X^{**}$) where $x \in X$ is identified with $x^{**} \in X^{**}$ such that $\chi_{x^{**}}(l) = \chi_l(x)$ for any upwards closed subset $l \subseteq X$. Any distributive lattice is isomorphic to its double dual ($L \cong L^{**}$) where $l \in L$ is identified with $l^{**} \in L^{**}$ such that $\chi_{l^{**}}(x) = \chi_x(l)$ for any prime filter $x \subseteq L$. Thus both units of the dual adjunction are morphisms whose underlying maps are the evaluation maps.

Stone's duality between boolean algebras and the Stone spaces relies similarly on the schizophrenic object $\mathbf{2}$, viewed as a boolean algebra and as a discrete topological space. More instructive in the context of this paper is the duality between spatial frames and sober spaces (see [28, 18]). Recall that a frame is a complete lattice with the generalized distributivity law (finite meets distribute over arbitrary joins: $f \wedge \bigvee_i f_i = \bigvee_i (f \wedge f_i)$). For any frame F, let $pt(F)$ be the set of *points* x of F defined as frame morphisms $x : F \to \mathbf{2}$. The dual F^* of F is the topological space $(pt(F), \Omega)$ whose open sets are the sets $O_f = \{x : F \to \mathbf{2} |\ x(f) = 1\}$ for f ranging over F. Conversely, the dual X^* of a topological space (X, Ω) is the frame of its open sets $O \in \Omega$, whose characteristic functions χ_O are the continuous maps from (X, Ω) to the Sierpinski space $\mathbf{2}$ (with open sets $\{0, 1\}$, $\{1\}$, and $\emptyset$). Frames and topological spaces are connected by a dual adjunction $\mathbf{Frame}(F, X^*) \cong \mathbf{Top}(X, F^*)$.

By restricting this adjunction at both sides on its kernel, one obtains a duality $\mathbf{Top}^* \overset{op}{\cong} \mathbf{Frame}^*$ between the subcategory $\mathbf{Top}^*$ of *spatial* frames and the subcategory **Frame*** of *sober* spaces. So, a frame F is isomorphic to its double dual F^{**} if and only if F is a spatial frame. Now, spatial frames are characterized by two conditions very similar to our separation conditions for automata, when regions are replaced by morphisms $x : F \to \mathbf{2}$. Namely, a frame F is spatial if and only if the following conditions are satisfied for all $f, f' \in F$, where $f \leq f' \Leftrightarrow f = f \wedge f'$:

$$\begin{array}{ll} (i) & f \neq f' \Rightarrow \exists x : F \to \mathbf{2} \ : \ x(f) \neq x(f') \\ (ii) & f \not\leq f' \Rightarrow \exists x : F \to \mathbf{2} \ : \ x(f) = 1 \wedge x(f') = 0 \end{array}$$

Condition *(i)* is the analogue of our state separation condition SSP. Condition *(ii)* is the counterpart of our event state separation condition ESSP, when the structure of labelled transition system is replaced by the structure of partial order.

The classical dualities recalled above are concerned with points x, properties p, and a binary relation of evaluation $ev(x)(p) = p(x)$ valued in the underlying set of the schizophrenic object, i.e. $\{0, 1\}$. When this relation is given a matrix form, duality appears as matrix transposition [37]. Now, dualities between transition systems and nets fit exactly in the same pattern: the points are the transitions $s \xrightarrow{e} s'$, the properties are the regions (σ, η), and the evaluation matrix given by $ev(s \xrightarrow{e} s', (\sigma, \eta)) = \sigma(s) \xrightarrow{\eta(e)} \sigma(s')$ describes the local effect of the transitions on the places (σ, η) of the dual net. The technical development presented in the remaining of the section is based on the material contained in [6].

9.1 Schizophrenic Objects and Dual Adjunctions

Definition 9.1 *A* Set-category *(or category over* **Set***) is a pair* $<\mathcal{C}, U>$ *where* $\mathcal{C}$ *is a category and* $U : \mathcal{C} \to \mathbf{Set}$ *is a functor called the underlying functor. It is a* concrete category *if* U *is faithful.*

In the sequel, the underlying functor is left implicit and we use the uniform notation $|C|$ and $|f|$ for respectively the underlying set of an object C and the underlying map of an arrow f. In a Set-category $\mathcal{C}$, a ***structured source*** is an indexed family of pairs $\{C_i; f_i : X \to |C_i|\}$, where the C_i's are objects of $\mathcal{C}$ and the f_i's are maps from a fixed set X to the underlying sets of the C_i's. A ***lift*** of a structured source is an indexed family $\tilde{f}_i : C \to C_i$ of arrows of $\mathcal{C}$ such that $|\tilde{f}_i| = f_i$, and hence $|C| = X$. An ***initial lift*** of a structured source is a lift such that, if $g_i : C' \to C_i$ is another lift and there exists a map $f : |C'| \to X$ such that $|g_i| = f_i \circ f$ for all indices, then there exists a unique arrow $\tilde{f} : C' \to C$ such that $|\tilde{f}| = f$ and $g_i = \tilde{f}_i \circ \tilde{f}$ for all indices. The following definition is an adaptation from [36].

Definition 9.2 *A schizophrenic object between two Set-categories* $\mathcal{A}$ *and* $\mathcal{B}$ *is a pair of objects* $<K_{\mathcal{A}}, K_{\mathcal{B}}> \in |\mathcal{A}| \times |\mathcal{B}|$ *with the same underlying set* $K = |K_{\mathcal{A}}| = |K_{\mathcal{B}}|$ *and such that*

1. *for every object* A *in* $\mathcal{A}$*, the family* $\{K_{\mathcal{B}}; ev_A(a) : \mathcal{A}(A, K_{\mathcal{A}}) \to K\}_{a \in |A|}$ *of evaluation maps* $ev_A(a)(f) = |f|(a)$ *has an initial lift* $\{\epsilon_A(a) : A^* \to K_{\mathcal{B}}\}_{a \in |A|}$

2. *for every object B in $\mathcal{B}$, the family $\{K_{\mathcal{A}}; ev_B(b) : \mathcal{B}(B, K_{\mathcal{B}}) \to K\}_{b \in |B|}$ has an initial lift $\{\epsilon_B(b) : B^* \to K_{\mathcal{A}}\}_{b \in |B|}$.*

A^*, called the *dual* of A, is therefore an object of the category $\mathcal{B}$ whose underlying set is the set of $\mathcal{A}$-morphisms from A to the *classifying object* $K_{\mathcal{A}}$. If $K = \{0, 1\}$ and $\mathcal{A}$ is concrete, then the elements of the underlying set of the dual of A can be identified with subsets of the underlying set of A: $|A^*| \subseteq 2^{|A|}$ and $|A^{**}| \subseteq 2^{2^{|A|}}$. In any case, A and A^{**} are linked by an evaluation morphism $Ev_A : A \to A^{**}$ according to the following statement.

Lemma 9.3 *Let $< K_{\mathcal{A}}, K_{\mathcal{B}} >$ be a schizophrenic object between two Set-categories $\mathcal{A}$ and $\mathcal{B}$. The initial lift $\{\epsilon_A(a) : A^* \to K_{\mathcal{B}}\}_{a \in |A|}$ of the evaluation maps, viewed as a mapping $\epsilon_A : |A| \to \mathcal{B}(A^*, K_{\mathcal{B}})$, is the underlying map of an arrow $Ev_A : A \to A^{**}$.*

As an initial lift, the dual A^* of A is only defined up to an isomorphism. However, once an arbitrary representative A^* is fixed for each class of isomorphic objects, the operator $(-)^*$ gives rise to a functor according to the following statement.

Lemma 9.4 *Let $< K_{\mathcal{A}}, K_{\mathcal{B}} >$ be a schizophrenic object between two Set-categories $\mathcal{A}$ and $\mathcal{B}$. For every morphism $f : A_1 \to A_2$ in $\mathcal{A}$, the map "composing with f" given by $f^{\bullet} : \mathcal{A}(A_2, K_{\mathcal{A}}) \to \mathcal{A}(A_1, K_{\mathcal{A}})$: $g \mapsto g \circ f$ is the underlying map of an arrow $f^* : A_2^* \to A_1^*$ in $\mathcal{B}$ such that the functoriality laws $(1_A)^* = 1_{A^*}$ and $(f \circ g)^* = g^* \circ f^*$ are satisfied.*

The following proposition tells us that the two functors $(-)^*$ induced from a schizophrenic object are in fact dual adjoints.

Proposition 9.5 *Let $< K_{\mathcal{A}}, K_{\mathcal{B}} >$ be a schizophrenic object between two Set-categories $\mathcal{A}$ and $\mathcal{B}$. The following identities, where $f : A \to B^*$ and $g : B \to A^*$, define a bijective correspondence $\mathcal{A}(A, B^*) \cong \mathcal{B}(B, A^*)$:*

$$g = f^* \circ Ev_B \text{ and } f = g^* \circ Ev_A$$

i.e. the functors $(-)^*$ are adjoint to the right with the evaluations as units.

In the particular case where $\mathcal{A}$ and $\mathcal{B}$ are concrete categories, the above correspondence may be presented as matrix transposition. Actually, in this special case, $\mathcal{A}(A, B^*) \cong \mathbf{Span}_K(A, B) \cong \mathcal{B}(B, A^*)$ where $\mathbf{Span}_K(A, B)$ is the set of matrices $|A| \times |B| \to K$ whose rows, resp. columns, are underlying maps of morphisms $\varphi_a : B \to K_{\mathcal{B}}$ (for $a \in |A|$), resp. of morphisms $\varphi^b : A \to K_{\mathcal{A}}$ (for $b \in |B|$). In such a matrix, the set of rows determines a unique morphism from A to B^*, and the set of columns φ^b determines a unique morphism from B to A^*.

9.2 Application to Automata and Nets

Let **Trans** be the category of deterministic and reduced transition systems (S, E, T) free of isolated states ($\forall s \in S\ \exists t \in T : s = \partial^0(t) \lor s = \partial^1(t)$), where a morphism $(\sigma, \eta) : (S, E, T) \to (S', E', T')$ is a pair of maps $\sigma : S \to S'$ and

$\eta : E \to E'$ such that $s \xrightarrow{e} s'$ (in T) entails $\sigma(s) \xrightarrow{\eta(e)} \sigma(s')$ (in T'). **Trans** is a concrete category with forgetful functor U : **Trans** $\to$ **Sets** given by $U(S, E, T) = T$ and $U(\sigma, \eta)(s \xrightarrow{e} s') = (\sigma(s) \xrightarrow{\eta(e)} \sigma(s'))$.

Let $\tau = (LS, LE, LT)$ be an arbitrary object of **Trans**, called the type of nets. Let **Nets** be the category of event-simple nets (P, E, W) of type τ, thus $W : P \times E \to LE$ has all columns distinct, where a morphism $(\beta, \eta) : (P, E, W) \to (P', E', W')$ is a pair of maps $\beta : P \to P'$ and $\eta : E' \to E$ such that $W(p, \eta(e')) = W(\beta(p), e')$. Owing to the assumption of event-simpleness, β determines η in any morphism (β, η), and **Nets** is a concrete category with forgetful functor U : **Nets** $\to$ **Sets** given by $U(P, E, W) = P$ and $U(\beta, \eta) = \beta$.

Let $\tau' = (LT, \{\bullet\}, W) \in$ **Nets** be the net with the unique event $\bullet$ such that $W(\ell s \xrightarrow{\ell e} \ell s') = \ell e$ for every place $\ell s \xrightarrow{\ell e} \ell s' \in LT$. Thus $U\tau' = LT = U\tau$. Figure 20 displays the net τ'_{EN} corresponding to the type τ_{EN} of elementary nets.

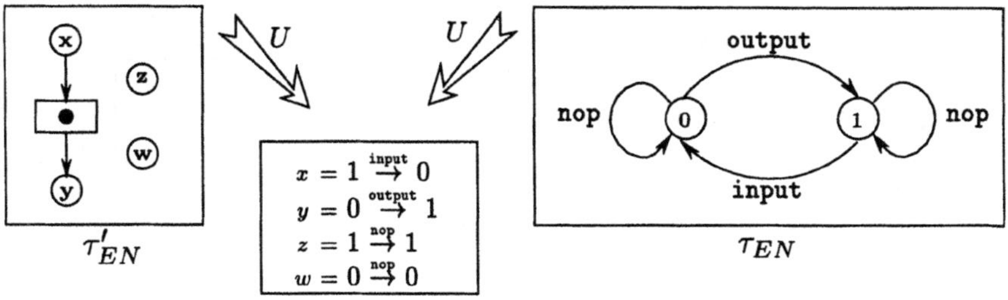

Fig. 20. the schizophrenic object for elementary nets

Proposition 9.6 *The pair* (τ, τ') *is a schizophrenic object between the categories* **Trans** *and* **Nets**, *inducing a dual adjunction* **Trans**$(A, N^*) \cong$ **Nets**(N, A^*).

It remains to interpret A^* and N^* in more familiar terms. For any transition system $A = (S, E, T)$, the homset **Trans**(A, τ) is the set $\mathcal{R}_\tau(A)$ of τ-regions of A. The evaluation $ev_A(s \xrightarrow{e} s')(\sigma, \eta) = (\sigma(s) \xrightarrow{\eta(e)} \sigma(s'))$ classifies therefore the transitions $t = (s \xrightarrow{e} s') \in T$ according to their local effect on each region. By definition, A^* is the net resulting from the initial lift of the family of evaluation maps $ev_A(t)$ for $t \in T$. The following proposition shows that A^* coincides with the net synthesized from the set of regions $\mathcal{R}_\tau(A)$ up to the confusion of indiscernible events.

Proposition 9.7 A^* *is isomorphic to the net* $(P, E_\equiv, W)$ *where* $P = \mathcal{R}_\tau(A)$ *is the set of regions of* A, $\equiv$ *is the equivalence relation on* E *such that* $e \equiv e'$ *when* $\eta(e) = \eta(e')$ *for every region* (σ, η), *and* $W((\sigma, \eta), [e]_\equiv) = \eta(e)$.

Now, for any net $N = (P, E, W)$, the homset **Nets**(N, τ') is in bijective correspondence with the set of transitions of the marking graph of N. In the sample case of elementary nets (see Fig. 21), each morphism $(\beta, \eta) : N \to \tau'_{EN}$ induces the transition $\beta^{-1}(\{x, z\}) \xrightarrow{\eta(e)} \beta^{-1}(\{y, z\})$, and conversely, each firing $M[e > M'$ induces the morphism (β, η) such that $\eta(\bullet) = e$ and for every place p,

$$\beta(p) = \begin{cases} x & \textbf{if} \quad p \in M \setminus M' \\ y & \textbf{if} \quad p \in M' \setminus M \\ z & \textbf{if} \quad p \in M \cap M' \\ w & \textbf{if} \quad p \notin M \cup M' \end{cases}$$

Therefore, the evaluation $ev_N(p)(\beta, \eta) = \beta(p)$ classifies the places of N ac-

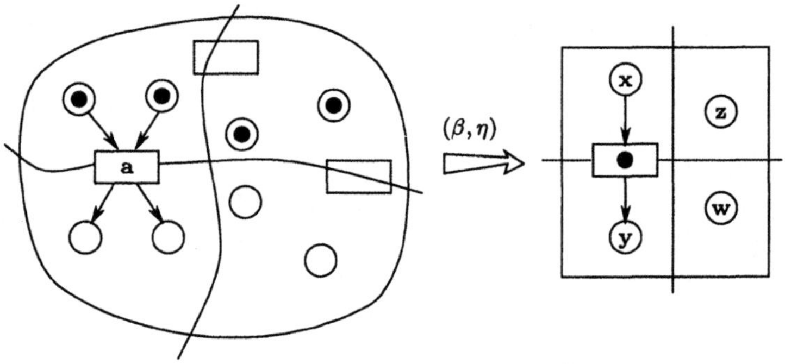

Fig. 21. firings as net morphisms

cording to the local transition they undergo in each global firing of the net. By definition, N^* is the transition system resulting from the initial lift of the family of evaluation maps $ev_N(p)$ for $p \in P$.

Proposition 9.8 *N^* is isomorphic to the marking graph of N.*

If one now specifies initial states for transition systems, and forward closed sets of markings for nets, the dual adjunction $\mathbf{Trans}(A, N^*) \cong \mathbf{Nets}(N, A^*)$ may be extended to a *Galois connection* $\mathbf{Aut}(\mathcal{A}, \mathcal{N}^*) \cong \mathbf{Netsys}(\mathcal{N}, \mathcal{A}^*)$ between automata and net systems, i.e. to a dual adjunction such that $\mathcal{A}^* \cong \mathcal{A}^{***}$ for every automaton $\mathcal{A}$, and $\mathcal{N}^* \cong \mathcal{N}^{***}$ for every net system $\mathcal{N}$. The details of the construction can be found in [6]. By restricting the Galois connection at both sides on its kernel, one finally obtains a duality $\mathbf{Netsys}^* \overset{op}{\cong} \mathbf{Aut}^*$ between separated automata and saturated net systems. As a consequence, the separated automata appear as a co-reflective subcategory of $\mathbf{Netsys}^{op}$. Similar co-reflections between separated automata and nets have been established in the literature for various categories of automata or concurrent automata, including elementary automata [34], asynchronous automata [35], automata with concurrency relations [20], and step automata [33].

10 Some Applications

Regions have come to use so far in two areas of application, asynchronous circuits and distributed protocols. A computer assisted solution to the state encoding problem for asynchronous circuits, based on elementary regions and supported by the tool `Petrify`, is described in [15]. A computer assisted solution to the distribution of protocols, based on Petri regions and supported by the tool SYNET, is described in [13]. Distributed and cooperative systems offer a wide range of problems to be solved prior to any successful application. The synthesis of stratified Petri nets [8], a weaker form of Valk's self-modifying nets [43, 44], may for instance be used for analysing cooperative systems in order to identify their normal and exceptional modes of operation, and possibly for simplifying the control of the transitions between these modes. Another goal of research is to derive systems from service specifications while decomposing large specifications into pieces. This might become feasible if one could solve the relaxed synthesis problem as follows: given a pair of rational languages L and L' such that $L \subseteq L'$, construct a (possibly not bounded) net system $\mathcal{N}$ such that $L \subseteq \mathcal{L}(\mathcal{N}) \subseteq L'$.

Acknowledgments. This work was partly supported by the H.C.M. Network *Express*.

References

[1] BADOUEL, E., *Splitting of Actions, Higher-Dimensional Automata and Net Synthesis.* Inria Research Report No 3013 (1996).

[2] BADOUEL, E., BERNARDINELLO, L. and DARONDEAU, PH., *Polynomial algorithms for the synthesis of bounded nets*, Proceedings Caap 95, Lecture Notes in Computer Science 915 (1995) 647–679.

[3] BADOUEL, E., BERNARDINELLO, L. and DARONDEAU, PH., *The synthesis problem for elementary net systems is NP-complete*, Inria Research Report 2558 (1995). To appear in Theoretical Computer Science.

[4] BADOUEL, E., and DARONDEAU, PH., *Trace Nets.* REX workshop, Beekbergen "Semantics: Foundation and Applications", Springer-Verlag Lecture Notes in Computer Science, vol. 666 (1993) 21–50.

[5] BADOUEL, E., and DARONDEAU, PH., *Trace nets and process automata*, Acta Informatica 32 (1995) 647–679.

[6] BADOUEL, E., and DARONDEAU, PH., *Dualities between Nets and Automata induced by Schizophrenic Objects*, 6th International Conference on Category Theory and Computer Science, Cambridge, Lecture Notes in Computer Science, vol. 953 (1995) 24–43.

[7] BADOUEL, E., and DARONDEAU, PH., *On the Synthesis of General Petri Nets.* Inria Research Report No 3025 (1996).

[8] BADOUEL, E., and DARONDEAU, PH., *Stratified Petri Nets.* Inria Research Report No 3128 (1997).

[9] BEDNARCZYK, M. A., *Categories of Asynchronous Systems.* Ph. D. Thesis, University of Sussex (1988).

[10] BERGE, C., *Graphes et hypergraphes.* Dunod, Paris (1970). English translation: *Graphs and Hypergraphs,* North Holland, Amsterdam (1973).

[11] BERNARDINELLO, L., *Synthesis of Net Systems.* Application and Theory of Petri Nets, Lecture Notes in Computer Science 691 (1993) 89–105.

[12] BERNARDINELLO, L., DE MICHELIS, G., PETRUNI, K., and VIGNA, S., *On the Synchronic Structure of Transitions Systems.* In "Structures in Concurrency Theory", J. Desel ed., Springer-Verlag (1996) 11–31.

[13] CAILLAUD, B., SYNET : *un outil de synthèse de réseaux de Petri bornés, applications* Irisa Research Report no 1101 (1997).

[14] CORTADELLA, J., KISHINEVSKY, M., LAVAGNO, L., and YAKOVLEV, A., *Synthesizing Petri Nets from State-Based Models.* Proceedings of ICCAD'95 (1995) 164–171.

[15] CORTADELLA, J., KISHINEVSKY, M., KONDRATYEV, A., LAVAGNO, L., and YAKOVLEV, A., *Complete state encoding based on the theory of regions.* Proceedings of the 2nd International Workshop on Advanced Research in Asynchronous Circuits and Systems (1996) 36–47.

[16] CHEN, W.K., *Applied Graph Theory.* North-Holland (1971).

[17] CHRISTOFIDES, N., *Graph Theory - An Algorithmic Approach.* Academic Press (1975).

[18] DAVEY, B.A., and PRIESTLEY, H.A., *Introduction to Lattices and Order.* Cambridge University Press, (1990).

[19] DESEL, J., and REISIG, W., *The Synthesis Problem of Petri Nets.* Acta Informatica vol. 33 (1996) 297–315.

[20] DROSTE, M., and SHORTT, R.M., *Petri Nets and Automata with Concurrency Relations – an Adjunction.* in "Semantics of Programming Languages and Model Theory", M. Droste and Y. Gurevich eds(1993) 69–87.

[21] DROSTE, M., and SHORTT, R.M., *From Petri Nets to Automata with Concurrency.* Draft communicated to the authors.

[22] EHRENFEUCHT, A., and ROZENBERG, G., *Partial 2-structures ;* Part I : *Basic Notions and the Representation Problem,* and Part II : *State Spaces of Concurrent Systems,* Acta Informatica, vol 27 (1990).

[23] GAREY, M.R., and JOHNSON, D.S., *Computer and Intractability. A guide to the theory of NP-Completeness.* W.H. Freeman and Company (1979).

[24] GONDRAN, M., and MINOUX, M., *Graphes et algorithmes.* Eyrolles, Paris (1979). English translation by Steven Vajda: *Graphs and Algorithms,* John Wiley (1984).

[25] HIRAISHI, K., *Some complexity results on transition systems and elementary net systems.* Theoretical Computer Science 135 (1994) 361–376.

[26] HOOGERS, P.W., KLEIJN, H.C.M., and THIAGARAJAN, P.S., *A Trace Semantics for Petri Nets.* Lecture Notes in Computer Science vol. 623 (1992) 595–604.

[27] HOOGERS, P.W., KLEIJN, H.C.M., and THIAGARAJAN, P.S., *An Event Structure Semantics for General Petri Nets.* Theoretical Computer Science, volume 153 (1996) 129–170.

[28] JOHNSTONE, P.T., *Stone Spaces.* Cambridge University Press, (1982).

[29] KISHINEVSKY, M., CORTADELLA, J., KONDRATYEV, A., LAVAGNO, L., TAUBIN, A., and YAKOVLEV, A., *Place Chart Nets and their Synthesis.* Technical Report 96-2-003 Department of Computer Hardware, University of Aizu (1996).

[30] KISHINEVSKY, M., CORTADELLA, J., KONDRATYEV, A., LAVAGNO, L., and YAKOVLEV, A., *Synthesis of General Petri Nets.* Technical Report 96-2-004, Department of Computer Hardware, University of Aizu (1996).

[31] LEFSCHETZ, S., *Applications of Algebraic Topology.* Applied Mathematical Science 16, Spinger-Verlag (1975).

[32] LIU, C.L., *Introduction to Combinatorial Mathematics.* Mac Graw Hill (1968).

[33] MUKUND, M., *Petri Nets and Step Transition Systems.* International Journal of Foundation of Computer Science, vol 3, n° 4 (1992) 443–478.

[34] NIELSEN, M., ROZENBERG, G., and THIAGARAJAN, P.S., *Elementary Transition Systems.* Theoretical Computer Science, vol. 96 (1992) 3–33.

[35] NIELSEN, M., and WINSKEL, G., *Models for Concurrency,* Handbook of Logic for Computer Science, vol. 4, Oxford University Press (1995).

[36] PORST, H.-E., and THOLEN, W., *Concrete Dualities.* In "Category Theory at Work", H. Herrlich, and H.-E. Porst (eds.), Heldermann Verlag Berlin (1991) 111-136.

[37] PRATT, V.R., *The Stone Gamut: A Coordinatization of Mathematics,* Proceedings of the 10th Symposium on Logics in Computer Science, IEEE Computer Society (1995) 444–454.

[38] SCHRIJVER, A., *Theory of Linear and Integer Programming.* John Wiley (1986).

[39] SCHMITT, V., *Flip-Flop Nets,* Proceedings of Stacs 96, Lecture Notes in Computer Science vol. 1046 (1996) 517–528.

[40] TUTTE, W.T., *Graph Theory.* Encyclopedia of Mathematics and its Application vol. 21, Addison-Wesley (1984).

[41] SHIELDS, M.W., *Concurrent machines,* The Computer Journal, vol. 28 (1985) 449–465.

[42] STARK, E.W., *Connections between a Concrete and an Abstract Model of Concurrent Systems.* 5th Mathematical Foundation of Programming Semantics (1989) 53–79.

[43] VALK, R., *Self-Modifying Nets, a Natural Extension of Petri Nets.* Proceedings of Icalp'78, Lecture Notes in Computer Science vol. 62 (1978) 464–476.

[44] VALK, R., *Generalizations of Petri Nets.* Proceedings of MFCS'81, Lecture Notes in Computer Science vol. 118 (1981) 140–155.

Petri Nets and Other Models of Concurrency

Mogens Nielsen and Vladimiro Sassone

ABSTRACT. This paper retraces, collects, and summarises contributions of the authors — in collaboration with others — on the theme of Petri nets and their categorical relationships to other models of concurrency.

CONTENTS

1991 *Mathematics Subject Classification.* Primary 68Q55, 68Q10, 68Q05.

Key words and phrases. Semantics and Models of Concurrency, Noninterleaving, Petri Nets.

Introduction

Concurrency theory is based on a number of different formal models of computation, with *Petri nets* [**66, 67**], or just *nets*, as a prominent example. Other models include the event structures of Winskel [**94**], the trace structures of Mazurkiewicz [**43**], the asynchronous and the concurrent transition systems of Bednarczyk [**4**], Shields [**83**] and Stark [**84**], just to name a few. Similarly, concurrency deals with an abundance of notions for behavioural equivalence, with the bisimulation of Milner [**50**], trace equivalence of Hoare [**30**], and pomset equivalence of Pratt [**69**] as prime examples.

During the past decade, attempts have been made in order to understand the relationships between the confusingly many different concepts within concurrency theory, and many of these are based on the language of *category theory*. Our main goal in this paper is to survey some of the main ideas behind this categorical approach to concurrency, and at the same time to present some particular categorical results for nets.

The first part of our paper is devoted to some categorical results on the *behaviour* of nets and their *relations* with other models, whereas the second part focuses on a categorical approach to the algebraic *structure* of net processes. In our presentation we have chosen to treat (only) three different classes of net systems: the *elementary net systems* of Thiagarajan [**87**], the *semiweighted net systems* [**48**], and *place/transition systems* [**72**], but the approaches presented here are, we claim, applicable to any class of net systems.

Let us start by a few general comments on the role of category theory in our treatment of the behaviour of net systems. Fist of all, how do we relate nets to other models for concurrency? Any model for concurrency is meant to model the behaviour of distributed systems at a certain level of abstraction, focusing on certain aspects of the behaviour, deliberately abstracting from others. Here, we shall attempt to classify models according to their 'level of abstraction', and in stating and proving such relationships we shall use the language of category theory — in particular the notion of *adjunction*. In many contexts this has proven to be a convenient language succinctly expressing such relationships, abstracting away from the details of the often very different mathematical formalisms of the individual models. As the reader will see, nets relate nicely to most of our chosen models, in the sense that one of the models '*embed*' into the other; 'embedding' is formalised here by special adjunctions called *coreflections*.

As the reader will see, adjunctions and hence coreflections between two categories of models, M_0 and M_1, consists of ways of translating from one model to the other, satisfying certain properties. Formally, an adjunction is expressed in terms of two functors $\mathsf{L}\colon \mathsf{M}_0 \to \mathsf{M}_1$ and $\mathsf{R}\colon \mathsf{M}_1 \to \mathsf{M}_0$, and a coreflection is a way of saying that M_0 embeds into M_1 — with L telling us how to *embed* M_0 into M_1, and R how to *project* M_1 onto M_0. This will be our formal way of saying that 'M_0 is an *abstract* version of M_1'.

In Part 1, we shall show examples of such embeddings between our classes of net systems and those of event structures, trace structures, domains, and transition systems. These result are part of a greater picture of relationships between models for concurrency, see e.g. [**97, 80, 79**]. We have chosen to present a few results in some detail, at the expense of the range of models covered.

It is important to notice that all our categories are based on notions of morphisms which should be thought of as '*simulations*'. This view is supported by the fact that they are all '*behaviour respecting*', as formalised in concrete theorems. This means that the existence of a morphism may be seen as a demonstration that one object (implementation) satisfies another (specification), and hence morphisms may also play a role in formal verification.

Once adjunctions are established between models, one may start comparing and transferring behavioural concepts from one model to another, formally via the adjoints L and R. In the final section of Part 1, we shall present on such example based on [**35, 63**], introducing a general way of understanding Milner's seminal notion of bisimulation [**50**] across a range of different models, including net systems.

It must be noted that there is more to the categorical view of models than we present here. For instance, universal constructs like products and coproducts serve as basis for giving semantics to process algebras. The reader is referred to [**97**] for more detail.

In Part 2 we restrict attention to the level of single nets in order to analyse the structure of their spaces of computations, i.e., the *algebraic structure* of their processes. Of course, we keep using categorical tools, following an approach that can be said '*in the small*', as opposed to the one in Part 1 that — dealing with the totality of nets — is '*in the large*'.

The idea is, given a net N, to describe in abstract terms its *concatenable processes*, a notion introduced in [**18**] to account for sequential composition of processes. The existence of an operation of concatenation leads easily to a category of concatenable processes of N, where objects are states (markings) and arrows are (concatenable) processes. It turns out that such a category is a *symmetric monoidal category* whose tensor product is the parallel composition of processes [**18**]. The relevance of this result is that it describes Petri net behaviours as *algebras* in a remarkable way.

Here we recall some of the results of [**77, 18, 45, 75**] providing, in particular, a construction that associates to each net N a symmetric monoidal category $\mathcal{P}(N)$ isomorphic to the category of concatenable processes of N. Such an approach is completely abstract, *axiomatic*, in that it is formulated in terms of *universal* constructions. Namely, as we shall see, $\mathcal{P}(N)$ is the *free symmetric strict monoidal* category on the net N modulo two simple additional axioms. The exposition is based on [**77**].

Most of the results presented here are based on work by the authors in co-operation with colleagues. Our main contribution here has been to collect and reformulate existing results, and to add a few new results as an attempt to obtain a uniform and coherent exposition. The results on elementary net systems and their relationships to other models in Part 1 is based on various works by G. Rozenberg, P.S. Thiagarajan, and G. Winskel in collaboration with Nielsen. The work on unfolding semiweighted nets and nets as monoidal categories are due to Sassone in collaboration with M. Meseguer, U. Montanari. And finally, Section 4 on nets and bisimulation is adopted from work A. Joyal, and G. Winskel and Nielsen.

Part 1. ON THE BEHAVIOUR OF NETS

1. Petri Nets, Hoare Structures, and Trace Structures

We start out be considering some fundamental and simple classes of Petri nets and their relationships to other models for concurrency. The theory of nets was originally a strong source of inspiration behind the introduction of traces by Mazurkiewicz in [**43**]. Also, the relationship between traces and nets have been extensively studied, see in particular the survey papers by Rozenberg and Thiagarajan in [**73, 87**]. The presentation here is based on joint work with Rozenberg and Thiagarajan, [**59**], in which proofs and details may be found.

1.1. Elementary net systems. Elementary net systems were introduced by Thiagarajan [**87**] as a fundamental class of nets. His definitions were as follows.

DEFINITION. A *condition/event net* (CE for short) is a triple (B, E, F) where B and E are disjoint sets of, respectively, *conditions* and *events*, $F \subseteq (B \times E) \cup (E \times B)$, the *flow relation*, admits no isolated elements, i.e.,

$$domain(F) \cup range(F) = B \cup E,$$

where $domain(F) = \{x \mid \exists y.\,(x,y) \in F\}$ and $range(F) = \{y \mid \exists x.\,(x,y) \in F\}$.

Let $N = (B, E, F)$ be a CE. Then $X_N = B \cup E$ is the set of *elements* of N. Let $x \in X_N$. It will be convenient to use the following notation.

$$^\bullet x \;=\; \{y \mid (y,x) \in F\} \quad \text{(the set of } \textit{pre-elements} \text{ of } x)$$
$$x^\bullet \;=\; \{y \mid (x,y) \in F\} \quad \text{(the set of } \textit{post-elements} \text{ of } x)$$

This 'dot' notation is extended to subsets of X_N in the obvious way. For $e \in E$ we shall call $^\bullet e$ the set of *pre-conditions* of e and we shall call $e^\bullet$ the set of *post-conditions* of e.

DEFINITION. A CE net is said to be *simple* if for all $x, y \in X_N$ such that $^\bullet x = {}^\bullet y$ and $x^\bullet = y^\bullet$, we have $x = y$.

DEFINITION. An *elementary net system* is a quadruple $N = (B, E, F, c_{in})$ where

▷ (B, E, F) is a simple net called the *underlying net* of N.
▷ $c_{in} \subseteq B$ is the *initial case* of N.

Thus a simple CE net may be viewed as a directed bipartite graph with no isolated or confused elements, and an elementary net system is a simple net together with a 'state' specified as subset of B-elements.

Presenting an elementary net system as a graph, following standard practise, the B-elements will be drawn as circles, the E-elements as boxes, the elements of the flow relation, F, as directed arcs, and the initial case will be indicated by dots (tokens) on its members. Figure 1 is an example of a net.

As a model for concurrency, B-elements are used to denote the (local) atomic states (or resources) called *conditions* and the E-elements are used to denote (local) atomic changes-of-states called *events*. The flow relation models the effect on conditions by an occurrence of an event, in the form of a *fixed* neighbourhood relation between the conditions and events of a system.

The dynamics of an elementary net system are simple. A state (usually called a *case*) of the system consists of a set of conditions holding concurrently. An event can occur at a case if all its pre-conditions and none of its post-conditions hold at the case. When an event occurs each of its pre-conditions ceases to hold and each of its post-conditions begins to hold. Let us formalise this dynamics of net systems.

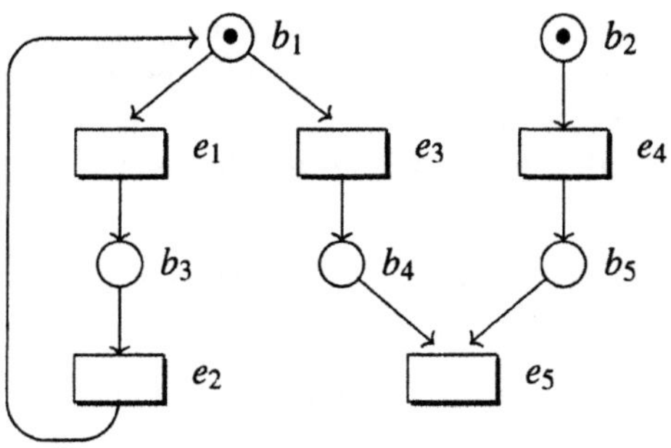

FIGURE 1

DEFINITION. Let $N = (B, E, F)$ be a net. Then $\longrightarrow_N \subseteq \mathcal{P}ow(B) \times E \times \mathcal{P}ow(B)$ is the *(elementary) transition relation* generated by N, and is given by

$$\longrightarrow_N = \{(k, e, k') \mid k \setminus k' = {}^\bullet e \quad \& \quad k' \setminus k = e^\bullet\}$$

DEFINITION. Let $N = (B, E, F, c_{in})$ be an elementary net system.

- ▷ C_N, the *state space of* N, is the least subset of $\mathcal{P}ow(B)$ containing c_{in} such that, if $c \in C_N$ and $(c, e, c') \in \longrightarrow_N$, then $c' \in C_N$. (Note that, whenever possible, we use N to denote both the net system and its underlying net.)
- ▷ $CG_N = (C_N, \longrightarrow_{C_N})$, where $\longrightarrow_{C_N} = \longrightarrow_N \cap (C_N \times E \times C_N)$, is the *case graph* associated with N.

The case graph of N describes the dynamics of N by giving, for any possible state, the diagram of the possible state-transitions.

Basic concepts concerning the behaviour of distributed systems such as causality, choice, concurrency, and confusion ('glitch') can now be cleanly defined — and separated from each other — with the help of net systems. The interested reader is referred to Thiagarajan [**87**] for details. Here we just bring out a few important behavioural concepts.

EXAMPLE. Let us illustrate by means of a few small examples how nets can be used to model concurrency, nondeterminism, and enabling.

(1) **Concurrency:**

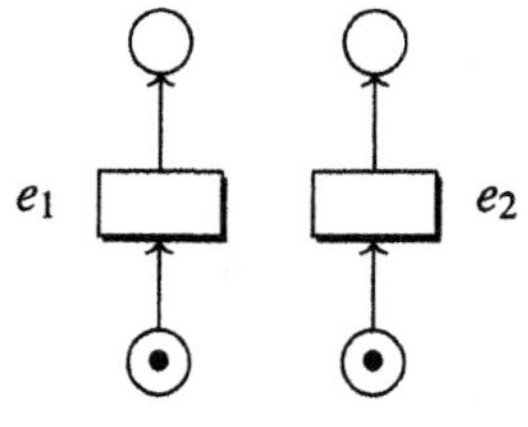

The events e_1 and e_2 can occur concurrently, in the sense that they both have concession and are independent in not having any pre or post conditions in common.

(2) **Conflict:**

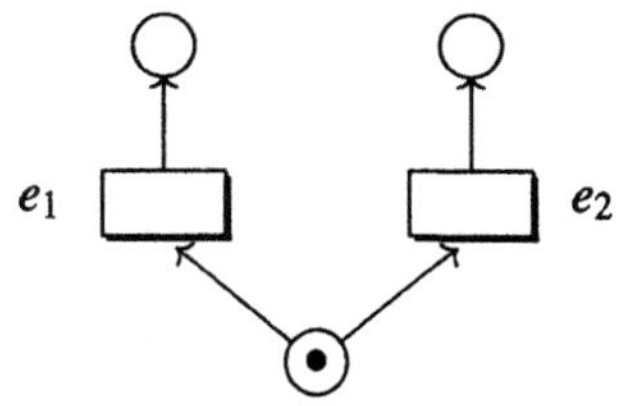

Either one of events e_1 and e_2 can occur, but not both. This shows how nondeterminism can be represented in a net.

(3) **Contact:**

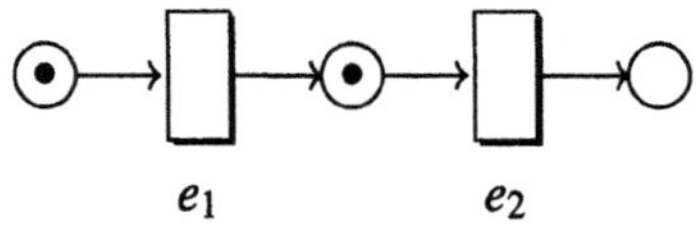

The event e_2 has concession. The event e_1 does not — its post condition holds — and it can only occur after e_2. This illustrates contact. In general, there is *contact* at a marking M when for some event e

$$^\bullet e \subseteq M \quad \& \quad e^\bullet \cap (M \setminus {}^\bullet e) \neq \varnothing.$$

As a further example, a critical region may be described as the elementary net system in Figure 2, where the condition in the center represents a kind of semaphore controlling the access (p's and v's events) to critical regions (c_0 and c_1) by the two processes.

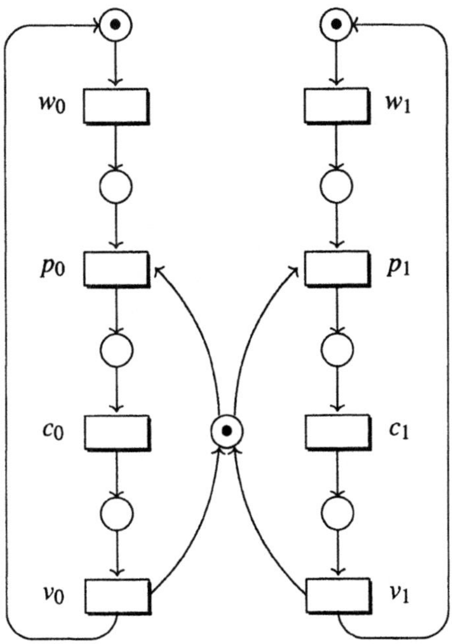

FIGURE 2

1.2. Trace structures. An traditional way to describe the behaviour of a system is to consider all the admissible sequences of event occurrences, the so-called *traces* of the system. Essentially, this amounts to giving a formal language whose alphabet is a set of events and whose strings represent the potential evolution of the system. Trace structures, introduced originally by Mazurkiewicz [**43**] as a model for concurrency, arose from a simple, yet powerful new idea: equip the alphabet of formal languages with an extra structure of *independence*, interpreted as computational independence between atomic *actions*. We recall this development starting with the following simpler notion.

DEFINITION. A *Hoare structure* is a pair (H,Σ) where Σ is an alphabet (of atomic actions), and H is a nonempty, prefix closed subset of the monoid Σ^*.

Actually, such structures are called traces in [**30**], but we prefer to reserve the word traces for the structures that will follow. Building on the definition of the transition relation we may associate an obvious Hoare structure with an elementary net system.

DEFINITION. The set FS_N of *firing sequences* of $N = (B,E,F,c_{in})$ is the subset of E^* defined inductively as follows.

▷ $\varepsilon \in FS_N$ and $c_{in}\,[\![\varepsilon\rangle\, c_{in}$, for ε the empty sequence;

▷ $$\frac{\rho \in FS_N \quad \text{and} \quad c_{in}\,[\![\rho\rangle\, c \quad \text{and} \quad c \xrightarrow{e}_N c'}{\rho e \in FS_N \quad \text{and} \quad c_{in}\,[\![\rho e\rangle\, c'}.$$

Observe that $[\![\ \rangle$ is the natural 'extension' of $\longrightarrow_N$ to $\{c_{in}\} \times E^* \times C_N$.

For the elementary net representation of the familiar example of mutual exclusion, we get the following Hoare structure

$$\{\varepsilon, w_0, w_1, w_0w_1, w_1w_0, w_0p_0, w_1p_1, w_0w_1p_0, \ldots\}.$$

One of the essential aspects of nets is that they allow an explicit representation of the distributed nature of computations. For instance, in the mutual exclusion example of Figure 2, the independence between actions w_0 and w_1 is represented, following our intuitive understanding of the net, by the disjointness of their local effects. However, as with Hoare structures in general, firing sequences 'hide' aspects of the behaviour of a net system to do with parallel or independent activities. To bring out this deficiency more clearly, we follow the original way of introducing independence between events of elementary net systems. In net theory this relation is most often referred to as the *concurrency relation*.

DEFINITION. For $e_1 \neq e_2 \in E$ and $c \in C_N$, say that e_1 and e_2 can occur *concurrently* at c — written $c[\![\{e_1,e_2\}\rangle$ or, when c can be omitted, also e_1 co e_2 — if $c[\![e_1\rangle$, $c[\![e_2\rangle$, and $({}^\bullet e_1 \cup e_1^\bullet) \cap ({}^\bullet e_2 \cup e_2^\bullet) = \varnothing$.

Thus e_1 and e_2 can occur concurrently at c iff they can occur individually and their neighbourhoods are disjoint. *Conflict* is clearly the 'dual' notion: e_1 and e_2 are in conflict at c if $c[\![e_1\rangle$, $c[\![e_2\rangle$, *but not* $c[\![\{e_1,e_2\}\rangle$, i.e., at c either e_1 may occur or e_2 may occur but not both. The choice as to whether e_1 or e_2 will occur is assumed to be resolved by the 'environment' of the system. For the system N of Figure 1, for instance, at the initial case e_1 and e_4 can occur concurrently. Consequently, the

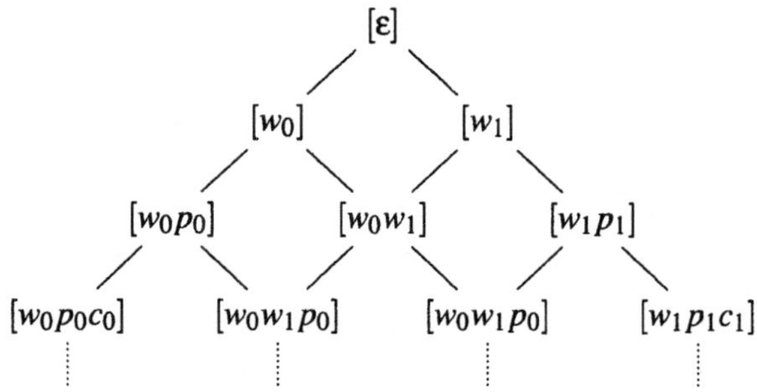

FIGURE 3

firing sequences $e_1e_2e_4$ and $e_4e_1e_2$ and $e_1e_4e_2$ all represent the *same* (non-sequential) stretch of behaviour of N. Also, e_1 and e_3 are in conflict at the initial case. Hence the firing sequences $e_1e_2e_4$ and $e_3e_4e_5$ represent two conflicting (alternative) stretches of behaviour of N.

The idea suggested by Mazurkiewicz [**43**] is to allow the modelling of such independent activities of components of system by introducing the extra structure of an independence relation I on the action alphabet. For nets, following our intuition we would relate two actions as independent if and only if they involve concurrent events. Based on an independence alphabet, the behaviour of a system will be modeled in terms of traces, i.e., of *equivalence classes* of

$$\eqsim_I \text{ : the least congruence on } E^* \text{ such that } \quad e_0e_1 \eqsim_I e_1e_0 \quad \text{whenever} \quad e_0 \, I \, e_1.$$

In our running example $w_0w_1p_0 \simeq_I w_0p_0w_1$ and the equivalence class of $w_0w_1p_0$ is $[w_0w_1p_0] = \{w_0w_1p_0, w_0p_0w_1, w_1w_0p_0\}$.

Now, Hoare structures generalise from subsets of Σ^* to subsets of the monoid of traces, denoted $M(\Sigma, I)$. The prefix ordering of Hoare structures generalise to a prefix ordering of traces, defined in terms of the following preorder on strings:

$$s \lesssim_I t \quad \text{if and only if} \quad \exists u.\, su \eqsim_I t$$

which induces the following partial order (*prefix order*) on traces:

$$\sqsubseteq_I = \lesssim_I / \eqsim_I,$$

that is,

$$[s] \sqsubseteq_I [t] \quad \text{if and only if} \quad \exists u, v.\, s \eqsim_I u \lesssim_I v \eqsim_I t.$$

In our example $[\varepsilon] \sqsubseteq_I [w_0] \sqsubseteq_I [w_0w_1] \sqsubseteq_I [w_0p_0w_1]$, and the initial traces and their prefix ordering are as shown in Figure 3. Notice that the extra modelling power boils down to the presence of traces like $[w_0w_1]$ in our example, representing actions w_0 and w_1 in any (unspecified) order, and interpreted as their concurrent or independent occurrences.

We are now ready for our formal definition of trace structures. Conceptually, we follow [**44**] where a trace structure is defined to be a prefix closed, proper subset of the monoid $M(\Sigma, I)$. However, only for technical reasons we prefer in our formal definition to work with such structures in terms of consistent subsets of Σ^* — with the traces a derived notion, as in Proposition 1.1 below.

DEFINITION. A *trace structure* is a triple $T = (M,\Sigma,I)$ where (Σ,I) is an *independence alphabet*, i.e. $I \subseteq \Sigma \times \Sigma$ is *irreflexive* and *symmetric*, and $M \subseteq \Sigma^*$ is such that for all $t,t' \in \Sigma^*$ and $a,b \in \Sigma$:

$$\begin{array}{lll} \textbf{\textit{consistency:}} & t \approx_I t' \in M & \Longrightarrow t \in M; \\ \textbf{\textit{prefix closure:}} & ta \in M & \Longrightarrow t \in M; \\ \textbf{\textit{properness:}} & ta, tb \in M \;\&\; a\,I\,b & \Longrightarrow tab \in M. \end{array}$$

We use the notation $M/\approx_I$ for the traces of T, i.e.,

$$M/\approx_I = \{[w] \mid w \in M\}.$$

We may think of a trace structure as a prefix closed set of traces, in the sense that from the axioms of consistency and prefix closure above, we get the following.

PROPOSITION 1.1. *Given a trace structure* $T = (M,\Sigma,I)$ *then* $(M/\approx_I, \sqsubseteq_I)$ *satisfies*

- ▷ $w \in M$ *if and only if* $[w] \in M/\approx_I$;
- ▷ $[w] \sqsubseteq_I [w'] \in M/\approx_I$ *implies* $[w] \in M/\approx_I$.

As will be expected by now, the information concerning concurrency and conflict-resolution hidden by Hoare structures may be retrieved by associating with a net a trace structure with concurrency as the appropriate independence relation.

THEOREM 1.2. *Let* $N = (B,E,F,c_{in})$ *be an elementary net system and let the independence relation associated with* N *be*

$$I = \{(e_1,e_2) \mid e_1,e_2 \in E \;\&\; ({}^\bullet e_1 \cup e_1^\bullet) \cap ({}^\bullet e_2 \cup e_2^\bullet) = \varnothing\}.$$

Then $nt(N) = (FS_N, E, I)$ *is a trace structure.*

PROOF. The required properties follow from definition. In particular, $nt(N)$ is consistent and proper by definition of the (elementary) transition relation. □

For the net system N from Figure 1, Figure 4 shows an initial portion of the associated poset of traces.

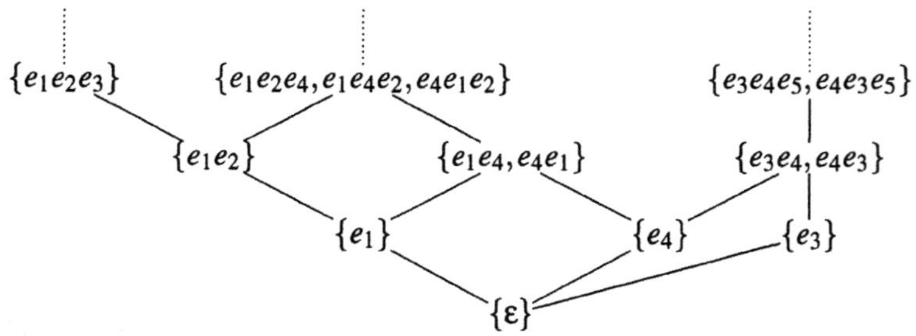

FIGURE 4

The beauty of the trace semantics is its simplicity. One of the classical results from concurrency theory is that the trace semantics is 'consistent' with an alternative way of defining the behaviour of net systems in terms of unfoldings into processes (or occurrence nets). Several results of this type have been shown **[9]**. The presentation that follows is adapted from [**59**]. For the sake of convenience we shall assume here that N is *contact-free*. In other words, we shall assume,

$$\forall c \in C_N.\ \forall e \in E.\ {}^\bullet e \subseteq c \implies e^\bullet \cap (c \setminus {}^\bullet e) = \varnothing.$$

As we shall see later, this does not involve any loss of generality, at least for the study of behavioural issues.

The theoretic development of Petri nets, focusing on the noninterleaving aspects of concurrency, brought to the foreground various notions of process, e.g. [**68, 26, 7, 45, 18**]. Generally speaking, these are structures accounting for the *causal relationships* which rule the occurrence of events in computations. Thus, ideally, processes are simply computations in which explicit information about such causal connections is added. Abstractly, the processes of a net N are ordered sets whose elements are labelled by events of N. Concretely, in order to describe exactly which sets of events give rise to processes, one takes a process of $N = (B, E, F, c_{in})$ will be a *labelled net* of the form $\tilde{N} = (\tilde{B}, \tilde{E}, \tilde{F}, \pi)$, where $(\tilde{B}, \tilde{E}, \tilde{F})$ is a restricted kind of a net (viz., finite, conflict-free, acyclic) called a *causal* or *process* net, and the labelling function $\pi\colon \tilde{B} \cup \tilde{E} \to B \cup E$ is required to connect the structure of $\tilde{N}$ to that of N in a suitable way. For a definition of a process along these lines see Part 2, or, e.g., [**73**].

Here we shall define processes with the help of firing sequences. This will enable us to build up the finite processes of N inductively. For a similar development of the process notion, see [**9**].

For each firing sequence ρ, we will define a process $N_\rho = (B_\rho, E_\rho, F_\rho, \pi_\rho)$. In doing so it will be convenient to keep track of the conditions that hold in N after the run represented by the firing sequence ρ. This set of conditions will be encoded as c_ρ.

Definition. Let N be (B, E, F, c_{in}). Then $N_\rho = (B_\rho, E_\rho, F_\rho, \pi_\rho)$ is defined inductively on the length of $\rho \in FS_N$ as follows.

Case $\rho = \varepsilon$: Then $N_\varepsilon = (\varnothing, \varnothing, \varnothing, \varnothing)$ and $c_\varepsilon = \{(b, \phi) \mid b \in c_{in}\}$.

Case $\rho = \rho' e$: Assume that $N_{\rho'} = (B_{\rho'}, E_{\rho'}, F_{\rho'}, \pi_{\rho'})$. Then $N_\rho = (B_\rho, E_\rho, F_\rho, \pi_\rho)$ where, for $X = \{(b, D) \mid b \in {}^\bullet e \ \&\ (b, D) \in c_{\rho'}\}$ and $Y = \{(b, \{(e, X)\}) \mid b \in e^\bullet\}$, we have

$$\begin{aligned} E_\rho &= E_{\rho'} \cup \{(e, X)\}, \\ B_\rho &= B_{\rho'} \cup X \cup Y, \\ F_\rho &= F_{\rho'} \cup (X \times \{(e, X)\}) \cup (\{(e, X)\} \times Y), \\ \pi_\rho &= \lambda(z, Z) \in B_\rho \cup E_\rho.\ z. \end{aligned}$$

Finally, $c_\rho = (c_{\rho'} \setminus X) \cup Y$.

It will turn out that N_ρ as defined above is a labelled net. For $\rho = e_1 e_2 e_4 e_3$ in the system N of Figure 1 we show N_ρ in Figure 5. For convenience we have displayed π_ρ by writing the value of $\pi_\rho(x)$ besides the graphical representation of x for each $x \in B_\rho \cup E_\rho$.

In order to establish a relationship between the traces of N and its processes it is necessary to define an ordering relation over the processes of N.

Definition. Let N be (B, E, F, c_{in}).

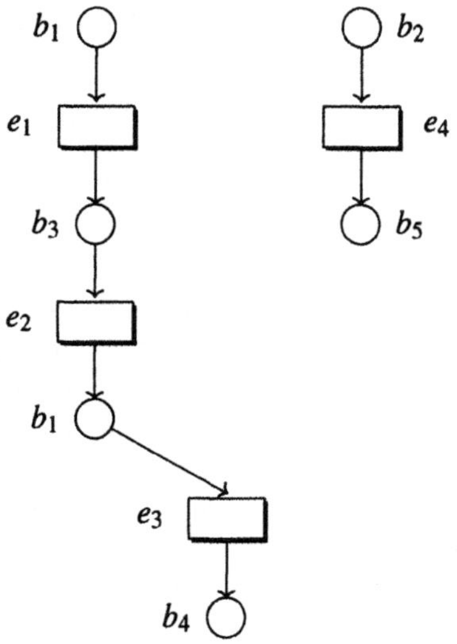

FIGURE 5

- ▷ The set of finite processes of N is $P_N = \{N_\rho \mid \rho \in FS_N\}$, for N_ρ as in the previous definition.
- ▷ $\sqsubseteq \subseteq P_N \times P_N$ is defined by

$$(B_\rho, E_\rho, F_\rho, \pi_\rho) \sqsubseteq (B_{\rho'}, E_{\rho'}, F_{\rho'}, \pi_{\rho'}) \quad \text{if} \quad B_\rho \subseteq B_{\rho'} \;\&\; E_\rho \subseteq E_{\rho'} \;\&\; F_\rho \subseteq F_{\rho'}.$$

Clearly $\sqsubseteq$ is a partial ordering relation. The main result relating trace semantics to processes is the following.

THEOREM 1.3. *For N any elementary net system* $(P_N, \sqsubseteq)$ *and the ordering of the traces from nt(N), i.e.,* $(FS_N/\approx_I, \sqsubseteq_I)$*, are isomorphic posets.*

PROOF. In **[59]** it is proved that $f\colon FS_N/\approx_I \to P$ given by $f([\rho]) = N_\rho$ is an isomorphism. □

1.3. A categorical way to relationships. In the last section we attempted to show connections between net systems and other structures. Although it is apparent that nets a more general, expressive, and powerful model, we lack at this stage a way to make precise any statement in this sense. Is there a formal way of saying that traces 'embeds' into nets, that 'nets generalise' them naturally? More generally, how can we relate nets to the other models? How do we establish relationships between models?

As we discussed in the introduction, we tend to classify models can for concurrency according to their '*level of abstraction*' (see, e.g., **[80, 97, 79]**), that is, according to those aspects of the behaviour of distributed systems they focus on and those they deliberately abstract from. In stating and proving relationships between models viewed under this perspective, the language of *category theory* as proven in many contexts to be very useful, as it is capable of *abstracting* away from unwanted details of the individual models and, therefore, of expressing the more essential aspects *succinctly* and in great *generality*. Let us review very briefly a few key steps behind these ideas.

First, all the models are introduced as a class of *objects*, e.g., the class of net systems or the class of trace structures, equipped with a notion of 'behaviour-preserving' (i.e., simulation) *morphism*, making each model into a category. The role of the morphisms is to make explicit (*if* and) *how* each single object relates to all the others. In particular, as behaviour is preserved, (*if* and) *how* it can be simulated. This makes explicit that central to our objects and, therefore, to the respective categories, is the dynamic notion of behaviour. Also, the very notion adopted for 'simulation' determines what aspects of behaviour are important, i.e., what can be ignored by a successful simulation (the aspects abstracted away) and what instead must be preserved (the aspects focused on). In other terms, the adopted notion of morphism define the abstraction level of the model.

From this standpoint, the notion of *functor* is the first tool category theory makes available to us in order to check the sanity of our translations from one model to another. Essentially, it requires us to map objects to objects *preserving* all the existing relationships, i.e., all the existing simulations. In other words, it requires to map also behaviours to behaviours.

Tools much more refined than functors are the notions of *adjunction* and *coreflection*, central to many papers on models of concurrency and, in particular, to our presentation here. Let us briefly comment on their formal definition and the intuitive way to understand them. Technically, an adjunction between categories M_0 and M_1, consists of ways of mapping from one to the other and back, satisfying certain properties. Formally, (see **[42]** for alternative characterisations) we shall express an adjunction in terms of two functors $\mathsf{L}\colon \mathsf{M}_0 \to \mathsf{M}_1$ (the *left adjoint*) and $\mathsf{R}\colon \mathsf{M}_1 \to \mathsf{M}_0$ (the *right adjoint* of the adjunction) satisfying (see Figure 6):

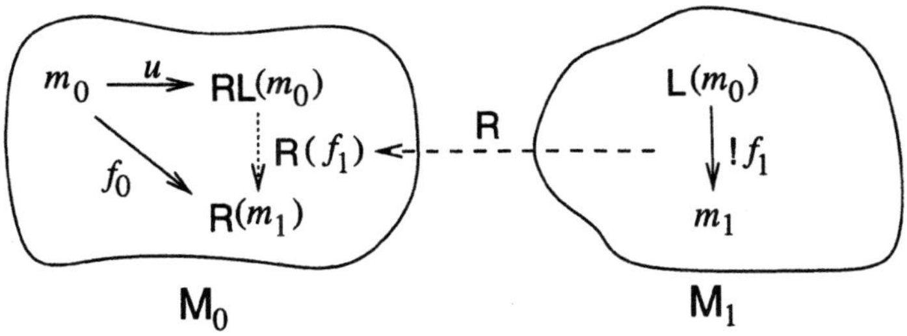

FIGURE 6

for *each* object m_0 of M_0, there is a morphism $u\colon m_0 \to \mathsf{R}\circ\mathsf{L}(m_0)$ (the *unit* at m_0) such that for *each* object m_1 of M_1 and *each* morphism $f_0\colon m_0 \to \mathsf{R}(m_1)$, then there is a *unique* morphism $f_1\colon \mathsf{L}(m_0) \to m_1$, such that $f_0 = \mathsf{R}(f_1)\circ u$.

In other terms, for each $m_0 \in \mathsf{M}_0$, $\mathsf{L}(m_0)$ is a 'special' object of M_1 in the sense that all the maps from m to objects of the kind $\mathsf{R}(m_1)$ in M_0 come exactly and unambiguously from maps from $L(m_0)$ to m_1 in M_1. Moreover, all such maps can be factored in the image of R via $u\colon m_0 \to \mathsf{R}\circ\mathsf{L}(m_0)$, a 'special' map that m_0 comes equipped with. Reading system for object and simulation for morphism, this definition has an evident significance in computational terms.

If all units of an adjunction are *isomorphisms*, then the adjunction is called a *coreflection*. This essentially means that no information is lost moving from M_0 to M_1, as the identity of objects is retained and recovered back by R. It follows from the definition that the left adjoint L of a coreflection is always *full* and *faithful*, i.e., an *embedding*. In other words, we may think of M_0 as a coreflective full subcategory of M_1 (the one identified by the image of L), and of L as the inclusion $\mathsf{M}_0 \hookrightarrow \mathsf{M}_1$, whereas R tells us how to project M_1 back onto M_0.

Paraphrasing this situation in terms of categories of models, behaviours and simulations, we can say that R selects for each $m \in \mathsf{M}_1$ its *best possible abstract 'approximation'* in M_0. That is, an object $\mathsf{R}(m) \in \mathsf{M}_0$ together with a simulation $\mathsf{R}(m) \to m$ such that any other $\mathsf{R}(m') \to m$ factors as $\mathsf{R}(m') \to \mathsf{R}(m) \to m$.

So much for the formal definition. In the following the existence of a coreflection of M_0 into M_1 will be our formal way of saying that 'M_0 is an *abstract* version of M_1'.

We therefore start by turning our models into a categories by defining appropriate notions of morphisms. Morphisms of languages are simply functions on their alphabets which send strings in one language to strings in another.

DEFINITION. A function $\lambda\colon \Sigma \to \Sigma'$ extends to strings by defining

$$\widehat{\lambda}(s\alpha) = \widehat{\lambda}(s)\lambda(\alpha).$$

A *morphism* of Hoare structures $(H,\Sigma) \to (H',\Sigma')$ consists of a function $\lambda\colon \Sigma \to \Sigma'$ such that $\forall s \in H.\, \widehat{\lambda}(s) \in H'$.
We write H for the category of Hoare structures with the above understanding of morphisms, where composition is our usual composition of functions.

Before we continue, let us comment briefly on our choice of morphisms — on Hoare structures as well as on all other models considered in this paper. In much of the literature, more liberal notions of morphisms are used, based on partial (rather than total) functions on the labelling sets. These more general morphisms have the advantage that many useful combinators (e.g., parallel composition) may be expressed as universal constructions in the corresponding categories of models. Furthermore, they may be thought of as specifying correctness properties: the 'correctness' of the mutual exclusion example, for instance, follows by the fact that the partial function λ from the alphabet of actions, which is undefined for $\{w_0, w_1\}$ and the identity function for all other action symbols, *is* a morphism from the Hoare structure of the mutual exclusion example to the Hoare structure consisting of all prefixes of the regular language $(p_0c_0v_0 + p_1c_1v_1)^*$. However, we have chosen here to restrict ourselves to morphisms based on total functions, purely as an attempt to simplify our presentation technically.

Similarly, morphisms between trace structures are morphisms between the underlying languages which preserve independence.

DEFINITION. A *morphism* of trace structures $(M,\Sigma,I) \to (M',\Sigma',I')$ consists of a function $\lambda\colon \Sigma \to \Sigma'$ which

preserves independence: $\alpha\, I\, \beta$ implies $\lambda(\alpha)\, I'\, \lambda(\beta)$, for all $\alpha,\beta \in \Sigma$;

preserves strings: $s \in M$ implies $\widehat{\lambda}(s) \in M'$, for all strings s.

This, with the usual composition of functions defines T, the category of trace structures.

It is easy to see that morphisms of trace structures preserve traces and the ordering between them.

PROPOSITION 1.4. *Let* $\lambda\colon (M,\Sigma,I) \to (M',\Sigma',I')$ *be a morphism of trace structures. If* $s \lesssim_I t$ *in the trace structure* (M,Σ,I) *then* $\widehat{\lambda}(s) \lesssim_{I'} \widehat{\lambda}(t)$ *in* (M',Σ',I').

It follows that $\widehat{\lambda}$ defines a monote function from $(M/\approx_I, \sqsubseteq_I)$ to $(M'/\approx_{I'}, \sqsubseteq_{I'})$. Concerning nets, we consider the following definition.

DEFINITION. Let $N = (B,E,F,c_{in})$ and $N' = (B',E',F',c'_{in})$ be elementary net systems. A *morphism* $(\beta,\eta)\colon N \to N'$ consists of a relation $\beta \subseteq B \times B'$, such that β^{op} is a partial function $B' \rightharpoonup B$, and a function $\eta\colon E \rightharpoonup E'$ such that

$$\forall (b,b') \in \beta.\ b \in c_{in} \iff b' \in c'_{in}, \qquad \beta({}^{\bullet}e) = {}^{\bullet}\eta(e), \qquad \beta(e^{\bullet}) = \eta(e)^{\bullet}.$$

Thus morphisms on nets preserve initial cases and events when defined. A morphism $(\beta,\eta)\colon N \to N'$ expresses how occurrences of events and conditions in N induce occurrences in N'. Morphisms on nets preserve behaviour.

PROPOSITION 1.5. *Let* $N = (B,E,F,c_{in})$ *and* $N' = (B',E',F',c'_{in})$ *be elementary nets and* $(\beta,\eta)\colon N \to N'$ *a morphism.*

- ▷ *If* $c\,[\![e\rangle\, c'$ *in* N, *then* $f_\beta(c)\,[\![\eta(e)\rangle f_\beta(c')$ *in* N', *for* $f_\beta(c) = \beta(c) \cup (c'_{in} \setminus \beta(c_{in}))$.
- ▷ *If* ${}^{\bullet}e_1^{\bullet} \cap {}^{\bullet}e_2^{\bullet} = \varnothing$ *in* N, *then* ${}^{\bullet}\eta(e_1)^{\bullet} \cap {}^{\bullet}\eta(e_2)^{\bullet} = \varnothing$ *in* N'.

PROOF. It is easily seen that ${}^{\bullet}\eta(e) = \beta({}^{\bullet}e)$ and that $\eta(e)^{\bullet} = \beta(e^{\bullet})$ for all events e of N. Observe too that because β^{op} is a partial function, β in addition preserves intersections and set differences. These observations mean that $\beta(c)\,[\![\eta(e)\rangle\, \beta(c')$ in N' follows from the assumption that $c\,[\![e\rangle\, c'$ in N, and that independence is preserved. □

DEFINITION. Let EN denote the category of elementary net systems and their morphisms under the obvious componentwise composition of morphisms, e.g., the composition of $(\beta_0,\eta_0)\colon N_0 \to N_1$ and $(\beta_1,\eta_1)\colon N_1 \to N_2$ is $(\beta_1 \circ \beta_0, \eta_1 \circ \eta_0)\colon N_0 \to N_2$

This choice of morphisms for elementary net systems may not be as obvious and intuitively clear as the those for the other models we consider. Indeed alternative categories of net systems have been studied — see, e.g., [**93, 45, 12, 48, 97**]. Here we just remark that Proposition 1.5 proves that these morphisms preserve behaviour (and concurrency), a fact that has been explored by, e.g. [**11**], where such morphism have been used to express correctness properties. Also, we note that the derived notion of isomorphism becomes identity up to names of conditions and events.

THEOREM 1.6. *The construction that maps* $N = (B,E,F,c_{in})$ *to the Hoare structure* (FS_N, E) *extends to a functor nh from* EN *to* H.

THEOREM 1.7. *The trace semantics nt extends to a functor nt from* EN *to* T.

However, these functors are *not* part of any adjunction. Following our discussion above, one would expect a formal result embedding T in EN, but for this to be the case it turns out that one needs a more abstract semantics. The reason why *nt* is too concrete is that it preserves information about event 'identities'. As we shall see in the next section, forgetting these will help yielding a 'nice' (read 'universal') unfolding of EN into event structures.

2. Petri Nets, Event Structures, and Domains

Consider again the prefix ordering of traces introduced above. What can be said about their structure and properties? In this section we shall provide a characterisation of such orderings in terms of a well-known class of *Scott domains* [**81, 6**]. Moreover, in the process of doing so, we shall also show that they arise exactly as the orderings associated with the dynamics of another well-known model for concurrency: the *event structures*, originally introduced in [**57**].

2.1. Event structures. The prefix ordering of the strings of a Hoare structure — which is in fact a *tree* ordering — may also be viewed as a structure over action occurrences, where individual occurrences may be either *ordered*, i.e., following each other in time in the same computation, or *not*, i.e., belong to different computations. Event structures may be seen as a generalisation of such structures, allowing a third relationship between occurrences, that of *concurrency*, i.e., belonging to the same computation, but without any *causal/temporal* ordering.

DEFINITION. Define an *event structure* to be a structure $(E,\leq,\#)$ consisting of a set E, of *events* which are partially ordered by $\leq$, the *causal dependency relation*, and a binary, symmetric, irreflexive relation $\# \subseteq E \times E$, the *conflict relation*, which satisfy for all $e, e', e'' \in E$

$$\{e' \mid e' \leq e\} \text{ is finite}, \qquad e \# e' \leq e'' \implies e \# e''.$$

Say two events $e, e' \in E$ are *concurrent*, and write $e \operatorname{co} e'$, if $\neg(e \leq e' \text{ or } e' \leq e \text{ or } e \# e')$. Write $\mathrel{W}$ for $\# \cup 1_E$, i.e., the reflexive closure of the conflict relation.

The finiteness assumption restricts attention to discrete processes where an event occurrence depends only on finitely many previous occurrences. The axiom on the conflict relation expresses that if two events causally depend on events in conflict then they too are in conflict.

Guided by our interpretation we can formulate a notion of computation state of an event structure $(E,\leq,\#)$. Taking a computation state of a process to be represented by the set x of events which have occurred in the computation, we expect that

$$e' \in x \quad \& \quad e \leq e' \quad \implies \quad e \in x,$$

i.e., if an event has occurred then all events on which it causally depends have occurred too, and also that

$$\forall e, e' \in x.\, \neg(e \# e'),$$

i.e., two conflicting events cannot occur together in the same computation.

DEFINITION. Let $(E,\leq,\#)$ be an event structure. Its *configurations*, $\mathcal{D}(E,\leq,\#)$, are those subsets $x \subseteq E$ which are

***conflict-free*:** $\forall e, e' \in x.\, \neg(e \# e')$;

***downwards-closed*:** $\forall e, e'.\, e' \leq e \in x \implies e' \in x$.

In particular, define $\lfloor e \rfloor = \{e' \in E \mid e' \leq e\}$, which is a configuration, as it is downward-closed and conflict-free. Write $\mathcal{D}^0(E,\leq,\#)$ for the set of finite configurations.

The important relations associated with an event structure can be recovered from its finite configurations (or indeed similarly from its configurations).

PROPOSITION 2.1. *Let* $(E,\leq,\#)$ *be an event structure. Then*

▷ $e \leq e'$ *if and only if* $\forall x \in \mathcal{D}^0(E,\leq,\#).\, e' \in x \Longrightarrow e \in x;$

▷ $e \# e'$ *if and only if* $\forall x \in \mathcal{D}^0(E,\leq,\#).\, e \in x \Longrightarrow e' \notin x;$

▷ $e \text{ co } e'$ *if and only if* $\exists x, x' \in \mathcal{D}^0(E,\leq,\#)$ *such that*

$$e \in x \setminus x' \;\&\; e' \in x' \setminus x \;\&\; x \cup x' \in \mathcal{D}^0(E,\leq,\#).$$

Events manifest themselves as atomic jumps from one configuration to another, and later it will follow that we can regard such jumps as transitions in the case graph associated with a net system.

DEFINITION. Let $(E,\leq,\#)$ be an event structure and x, x' be configurations. Write

$$x \xrightarrow{e} x' \quad \text{if and only if} \quad e \notin x \;\&\; x' = x \cup \{e\}.$$

PROPOSITION 2.2. *Two events* e_0, e_1 *of an event structure are in the concurrency relation* co *if and only if there exist configurations* x, x_0, x_1, x' *such that*

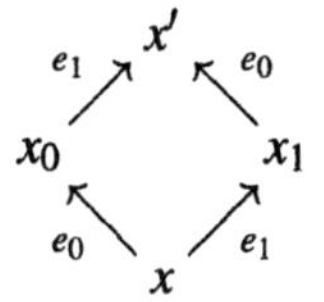

Morphisms on event structures are defined as follows **[92, 91]**:

DEFINITION. Let $ES = (E,\leq,\#)$ and $ES' = (E',\leq',\#')$ be event structures. A *morphism* from ES to ES' consists of a function $\eta : E \to E'$ on events which satisfies

$$x \in \mathcal{D}(ES) \quad \Longrightarrow \quad \eta(x) \in \mathcal{D}(ES')$$

$$\forall e_0, e_1 \in x.\, \eta(e_0) = \eta(e_1) \Longrightarrow e_0 = e_1.$$

A morphism $\eta : ES \to ES'$ between event structures expresses how behaviour in ES determines behaviour in ES'. The function η expresses how the occurrence of events in ES implies the simultaneous occurrence of events in ES'; the fact that $\eta(e) = e'$ can be understood as expressing that the event e' is a 'component' of the event e and, in this sense, that the occurrence of e implies the simultaneous occurrence of e'. If two distinct events in ES have the same image in ES' under η then they cannot belong to the same configuration.

Morphisms of event structures preserve the concurrency relation. This is a simple consequence of Proposition 2.2, showing how the concurrency relation holding between events appears as a 'little square' of configurations.

PROPOSITION 2.3. *Let* E *be an event structure with concurrency relation* co *and* E' *an event structure with concurrency relation* co'. *Let* $\eta : E \to E'$ *be a morphism of event structures. Then, for any events* e_0, e_1 *of* E,

$$e_0 \text{ co } e_1 \quad \Longrightarrow \quad \eta(e_0) \text{ co}' \eta(e_1).$$

Morphisms between event structures can be described more directly in terms of the causality and conflict relations of the event structure.

PROPOSITION 2.4. *A morphism of event structures from* $(E,\leq,\#)$ *to* $(E',\leq',\#')$ *is a function* $\eta: E \to E'$ *such that*

- ▷ $\lfloor \eta(e) \rfloor \subseteq \eta(\lfloor e \rfloor)$,
- ▷ $\eta(e_0) \,\mathrm{W}'\, \eta(e_1) \quad \Longrightarrow \quad e_0 \,\mathrm{W}\, e_1$.

Let E denote the category of event structures with morphism as described above and composition named composition of functions.

2.2. Event structures and domains. Let us turn our attention to the class of partial orders corresponding with the orderings of configurations of event structures. The characterisation given below in terms of special Scott domains has been originally formulated in [**57**].

In the following, we shall need a few standard definitions from domain theory. For $(D,\sqsubseteq)$ a partial order and X a subset of D, we write as usual $\bigsqcup X$ for the least upper bound of X, when it exists.

DEFINITION. Let $(D,\sqsubseteq)$ be a partial order. A *complete prime* of D is an element $p \in D$ such that

$$p \sqsubseteq \bigsqcup X \Longrightarrow \exists x \in X.\, p \sqsubseteq x$$

for any set X for which $\bigsqcup X$ exists.

DEFINITION. For $(D,\sqsubseteq)$ a partial and $d_0, d_1 \in D$, we say that d_1 *covers* d_0, in symbols $d_0 < d_1$, if and only if $d_0 \sqsubset d_1$ and, for every d,

$$d_0 \sqsubseteq d \sqsubseteq d_1 \Longrightarrow d = d_0 \text{ or } d = d_1.$$

DEFINITION. Let $(D,\sqsubseteq)$ be a partial order. We say that D is

- ▷ *bounded complete* if all subsets $X \subseteq D$ which have an upper bound in D have a *least* upper bound $\bigsqcup X$ in D.
- ▷ *coherent* if all subsets $X \subseteq D$ which are *pairwise* bounded (i.e., such that all pairs of elements $d_0, d_1 \in X$ have upper bounds in D), have *least* upper bounds $\bigsqcup X$ in D. (Note that coherence implies bounded completeness).
- ▷ *prime algebraic* if

  $$x = \bigsqcup \{p \sqsubseteq x \mid p \text{ is a complete prime}\},$$

  for all $x \in D$. If furthermore the sets

  $$\{p \sqsubseteq q \mid p \text{is a complete prime}\}$$

  are always finite when q is a complete prime, then D is said to be *finitary*.

A *prime algebraic domain* domain is a bounded complete and prime algebraic partial order.

THEOREM 2.5. *Let* $(E,\leq,\#)$ *be an event structure. The partial order* $(\mathcal{D}(E,\leq,\#),\subseteq)$, *that we shall indicate simply as* $\mathcal{D}(E,\leq,\#)$, *is a coherent, finitary, prime algebraic domain whose complete primes are the* $\{\lfloor e \rfloor \mid e \in E\}$.

PROOF. See [**57**, **95**]. □

Conversely, any coherent, finitary, prime algebraic domain is associated with the partial order of configurations of event structures.

THEOREM 2.6. *Let $(D,\sqsubseteq)$ be a coherent, finitary, prime algebraic domain. Define $\mathcal{P}r(D,\sqsubseteq)$ as the event structure $(E,\leq,\#)$*

$$\begin{array}{rcl} E & = & \textit{the complete primes of } (D,\sqsubseteq) \\ \leq & = & \textit{is the restriction of } \sqsubseteq \textit{ to } E \\ \# & = & \{(x,y) \in E \times E \mid x \sqcup y \textit{ does not exist in } D\} \end{array}$$

Then $(D,\sqsubseteq)$ and $\mathcal{D}(E,\leq,\#)$ are isomorphic partial orders.

PROOF. See [**57**]. □

Actually, the relationship between event structures and coherent, finitary prime algebraic domains is very strong, in that they are *equivalent*: one can be used to represent the other. This may be formalised also in terms of a categorical equivalence between D and a category of coherent, finitary prime algebraic domains equipped with stable functions as morphisms.

THEOREM 2.7. *Let* D *denote the category of coherent, finitary prime algebraic domains with morphism functions $f\colon (D_0,\sqsubseteq_0) \to (D_1,\sqsubseteq_1)$ satisfying:*

additivity: *for all $x,y \in D_0$ such that $x \sqcup y$ exists, $f(x \sqcup y) = f(x) \sqcup f(y)$;*

stability: *for all $x,y \in D_0$ such that $x \sqcup y$ exists, $f(x \sqcap y) = f(x) \sqcap f(y)$;*

covering preserving: *for all $x,y \in D_0$. if $x <_0 y$ then $f(x) <_1 f(y)$.*

Then D *and* E *are equivalent categories.*

PROOF. One can prove that $\mathcal{D}$ and $\mathcal{P}r$ can be extended to functors that form an equivalence of categories. See [**95**]. □

Getting back to trace structures, we may now formulate a functor, which in essence performs the abstraction from identities of events mentioned previously, and based on this a universal form of unfolding elementary nets into D (and hence E).

THEOREM 2.8. *Given a trace structure $T = (M,\Sigma,I)$ then $td(T) = (M/\!\approx_I,\sqsubseteq_I)$ is a coherent, finitary prime algebraic domain, and td extends to a functor from* T *to* D.

PROOF. See [**97**]. □

The following is the result announced at the end of the previous section.

THEOREM 2.9. *$td \circ nt$ is the right adjoint of a coreflection between* EN *and* D.

PROOF. The proof of this is rather involved, but may be found in [**61**], and for more general forms of net systems in [**97**] and [**33**]. □

2.3. Semiweighted nets. Having introduced event structures and used them as a 'bridge' across elementary net systems, trace structures and domains, we now set out to study the relationships between nets and event structures directly, by means of a so-called *unfolding* construction.

Such an approach to net systems was devised in [**57, 94**] for the category *safe nets* and extended in [**48, 46**] to the more generous category of *semiweighted nets*. Informally, it consists of a coreflection of event structures into nets to whose right adjoint, the 'unfolding', is obtained by *'unrolling'* the *'dynamic'* structure of nets to the *'static'* structure of event structures. In other words, it amounts to 'compiling' transitions to events, so yielding a fine-grain description of the *causal* interaction between the components of a computation, as such interactions must be resolved to the global, static relations of causality and conflict of event structures. Under this translation events are to be thought of as unique occurrences of transitions which bear unique, static causal links to each other.

The unfolding construction factors via a coreflection through Occ, the category of *occurrence nets* [**57**], so yielding the following global picture, where $\hookleftarrow$ is the inclusion functor, and the lower arrows are left adjoints.

$$\mathsf{SWNets}\ \underset{\mathcal{U}(\_)}{\overset{\hookleftarrow}{\longrightarrow}}\ \mathsf{Occ}\ \underset{\mathcal{E}(\_)}{\overset{\mathcal{N}(\_)}{\hookleftarrow\!\!\longrightarrow}}\ \mathsf{E}\ \underset{\mathcal{D}(\_)}{\overset{\mathcal{P}r(\_)}{\hookleftarrow\!\!\longrightarrow}}\ \mathsf{D}$$

In presenting these results, we shall follow closely [**48**]. Let us start by generalising the class of nets we consider.

The first generalisation is to pass from condition/event to place/transition nets, i.e., to allow markings to be *multisets* (rather than sets) of places. The state of a net is now thought as a distribution of resources ('tokens') in places. Differently from conditions, that may simply hold or not, resources may be absent, but also present in multiple copies.

The flow relation F of elementary nets can be equivalently formalised by a pair of functions ${}^\bullet(\_),(\_)^\bullet\colon E \to \mathcal{P}ow(B)$ assigning to each event its pre- and post-set of conditions. In the same way, the structure of a place/transition net can be equivalently described by generalising F to a *multi*relation or by a pair of functions assigning pre- and post-*multi*sets to transitions. Following [**45**], we choose here this second way.

In the following we shall denote by $\mu(S)$ the monoid of multisets of S. Recall that a multiset is a function from S to ω and that the sum $\mu_1+\mu_2$ is the multiset μ such that $\mu(s)=\mu_1(s)+\mu_2(s)$, for all $s\in S$. Often, we represent $\mu\in\mu(S)$ as a formal sum $\sum\mu(s_i)\cdot s_i$ where only the $s_i\in S$ such that $\mu(s_i)>0$ appear; the empty multiset will be denoted by 0.

DEFINITION. A *place/transition net* (PT for short) is a structure

$$N = (pre_N, post_N\colon T_N \to \mu(S_N))$$

where S_N is a set whose elements are called *places*, T_N is a set whose elements are called *transitions*, pre_N and $post_N$ are functions which assign to each transition, respectively, a *source* (or *pre-set*) and a *target* (or *post-set*) multiset of places, For $t\in T_N$, we write $t\colon u\to v$ to indicate that $pre_N(t)=u$ and $post_N(t)=v$.

A *morphism* of nets $f\colon N_0 \to N_1$ consists of a pair of functions

$$\langle f_t\colon T_{N_0} \to T_{N_1}, f_p\colon \mu(S_{N_0}) \to \mu(S_{N_1})\rangle,$$

where the place component f_p is a *monoid homomorphism*, which respect initial marking and source and target, i.e., the two diagrams below commute.

$$\begin{array}{ccc} T_{N_0} & \xrightarrow{pre_{N_0}} & \mu(S_{N_0}) \\ {\scriptstyle f_t}\downarrow & & \downarrow{\scriptstyle f_p} \\ T_{N_1} & \xrightarrow[pre_{N_1}]{} & \mu(S_{N_1}) \end{array} \qquad \begin{array}{ccc} T_{N_0} & \xrightarrow{post_{N_0}} & \mu(S_{N_0}) \\ {\scriptstyle f_t}\downarrow & & \downarrow{\scriptstyle f_p} \\ T_{N_1} & \xrightarrow[post_{N_1}]{} & \mu(S_{N_1}) \end{array}$$

Explicitly, f_t and f_p are such that:

$$pre_{N_1} \circ f_t = f_p \circ pre_{N_0}, \quad \text{and} \quad post_{N_1} \circ f_t = f_p \circ post_{N_0}$$

This, with the obvious componentwise composition of morphisms, defines the category PTNets.

A PT net is thus a graph whose arcs are the transitions and whose nodes are the multisets on the set of places, i.e., *markings* of the net. As usual, transitions have pre- and post-sets, i.e., sources and targets, in which each place has only finitely many tokens, i.e., finite multiplicity. The same applies to markings. Finally, morphisms of PT nets are graph morphisms in the precise sense of preserving source and target of transitions. In addition, they respect the monoidal structure of multisets, which simply boils down to saying that $f_p(0) = 0$ and that $f_p(\mu_1 + \mu_2) = f_p(\mu_1) + f_p(\mu_2)$ for each pair of multisets $\mu_1, \mu_2 \in \mu(S_{N_0})$.

In the following we shall consider the category of those PT nets whose initial markings and whose post-sets are sets, as opposed to multisets. Since weights are allowed only on the arcs from places to transitions, they are referred to as *semiweighted* nets [**48**].

DEFINITION. A PT net N is *semiweighted* (SW for short) is for all $t \in T_N$, $post_N(t)$ is a set. Moreover, we assume the standard constraint that $pre_N(t) \neq 0$. A *semiweighted net systems* is a semiweighted net N together with an initial marking u_N that is a *set*. A *morphism* of semiweighted net systems $f\colon N_0 \to N_1$ is a morphism of the underlying place/transition nets that, in addition, preserves the initial marking, i.e., $f_p(u_{N_0}) = u_{N_1}$. Semiweighted net systems and their morphisms define the category SWNets.

Notice that disallowing transitions with empty pre-set is a step necessary in any behavioural construction involving nets, as such transitions are highly degenerated; in particular, any number of parallel copies of them can fire at any marking.

Starting from the primitive $t\colon u \to v$ — to be read as t performs a computation *consuming* the tokens in u and *producing* the tokens in v — the notion of *firing* and *state space* is extended to PT nets as follows. A finite number of transitions can be composed in parallel to form a *step*, which, therefore, is a finite multiset of transitions. We write $u\,[\![\alpha\rangle\, v$ to denote a step α with source u and target v. The set $\mathcal{S}(N)$ of steps of N is generated by the rules:

$$\frac{u \in \mu(S)}{u\,[\![0\rangle\, u} \qquad \frac{t\colon u \to v \text{ in } N \text{ and } w \text{ in } \mu(S)}{(u+w)\,[\![t\rangle\,(v+w) \text{ in } \mathcal{S}(N)} \qquad \frac{u\,[\![\alpha\rangle\, v \text{ and } u'\,[\![\beta\rangle\, v' \text{ in } \mathcal{S}(N)}{(u+u')\,[\![\alpha+\beta\rangle\,(v+v') \text{ in } \mathcal{S}(N)}.$$

A finite number of steps from the initial marking can be sequentially composed thus yielding a *step sequence*. The set of step sequences, denoted $\mathcal{SS}(N)$, is given by:

$$\frac{u_N \,[\![\alpha_0\rangle\, v_0, \ldots, u_n \,[\![\alpha_n\rangle\, v_n \text{ in } \mathcal{S}(N) \text{ and } u_i = v_{i-1},\ i = 1, \ldots, n}{u_N \,[\![\alpha_0\rangle[\![\alpha_1\rangle \cdots [\![\alpha_n\rangle\, v_n \text{ in } \mathcal{SS}(N)}.$$

The set $\mathcal{R}(N)$ of *reachable markings* of N is the set of markings which are target of some step sequence, i.e.,

$$\mathcal{R}(N) = \{v \mid \exists (u_N \,[\![\alpha_0\rangle \cdots [\![\alpha_n\rangle\, v) \text{ in } \mathcal{SS}(N)\}.$$

An important class of nets is that of *occurrence nets*, introduced originally in [57] by 'unfolding' safe nets into a suitable 'collection' of their processes (as defined for elementary net systems in Section 1.2). Occurrence nets are elementary nets with a nicely stratified structure whose minimal elements constitute the initial marking.

DEFINITION. An *occurrence net* is an semiweighted net Θ such that $pre_\Theta(t)$ is a set for all $t \in T_\Theta$, and

i) $\forall a \in S_\Theta$, $|{}^\bullet a| \leq 1$, and $a \in u_\Theta$ if and only if ${}^\bullet a = \varnothing$;

ii) $\prec$ is irreflexive, where $\prec$ is the transitive closure of the (flow) relation

$$\prec^1 = \{(a,t) \mid a \in S_\Theta, t \in T_\Theta, t \in a^\bullet\} \cup \{(t,a) \mid a \in S_\Theta, t \in T_\Theta, t \in {}^\bullet a\};$$

moreover, $\forall t \in T_\Theta$, $\{t' \in T_\Theta \mid t' \prec t\}$ is finite;

iii) the binary '*conflict*' relation # on $T_\Theta \cup S_\Theta$ is irreflexive, where

$$\forall t_1, t_2 \in T_\Theta,\ t_1 \#_m t_2 \iff pre_\Theta(t_1) \cap pre_\Theta(t_2) \neq \varnothing \ \&\ t_1 \neq t_2,$$

$$\forall x, y \in T_\Theta \cup S_\Theta,\ x \# y \iff \exists t_1, t_2 \in T_\Theta,\ t_1 \#_m t_2 \ \&\ t_1 \preceq x \ \&\ t_2 \preceq y,$$

and $\preceq$ is the reflexive closure of $\prec$.

This defines the category Occ as a full subcategory of SWNets.

Elements x and y of Θ by are *concurrent*, x co y, if they are related neither by $\preceq$ nor by #. For X a set of elements, we write Co(X) to mean that x co y for all $x, y \in X$.

Thanks to the stratified structure of the nets in Occ, for them we can define the concepts of *depth* of elements and, consequently, of *subnet of depth n*. Essentially, this will allow us to work on such nets by induction.

DEFINITION. Let Θ be a net in Occ. The *depth* of elements in $T_\Theta \cup S_\Theta$ is defined inductively by:

- ▷ $\text{depth}(b) = \begin{cases} 0 & \text{if } b \in u_\Theta; \\ \text{depth}(t) & \text{if } {}^\bullet b = \{t\}; \end{cases}$
- ▷ $\text{depth}(t) = \max\{\text{depth}(b) \mid b \prec t\} + 1.$

DEFINITION. Given a net Θ in Occ define its *subnet* of depth n, $\Theta^{(n)}$, as

- ▷ $T_{\Theta^{(n)}} = \{t \in T_\Theta \mid \text{depth}(t) \leq n\}$;
- ▷ $S_{\Theta^{(n)}} = \{b \in S_\Theta \mid \text{depth}(b) \leq n\}$;
- ▷ $pre_{\Theta^{(n)}}$ and $post_{\Theta^{(n)}}$ are the restrictions of pre_Θ and $post_\Theta$ to $T_{\Theta^{(n)}}$;
- ▷ $u_{\Theta^{(n)}} = u_\Theta$.

Clearly, $\Theta^{(n)}$ is a net in Occ, whenever Θ is such. For each $n \leq m$ there exists a morphism $in_{n,m}\colon \Theta^{(n)} \to \Theta^{(m)}$ whose components are both set inclusions. In the following we shall call such net morphisms simply *inclusions*. Observe that, if $\langle f,g\rangle\colon \Theta_0 \to \Theta_1$ is an inclusion, we obviously have $u_{\Theta_0} = u_{\Theta_1}$ and, for each $t \in T_{\Theta_0}$, $pre_{\Theta_0}(t) = pre_{\Theta_1}(t)$ and $post_{\Theta_0}(t) = post_{\Theta_1}(t)$.

The sequence of nets $\Theta^{(n)}$, $n \in \omega$, can be seen as a sequence of finite approximations which, together with the corresponding inclusions, determines Θ uniquely (up to isomorphisms). We shall formalise this intuition by means of the categorical notion of colimit. The following results will allow us to define the unfolding of a net in terms of finite unfoldings, viz., its subnets of depth n. We first need to show that Occ possesses the required colimits. Consider the category $\underline{\omega} = \{0 \to 1 \to 2 \to 3 \cdots\}$ and the class $\mathcal{D}$ of diagrams $D\colon \underline{\omega} \to \mathrm{Occ}$ such that $D(n \to n+1) = in_n\colon D(n) \to D(n+1)$ is an inclusion. For such a class we have the following results. The reader is referred to [**42**] for the definition of the categorical concepts involved.

PROPOSITION 2.10. *For any $D \in \mathcal{D}$, the colimit of D in* Occ *exists.*

PROOF. Consider the net $\Theta = (pre_\Theta, post_\Theta\colon T_\Theta \to \mu(S_\Theta), u_\Theta)$ where

$$T_\Theta = \textstyle\bigcup_n T_{D(n)}, \qquad S_\Theta = \bigcup_n S_{D(n)}, \qquad u_\Theta = u_{D(0)},$$

$$pre^i_\Theta(t) = pre^i_{D(n_t)}(t), \qquad post^i_\Theta(t) = post^i_{D(n_t)}(t),$$

where n_t above denotes any $n \in \omega$ such that $t \in T_{D(n)}$.

Clearly, Θ is well-defined, is a net, and belongs to Occ. Then taking for each $n \in \omega$ $\mu_n\colon D(n) \to \Theta$ to be the obvious inclusion, we have that μ is the limiting cocone. □

PROPOSITION 2.11. *Given a net Θ in* Occ, *let $D_\Theta\colon \underline{\omega} \to \mathrm{Occ}$ be the functor such that $D_\Theta(n) = \Theta^{(n)}$ and $D_\Theta(n \to n+1) = in_{n,n+1}\colon \Theta^{(n)} \to \Theta^{(n+1)}$. Then $\Theta = \mathrm{Colim}(D_\Theta)$.*

PROOF. It is enough to observe that the colimit construction for diagrams in $\mathcal{D}$ in the proof of the previous proposition gives a family $\mu_n\colon D(n) \to \Theta$, $n \in \omega$, where $\mu_n\colon \Theta^{(n)} \to \Theta$ is the inclusion of $\Theta^{(n)}$ in Θ. □

2.4. Unfolding semiweighted nets. We are now ready to define the *unfolding* of SW nets in terms of occurrence nets and show that it is a functor from SWNets to Occ which is right adjoint to the inclusion of Occ in SWNets.

We start by giving the object component of such a functor. To this end, given a net N, we define a sequence occurrence nets, whose nth element approximates the unfolding of N up to depth n, i.e., it reflects *all* the possible behaviours of N up to (step) sequences of length at *most n*. Clearly, the unfolding of N will be defined as the colimit of an appropriate $\underline{\omega}$-diagram built on the sequence of approximating nets.

The purpose of the following inductive definition is to generate all the possible instances of places and transitions of N by decorating them with their '*history*'. Places in the approximating nets represent instances of places of N: precisely, they are pairs (x,b), where $b \in S_N$ and x is a set encoding the history of this particular instance of b. Analogously, the transitions are pairs (B,t) where $t \in T_N$ and the set B represents the history of the instance of t.

DEFINITION. Let $N = (pre_N, post_N \colon T_N \to \mu(S_N), u_N)$ be a net in SWNets. We define the nets $\mathcal{U}(N)^{(k)} = (pre_k, post_k \colon T_k \to \mu(S_k), u_k)$, for $k \in \omega$, where (cf. Figure 7)

- ▷ $S_0 = \{(\varnothing, b) \mid b \in u_N\}$;
- ▷ $T_0 = \varnothing$, and pre_0 and $post_0$ with the obvious definitions;
- ▷ $u_0 = \sum S_0$;

and for $k > 0$,

▷ $$\frac{B = \{(x_j, b_j) \mid j \in J\} \subseteq S_{k-1}, \quad \mathrm{Co}(B), \quad \sum_{j \in J} b_j = pre_N(t) \text{ for } t \in T_N}{(B,t) \in T_k \quad \text{and} \quad pre_k(B,t) = \sum B}$$

▷ $$\frac{t_0 = (B,t) \in T_k, \quad post_N(t) = \sum_{j \in J} b_j}{(\{t_0\}, b_j) \in S_k, \ \forall j \in J, \quad \text{and} \quad post_k(t_0) = \sum_j (\{t_0\}, b_j)}$$

▷ $u_k = \sum_j (\varnothing, b_j) = \sum S_0 = u_0$.

Therefore $\mathcal{U}(N)^{(0)}$ consists of the initial marking of N, and, informally speaking, $\mathcal{U}(N)^{(n+1)}$ is obtained, inductively, by generating a new transition for each possible subset of *concurrent* places of $\mathcal{U}(N)^{(n)}$ whose corresponding multiset of places of N is the source of some transition t of N; the target of t is then decorated with its history and added to $\mathcal{U}(N)^{(n+1)}$.

Clearly, we shall take $\mathcal{U}(N)$ to be the colimit of the sequence of the $\mathcal{U}(N)^{(n)}$, $n \in \omega$. To do that, we first need to prove the following lemma.

LEMMA 2.12. *For all $n \in \omega$, $\mathcal{U}(N)^{(n)}$ is an occurrence net of depth n. Moreover, for each $n \in \omega$ there is an inclusion $in_n \colon \mathcal{U}(N)^{(n)} \to \mathcal{U}(N)^{(n+1)}$.*

PROOF. That $\mathcal{U}(N)^{(n)}$ has depth n and that there exists an inclusion from $\mathcal{U}(N)^{(n)}$ to $\mathcal{U}(N)^{(n+1)}$ is obvious from the definition. We have to show that $\mathcal{U}(N)^{(n)}$ is an occurrence net. For each $t \in T_n$, $pre_n(t)$ and $post_n(t)$ are multisets where all the elements have multiplicity one, i.e., sets. The same happens for u_n.

i) For each $(x,b) \in S_n$, ${}^\bullet(x,b) = x$ which is either the empty set or a singleton. So $|{}^\bullet(x,b)| \leq 1$. Moreover, $(x,b) \in u_n$ if and only if $x = \varnothing$ if and only if ${}^\bullet(x,b) = \varnothing$.

ii) By definition of $\mathcal{U}(N)^{(n)}$, whenever $x \prec^1 y \prec^1 z$, $\mathrm{depth}(z) = \mathrm{depth}(x) + 1$. Since $x, z \in T_n$ or $x, z \in S_n$ implies that there exists at least one y such that $x \prec y \prec^1 z$ we have $\mathrm{depth}(x) < \mathrm{depth}(z)$. So $x \neq z$ and $\prec$ is irreflexive. Observe that this, together with (*i*), implies that, in each reachable marking, every place has multiplicity at most one. In fact, since that happens in u_n, since each place has only one pre-event and each transition occurs at most once in any computation, there is no way to generate multiple tokens in a place. Moreover, $\{t' \in T_n \mid t' \prec t\}$ is obviously finite for all $t \in T_n$.

iii) Recall that $x \# x$ if and only if $\exists t, t' \in T_n$, $t \neq t'$ and $t \#_m t'$ such that $t \preceq x$ and $t' \preceq x$. So, by (*i*), x cannot be a place, otherwise we would have backward branching. This means that there exist $b, b' \in pre_n(x)$, $b \neq b'$ such that b co b', i.e., $x = (B,t)$ with *not* $\mathrm{Co}(B)$, that is impossible.

The other conditions of occurrence nets obviously hold. □

DEFINITION. We define $\mathcal{U}(N)$ to be the colimit of the diagram $D\colon \underline{\omega} \to$ DecOcc such that $D(n) = \mathcal{U}(N)^{(n)}$ and $D(n \to n+1) = in_n$. By Lemma 2.12 D belongs to $\mathcal{D}$ and so, by Proposition 2.10, the colimit exists and is a decorated occurrence net.

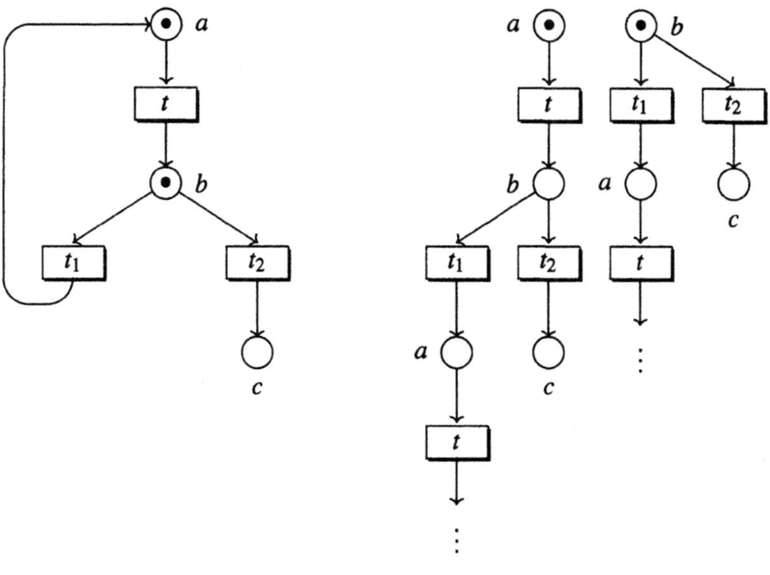

FIGURE 7. An SW net N and (part of) its unfolding $\mathcal{U}(N)$

The correspondence between elements of the unfolding and elements of the original net is formalised by the folding morphism, which will also be the counit of the adjunction.

PROPOSITION 2.13. *Consider the map* $\varepsilon\colon \mathcal{U}(N) \to N$ *defined by*

- ▷ $\varepsilon_t(B,t) = t$;
- ▷ $\varepsilon_p(0) = 0$;
- ▷ $\varepsilon_p(\sum_i (x_i, y_i)) = \sum_i y_i$.

Then, ε_N *is a morphism in* SWNets, *called the folding of* $\mathcal{U}(N)$ *into* N.

PROOF. Since the transitions of $\mathcal{U}(N)$ are of the form $t_0 = (B,t)\colon \sum B \to \sum C$, where $B = \{(x_j, b_j) \mid j \in J\} \subseteq S_{\mathcal{U}(N)}$, $C = \{(\{t_0\}, c_k) \mid k \in K\}$, $t \in T_N$, $\sum_{j\in J} b_j = pre_N(t)$, and $\sum_{k\in K} c_k = post_N(t)$, we immediately obtain

$$\varepsilon_p(pre_{\mathcal{U}(N)}(B,t)) = pre_N(\varepsilon_t(B,t)),$$

and analogously for *post*. Since $u_{\mathcal{U}(N)} = \sum_{b\in S_N} u_N(b)\cdot(\varnothing, b)$, we have $\varepsilon_p(u_{\mathcal{U}(N)}) = \sum_{b\in S_N} u_N(b)\cdot b = u_N$. □

The next lemma is the final ingredient we need to prove that $\mathcal{U}(\_)$ is right adjoint to the inclusion. The missing details can be found in [**48**].

LEMMA 2.14. *Let* Θ_0 *and* Θ_1 *be occurrence nets and let* $f\colon \Theta_0 \to \Theta_1$ *be a morphism. Then, for each* $t_0 \in T_{\Theta_0}$, *we have* $\mathrm{Co}\left(pre_{\Theta_0}(t_0)\right)$ *and* $\mathrm{Co}\left(f_p(pre_{\Theta_0}(t_0))\right)$.

PROOF. Since, by definition of occurrence nets, $\{t' \preceq t\}$ is finite, we have *not* $\mathrm{Co}(pre_{\Theta_0}(t_0))$ iff $\exists b, b' \in pre_{\Theta_0}(t_0)$ such that $b \# b'$. This would mean that $\exists t, t' \in T_{\Theta_0}$, $t \neq t'$ and $t \#_m t'$ such that $t \preceq b$ and $t' \preceq b'$. Thus, since $t \preceq t_0$ and $t' \preceq t_0$, we would have $t_0 \# t_0$ which is impossible since Θ_0 is a occurrence net. Furthermore, $f_p(pre_{\Theta_0}(t_0)) = pre_{\Theta_1}(f_t(t_0))$, which is the pre-set of a transition of a occurrence net and so, by the first part of this proposition, $\mathrm{Co}(f_p(pre_{\Theta_0}(t_0)))$. □

THEOREM 2.15. *The pair* $\langle \hookrightarrow, \mathcal{U}(\_) \rangle : \mathsf{Occ} \rightharpoonup \mathsf{SWNets}$ *constitutes an adjunction.*

PROOF. Let N be a SW net and $\mathcal{U}(N)$ its unfolding. We show that the folding $\varepsilon\colon \mathcal{U}(N) \to N$ is universal, i.e., for any occurrence net Θ and any morphism $k\colon \Theta \to N$ in SWNets, there exists a unique $h\colon \Theta \to \mathcal{U}(N)$ in Occ such that $k = \varepsilon \circ h$.

$$\begin{array}{ccccc} N & & \mathcal{U}(N) & & \mathcal{U}(N) \xrightarrow{\ \varepsilon\ } N \\ \uparrow{\scriptstyle \forall k} & & \uparrow{\scriptstyle \exists ! h} & \text{s.t.} & {\scriptstyle h}\uparrow \quad \nearrow{\scriptstyle k} \qquad \text{commutes.} \\ \Theta & & \Theta & & \Theta \end{array}$$

Consider the diagram in Occ given by $D_\Theta(n) = \Theta^{(n)}$, the subnet of Θ of depth n and $D_\Theta(n \to n+1) = in_n\colon \Theta^{(n)} \to \Theta^{(n+1)}$. We define a sequence of morphisms of nets $h_n\colon \Theta^{(n)} \to \mathcal{U}(N)$, such that for each n, $h_n = h_{n+1} \circ in_n$. Since $\Theta = \mathrm{Colim}(D_\Theta)$, there is a unique $h\colon \Theta \to \mathcal{U}(N)$ such that $h \circ \mu_n = h_n$ for each n. At the same time, we show that

$$\forall n \in \omega,\ k \circ \mu_n = \varepsilon \circ h_n$$

and that the h_n form the unique sequence of morphisms $h_n\colon \Theta^{(n)} \to \mathcal{U}(N)$ such that this holds. Thus we have

$$\forall n \in \omega,\ k \circ \mu_n = \varepsilon \circ h \circ \mu_n$$

and, by the universal property of the colimit, $k = \varepsilon \circ h$. To show the uniqueness of h, let h' be such that $k = \varepsilon \circ h'$. Then we have $k \circ \mu_n = \varepsilon \circ h' \circ \mu_n$. But h_n is the unique morphism for which this happens. Therefore, for each n, $h_n = h' \circ \mu_n$ and so, again by the universal property of the colimit, $h = h'$.

Let us now define h_n and therefore $h\colon \Theta \to \mathcal{U}(N)$, and show that the h_n, $n \in \omega$, form the unique sequence of morphisms for which (1) above holds.

depth 0. This is a special case of the inductive step, and we omit it (see [**48**].)

depth n+1. Let us suppose that we have defined $h_n\colon \Theta^{(n)} \to \mathcal{U}(N)$ and that it is a morphism. Suppose that for each $m \leq n$, h_m is the unique morphism such that $\varepsilon \circ h_m = k \circ \mu_m$. Let h_{n+1} be h_n on the elements of depth less or equal to n. Now, we define h_{n+1} on the elements of depth $n+1$. Let $t_1 \in T_\Theta$ such that $\mathrm{depth}(t_1) = n+1$ and $k(t_1) = t$. Since $pre_\Theta(t_1)$ is a set of elements of depth less or equal to n, $h_n(pre_\Theta(t_1))$ is defined. Since h_n is a morphism, by Lemma 2.14, we have $\mathrm{Co}(h_n(pre_\Theta(t_1)))$. Moreover, since $\varepsilon \circ h_n = k \circ \mu_n$, we have that

$$pre_N(t) = k(pre_\Theta(t_1)) = \varepsilon \circ h_n(pre_\Theta(t_1)) = \textstyle\sum_{j \in J} b_j,$$
$$\text{for } J \text{ such that } \{(x_j, b_j) \mid j \in J\} = h_n(pre_\Theta(t_1)).$$

Therefore $t_0 = \big(h_n(pre_\Theta(t_1)), t\big) = \big(h_{n+1}(pre_\Theta(t_1)), t\big) \in T_{\mathcal{U}(N)}$. Now, since h_{n+1} has to make the diagram commute, $h_{n+1}(t_1)$ must be of the form (B, t) and, since it has to be a morphism, it must be $pre_{\mathcal{U}(N)}((B,t)) = \sum B = h_{n+1}(pre_\Theta(t_1))$. Therefore $h_{n+1}(t_1) = t_0$. Observe that there is only one choice for $h_{n+1}(t_1)$, given k and h_n by inductive hypothesis. Obviously, $\varepsilon \circ h_{n+1}(t_1) = t = k(t_1) = k \circ \mu_{n+1}(t_1)$. Now, let $post_\Theta(t_1) = \sum_i a_i$. Suppose that $k(a_i) = \sum_j m^i_j b^i_j$. Since $k(post_\Theta(t_1)) = post_N(k(t_1))$, we have $post_N(k(t_1)) = \sum_{i,j} m^i_j b^i_j$, with all b^i_j distinct. It follows that $m^i_j = 1$ and thus in $\mathcal{U}(N)$ we have the places $\bigcup_{i,j}\{(\{t_0\}, b^j_l)\}$. We define

$$h_{n+1}(a_i) = \textstyle\sum_j \{(\{t_0\}, b^j_l)\}$$

and, as before, conclude that $\varepsilon \circ h_{n+1}(a_i) = \sum_j b^i_j = k(a_i) = k \circ \mu_{n+1}(a_i)$.

Observe that $h_{n+1}(a_i)$ is completely determined by k and by the conditions of decorated occurrence net morphisms.

Finally, we have to show that h_{n+1} is a morphism $\Theta^{(n+1)} \to \mathcal{U}(N)$. But this task is really trivial because, by its own construction, h_{n+1} preserves source, target and initial marking. □

THEOREM 2.16. $\langle \hookrightarrow, \mathcal{U}(\_) \rangle$ *is a coreflection* Occ $\rightharpoonup$ SWNets.

It is worth observing that when N is a *safe* net, $\mathcal{U}(N)$ is (isomorphic to) the unfolding of N defined in [**57, 94**]. In other words, $\langle \hookrightarrow, \mathcal{U}(\_) \rangle$ restricts to the coreflection Occ $\rightharpoonup$ Safe presented in *loc. cit.*

Our final step in relating SW nets to event structures and domains is to fill the gap between occurrence nets and event structures. To this aim, we conclude this section recalling the definitions of the functors forming the coreflection $\langle \mathcal{N}, \mathcal{E} \rangle$: E $\rightharpoonup$ Occ as studied in [**57, 94**].

DEFINITION. Let Θ be an occurrence net. Then, $\mathcal{E}(\Theta)$ is the event structure

$$(T_\Theta, \preceq, \#),$$

where $\preceq$ and $\#$ are the restriction to T_Θ, the set of transitions of Θ, of, respectively, the *flow* ordering $\preceq_\Theta$ and the *conflict* relation $\#_\Theta$ implicitly defined by Θ.
For $f\colon N_0 \to N_1$ a morphism in Occ, we take $\mathcal{E}(f)$ to be $f_t\colon \mathcal{E}(N_0) \to \mathcal{E}(N_1)$, which clearly gives a functor $\mathcal{E}\colon$ Occ $\to$ E.

Consider now an event structure $(E, \leq, \#)$. As a notation, for a subset A of E, we write $\#A$ to mean that $a \# a'$ for all $a \neq a' \in A$ Similarly, $e < A$ means that $e < e'$ for all $e' \in A$. Then, we can define define

$$\mathcal{N}(E, \leq, \#) = (pre, post\colon E \to \mu(M \cup B), \textstyle\sum M),$$

where

$$\begin{aligned} M &= \{(\varnothing, A) \mid A \subseteq E \text{ and } \#A\}; \\ B &= \{(e, A) \mid e \in E,\ \#A \text{ and } e < A\}; \\ pre(e) &= \{(c, A) \in B \cup M \mid e \in A\}; \\ post(e) &= \{(e, A) \in B\}. \end{aligned}$$

Then, we have the following.

THEOREM 2.17. *For each event structure E, $\mathcal{N}(E)$ is an occurrence net such that $\mathcal{E}\mathcal{N}(E) = E$. Moreover, $\mathcal{N}$ extends to a functor that is left adjoint to $\mathcal{E}$, so yielding a coreflection whose unit is the identity function $E \to \mathcal{E}\mathcal{N}(E) = E$.*

PROOF. See [95]. □

Although $\mathcal{N}\mathcal{E}(\Theta)$ and Θ are not isomorphic in Occ, it is worth observing that they are 'behaviourally' such. In particular, an inspection of the definition will prove that $\mathcal{N}\mathcal{E}(N)$ is *place-saturated* version of N, i.e., that is obtained by adding to N all the places that it is possible to add without duplicating any or altering the behaviour.

The coreflection between Occ and SWNets can of course be composed with the coreflection between Occ and E and with the equivalence (*a fortiori* a coreflection!) of D and E. The following example shows the structures associated by this chain of coreflections to a simple, well-known, (non-safe) semiweighted net.

EXAMPLE.

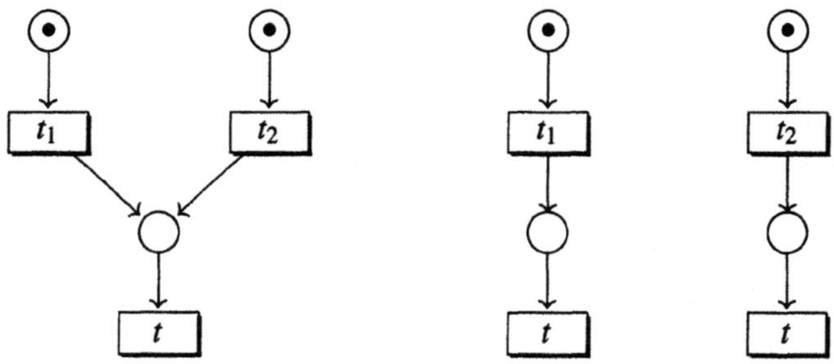

N $\mathcal{U}(N)$

Observe that the unfolding contains two concurrent copies of t. These correspond to the occurrences of t in two possible 'causal contexts', namely t caused by t_1 and t caused by t_2. In the picture below, which shows the event structure and the prime algebraic domain associated to N, the four events so arising are labelled by the transition they correspond to.

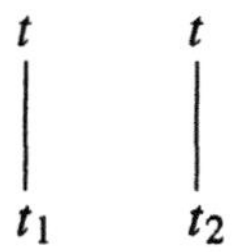

$\mathcal{E}\mathcal{U}(N)$

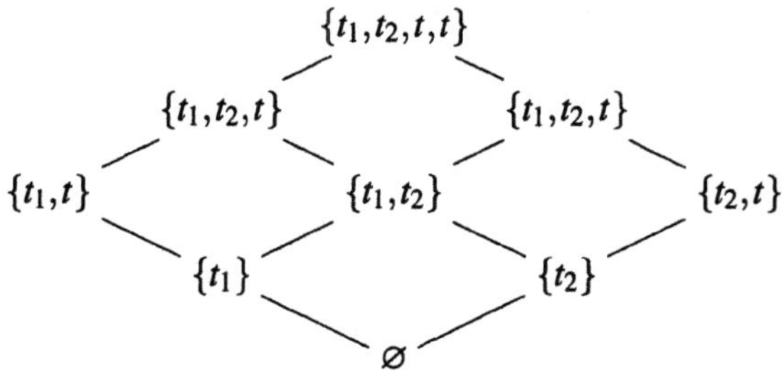

$\mathcal{D}\mathcal{E}\mathcal{U}(N)$

3. Petri Nets and Transition Systems

Looking back at the notion of case graph CG_N in Section 1.1, we certainly know that — to a certain extent and with a certain degree of precision — nets can be described by some sort of transition system. The details, however, are far from trivial and in this section we set out to study them by presenting a coreflection between elementary net systems and so-called *elementary transition systems* due to **[59, 60]**.

Intuitively, this result identifies a category of transition systems which may be seen as an abstract behavioural characterisation of elementary net systems, formally stated as a coreflection 'embedding' a category of certain transition systems into EN. This was the first of a series of behavioural characterisation results for nets, in the sense that several similar coreflections have been shown for more general classes of nets, see e.g. **[97, 53]**. We restrict ourselves here to elementary nets, though we believe that the ideas from **[53]** generalise smoothly to provide a corresponding result for SWNets.

3.1. Transition systems. Transition systems are a frequently used model of parallel processes. They consist of a set of states, with an initial state, together with transitions between states which are labelled to specify the kind of events they represent.

DEFINITION. A *transition system* is a structure $(S,i,L,tran)$ where

- ▷ S is a set of *states* with *initial state* i,
- ▷ L is a set of *labels*,
- ▷ $tran \subseteq S \times L \times S$ is the *transition relation.*

A transition $(s,a,s') \in tran$ is often indicated as $s \xrightarrow{a} s'$.

DEFINITION. Let $T_0 = (S_0,i_0,L_0,tran_0)$ and $T_1 = (S_1,i_1,L_1,tran_1)$ be transition systems. A *morphism* $f\colon T_0 \to T_1$ is a pair $f = (\sigma,\lambda)$ where

- ▷ $\sigma\colon S_0 \to S_1$, a function between sets of states,
- ▷ $\lambda\colon L_0 \to L_1$, a function between sets of labels,

are such that $\sigma(i_0) = i_1$ and $(s,\alpha,s') \in tran_0$ implies $(\sigma(s),\lambda(\alpha),\sigma(s')) \in tran_1$.

The intention behind the definition of morphism is, as usual in this paper, to guarantee that the relevant behavioural notions are preserved. In case of transition systems this amounts to saying that the effect of a transition $s \xrightarrow{\alpha} s'$ in T_0 is matched in T_1 by the effect of a corresponding transition $\sigma(s) \xrightarrow{\lambda(\alpha)} \sigma(s')$. To complete the definition, morphism are required to preserve initial states.

Transition system and their morphisms form a *category* TS in which composition is defined componentwise — i.e., composing $(\sigma,\lambda)\colon T_0 \to T_1$ and $(\sigma',\lambda')\colon T_1 \to T_2$ yields $(\sigma' \circ \sigma, \lambda' \circ \lambda)\colon T_0 \to T_2$ — and the identity morphism of T is $(1_{T_S}, 1_{T_L})$, where 1_X is the identity function on the set X.

The definition below refines the notion of case graph of an elementary net systems by recasting it in terms of transition systems. In the following, for $N = (B,E,F,c_{in})$ an elementary net system, we shall say that an event $e \in E$ is '*active*' if it has an occurrence $(c,e,c') \in \longrightarrow_{C_N}$.

DEFINITION. Let $N = (B,E,F,c_{in})$ an elementary net system. The *reachable case graph* associated with N is the transition system $TS_N = (C_N, c_{in}, E_N, \longrightarrow_{C_N})$, where E_N is the set of *active events* of N.

The reachable case graph, or simply the transition system, associated with the net system of Figure 1 is presented in Figure 8.

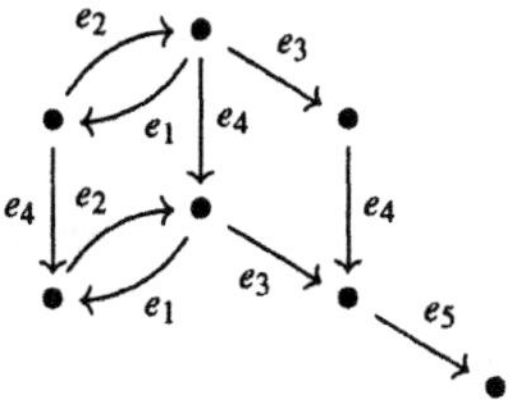

FIGURE 8

With the definition of TS given above, it is quite easy to see that the case graph construction of elementary net systems extends to a *functor* from EN to TS (see below). In order for this functor to be part (left adjoint) of a coreflection, however, we first need to identify those transition systems that arise as reachable case graphs of elementary nets. Such transition systems can be characterised in terms of *regions* by Ehrenfeucht and Rozenberg [**20**].

DEFINITION. Let $T = (S, i, L, tran)$ be a transition system. Then $r \subseteq S$ is a *region* of T if and only if the following two conditions are satisfied for all $(s_0, e, t_0), (s_1, e, t_1) \in tran$:

▷ if $s_0 \in r \;\&\; t_0 \notin r$, then $s_1 \in r \;\&\; t_1 \notin r$;

▷ if $s_0 \notin r \;\&\; t_0 \in r$, then $s_1 \notin r \;\&\; t_1 \in r$.

So, a region is a subset of states, such that for all $e \in L$, *all* occurrences of an e-labelled transitions have the same '*crossing*' relationship with respect to r (leaving or entering). As an example, consider again the transition system from Figure 8. It is easy to check, for instance, that the singleton set consisting of the state entered by the e_5-labelled transition is a region, whereas the singleton set consisting of the initial state is *not*.

It follows trivially that both $\varnothing$ and S are regions of T. They are called the *trivial regions*, and we use $\mathcal{R}_T$ as notation for the set of *non-trivial* regions of T. Also, for $s \in S$, we use $\mathcal{R}_s$ to denote the set of non-trivial regions of T containing s. More precisely

$$\mathcal{R}_s = \{r \mid s \in r \text{ and } r \in \mathcal{R}_T\}.$$

Yet another crucial notation concerns the set of *pre-* and *post-regions* of a label. Formally, for $e \in L$, we define

the *pre-regions* of e: ${}^\circ e = \{r \in \mathcal{R}_T \mid \exists (s, e, s') \in tran.\ s \in r \;\&\; s' \notin r\}$;

the *post-regions* of e: $e^\circ = \{r \in \mathcal{R}_T \mid \exists (s, e, s') \in tran.\ s \notin r \;\&\; s' \in r\}$.

Some useful properties of regions can now be stated.

PROPOSITION 3.1. *Let $T = (S, i, L, tran)$ be a transition system. Then*

i) $r \subseteq S$ *is a region if and only if* $\bar{r} = S \setminus r$ *is a region.*

ii) *Let* $e \in L$. *Then* $e^\circ = \{\bar{r} \mid r \in {}^\circ e\}$.

iii) *Let* $(s, e, s') \in tran$. *Then* $\mathcal{R}_s \setminus \mathcal{R}_{s'} = {}^\circ e$ *and* $\mathcal{R}_{s'} \setminus \mathcal{R}_s = e^\circ$.

Another important property of regions is that they are preserved by inverse morphism.

PROPOSITION 3.2. *Let $f = (\sigma, \lambda)$ be a morphism from T_0 to T_1. If r is a region of T_1 then $\sigma^{-1}(r)$ is a region of T_0. Furthermore, for every $e \in L_{T_0}$, we have $\sigma^{-1}(r) \in {}^\circ e$ (respectively $\sigma^{-1}(r) \in e^\circ$) if and only if $r \in {}^\circ\lambda(e)$ (respectively to $r \in \lambda(e)^\circ$).*

Using the notion of regions, we may now define the transition systems which arise as case graphs of elementary nets, following **[20]**.

DEFINITION. A transition system $T = (S, i, L, tran)$ is said to be *elementary* if it satisfies the following conditions:

(A1) $\forall e \in L.\ \exists (s, e, s') \in tran$, where $s \neq s'$;

(A2) $\forall s \in S \setminus \{i\}.\ \exists s_0, s_1, \ldots, s_{n+1} \in S$ and $e_0, e_1, \ldots, e_n \in L$ such that

$$i = s_0, \quad s_{n+1} = s, \quad \text{and} \quad (s_i, e_i, s_{i+1}) \in tran \text{ for } 0 \leq i \leq n;$$

(A3) $\forall e, e' \in L.\ {}^\circ e = {}^\circ e' \implies e = e'$;

(A4) $\forall s, s' \in S.\ \mathcal{R}_s = \mathcal{R}_{s'} \implies s = s'$;

(A5) $\forall s \in S.\ \forall e \in L.\ {}^\circ e \subseteq \mathcal{R}_s \ \&\ e^\circ \cap \mathcal{R}_s = \varnothing \implies \exists (s, e, s') \in tran.$

We let ETS denote the full subcategory of TS with elementary transition systems as objects.

Notice that the axioms presented here are a cleaned up version of the axioms presented in **[59]**, but in this context, the two sets of axioms can be proved to be equivalent. The two first axioms just state that each label and each state can be *reached* form the initial state by a finite number of transitions. The next two enforce certain obvious regional *separation* properties, for labels and states respectively. And the final axiom enforces a relationship between regional properties of labels and the transition relation.

3.2. Elementary nets and transition systems. It turns out that ETS is exactly the category of case graphs associated with EN, a result which we shall formalise in this section. For full proofs of the theorems quoted, we refer to **[59]**

THEOREM 3.3. *The construction that maps $N = (B, E, F, c_{in})$ to the its reachable case graph $TS_N = (C_N, c_{in}, E_N, \longrightarrow_{C_N})$ extends to a functor nt from* EN *to* ETS.

PROOF. Easy. It follows essentially from Proposition 1.5. □

More importantly, *nt* is the left adjoint component of a coreflection whose a right adjoint, *tn*: ETS $\to$ EN, operates on objects as detailed below.

DEFINITION. Let $T = (S, i, L, tran)$ be a transition system. The *net structure* associated to T is the net $N_T = (\mathcal{R}_T, L, F_T, \mathcal{R}_i)$, where

- ▹ $(r, e) \in F_T$ if and only if $r \in {}^\circ e$;
- ▹ $(e, r) \in F_T$ if and only if $r \in e^\circ$.

Essentially, this definition 'reads' a transition systems as the case graph of some net. Thus, as expected, the labels of T become events of N_T and the regions become the conditions, with the pre- and post-sets obviously given by pre- and post-regions.

PROPOSITION 3.4. *The net structure $N_T = (R_T, L, F_T, R_i)$ associated to the transition system $T = (S, i, L, tran)$ is an elementary net system.*

PROOF. Easy. □

As anticipated, N_T is the object part of the right adjoint of *nt*.

THEOREM 3.5. *The construction that maps a transiton system T to its net structure N_T extends to a functor tn from* ETS *to* EN.

PROOF. A morphism $(\sigma, \lambda): T_0 \to T_1$ is mapped to $tn(\sigma, \lambda) = (\beta, \lambda): N_{T_0} \to N_{T_1}$, where $\beta^{-1}(r_1) = \sigma^{-1}(r_1)$. With this definition, the proof essentially follows from Proposition 3.2. □

THEOREM 3.6. *Functors tn and nt form a coreflection* ETS $\rightharpoonup$ EN *with tn as left adjoint.*

PROOF. This proof is rather involved. We refer the interested reader to [**59**]. □

One interesting part of the construction is that, exactly as in the case of event structures, $tn \circ nt(N)$ is a '*condition-saturated*' version of N, in the formal sense that any further addition of a condition will change the behaviour (case graph) of the net. As a consequence, the saturated net will have, for each condition b, a condition representing the *complement* of b — formally the set complement $\bar{r}$ of the region r associated with b (cf. Proposition 3.1*(i)*). Hence, as a corollary of this construction, we have that any elementary net system is behaviourally equivalent to a contact-free system.

4. Petri Nets and Bisimulations

The coreflections that we studied in the previous sections establish a web of formal relationships between nets and other models, enabling us to place nets in a broader picture of models for concurrency and, often, allowing the translation of concepts to between models. This section aims at providing an illustration of this point. It is an exposition of a the categorical approach to bisimulation obtained from spans of open maps as defined in [**35**] — to which we refer for the missing proofs — with an additional treatment of SWNets in the general picture. The open map approach presented here has also been applied successfully to capture other familiar behavioural equivalences on nets, e.g., Hoare's trace equivalence [**30**] and Milner's weak bisimulation [**49, 50**], both of which may be obtained by slightly changing the notion of path extension from the one presented here [**55**]. Also, the open morphism approach has been applied successfully to different categories of models, e.g., probabilistic systems [**55**], and timed systems [**34**].

Once again, the idea here is to put forward that the categorical view of models for concurrency provides guidelines for definitions of concepts, like behavioural equivalences, consistent across a range of models. In particular, the notion of bisimulation derived for nets comes automatically equipped with a number of essential properties. The categorical approach here contrasts with the more common alternative of searching for a sensible candidate for bisimulation on nets and, having found one of then checking it possesses these essential properties.

4.1. Labelled models and their relationship. Like most models for concurrency, nets [**65**] and asynchronous transition systems [**54**], or more precisely their labelled versions, have been used as models for process languages like CCS, [**50**]. As an illustration, following [**65**], the (strongly bisimilar) CCS expressions *a.nil* | *b.nil* and *a.b.nil* + *b.a.nil* are represented by the following rather different nets.

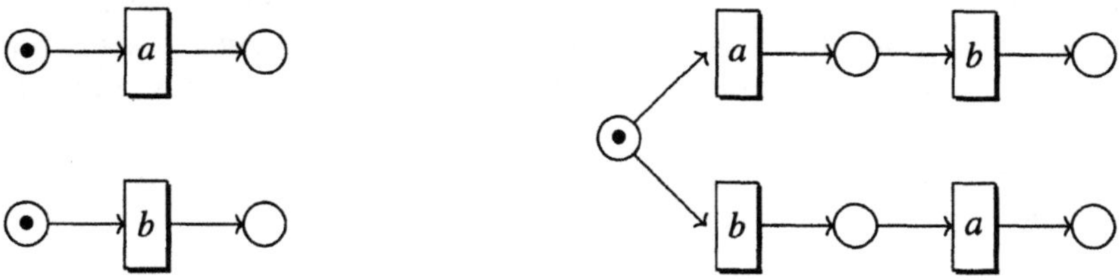

a.nil | *b.nil* *a.b.nil* + *b.a.nil*

There is a general way of introducing labels to models in such a way that one may carry over adjunctions between unlabelled models to their labelled counterparts, following [**35**]. Here we sketch the idea, applicable to the categories of nets and event structures. We assume a category X of structures each of which possesses a distinguished set of events and where morphisms have as a component a function between sets of events. In any such setting, morphisms may be lifted uniformly to a category X_L with labels.

- ▷ The objects of X_L consist of structures (X, l) where X is an object of X, and $l: E \to L$ is a (total) labelling function from E the events of X to the labelling set L.
- ▷ The morphisms of X_L from (X, l) to (X', l') correspond to morphisms $f: X \to X'$ of X of which the event component η preserves labels, i.e. $l' \circ \eta = l$.

Correspondingly, for a set of labels L, we denote the fibres over L in the labelled versions of our categories of nets and event structures by EN_L, SWNets_L and E_L, respectively. Similarly the category of transition systems over label set L, with morphisms having the identity as label component, will be denoted TS_L, and its full subcategory of *synchronisation trees* — that are precisely the tree-'shaped' transition systems — by S_L.

It follows for general reasons [**97**] (and is easy to see) that the coreflection between nets and event structures lift to a coreflection between the labelled versions. The modified adjoints are essentially the adjoints presented in the previous sections, simply carrying the label parts across from one model to the other. Furthermore, this coreflection is part of greater collection [**97, 80, 79**] of which here we are interested in the small portion shown in the diagram below.

$$\mathsf{S}_L \underset{es}{\overset{se}{\rightleftarrows}} \mathsf{E}_L \underset{\mathcal{EU}(\_)}{\overset{\mathcal{N}(\_)}{\rightleftarrows}} \mathsf{SWNets}_L$$

Concerning synchronisation trees, we remark that, as hinted at in Section 2.1, they can be identified with those labelled event structures having *empty* co-relation, i.e., those whose every two events are either causally dependent or conflicting. This gives rise to an embedding $se\colon \mathsf{S}_L \hookrightarrow \mathsf{E}_L$ which, actually, admits a coreflection right adjoint whose action on objects yields the tree of event sequences ordered by prefix.

4.2. Path-lifting morphisms. Following [35], a computation path represents a particular run or history of a process. For transition systems, a computation path is traditionally taken to be a finite sequence of transitions. For a labelling set L, define the category of *branches* Bran_L to be the full subcategory of transition systems, with labelling set L, with objects those finite synchronisation trees with exactly one (maximal) branch; so the objects of Bran_L are essentially *strings* over alphabet L. A computation path in a transition system T, with labelling set L, can then be represented by a morphism

$$p\colon P \to T$$

in TS_L from an object P of Bran_L.

How should we represent a computation path of a net or an event structure? To take into account the explicit concurrency exhibited by an event structure, it is reasonable to represent a computation path as a morphism from a partial order of labelled events, that is from a *pomset* [69]. Observe that pomsets with labels in L can be identified with special kinds of labelled event structures in E_L: those with *empty* conflict relation. Then, define the category of pomsets Pom_L, with respect to a labelling set L, to be the full subcategory of E_L whose objects consist exclusively of *finite* pomsets. A computation path in an event structure E, with labelling set L, is a morphism

$$p\colon P \to E$$

in E_L from an object P of Pom_L.

What about computation paths in nets? The answer is that our embeddings of event structures into nets allow us to view *pomsets as labelled nets*! Let us illustrate the point more concretely. The left adjoint $\mathcal{N}(\_)$ of the coreflection $\mathsf{E}_L \to \mathsf{SWNets}_L$ embeds labelled event structures, and so pomsets, in labelled SW nets. This enables us to identify pomsets P in Pom_L with their images $\mathcal{N}(P)$ as labelled saturated nets in SWNets_L. Now, we can take a computation path in a net N, with labelling set L, to be a morphism

$$p\colon P \to N$$

in SWNets_L from a 'pomset' P, with labelling set L — where P actually stands for labelled saturated net in SWNets_L corresponding to a pomset. In future, when discussing nets, we will deliberately confuse pomsets with their image in SWNets_L under the embedding.

Generally, assume a category of models M (this can be any of the categories of labelled structures we are considering) and a choice of path category, a subcategory $\mathsf{P} \hookrightarrow \mathsf{M}$ consisting of path objects (these could be branches, or pomsets) together with morphisms expressing how they can be extended. Define a *computation path* in an object X of M to be a morphism

$$p\colon P \to X,$$

in M, where P is an object in P. A morphism $f: X \to Y$ in M takes such a path p in X to the path $f \circ p: P \to Y$ in Y. The morphism f expresses the sense in which Y simulates X: any computation path in X is matched by the computation path $f \circ p$ in Y.

We might demand a stronger condition of a morphism $f: X \to Y$ expressed succinctly in the following *path-lifting condition*: whenever, for $m: P \to Q$ a morphism in P, a 'square'

$$\begin{array}{ccc} P & \xrightarrow{p} & X \\ {\scriptstyle m}\downarrow & & \downarrow{\scriptstyle f} \\ Q & \xrightarrow[q]{} & Y \end{array}$$

in M commutes, i.e., $q \circ m = f \circ p$, meaning the path $f \circ p$ in Y can be extended via m to a path q in Y, then there is a morphism p' such that in the diagram

$$\begin{array}{ccc} P & \xrightarrow{p} & X \\ {\scriptstyle m}\downarrow & \overset{p'}{\nearrow} & \downarrow{\scriptstyle f} \\ Q & \xrightarrow[q]{} & Y \end{array}$$

the two 'triangles' commute, i.e., $p' \circ m = p$ and $f \circ p' = q$, meaning the path p can be extended via m to a path p' in X which matches q. When the morphism f satisfies this condition we shall say it is P-*open*.

It is easily checked that P-open morphisms include all the identity morphisms (in fact, all isomorphisms) of M and are closed under composition there; in other words they form a subcategory of M.

For transition systems, Bran_L-open morphisms are already familiar.

PROPOSITION 4.1. *With respect to a labelling set L, the* Bran_L*-open morphisms of* TS_L *are the 'zig-zag morphisms' of* **[88]**, *the 'p-morphism' of* **[82]**, *the 'abstraction homomorphisms' of* **[14]**, *and the 'pure morphisms' of* **[5]**, *the 'transition-preserving homomorphisms' of* **[23]**, *i.e., those label-preserving morphisms* $(\sigma, 1_L): T \to T'$ *on transition systems over labelling set L with the property that for all reachable states s of T*

$$\textit{if } \sigma(s) \xrightarrow{a} s' \textit{ in } T' \textit{ then } s \xrightarrow{a} u \textit{ in } T \textit{ and } \sigma(u) = s', \textit{for } u \textit{ a state of } T.$$

DEFINITION. Let P be a category in a category of models M. Objects X_1, X_2 of M are P-*bisimilar* iff they are in the equivalence generated by being related by a P-open map.

For the interleaving models of transition systems and synchronisation trees with path category P taken to be branches, P-bisimulation coincides with Milner's *strong bisimulation* **[50, 49]**.

THEOREM 4.2. *Two transition systems (and so synchronisation trees), over the same labelling set L, are* Bran_L*-bisimilar iff they are strongly bisimilar in the sense of* **[50]**.

In many cases, including the ones considered here, P-bisimilarity between two objects have a particularly simple presentation.

THEOREM 4.3. *If the category* M *has pullbacks, then* M_1 *and* M_2 *are* P*-bisimilar iff there is a span of* P*-open morphisms* f_1, f_2*:*

$$M_1 \xleftarrow{\;f_1\;} M \xrightarrow{\;f_2\;} M_2$$

PROOF. It follows since pullbacks of P-open morphisms are P-open. □

PROPOSITION 4.4. *The categories* S_L, E_L *and* Occ_L *have pullbacks.*

PROOF. One shows that Occ_L has pullbacks. Then, using the facts that right adjoints preserve limits, and pullbacks in particular, and that there are coreflections from categories S_L and E_L to Occ_L, we obtain pullbacks in any of these as images under the right adjoints of the pullback in Occ_L of diagrams transported into Occ_L by the left adjoints. □

We conclude this section presenting a few general facts from [**35**] about how open morphisms and bisimilarity are preserved and reflected by functors, especially when part of a coreflection. For notational simplicity we shall assume the left adjoints of the coreflections are inclusions. It follows that for the coreflections of Section 4.1, in which the categories of models share the same choice of path category, open morphisms and bisimilarity are preserved in both directions of the adjunction.

LEMMA 4.5. *Let* M *be a coreflective subcategory of* X *with* R *right adjoint to the inclusion function* $\mathsf{M} \hookrightarrow \mathsf{X}$ *and* P *a subcategory of* M.

i) A morphism f *of* M *is* P*-open in* M *if and only if* f *is* P*-open in* X.
ii) The components of the counit $\varepsilon_X : \mathsf{R}(X) \to X$ *are* P*-open in* X.
iii) A morphism f *is* P*-open in* X *if and only if* $\mathsf{R}(f)$ *is* P*-open in* M.

COROLLARY 4.6. *Let* M *be a coreflective subcategory of* X *with* R *right adjoint to the inclusion functor* $\mathsf{M} \hookrightarrow \mathsf{X}$ *and* P *a subcategory of* M.

i) M_1 *and* M_2 *are* P*-bisimilar in* M *if and only if they are* P*-bisimilar in* X.
ii) M_1 *and* M_2 *are* P*-bisimilar in* X *if and only if* $\mathsf{R}(M_1)$ *and* $\mathsf{R}(M_2)$ *are* P*-bisimilar in* M.

PROOF. *(i)* Directly from *(i)* of Lemma 4.5.

(ii) '*only if*': By Lemma 4.5*(iii)*, a span of open morphisms in X has, as image under R, a span of open morphisms in M. Thus P-bisimilarity of M_1 and M_2 in X implies P-bisimilarity of $\mathsf{R}(M_1)$ and $\mathsf{R}(M_2)$ in M.

'*if*': Suppose $\mathsf{R}(M_1)$ and $\mathsf{R}(M_2)$ in M are P-bisimilar in M via a span of open morphisms $f_1 : M \to \mathsf{R}(M_1)$, $f_2 : M \to \mathsf{R}(M_2)$ in M. By Lemma 4.5*(i)*, f_1 and f_2 form a span of open morphisms in X. The components of the counits of the coreflection $\varepsilon_1 : \mathsf{R}(M_1) \to M_1$ and $\varepsilon_2 : \mathsf{R}(M_2) \to M_2$ are open by Lemma 4.5*(ii)*. Hence the compositions $\varepsilon_1 \circ f_1$ and $\varepsilon_2 \circ f_2$ form a span of open morphisms in X showing the P-bisimilarity of M_1 and M_2 in X. □

4.3. Pom_L-bisimulation for nets. As seen in Lemma 4.1 and Theorem 4.2, for transition systems the general definition of P-open morphism and P-bisimilarity coincide with familiar notions: the equivalence of strong bisimilarity central to Milner's work. Here we explore how the general definitions specialise to the models of event structures and nets, with nonsequential observations in the form of pomsets. We focus our attention on SWNets, but the answers provided follow closely those obtained for other classes of net systems in [**63**].

We start by characterising Pom_L-open morphisms of SWNets_L.

PROPOSITION 4.7. *The Pom_L-open morphisms of SWNets_L are precisely those which satisfy the 'zig-zag' condition of Proposition 4.1 and which, in addition, reflect consecutive independence, i.e., those $f\colon N_1 \to N_2$ satisfying:*

- ▷ *f_t is total and label preserving;*
- ▷ *whenever $f_p(\mu) \xrightarrow{e'} \nu'$ in N_2, for μ reachable, then there exists $\mu \xrightarrow{e} \nu$ in N_1 such that $f_t(e) = e'$ and $f_p(\nu) = \nu'$;*
- ▷ *whenever $\mu \xrightarrow{e} \mu'$ and $\mu' \xrightarrow{e'} \mu''$ in N_1, for μ reachable, and $f_t(e)$ co $f_t(e')$ in N_2, then e co e' in N_1.*

PROOF. Let $f\colon N_1 \to N_2$ be an open morphism in SWNets_L. The function f_t is total and label preserving from definition of morphisms in SWNets_L, and by considering linear pomsets, where causal dependency is a total order, it is clear as in Proposition 4.1 that f satisfies the 'zig-zag' condition. The only nontrivial part is the reflection of consecutive independence. Let μ be a reachable marking and let

$$\mu \xrightarrow{e} \mu' \quad \text{and} \quad \mu' \xrightarrow{e'} \mu'',$$

be two consecutive transitions in N_1. Consider the corresponding transitions

$$f_P(\mu) \xrightarrow{f_t(e)} f_P(\mu') \quad \text{and} \quad f_P(\mu') \xrightarrow{f_t(e')} f_P(\mu'')$$

of N_2, and assume that $f_t(e)$ and $f_t(e')$ are independent in N_2. Assume further that $l(e) = l(f_t(e)) = a$ and $l(e') = l(f_t(e')) = a'$.

Because μ is reachable there is a chain of transitions

$$c_{in} = \mu_0 \xrightarrow{e_1} \mu_1 \xrightarrow{e_2} \cdots \xrightarrow{e_n} \mu_n = \mu$$

in N_1 from its initial marking c_{in}. Let $l(e_i) = a_i$, and take P to be the linear pomset with $n+2$ elements, ordered and labelled as indicated in the following pomset

$$a_1 \leq a_2 \leq \cdots \leq a_n \leq a \leq a'$$

Let $p\colon P \to N_1$ be that morphism in SWNets_L which maps this chain of transitions to

$$\mu_0 \xrightarrow{e_1} \mu_1 \xrightarrow{e_2} \cdots \xrightarrow{e_n} \mu \xrightarrow{e} \mu' \xrightarrow{e'} \mu''$$

in N_1. Let Q be the pomset differing from P only in that the a and a' labelled elements are unordered, i.e., the pomset corresponding to the following graph.

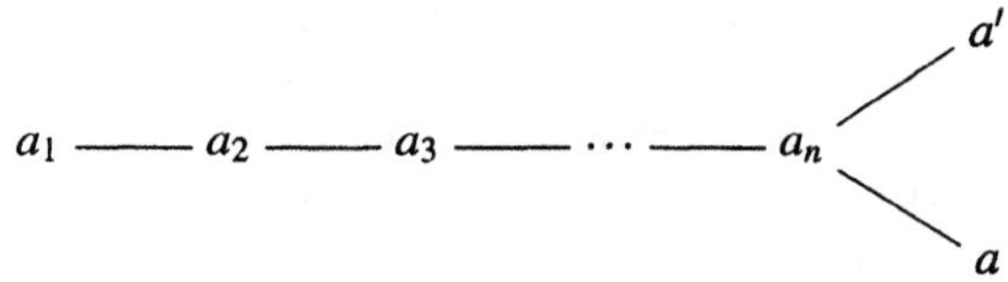

Let $q\colon Q \to N_2$ be that morphism in SWNets_L mapping these transitions to

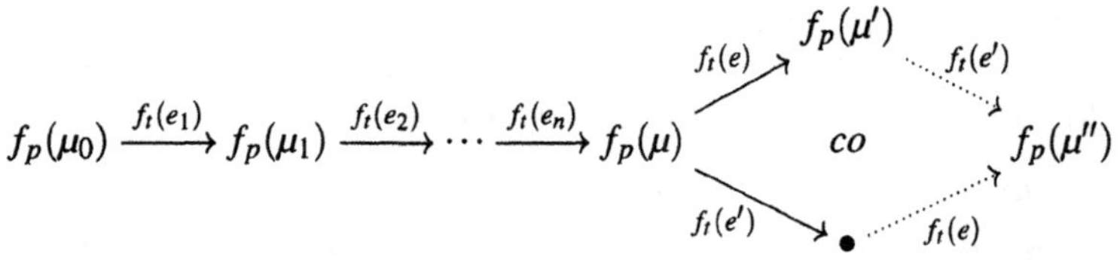

in N_2, where the dotted arrows represent the concurrency diamond. Letting $m\colon P \to Q$ be the obvious morphism of pomsets, we observe the commuting diagram:

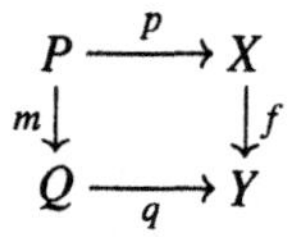

But f is open, so we obtain a morphism $p'\colon Q \to T$ such that the two 'triangles' commute in the following diagram.

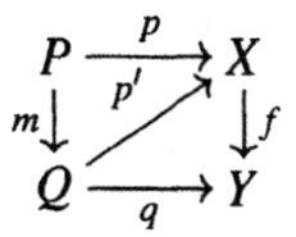

Because p' preserves independence, we see that e and e' are independent in T. So because f is open it satisfies the 'zig-zag' condition and reflects consecutive independence.

The proof in the other direction is omitted; we refer to [**35**] for a similar proof involving asynchronous transitions systems [**4**]. □

We now turn to the question of bisimulation. As shown in [**35**], taking *pomsets* as the path category P yields in the case of *event structures* a reasonable strengthening of a previously studied equivalence: the *history-preserving bisimulation* [**71, 24**]. Its definition below depends on the simple but important remark that a *configuration* of an event structure can be regarded as a *pomset*, with causal dependency relation and labelling got by restricting those of the event structure.

DEFINITION. A *history-preserving bisimulation* between event structures E_1 and E_2 consists of a set $\mathcal{H}$ of triples (x_1, f, x_2), where x_i is a configuration of E_i and $f\colon x_1 \cong x_2$ is an isomorphism of pomsets, such that $(\varnothing,\varnothing,\varnothing) \in \mathcal{H}$ and, whenever $(x_1, f, x_2) \in \mathcal{H}$,

i) if $x_1 \xrightarrow{a} x_1'$ in E_1, then $\exists\, x_2 \xrightarrow{a} x_2'$ in E_2 and $(x_1', f', x_2') \in \mathcal{H}$ with $f \subseteq f'$;

ii) if $x_2 \xrightarrow{a} x_2'$ in E_2 then $\exists\, x_1 \xrightarrow{a} x_1'$ in E_1 and $(x_1', f', x_2') \in \mathcal{H}$ with $f \subseteq f'$.

We say a history-preserving bisimulation $\mathcal{H}$ is *strong* if whenever $(x, f, y) \in \mathcal{H}$

I) if $x' \subseteq x$, for x' a configuration of E_1, then $(x', f', y') \in \mathcal{H}$, for $f' \subseteq f$ and $y' \subseteq y$;

II) if $y' \subseteq y$, for y' a configuration of E_2, then $(x', f', y') \in \mathcal{H}$, for $f' \subseteq f$ and $x' \subseteq x$.

THEOREM 4.8. *Let E_1, E_2 be event structures with labelling sets L. The following are equivalent:*

i) E_1 and E_2 are Pom_L*-bisimilar in* E_L.
ii) E_1 and E_2 are strong history-preserving bisimilar.

PROOF. See [**63**]. □

Via the coreflection between event structures and Petri nets, from Corollary 4.6 we can draw characterisations of Pom_L-bisimilarity on nets.

THEOREM 4.9. *Let N_1 and N_2 be nets with labelling sets L. The following are equivalent.*

i) The nets N_1 and N_2 are Pom_L*-bisimilar in* SWNets_L.
ii) The unfoldings to event structures $\mathcal{EU}(N_1)$ and $\mathcal{EU}(N_2)$ are strong history-preserving bisimilar.

So, for general reasons, the notion of bisimilarity for nets agrees with the notion of bisimilarity for the associated case graphs and unfoldings (where it amounts to strong history-preserving bisimilarity). Results expressing agreements of this kind would probably be required of any notion of bisimilarity, and, without the help of some categorical machinery, would seem to require separate proofs. Of course, having characterised Pom_L-bisimilarity on nets as strong history-preserving bisimilarity of their unfoldings to event structures, it is possible to produce a characterisation in terms of nets and their 'processes' along the lines of [**89**].

Many attempts have been made to define bisimilarity for noninterleaving models like Petri nets, and the idea of parameterising the definition on a notion of observation has been used in other attempts, e.g., [**16**]. One of the advantages of Pom_L-bisimilarity is, as shown, e.g., by Theorem 4.9, its *robustness* across a range of models. Another issue is the the *sensitivity* of Pom_L-bisimilarity for nets to the particular choice of path category Pom_L, as the notion of Pom_L-bisimilarity might in fact seem questionable to those who view general pomsets as not observable. To answer such a question, let us define a pomset to be an *almost totally ordered multiset* if it is of one of the two simple forms considered in the proof of Proposition 4.7, i.e., allowing at *most* two (maximal) elements to be unordered. Note that in the range of subclasses of *pomsets* considered in the literature (see [**69**]), this one is as close to Bran_L as one can get! The following result shows that restricting ourselves to such pomsets does not change the notion of P-bisimilarity.

COROLLARY 4.10. *Let* Atom_L *denote the full subcategory of* Pom_L *consisting of the almost totally ordered multisets.*

i) A morphism in SWNets_L *is* Pom_L*-open if and only if it is* Atom_L*-open.*
ii) Two nets are Pom_L*-bisimilar iff they are* Atom_L*-bisimilar.*

PROOF. Clearly (ii) follows from (i), so we concentrate on a proof of (i).

The *'only if'* part of (i) follows immediately from definition of open maps. By inspecting the proof of Proposition 4.7, we observe that a morphism in SWNets_L is Pom_L-open if it is Atom_L-open. □

Here, we have illustrated how to introduce bisimilarity for SW nets, using open maps as in [**35**]. Much of the theory developed since then on this approach to concurrency may be transferred to the setting of nets. As examples, we mention the logical and game theoretic characterisations from [**56**], and the treatment of higher order models in [**96**]. But many questions are still left open.

Part 2. ON THE STRUCTURE OF NETS

5. Petri nets as monoids

A very prominent role in the semantic theory of Petri nets is played by various notions of process, as, e.g. [**68, 26, 7, 45, 18**]. This, as we already pointed out in Section 1.2, is because processes are structures capable of accounting, not only for the mere occurrence of events in a computation, but also for the *causal relationships* which ruled such occurrences. In other terms, processes are *noninterleaving* structures and, as such, very suited to describe Petri nets.

A parallel and extremely successful line of research in concurrency, rooted in the very ideas of denotational semantics, is the one following the *algebraic* approach. Here the focus is on *structural* and *compositional* aspects of systems and behaviours, and the leading idea is to describing them by means of a few basic building blocks and a small number of *combinators* [**30, 49, 29, 50**]. The appeal of this approach is that it tends to devise neat algebraic structures that capture the *essential* nature of the class of systems considered.

In this section, we shall focus on a line of research — detailed, e.g., in [**45, 18, 47, 77, 78, 75, 76**] — aimed at recasting Petri *net processes* in lieu of ideas from *process algebras* and *categorical algebra*. In particular, we shall focus on Petri net *concatenable processes*, introduced in [**18**] to account, as their name indicates, for the issue of process concatenation. We start by briefly reconsidering the ideas that lead to their definition. The exposition will follow [**77**] closely.

5.1. Concatenable processes. Ideally, Petri net processes are simply computations in which explicit information about cause/effect relationship between event occurrences is added. More precisely, as we describe causality by means of partial orderings, the processes of a net N are ordered sets whose elements are labelled by transitions of N. In order to describe exactly which multisets of transitions form 'legal' processes, it is very convenient to define a process of N to be a map $\pi: \Theta \to N$ which maps transitions to transitions and places to places respecting the 'bipartite graph structure' of nets. Here Θ is a *finite deterministic occurrence net*, i.e., roughly speaking, a finite, conflict-free, 1-safe, acyclic net. The role of π is to 'label' the places and the (partially ordered) transitions of Θ with places and transitions of N in a way compatible with the structure of N.

Given this definition, one can assign the correct *source* and *target* states to a process $\pi: \Theta \to N$ by considering the multisets of places of N which are the image via π of the places of Θ with, respectively, empty pre-set and empty post-set (henceforth referred to as *minimal* and *maximal* places of Θ). Now, the simple minded attempt to concatenate a process $\pi_1 : \Theta_1 \to N$ with source u to a process $\pi_0 : \Theta_0 \to N$ with target u by *gluing* the maximal places of Θ_0 with the minimal places of Θ_1 in a way which preserves the labellings fails immediately. In fact, if more than one place of u is labelled by a single place of N, there are many ways to put in one-to-one correspondence the maximal places of Θ_0 and the minimal places of Θ_1 respecting the labels, i.e., there are many possible concatenations of π_0 and π_1, each of which gives a possibly different process of N. In other words, as the above argument shows, process concatenation has to do with *gluing tokens*, i.e., instances of places, rather than *gluing places*.

Therefore, to deal with process concatenation one must disambiguate the *identity* of each token in a process. This is exactly the idea of *concatenable processes*, which are simply Goltz-Reisig [**26**] processes in which the minimal and maximal places carrying the same label are linearly ordered. This yields immediately an operation of concatenation, since the ambiguity about the identity of tokens is resolved using the additional information given by the orderings. Moreover, the existence of concatenation leads easily to the definition of the category of concatenable processes of N.

It turns out that such a category is a *symmetric monoidal category* whose tensor product is the parallel composition of processes [**18**]. We shall now recall this result, whose relevance is that it describes net behaviours as *algebras* in a remarkably abstract and smooth way, showing how to describe the concatenable processes of N as a symmetric monoidal category $\mathcal{P}(N)$ defined axiomatically by means of *universal* constructions. Namely, $\mathcal{P}(N)$ is the *free symmetric strict monoidal* category on the net N modulo two simple additional axioms [**77**].

5.2. Monoidal categories and concatenable processes. The notion of *monoidal category* dates back to [**3**] (see [**42**] for an easy thorough introduction and [**21**] for advanced topics). In this paper we shall be concerned only with a particular kind of symmetric monoidal categories, namely those which are *strict monoidal* and whose objects form a *free commutative monoid.*

A *symmetric strict monoidal category* (ssmc for short) is a structure $(\mathsf{C}, \otimes, e, \gamma)$, where C is a category, e is an object of C, called the *unit* object, $\otimes : \mathsf{C} \times \mathsf{C} \to \mathsf{C}$ is a functor, called the *tensor product*, subject to the following equations

$$\otimes \circ \langle \otimes \times 1_{\mathsf{C}} \rangle = \otimes \circ \langle 1_{\mathsf{C}} \times \otimes \rangle, \tag{1}$$

$$\otimes \circ \langle \underline{e}, 1_{\mathsf{C}} \rangle = 1_{\mathsf{C}}, \tag{2}$$

$$\otimes \circ \langle 1_{\mathsf{C}}, \underline{e} \rangle = 1_{\mathsf{C}}, \tag{3}$$

where $\underline{e} : \mathsf{C} \to \mathsf{C}$ is the constant functor which associate e and id_e respectively to each object and each morphism of C, $\langle \_, \_ \rangle$ is the pairing of functors induced by the cartesian product, and $\gamma : x_1 \otimes x_2 \xrightarrow{\sim} x_2 \otimes x_1$ is a natural isomorphism, called the *symmetry* of C, subject to the following Kelly-MacLane *coherence* axioms [**41, 39**]:

$$(\gamma_{x,z} \otimes id_y) \circ (id_x \otimes \gamma_{y,z}) = \gamma_{x \otimes y, z}, \tag{4}$$

$$\gamma_{y,x} \circ \gamma_{x,y} = id_{x \otimes y}. \tag{5}$$

Clearly, equation (1) states that the tensor is associative on both objects and arrows, while (2) and (3) state that e and id_e are, respectively, the unit object and the unit arrow

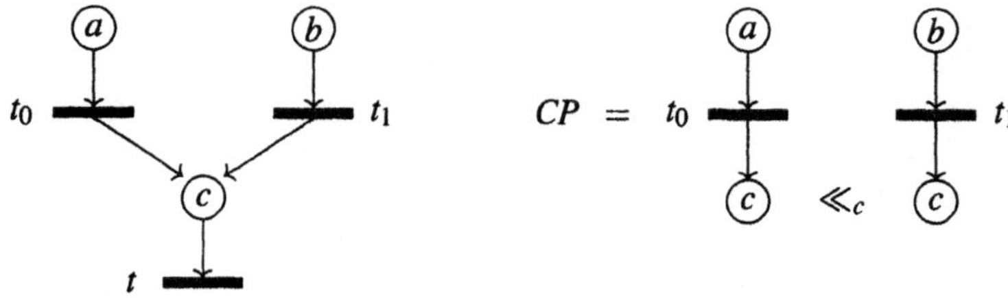

FIGURE 9. A net and one of its two concatenable processes $CP: a+b \to 2c$

for $\otimes$. Concerning the coherence axioms, axiom (5) says that $\gamma_{y,x}$ is the inverse of $\gamma_{x,y}$, while (4), the *real key* of symmetric monoidal categories, links the symmetry at composed objects to the symmetry at the components.

As an example easy enough to ponder the above equations, one might think of C as the category of sets, with $\otimes$ being the cartesian product, e the singleton set $\{\bullet\}$, and $\gamma_{X,Y}: X \times Y \to Y \times X$ as the map $(x,y) \mapsto (y,x)$.

A *symmetry* s in a symmetric monoidal category C is any arrow obtained as composition and tensor of *identities* and *components* of γ. We use Sym_{C} to denote the subcategory of the symmetries of C.

A *symmetric strict monoidal functor* from $(\mathsf{C},\otimes,e,\gamma)$ to $(\mathsf{D},\otimes',e',\gamma')$, is a functor $\mathsf{F}: \mathsf{C} \to \mathsf{D}$ which preserves the monoidal structure, i.e., such that

$$\mathsf{F}(e) = e', \tag{6}$$

$$\mathsf{F}(x \otimes y) = \mathsf{F}(x) \otimes' \mathsf{F}(y), \tag{7}$$

$$\mathsf{F}(\gamma_{x,y}) = \gamma'_{\mathsf{F}x,\mathsf{F}y}. \tag{8}$$

Let SSMC be the category of ssmc's and symmetric strict monoidal functors and let $\mathsf{SSMC}^{\oplus}$ be the full subcategory consisting of the monoidal categories whose objects form *free* commutative monoids.

In this section, we consider only PT nets with *finite markings*, but release all remaining restrictions. Let $S^{\oplus}$ denote the submonoid of $\mu(S)$ consisting of the *finite* multisets of S, i.e., the functions $S \to \omega$ which yield nonzero values at most on finitely many arguments. Then, define Petri to be the subcategory of PTNets consisting of those N whose transitions have source and target in $S_N^{\oplus}$, and of those $f: N_0 \to N_1$ whose place components map $S_{N_0}^{\oplus}$ to $S_{N_1}^{\oplus}$.

Recall that $S^{\oplus}$ can be characterised as the *free commutative monoid* on S. In particular this means that the place component f_p of a morphism $f: N_0 \to N_1$ in Petri is determined by assigning a multiset $f_p(a) \in S_{N_1}^{\oplus}$ for each place $a \in S_{N_0}$, and then freely extending it to the entire $S_{N_0}^{\oplus}$.

DEFINITION. A *process net* is a finite, acyclic net Θ such that for all $t \in T_\Theta$, $pre_\Theta(t)$ and $post_\Theta(t)$ are sets (as opposed to multisets), and for all $t_0 \neq t_1 \in T_\Theta$,

$$pre_\Theta(t_0) \cap pre_\Theta(t_1) = \varnothing \qquad \text{and} \qquad post_\Theta(t_0) \cap post_\Theta(t_1) = \varnothing.$$

Given $N \in$ Petri, a *process* of N is a morphism $\pi: \Theta \to N$, where Θ is a process net and π is a net morphism which maps places to places (as opposed to morphisms which map places to markings).

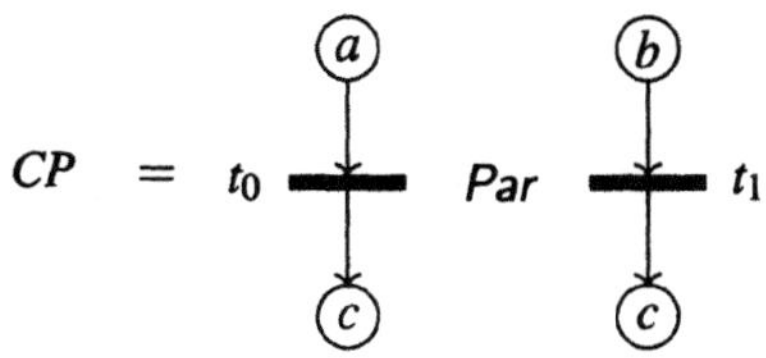

FIGURE 10. *CP* of Figure 9 as the parallel composition of two simpler processes

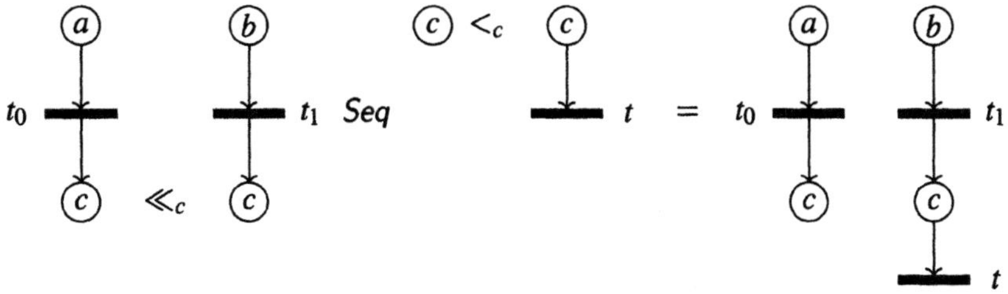

FIGURE 11. Sequential composition (concatenation) of concatenable processes

DEFINITION. A *concatenable process* of N is a triple

$$(\pi\colon \Theta \to N, \{<_a\}_{a \in S_N}, \{\ll_a\}_{a \in S_N}),$$

where π is a process, and $<_a$ and $\ll_a$ are linear orderings of, respectively, the set of minimal and the set of maximal places of Θ contained in $\pi_p^{-1}(a)$ (cf. Figure 9).

In order to abstract from the details concerning the underlying process nets, concatenable processes are considered up to isomorphisms. Formally, two concatenable processes, say with underlying processes $\pi_0 : \Theta_0 \to N$ and $\pi_1 : \Theta_1 \to N$, are identified if there exists an isomorphism $\varphi\colon \Theta_0 \to \Theta_1$ which preserves all the orderings and such that $\pi_1 \circ \varphi = \pi_0$.

Concatenable processes allow the operations of *sequential* and *parallel* composition (see Figures 10 and 11, and consult **[18]** for further examples). Let CP_0 and CP_1 be concatenable processes of N, and let $\pi_0\colon \Theta_0 \to N$ and $\pi_1 : \Theta_1 \to N$ denote their underlying processes.

DEFINITION. The *parallel composition* CP_0 *Par* CP_1 is the concatenable process of N whose underlying process is the disjoint union of π_0 and π_1, i.e., $\pi_0 + \pi_1 : \Theta_0 + \Theta_1 \to N$, where $+$ denotes the coproduct in Petri, and whose orderings extend those of CP_0 and CP_1 by making all the places of Θ_0 precede all the places of Θ_1.

DEFINITION. The *sequential composition*, or concatenation, $CP = CP_0$ *Seq* CP_1 is defined if and only if the state reached by CP_0 coincide with the source state of CP_1. In this case, CP is obtained by gluing together π_0 and π_1, identifying injectively each maximal place of Θ_0 with a minimal place of Θ_1 in the *unique* way compatible with the orderings $\ll_a$ on Θ_0 and $<_a$ on Θ_1 for all $a \in S_N$.

Next, we recall the construction of the symmetric strict monoidal category $\mathcal{P}(N)$. We start by introducing the *vectors of permutations* (*vperms*) of N, which will provide the symmetry isomorphism of $\mathcal{P}(N)$.

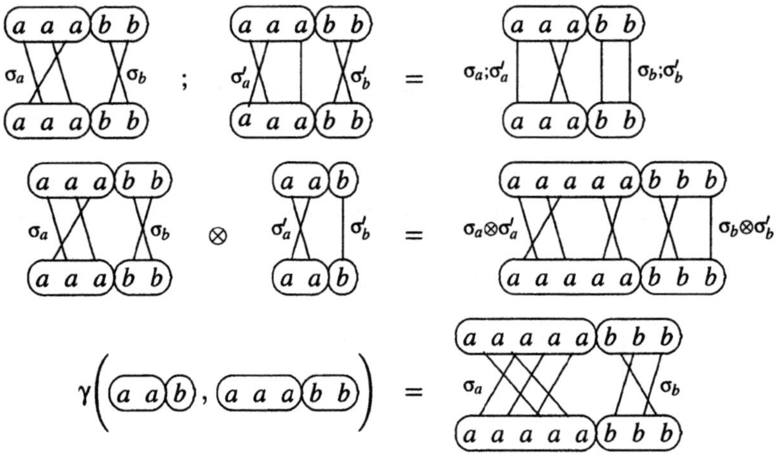

FIGURE 12. The monoidal structure of vperms

For $u \in S^{\oplus}$, a *vperm* $s\colon u \to u$ is a function which assigns to each $a \in S$ a permutation $s(a) \in \Pi(u(a))$. Given $u = n_1 \cdot a_1 + \ldots + n_k \cdot a_k$ in $S_N^{\oplus}$, we shall represent a vperm s on u as a vector of permutations, $\langle \sigma_{a_1}, \ldots, \sigma_{a_k} \rangle$, where $s(a_j) = \sigma_{a_j}$, whence their name. One can define the operations of sequential and parallel composition of vperms, so that they can be organised as the arrows of a ssmc. The details follow (see also Figure 12).

Given the vperms $s = \langle \sigma_{a_1}, \ldots, \sigma_{a_k} \rangle\colon u \to u$ and $s' = \langle \sigma'_{a_1}, \ldots, \sigma'_{a_k} \rangle\colon u \to u$ their *sequential composition* $s;s'\colon u \to u$ is the vperm

$$\langle \sigma_{a_1};\sigma'_{a_1}, \ldots, \sigma_{a_k};\sigma'_{a_k} \rangle,$$

where $\sigma;\sigma'$ is the composition of permutation which we write in the diagrammatic order from left to right.

Given the vperms $s = \langle \sigma_{a_1}, \ldots, \sigma_{a_k} \rangle\colon u \to u$ and $s' = \langle \sigma'_{a_1}, \ldots, \sigma'_{a_k} \rangle\colon v \to v$ (where possibly $\sigma_{a_j} = \varnothing$ for some j), their *parallel composition* $s \otimes s'\colon u+v \to u+v$ is the vperm

$$\langle \sigma_{a_1} \otimes \sigma'_{a_1}, \ldots, \sigma_{a_k} \otimes \sigma'_{a_k} \rangle,$$

where

$$(\sigma \otimes \sigma')(x) = \begin{cases} \sigma(x), & \text{if } 0 < x \leq |\sigma| \\ \sigma'(x - |\sigma|) + |\sigma|, & \text{if } |\sigma| < x \leq |\sigma| + |\sigma'| \end{cases}$$

Let γ be $(1\ 2) \in \Pi(2)$ and consider $u_i = n_1^i \cdot a_1 + \ldots + n_k^i \cdot a_k$, $i = 1,2$, in $S^{\oplus}$. The *interchange vperm* $\gamma(u_1, u_2)$ is the vperm $\langle \sigma_{a_1}, \ldots, \sigma_{a_k} \rangle\colon u_1 + u_2 \to u_1 + u_2$ where

$$\sigma_{a_j}(x) = \begin{cases} x + n_j^2, & \text{if } 0 < x \leq n_j^1 \\ x - n_j^1, & \text{if } n_j^1 < x \leq n_j^1 + n_j^2 \end{cases}$$

It is immediate to verify that $\_;\_$ is associative. Moreover, for each $u \in S^{\oplus}$, the vperm $u = \langle id_{a_1}, \ldots, id_{a_n} \rangle\colon u \to u$, where id_{a_j} is the identity permutation, is an identity for sequential composition. Finally, writing 0 for the empty multiset on S, the (unique) vperm $s\colon 0 \to 0$, is a unit for parallel composition.

Now, for N a net, let Sym_N be the category whose objects are the elements of $S_N^{\oplus}$ and whose arrows are the vperms $s\colon u \to u$ for $u \in S_N^{\oplus}$. It is easy to show that Sym_N is a ssmc with respect to the given composition and tensor product, with identities and unit element as explained above, and with the symmetry natural isomorphism given by the collection $\gamma = \{\gamma(u,v)\}_{u,v \in Sym_N}$ of the interchange vperms. Observe that, although Sym_N is not strictly symmetric, it is so on the objects. More strongly, the objects form a free commutative monoid, i.e., $Sym_N \in \mathsf{SSMC}^{\oplus}$.

We can now define $\mathcal{P}(N)$ as the category which includes Sym_N as a subcategory and has as additional arrows those defined by the following rules:

$$\frac{t\colon u \to v \text{ in } T_N}{t\colon u \to v \text{ in } \mathcal{P}(N)}$$

$$\frac{\alpha\colon u \to v \text{ and } \beta\colon u' \to v' \text{ in } \mathcal{P}(N)}{\alpha \otimes \beta\colon u + u' \to v + v' \text{ in } \mathcal{P}(N)} \qquad \frac{\alpha\colon u \to v \text{ and } \beta\colon v \to w \text{ in } \mathcal{P}(N)}{\alpha;\beta\colon u \to w \text{ in } \mathcal{P}(N)}$$

plus axioms expressing the fact that $\mathcal{P}(N)$ is a ssmc with composition $\_;\_$, tensor $\_\otimes\_$ (extending those already defined on vperms) and symmetry isomorphism γ, and the following axioms involving transitions and vperms

$$\begin{array}{ll} t;s = t & \text{where } t\colon u \to v \text{ in } T_N \text{ and } s\colon v \to v \text{ in } Sym_N, \\ s;t = t & \text{where } t\colon u \to v \text{ in } T_N \text{ and } s\colon u \to u \text{ in } Sym_N. \end{array} \qquad (\Psi)$$

In other words, $\mathcal{P}(N)$ is built on the category Sym_N by adding the transitions of N and freely closing with respect to sequential and parallel composition of arrows, so that $\mathcal{P}(N)$ is made symmetric strict monoidal and axioms (Ψ) hold.

The relevant fact about $\mathcal{P}(N)$ is that its arrows represent exactly the concatenable processes of N, i.e., $\mathcal{P}(N)$ represents the noninterleaving behaviour of N, including its algebraic structure. (See **[18]** for proof and details.)

Theorem 5.1. *For any net N there exists a one-to-one correspondence between the arrows of $\mathcal{P}(N)$ and the concatenable processes of N such that, for each $u, v \in S_N^{\oplus}$, the arrows of type $u \to v$ correspond to the processes enabled by u and producing v, and such that sequential and parallel composition (tensor product) of processes (arrows) are respected.*

Vperms play in this correspondence an absolutely fundamental role: Sym_N accounts for the families of orderings $\{<_a\}_{a \in S_N}$ and $\{\ll_a\}_{a \in S_N}$, which are the key to concatenable processes, guaranteeing a correct treatment of sequential composition. In other words, Sym_N is an algebraic representation of the 'threads of causality' in process concatenation.

5.3. Axiomatising concatenable processes. Unfortunately, the concrete definition of vperms weakens considerably the essentially axiomatic character of $\mathcal{P}(N)$, as the laws which rule it remain partly concealed in Sym_N. The aim of this section is to provide a fully axiomatic description of the concatenable processes of N obtained by proving that $\mathcal{P}(N)$ is a quotient of the free ssmc on N **[77]**. The key to this result will be an axiomatisation of the category of vperms Sym_N. We start by showing that we can associate a free ssmc to each net N.

PROPOSITION 5.2. *The forgetful functor* $\mathcal{U}\colon \mathsf{SSMC}^{\oplus} \to \mathsf{Petri}$ *admits a left adjoint* $\mathcal{F}\colon \mathsf{Petri} \to \mathsf{SSMC}^{\oplus}$.

PROOF. Consider the category $\mathcal{F}(N)$ whose objects are the elements of $S_N^{\oplus}$ and whose arrows are generated by the inference rules

$$\frac{u \in S_N^{\oplus}}{id_u\colon u \to u \text{ in } \mathcal{F}(N)} \qquad \frac{a \text{ and } b \text{ in } S_N}{c_{a,b}\colon a+b \to b+a \text{ in } \mathcal{F}(N)} \qquad \frac{t\colon u \to v \text{ in } T_N}{t\colon u \to v \text{ in } \mathcal{F}(N)}$$

$$\frac{\alpha\colon u \to v \text{ and } \beta\colon u' \to v' \text{ in } \mathcal{F}(N)}{\alpha \otimes \beta\colon u+u' \to v+v' \text{ in } \mathcal{F}(N)} \qquad \frac{\alpha\colon u \to v \text{ and } \beta\colon v \to w \text{ in } \mathcal{F}(N)}{\alpha;\beta\colon u \to w \text{ in } \mathcal{F}(N)}$$

modulo the axioms expressing that $\mathcal{F}(N)$ is a strict monoidal category, namely,

$$\begin{array}{rcl} \alpha; id_v = \alpha = id_u;\alpha, & \text{and} & (\alpha;\beta);\gamma = \alpha;(\beta;\gamma), \\ (\alpha\otimes\beta)\otimes\gamma = \alpha\otimes(\beta\otimes\gamma), & \text{and} & id_0\otimes\alpha = \alpha = \alpha\otimes id_0, \\ id_u\otimes id_v = id_{u+v}, & \text{and} & (\alpha\otimes\alpha');(\beta\otimes\beta') = (\alpha;\beta)\otimes(\alpha';\beta'), \end{array} \tag{9}$$

the latter whenever the right-hand term is defined, and the following axioms

$$c_{a,b};c_{b,a} = id_{a+b}, \tag{10}$$

$$c_{u,u'};(\beta\otimes\alpha) = (\alpha\otimes\beta);c_{v,v'}, \quad \text{for } \alpha\colon u \to v,\ \beta\colon u' \to v', \tag{11}$$

where $c_{u,v}$ for $u,v \in S_N^{\oplus}$ denote *any* term obtained from $c_{a,b}$ for $a,b \in S_N$ by applying recursively the following rules (compare with axiom (4)):

$$\begin{array}{rcl} c_{0,u} & = & c_{0,u} \;=\; id_u, \\ c_{a+u,v} & = & (id_a\otimes c_{u,v});(c_{a,v}\otimes id_u), \\ c_{u,v+a} & = & (c_{u,v}\otimes id_a);(id_v\otimes c_{u,a}). \end{array} \tag{12}$$

Observe that equation (11), in particular, equalises all the terms obtained from (12) for fixed u and v. In fact, let $c_{u,v}$ and $c'_{u,v}$ be two such terms and take α and β to be, respectively, the identities of u and v. Now, since $id_u\otimes id_v = id_{u\oplus v} = id_v\otimes id_u$, from (11) we have that $c_{u,v} = c'_{u,v}$ in $\mathcal{F}(N)$. Then, we claim that the collection $\{c_{u,v}\}_{u,v\in S_N^{\oplus}}$ is a symmetry natural isomorphism which makes $\mathcal{F}(N)$ into a ssmc and that, in addition, $\mathcal{F}(N)$ is the free ssmc on N.

In order to show the first claim, observe that the naturality of c is expressed directly from axiom (11). We need to check that for any u and v we have $c_{u,v};c_{v,u} = id_{u\oplus v}$, which follows from (10) by induction on the sum of the sizes of u and v.

As for the second, for C in $\mathsf{SSMC}^{\oplus}$, the net $\mathcal{U}(\mathsf{C})$ is obtained by forgetting the categorical structure of C. The markings and the transitions of $\mathcal{U}(\mathsf{C})$ are, respectively, the objects and the arrows of C with the given sources and targets. Similarly, for F a symmetric strict monoidal functor in $\mathsf{SSMC}^{\oplus}$, $\mathcal{U}(\mathsf{F})$ is the net morphism whose components are the restrictions of F to, respectively, arrows and objects. Consider the net $\mathcal{UF}(N)$ and the net morphism $\eta\colon N \to \mathcal{UF}(N)$, where η_p is the identity homomorphism and η_t is the obvious injection of T_N in $T_{\mathcal{UF}(N)}$. One can then show (cf. [**77**] for the details) that η is universal, i.e., that for any C in $\mathsf{SSMC}^{\oplus}$ and for any net morphism $f\colon N \to \mathcal{U}(\mathsf{C})$, there is a unique symmetric strict monoidal functor $\mathsf{F}\colon \mathcal{F}(N) \to \mathsf{C}$ which

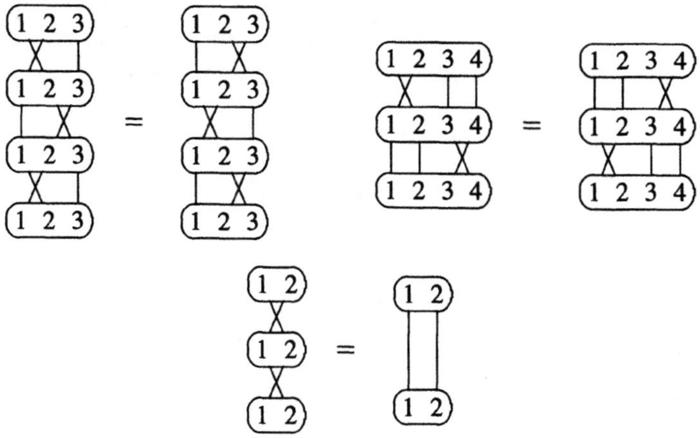

FIGURE 13. Some instances of the axioms of permutations

makes the following diagram commute.

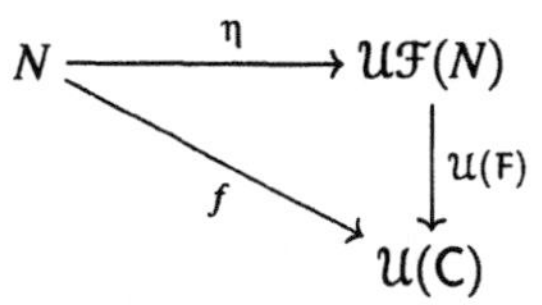

□

Thus, establishing the adjunction $\mathcal{F} \dashv \mathcal{U}$: Petri $\rightharpoonup$ SSMC$^{\oplus}$, we have identified $\mathcal{F}(N)$, the *free ssmc on N*, as a category generated, modulo appropriate equations, from the net N viewed as a graph enriched with formal arrows id_u, which play the role of the identities, and $c_{a,b}$ for $a, b \in S_N$, which generate all the needed symmetries.

Our aim is to relate $\mathcal{F}(N)$ and $\mathcal{P}(N)$. As a matter of fact, $\mathcal{F}(N)$ is positively more concrete than $\mathcal{P}(N)$ and far from being isomorphic (or equivalent) to it. For example, for $a \neq b$ in S_N, we have $c_{a,b} \neq id_{a \oplus b}$ in $\mathcal{F}(N)$, whilst $\gamma(a,b) = id_{a \oplus b}$ in $\mathcal{P}(N)$. Therefore, no symmetric monoidal functor $\mathsf{Q}: \mathcal{F}(N) \to \mathcal{P}(N)$ can be mono. Also, $\mathcal{F}(N)$ possesses no counterpart of axioms (Ψ). We shall see that these are precisely the differences between $\mathcal{F}(N)$ and $\mathcal{P}(N)$. Namely, we shall obtain $\mathcal{P}(N)$ as a quotient of $\mathcal{F}(N)$ by enforcing the axioms outlined above. The next proposition, which is the adaptation to ssmc's of the usual notion of quotient algebras, provides the tool we shall use for this purpose.

PROPOSITION 5.3. *For C a ssmc, let $\mathcal{R}$ be a function which assigns to each pair of objects a and b of C a binary relation $\mathcal{R}_{a,b}$ on the homset $\mathsf{C}(a,b)$. Then, there exist a ssmc $\mathsf{C}/\mathcal{R}$ and a symmetric strict monoidal functor $[\_]_{\mathcal{R}}: \mathsf{C} \to \mathsf{C}/\mathcal{R}$ such that*

i) If $f \, \mathcal{R}_{a,b} \, f'$ then $[f]_{\mathcal{R}} = [f']_{\mathcal{R}}$;

ii) For each symmetric strict monoidal $\mathsf{H}\colon \mathsf{C} \to \mathsf{D}$ *such that* $\mathsf{H}(f) = \mathsf{H}(f')$ *whenever* $f\mathcal{R}_{a,b}f'$, *there exists a unique* $\mathsf{K}\colon \mathsf{C}/\mathcal{R} \to \mathsf{D}$, *which is necessarily symmetric strict monoidal, such that the following diagram commutes.*

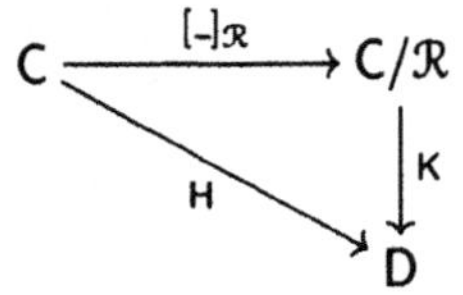

PROOF. Take $\mathsf{C}/\mathcal{R}$ to be the category whose objects are those of C, whose homset $\mathsf{C}/\mathcal{R}(a,b)$ is $\mathsf{C}(a,b)/\mathcal{R}_{a,b}$, with composition of arrows given by $[g]_{\mathcal{R}} \circ [f]_{\mathcal{R}} = [g \circ f]_{\mathcal{R}}$, and define $[f]_{\mathcal{R}} \otimes [g]_{\mathcal{R}} = [f \otimes g]_{\mathcal{R}}$.

Say that $\mathcal{R}$ is a *congruence* if $\mathcal{R}_{a,b}$ is an equivalence for each a and b and if $\mathcal{R}$ respects composition, i.e., whenever $f\mathcal{R}_{a,b}f'$ then, for all $h\colon a' \to a$ and $k\colon b \to b'$, we have $(k \circ f \circ h)\mathcal{R}_{a',b'}(k \circ f' \circ h)$. Call $\mathcal{R}$ a $\otimes$*-congruence* if it is a congruence and it similarly respects tensor. It is easy to check that, if $\mathcal{R}$ is a $\otimes$-congruence, then the above definition makes the quotient category $\mathsf{C}/\mathcal{R}$ into a ssmc with symmetry isomorphisms $[\gamma_{u,v}]_{\mathcal{R}}$ and unit object e.

Observe now that, given $\mathcal{R}$ as in the hypothesis, it is always possible to find the least $\otimes$-congruence $\mathcal{R}'$ which includes (componentwise) $\mathcal{R}$. Then, take $\mathsf{C}/\mathcal{R}$ to be $\mathsf{C}/\mathcal{R}'$ and $[\_]_{\mathcal{R}}$ to be the obvious projection of C into $\mathsf{C}/\mathcal{R}$, which is clearly a symmetric strict monoidal functor. □

In order to show that $\mathcal{P}(N)$ is a monoidal quotient of $\mathcal{F}(N)$, we need a more abstract understanding of the structure of the vperms of $\mathcal{P}(N)$. To this aim, we shall make use of the following lemma, originally proved in [**52**].

LEMMA 5.4. *The symmetric group* $\Pi(n)$ *is (isomorphic to) the group G freely generated from the set* $\{\tau_i \mid 1 \le i < n\}$, *modulo the equations (see also Figure 13)*

$$\begin{aligned} \tau_i\tau_{i+1}\tau_i &= \tau_{i+1}\tau_i\tau_{i+1}, \\ \tau_i\tau_j &= \tau_j\tau_i \quad \textit{if}\ |i-j| \ge 1, \\ \tau_i\tau_i &= e, \end{aligned} \tag{13}$$

where e is the unit element of G.

The previous lemma is easily adapted to vperms by translating axioms (13) as follows.

LEMMA 5.5. *The arrows of* Sym_N *are freely generated by composition and tensor from the vperms* $\gamma(a,a)\colon 2\cdot a \to 2\cdot a$, *for* $a \in S_N$, *modulo the axioms (9) of strict monoidal categories and the following additional axioms*

$$\begin{aligned} ((id_a \otimes \gamma(a,a));(\gamma(a,a) \otimes id_a))^3 &= id_{3\cdot a}, \\ \gamma(a,a)^2 &= id_{2\cdot a}, \\ (id_b \otimes \gamma(a,a));(\gamma(a,a) \otimes id_b) &= id_{2\cdot a+b}, \quad \textit{if}\ a \neq b \in S_N, \end{aligned} \tag{14}$$

where f^n *indicates the composition of f with itself n times.*

PROOF. See [**77**]. □

We are now ready to give the promised characterisation of $\mathcal{P}(N)$.

PROPOSITION 5.6. *$\mathcal{P}(N)$ is the monoidal quotient of the free ssmc on N modulo the axioms*

$$(15)\qquad c_{a,b} = id_{a+b}, \quad \textit{if } a,b \in S_N \textit{ and } a \neq b,$$
$$(16)\qquad s;t;s' = t, \quad \textit{if } t \in T_N \textit{ and } s,s' \textit{ are symmetries.}$$

PROOF. We prove that $\mathcal{P}(N)$ is isomorphic to $\mathcal{F}(N)/\mathcal{R}$, where $\mathcal{R}$ is the congruence for $\otimes$ and $\_;\_$ generated from equations (15) and (16).

Since $\mathcal{P}(N)$ belongs to $\mathsf{SSMC}^{\oplus}$, it follows from Proposition 5.2 that, corresponding to the net *inclusion* morphism $N \to \mathcal{U}\mathcal{P}(N)$, there is a *unique* symmetric strict monoidal functor $\mathsf{Q}\colon \mathcal{F}(N) \to \mathcal{P}(N)$ which is the identity on the places and on the transitions of N. In particular, Q is such that

$$\mathsf{Q}(c_{a,b}) = \gamma(a,b), \quad \text{for } a,b \in S_N.$$

For $a \neq b \in S_N$, since $\gamma(a,b) = id_{a\oplus b}$, we have that $\mathsf{Q}(c_{a,b}) = \mathsf{Q}(id_{a\oplus b})$. Moreover, since symmetric monoidal functors map symmetries to symmetries, and since (16) holds in $\mathcal{P}(N)$, we have that $\mathsf{Q}(s;t;s') = \mathsf{Q}(s);t;\mathsf{Q}(s') = t = \mathsf{Q}(t)$ for s and s' in $Sym_{\mathcal{F}(N)}$ and $t \in T_N$. In other words, Q equalises the pairs $\langle c_{a,b}, id_{a\oplus b}\rangle$ with $a \neq b \in S_N$ and the pairs $\langle s;t;s',t\rangle$ with s and s' symmetries and $t \in T_N$. Then, in force of Proposition 5.3 applied to Q, there is a (unique) symmetric strict monoidal functor $\mathsf{H}\colon \mathcal{F}(N)/\mathcal{R} \to \mathcal{P}(N)$ which is the identity on the objects and is such that

$$\mathsf{H}([t]_{\mathcal{R}}) = t, \quad \text{for } t \in T_N.$$

One can then prove that H is an isomorphism by producing its inverse $\mathcal{P}(N) \to \mathcal{F}(N)/\mathcal{R}$ as the functor G which acts identically on the objects and is defined on the arrows by

$$\begin{aligned} \mathsf{G}(t) &= [t]_{\mathcal{R}}, && \text{if } t \in T_N, \\ \mathsf{G}(\gamma(a,a)) &= [c_{a,a}]_{\mathcal{R}}, && \text{if } a \in S_N, \end{aligned}$$

extended to identities, composition and tensor as usual: $\mathsf{G}(id_u) = [id_u]_{\mathcal{R}}$, $\mathsf{G}(\alpha;\beta) = \mathsf{G}(\alpha)\,;\mathsf{G}(\beta)$, and $\mathsf{G}(\alpha\otimes\beta) = \mathsf{G}(\alpha)\otimes\mathsf{G}(\beta)$. Notice that it follows from the definition of $\mathcal{P}(N)$ and from Lemma 5.5 that the equations above define G uniquely. □

The merit of this result is to describe the algebraic structure of $\mathcal{P}(N)$, and thus of the concatenable processes of N, in terms of *universal* constructions, namely the construction on the free ssmc on Petri and a quotient construction on $\mathsf{SSMC}^{\oplus}$, providing in this way a completely abstract view of $\mathcal{P}(N)$. It may be worth noticing in this context that (15) is actually a problematic axiom: because of its negative premise, viz., $a \neq b$, it invalidates the freeness of $\mathcal{F}(N)$ on Petri. Even worse, $\mathcal{F}(\_)/\mathcal{R}$ and $\mathcal{P}(\_)$ fail to be functors from Petri to SSMC. On the other hand, axiom (15) plays very relevant a role in capturing algebraically the essence of concatenable process, and it cannot be dispensed with easily. A detailed study of this issue and a possible solution is provided in [**78, 76**]. In particular, in *loc. cit.*, a functorial and universal construction for net computations is devised, based on a refinement of the notion of concatenable processes called *strongly concatenable processes.*

Resuming our work, we give an alternative form of axiom (16).

COROLLARY 5.7. *Axiom (16) in Proposition 5.6 can be replaced by the axioms*

$$
(17)\qquad
\begin{array}{rcll}
t;(id_u \otimes c_{a,a} \otimes id_v) & = & t, & \textit{if } t \in T_N \textit{ and } a \in S_N, \\
(id_u \otimes c_{a,a} \otimes id_v);t & = & t, & \textit{if } t \in T_N \textit{ and } a \in S_N.
\end{array}
$$

PROOF. Since $(id_u \otimes \gamma_{a,a} \otimes id_v)$ and all the identities are symmetries, axiom (16) implies the present ones. It is easy to see that, on the other hand, the axioms above, together with axiom (15), imply (16).

Let $s\colon u \to u$ by a symmetry of $\mathcal{F}(N)$ and suppose $s \neq id_u$. By repeated applications of (12), together with the functoriality of $\otimes$, we obtain the following equality:

$$s = (id_{u_1} \otimes c_{a_1,b_1} \otimes id_{v_1});\ldots;(id_{u_h} \otimes c_{a_h,b_h} \otimes id_{v_h})$$

for some $h \in \omega$. Moreover, by exploiting axiom (15), we can drop every term in which $a_i \neq b_i$. Thus we have

$$s = (id_{u_1} \otimes c_{a_1,a_1} \otimes id_{v_1});\ldots;(id_{u_k} \otimes c_{a_k,a_k} \otimes id_{v_k})$$

for some $k \leq h$. Then, by this equation and by repeated applications of axioms (17), one can prove $s;t;s' = t$. □

Finally, the next corollary sums up the purely algebraic characterisation of the category of concatenable processes that we illustrated here. In particular, it identifies in algebraic terms the essential components of concatenable processes and the laws which rule their sequential and parallel composition.

COROLLARY 5.8. *The category $\mathcal{P}(N)$ of concatenable processes of N is the category whose objects are the elements of $S_N^{\oplus}$ and whose arrows are generated by the inference rules*

$$
\frac{u \in S_N^{\oplus}}{id_u\colon u \to u \textit{ in } \mathcal{P}(N)}
\qquad
\frac{a \textit{ in } S_N}{c_{a,a}\colon a+a \to a+a \textit{ in } \mathcal{P}(N)}
\qquad
\frac{t\colon u \to v \textit{ in } T_N}{t\colon u \to v \textit{ in } \mathcal{P}(N)}
$$

$$
\frac{\alpha\colon u \to v \textit{ and } \beta\colon u' \to v' \textit{ in } \mathcal{P}(N)}{\alpha \otimes \beta\colon u+u' \to v+v' \textit{ in } \mathcal{P}(N)}
\qquad
\frac{\alpha\colon u \to v \textit{ and } \beta\colon v \to w \textit{ in } \mathcal{P}(N)}{\alpha;\beta\colon u \to w \textit{ in } \mathcal{P}(N)}
$$

modulo the axioms expressing that $\mathcal{P}(N)$ is a strict monoidal category, namely,

$$
\begin{array}{rcl}
\alpha;id_v = \alpha = id_u;\alpha, & \textit{and} & (\alpha;\beta);\gamma = \alpha;(\beta;\gamma), \\
(\alpha \otimes \beta) \otimes \gamma = \alpha \otimes (\beta \otimes \gamma), & \textit{and} & id_0 \otimes \alpha = \alpha = \alpha \otimes id_0, \\
id_u \otimes id_v = id_{u \oplus v}, & \textit{and} & (\alpha \otimes \alpha');(\beta \otimes \beta') = (\alpha;\beta) \otimes (\alpha';\beta'),
\end{array}
$$

the latter whenever the right-hand term is defined, and the following axioms

$$
\begin{array}{rcll}
c_{a,a};c_{a,a} & = & id_{a+a}, & \\
t;(id_u \otimes c_{a,a} \otimes id_v) & = & t, & \textit{if } t \in T_N, \\
(id_u \otimes c_{a,a} \otimes id_v);t & = & t, & \textit{if } t \in T_N, \\
c_{u,u'};(\beta \otimes \alpha) & = & (\alpha \otimes \beta);c_{v,v'}, & \textit{for } \alpha\colon u \to v,\ \beta\colon u' \to v',
\end{array}
$$

where $c_{u,v}$, for $u,v \in S_N^{\oplus}$, is obtained from $c_{a,a}$ by applying recursively the rules:

$$\begin{aligned} c_{a,b} &= id_{a+b}, \quad \text{if } a=0 \text{ or } b=0 \text{ or } (a,b \in S_N \text{ and } a \neq b), \\ c_{a+u,v} &= (id_a \otimes c_{u,v}); (c_{a,v} \otimes id_u), \\ c_{u,v+a} &= (c_{u,v} \otimes id_a); (id_v \otimes c_{u,a}). \end{aligned}$$

PROOF. Observe that the terms and the axioms above are obtained normalising those of $\mathcal{F}(N)$ with respect to $c_{a,b} = id_{a+b}$, for $a \neq b \in S_N$, and then adding axioms (15) and (17). The claim then follows immediately from Proposition 5.2, Proposition 5.6, and Corollary 5.7. □

6. Conclusions and Related Work

We have presented some examples of the use of category theory in understanding the behaviour and structure of Petri nets and their relationships to other models for concurrency. However, there are many further examples of applications of categorical ideas in concurrency.

The results on categorical relationships between models is a small part of a general picture, as illustrated in **[80]**, in which a number of important constructions from concurrency theory emerges as parts of (coreflective) 'unfoldings' and 'sequentialisations', and (reflective) 'determinizations'. Also, several results in the literature **[74, 79, 62]** concern trace structures and other models for concurrency — including the pomsets of Pratt **[69]** and the partial words of Grabowsky **[28]**.

We have illustrated how to introduce bisimilarity for Petri nets following a general pattern which automatically guarantees consistency with bisimilarity on a number of related models. From the work on open maps, it was suggested in **[35]** to study *presheaves* as models derived directly from path categories. Intuitively, a presheaf represents a system by 'gluing' together (an abstract set of) computation paths, and the advantage of this approach is that a number of categorical concepts provide a *uniform* notion of model of computation, as an alternative to the often ad hoc, concrete constructions adopted in the literature. Formally, given a path category P, the category $\widehat{\mathsf{P}}$ of presheaves over P consists of functors $\mathsf{P}^{op} \to \mathsf{Set}$ (where Set is the category of sets with functions) as objects, and natural transformations between them as morphisms. A presheaf $\mathsf{F}\colon \mathsf{P}^{op} \to \mathsf{Set}$ can be thought of as specifying for each path object P, the set $\mathsf{F}(P)$ of paths from P. It acts on a morphism $m\colon P \to Q$ in P to give a function $\mathsf{F}(m)\colon \mathsf{F}(Q) \to \mathsf{F}(P)$ saying how Q-paths restrict to P-paths.

Presheaves may thus be looked upon as labelled transition systems, where states are (abstract representations of) sets of possible runs of the paths in P, labels are path extensions, and the transitions describe how runs extend each other. Based on this view of presheaves, **[98]** provided logical and game-theoretic characterisations of open maps and their bisimulations on presheaves, which in turn may be specialised to concrete models like Petri nets via uniform representations as presheaves. Also, i n recent work by G. Winskel and others, these presheaf models have been used successfully in dealing with higher-order models in concurrency **[96, 15]**.

We then illustrated how to use categorical tools 'in the small', focusing our study at level of single nets. Building on **[45, 18, 77]**, we described the concatenable processes of a Petri net N in terms of *universal* constructions, providing in such a way an abstract, fully axiomatic presentation of their algebraic structure. In particular, Corollary 5.8 provides a *term algebra* and an *equational theory* of the concatenable processes of N. Technically, relying on the characterisation of the concatenable processes of N as the arrows of the symmetric strict monoidal category $\mathcal{P}(N)$, the result was illustrated by showing (cf. Proposition 5.6) that $\mathcal{P}(N)$ is the quotient of the free symmetric strict monoidal category on N modulo two simple axioms. The proof of this fact makes an essential use of the axiomatisation of Sym_N, the category of *symmetries* of $\mathcal{P}(N)$, provided by Lemma 5.5.

It is worth noticing that lifting these result to the totality of nets is rather problematic, as the negative premise of axiom (15) — essential from the computational viewpoint — breaks the freeness of $\mathcal{F}(N)$ on Petri and makes $\mathcal{P}(\_)$ fail to be functor from Petri to SSMC. The interested reader is referred to [**78**] for a detailed study of the problem, and for a suggested solution based on a refined notion of so-called *strongly* concatenable processes.

An aspect of Petri nets we did not touch in this paper is their use as a semantic basis to interpret concurrent languages (see for example [**90, 64, 25, 17**]), an application that clearly calls for a '*process algebra-like*' description of nets and, possibly, for a suitable abstract characterisation of it. And in fact, the literature is rich of examples of process algebras over nets, as, e.g., [**54, 27, 8, 51, 58**] (see also the early [**19, 13, 99, 38, 70**] on compositionality issues). Observe that category theory can clearly play an interesting role in this, as we are called to consider the totality of nets, as in Part 1, focusing this time — as in Part 2 — on algebraic and compositional aspects. We conclude this paper explaining the basic ideas underlying the algebra of nets presented in [**58**], and how it may be related to the categorical approach we have taken here.

The approach is entirely based on a notion of *interface* for Petri nets. Informally, an interface for a net N is a selection of places and transitions of N which specifies what parts of N are *public*, i.e., accessible to the environment, and what parts are *private* to N. The private places and transitions cannot be accessed and, therefore, they cannot be used for connecting N with other nets. Net interfaces are built out of two components: an '*input*' interface, consisting of places, and an '*output*' interface, consisting of transitions. Intuitively, the input interface provides the buffers in which the tokens arriving from the environment are gathered, whilst the output interface sends tokens out to the environment.

Drawing on the experience of developments in concurrency theory, one aims at defining a minimal set of net combinators expressive enough to form a rudimentary calculus of nets. This should certainly include operations allowing (forms of) *communication* and *parallel composition* (and, to make easier the description of (large) modular systems, also operations as *relabelling* and *hiding*). However, in order to avoid a chaotic 'structural' calculus where everything is permitted, it is obvious that some restrictions on the allowed connections via places and transitions must be imposed. Interfaces readily suggest a reasonable discipline of interaction: connections between nets should go from outputs to inputs, involving only public components. This formalises the well-motivated and solid intuition that the only allowed interactions are achieved by *sending* and *receiving* along interfaces, to be thought of as communication channels.

The main way of combining nets provided in [**58**] therefore consists of connecting the outputs of one net to the inputs of another net and, possibly, vice versa, as schematically shown by the following picture.

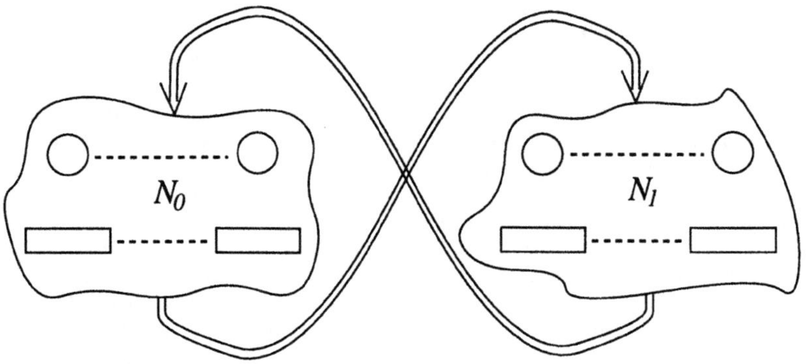

Following the principle of considering as simple operations as possible, this is realised by two more basic combinators: *par*(_,_), which puts its two arguments side by side, and *add*(_), which augments its argument by a new arc from an interface-transition to an interface-place. The operation above is then obtained by repeatedly applying *add*(_) to $par(N_0,N_1)$.

Observe that *add*(_) in isolation provides an interesting form of *recursion* consisting of *feeding back* outputs to inputs, and represents a bridge to structures of recent common interest in category theory and in computer science: the *traced monoidal categories* [**36, 40**], i.e., monoidal categories equipped with an *feedback* operation completely analogous to the one discussed above. Algebraic structures based on a central operation of *iteration*, or *feedback* — inspired by flowcharts and program schemata — have appeared rather early in computer science, see, e.g., [**22, 2, 85, 86, 51**] and [**10**], that offers for a thorough exposition of so-called '*iteration theory*' and more references. The advent of traced monoidal categories, though, has recently revived interest in using such abstract structures in semantics of computation, as e.g., in [**1, 37, 31, 32**]. Obviously, the calculus of [**58**] fits nets into this framework very nicely, although some of the details still need to be clarified.

Acknowledgements

We would like to thank many colleagues who have been part of developing the material behind this chapter. In particular, we acknowledge our close collaboration with A. Joyal, J. Meseguer, U. Montanari, L. Priese, G. Rozenberg, P.S. Thiagarajan, and G. Winskel on many parts of this work.

References

[1] S. Abramsky (1996), Retracing Some Paths in Process Algebra, in *Proceedings of CONCUR 96*, U. Montanari and V. Sassone (Eds.), *Lecture Notes in Computer Science* **1119**, 1–17, Springer-Verlag.

[2] E.S. Bainbridge (1976), Feedback and Generalized Logic. *Information and Control* **31**, 75–96.

[3] J. Bénabou (1963), Categories with Multiplication. *Comptes Rendus Académie Science Paris* **256**, 1887–1890.

[4] M.A. BEDNARCZYK (1987), *Categories of Asynchronous Systems*. PhD thesis, University of Sussex, Report **1/88**, School of Cognitive and Computing Sciences, University of Sussex.

[5] D. BENSON AND O. BEN-SHACHAR (1988), Bisimulation of Automata. *Information and Computation* **79**, 60–83, Academic Press.

[6] G. BERRY (1979), *Modèles completement adéquats et stables des λ-calculs typèes*. Thèse de Doctorat d'Etat, Université Paris VII.

[7] E. BEST AND R. DEVILLERS (1987), Sequential and Concurrent Behaviour in Petri Net Theory. *Theoretical Computer Science* **55**, 87–136, Elsevier.

[8] E. BEST, R. DEVILLERS, AND J. HALL (1992), The Petri Box Calculus: a New Causal Algebra with Multilabel Communication, in *Advances in Petri Nets 92*, G. Rozenberg (Ed.), *Lecture Notes in Computer Science* **609**, 21–69, Springer-Verlag.

[9] E. BEST AND C. FERNANDEZ (1988), *Non-sequential Processes: A Petri Net View*. EATCS Monographs on Theoretical Computer Science **13**, Springer-Verlag.

[10] S.L. BLOOM AND Z. ÉSIK (1991), *Iteration Theories: the Equational Logic of Iterative Processes*. EATCS Monographs on Theoretical Computer Science **30**, Springer-Verlag.

[11] C. BROWN AND D. GURR (1992), Temporal Logic and Categories of Petri Nets, in *Proceedings of ICALP 93*, A. Lingas *et al.* (Eds.), *Lecture Notes in Computer Science* **700**, 570–581, Springer-Verlag.

[12] C. BROWN, D. GURR, AND V. DE PAIVA (1991), *A Linear Specification Language for Petri Nets*. Technical Report DAIMI PB-**363**, Computer Science Dept., University of Aarhus.

[13] J. BRUNO AND S.M. ALTMAN (1971), A Theory of Asynchronous Control Networks. *IEEE Transactions on Computers*, Vol. C-**20**, 629–638, IEEE Press.

[14] I. CASTELLANI (1985), Bisimulation and Abstraction Homomorphisms, in *Proceedings of CAAP 85*, H. Ehrig *et al.* (Eds.), *Lecture Notes in Computer Science* **185**, 223–238, Springer-Verlag.

[15] G.L. CATTANI, M. FIORE, AND G. WINSKEL (1998), A Theory of Recursive Domains with Applications to Concurrency, in *Proceedings of the 13th LICS Symposium*, IEEE Press. To appear.

[16] P. DEGANO, R. DE NICOLA, AND U. MONTANARI (1987). Observational Equivalences for Concurrency Models, in *Formal Description of Programming Concepts – III, IFIP*, M. Wirsing (Ed.), 105–132, Elsevier.

[17] P. DEGANO, R. DE NICOLA, AND U. MONTANARI (1988), A Distributed Operational Semantics for CCS based on Condition/Event Systems. *Acta Informatica* **26**, 59–91, Springer-Verlag.

[18] P. DEGANO, J. MESEGUER, AND U. MONTANARI (1989), Axiomatizing Net Computations and Processes, in *Proceedings of the 4th LICS Symposium*, 175–185, IEEE Press.

[19] J.B. DENNIS (1970), Modular, Asynchronous Control Structures for a High Performance Processor, in *Record of the Project MAC Conference on Concurrent Systems and Parallel Computation*, ACM Press.

[20] A. EHRENFEUCHT AND G. ROZENBERG (1987), On the Structure of Dependence Graphs, in *Concurrency and Nets*, K. Voss *et al.* (Eds.), 141–170, Springer-Verlag.

[21] S. EILENBERG, AND G.M. KELLY (1966), Closed Categories, in *Proceedings of the Conference on Categorical Algebra*, S. Eilenberg *et al.* (Eds.), 421–562, Springer-Verlag.

[22] C. ELGOT (1975), Monadic Computation and Iterative Algebraic Theories, in *Logic Colloquium '73*, H.E. Rose and J.C. Shepherdson (Eds.), 175–230, North-Holland.

[23] G. FERRARI, U. MONTANARI, AND M. MOWBRAY (1997), Structured Transition Systems with Parametric Observations: Observational Congruences and Minimal Realizations. *Mathematical Structures in Computer Science* **7**, 241–282, Cambridge University Press.

[24] R.J. VAN GLABBEEK AND U. GOLTZ (1989), Equivalence Notions for Concurrent Systems and Refinement of Actions, in *Proceedings of MFCS 89*, A. Kreczmar and G. Mirkowska (Eds.), *Lecture Notes in Computer Science* **379**, 237–248, Springer-Verlag.

[25] R.J. VAN GLABBEEK AND F. VAANDRAGER (1987), Petri Net Models for Algebraic Theories of Concurrency, in *Proceedings of PARLE*, J.W. de Bakker *et al.* (Eds.), *Lecture Notes in Computer Science* **259**, 224–242, Springer-Verlag.

[26] U. GOLTZ AND W. REISIG (1983), The Non-Sequential Behaviour of Petri Nets. *Information and Computation* **57**, 125–147, Academic Press.

[27] R. GORRIERI AND U. MONTANARI (1990), Scone: A Simple Calculus of Nets, in *Proceedings of CONCUR 90*, J.C.M. Baeten and J.W. Klop (Eds.), *Lecture Notes in Computer Science* **458**, 2–31, Springer-Verlag.

[28] J. GRABOWSKY (1981), On partial languages. *Annales Societatis Mathematicae Polonae* **IV.2**, 428–498, North-Holland.

[29] M. HENNESSY (1988), *Algebraic Theory of Processes*. MIT Press.

[30] C.A.R. HOARE (1985), *Communicating Sequential Processes*. Prentice Hall.

[31] M. HASEGAWA (1997), Recursion from Cyclic Sharing: Traced Monoidal Categories, in *Proceedings of TLCA 97*, Ph. de Groote and J.R. Hindley (Eds.), *Lecture Notes in Computer Science* **1210**, 196–213, Springer-Verlag.

[32] T.T. HILDEBRANDT, P. PANANGADEN, AND G. WINSKEL (1998), Relational Semantics of Non-Deterministic Dataflow, in *Proceedings of CONCUR 98*, D. Sangiorgi and R. de Simone (Eds.), *Lecture Notes in Computer Science*, Springer-Verlag. To appear.

[33] P.W. HOOGERS, H.C.M. KLEIJN, AND P.S. THIAGARAJAN (1992), A Trace Semantics for Petri Nets, in *Proceedings of ICALP 92*, W. Kuich (Ed.), *Lecture Notes in Computer Science* **623**, 595–604, Springer-Verlag.

[34] T. HUNE AND M. NIELSEN (1998), Timed Bisimulation and Open Maps, in *Proceedings of MFCS 98*, J. Gruska *et al.* (Eds.), *Lecture Notes in Computer Science*, Springer-Verlag. To appear.

[35] A. JOYAL, M. NIELSEN, AND G. WINSKEL (1996), Bisimulation from Open Maps. *Information and Computation* **127**, 164–185, Academic Press.

[36] A. JOYAL, R. STREET, AND D. VERITY (1996), Traced Monoidal Categories. *Mathematical Proceedings of the Cambridge Philosophical Society* **119**, 447–468, Cambridge University Press.

[37] P. KATIS, N. SABADINI, AND R. WALTERS (1997), Bicategories of Processes. *Journal of Pure and Applied Algebra* **115**, 141–178, North-Holland.

[38] R.M. KELLER (1974), Towards a Theory of Universal Speed-Independent Modules. *IEEE Transactions on Computers* vol. **C-23**, 21–33, IEEE Press.

[39] G.M. KELLY (1964), On MacLane's Conditions for Coherence of Natural Associativities, Commutativities, etc. *Journal of Algebra* **1**, 397–402, Acamedic Press.

[40] G.M. KELLY AND M. LAPLAZA (1980), Coherence for Compact Closed Categories. *Journal of Pure and Applied Algebra* **19**, 193–213, North-Holland.

[41] S. MACLANE (1963), Natural Associativity and Commutativity. *Rice University Studies* **49**, pp. 28–46, Rice University Press.

[42] S. MACLANE (1971), *Categories for the Working Mathematician*. Graduate Texts in Mathematics, Springer-Verlag.

[43] A. MAZURKIEWICZ (1977), *Concurrent Program Schemes and their Interpretations*. Technical Report DAIMI PB-**78**, Computer Science Dept., University of Aarhus.

[44] A. MAZURKIEWICZ (1987), Trace theory, in *Petri Nets, Applications and Relationship to other Models of Concurrency*, W. Brauer *et al.* (Eds.), *Lecture Notes in Computer Science* **255**, 279–324, Springer-Verlag.

[45] J. MESEGUER AND U. MONTANARI (1990), Petri Nets are Monoids. *Information and Computation* **88**, 105- 155, Academic Press.

[46] J. MESEGUER, U. MONTANARI, AND V. SASSONE (1992), On the Semantics of Petri Nets, in *Proceedings of CONCUR 92*, R. Cleaveland (Ed.), *Lecture Notes in Computer Science* **630**, 286–301, Springer-Verlag.

[47] J. MESEGUER, U. MONTANARI, AND V. SASSONE (1996), Process versus Unfolding Semantics for Place/Transition Petri Nets. *Theoretical Computer Science* **153**, 171–210, Elsevier.

[48] J. MESEGUER, U. MONTANARI, AND V. SASSONE (1997), On the Semantics of Place/Transition Petri Nets. *Mathematical Structures in Computer Science* **7**, 359–397, Cambridge University Press.

[49] R. MILNER (1980), *A Calculus of Communicating Systems*. Lecture Notes in Computer Science **92**, Springer-Verlag.

[50] R. MILNER (1989), *Communication and Concurrency*. Prentice-Hall.

[51] R. MILNER (1993), Action Calculi or Syntactic Action Structures, in *Proceedings of MFCS 93*, A. Borzyszkowski, S. Sokolowski (Eds.), *Lecture Notes in Computer Science* **711**, 105–121, Springer-Verlag.

[52] E.H. MOORE (1897), Concerning the abstract group of order $k!$ isomorphic with the symmetric substitution group on k letters. *Proceedings of the London Mathematical Society* **28**, 357–366, Claredon Press.

[53] M. MUKUND (1992), Petri Nets and Step Transition Systems. *International Journal of Foundations of Computer Science* **3**, 443- 478, World Scientific.

[54] M. MUKUND AND M. NIELSEN (1992), Locations and Asynchronous Transition Systems, in *Proceedings of FST & TCS 92*, R. Shyamasundar (Ed.), *Lecture Notes in Computer Science* **652**, 328–341, Springer-Verlag.

[55] M. NIELSEN AND A. CHENG (1995), Observing Behaviour Categorically, in *Proceedings of FST & TCS 95*, P.S. Thiagarajan (Ed.), *Lecture Notes in Computer Science* **1026**, 263–278, Springer-Verlag.

[56] M. NIELSEN AND C. CLAUSEN (1995), Games and Logics for a Noninterleaving Bisimulation. *Nordic Journal of Computing* **2**, 222–250, Publishing Association NJC.

[57] M. NIELSEN, G. PLOTKIN, AND G. WINSKEL (1981), Petri Nets, Event Structures and Domains, Part 1. *Theoretical Computer Science* **13**, 85–108, Elsevier.

[58] M. NIELSEN, L. PRIESE, AND V. SASSONE (1995), Characterizing Behavioural Congruences for Petri Nets, in *Proceedings of CONCUR 95*, I. Lee and S. Smolka (Eds.), *Lecture Notes in Computer Science* **962**, 175–189, Springer-Verlag.

[59] M. NIELSEN, G. ROZENBERG, AND P.S. THIAGARAJAN (1990), Behavioural Notions for Elementary Net Systems. *Distributed Computing* **4**, 45–57, Springer-Verlag.

[60] M. NIELSEN, G. ROZENBERG, AND P.S. THIAGARAJAN (1992), Elementary Transition Systems. *Theoretical Computer Science* **96**, 3–33, Elsevier.

[61] M. NIELSEN, G. ROZENBERG, AND P.S. THIAGARAJAN (1995), Transition Systems, Event Structures and Unfoldings. *Information and Computation* **118**, 191–207, Academic Press.

[62] M. NIELSEN AND G. WINSKEL (1995), Trace Structures and other Models for Concurrency, in *The Book of Traces*, V. Diekert and G. Rozenberg (Eds.), 271–306, World Scientific.

[63] M. Nielsen and G. Winskel (1996), Petri Nets and Bisimulation. *Theoretical Computer Science* **153**, 211–244, Elsevier.

[64] E.R. OLDEROG (1987), A Petri Net Semantics for CCSP, in *Advances in Petri Nets 86*, W. Brauer *et al.* (Eds.), *Lecture Notes in Computer Science* **255**, 196–223, Springer-Verlag.

[65] E.R. OLDEROG (1991), *Nets, Terms and Formulas*. Cambridge Tracts in Theoretical Computer Science, Cambridge University Press.

[66] C.A. PETRI (1962), *Kommunikation mit Automaten*. PhD thesis, Institut für Instrumentelle Mathematik, Bonn.

[67] C.A. PETRI (1973), Concepts of Net Theory, in *Proceedings of MFCS 73*, 137–146, Mathematics Institute of the Slovak Academy of Science.

[68] C.A. PETRI (1977), *Non-Sequential Processes*. Interner Bericht ISF–**77–5**, Gesellschaft für Mathematik und Datenverarbeitung, Bonn.

[69] V. PRATT (1986), Modelling Concurrency with Partial Orders. *International Journal of Parallel Programming* **15**, 33–71, Plenum.

[70] L. PRIESE (1983), Automata and Concurrency. *Theoretical Computer Science* **25**, 221–265, Elsevier.

[71] A. RABINOVITCH AND B. TRAKHTENBROT (1988), Behaviour structures and nets. *Fundamenta Informatica* **11**, 357–404, North-Holland.

[72] W. REISIG (1985), *Petri Nets (an Introduction)*. EATCS Monographs on Theoretical Computer Science **4**, Springer-Verlag.

[73] G. ROZENBERG (1987), Behaviour of Elementary Net Systems, in *Advances in Petri Nets 86*, W. Brauer *et al.* (Eds.), *Lecture Notes in Computer Science* **254**, 60–94, Springer-Verlag.

[74] B. ROZOY AND P.S. THIAGARAJAN (1991), Event Structures and Trace Monoids. *Theoretical Computer Science* **91**, 285–313, Elsevier.

[75] V. SASSONE (1995), Axiomatizing Petri Net Concatenable Processes, in *Proceedings of FCT 95*, H. Reichel (Ed.), *Lecture Notes in Computer Science* **962**, 414–423, Springer-Verlag

[76] V. SASSONE (1995), On the Category of Petri Net Computations, in *Proceedings of TAPSOFT 95*, P. Mosses *et al.* (Eds,), *Lecture Notes in Computer Science* **915**, 334–348, Springer-Verlag.

[77] V. SASSONE (1996), An Axiomatization of the Algebra of Petri Net Concatenable Processes. *Theoretical Computer Science* **170**, 277–296, Elsevier.

[78] V. SASSONE (1998), An Axiomatization of the Category of Petri Net Computations. *Mathematical Structures in Computer Science* **8**, 117–151, Cambridge University Press.

[79] V. SASSONE, M. NIELSEN, AND G. WINSKEL (1993), Deterministic Behavioural Models for Concurrency, in *Proceedings of MFCS 93* A. Borzyszkowski, S. Sokolowski (Eds.), *Lecture Notes in Computer Science* **711**, 682–692, Springer-Verlag.

[80] V. SASSONE, M. NIELSEN, AND G. WINSKEL (1996), Models for Concurrency: Towards a Classification. *Theoretical Computer Science* **170**, 297–348, Elsevier.

[81] D. SCOTT (1970), Outline of a Mathematical Theory of Computation, in *Proceedings of 4th Annual Princeton Conference on Information Science and Systems*, 169–176, Princeton University Press.

[82] K. SEGERBERG (1968), Decidability of S4.1. *Theoria* **34**,7–20, Gordon and Breach.

[83] M.W. SHIELDS (1985), Concurrent machines. *Computer Journal* **28**, 449–465, Cambridge University Press.

[84] E.W. STARK (1989), Concurrent Transition Systems. *Theoretical Computer Science* **64**, 221–269, Elsevier.

[85] G. ŞTEFANĔSCU (1987), On Flowchart Theories: Part I. The Deterministic Case. *Journal of Computer and System Sciences* **35**, 163–191, Academic Press.

[86] G. ŞTEFANĔSCU (1987), On Flowchart Theories: Part I. The Nondeterministic Case. *Theoretical Computer Science* **52**, 307–340, Elsevier.

[87] P.S. THIAGARAJAN (1987), Elementary Net Systems, in *Advances in Petri Nets 86*, W. Brauer *et al.* (Eds.), *Lecture Notes in Computer Science* **254**, 26–59, Springer-Verlag.

[88] J. VAN BENTHAM (1984), Correspondence Theory, in *Handbook of Philosophical Logic* vol. **2**, P. Gabbay and F. Guenthner (Eds.), 167–247, Reidel.

[89] W. VOGLER (1991), Deciding History Preserving Bisimilarity. in *Proceedings of ICALP 91*, J. Leach Albert *et al.* (Eds.), *Lecture Notes in Computer Science* **510**, 495–505, Springer-Verlag.

[90] G. WINSKEL (1982), Event Structure Semantics for CCS and related languages, in *Proceedings of the 9th ICALP*, M. Nielsen and E.M. Schmidt (Eds.), *Lecture Notes in Computer Science* **140**, 561–576, Springer-Verlag.

[91] G. WINSKEL (1984), A New Definition of Morphism on Petri Nets, in *Proceedings of STACS 84*, M. Fontet and K. Mehlhorn (Eds.), *Lecture Notes in Computer Science* **166**, 140–150, Springer-Verlag.

[92] G. WINSKEL (1984), Categories of Model for Concurrency, in *Seminar on Concurrency*, S. Brookes *et al.* (Eds.), *Lecture Notes in Computer Science* **197**, 246–267, Springer-Verlag.

[93] G. WINSKEL (1987), Petri Nets, Algebras, Morphisms and Compositionality. *Information and Computation* **72**, 197- 238, Academic Press.

[94] G. WINSKEL (1986), Event Structures, in *Advances in Petri Nets 86*, W. Brauer *et al.* (Eds.), *Lecture Notes in Computer Science* **255**, 325–392, Springer-Verlag.

[95] G. WINSKEL (1988), An Introduction to Event Structures, in *Linear time, branching time, and partial order in logics and models for concurrency*, J.W. de Bakker *et al.* (Eds.), *Lecture Notes in Computer Science* **354**, 365–397, Springer-Verlag.

[96] G. WINSKEL (1996), A Presheaf Semantics of Value-Passing Processes, in *Proceedings of CONCUR 96*, U. Montanari and V. Sassone (Eds.), *Lecture Notes in Computer Science* **1119**, 98–114, Springer-Verlag.

[97] G. WINSKEL AND M. NIELSEN (1995), Models for Concurrency, in *Handbook of Logic in Computer Science* vol. **4**, S. Abramsky *et al.* (Eds), 1–148, Oxford University Press.

[98] G. WINSKEL AND M. NIELSEN (1997), Presheaves as Transition Systems, in *Proceedings of POMIV 96*, D. Peled *et al.* (Eds.), *DIMACS Series in Discrete Mathematics and Theoretical Computer Science* **29**, 129–140, American Mathematical Society Press.

[99] M. YOELI (1973), *Petri Nets and Asynchronous Control Networks*. Research Report CS–73–07, Department of Applied Mathematics and Computer Science, University of Waterloo.

BRICS, BASIC RESEARCH IN COMPUTER SCIENCE, DEPT. OF COMPUTER SCIENCE, UNIVERSITY OF AARHUS, NY MUNKEGADE BLD. 540, DK-8000 ÅRHUS C, DENMARK
E-mail address: `mnielsen@brics.dk`

QUEEN MARY AND WESTFIELD COLLEGE, UNIVERSITY OF LONDON, MILE END ROAD, LONDON E1 4NS, UK
E-mail address: `vs@dcs.qmw.ac.uk`

Distributed Versions of Linear Time Temporal Logic: A Trace Perspective

P. S. Thiagarajan[1*] and Jesper G. Henriksen[2]

[1] SPIC Mathematical Institute,
92 G.N. Chetty Road, T. Nagar, Chennai 600 017, India
Email: pst@smi.ernet.in

[2] BRICS**, Department of Computer Science,
University of Aarhus, Ny Munkegade, 8000 Aarhus C, Denmark
Email: gulmann@brics.dk

1 Introduction

Linear time Temporal Logic (LTL) as proposed by Pnueli [37] has become a well established tool for specifying the dynamic behaviour of distributed systems. A basic feature of LTL is that its formulas are interpreted over sequences. Typically, such a sequence will model a computation of a system; a sequence of states visited by the system or a sequence of actions executed by the system during the course of the computation. A system is said to satisfy a specification expressed as an LTL formula in case every computation of the system is a model of the formula. A rich theory of LTL is now available using which one can effectively verify whether a finite state system meets its specification [51]. Indeed, the verfication task can be automated (for instance using the software packages SPIN [21] and FormalCheck [2]) to handle large systems of practical interest.

In many applications the computations of a distributed system will constitute interleavings of the occurrences of causally independent actions. Consequently, the computations can be naturally grouped together into equivalence classes where two computations are equated in case they are two different interleavings of the same partially ordered stretch of behaviour. It turns out that many of the properties expressed as LTL-formulas happen to have the so called "all-or-none" property. Either all members of an equivalence class of computations will have the desired property or none will do ("leads to deadlock" is one such property). For verifying such properties one has to check the property for just one member of each equivalence class. This is the insight underlying many of the partial-order based verification methods [17, 35, 50]. As may be guessed, the importance of these methods lies in the fact that via these methods the computational resources required for the verification task can often be dramatically reduced.

It is often the case that the equivalence classes of computations generated by a distributed system constitute objects called Mazurkiewicz traces. They can be

* This work has been supported by BRICS and IFCPAR Project 1502-1.
** Basic Research in Computer Science,
Centre of the Danish National Research Foundation.

canonically represented as restricted labelled partial orders. This opens up an alternative way of exploiting the non-sequential nature of the computations of a distributed systems and the attendant partial-order based methods. It consists of developing linear time temporal logics that can be directly interpreted over Mazurkiewicz traces. In these logics, every specification is guaranteed to have the "all-or-none" property and hence can take advantage of the partial-order based reduction methods during the verification process. The study of these logics also exposes the richness of the partial-order settings from a logical standpoint and the complications that can arise as a consequence.

Our aim here is to present an overview of linear time temporal logics whose models can be viewed as Mazurkiewicz traces. The presentation is, in principle, self-contained though previous exposure to temporal logics [12] and automata over infinite objects [49] will be very helpful. We have provided net-theoretic examples whenever possible in order to emphasize the broad scope of applicability of the material.

In the next section we introduce linear time temporal logic and sketch the automata-theoretic solutions to the satisfiability problem (does a formula have a model?) and the model checking problem (do all computations of a system constitute models of a given specification formula?). In Section 3 we introduce Mazurkiewicz traces viewed as equivalence classes of sequences. This leads to the precise formualtion of the notion "all-or-none" LTL properties.

Next we introduce a well-understood class of trace languages called product languages. The automata that recognize these languages are called product automata and they incorporate a simple and yet useful method of forming distributed systems. The system consists of a network of sequential agents, each with its own alphabet of actions. In the interesting instances the alphabets are not pair-wise disjoint. One then imposes a synchronization regime under which the agents are forced to carry out common actions together. After presenting a theory of product languages and automata, we formulate in Section 5 a simple version of a trace-based version of LTL called product LTL. The formulas of this logic have a natural semantics in terms of the computations generated by a network of sequential agents as introduced in the previous section. Using the theory of product automata we then provide solutions to the satisfiability and model checking problems for product LTL.

In Section 6 we introduce the representation of Mazurkiewicz traces as restricted labelled partial orders. We then provide a rapid introduction to the theory of trace languages and automata that we call asynchronous automata for recognizing trace languages. In the subsequent section we introduce the logic TrPTL which is a trace-based logic with much richer possibilities than product LTL. We then provide solutions to the satisfiability and model checking problems for TrPTL using asynchronous automata. This is followed by a brief survey of other trace-based linear time temporal logics available in the literature. Section 8 is devoted to considering various expressiveness issues associated with our temporal logics. We conclude in the final section with remarks about branching time temporal logics based on traces.

2 Linear Time Temporal Logic

In our formulation of linear time temporal logics it will be convenient to treat *actions* as first class objects both at the syntactic and semantic levels. As a first step we shall consider a version of LTL (linear time temporal logic) in which the next-state modality is indexed by actions.

Through the rest of the paper we fix a finite non-empty alphabet of actions Σ. We let a, b range over Σ and refer to members of Σ as actions. Σ^* is the set of finite words and Σ^ω is the set of infinite words generated by Σ with $\omega = \{0, 1, 2, \ldots\}$. We set $\Sigma^\infty = \Sigma^* \cup \Sigma^\omega$ and denote the null word by ε. We let σ, σ' range over Σ^ω and τ, τ', τ'' range over Σ^*. Finally $\preceq$ is the usual prefix ordering defined over Σ^* and for $u \in \Sigma^\infty$, we let $\mathrm{prf}(u)$ be the set of finite prefixes of u.

Next we fix a finite non-empty set of atomic propositions $P = \{p_1, p_2, \ldots\}$ and let p, q range over P. The set of formulas of LTL(Σ) is then given by the syntax:

$$\mathrm{LTL}(\Sigma) ::= p \mid \sim\alpha \mid \alpha \vee \beta \mid \langle a \rangle \alpha \mid \alpha \, U \, \beta.$$

Through the rest of this section α, β will range over LTL(Σ).

A model of LTL(Σ) is a pair $M = (\sigma, V)$ where $\sigma \in \Sigma^\omega$ and $V : \mathrm{prf}(\sigma) \to 2^P$ is a valuation function. Let $M = (\sigma, V)$ be a model, $\tau \in \mathrm{prf}(\sigma)$ and α be a formula. Then $M, \tau \models \alpha$ will stand for α being satisfied at τ in M. This notion is defined inductively in the expected manner.

- $M, \tau \models p$ iff $p \in V(\tau)$.
- $M, \tau \models \sim\alpha$ iff $M, \tau \not\models \alpha$.
- $M, \tau \models \alpha \vee \beta$ iff $M, \tau \models \alpha$ or $M, \tau \models \beta$.
- $M, \tau \models \langle a \rangle \alpha$ iff $\tau a \in \mathrm{prf}(\sigma)$ and $M, \tau a \models \alpha$.
- $M, \tau \models \alpha \, U \, \beta$ iff there exists τ' such that $\tau\tau' \in \mathrm{prf}(\sigma)$ and $M, \tau\tau' \models \beta$. Moreover for every τ'' such that $\varepsilon \preceq \tau'' \prec \tau'$, it is the case that $M, \tau\tau'' \models \alpha$.

Along with the usual propositional connectives $\wedge, \supset$ and $\equiv$ we will also use the propositional constants, $\top \overset{\Delta}{\Longleftrightarrow} p_1 \vee \sim p_1$ and $\bot \overset{\Delta}{\Longleftrightarrow} \sim \top$. Some useful derived modalities are:

- $O\alpha \overset{\Delta}{\Longleftrightarrow} \bigvee_{a \in \Sigma} \langle a \rangle \alpha$.
- $\Diamond\alpha \overset{\Delta}{\Longleftrightarrow} \top U \alpha$.
- $\Box\alpha \overset{\Delta}{\Longleftrightarrow} \sim\Diamond\sim\alpha$.

Let $M = (\sigma, V)$ be a model and $\tau \in \mathrm{prf}(\sigma)$. Then it is easy to check the following assertions.

- $M, \tau \models O\alpha$ iff $M, \tau' \models \alpha$ where $\tau' \in \mathrm{prf}(\sigma)$ is such that $|\tau'| = |\tau| + 1$.
- $M, \tau \models \Diamond\alpha$ iff there exists a $\tau' \in \Sigma^*$ with $\tau\tau' \in \mathrm{prf}(\sigma)$ such that $M, \tau\tau' \models \alpha$.
- $M, \tau \models \Box\alpha$ iff for each $\tau' \in \Sigma^*$, $\tau\tau' \in \mathrm{prf}(\sigma)$ implies $M, \tau\tau' \models \alpha$.

Note that $O\alpha$ is the usual next-state operator of LTL.

We say that a formula $\alpha \in \mathrm{LTL}(\Sigma)$ is ***satisfiable*** iff there exist a model $M = (\sigma, V)$ and $\tau \in \mathrm{prf}(\sigma)$ such that $M, \tau \models \alpha$. This logic does not refer to the past either in the syntax or in the semantics. Hence the formula α is satisfiable iff there exists a model M such that $M, \varepsilon \models \alpha$. This is easy to check. The ***satisfiability problem*** for LTL is to develop a decision procedure which will determine whether a given formula α is satisfiable. We will later in this section describe such a decision procedure.

We now wish to formulate the model checking problem for $\mathrm{LTL}(\Sigma)$. A *finite-state program* over Σ is a structure $Pr = (S, \longrightarrow, S_{in}, V_{Pr})$ where:

- S is a finite set of states.
- $\longrightarrow \subseteq S \times \Sigma \times S$ is a transition relation.
- $S_{in} \subseteq S$ is a set of initial states of the program.
- $V_{Pr} : S \to 2^P$ assigns a subset of P to each state of the program.

The members of P capture a finite set of basic assertions concerning the program which can usually be "read off" by examining the states of Pr and this is described by V_{Pr}. It will often be the case that the set of initial states is a singleton.

It is easy to arrange matters so that at each reachable state of the program at least one transition can be performed. We will assume that this is indeed the case for all program models we consider in this paper. Further we will say "program" instead "finite-state program" from now on.

A *computation* of the program Pr is a pair (σ, ρ) where $\sigma \in \Sigma^\omega$ and $\rho : \mathrm{prf}(\sigma) \to S$ is a map which satisfies:

- $\rho(\varepsilon) \in S_{in}$.
- $\rho(\tau) \xrightarrow{a} \rho(\tau a)$ for each $\tau a \in \mathrm{prf}(\sigma)$.

Let (σ, ρ) be a computation of the program Pr. Then this computation canonically induces the model $M_{\sigma,\rho} = (\sigma, V_\rho)$ where V_ρ is given by: $V_\rho(\tau) = V_{Pr}(\rho(\tau))$ for each $\tau \in \mathrm{prf}(\sigma)$.

Let Pr be a program and α be a formula of $\mathrm{LTL}(\Sigma)$. We say that Pr *meets the specification* α — denoted $Pr \models \alpha$ — if for every computation (σ, ρ) of Pr, it is the case that $M, \varepsilon \models \alpha$ where M is the model induced by the computation (σ, ρ). The *model checking problem* is to decide for a given program Pr and a given formula α whether or not $Pr \models \alpha$. We will sketch a solution to the model checking problem later in this section.

Let $\mathcal{N} = (B, E, F, c_{in})$ be a finite elementary net system. In other words, it is an elementary net system in which both B, the set of conditions and E, the set of events are finite sets. We can associate the program $Pr_{\mathcal{N}} = (S, \longrightarrow, S_{in}, V_{Pr})$ with $\mathcal{N}$ as follows:

- $\Sigma = E$ and $P = B$.
- S is the least subset of 2^B and $\longrightarrow$ is the least subset of $S \times \Sigma \times S$ satisfying:
  - $c_{in} \in S$.

- Suppose $c \in S$ and $e \in E$ such that ${}^\bullet e \subseteq c$ and $e^\bullet \cap c = \emptyset$. Then $c' \in S$ and $(c, e, c') \in \longrightarrow$ where $c' = (c - {}^\bullet e) \cup e^\bullet$.

– $S_{in} = \{c_{in}\}$.
– $V_{Pr}(c) = c$ for every $c \in S$.

Thus the so called case graph — or the sequential configuration graph as called in the chapter on elementary net systems in this volume — is the underlying transition system of the program. The conditions serve as the atomic propositions.

For $c \subseteq B$, let α_c be the formula $\bigwedge_{b \in c} b$. Now consider the specification $\Box \sim \alpha_c$ for some $c \subseteq B$. Then $Pr_{\mathcal{N}} \not\models \Box \sim \alpha_c$ iff c is a reachable state (i.e. $c \in S$) in $\mathcal{N}$. Next suppose e and e' are two events. Then $Pr_{\mathcal{N}} \models \Box\Diamond\langle e\rangle\top \supset \Box\Diamond\langle e'\rangle\top$ captures the fact that in $\mathcal{N}$, along every computation, if e occurs infinitely often then so does e'. A rich variety of liveness and safety properties can be expressed in LTL(Σ). For a substantial collection of examples the reader should see [26].

It turns out that both the satisfiability and model checking problems for LTL can be solved elegantly using Büchi automata [51]. We start with a brief introduction to these automata. A *Büchi automaton* over Σ is a tuple $\mathcal{B} = (Q, \longrightarrow, Q_{in}, F)$ where:

– Q is a finite non-empty set of states.
– $\longrightarrow \subseteq Q \times \Sigma \times Q$ is a transition relation.
– $Q_{in} \subseteq Q$ is a set of initial states.
– $F \subseteq Q$ is a set of accepting states.

Let $\sigma \in \Sigma^\omega$. Then a *run* of $\mathcal{B}$ over σ is a map $\rho : \text{prf}(\sigma) \longrightarrow Q$ such that:

– $\rho(\varepsilon) \in Q_{in}$.
– $\rho(\tau) \stackrel{a}{\longrightarrow} \rho(\tau a)$ for each $\tau a \in \text{prf}(\sigma)$.

The run ρ is *accepting* iff $\inf(\rho) \cap F \neq \emptyset$ where $\inf(\rho) \subseteq Q$ is given by $q \in \inf(\rho)$ iff $\rho(\tau) = q$ for infinitely many $\tau \in \text{prf}(\sigma)$. Finally $\mathcal{L}(\mathcal{B})$, the *language of ω-words accepted by* $\mathcal{B}$, is:

$$\mathcal{L}(\mathcal{B}) = \{\sigma \mid \exists \text{ an accepting run of } \mathcal{B} \text{ over } \sigma\}.$$

The languages recognized by Büchi automata are called the *ω-regular languages*. For an excellent survey of regular languages and automata over infinite objects, the reader is referred to [49].

It is easy to solve the *emptiness problem* for Büchi automata; to determine whether or not the language accepted by a Büchi automaton is empty. This can be done in time linear in the size of the automaton where the size of a Büchi automaton is the number of states of the automaton [49].

We will now show how one can effectively construct for each $\alpha \in$ LTL(Σ), a Büchi automaton $\mathcal{B}_\alpha$ such that the language of ω-words accepted by $\mathcal{B}_\alpha$ is non-empty iff α is satisfiable. This is an action-based version of the elegant solution presented in [51] for LTL.

Through the rest of the section we fix a formula α_0. To construct $\mathcal{B}_{\alpha_0}$ we first define the (Fischer-Ladner) closure of α_0. For convenience we will assume that the derived next-state modality modality O is included in the syntax of LTL(Σ). We take $cl(\alpha_0)$ to be the least set of formulas that satisfies:

- $\alpha_0 \in cl(\alpha_0)$.
- If $\sim\beta \in cl(\alpha_0)$ then $\beta \in cl(\alpha_0)$.
- If $\alpha \vee \beta \in cl(\alpha_0)$ then $\alpha, \beta \in cl(\alpha_0)$.
- If $\langle a \rangle \alpha \in cl(\alpha_0)$ then $\alpha \in cl(\alpha_0)$.
- If $\alpha\ U\ \beta \in cl(\alpha_0)$ then $\alpha, \beta \in cl(\alpha_0)$. In addition, $O(\alpha\ U\ \beta) \in cl(\alpha_0)$.

Now $CL(\alpha_0)$, the *closure* of α_0, is defined to be:

$$CL(\alpha_0) = cl(\alpha_0) \cup \{\sim\beta \mid \beta \in cl(\alpha_0)\}.$$

In what follows $\sim\sim\beta$ will be identified with β. Moreover, throughout the section, all the formulas that we encounter will be assumed to be members of $CL(\alpha_0)$. For convenience, we shall often write CL instead of $CL(\alpha_0)$.

$A \subseteq CL$ is called an *atom* iff it satisfies :

- $\beta \in A$ iff $\sim\beta \notin A$.
- $\alpha \vee \beta \in A$ iff $\alpha \in A$ or $\beta \in A$.
- $\alpha\ U\ \beta \in A$ iff $\beta \in A$ or $\alpha, O(\alpha\ U\ \beta) \in A$.
- If $\langle a \rangle \alpha \in A$ and $\langle b \rangle \beta \in A$ then $a = b$.

$AT(\alpha_0)$ is the set of atoms and again we shall often write AT instead of $AT(\alpha_0)$. Finally we set U_{α_0}, the set of *until requirements* of α_0, to be the given by $U_{\alpha_0} = \{\alpha\ U\ \beta \mid \alpha\ U\ \beta \in CL\}$. We will often write U_0 instead of U_{α_0}.

The Büchi automaton $\mathcal{B}_{\alpha_0}$ (from now on denoted as $\mathcal{B}$) is now defined as $\mathcal{B} = (Q, \longrightarrow, Q_{in}, F)$, where the various components of $\mathcal{B}$ are specified as follows.

- $Q = AT \times 2^{U_0}$ is the set of states.
- The transition relation $\longrightarrow \subseteq Q \times \Sigma \times Q$ is given by $(A, x) \xrightarrow{a} (B, y)$ iff the following requirements are met:
  - For every $\langle a \rangle \alpha \in CL$, $\langle a \rangle \alpha \in A$ iff $\alpha \in B$ and for every $O(\alpha) \in CL$, $O(\alpha) \in A$ iff $\alpha \in B$.
  - if $\langle b \rangle \beta \in A$ then $b = a$.
  - if $x \neq \emptyset$ then $y = \{\alpha\ U\ \beta \mid \alpha\ U\ \beta \in x$ and $\beta \notin B\}$. If $x = \emptyset$ then $y = \{\alpha\ U\ \beta \mid \alpha\ U\ \beta \in B$ and $\beta \notin B\}$.
- $Q_{in} \subseteq Q$ is given by $(A, x) \in Q_{in}$ iff $\alpha_0 \in A$ and $x = \emptyset$.
- $F \subseteq Q$ is given by $(A, x) \in F$ iff $x = \emptyset$.

It is easy to show that $\mathcal{L}(\mathcal{B}) \neq \emptyset$ iff α_0 is satisfiable. It is also easy to check that the size of $\mathcal{B}$ is at most exponential in the size of α_0. As observed earlier the emptiness problem for a Büchi automaton can be solved in time linear in the size of the automaton. Thus we arrive at:

Theorem 1. *The satisfiability problem for* LTL(Σ) *is decidable in exponential time.*

Turning now to the model checking problem we first recall that the *intersection problem* for Büchi automata can be easily solved. In other words, let $\mathcal{B}_1, \mathcal{B}_2$ be two Büchi automata both operating over Σ. Then one can effectively construct a Büchi automaton $\mathcal{B}$ over the same alphabet such that the language accepted by $\mathcal{B}$ is the intersection of the languages accepted by $\mathcal{B}_1$ and $\mathcal{B}_2$. Moreover, the size of $\mathcal{B}$ can be assumed to be bounded by $2n_1n_2$ where n_1 is the size of $\mathcal{B}_1$ and n_2 is the size of $\mathcal{B}_2$ [49].

Now let $Pr = (S, \longrightarrow, S_{in}, V_{Pr})$ be a program. We associate the Büchi automaton $\mathcal{B}_{Pr} = (S, \rightsquigarrow, S_{in}, S)$ over the alphabet $\Sigma \times 2^P$ with Pr where $\rightsquigarrow$ is given by: $(s, (a, R), s') \in \rightsquigarrow$ iff $(s, a, s') \in \longrightarrow$ and $V_{Pr}(s) = R$.

Let α be a specification. Then we construct the Büchi automaton $\mathcal{B}_{\sim\alpha}$ corresponding to the *negation* of α. Let $\mathcal{B}_{\sim\alpha} = (Q, \Longrightarrow, Q_{in}, F)$. Recall that each state in Q is of the form (A, x) where A is an atom. We now convert this automaton into the automaton $\widehat{\mathcal{B}} = (Q, \Rrightarrow, Q_{in}, F)$ over the alphabet $\Sigma \times 2^P$ by defining $\Rrightarrow$ as: $((A, x), (a, R), (B, y)) \in \Rrightarrow$ iff $((A, x), a, (B, y)) \in \Longrightarrow$ and $A \cap P = R$. Finally, let $\mathcal{B}$ be the Büchi automaton which accepts the intersection of the languages accepted by $\mathcal{B}_{Pr}$ and $\widehat{\mathcal{B}}$. It is straightforward to check that $Pr \models \alpha$ iff the language accepted by $\mathcal{B}$ is *empty*. An easy analysis of the size of $\mathcal{B}$ leads to:

Theorem 2. *The model checking problem for* LTL(Σ) *is decidable in time* $O(|Pr| \cdot 2^{|\alpha|})$.

In what follows, automata-theoretic constructions and expressiveness issues will play a considerable role. These topics can be treated in a simpler fashion if we eliminate atomic propositions. Most of the material we present can easily accomodate atomic propositions with some notational overhead. Hence from now on, we will not — except for some passing remarks — deal with atomic propositions. To be specific, the syntax of LTL(Σ) will be assumed to be:

$$\mathrm{LTL}(\Sigma) ::= \top \mid \sim\alpha \mid \alpha \vee \beta \mid \langle a \rangle \alpha \mid \alpha\, U\, \beta.$$

Notice that a model is now just a member of Σ^ω with the semantics being the obvious one ($\top$ is always true). The set of models of a formula constitute a language of infinite words. More precisely, each α induces the language L_α given by:

$$L_\alpha = \{\sigma \mid \sigma, \varepsilon \models \alpha\}.$$

A program is now just a finite-state transition system $Pr = (S, \longrightarrow, S_{in})$ over Σ. Each such program Pr has the language L_{Pr} associated with it. This is just the language accepted by the Büchi automaton $(S, \longrightarrow, S_{in}, S)$. It is also easy to see that $Pr \models \alpha$ iff $L_{Pr} \subseteq L_\alpha$ iff $L_{Pr} \cap L_{\sim\alpha} = \emptyset$.

3 Mazurkiewicz Traces and Trace Consistent Properties

Here we wish to introduce the notion of traces from the standpoint of sequences. This will enable us to define the notion of a trace consistent property. This notion plays an important role in partial order based reducion methods. As pointed out in the introduction, it also provides the motivation for studying trace based linear time temporal logics.

A *(Mazurkiewicz) trace alphabet* is a pair (Σ, I), where Σ, the alphabet, is a finite set and $I \subseteq \Sigma \times \Sigma$ is an irreflexive and symmetric ***independence relation.*** In most applications, Σ consists of the actions performed by a distributed system while I captures a static notion of causal independence between actions. The idea is that contiguous independent actions occur with no causal order between them. Thus, every sequence of actions from Σ corresponds to an interleaved observation of a partially-ordered stretch of system behaviour. This leads to a natural equivalence relation over execution sequences: two sequences are equated iff they correspond to different interleavings of the same partially-ordered stretch of behaviour.

For the rest of the section we fix a trace alphabet (Σ, I) and assume the terminology developed in the previous section for objects derived from Σ. We define $D = (\Sigma \times \Sigma) - I$ to be the *dependency relation.* Note that D is reflexive and symmetric. A set $p \subseteq \Sigma$ is called a *D-clique* iff $p \times p \subseteq D$. The equivalence relation $\approx_I \subseteq \Sigma^\infty \times \Sigma^\infty$ induced by I is given by:

$$\sigma \approx_I \sigma' \text{ iff } \sigma \restriction p = \sigma' \restriction p \text{ for every } D\text{-clique } p.$$

Here and elsewhere, if A is a finite set, $\rho \in A^\infty$ and $B \subseteq A$ then $\rho \restriction B$ is the sequence obtained by erasing from ρ all occurrences of letters in $A - B$.

Clearly $\approx_I$ is an equivalence relation. Notice that if $\sigma = \tau a b \sigma_1$ and $\sigma' = \tau b a \sigma_1$ with $(a, b) \in I$ then $\sigma \approx_I \sigma'$. Thus σ and σ' are identified if they differ only in the order of appearance of a pair of adjacent independent actions. In fact, for finite words, an alternative way to characterize $\approx_I$ is to say that $\sigma \approx_I \sigma'$ iff σ' can be obtained from σ by a finite sequence of permutations of adjacent independent actions. However the definition of $\approx_I$ in terms of permutations can not be directly transported to infinite words, which is why we work with the definition presented here.

The equivalence classes generated by $\approx_I$ are called *(Mazurkiewicz) traces.* A set of traces is called a *trace language.* The theory of traces is well developed and documented—see [6, 7] for basic material as well as a substantial number of references to related work.

A variety of models of distributed systems naturally have a trace alphabet associated with them [55]. It also turns out that many interesting properties of distributed systems respect the equivalence relation induced by these trace alphabets. This has important consequences for the practical verification of such properties.

The key notion in this context is that of a trace consistent property. To bring this out, we start with a trace alphabet (Σ, I) and recall the remarks concerning the abolition of atomic propositions at the end of Section 2. Let $L \subseteq \Sigma^\omega$. We say that L is *trace consistent* in case $\sigma \in L$ and $\sigma \approx_I \sigma'$ implies $\sigma' \in L$; for every $\sigma, \sigma' \in \Sigma^\omega$. In other words, either *all* members of a trace are in L or *none* of them are. We say that the formula α in LTL(Σ) is trace consistent in case L_α is trace consistent. It is not hard to see that there is a one-to-one correspondence between trace languages and trace consistent languages of strings.

Now suppose Pr is a program over Σ which has a trace alphabet (Σ, I) associated with it in some natural manner. Suppose further that L_{Pr}, the linear time behaviour of Pr, is trace consistent (we will see a number of models of distributed programs that possess these features in the material to follow). Now consider a specification α which happens to be trace consistent. Then, as remarked at the end of Section 2, verifying $Pr \models \alpha$ boils down to verifying $L_{Pr} \subseteq L_\alpha$. Instead of checking $L_{Pr} \subseteq L_\alpha$ we can choose to check $L' \subseteq L_\alpha$ where L' is designed to be such that $L' \subseteq L_{Pr}$ and for every $\sigma \in L_{Pr}$, $[\sigma] \cap L' \neq \emptyset$. The key point is, the finite representation of L' can be often substantially smaller than the representation of Pr. This is the insight underlying many of the so called partial-order methods deployed in the model checking world [17, 35, 50].

As pointed out in the introduction this is also the main motivation for considering the trace-based linear time temporal logics that we will encounter later. We shall conclude this section with some examples.

Recall the material on elementary net systems introduced in Section 2. Suppose $\mathcal{N} = (B, E, F, c_{in})$ is an elementary net system. Each such system induces the independence relation $I_{\mathcal{N}}$ given by:

$$I_{\mathcal{N}} = \{(e_1, e_2) \mid ({}^\bullet e_1 \cup e_1^\bullet) \cap ({}^\bullet e_2 \cup e_2^\bullet) = \emptyset\}.$$

Let $e \in E$ and consider the formula $\Box\Diamond\langle e\rangle\top$. The property captured by this formula says that (along every computation) the event e occurs infinitely often. It is easy to see that this is a trace consistent property with respect to the trace alphabet $(E, I_{\mathcal{N}})$. Next consider the net system of Figure 1.

Consider the formula $\beta = \Box\Diamond(\langle e\rangle\top \wedge \langle e'\rangle\top)$. Suppose $\sigma = (e_1 e_2 e e')^\omega$ and $\sigma' = (e_1 e' e_2 e)^\omega$. Then $\sigma, \varepsilon \models \beta$ and $\sigma \approx_{I_{\mathcal{N}}} \sigma'$ but $\sigma', \varepsilon \not\models \beta$. Thus this property is not trace consistent with respect to the trace alphabet induced by this net system.

4 Product Languages and Automata

We will now exhibit a restricted but useful class of distributed behaviours that we call product behaviours. Such behaviours are generated by a network of sequential agents that coordinate their activities by performing common actions together. It will turn out that product behaviours are naturally trace consistent. They also constitute a clean and yet non-trivial subset of the class of trace behaviours considered later.

We first study product Büchi automata. We then formulate in Section 5 the product version of LTL(Σ). We will then use product Büchi automata to solve the satisfiability and model checking problems for the product version of LTL(Σ). The technical details — which we suppress here — can be found in [47]. The key notion underlying product behaviours is that of a distributed alphabet. It can be viewed as an "implementation" of a trace alphabet. As a result, distributed alphabets play a fundamental role in the automata-theoretic aspects of trace

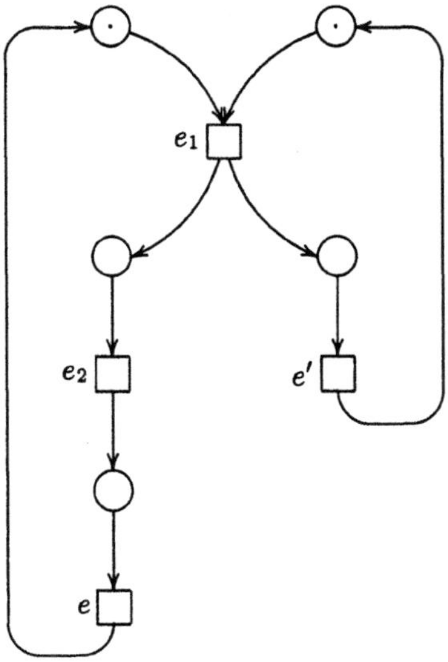

Fig. 1. Example elementary net system

languages [15, 58]. This will become more clear when the material in Section 6 is encountered.

A *distributed alphabet* is a family $\{\Sigma_p\}_{p \in \mathcal{P}}$ where $\mathcal{P}$ is a finite non-empty set of agents (also referred to as processes in the sequel) and Σ_p is a finite non-empty alphabet for each $p \in \mathcal{P}$. The idea is that whenever an action from Σ_p occurs, the agent p must participate in it. Hence the agents can constrain each other's behaviour, both directly and indirectly.

Trace alphabets and distributed alphabets are closely related to each other. Let $\widetilde{\Sigma} = \{\Sigma_p\}_{p \in \mathcal{P}}$ be a distributed alphabet. Then $\Sigma_{\mathcal{P}}$, the global alphabet associated with $\widetilde{\Sigma}$, is the collection $\bigcup_{p \in \mathcal{P}} \Sigma_p$. The distribution of $\Sigma_{\mathcal{P}}$ over $\mathcal{P}$ can be described using a *location function* $\mathrm{loc}_{\widetilde{\Sigma}} : \Sigma_{\mathcal{P}} \to 2^{\mathcal{P}}$ defined as follows:

$$\mathrm{loc}_{\widetilde{\Sigma}}(a) = \{p \mid a \in \Sigma_p\}.$$

This in turn induces the relation $I_{\widetilde{\Sigma}} \subseteq \Sigma_{\mathcal{P}} \times \Sigma_{\mathcal{P}}$ given by:

$$(a, b) \in I_{\widetilde{\Sigma}} \text{ iff } \mathrm{loc}_{\widetilde{\Sigma}}(a) \cap \mathrm{loc}_{\widetilde{\Sigma}}(b) = \emptyset.$$

Clearly $I_{\widetilde{\Sigma}}$ is irreflexive and symmetric and hence $(\Sigma_{\mathcal{P}}, I_{\widetilde{\Sigma}})$ is a trace alphabet. Thus every distributed alphabet canonically induces a trace alphabet. Two actions are independent according to $\widetilde{\Sigma}$ if they are executed by disjoint sets of processes. Henceforth, we write loc for $\mathrm{loc}_{\widetilde{\Sigma}}$ whenever $\widetilde{\Sigma}$ is clear from the context.

Going in the other direction there are, in general, many different ways to implement a trace alphabet as a distributed alphabet. A standard approach is to create a separate agent for each maximal D-clique generated by (Σ, I). Recall that a D-clique of (Σ, I) is a non-empty subset $p \subseteq \Sigma$ such that $p \times p \subseteq D$. Let $\mathcal{P}$ be the set of maximal D-cliques of (Σ, I). This set of processes induces the distributed alphabet $\widetilde{\Sigma} = \{\Sigma_p\}_{p \in \mathcal{P}}$ where $\Sigma_p = p$ for every process p. The alphabet $\widetilde{\Sigma}$ implements (Σ, I) in the sense that the canonical trace alphabet induced by it is exactly (Σ, I). In other words, $\Sigma_{\mathcal{P}} = \Sigma$ and $I_{\widetilde{\Sigma}} = I$.

For example, consider the trace alphabet (Σ, I) where $\Sigma = \{a, b, d\}$ and $I = \{(a, b), (b, a)\}$. The canonical D-clique implementation of (Σ, I) yields the distributed alphabet $\widetilde{\Sigma} = \{\{a, d\}, \{d, b\}\}$.

Through the rest of the section we fix a distributed alphabet $\{\Sigma_p\}_{p \in \mathcal{P}}$ and set $\Sigma = \Sigma_{\mathcal{P}}$. It will be convenient to assume that $\mathcal{P} = \{1, 2, \ldots, K\}$. Further, the ith component of a K-tuple $x = (x_1, x_2, \ldots, x_K)$ will be written as $x[i]$. In other words, $x[i] = x_i$.

A *product Büchi automaton* over $\widetilde{\Sigma}$ is a structure $\mathcal{A} = (\{\mathcal{A}_i\}_{i=1}^K, Q_{in})$ where $\mathcal{A}_i = (Q_i, \longrightarrow_i, F_i, F_i^\omega)$ for each i such that :

- Q_i is a finite set of i-local states.
- $\longrightarrow_i \subseteq Q_i \times \Sigma_i \times Q_i$ is the transition relation of the ith component.
- $F_i \subseteq Q_i$ is a set of finitary accepting states.
- $F_i^\omega \subseteq Q_i$ is a set of infinitary accepting states.
- $Q_{in} \subseteq Q_1 \times Q_2 \times \cdots \times Q_K$ is a set of global initial states.

We use two types of accepting states for the components in order to be able to handle both finite and infinite behaviours. Even if one is interested only in global infinite behaviours, finite behaviours at the component level must be treated; a component might quit after engaging in a finite number of actions while a part of the network runs forever. We use global initial states to obtain the required expressive power. In general, the automaton will not be able to branch off into different parts of the state space, starting from a single global initial state. This will be brought out through a simple example after we define the language behaviour of product automata. The same example will also illustrate why using the cartesian product of local initial state sets as global initial states will result in a loss of expressive power.

Let $\mathcal{A} = (\{\mathcal{A}_i\}_{i=1}^K, Q_{in})$ be a product Büchi automaton over $\widetilde{\Sigma}$. From now on we will say just "product automata". Also, we shall often suppress the mention of $\widetilde{\Sigma}$. We will also write $\{\mathcal{A}_i\}$ instead of $\{\mathcal{A}_i\}_{i=1}^K$. Let $\mathcal{A}_i = (Q_i, \longrightarrow_i, F_i, F_i^\omega)$. Then we set $Q_G^{\mathcal{A}} = Q_1 \times Q_2 \times \ldots \times Q_K$. When $\mathcal{A}$ is clear from the context, we will write Q_G instead of $Q_G^{\mathcal{A}}$. The global transition relation of $\mathcal{A}$ is denoted as $\longrightarrow_{\mathcal{A}}$ and it is the subset of $Q_G \times \Sigma \times Q_G$ given by:

$$q \xrightarrow{a}_{\mathcal{A}} q' \text{ iff } \forall\, i \in \text{loc}(a) : q[i] \xrightarrow{a}_i q'[i] \text{ and } \forall\, i \notin \text{loc}(a) : q[i] = q'[i].$$

Let $\sigma \in \Sigma^\infty$. A *run* of $\mathcal{A}$ over σ is a map $\rho : \text{Prf}(\sigma) \longrightarrow Q_G$ which satisfies:

- $\rho(\varepsilon) \in Q_{in}$.
- $\forall\, \tau a \in \text{prf}(\sigma).\ \rho(\tau) \xrightarrow{a}_{\mathcal{A}} \rho(\tau a)$.

A simple but useful property of runs is the following. Suppose ρ is a run of the product automaton $\mathcal{A}$ over σ. Further suppose that $\tau, \tau' \in \mathrm{Prf}(\sigma)$ such that $\tau \upharpoonright i = \tau' \upharpoonright i$ for some i. Then $\rho(\tau)[i] = \rho(\tau')[i]$.

Let ρ be a run of the product automaton $\mathcal{A}$ over σ. Then ρ is *accepting* iff for each i, the following condition is satisfied:

- If $\sigma \upharpoonright i$ is finite then $\rho(\tau)[i] \in F_i$ where $\tau \in \mathrm{prf}(\sigma)$ such that $\tau \upharpoonright i = \sigma \upharpoonright i$.
- If $\sigma \upharpoonright i$ is infinite then $\rho(\tau a)[i] \in F_i^{\omega}$ for infinitely many $\tau a \in \mathrm{prf}(\sigma)$ with $a \in \Sigma_i$.

If $\sigma \upharpoonright i$ is finite then clearly there exists $\tau \in \mathrm{prf}(\sigma)$ such that $\tau \upharpoonright i = \sigma \upharpoonright i$. Now the above property of runs assures us that the notion of an accepting run is well-defined. In case $\sigma \upharpoonright i$ is infinite the acceptance condition can also be phrased as:

- $\rho(\tau)[i] \in F_i^{\omega}$ for infinitely many $\tau \in \mathrm{prf}(\sigma)$.

This once again follows easily from the definition of a run. We now define $\mathcal{L}(\mathcal{A})$, the *language accepted by the product automaton* $\mathcal{A}$ as,

$$\mathcal{L}(\mathcal{A}) = \{\sigma \mid \exists \text{ an accepting run of } \mathcal{A} \text{ over } \sigma\}.$$

Now consider the alphabet $(\{a,d\},\{d,b\})$ and the language $L = \{ad, bd\}$. Figure 2 shows a product automaton over this alphabet which accepts L. It is

$$F_1 = \{q_3, q_5\} \qquad \widehat{F_1} = \emptyset = \widehat{F_2} \qquad F_2 = \{p_3, p_5\}$$

$$Q_{in} = \{(q_1, p_4), (q_4, p_1)\}$$

Fig. 2. Product automaton accepting $L = \{ad, bd\}$

easy to verify that *no* product automaton over this alphabet with a *single* global initial state can accept L. It is also easy to verify that no product automaton whose set of initial states is a cartesian product of component initial state sets can accept this language.

A crucial property of product automata is that they accept $\approx$-consistent languages.

Lemma 3. *Let $\mathcal{A} = (\{\mathcal{A}_i\}, Q_{in})$ be a product automaton over $\widetilde{\Sigma}$. Then $\mathcal{L}(\mathcal{A})$ is trace consistent.*

The class of languages accepted by product automata can now be characterized. To this end we define the K-ary operation $\otimes : 2^{\Sigma_1^\infty} \times 2^{\Sigma_2^\infty} \times \cdots \times 2^{\Sigma_K^\infty} \rightarrow 2^{\Sigma^\infty}$ via $\otimes(L_1, \ldots, L_K) = \{\sigma \mid \sigma \restriction i \in L_i \text{ for each } i\}$.

In what follows we will write $L = L_1 \otimes L_2 \cdots \otimes L_K$ to denote the fact $\otimes(L_1, \ldots, L_K) = L$. We say that $L \subseteq \Sigma^\infty$ is a *direct product language* over $\widetilde{\Sigma}$ iff $\exists\, L_i \subseteq \Sigma_i^\infty$ for each i such that $L = L_1 \otimes L_2 \otimes \cdots \otimes L_K$. Here are two useful properties of direct product languages. In stating this result and elsewhere we will say "product language" instead of "product language over $\widetilde{\Sigma}$" etc.

Proposition 4.

1. *Let L be a direct product language and $\sigma \in \Sigma^\infty$. Then $\sigma \in L$ iff for each i there exists $\sigma_i \in L$ such that $\sigma \restriction i = \sigma_i \restriction i$.*
2. *Let $L \subseteq \Sigma^\infty$. Then L is a direct product language iff $L = \widehat{L}_1 \otimes \widehat{L}_2 \otimes \cdots \otimes \widehat{L}_K$ where $\widehat{L}_i = \{\sigma \restriction i \mid \sigma \in L\}$ for each i.*

As usual, for an alphabet Σ and $L \subseteq \Sigma^\infty$ we say that L is *regular* iff $L \cap \Sigma^*$ is a regular subset of Σ^* and $L \subseteq \Sigma^\omega$ is an ω-regular subset of Σ^ω as described in Section 2. We can now define the class of languages accepted by product automata.

Definition 5.

- *$\mathcal{R}_0^\otimes(\widetilde{\Sigma})$ is the subset of 2^{Σ^∞} given by $L \in \mathcal{R}_0^\otimes(\widetilde{\Sigma})$ iff $L = L_1 \otimes L_2 \otimes \cdots \otimes L_K$ with each L_i a regular subset of Σ_i^∞.*
- *$\mathcal{R}^\otimes(\widetilde{\Sigma})$ is the least subset of 2^{Σ^∞} which contains $\mathcal{R}_0^\otimes$ and is closed under finite unions.*

The class $\mathcal{R}^\otimes(\widetilde{\Sigma})$ defined above will be called the *regular product languages* over $\widetilde{\Sigma}$. As usual, we shall often write $\mathcal{R}_0^\otimes$ instead of $\mathcal{R}_0^\otimes(\widetilde{\Sigma})$ and write $\mathcal{R}^\otimes$ instead of $\mathcal{R}^\otimes(\widetilde{\Sigma})$. An interesting observation concerning $\mathcal{R}^\otimes$ is the following:

Proposition 6. *$\mathcal{R}^\otimes$ is closed under boolean operations.*

It turns out that $\mathcal{R}^\otimes$ is precisely the class of languages accepted by product automata.

Theorem 7 ([47]). *Let $L \subseteq \Sigma^\infty$. Then $L \in \mathcal{R}^\otimes$ iff there exists a product automaton $\mathcal{A}$ such that $L = \mathcal{L}(\mathcal{A})$.*

We shall be using product automata to settle the decidability and model checking problems for the logic LTL$^\otimes$ to be introduced in the next section. In anticipation of this, we shall put down two more results concerning product automata. While doing so and elsewhere the *size* of the product automaton $\mathcal{A}$ will be understood to be $|Q_G|$.

Theorem 8. *Let $\mathcal{A}$ be a product automaton. Then the question $\mathcal{L}(\mathcal{A}) \stackrel{?}{=} \emptyset$ can be settled in time $O(2^{2K} \cdot n^2)$ where n is the size of $\mathcal{A}$.*

Theorem 9. *Let $\mathcal{A}^1$ and $\mathcal{A}^2$ be two product automata. Then one can effectively construct a product automaton $\mathcal{A}$ such that $\mathcal{L}(\mathcal{A}) = \mathcal{L}(\mathcal{A}^1) \cap \mathcal{L}(\mathcal{A}^2)$ and moreover $n = O(2^K \cdot n_1 \cdot n_2)$ where n is the size of $\mathcal{A}$ and n_ℓ is the size of $\mathcal{A}^\ell$ for $\ell = 1, 2$.*

5 A Product Version of LTL

We now wish to design a product version of LTL denoted $\mathrm{LTL}^{\otimes}(\widetilde{\Sigma})$. The set of formulas and their *locations* are given by:

- $\top$ is a formula and $\mathrm{loc}(\top) = \emptyset$.
- Suppose α and β are formulas. Then so are $\sim\alpha$ and $\alpha \vee \beta$. Furthermore, $\mathrm{loc}(\sim\alpha) = \mathrm{loc}(\alpha)$ and $\mathrm{loc}(\alpha \vee \beta) = \mathrm{loc}(\alpha) \cup \mathrm{loc}(\beta)$.
- Suppose $a \in \Sigma_i$ and α is a formula with $loc(\alpha) \subseteq \{i\}$. Then $\langle a \rangle_i \alpha$ is a formula and $loc(\langle a \rangle_i \alpha) = \{i\}$.
- Suppose α and β are formulas such that $\mathrm{loc}(\alpha), \mathrm{loc}(\beta) \subseteq \{i\}$. Then $\alpha \mathcal{U}_i \beta$ is a formula. Moreover, $\mathrm{loc}(\alpha \mathcal{U}_i \beta) = \{i\}$.

We note that each formula in $\mathrm{LTL}^{\otimes}(\widetilde{\Sigma})$ is a boolean combination of formulas taken from the set $\bigcup_{i \in Loc} \mathrm{LTL}_i^{\otimes}(\widetilde{\Sigma})$ where, for each i,

$$\mathrm{LTL}_i^{\otimes}(\widetilde{\Sigma}) = \{\alpha \mid \alpha \in \mathrm{LTL}^{\otimes}(\widetilde{\Sigma}) \text{ and } \mathrm{loc}(\alpha) \subseteq \{i\}\ \}.$$

Stated differently, the syntax of $\mathrm{LTL}_i^{\otimes}(\widetilde{\Sigma})$ is given inductively by:

- $\top \in \mathrm{LTL}_i^{\otimes}(\widetilde{\Sigma})$.
- If α and β are in $\mathrm{LTL}_i^{\otimes}(\widetilde{\Sigma})$ then $\sim\alpha$ and $\alpha \vee \beta$ are in $\mathrm{LTL}_i^{\otimes}(\widetilde{\Sigma})$.
- If α is in $\mathrm{LTL}_i^{\otimes}(\widetilde{\Sigma})$ and $a \in \Sigma_i$ then $\langle a \rangle_i \alpha$ is in $\mathrm{LTL}_i^{\otimes}(\widetilde{\Sigma})$.
- If α and β are in $\mathrm{LTL}_i^{\otimes}(\widetilde{\Sigma})$ then $\alpha \mathcal{U}_i \beta$ is in $\mathrm{LTL}_i^{\otimes}(\widetilde{\Sigma})$.

Once again, we have chosen to avoid dealing with atomic propositions for the sake of convenience. They can be introduced in a local fashion as done in [47]. The decidability result to be presented will go through with minor notational overheads.

As before, we will often suppress the mention of $\widetilde{\Sigma}$. We will also often write τ_i, τ_i' and τ_i'' instead of $\tau \restriction i$, $\tau' \restriction i$ and $\tau'' \restriction i$, respectively with $\tau, \tau', \tau'' \in \Sigma^*$.

A model is a sequence $\sigma \in \Sigma^\infty$ and the semantics of this logic is given, as before, with $\tau \in \mathrm{prf}(\sigma)$.

- $\sigma, \tau \models \top$.
- $\sigma, \tau \models \sim\alpha$ iff $\sigma, \tau \not\models \alpha$.
- $\sigma, \tau \models \alpha \vee \beta$ iff $\sigma, \tau \models \alpha$ or $\sigma, \tau \models \beta$.
- $\sigma, \tau \models \langle a \rangle_i \alpha$ iff there exists $\tau' \in \mathrm{prf}(\sigma)$ such that $\sigma, \tau' \models \alpha$ and $\tau_i' = \tau_i a$. (recall that $\tau_i' = \tau' \restriction i$.)

- $\sigma, \tau \models \alpha \mathcal{U}_i \beta$ iff there exists τ' such that $\tau\tau' \in \mathrm{prf}(\sigma)$ and $\sigma, \tau\tau' \models \beta$. Further, for every $\tau'' \in \mathrm{prf}(\tau')$, if $\varepsilon \preceq \tau''_i \prec \tau'_i$ then $\sigma, \tau\tau'' \models \alpha$.

As before we derive some useful modalities:

- $O_i\alpha \overset{\Delta}{\Longleftrightarrow} \bigvee_{a \in \Sigma_i} \langle a \rangle_i \alpha$.
- $\Diamond_i \alpha \overset{\Delta}{\Longleftrightarrow} \top \mathcal{U}_i \alpha$.
- $\Box_i \alpha \overset{\Delta}{\Longleftrightarrow} \sim \Diamond_i \sim \alpha$.

Let $M = \sigma$ be a model and $\tau \in \mathrm{prf}(\sigma)$. The following assertions can now easily be checked.

- $\sigma, \tau \models O_i\alpha$ iff there exists $\tau' \in prf(\sigma)$ such that $\sigma, \tau' \models \alpha$ and $|\tau'_i| = |\tau_i| + 1$.
- $\sigma, \tau \models \Diamond_i \alpha$ iff there exists τ' with $\tau\tau' \in \mathrm{prf}(\sigma)$ such that $\sigma, \tau\tau' \models \alpha$.
- $\sigma, \tau \models \Box_i \alpha$ iff for each τ', $\tau\tau' \in \mathrm{prf}(\sigma)$ implies $\sigma, \tau\tau' \models \alpha$.

Note that $O_i\alpha$ is the i-local version of the usual next-state operator of LTL.

We will say that a formula $\alpha \in \mathrm{LTL}^{\otimes}(\widetilde{\Sigma})$ is *satisfiable* if there exist $\sigma \in \Sigma^{\infty}$ and $\tau \in \mathrm{prf}(\sigma)$ such that $\sigma, \tau \models \alpha$. The *language defined by* α is given by

$$L_\alpha = \{\sigma \in \Sigma^{\infty} \mid \sigma, \varepsilon \models \alpha\}.$$

We will show the satisfiability problem for $\mathrm{LTL}^{\otimes}(\widetilde{\Sigma})$ is solvable in deterministic exponential time. This will be achieved by effectively constructing a product automaton $\mathcal{A}_\alpha$ for each $\alpha \in \mathrm{LTL}^{\otimes}(\widetilde{\Sigma})$ such that the language accepted by $\mathcal{A}_\alpha$ is non-empty iff α is satisfiable. Our construction is a generalization of the one for LTL in Section 2. The solution to the satisfiability problem will at once lead to a solution to the model checking problem for programs modelled as a product of sequential agents.

Through the rest of the section we fix a formula $\alpha_0 \in \mathrm{LTL}^{\otimes}(\widetilde{\Sigma})$. As before we will for convenience assume that the derived local next-state modality O_i is included in the syntax of $\mathrm{LTL}^{\otimes}$. In order to construct $\mathcal{A}_{\alpha_0}$ we first define the (Fischer-Ladner) closure of α_0. As a first step let $cl(\alpha_0)$ be the least set of formulas satisfying:

- $\alpha_0 \in cl(\alpha_0)$.
- $\sim\alpha \in cl(\alpha_0)$ implies $\alpha \in cl(\alpha_0)$.
- $\alpha \vee \beta \in cl(\alpha_0)$ implies $\alpha, \beta \in cl(\alpha_0)$.
- $\langle a \rangle_i \alpha \in cl(\alpha_0)$ implies $\alpha \in cl(\alpha_0)$.
- $\alpha \mathcal{U}_i \beta \in cl(\alpha_0)$ implies $\alpha, \beta \in cl(\alpha_0)$. In addition, $O_i(\alpha \mathcal{U}_i \beta) \in cl(\alpha_0)$.

We will now take the *closure* of α_0 to be $CL(\alpha_0) = cl(\alpha_0) \cup \{\sim\alpha \mid \alpha \in cl(\alpha_0)\}$. From now on we shall identify $\sim\sim\alpha$ with α. Set $CL_i(\alpha_0) = CL(\alpha_0) \cap \mathrm{LTL}_i^{\otimes}$ for each i. We will often write CL instead of $CL(\alpha_0)$ and CL_i instead of $CL_i(\alpha_0)$. All formulas considered from now on will be assumed to belong to CL unless otherwise stated.

An *i-type atom* is a subset $A \subseteq CL_i$ which satisfies:

- $\top \in A$.
- $\alpha \in A$ iff $\sim\alpha \notin A$.
- $\alpha \vee \beta \in A$ iff $\alpha \in A$ or $\beta \in A$.
- $\alpha \mathcal{U}_i \beta \in A$ iff $\beta \in A$ or $\alpha, O_i(\alpha \mathcal{U}_i \beta) \in A$.

The set of i-type atoms is denoted AT_i. We next define, for each $\alpha \in CL(\alpha_0)$ and $(A_1, \ldots, A_K) \in AT_1 \times \ldots \times AT_K$, the predicate Member$(\alpha, (A_1, \ldots, A_K))$. For convenience this predicate will be denoted as $\alpha \in (A_1, \ldots, A_K)$ and is given inductively by:

- Let $\alpha \in CL_i$. Then $\alpha \in (A_1, \ldots, A_K)$ iff $\alpha \in A_i$.
- Let $\alpha = \sim\beta$. Then $\alpha \in (A_1, \ldots, A_K)$ iff $\beta \notin (A_1, \ldots, A_K)$.
- Let $\alpha = \beta \vee \gamma$. Then $\alpha \in (A_1, \ldots, A_K)$ iff $\beta \in (A_1, \ldots, A_K)$ or $\gamma \in (A_1, \ldots, A_K)$.

Finally, we set $U_i = \{\alpha \mathcal{U}_i \beta \mid \alpha \mathcal{U}_i \beta \in CL_i(\alpha_0)\}$ for each i. The product automaton $\mathcal{A}_{\alpha_0}$ associated with α_0 is now defined to be $\mathcal{A}_{\alpha_0} = (\{\mathcal{A}_i\}, Q_{in})$ where, for each i, $\mathcal{A}_i = (Q_i, \longrightarrow_i, F_i, F_i^\omega)$ is specified as follows:

- $Q_i = AT_i \times \{\mathsf{off}, \mathsf{on}\} \times 2^{U_i}$
- $\longrightarrow_i \subseteq Q_i \times \Sigma_i \times Q_i$ is given by, $(A, x, u) \xrightarrow{a}_i (B, y, v)$ iff the following conditions are met.
  1. $x = \mathsf{on}$ and for all $\langle a \rangle_i \alpha \in CL_i(\alpha_0)$, $\langle a \rangle_i \alpha \in A$ iff $\alpha \in B$ and for all $O_i \alpha \in CL_i(\alpha_0)$, $O_i \alpha \in A$ iff $\alpha \in B$. Moreover, if $\langle b \rangle_i \beta \in A$ then $b = a$.
  2. If $u \neq \emptyset$ then $v = \{\alpha \mathcal{U}_i \beta \mid \alpha \mathcal{U}_i \beta \in u \text{ and } \beta \notin B\}$. If $u = \emptyset$ then $v = \{\alpha \mathcal{U}_i \beta \mid \alpha \mathcal{U}_i \beta \in B \text{ and } \beta \notin B\}$.
- $F_i \subseteq Q_i$ is given by: $(A, x, u) \in F_i$ iff $x = \mathsf{off}$ and for all $\langle a \rangle_i \alpha \in CL_i(\alpha_0)$, $\langle a \rangle_i \alpha \notin A$ and for all $O_i \alpha \in CL_i(\alpha_0)$, $O_i \alpha \notin A$.
- $F_i^\omega \subseteq Q_i$ is given by: $(A, x, u) \in F_i^\omega$ iff $u = \emptyset$.
- $Q_{in} \subseteq Q_1 \times Q_2 \times \ldots \times Q_K$ is given by: $((A_1, x_1, u_1), \ldots, (A_K, x_K, u_K)) \in Q_{in}$ iff $\alpha_0 \in (A_1, \ldots, A_K)$ and $u_i = \emptyset$ for every i.

It is not difficult to now establish the next result by an application of Theorem 8.

Theorem 10. *α_0 is satisfiable iff $\mathcal{L}(\mathcal{A}_{\alpha_0}) \neq \emptyset$. Hence the satisfiability problem for* LTL$^\otimes$ *is decidable in exponential time.*

We now turn to the model checking problem for LTL$^\otimes$. A product program (over $\widetilde{\Sigma}$) is a structure $Pr = (\{Pr_i\}_{i=1}^K, Q_{in}^{Pr})$ where, for each i, $Pr_i = (Q_i, \longrightarrow_i)$ with Q_i a finite set and $\longrightarrow_i \subseteq Q_i \times \Sigma_i \times Q_i$. Since we have agreed to drop atomic propositions there is no need for (local) interpretations for the atomic propositions. Let us further assume for convenience that Q_{in}^{Pr} is a singleton with q_{in} as its sole member and with $q_{in}[i] = q_{in}^i$ for each i. With each such program we can associate the product automaton $\mathcal{A}_{Pr} = (\{\mathcal{A}_i\}_{i=1}^K, \{q_{in}\})$ where $\mathcal{A}_i = (Q_i, \longrightarrow_i, Q_i, Q_i)$ for each i.

Now let Pr be a product program and α_0 be a formula of LTL$^\otimes$. As in the case for LTL, we say that Pr *meets the specification* α_0 — again denoted $Pr \models \alpha_0$ — iff $\sigma, \varepsilon \models \alpha_0$ for every $\sigma \in \mathcal{L}(\mathcal{A}_{Pr})$. Once again, using Theorem 9 it is not difficult to prove the following.

Theorem 11. *The model checking problem for* LTL$^{\otimes}$ *is decidable in time* $O(|Pr| \cdot 2^{|\alpha_0|})$.

We wish to observe that each product program can be represented as a Σ-labelled 1-safe net system. To see this let $Pr = (\{Pr_i\}_{i=1}^{K}, \{q_{in}\})$ be a product program. Let's assume without loss of generality that the family of local states $\{Q_i\}$ is pairwise disjoint. We set $Q = \bigcup_{i \in \mathcal{P}} Q_i$ and define an a-state to be a map $q_a : \text{loc}(a) \to Q$ which satisfies $q_a(i) \in Q_i$ for each i in loc(a). (A more elaborate development of these notions will appear in the next section). An a-event is a pair of a-states (q_a, q'_a) which satisfies $q_a(i) \xrightarrow{a}_i q'_a(i)$ for each i in loc(a). We let E_a be the set of a-events. We can now define the Σ-labelled 1-safe net system representing Pr to be $\mathcal{N} = (B, E, F, c_{in}, \phi)$ where:

- $B = Q$
- $E = \bigcup_{a \in \Sigma} E_a$
- Let $q_i \in Q_i$ and $e = (q_a, q'_a) \in E_a$. Then $(q_i, e) \in F$ iff $i \in \text{loc}(a)$ and $q_a(i) = q_i$. Similarly $(e, q_i) \in F$ iff $i \in \text{loc}(a)$ and $q'_a(i) = q_i$.
- Let $e \in E$. Then $\phi(e) = a$ iff e is an a-event.

On the other hand each 1-safe net system which is covered by a set of S-components can be viewed as a (deterministic) product program; the alphabet of each component is its set of events. If necessary, S-complementation can be performed to ensure that the system is covered by a set of S-components. We do not wish to enter into details here. Instead we show on Figure 3 an example of a 1-safe net system composed out of three components.

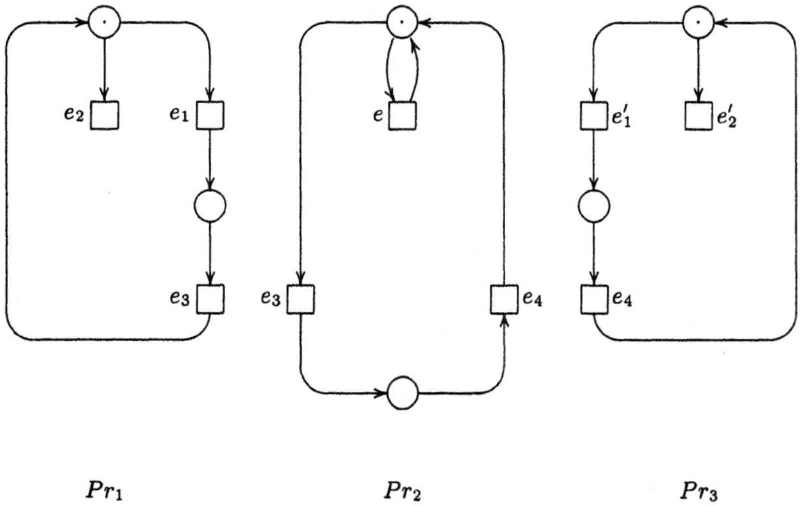

Fig. 3. 1-safe net with three components

Let Pr denote the associated product program over the distributed alphabet $\{\{e_1, e_2, e_3\}, \{e_3, e_4\}, \{e'_1, e'_2, e_4\}\}$. Then it is easy to check that

$$Pr \models \Box_1 O_1 \top \supset \Box_3 O_3 \top .$$

This property says that along every computation, if the first component executes infinitely often then so does the third component. The point to note is that the first component and the third component do not have any common events and hence there is no direct communication between them. Nevertheless through the power of the boolean connectives alone the logic can make assertions about the way components that are "far apart" are required to influence each other's behaviour.

6 Trace Languages and Automata

Traces have many equivalent representations. Here we shall view them as restricted Σ-labelled partial orders. Abusing terminology we shall call these objects also *traces*. We will then argue that these objects are in a rather precise sense the same as the objects called traces defined in Section 3 in terms of equivalence classes of sequences.

Let T be a Σ-labelled poset. In other words, $(E, \leq)$ is a poset and $\lambda : E \to \Sigma$ is a labelling function. For $Y \subseteq E$ we define $\downarrow Y = \{x \mid \exists y \in Y : \ x \leq y\}$ and $\uparrow Y = \{x \mid \exists y \in Y : \ y \leq x\}$. In case $Y = \{y\}$ is a singleton we shall write $\downarrow y$ ($\uparrow y$) instead of $\downarrow \{y\}$ ($\uparrow \{y\}$). We also let $\lessdot$ be the relation: $x \lessdot y$ iff $x < y$ and for all $z \in E$, $x \leq z \leq y$ implies $x = z$ or $z = y$.

A *trace* (over (Σ, I)) is a Σ-labelled poset $T = (E, \leq, \lambda)$ satisfying:

(T1) $\forall e \in E.$ $\quad \downarrow e$ is a finite set
(T2) $\forall e, e' \in E.$ $\quad e \lessdot e'$ implies $\lambda(e) \ D \ \lambda(e')$.
(T3) $\forall e, e' \in E.$ $\quad \lambda(e) \ D \ \lambda(e')$ implies $e \leq e'$ or $e' \leq e$.

We shall refer to members of E as *events*. The trace $T = (E, \leq, \lambda)$ is said to be *finite* if E is a finite set. Otherwise it is an *infinite* trace. Note that E is always a countable set. T is said to be *non-empty* in case $E \neq \emptyset$. We let $TR^{fin}(\Sigma, I)$ be the set of finite traces and $TR^{\omega}(\Sigma, I)$ be the set of infinite traces over (Σ, I) and set $TR(\Sigma, I) = TR^{fin}(\Sigma, I) \cup TR^{\omega}(\Sigma, I)$. Often we will write TR^{fin} instead of $TR^{fin}(\Sigma, I)$ etc. As before, a subset of traces $L_{Tr} \subseteq TR$ will be called a *trace language*.

Let $T = (E, \leq, \lambda)$ be a trace. The finite prefixes of T, to be called configurations, will play a crucial role in what follows. A *configuration* of T is a finite subset $c \subseteq E$ such that $c = \downarrow c$. We let C_T be the set of configurations of T and let c, c', c'' range over C_T. Note that $\emptyset$, the empty set, is a configuration and $\downarrow e$ is a configuration for every $e \in E$. Finally, the transition relation $\longrightarrow_T \subseteq C_T \times \Sigma \times C_T$ is given by: $c \xrightarrow{a}_T c'$ iff there exists $e \in E$ such that $\lambda(e) = a$ and $e \notin c$ and $c' = c \cup \{e\}$. It is easy to see that if $c \xrightarrow{a}_T c'$ and $c \xrightarrow{a}_T c''$ then $c' = c''$.

Note that we have now introduced two different notions of traces; one in terms of equivalence classes of strings as in Section 3 and the other in terms of Σ-labelled partial orders as in this section. We now sketch briefly the constructions that show that $\Sigma^\infty/\approx_I$ and $TR(\Sigma, I)$ represent the same class of objects. We shall construct representation maps $\mathsf{str} : \Sigma^\infty/\approx_I \to TR(\Sigma, I)$ and $\mathsf{trs} : TR(\Sigma, I) \to \Sigma^\infty/\approx_I$ and state some results which show that these maps are "inverses" of each other. We shall not prove these results. The details can be easily obtained using the constructions developed in [55] for relating traces and event structures.

Henceforth, we will not distinguish between isomorphic elements in $TR(\Sigma, I)$. In other words, whenever we write $T = T'$ for traces $T = (E, \leq, \lambda)$ and $T' = (E', \leq', \lambda')$, we mean that there is a label-preserving isomorphism between T and T'.

Recall that for $\sigma \in \Sigma^\infty$, $[\sigma]$ stands for the $\approx_I$-equivalence class containing σ. We now define $\mathsf{str} : \Sigma^\infty \to TR(\Sigma, I)$. Let $\sigma \in \Sigma^\infty$. Then $\mathsf{str}(\sigma) = (E, \leq, \lambda)$ where:

- $E = \{\tau a \mid \tau a \in \mathrm{prf}(\sigma)\}$. Recall that $\tau \in \Sigma^*$ and $a \in \Sigma$. Thus $E = \mathrm{prf}(\sigma) - \{\varepsilon\}$, where ε is the null string.
- $\leq \subseteq E \times E$ is the least partial order which satisfies: For all $\tau a, \tau' b \in E$, if $\tau a \preceq \tau' b$ and $(a, b) \in D$ then $\tau a \leq \tau' b$.
- For $\tau a \in E$, $\lambda(\tau a) = a$.

The map str induces a natural map str' from $\Sigma^\infty/\approx_I$ to $TR(\Sigma, I)$ defined by $\mathsf{str}'([\sigma]) = \mathsf{str}(\sigma)$. One can show that if $\sigma, \sigma' \in \Sigma^\infty$, then $\sigma \approx_I \sigma'$ iff $\mathsf{str}(\sigma) = \mathsf{str}(\sigma')$. This observation guarantees that str' is well-defined. In fact, henceforth we shall write str to denote both str and str'.

Next, let $T = (E, \leq, \lambda) \in TR(\Sigma, I)$. Then $\sigma \in \Sigma^\infty$ is a *linearization* of T iff there exists a map $\rho : \mathrm{prf}(\sigma) \to \mathcal{C}_T$, such that the following conditions are met:

- $\rho(\varepsilon) = \emptyset$.
- $\forall \tau a \in \mathrm{prf}(\sigma)$ with $\tau \in \Sigma^*$, $\rho(\tau) \xrightarrow{a}_T \rho(\tau a)$.
- $\forall e \in E\ \exists \tau \in \mathrm{prf}(\sigma).\ e \in \rho(\tau)$.

The function ρ will be called a *run map* of the linearization σ. Note that the run map of a linearization is unique. In what follows, we shall let $\mathrm{lin}(T)$ to be the set of linearizations of the trace T.

We can now define the map $\mathsf{trs} : TR(\Sigma, I) \to \Sigma^\infty/\approx_I$ as: $\mathsf{trs}(T) = \mathrm{lin}(T)$. One can now show that for every $\sigma \in \Sigma^\infty$, $\mathsf{trs}(\mathsf{str}(\sigma)) = [\sigma]$ and for every $T \in TR(\Sigma, I)$, $\mathsf{str}(\mathsf{trs}(T)) = T$. This justify our claim that $\Sigma^\infty/\approx_I$ and $TR(\Sigma, I)$ are indeed two equivalent ways of talking about the same class of objects.

We note that every trace consistent subset L of Σ^∞ defines a trace language L_{Tr} given by $L_{Tr} = \{\mathsf{str}(\sigma) \mid \sigma \in L\}$ which has the property $\mathsf{trs}(L_{Tr}) = L$. In this sense every product language defines a trace language. We say that a trace language L_{Tr} is *regular* iff $\mathsf{trs}(L_{Tr})$ is a regular subset of Σ^∞. As we will see later not every (regular) trace language is a (regular) product language. Hence in order to recognize regular trace languages one will have to use strengthened versions of product automata. Such automata called asynchronous automata were formulated by Zielonka for recognizing regular languages of finite traces. These were then generalized for handling infinite traces by Gastin and Petit [15]. We will use a combination of these two types of automata for solving the satisfiability and model checking problems for the trace-based temporal logic called TrPTL to be considered in the next section.

Let $\widetilde{\Sigma}$ be a distributed alphabet with $\mathcal{P}$ as the associated set of agents. In an asynchronous automaton, each process $p \in \mathcal{P}$ is equipped with a finite non-empty set of local p-states, denoted S_p. It will be convenient to develop some notations for talking about "more global" states before defining these automata.

First we set $S = \bigcup_{p \in \mathcal{P}} S_p$ and call S the set of *local states*. We let P, Q range over non-empty subsets of $\mathcal{P}$ and let p, q range over $\mathcal{P}$. A Q-state is a map $s : Q \to S$ such that $s(q) \in S_q$ for every $q \in Q$. We let S_Q denote the set Q-states. We call $S_{\mathcal{P}}$ the set of *global states*.

We use a to abbreviate loc(a) when talking about states (recall that loc$(a) = \{ p \mid a \in \Sigma_p \}$). Thus an a-state is just a loc(a)-state and S_a denotes the set of all loc(a)-states.

A *distributed transition system* TS over $\widetilde{\Sigma}$ is a structure $(\{S_p\}, \{\longrightarrow_a\}, S_{in})$ where

- S_p is a finite non-empty set of p-states for each process p.
- For $a \in \Sigma$, $\longrightarrow_a \subseteq S_a \times S_a$ is a transition relation between a-states.
- $S_{in} \subseteq S_{\mathcal{P}}$ is a set of initial global states.

The idea is that an a-move by TS involves only the local states of the agents which participate in the execution a. This is reflected in the global transition relation $\longrightarrow_{TS} \subseteq S_{\mathcal{P}} \times \Sigma \times S_{\mathcal{P}}$ which is defined as follows: Suppose s and s' are two global states and s_a and s'_a are the two corresponding a-states. In other words, $s_a(i) = s(i)$ and $s'_a(i) = s'(i)$ for each i in loc(a). Then

$$s \overset{a}{\longrightarrow}_{TS} s' \text{ iff } (s_a, s'_a) \in \longrightarrow_a \text{ and } s(j) = s'(j) \text{ for every } j \notin \text{loc}(a).$$

From the definition of $\longrightarrow_{TS}$, it is clear that actions which are executed by disjoint sets of agents are processed independently by TS.

An *asynchronous automaton* over $\widetilde{\Sigma}$ is then a distributed transition system equipped with a set of global accepting states. More precisely, it is a structure $\mathcal{A} = (\{S_p\}, \{\longrightarrow_a\}, S_{in}, F)$ where

- $F \subseteq S_{\mathcal{P}}$ is a set of accepting global states.

A *trace run* of $\mathcal{A}$ over the finite trace $T = (E, \leq, \lambda)$ is a map $\rho : \mathcal{C}_T \to S_{\mathcal{P}}$ such that $\rho(\emptyset) \in S_{in}$ and for every $(c, a, c') \in \longrightarrow_T$, $\rho(c) \overset{a}{\longrightarrow}_{TS} \rho(c')$. We say that ρ is an *accepting run* whenever $\rho(E) \in F$. The *language of finite traces accepted by* $\mathcal{A}$ is given by

$$\mathcal{L}_{Tr}(\mathcal{A}) = \{ T \in TR^{fin} \mid \exists \text{ an accepting run of } \mathcal{A} \text{ over } T \}.$$

In the present setting Zielonka's fundamental result can now be formulated as

Theorem 12 ([58]). *$L \subseteq TR^{fin}(\Sigma, I)$ is regular iff $L = L_{Tr}(\mathcal{A})$ for some asynchronous automaton $\mathcal{A}$ over some $\widetilde{\Sigma}$ where $\widetilde{\Sigma}$ is a distributed alphabet whose induced trace alphabet is (Σ, I). Further, one may assume $\mathcal{A}$ to be deterministic and one may assume $\widetilde{\Sigma}$ to be the distributed alphabet induced by the maximal D-cliques of (Σ, I).*

This result has been generalized to the set of ω-regular trace languages by Gastin and Petit [15] in terms of asynchronous automata with Büchi acceptance conditions. Since we will treat both finite and infinite traces on an equal footing we will present a class of automata capable of accepting both finite and infinite traces. Hence our automata are essentially distributed transition systems augmented with *both* finite and infinite accepting states.

An *asynchronous Büchi automaton* over $\widetilde{\Sigma}$ is a structure

$$\mathcal{A} = (\{S_p\}, \{\longrightarrow_a\}, S_{in}, \{(F_p, F_p^\omega)\}),$$

where:

- $(\{S_p\}, \{\longrightarrow_a\}, S_{in})$ is a distributed transition system.
- $F_p \subseteq S_p$ is a set of local finitary accepting states of process p.
- $F_p^\omega \subseteq S_p$ is a set of local infinitary accepting states of process p.

For convenience we will from now on denote this class of automata just "asynchronous automata".

To define acceptance we must now compute $\text{Inf}_p(\rho)$, the set of p-states that are encountered infinitely often along ρ. When incorporating both finite and infinite behaviour in this richer domain we have to take care in defining the set of infinitely occuring states of process p. The obvious definition, namely $\text{Inf}_p(\rho) = \{s_p \mid \rho(c)(p) = s_p$ for infinitely many $c \in \mathcal{C}_T\}$, will not work. The complication arises because some processes may make only finitely many moves, even though the overall trace consists of an infinite number of events.

For instance, consider the distributed alphabet $\widetilde{\Sigma}_0 = \{\{a\}, \{b\}\}$. In the corresponding distributed transition system, there are two processes p and q which execute a's and b's completely independently. Consider the trace $T = (E, \leq, \lambda)$ where $|E_p| = 1$ and E_q is infinite — i.e., all the infinite words in $\text{trs}(T)$ contain one a and infinitely many b's. Let s_p be the state of p after executing a. Then, there will be infinitely many configurations whose p-state is s_p, even though p only moves a finite number of times.

Continuing with the same example, consider another infinite trace $T' = (E', \leq', \lambda')$ over the same alphabet where both E_p and E_q are infinite. Once again, let s_p be the local state of p after reading one a. Further, let us suppose that after reading the second a, p never returns to the state s_p. It will still be the case that there are infinitely many configurations whose p-state is s_p: consider the configurations $c_0, c_1, c_2, \ldots$ where c_j is the finite configuration after one a and j b's have occurred.

So, we have to define $\text{Inf}_p(\rho)$ so as to detect whether or not process p is making progress. The appropriate formulation is as follows:

Case 1 *E_p is finite:* $\mathrm{Inf}_p(\rho) = \{s_p\}$, where $\rho(\downarrow E_p) = s$ and $s_p = s(p)$.
Case 2 *E_p is an infinite set:*

$$\mathrm{Inf}_p(\rho) = \{s_p \mid \text{for infinitely many } e \in E_p, s_e(p) = s_p, \text{ where } \rho(\downarrow e) = s_e\}.$$

A *trace run* of an asynchronous automaton over the (possibly infinite) trace $T = (E, \leq, \lambda) \in TR$ is now defined in the obvious way. A run ρ of $\mathcal{A}$ over the (possibly infinite) trace $T = (E, \leq, \lambda)$ is *accepting* iff for each process p the following conditions are met:

- If E_p is *finite* then $\mathrm{Inf}_p(\rho) \cap F_p \neq \emptyset$.
- If E_p is *infinite* then $\mathrm{Inf}_p(\rho) \cap F_p^\omega \neq \emptyset$.

We then have the following characterization extending Theorem 12.

Theorem 13. *A trace language $L \subseteq TR(\Sigma, I)$ is regular iff $L = L_{Tr}(\mathcal{A})$ for an asynchronous automaton over $\widetilde{\Sigma}$ where $\widetilde{\Sigma}$ is a distributed alphabet whose induced trace alphabet is (Σ, I).*

It should be noted however that deterministic automata no longer suffice for accepting *all* regular languages.

We say that $\mathcal{A}$ is *in standard form* if

- For each p, $F_p \cap F_p^\omega = \emptyset$.
- For each $(s_a, t_a) \in \longrightarrow_a$ and $p \in \mathrm{loc}(a)$ we have that $s_a(p) \notin F_p$.

Thus, $\mathcal{A}$ is in standard form if the p-states in F_p are all "dead" and disjoint from F_p^ω. It is easy to convert every asynchronous automaton into standard form. All our asynchronous automata will be in standard form.

We conclude with a result concerning the emptiness problem for asynchronous automata.

Proposition 14 ([30]). *Let $\mathcal{A}$ be an asynchronous automaton in standard form. The emptiness problem is decidable in time $O(n^{2|\mathcal{P}|})$, where n is the largest of the local state spaces, S_p.*

We have defined here the languages defined by asynchronous automata in terms of traces. We note that these automata can be viewed — and this is the conventional approach — as automata running over Σ-sequences. Using the global transition relations of these automata one can easily define the string languages accepted by these automata. These languages will be naturally trace consistent w.r.t. the trace alphabets induced by the associated distributed alphabets. The resulting trace languages will be precisely the trace languages accepted by these automata according to the definitions we have provided here.

7 TrPTL

We present here the linear time temporal logic over traces called TrPTL. This is the first such logic patterned after PTL (i.e. LTL) formulated for traces. For a detailed treatment of this logic the reader is referred to [44, 45].

As before, it will be notationally convenient to deal with distributed alphabets in which the names of the processes are positive integers. Through this section and the next, we fix a distributed alphabet $\widetilde{\Sigma} = \{\Sigma_i\}_{i\in\mathcal{P}}$ with $\mathcal{P} = \{1, 2, \ldots, K\}$ and $K \geq 1$. We let i, j and k range over $\mathcal{P}$. As before, let P, Q range over non-empty subsets of $\mathcal{P}$. The trace alphabet induced by $\widetilde{\Sigma}$ is denoted (Σ, I). We assume the terminology and notations developed in the previous sections. In particular, when dealing with a $\mathcal{P}$-indexed family $\{X_i\}_{i\in\mathcal{P}}$ we will often write just $\{X_i\}$.

The logic TrPTL is parameterized by the class of distributed alphabets. Having fixed $\widetilde{\Sigma}$ we shall often almost always write TrPTL to mean TrPTL($\widetilde{\Sigma}$), the logic associated with $\widetilde{\Sigma}$. In order to better illustrate the main features of the logic we will first include atomic propositions. They will be dropped once we return to considering the technical aspects of the logic. We fix a finite non-empty set of atomic propositions P with p, q ranging over P. Then $\Phi_{\mathrm{TrPTL}(\widetilde{\Sigma})}$, the set of formulas of TrPTL($\widetilde{\Sigma}$), is defined inductively via:

- For $p \in P$ and $i \in \mathcal{P}$, $p(i)$ is a formula (which is to be read "p at i").
- If α and β are formulas, so are $\sim\alpha$ and $\alpha \vee \beta$.
- If α is a formula and $a \in \Sigma_i$ then $\langle a\rangle_i\alpha$ is a formula.
- If α and β are formulas so is $\alpha\mathcal{U}_i\beta$.

Throughout this section, we denote $\Phi_{\mathrm{TrPTL}(\widetilde{\Sigma})}$ as just Φ. In the semantics of the logic, which will be based on infinite traces, the i-view of a configuration will play a crucial role. Let $T \in TR^\omega$ with $T = (E, \leq, \lambda)$. Recall that $E_i = \{e \mid e \in E$ and $\lambda(e) \in \Sigma_i\}$. Let $c \in \mathcal{C}_T$ and $i \in \mathcal{P}$. Then $\downarrow^i(c)$ is the i-view of c and it is defined as:

$$\downarrow^i(c) = \downarrow(c \cap E_i).$$

We note that $\downarrow^i(c)$ is also a configuration. It is the "best" configuration that the agent i is aware of at c. We say that $\downarrow^i(c)$ is an *i-local configuration.* Let $\mathcal{C}_T^i = \{\downarrow^i(c) \mid c \in \mathcal{C}_T\}$ be the set of i-local configurations. For $Q \subseteq \mathcal{P}$ and $c \in \mathcal{C}_T$, we let $\downarrow^Q(c)$ denote the set $\bigcup\{\downarrow^i(c) \mid i \in Q\}$. Once again, $\downarrow^Q(c)$ is a configuration. It represents the collective knowledge of the processes in Q about the configuration c.

The following basic properties of traces follow directly from the definitions.

Proposition 15. *Let $T = (E, \leq, \lambda)$ be an infinite trace. The following statements hold.*

1. *Let $\leq_i = \leq \cap (E_i \times E_i)$. Then $(E_i, \leq_i)$ is a linear order isomorphic to ω if E_i is infinite and isomorphic to a finite initial segment of ω if E_i is finite.*
2. *$(\mathcal{C}_T^i, \subseteq)$ is a linear order. In fact $(\mathcal{C}_T^i - \{\emptyset\}, \subseteq)$ is isomorphic to $(E_i, \leq_i)$.*
3. *Suppose $\downarrow^i(c) \neq \emptyset$ where $c \in \mathcal{C}_T$. Then there exists $e \in E_i$ such that $\downarrow^i(c) = \downarrow e$. In fact e is the $\leq_i$-maximum event in $(c \cap E_i)$.*
4. *Suppose $Q \subseteq Q' \subseteq \mathcal{P}$ and $c \in \mathcal{C}_T$. Then $\downarrow^Q(c) = \downarrow^Q(\downarrow^{Q'}(c))$. In particular, for a single process i, $\downarrow^i(c) = \downarrow^i(\downarrow^i(c))$.*

We can now present the semantics of TrPTL. A model is a pair $M = (T, \{V_i\}_{i \in \mathcal{P}})$ where $T = (E, \leq, \lambda) \in TR^\omega$ and $V_i : \mathcal{C}_T^i \to 2^P$ is a valuation function which assigns a set of atomic propositions to i-local configurations for each process i. Let $c \in \mathcal{C}_T$ and $\alpha \in \Phi$. Then $M, c \models \alpha$ denotes that α is satisfied at c in M and it is defined inductively as follows:

- $M, c \models p(i)$ for $p \in P$ iff $p \in V_i(\downarrow^i(c))$.
- $M, c \models \sim\alpha$ iff $M, c \not\models \alpha$.
- $M, c \models \alpha \vee \beta$ iff $M, c \models \alpha$ or $M, c \models \beta$.
- $M, c \models \langle a \rangle_i \alpha$ iff there exists $e \in E_i - c$ such that $\lambda(e) = a$ and $M, \downarrow e \models \alpha$. Moreover, for every $e' \in E_i$, $e' < e$ iff $e' \in c$.
- $M, c \models \alpha \mathcal{U}_i \beta$ iff there exists $c' \in \mathcal{C}_T$ such that $c \subseteq c'$ and $M, \downarrow^i(c') \models \beta$. Moreover, for every $c'' \in \mathcal{C}_T$, if $\downarrow^i(c) \subseteq \downarrow^i(c'') \subset \downarrow^i(c')$ then $M, \downarrow^i(c'') \models \alpha$.

Thus TrPTL is an action based multi-agent version of LTL. Indeed both in terms of its syntax and semantics, LTL(Σ) corresponds to the case where there is only one agent. The semantics of TrPTL when specialized down to this case yields the previous LTL(Σ) semantics.

Returning to TrPTL, the assertion $p(i)$ says that the i-view of c satisfies the atomic proposition p. Observe that we could well have $p(i)$ satisfied at c but not $p(j)$ (with $i \neq j$). It is interesting to note that all atomic assertions (that we know of) concerning distributed behaviours are local in nature. Indeed, it is well-known that global atomic propositions will at once lead to an undecidable logic in the current setting [25, 36].

Suppose $M = (T, \{V_i\})$ is a model and $c \xrightarrow{a}_T c'$ with $j \notin \text{loc}(a)$. Then $M, c \models p(j)$ iff $M, c' \models p(j)$. In this sense the valuation functions are local. There are, of course, a number of equivalent ways of formulating this idea which we will not get into here.

The assertion $\langle a \rangle_i \alpha$ says that the agent i will next participate in an a-event. Moreover, at the resulting i-view, the assertion α will hold. The assertion $\alpha \mathcal{U}_i \beta$ says that there is a future i-view (including the present i-view) at which β will hold and for all the intermediate i-views (if any) starting from the current i-view, the assertion α will hold.

Before considering examples of TrPTL specifications, we will introduce some notation. We let α, β with or without subscripts range over Φ. Abusing notation, we will use loc to denote the map which associates a set of *locations* with each formula.

- $\text{loc}(p(i)) = \text{loc}(\langle a \rangle_i \alpha) = \text{loc}(\alpha \mathcal{U}_i \beta) = \{i\}$.
- $\text{loc}(\sim\alpha) = \text{loc}(\alpha)$.
- $\text{loc}(\alpha \vee \beta) = \text{loc}(\alpha) \cup \text{loc}(\beta)$.

In what follows, $\Phi^i = \{\alpha \mid \text{loc}(\alpha) = \{i\}\}$ is the set of i-type formulas. We note that unlike LTL$^\otimes$, a TrPTL formula of the form $\langle a \rangle_i \alpha$ could have $j \in \text{loc}(\alpha)$ with $j \neq i$. A similar remark applies to the indexed until-operators.

A basic observation concerning the semantics of TrPTL can be phrased as follows:

Proposition 16. *Let* $M = (T, \{V_i\})$ *be a model,* $c \in \mathcal{C}_T$ *and* α *a formula such that* $\mathrm{loc}(\alpha) \subseteq Q$*. Then* $M, c \models \alpha$ *iff* $M, \downarrow^Q(c) \models \alpha$.

A corollary to this result is that in case $\alpha \in \Phi^i$ then $M, c \models \alpha$ if and only if $M, \downarrow^i(c) \models \alpha$. As a result, the formulas in Φ^i can be used in exactly the same manner as one would use LTL$^{\otimes}$ to express properties of the agent i. Boolean combinations of such local assertions can be used to capture various interaction patterns between the agents implied by the logical connectives as well as the coordination enforced by the distributed alphabet $\widetilde{\Sigma}$. For writing specifications, apart from the usual derived connectives that we already introduced in Section 2 for LTL, the following operators are also available:

- $\top \stackrel{\Delta}{\Longleftrightarrow} p_1(1) \vee {\sim} p_1(1)$ denotes the constant "True", where $P = \{p_1, p_2, \ldots\}$. We use $\bot = {\sim}\top$ to denote "False".
- $\Diamond_i \alpha \stackrel{\Delta}{\Longleftrightarrow} \top \mathcal{U}_i \alpha$ is a local version of the $\Diamond$ modality of LTL.
- $\Box_i \alpha \stackrel{\Delta}{\Longleftrightarrow} {\sim}\Diamond_i {\sim}\alpha$ is a local version of the $\Box$ modality of LTL.
- Let $X \subseteq \Sigma_i$ and $\overline{X} = \Sigma_i - X$. Then $\alpha \mathcal{U}_i^X \beta \stackrel{\Delta}{\Longleftrightarrow} (\alpha \wedge \bigwedge_{a \in \overline{X}} [a]_i \bot) \mathcal{U}_i \beta$. In other words $\alpha \mathcal{U}_i^X \beta$ is fulfilled using (at most) actions taken from X. We set $\Diamond_i^X \alpha \stackrel{\Delta}{\Longleftrightarrow} \top \mathcal{U}_i^X \alpha$ and $\Box_i^X \alpha \stackrel{\Delta}{\Longleftrightarrow} {\sim}\Diamond_i^X {\sim}\alpha$.
- $\alpha(i) \stackrel{\Delta}{\Longleftrightarrow} \alpha \mathcal{U}_i \alpha$ (or equivalently $\bot \mathcal{U}_i \alpha$). $\alpha(i)$ is to be read as "α at i". If $M = (T, \{V_i\})$ is a model and $c \in \mathcal{C}_T$ then $M, c \models \alpha(i)$ iff $M, \downarrow^i(c) \models \alpha$. It could of course be the case that $\mathrm{loc}(\alpha) \neq \{i\}$.

A simple but important observation is that every formula is a boolean combination of formulas taken from $\bigcup_{i \in \mathcal{P}} \Phi^i$. In TrPTL we can say that a specific global configuration is reachable from the initial configuration. Let $\{\alpha_i\}_{i \in \mathcal{P}}$ be a family with $\alpha_i \in \Phi^i$ for each i. Then we can define a derived connective $\Diamond(\alpha_1, \alpha_2, \ldots, \alpha_K)$ which has the following semantics at the empty configuration. Let $M = (T, \{V_i\})$ be a model. Then $M, \emptyset \models \Diamond(\alpha_1, \alpha_2, \ldots, \alpha_k)$ iff there exists $c \in \mathcal{C}_T$ such that $M, c \models \alpha_1 \wedge \alpha_2 \wedge \cdots \wedge \alpha_K$.

To define this derived connective set $\Sigma_1' = \Sigma_1$ and, for $1 < i \leq K$, set $\Sigma_i' = \Sigma_i - \cup\{\Sigma_j \mid 1 \leq j < i\}$. Then $\Diamond(\alpha_1, \alpha_2, \ldots, \alpha_K)$ is the formula:

$$\Diamond_1^{\Sigma_1'}(\alpha_1 \wedge \Diamond_2^{\Sigma_2'}(\alpha_2 \wedge \Diamond_3^{\Sigma_3'}(\alpha_3 \wedge \cdots \Diamond_K^{\Sigma_K'} \alpha_K)) \cdots).$$

The idea is that the sequence of actions leading up to the required configuration can be reordered so that one first performs all the actions in Σ_1, then all the actions in $\Sigma_2 - \Sigma_1$ etc. Hence, if now is an atomic proposition, the formula $\Diamond(\mathsf{now}(1), \mathsf{now}(2), \ldots, \mathsf{now}(K))$ is satisfied at the empty configuration iff there is a reachable configuration at which all the agents assert now.

Dually, safety properties that hold at the initial configuration can also be expressed. For example, let crt_i be the atomic assertion declaring that the agent i is currently in its critical section. Then it is possible to write a formula φ_{ME} which asserts that at all reachable configurations at most one agent is in its critical section, thereby guaranteeing that the system satisfies the mutual exclusion property. We omit the details of how to specify φ_{ME}.

On the other hand, it seems difficult to express nested global and safety properties in TrPTL. It is also the case that due to the local nature of the modalities, information about the past sneaks into the semantics even though there are no explicit past operators in the logic.

A formula α is said to be *root-satisfiable* iff there exists a model M such that $M, \emptyset \models \alpha$. On the other hand, α is said to be *satisfiable* iff there exists a model $M = (T, \{V_i\})$ and $c \in \mathcal{C}_T$ such that $M, c \models \alpha$. It turns out that these two notions are *not* equivalent. Consider the distributed alphabet $\widetilde{\Sigma}_0 = \{\Sigma_1, \Sigma_2\}$ with $\Sigma_1 = \{a, d\}$ and $\Sigma_2 = \{b, d\}$. Then it is not difficult to verify that the formula $p(2)(1) \wedge \Box_2 \sim p(2)$ is satisfiable but not root-satisfiable. (Recall that $p(2)(1)$ abbreviates $\bot \mathcal{U}_1 p(2)$). One can however transform every formula α into a formula α' such that α is satisfiable iff α' is root satisfiable.

This follows from the observation that every α can be expressed as a boolean combination of formulas taken from the set $\bigcup_{i \in \mathcal{P}} \Phi^i$. Hence the given formula α can be assumed to be of the form $\alpha = \bigvee_{j=1}^{m} (\alpha_{j1} \wedge \alpha_{j2} \wedge \cdots \wedge \alpha_{jK})$ where $\alpha_{ji} \in \Phi^i$ for each $j \in \{1, 2, \ldots, m\}$ and each $i \in \mathcal{P}$. Now convert α to the formula α' where $\alpha' = \bigvee_{j=1}^{m} \Diamond(\alpha_{j1}, \alpha_{j2}, \cdots, \alpha_{jK})$. (Recall the derived modality $\Diamond(\alpha_1, \alpha_2, \ldots, \alpha_K)$ introduced earlier.) From the semantics of $\Diamond(\alpha_1, \alpha_2, \ldots, \alpha_K)$ it follows that α is satisfiable iff α' is root-satisfiable.

Hence, in principle, it suffices to consider only root-satisfiability in developing a decision procedure for TrPTL. There is of course a blow-up involved in converting satisfiable formulas to root-satisfiable formulas. If one wants to avoid this blow-up then the decision procedure for checking root-satisfiability can be suitably modified to yield a direct decision procedure for checking satisfiability as done in [44]. In any case, it is root satisfiability which is of importance from the standpoint of model checking. Hence here we shall only develop a procedure for deciding if a given formula of TrPTL is root-satisfiable.

As a first step we augment the syntax of our logic by one more construct.

- If α is a formula, so is $O_i\alpha$. In the model $M = (T, \{V_i\})$, at the configuration $c \in \mathcal{C}_T$, $M, c \models O_i\alpha$ iff $M, c \models \langle a \rangle_i \alpha$ for some $a \in \Sigma_i$. We also define $\mathrm{loc}(O_i\alpha) = \{i\}$.

Secondly, we will from now on drop the atomic propositions and instead work with the constant $\top$ and its negation $\bot$ as done earlier. The semantic definitions are assumed to be suitably modified.

Thus $O_i\alpha \equiv \bigvee_{a \in \Sigma_i} \langle a \rangle_i \alpha$ is a valid formula and O_i is expressible in the former syntax. It will be however more efficient to admit O_i as a first class modality as we did in Section 2.

Fix a formula α_0. Our aim is to effectively associate an asynchronous automaton $\mathcal{A}_{\alpha_0}$ with α_0 such that α_0 is root-satisfiable iff $L_{Tr}(\mathcal{A}_{\alpha_0}) \neq \emptyset$. Since the emptiness problem for asynchronous automata is decidable (Proposition 14), this will yield the desired decision procedure. Let $cl(\alpha_0)$ be the least set of formulas containing α_0 which satisfies:

- $\sim\alpha \in cl(\alpha_0)$ implies $\alpha \in cl(\alpha_0)$.
- $\alpha \vee \beta \in cl(\alpha_0)$ implies $\alpha, \beta \in cl(\alpha_0)$.
- $\langle a \rangle_i \alpha \in cl(\alpha_0)$ implies $\alpha \in cl(\alpha_0)$.
- $O_i \alpha \in cl(\alpha_0)$ implies $\alpha \in cl(\alpha_0)$.
- $\alpha \mathcal{U}_i \beta \in cl(\alpha_0)$ implies $\alpha, \beta \in cl(\alpha_0)$. In addition, $O_i(\alpha \mathcal{U}_i \beta) \in cl(\alpha_0)$.

We then define $CL(\alpha_0)$ to be the set $cl(\alpha_0) \cup \{\sim\beta \mid \beta \in cl(\alpha_0)\}$.

Thus $CL(\alpha_0)$, sometimes called the Fisher-Ladner closure of α_0, is closed under negation with the convention that $\sim\sim\beta$ is identified with β. Moreover, throughout the remainder of the section all formulas that we encounter will be assumed to be members of $CL(\alpha_0)$. From now we shall write CL instead of $CL(\alpha_0)$.

$A \subseteq CL$ is called an *i-type atom* iff it satisfies:

- $\top \in A$.
- $\alpha \in A$ iff $\sim\alpha \notin A$.
- $\alpha \vee \beta \in A$ iff $\alpha \in A$ or $\beta \in A$.
- $\alpha \mathcal{U}_i \beta \in A$ iff $\beta \in A$ or ($\alpha \in A$ and $O_i(\alpha \mathcal{U}_i \beta) \in A$).
- If $\langle a \rangle_i \alpha$, $\langle b \rangle_i \beta \in A_i$ then $a = b$.

AT_i denotes the set of i-type atoms. We now need to define the notion of a formula in CL being a member of a collection of atoms. Let $\alpha \in CL$ and $\{A_i\}_{i \in Q}$ be a family of atoms with $\mathrm{loc}(\alpha) \subseteq Q$ and $A_i \in AT_i$ for each $i \in Q$. We'll define the predicate Member(α,$\{A_i\}_{i \in Q}$), which for convenience will be denoted by $\alpha \in \{A_i\}_{i \in Q}$. It is defined inductively as:

- If $\mathrm{loc}(\alpha) = \{j\}$ then $\alpha \in \{A_i\}_{i \in Q}$ iff $\alpha \in A_j$.
- If $\alpha = \sim\beta$ then $\alpha \in \{A_i\}_{i \in Q}$ iff $\beta \notin \{A_i\}_{i \in Q}$.
- If $\alpha = \alpha_1 \vee \alpha_2$ then $\alpha_1 \vee \alpha_2 \in \{A_i\}_{i \in Q}$ iff $\alpha_1 \in \{A_i\}_{i \in Q}$ or $\alpha_2 \in \{A_i\}_{i \in Q}$.

The construction of the asynchronous automaton $\mathcal{A}_{\alpha_0}$ is guided by the construction developed for LTL in Section 2. However in the much richer setting of traces it turns out that one must make crucial use of the latest information that the agents have about each other when defining the transitions of $\mathcal{A}_{\alpha_0}$. It has been shown by Mukund and Sohoni [29] that this information can be kept track of by a deterministic asynchronous automaton whose size depends only on $\tilde{\Sigma}$. (Actually the automaton described in [29] operates over finite traces but it is a trivial task to convert it into an asynchronous automaton having the desired properties). To bring out the relevant properties of this automaton, let $T \in TR^\omega$ with $T = (E, \leq, \lambda)$. For each subset Q of processes, the function $\mathsf{latest}_{T,Q} : C_T \times \mathcal{P} \to Q$ is given by $\mathsf{latest}_{T,Q}(c, j) = \ell$ iff ℓ is the least member of Q (under the usual ordering over the integers) with the property $\downarrow^j(\downarrow^q(c)) \subseteq \downarrow^j(\downarrow^\ell(c))$ for every $q \in Q$. In other words, among the agents in Q, ℓ has the best information about j at c, with ties being broken by the usual ordering over integers.

Theorem 17 (**[29]**)**.** *There exists an effectively constructible deterministic asynchronous automaton* $\mathcal{A}_\Gamma = (\{\Gamma_i\}, \{\Longrightarrow_a\}, \Gamma_{in}, \{(F_i, F_i^\omega)\})$ *such that:*

1. $L_{Tr}(\mathcal{A}_\Gamma) = TR^\omega$.
2. *For each* $Q = \{i_1, i_2, \ldots, i_n\}$, *there exists an effectively computable function* $\mathsf{gossip}_Q : \Gamma_{i_1} \times \Gamma_{i_2} \times \cdots \times \Gamma_{i_n} \times \mathcal{P} \to Q$ *such that for every* $T \in TR^\omega$, *every* $c \in \mathcal{C}_T$ *and every* $j \in \mathcal{P}$, $\mathsf{latest}_{T,Q}(c, j) = \mathsf{gossip}_Q(\gamma(i_1), \ldots, \gamma(i_n), j)$ *where* $\rho_T(c) = \gamma$ *and* ρ_T *is the unique (accepting) run of* $\mathcal{A}_\Gamma$ *over* T.

Henceforth, we refer to $\mathcal{A}_\Gamma$ as the *gossip automaton.* Each process in the gossip automaton has $2^{O(K^2 \log K)}$ local states, where $K = |\mathcal{P}|$. Moreover the function gossip_Q can be computed in time which is polynomial in the size of K.

Each i-state of the automaton $\mathcal{A}_{\alpha_0}$ will consist of an i-type atom together with an appropriate i-state of the gossip automaton. Two additional components will be used to check for liveness requirements. One component will take values from the set $N_i = \{0, 1, 2, \ldots, |U_i|\}$ where $U_i = \{\alpha \mathcal{U}_i \beta \mid \alpha \mathcal{U}_i \beta \in CL\}$. This component will be used to ensure that all "until" requirements are met. The other component will take values from the set {on,off}. This will be used to detect when an agent has quit.

The automaton $\mathcal{A}_{\alpha_0}$ can now be defined as:

$$\mathcal{A}_{\alpha_0} = (\{S_i\}, \{\longrightarrow_a\}, S_{in}, \{(F_i, F_i^\omega)\}),$$

where:

- For each i, $S_i = AT_i \times \Gamma_i \times N_i \times$ {on,off}. Recall that Γ_i is the set of i-states of the gossip automaton and $N_i = \{0, 1, 2, \ldots, |U_i|\}$ with $U_i = \{\alpha \mathcal{U}_i \beta \mid \alpha \mathcal{U}_i \beta \in CL\}$.
- Let $s_a, s'_a \in S_a$ with $s_a(i) = (A_i, \gamma_i, n_i, v_i)$ and $s'_a(i) = (A'_i, \gamma'_i, n'_i, v'_i)$ for each $i \in \mathrm{loc}(a)$. Then $(s_a, s'_a) \in \longrightarrow_a$ iff the following conditions are met.
  - $(\gamma_a, \gamma'_a) \in \Longrightarrow_a$ (recall that $\{\Longrightarrow_a\}$ is the family of transition relations of the gossip automaton) where $\gamma_a, \gamma'_a \in \Gamma_a$ such that $\gamma_a(i) = \gamma_i$ and $\gamma'_a(i) = \gamma'_i$ for each $i \in \mathrm{loc}(a)$.
  - $\forall i, j \in \mathrm{loc}(a)$, $A'_i = A'_j$.
  - $\forall i \in \mathrm{loc}(a)\ \forall \langle a \rangle_i \alpha \in CL$. $\langle a \rangle_i \alpha \in A_i$ iff $\alpha \in A'_i$.
  - $\forall i \in \mathrm{loc}(a)\ \forall O_i \alpha \in CL$. $O_i \alpha \in A$ iff $\alpha \in A'_i$.
  - $\forall i \in \mathrm{loc}(a) \forall \langle b \rangle_i \beta \in CL$. If $\langle b \rangle_i \beta \in A_i$ then $b = a$.
  - Suppose $j \notin \mathrm{loc}(a)$ and $\beta \in CL$ with $\mathrm{loc}(\beta) = \{j\}$. Further suppose that $\mathrm{loc}(a) = \{i_1, i_2, \ldots, i_n\}$. Then $\beta \in A'_i$ iff $\beta \in A_\ell$ where $\ell = \mathsf{gossip}_{\mathrm{loc}(a)}(\gamma_{i_1}, \gamma_{i_2}, \ldots, \gamma_{i_n}, j)$.
  - Let $i \in \mathrm{loc}(a)$, $U_i = \{\alpha_1 \mathcal{U}_i \beta_1, \alpha_2 \mathcal{U}_i \beta_2, \ldots, \alpha_{n_i} \mathcal{U}_i \beta_{n_i}\}$. Then u'_i and u_i are related to each other via:

$$u'_i = \begin{cases} (u_i{+}1) \bmod (n_i{+}1), & \text{if } u_i = 0 \text{ or } \beta_{u_i} \in A_i \text{ or } \alpha_{u_i} \mathcal{U}_i \beta_{u_i} \notin A_i \\ u_i, & \text{otherwise} \end{cases}$$

  - For each $i \in \mathrm{loc}(a)$, $v_i =$ on. Moreover, if $v'_i =$ off then $\langle a \rangle_i \alpha \notin A'_i$ for every $i \in \mathrm{loc}(a)$ and every $\langle a \rangle_i \alpha \in CL$.
- Let $s \in S_{\mathcal{P}}$ with $s(i) = (A_i, \gamma_i, u_i, v_i)$ for every i. Then $s \in S_{in}$ iff $\alpha_0 \in \{A_i\}_{i \in \mathcal{P}}$ and $\gamma \in \Gamma_{in}$ where $\gamma \in \Gamma_{\mathcal{P}}$ satisfies $\gamma(i) = \gamma_i$ for every i. Furthermore, $u_i = 0$ for every i. Finally, for every i, $v_i =$ off implies that $\langle a \rangle_i \alpha \notin A_i$ for every $\langle a \rangle_i \alpha \in CL$.

- For each i, $F_i^\omega \subseteq S_i$ is given by $F_i^\omega = \{(A_i, \gamma_i, u_i, v_i) \mid u_i = 0 \text{ and } v_i = \mathsf{on}\}$ and $F_i \subseteq S_i$ is given by $F_i = \{(A_i, \gamma_i, u_i, v_i) \mid v_i = \mathsf{off}\}$.

This construction is an optimized version of the original construction for TrPTL presented in [44, 45]. Note that $\mathcal{A}_{\alpha_0}$ is indeed in standard form. Arguments similar to those presented in [44, 45] lead to the next set of results.

Theorem 18.

1. *α_0 is root-satisfiable iff $L_{Tr}(\mathcal{A}_{\alpha_0}) \neq \emptyset$.*
2. *The number of local states of $\mathcal{A}_{\alpha_0}$ is bounded by $2^{O(\max(n, m^2 \log m))}$ where $n = |\alpha_0|$ and m is the number of agents mentioned in α_0. Clearly, $m \leq n$. It follows that the root-satisfiability problem (and in fact the satisfiability problem) for* TrPTL *is solvable in time $2^{O(\max(n, m^2 \log m) \cdot m)}$.*

The number of local states of each process in $\mathcal{A}_{\alpha_0}$ is determined by two quantities: the length of α_0 and the size of the gossip automaton $\mathcal{A}_\Gamma$. As far as the size of $\mathcal{A}_\Gamma$ is concerned, it is easy to verify that we need to consider only those agents in $\mathcal{P}$ that are mentioned in $\mathrm{loc}(\alpha_0)$, rather than all agents in the system.

The *model checking problem* for TrPTL can be phrased as follows. A finite state distributed program Pr over $\widetilde{\Sigma}$ is an asynchronous automaton $\mathcal{A}_{Pr} = (\{S_i^{Pr}\}, \{\Longrightarrow_a^{Pr}\}, S_{in}^{Pr}, \{(S_i^{Pr}, S_i^{Pr})\})$ modelling the state space of Pr.

Viewing a formula α_0 as a specification, we say that Pr *meets the specification* α_0 — denoted $Pr \models \alpha_0$ — if for every $T \in TR^\omega$, if $\mathcal{A}_{Pr}$ has a run over T then $T, \emptyset \models \alpha_0$.

The model checking problem for TrPTL can be solved by "intersecting" the program automaton $\mathcal{A}_{Pr}$ with the formula automaton $\mathcal{A}_{\sim\alpha_0}$ to yield an automaton $\mathcal{A}$ such that $L_{Tr}(\mathcal{A}) = L_{Tr}(\mathcal{A}_{Pr}) \cap L_{Tr}(\mathcal{A}_{\sim\alpha_0})$. As before, $L_{Tr}(\mathcal{A}) = \emptyset$ iff $Pr \models \alpha_0$.

It turns out that this model checking problem has time complexity $O(|\mathcal{A}_{Pr}| \cdot 2^{O(\max(n, m^2 \log m) \cdot m)})$ where $|\mathcal{A}_{Pr}|$ is the size of the global state space of the asynchronous automaton modelling the behaviour of the given program Pr and, as before, $n = |\alpha_0|$ and m is the number of agents mentioned in α_0, where α_0 is the specification formula.

We now take a brief look at some related agent-based linear time temporal logics over traces. The first one is the sublogic of TrPTL denoted which consists of the so called connected formulas of TrPTL. We define $\Phi_{\mathrm{TrPTL}}^{\mathrm{con}}$ (from now on written as Φ^{con}) to be the least subset of Φ satisfying the following conditions:

- $\top \in \Phi^{\mathrm{con}}$ and as before $\mathrm{Loc}(\top) = \emptyset$
- If $\alpha, \beta \in \Phi^{\mathrm{con}}$, so are $\sim\alpha$ and $\alpha \vee \beta$.
- If $\alpha \in \Phi^{\mathrm{con}}$ and $a \in \Sigma_i$ such that $\mathrm{loc}(\alpha) \subseteq \mathrm{loc}(a)$ then $\langle a \rangle_i \alpha \in \Phi^{\mathrm{con}}$.
- If $\alpha, \beta \in \Phi^{\mathrm{con}}$ with $\mathrm{loc}(\alpha) = \mathrm{loc}(\beta) = \{i\}$ then $\alpha \mathcal{U}_i \beta \in \Phi^{\mathrm{con}}$. Actually one need only demand that $\mathrm{loc}(\alpha), \mathrm{loc}(\beta) \subseteq \bigcap\{\mathrm{loc}(a) \mid a \in \Sigma_i\}$ but this leads to notational complications that we wish to avoid here.
- If $\alpha \in \Phi^{\mathrm{con}}$ and $\mathrm{loc}(\alpha) = \{i\}$ then $O_i \alpha \in \Phi^{\mathrm{con}}$. (Once again one needs to just demand that $\alpha \subseteq \bigcap\{\mathrm{loc}(a) \mid a \in \Sigma_i\}$.)

Connected formulas were first identified by Niebert and used by Huhn [22]. They have also been independently identified by Ramanujam [38]. Thanks to the syntactic restrictions imposed on the next state and until formulas, past information is not allowed to creep in. Indeed one can prove the following:

Proposition 19. *Let $\alpha \in \Phi^{con}$. Then α is satisfiable iff α is root-satisfiable.*

Yet another pleasing feature of TrPTL^{con} is that the gossip automaton can be eliminated in the construction of the automaton $\mathcal{A}_{\alpha_0}$ whenever $\alpha_0 \in \Phi^{con}$. In fact one can prove the following.

Theorem 20. *The satisfiability problem for* TrPTL^{con} *is solvable in time* $2^{O(|\alpha_0|)}$.

Once again, a suitably modified statement can be made about the associated model checking problem. At present we do not know whether or not TrPTL is strictly more expressive than TrPTL^{con}, but it is clear that $\mathrm{LTL}^{\otimes}$ is a strict sublogic of TrPTL^{con}. We shall deal with the relative strengths of these logics in the next section. Two of the four logics considered by Ramanujam [38] in a closely related setting turn out to be $\mathrm{LTL}^{\otimes}$ and TrPTL^{con}. We conjecture that the other two logics are also expressible within TrPTL.

Katz and Peled introduced the logic ISTL [24] whose semantics has a trace-theoretic flavour. In a subsequent paper by Peled and Pnueli [34] on ISTL, the connection to traces was made more directly. Indeed this is one of the first instances of the explicit use of traces in a temporal logical setting that we know of. However, it has branching time modalities which permit quantification over the so called observations of a trace. ISTL uses global atomic propositions rather than local atomic propositions. Penczek has also studied a number of temporal logics (including a version of ISTL) with branching time modalities and global atomic propositions [36]. His logics are interpreted directly over the space of configurations of a trace resulting in a variety of axiomatizations and undecidability results. We feel that local atomic propositions (as used in TrPTL) are crucial for obtaining tractable partial order based temporal logics. Niebert has considered several μ-calculus versions of TrPTL [31, 32] and has obtained various decidability results using a variant of asynchronous Büchi automata.

The temporal logic of causality (TLC) proposed by Alur, Peled and Penczek is basically a temporal logic over traces [1]. The concurrent structures used in [1] as frames for TLC can be easily represented as traces over an appropriately chosen trace alphabet. The interesting feature of TLC is that its branching time modalities are interpreted over causal paths. In a trace $(E, \leq, \lambda)$, the sequence $e_0 e_1 \cdots \in E^{\infty}$ is a causal path if $e_0 \lessdot e_1 \lessdot e_2 \cdots$. This logic admits an essentially exponential time decision procedure for checking satisfiablity in terms of a variant of Büchi automata called Street automata.

8 Expressiveness Issues

Our aim here is to discuss some expressiveness issues concerning trace-based linear time temporal logics. To set the stage we first quickly review the classical case of sequences.

The *monadic second-order theory of infinite sequences over* Σ is denoted $\mathrm{MSO}(\Sigma)$. Its vocabulary consists of a family of unary predicates $\{R_a\}_{a\in\Sigma}$, one for each $a \in \Sigma$; a binary predicate $\leq$; a binary predicate $\in$; a countable supply of individual variables $Var = \{x, y, z, \ldots\}$; a countable supply of set variables (i.e. monadic predicate variables) $SVar = \{X, Y, Z, \ldots\}$. The formulas of $\mathrm{MSO}(\Sigma)$ are then built up by:

- $R_a(x)$, $x \leq y$ and $x \in X$ are atomic formulas.
- If ϕ and ϕ' are formulas then so are $\sim\phi$, $\phi \vee \phi'$, $(\exists x)\phi$ and $(\exists X)\phi$.

A structure for $\mathrm{MSO}(\Sigma)$ is a ω-sequence $\sigma \in \Sigma^\omega$. Let $\mathcal{I}$ be an interpretation of the variables with $\mathcal{I} : Var \longrightarrow \omega$ and $\mathcal{I} : SVar \longrightarrow 2^\omega$. Then the notion of σ being a model of ϕ under the interpretation $\mathcal{I}$, denoted $\sigma \models_{\mathcal{I}} \phi$, is defined in the expected manner. In particular, $\sigma \models_{\mathcal{I}} R_a(x)$ iff $\sigma(\mathcal{I}(x)) = a$ (note that $\sigma \in \Sigma^\omega$ is viewed as $\sigma : \omega \longrightarrow \Sigma$); $\sigma \models_{\mathcal{I}} x \leq y$ iff $\mathcal{I}(x) \leq \mathcal{I}(y)$ (here $\leq$ is the usual ordering over ω); $\sigma \models_{\mathcal{I}} x \in X$ iff $\mathcal{I}(x) \in \mathcal{I}(X)$.

As usual, a sentence is a formula with no free variables. Each sentence ϕ defines an ω-language, denoted L_ϕ, where:

$$L_\phi = \{\sigma \mid \sigma \models \phi\}.$$

We say that $L \subseteq \Sigma^\omega$ is $\mathrm{MSO}(\Sigma)$*-definable* iff there exists a sentence $\phi \in \mathrm{MSO}(\Sigma)$ such that $L = L_\phi$. A celebrated result of Büchi [4] shows that the class of languages expressible by sentences in $\mathrm{MSO}(\Sigma)$ coincides with the class of languages recognized by Büchi automata over Σ. This class is the ω*-regular languages* over Σ.

The *first-order theory of infinite sequences over* Σ is denoted $\mathrm{FO}(\Sigma)$ and is obtained from $\mathrm{MSO}(\Sigma)$ by abolishing the monadic second-order quantifications from the logic. The semantics and notions of first-order definability are carried over in the obvious manner.

A fundamental result in the theory of temporal logic is Kamp's Theorem [23] which was later strengthened in [14] to establish that $\mathrm{LTL}(\Sigma)$ is expressively equivalent to the $\mathrm{FO}(\Sigma)$. The surprise here being that $\mathrm{LTL}(\Sigma)$ admits only a bounded number of operators (one unary and one binary as we have formulated it) whereas infinitely many operators of increasing arities can be defined in $\mathrm{FO}(\Sigma)$. Secondly, as we saw in Section 2, the satisfiability problem for $\mathrm{LTL}(\Sigma)$ can be solved in deterministic exponential time. The satisfiability problem for $\mathrm{FO}(\Sigma)$ on the other hand, even when the sentences are interpreted over *finite* words, is known to be non-elementary hard [43]. It is quite easy to see that $\mathrm{FO}(\Sigma)$ — and hence $\mathrm{LTL}(\Sigma)$ — is strictly less expressive than $\mathrm{MSO}(\Sigma)$ in the sense that there is a language which is $\mathrm{MSO}(\Sigma)$-definable but not $\mathrm{FO}(\Sigma)$-definable. (Indeed this is the sense in which we shall compare the expressive power of various logics in what follows.) For instance, as pointed out by Wolper in a state-based setting [56], the language $L \subseteq \{a,b\}^\omega$ given by "a is executed at every even position" is not definable in this logic. On the other hand, it is easy to come up with a formula of MSO(Σ) defining L.

The expressive power of LTL can be extended to obtain the expressive power of MSO while still guaranteeing an exponential time decidable satisfiability problem as demonstrated first in [57]. Here we sketch how the regular programs over Σ can be used to achieve this goal [19].

The syntax of regular programs over Σ is given by:

$$\mathrm{Prg}(\Sigma) ::= a \mid \pi_0 + \pi_1 \mid \pi_0;\pi_1 \mid \pi^*.$$

With each program we associate a set of finite words via the map $||\cdot|| : \mathrm{Prg}(\Sigma) \longrightarrow 2^{\Sigma^*}$. This map is defined in the standard fashion:

- $||a|| = \{a\}$.
- $||\pi_0 + \pi_1|| = ||\pi_0|| \cup ||\pi_1||$.
- $||\pi_0;\pi_1|| = \{\tau_0\tau_1 \mid \tau_0 \in ||\pi_0|| \text{ and } \tau_1 \in ||\pi_1||\}$.
- $||\pi^*|| = \bigcup_{i\in\omega} ||\pi^i||$, where
  - $||\pi^0|| = \{\varepsilon\}$ and
  - $||\pi^{i+1}|| = \{\tau_0\tau_1 \mid \tau_0 \in ||\pi|| \text{ and } \tau_1 \in ||\pi^i||\}$ for every $i \in \omega$.

The set of formulas of DLTL(Σ) is given by the following syntax.

$$\mathrm{DLTL}(\Sigma) ::= \top \mid {\sim}\alpha \mid \alpha \vee \beta \mid \alpha\, U^\pi \beta, \quad \pi \in \mathrm{Prg}(\Sigma)$$

A model is a ω-sequence $\sigma \in \Sigma^\omega$. For $\tau \in \mathrm{prf}(\sigma)$ we define $\sigma,\tau \models \alpha$ just as we did for LTL(Σ) in the case of the first three clauses. As for the last one,

- $\sigma,\tau \models \alpha\, U^\pi \beta$ iff there exists $\tau' \in ||\pi||$ such that $\tau\tau' \in \mathrm{prf}(\sigma)$ and $\sigma,\tau\tau' \models \beta$. Moreover, for every τ'' such that $\varepsilon \preceq \tau'' \prec \tau'$, it is the case that $\sigma,\tau\tau'' \models \alpha$.

Thus DLTL(Σ) adds to LTL(Σ) by strengthening the until-operator. To satisfy $\alpha\, U^\pi \beta$, one must satisfy $\alpha\, U \beta$ along some finite stretch of behaviour which is required to be in the (linear time) behaviour of the program π. We associate with a formula α of DLTL(Σ) the ω-language L_α in the obvious manner.

A useful derived operator of DLTL is:

- $\langle\pi\rangle\alpha \stackrel{\Delta}{\Longleftrightarrow} \top\, \mathcal{U}^\pi \alpha$.

By replacing the until-modality of DLTL with the above derived operator we obtain the sublogic DLTL$^-$(Σ), which is essentially Propositional Dynamic Logic [13] equipped with a linear time semantics. It turns out that DLTL(Σ) and DLTL$^-$(Σ) both have the same expressive power as MSO(Σ).

Theorem 21. *Let $L \subseteq \Sigma^\omega$. Then the following statements are equivalent.*

1. *L is ω-regular (i.e. definable in* MSO(Σ)*).*
2. *L is* DLTL(Σ)*-definable.*
3. *L is* DLTL$^-$(Σ)*-definable.*

Both the satisfiablity and model checking problems for DLTL(Σ) are decidable with the same time complexity as for LTL(Σ).

Let (Σ, I) be trace alphabet. Then MSO(Σ, I), the *monadic second-order theory of infinite traces (over Σ, I)*, has the same syntax as MSO(Σ). The structures are elements of $TR^\omega(\Sigma, I)$. Let $T \in TR^\omega(\Sigma, I)$ with $T = (E, \leq, \lambda)$ and let $\mathcal{I} : X \to E$ be an interpretation. Then $T \models_{\mathcal{I}}^{\mathrm{MSO}} R_a(x)$ iff $\lambda(\mathcal{I}(x)) = a$ and $T \models_{\mathcal{I}}^{\mathrm{MSO}} x \leq y$ iff $\mathcal{I}(x) \leq \mathcal{I}(y)$. Hence, the essential difference is that the binary predicate symbols is now interpreted as the causal partial order of the trace. The remaining semantic definitions go along the expected lines. Each sentence φ (i.e., a formula with no free occurrences of variables) defines the ω-trace language

$$L_\varphi = \{T \mid T \models^{\mathrm{MSO}} \varphi\}.$$

We say that $L \subseteq TR^\omega$ is MSO-definable iff there exists a sentence φ in MSO(Σ, I) such that $L = L_\varphi$. It is known that MSO-definable languages are precisely the regular trace languages; i.e. those recognized by asynchronous automata [11].

FO(Σ, I), *the first-order theory of traces*, is defined in the obvious way. Clearly it will be strictly weaker than MSO(Σ, I). For more information the reader is referred to [7]. Naturally both these theories can be made to handle finite traces as well.

Through the rest of this section we fix a distributed alphabet $\tilde{\Sigma}$ and let (Σ, I) be the induced trace alphabet. By MSO($\tilde{\Sigma}$) we shall mean the theory MSO(Σ, I) and similarly for FO($\tilde{\Sigma}$), the first-order fragment of MSO($\tilde{\Sigma}$). In what follows we shall often supress the mention of $\tilde{\Sigma}$ as well as the induced (Σ, I).

We first consider the logic LTL$^\otimes$. Recall that product languages are trace consistent and hence they induce trace languages via the map **str**. The resulting trace languages will be called product trace languages. As might be expected, the regular product trace languages are the ones obtained from regular product languages via the map **str**. It is easy to show that not every (regular) trace language is a product trace language [47]. It is also easy to see that LTL$^\otimes$-definable trace languages constitute a strict subclass of regular product trace languages. It has been shown that a product version of DLTL denoted DLTL$^\otimes$ captures exactly the class of regular product trace languages [20]. We also claim that it is an easy exercise to formulate a product version of MSO($\tilde{\Sigma}$) and show that it captures exactly the regular product trace languages. Let us denote this product version of MSO($\tilde{\Sigma}$) as MSO$^\otimes$($\tilde{\Sigma}$) and its first-order fragment as FO$^\otimes$($\tilde{\Sigma}$). It is easy to show — using Kamp's theorem — that LTL$^\otimes$($\tilde{\Sigma}$) has exactly the same expressive power as FO$^\otimes$($\tilde{\Sigma}$).

We also know that LTL$^\otimes$ is strictly weaker than TrPTL. First note that each formula (say α of TrPTL) defines a trace language L_α via :

$$L_\alpha = \{\, T \mid T, \emptyset \models \alpha\}.$$

Hence we can compare the relative expressive powers of LTL$^\otimes$ and TrPTL. It is known that ([30, 47]):

$$\mathrm{LTL}^{\otimes} \subset \mathrm{TrPTL}^{\mathrm{con}} \subseteq \mathrm{TrPTL}.$$

It is still open whether $\mathrm{TrPTL}^{\mathrm{con}}$ is equal to TrPTL in expressive power.

It is not difficult to show that TrPTL is no more expressive than the first-order theory of traces but it is not known whether the converse also holds. It would be nice to have a linear time temporal logic over traces patterned after LTL which has the same expressive power as the first-order theory of traces. The motivation is provided by the next result [11]:

Proposition 22. *Let $L \subseteq \Sigma^{\omega}$. Then the following statements are equivalent.*

1. *L is trace consistent and* $\mathrm{LTL}(\Sigma)$*-definable.*
2. $\{\mathsf{str}(\sigma) \mid \sigma \in L\}$ *is* $\mathrm{FO}(\Sigma, I)$*-definable.*

Egged on by this result, recently a different kind of trace-based linear time temporal logic called LTrL has been proposed [48]. This logic works directly with a trace alphabet (i.e. it is not based on agents). It is interpreted over the configurations of a trace and its syntax is given by:

$$\mathrm{LTrL}(\Sigma, I) ::= \top \mid {\sim}\alpha \mid \alpha \vee \beta \mid \langle a\rangle\alpha \mid \alpha\ U\ \beta \mid \langle a^{-1}\rangle\top.$$

Thus the syntax is very close to LTL except for the addition of a very restricted past-operator. In fact, just a *constant* number of past-operators are present in the logic; one for each action.

A model of $\mathrm{LTrL}(\Sigma, I)$ is a trace $T = (E, \leq, \lambda)$. Let $c \in \mathcal{C}_T$ be a configuration of T. Then $T, c \models \alpha$ will stand for α being satisfied at c in T. This notion is defined inductively as follows:

- $T, c \models \top$.
- $T, c \models {\sim}\alpha$ and $T, c \models \alpha \vee \beta$ are defined in the expected manner.
- $T, c \models \langle a\rangle\alpha$ iff there exists $c' \in \mathcal{C}_T$ with $c \xrightarrow{a}_T c'$ with $T, c' \models \alpha$.
- $T, c \models \alpha\ U\ \beta$ iff there exists $c' \in \mathcal{C}_T$ with $c \subseteq c'$ such that $T, c' \models \beta$. Moreover, for every $c'' \in \mathcal{C}_T$, $c \subseteq c'' \subset c'$ implies $T, c'' \models \alpha$.
- $T, c \models \langle a^{-1}\rangle\alpha$ iff there exists $c' \in \mathcal{C}_T$ with $c' \xrightarrow{a}_T c$.

The major result concerning LTrL is the following:

Theorem 23 ([48]). *Let $L \subseteq TR^{\omega}(\Sigma, I)$. Then the following statements are equivalent.*

1. *L is* $\mathrm{FO}(\Sigma, I)$*-definable.*
2. *L is* $\mathrm{LTrL}(\Sigma, I)$*-definable.*

Thus — except for the addition of the restricted past-operators — LTrL is a generalization of Kamp's Theorem to the much richer setting of traces. Meyer and Petit have shown that the past-operators can be eliminated without loss of expressive power when the logic is interpreted over *finite* traces [28]. A similar result for infinite traces is not known at present. Unfortunately this logic does not have a matching time complexity in relation to LTL. Recently Walukiewicz has shown that the satisfiability problem for LTrL is non-elementary hard [53]. A related result concerns the logic TLPO formulated by Ebinger [10]. This is also a linear time temporal logic interpreted over traces but with full-fledged past-operators. TLPO is claimed to be expressively complete when interpreted over *finite* traces but nothing is known about the complexity of the satisfiability problem nor about its expressive power in relation to infinite traces.

At present we do not know much about the relationship between TLC and the logics we have mentioned so far, except that it is strictly weaker than the monadic second-order theory of traces.

In an interesting recent development Niebert [32] has formulated a fixed point based linear time temporal logic for traces in the setting of distributed alphabets. This logic is denoted as νTrPTL. It is equal in expressive power to the monadic second-order theory of traces *and* it has decision procedure of essentially exponential time complexity. However, the formulas of this logic are required to satisfy what appears to be awkward syntactic restrictions and it is not clear how one could express global properties of interest in this formalism.

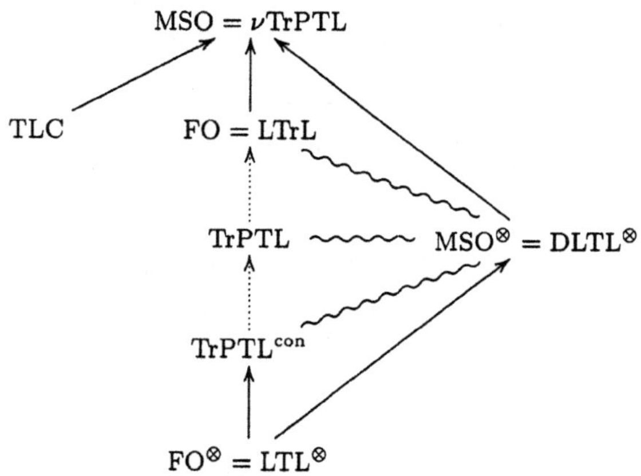

Fig. 4. Relative expressive power of the logics

The relative strengths of the various linear time temporal logics over traces mentioned in this section are displayed in Figure 4. A dotted (solid) arrow from A to B indicates that B is at least as expressive as (strictly more expressive than) A. Squiggled lines denote that the logics are incomparable to each other.

To conclude this section, a lot is known about linear time temporal logics for traces but at present we still do not have — unlike the case of sequences — pleasing counterparts to the first-order and monadic second-order theories of traces.

9 Conclusion

In this paper we have attempted an overview of linear time temporal logics interpreted over traces. We have mainly concentrated on the satisfiability and model checking problems as well as expressiveness issues. The problem of axiomatizing these logics seems to be a non-trivial task. Some partial results may be found in [39]. In [34] the authors present proof rules for the logic ISTL with a trace semantics together with a relative expressive completeness result. Reisig has also developed a kit of proof rules for a version of UNITY logic [40,41]. The models of this logic are the non-sequential processes of a net system and the proof rules are mainly designed to help reason about distributed algorithms modelled using net systems.

At present not much is known about corresponding logics in a branching time setting. Most of the attempts in this direction have lead to logics whose satisfiablity problems are undecidable [5,25,36]. It is however the case that the model checking problem often remains tractable [5,36]. We do not know at present whether the properties expressible in such logics have any type of "all-or-none" flavour and if so whether one can develop some reduction techniques for verifying such properties. Some preliminary attempts in this direction have been made in [16,54].

References

1. Alur, R., Peled, D., Penczek, W.: Model checking of causality properties. Proceedings of the 10th Annual IEEE Symposium on Logic in Computer Science, IEEE Computer Society Press (1995) 90–100
2. Bell Labs Design Automation: FormalCheck^{tm}. Further information can be obtained at `http://www.bell-labs.com/formalcheck/`
3. Bracho, F., Droste, M., Kuske, D.: Representation of computations in concurrent automata by dependence orders. Theoretical Computer Science **174**(1-2) (1997) 67–96
4. Büchi, J. R.: On a decision method in restricted second order arithmetic. Proceedings of the International Congress on Logic, Methodology and Philosophy of Science, Stanford University Press (1960) 1–11
5. Cheng, A.: Petri nets, traces, and local model checking. Proceedings of the 4th International Conference on Algebraic Methodology and Software Technology, Lecture Notes in Computer Science **936**, Springer-Verlag (1995) 322–337
6. Diekert, V.: Combinatorics of traces. Lecture Notes in Computer Science **454**, Springer-Verlag (1990)
7. Diekert, V., Rozenberg, G. (eds.): The book of traces. World Scientific (1995)
8. Droste, M.: Recognizable languages in concurrency monoids. Theoretical Computer Science **150**(1) (1995) 77–109
9. Droste, M., Gastin, P.: Asynchronous cellular automata for pomsets without autoconcurrency. Proceedings of the 7th International Conference on Concurrency Theory, Lecture Notes in Computer Science **1119**, Springer-Verlag (1996) 627–638
10. Ebinger, W.: Charakterisierung von sprachklassen unendlicher spuren durch logiken. Dissertation, Institut für Informatik, Universität Stuttgart (1994)

11. Ebinger, W., Muscholl, A.: Logical definability on infinite traces. Theoretical Computer Science **154**(1) (1996) 67–84
12. Emerson, A. E.: Temporal and modal logic. In Handbook of Theoretical Computer Science, volume B: Formal Models and Semantics, Elsevier Science Publishers (1990) 996–1072
13. Fischer, M. J., Ladner, R. E.: Propositional dynamic logic of regular programs. Journal of Computer and System Sciences **18**(2) (1979) 194–211
14. Gabbay, A., Pnueli, A., Shelah, S., Stavi, J.: On the temporal analysis of fairness. Proceedings of the 7th Annual Symposium on Principles of Programming Languages, ACM (1980) 163–173
15. Gastin, P., Petit, A.: Asynchronous cellular automata for infinite traces. Proceedings of the 19th International Colloquium on Automata, Languages and Programming. Lecture Notes in Computer Science **623**, Springer-Verlag (1992) 583–594
16. Gerth, R., Kuiper, R., Peled, D., Penczek, W.: A partial-order approach to branching time model checking. Proceedings of the 3rd Israeli Symposium on Theory of Computing and Systems, IEEE Computer Society Press (1995) 130–139
17. Godefroid, P.: Partial-order methods for the verification of concurrent systems. Lecture Notes in Computer Science **1032**, Springer-Verlag (1996)
18. Gerth, R., Peled, D., Vardi, M., Wolper, P.: Simple on-the-fly automatic verification of linear time temporal logic. Proceedings of the 15th IFIP WG 6.1 International Workshop on Protocol Specification, Testing, and Verification, North-Holland (1995)
19. Henriksen, J. G., Thiagarajan, P. S.: Dynamic linear time temporal logic. Journal of Pure and Applied Logic, Elsevier (to appear)
20. Henriksen, J. G., Thiagarajan, P. S.: A product version of dynamic linear time temporal logic. Proceedings of the 8th International Conference on Concurrency Theory, Lecture Notes in Computer Science **1243**, Springer-Verlag (1997) 45–58
21. Holzmann, G. J.: An overview of the SPIN model checker. In "On-the-fly Model Checking Tutorial", BRICS Autumn School on Verification, Note NS-96-6, BRICS, Department of Computer Science, University of Aarhus (1996)
22. Huhn, M.: On semantic and logical refinement of actions. Technical Report, Institut für Informatik, Universität Hildesheim, Germany (1996)
23. Kamp, H. R.: Tense logic and the theory of linear order. Ph.D. thesis, University of California (1968)
24. Katz, S., Peled, D.: Interleaving set temporal logic. Theoretical Computer Science **73**(3) (1992) 21–43
25. Lodaya, K., Parikh, R., Ramanujam, R., Thiagarajan, P. S.: A logical study of distributed transition systems. Information and Computation **119**(1) (1995) 91–118
26. Manna, Z., Pnueli, A.: The temporal logic of reactive and concurrent systems (specification), Springer-Verlag (1991)
27. Mazurkiewicz, A.: Concurrent program schemes and their interpretations. Technical report DAIMI PB-78, Department of Computer Science, University of Aarhus, Denmark (1977)
28. Meyer, R., Petit, A.: Expressive completeness of LTrL on finite traces: an algebraic proof. Proceedings of the 15th Annual Symposium on Theoretical Aspects of Computer Science **1373**, Lecture Notes in Computer Science, Springer-Verlag (1998) 533–543
29. Mukund, M., Sohoni, M.: Keeping track of the latest gossip in a distributed system. Distributed Computing **10**(3) (1997) 117-127

30. Mukund, M., Thiagarajan, P. S.: Linear time temporal logics over Mazurkiewicz traces. Proceedings of the 21st International Symposium on Mathematical Foundations of Computer Science, Lecture Notes in Computer Science **1113**, Springer-Verlag (1996) 62–92
31. Niebert, P.: A ν-calculus with local views for systems of sequential agents. Proceedings of the 20th International Symposium on Mathematical Foundations of Computer Science, Lecture Notes in Computer Science **969**, Springer-Verlag (1995) 563–573
32. Niebert, P.: A temporal logic for the specification and validation of distributed behaviour. Ph.D. thesis, University of Hildesheim (1997)
33. Nielsen, M., Plotkin, G., Winskel, G.: Petri nets, event structures and domains, part I. Theoretical Computer Science **13**(1) (1981) 85–108
34. Peled, D., Pnueli, A.: Proving partial order properties. Theoretical Computer Science **126**(2) (1994) 143–182
35. Peled, D.: Partial order reduction: model checking using representatives. Proceedings of the 21st International Symposium on Mathematical Foundations of Computer Science, Lecture Notes in Computer Science **1113**, Springer-Verlag (1996) 93–112
36. Penczek, W.: Temporal logics for trace systems: on automated verification. International Journal of the Foundations of Computer Science **4**(1) (1993) 31–68
37. Pnueli, A.: The temporal logic of programs. Proceedings of the 18th Annual Symposium on Foundations of Computer Science, IEEE Computer Society Press (1977) 46–57
38. Ramanujam, R.: Locally linear time temporal logic. Proceedings of the 11th Annual IEEE Symposium on Logic in Computer Science, IEEE Computer Society Press (1996) 118–127
39. Ramanujam, R.: Rules for trace consistent reasoning. Proceedings of the 3rd Asian Computing Science Conference, Lecture Notes in Computer Science **1345**, Springer-Verlag (1997) 57–71
40. Reisig, W.: Temporal logic and causality in concurrent systems. Proceedings of CONCURRENCY'88, Lecture Notes in Computer Science **335**, Springer-Verlag (1988) 121–139
41. Reisig, W.: Petri net models for distributed algorithms. In Computer Science Today — Recent Trends and Developments, Lecture Notes in Computer Science **1000**, Springer-Verlag (1995) 441–454
42. Sistla, A. P., Clarke, E.: The complexity of propositional linear temporal logics. Journal of the ACM **32**(3) (1985) 733–749
43. Stockmeyer, L. J.: The complexity of decision problems in automata theory and logic. Ph.D. thesis, MIT, Cambridge, Massachusetts (1974)
44. Thiagarajan, P. S.: A trace based extension of linear time temporal logic. Proceedings of the 9th Annual IEEE Symposium on Logic in Computer Science, IEEE Computer Society Press (1994) 438–447
45. Thiagarajan, P. S.: TrPTL: a trace based extension of linear time temporal logic. Technical report TCS-93-6, School of Mathematics, SPIC Science Foundation, Madras (1993)
46. Thiagarajan, P. S.: A trace consistent subset of PTL. Proceedings of the 6th International Conference on Concurrency Theory, Lecture Notes in Computer Science **962**, Springer-Verlag (1995) 438–452
47. Thiagarajan, P. S.: PTL over product state spaces. Technical report TCS-95-4, School of Mathematics, SPIC Science Foundation, Madras (1995)

48. Thiagarajan, P. S., Walukiewicz, I.: An expressively complete linear time temporal logic for Mazurkiewicz traces. Proceedings of the 12th Annual IEEE Symposium on Logic in Computer Science, IEEE Computer Society Press (1997) 183–194
49. Thomas, W.: Automata on infinite objects. In Handbook of Theoretical Computer Science, volume B: Formal Models and Semantics, Elsevier Science Publishers (1990) 133–191
50. Valmari, A.: A stubborn attack on state explosion. Formal Methods in Systems Design **1** (1992) 285–313
51. Vardi, M. Y., Wolper, P.: An automata-theoretic approach to automatic program verification. Proceedings of the 1st Annual IEEE Symposium on Logic in Computer Science, IEEE Computer Society Press (1986) 332-345
52. Vardi, M. Y.: An automata-theoretic approach to linear time temporal logic. In Logics for Concurrency - Structure vs. Automata, Lecture Notes in Computer Science **1043**, Springer-Verlag (1996) 238–266
53. Walukiewicz, I.: Difficult configurations — on the complexity of LTrL (extended abstract). Proceedings of the 25th International Colloquium on Automata, Languages and Programming. Lecture Notes in Computer Science, Springer-Verlag (to appear)
54. Willems, B., Wolper, P.: Partial-order methods for model checking: from linear time to branching time. Proceedings of the 11th Annual IEEE Symposium on Logic in Computer Science, IEEE Computer Society Press (1996) 294–303
55. Winskel, G., Nielsen, M.: Models for concurrency. In Handbook of Logic and the Foundations of Computer Science, volume 4, Oxford University Press (1995) 1–148
56. Wolper, P.: Temporal logic can be more expressive. Proceedings of the 22nd Annual IEEE Symposium on Foundations of Computer Science, IEEE Computer Society Press (1981) 340–348
57. Wolper, P., Vardi, M. Y., Sistla, A. P.: Reasoning about infinite computation paths. Proceedings of the 24th Annual IEEE Symposium on Foundations of Computer Science, IEEE Computer Society Press (1983) 185–194.
58. Zielonka, W.: Notes on finite asynchronous automata. R.A.I.R.O. Informatique Théorique et Applications **21** (1987) 99–135

Author Index

Springer and the environment

Lecture Notes in Computer Science

For information about Vols. 1–1447

please contact your bookseller or Springer-Verlag

Vol. 1448: M. Farach-Colton (Ed.), Combinatorial Pattern Matching. Proceedings, 1998. VIII, 251 pages. 1998.

Vol. 1449: W.-L. Hsu, M.-Y. Kao (Eds.), Computing and Combinatorics. Proceedings, 1998. XII, 372 pages. 1998.

Vol. 1450: L. Brim, F. Gruska, J. Zlatuška (Eds.), Mathematical Foundations of Computer Science 1998. Proceedings, 1998. XVII, 846 pages. 1998.

Vol. 1451: A. Amin, D. Dori, P. Pudil, H. Freeman (Eds.), Advances in Pattern Recognition. Proceedings, 1998. XXI, 1048 pages. 1998.

Vol. 1452: B.P. Goettl, H.M. Halff, C.L. Redfield, V.J. Shute (Eds.), Intelligent Tutoring Systems. Proceedings, 1998. XIX, 629 pages. 1998.

Vol. 1453: M.-L. Mugnier, M. Chein (Eds.), Conceptual Structures: Theory, Tools and Applications. Proceedings, 1998. XIII, 439 pages. (Subseries LNAI).

Vol. 1454: I. Smith (Ed.), Artificial Intelligence in Structural Engineering. XI, 497 pages. 1998. (Subseries LNAI).

Vol. 1455: A. Hunter, S. Parsons (Eds.), Applications of Uncertainty Formalisms. VIII, 474 pages. 1998. (Subseries LNAI).

Vol. 1456: A. Drogoul, M. Tambe, T. Fukuda (Eds.), Collective Robotics. Proceedings, 1998. VII, 161 pages. 1998. (Subseries LNAI).

Vol. 1457: A. Ferreira, J. Rolim, H. Simon, S.-H. Teng (Eds.), Solving Irregularly Structured Problems in Prallel. Proceedings, 1998. X, 408 pages. 1998.

Vol. 1458: V.O. Mittal, H.A. Yanco, J. Aronis, R-. Simpson (Eds.), Assistive Technology in Artificial Intelligence. X, 273 pages. 1998. (Subseries LNAI).

Vol. 1459: D.G. Feitelson, L. Rudolph (Eds.), Job Scheduling Strategies for Parallel Processing. Proceedings, 1998. VII, 257 pages. 1998.

Vol. 1460: G. Quirchmayr, E. Schweighofer, T.J.M. Bench-Capon (Eds.), Database and Expert Systems Applications. Proceedings, 1998. XVI, 905 pages. 1998.

Vol. 1461: G. Bilardi, G.F. Italiano, A. Pietracaprina, G. Pucci (Eds.), Algorithms – ESA'98. Proceedings, 1998. XII, 516 pages. 1998.

Vol. 1462: H. Krawczyk (Ed.), Advances in Cryptology - CRYPTO '98. Proceedings, 1998. XII, 519 pages. 1998.

Vol. 1463: N.E. Fuchs (Ed.), Logic Program Synthesis and Transformation. Proceedings, 1997. X, 343 pages. 1998.

Vol. 1464: H.H.S. Ip, A.W.M. Smeulders (Eds.), Multimedia Information Analysis and Retrieval. Proceedings, 1998. VIII, 264 pages. 1998.

Vol. 1465: R. Hirschfeld (Ed.), Financial Cryptography. Proceedings, 1998. VIII, 311 pages. 1998.

Vol. 1466: D. Sangiorgi, R. de Simone (Eds.), CONCUR'98: Concurrency Theory. Proceedings, 1998. XI, 657 pages. 1998.

Vol. 1467: C. Clack, K. Hammond, T. Davie (Eds.), Implementation of Functional Languages. Proceedings, 1997. X, 375 pages. 1998.

Vol. 1468: P. Husbands, J.-A. Meyer (Eds.), Evolutionary Robotics. Proceedings, 1998. VIII, 247 pages. 1998.

Vol. 1469: R. Puigjaner, N.N. Savino, B. Serra (Eds.), Computer Performance Evaluation. Proceedings, 1998. XIII, 376 pages. 1998.

Vol. 1470: D. Pritchard, J. Reeve (Eds.), Euro-Par'98: Parallel Processing. Proceedings, 1998. XXII, 1157 pages. 1998.

Vol. 1471: J. Dix, L. Moniz Pereira, T.C. Przymusinski (Eds.), Logic Programming and Knowledge Representation. Proceedings, 1997. IX, 246 pages. 1998. (Subseries LNAI).

Vol. 1472: B. Freitag, H. Decker, M. Kifer, A. Voronkov (Eds.), Transactions and Change in Logic Databases. Proceedings, 1996, 1997. X, 396 pages. 1998.

Vol. 1473: X. Leroy, A. Ohori (Eds.), Types in Compilation. Proceedings, 1998. VIII, 299 pages. 1998.

Vol. 1474: F. Mueller, A. Bestavros (Eds.), Languages, Compilers, and Tools for Embedded Systems. Proceedings, 1998. XIV, 261 pages. 1998.

Vol. 1475: W. Litwin, T. Morzy, G. Vossen (Eds.), Advances in Databases and Information Systems. Proceedings, 1998. XIV, 369 pages. 1998.

Vol. 1476: J. Calmet, J. Plaza (Eds.), Artificial Intelligence and Symbolic Computation. Proceedings, 1998. XI, 309 pages. 1998. (Subseries LNAI).

Vol. 1477: K. Rothermel, F. Hohl (Eds.), Mobile Agents. Proceedings, 1998. VIII, 285 pages. 1998.

Vol. 1478: M. Sipper, D. Mange, A. Pérez-Uribe (Eds.), Evolvable Systems: From Biology to Hardware. Proceedings, 1998. IX, 382 pages. 1998.

Vol. 1479: J. Grundy, M. Newey (Eds.), Theorem Proving in Higher Order Logics. Proceedings, 1998. VIII, 497 pages. 1998.

Vol. 1480: F. Giunchiglia (Ed.), Artificial Intelligence: Methodology, Systems, and Applications. Proceedings, 1998. IX, 502 pages. 1998. (Subseries LNAI).

Vol. 1481: E.V. Munson, C. Nicholas, D. Wood (Eds.), Principles of Digital Document Processing. Proceedings, 1998. VII, 152 pages. 1998.

Vol. 1482: R.W. Hartenstein, A. Keevallik (Eds.), Field-Programmable Logic and Applications. Proceedings, 1998. XI, 533 pages. 1998.

Vol. 1483: T. Plagemann, V. Goebel (Eds.), Interactive Distributed Multimedia Systems and Telecommunication Services. Proceedings, 1998. XV, 326 pages. 1998.

Vol. 1484: H. Coelho (Ed.), Progress in Artificial Intelligence – IBERAMIA 98. Proceedings, 1998. XIII, 421 pages. 1998. (Subseries LNAI).

Vol. 1485: J.-J. Quisquater, Y. Deswarte, C. Meadows, D. Gollmann (Eds.), Computer Security – ESORICS 98. Proceedings, 1998. X, 377 pages. 1998.

Vol. 1486: A.P. Ravn, H. Rischel (Eds.), Formal Techniques in Real-Time and Fault-Tolerant Systems. Proceedings, 1998. VIII, 339 pages. 1998.

Vol. 1487: V. Gruhn (Ed.), Software Process Technology. Proceedings, 1998. VIII, 157 pages. 1998.

Vol. 1488: B. Smyth, P. Cunningham (Eds.), Advances in Case-Based Reasoning. Proceedings, 1998. XI, 482 pages. 1998. (Subseries LNAI).

Vol. 1489: J. Dix, L. Fariñas del Cerro, U. Furbach (Eds.), Logics in Artificial Intelligence. Proceedings, 1998. X, 391 pages. 1998. (Subseries LNAI).

Vol. 1490: C. Palamidessi, H. Glaser, K. Meinke (Eds.), Principles of Declarative Programming. Proceedings, 1998. XI, 497 pages. 1998.

Vol. 1491: W. Reisig, G. Rozenberg (Eds.), Lectures on Petri Nets I: Basic Models. XII, 683 pages. 1998.

Vol. 1492: W. Reisig, G. Rozenberg (Eds.), Lectures on Petri Nets II: Applications. XII, 479 pages. 1998.

Vol. 1493: J.P. Bowen, A. Fett, M.G. Hinchey (Eds.), ZUM '98: The Z Formal Specification Notation. Proceedings, 1998. XV, 417 pages. 1998.

Vol. 1494: G. Rozenberg, F. Vaandrager (Eds.), Lectures on Embedded Systems. Proceedings, 1996. VIII, 423 pages. 1998.

Vol. 1495: T. Andreasen, H. Christiansen, H.L. Larsen (Eds.), Flexible Query Answering Systems. IX, 393 pages. 1998. (Subseries LNAI).

Vol. 1496: W.M. Wells, A. Colchester, S. Delp (Eds.), Medical Image Computing and Computer-Assisted Intervention – MICCAI'98. Proceedings, 1998. XXII, 1256 pages. 1998.

Vol. 1497: V. Alexandrov, J. Dongarra (Eds.), Recent Advances in Parallel Virtual Machine and Message Passing Interface. Proceedings, 1998. XII, 412 pages. 1998.

Vol. 1498: A.E. Eiben, T. Bäck, M. Schoenauer, H.-P. Schwefel (Eds.), Parallel Problem Solving from Nature – PPSN V. Proceedings, 1998. XXIII, 1041 pages. 1998.

Vol. 1499: S. Kutten (Ed.), Distributed Computing. Proceedings, 1998. XII, 419 pages. 1998.

Vol. 1501: M.M. Richter, C.H. Smith, R. Wiehagen, T. Zeugmann (Eds.), Algorithmic Learning Theory. Proceedings, 1998. XI, 439 pages. 1998. (Subseries LNAI).

Vol. 1502: G. Antoniou, J. Slaney (Eds.), Advanced Topics in Artificial Intelligence. Proceedings, 1998. XI, 333 pages. 1998. (Subseries LNAI).

Vol. 1503: G. Levi (Ed.), Static Analysis. Proceedings, 1998. IX, 383 pages. 1998.

Vol. 1504: O. Herzog, A. Günter (Eds.), KI-98: Advances in Artificial Intelligence. Proceedings, 1998. XI, 355 pages. 1998. (Subseries LNAI).

Vol. 1506: R. Koch, L. Van Gool (Eds.), 3D Structure from Multiple Images of Large-Scale Environments. Proceedings, 1998. VIII, 347 pages. 1998.

Vol. 1507: T.W. Ling, S. Ram, M.L. Lee (Eds.), Conceptual Modeling – ER '98. Proceedings, 1998. XVI, 482 pages. 1998.

Vol. 1508: S. Jajodia, M.T. Özsu, A. Dogac (Eds.), Advances in Multimedia Information Systems. Proceedings, 1998. VIII, 207 pages. 1998.

Vol. 1510: J.M. Zytkow, M. Quafafou (Eds.), Principles of Data Mining and Knowledge Discovery. Proceedings, 1998. XI, 482 pages. 1998. (Subseries LNAI).

Vol. 1511: D. O'Hallaron (Ed.), Languages, Compilers, and Run-Time Systems for Scalable Computers. Proceedings, 1998. IX, 412 pages. 1998.

Vol. 1512: E. Giménez, C. Paulin-Mohring (Eds.), Types for Proofs and Programs. Proceedings, 1996. VIII, 373 pages. 1998.

Vol. 1513: C. Nikolaou, C. Stephanidis (Eds.), Research and Advanced Technology for Digital Libraries. Proceedings, 1998. XV, 912 pages. 1998.

Vol. 1514: K. Ohta,, D. Pei (Eds.), Advances in Cryptology – ASIACRYPT'98. Proceedings, 1998. XII, 436 pages. 1998.

Vol. 1515: F. Moreira de Oliveira (Ed.), Advances in Artificial Intelligence. Proceedings, 1998. X, 259 pages. 1998. (Subseries LNAI).

Vol. 1516: W. Ehrenberger (Ed.), Computer Safety, Reliability and Security. Proceedings, 1998. XVI, 392 pages. 1998.

Vol. 1517: J. Hromkovič, O. Sýkora (Eds.), Graph-Theoretic Concepts in Computer Science. Proceedings, 1998. X, 385 pages. 1998.

Vol. 1518: M. Luby, J. Rolim, M. Serna (Eds.), Randomization and Approximation Techniques in Computer Science. Proceedings, 1998. IX, 385 pages. 1998.

Vol. 1520: M. Maher, J.-F. Puget (Eds.), Principles and Practice of Constraint Programming - CP98. Proceedings, 1998. XI, 482 pages. 1998.

Vol. 1521: B. Rovan (Ed.), SOFSEM'98: Theory and Practice of Informatics. Proceedings, 1998. XI, 453 pages. 1998.

Vol. 1522: G. Gopalakrishnan, P. Windley (Eds.), Formal Methods in Computer-Aided Design. Proceedings, 1998. IX, 529 pages. 1998.

Vol. 1524: G.B. Orr, K.-R. Müller (Eds.), Neural Networks: Tricks of the Trade. VI, 432 pages. 1998.

Vol. 1526: M. Broy, B. Rumpe (Eds.), Requirements Targeting Software and Systems Engineering. Proceedings, 1997. VIII, 357 pages. 1998.

Vol. 1529: D. Farwell, L. Gerber, E. Hovy (Eds.), Machine Translation and the Information Soup. Proceedings, 1998. XIX, 532 pages. 1998. (Subseries LNAI).

Vol. 1531: H.-Y. Lee, H. Motoda (Eds.), PRICAI'98: Topics in Artificial Intelligence. XIX, 646 pages. 1998. (Subseries LNAI).

Vol. 1096: T. Schael, Workflow Management Systems for Process Organisations. Second Edition. XII, 229 pages. 1998.

GPSR Compliance

The European Union's (EU) General Product Safety Regulation (GPSR) is a set of rules that requires consumer products to be safe and our obligations to ensure this.

If you have any concerns about our products, you can contact us on ProductSafety@springernature.com

In case Publisher is established outside the EU, the EU authorized representative is:

Springer Nature Customer Service Center GmbH
Europaplatz 3
69115 Heidelberg, Germany

Batch number: 10682269

Printed by Printforce, the Netherlands